The R Book

The R Book

Second Edition

Michael J. Crawley
Imperial College London at Silwood Park, UK
http://www.bio.ic.ac.uk/research/mjcraw/**therbook/index.htm**

A John Wiley & Sons, Ltd., Publication

Registered office

John Wiley & Sons Ltd, The Atrium, Southern Gate, Chichester, West Sussex PO19 8SQ, United Kingdom

For details of our global editorial offices, for customer services and for information about how to apply for permission to reuse the copyright material in this book please see our website at www.wiley.com.

Library of Congress Cataloging-in-Publication Data

Crawley, Michael J.
  The R book / Michael J. Crawley. – 2e.
    pages cm
  Includes bibliographical references and index.
  ISBN 978-0-470-97392-9 (hardback)
 1. R (Computer program language)   2. Mathematical statistics–Data processing.   I. Title.
  QA276.45.R3C73 2013
  519.50285′5133–dc23

                                                              2012027339

A catalogue record for this book is available from the British Library.

ISBN: 978-0-470-97392-9

Set in 10/12pt Times by Aptara Inc., New Delhi, India.

Printed and bound in Singapore by Markono Print Media Pte Ltd

Chapters

Detailed Contents

Preface

R is a high-level language and an environment for data analysis and graphics. The design of R was heavily influenced by two existing languages: Becker, Chambers and Wilks' S and Sussman's Scheme. The resulting language is very similar in appearance to S, but the underlying implementation and semantics are derived from Scheme. This book is intended as an introduction to the riches of the R environment, aimed at beginners and intermediate users in disciplines ranging from science to economics and from medicine to engineering. I hope that the book can be read as a text as well as dipped into as a reference manual. The early chapters assume absolutely no background in statistics or computing, but the later chapters assume that the material in the earlier chapters has been studied. The book covers data handling, graphics, mathematical functions, and a wide range of statistical techniques all the way from elementary classical tests, through regression and analysis of variance and generalized linear modelling, up to more specialized topics such as Bayesian analysis, spatial statistics, multivariate methods, tree models, mixed-effects models and time series analysis. The idea is to introduce users to the assumptions that lie behind the tests, fostering a critical approach to statistical modelling, but involving little or no statistical theory and assuming no background in mathematics or statistics.

Why should you switch to using R when you have mastered a perfectly adequate statistical package already? At one level, there is no point in switching. If you only carry out a very limited range of statistical tests, and you do not intend to do more (or different) in the future, then fine. The main reason for switching to R is to take advantage of its unrivalled coverage and the availability of new, cutting-edge applications in fields such as generalized mixed-effects modelling and generalized additive models. The next reason for learning R is that you want to be able to understand the literature. More and more people are reporting their results in the context of R, and it is important to know what they are talking about. Third, look around your discipline to see who else is using R: many of the top people will have switched to R already. A large proportion of the world's leading statisticians use R, and this should tell you something (many, indeed, contribute to R, as you can see below). Another reason for changing to R is the quality of back-up and support available. There is a superb network of dedicated R wizards out there on the web, eager to answer your questions. If you intend to invest sufficient effort to become good at statistical computing, then the structure of R and the ease with which you can write your own functions are major attractions. Last, and certainly not least, the product is free. This is some of the finest integrated software in the world, and yet it is yours for absolutely nothing.

Although much of the text will equally apply to S-PLUS, there are some substantial differences, so in order not to confuse things I concentrate on describing R. I have made no attempt to show where S-PLUS is different from R, but if you have to work in S-PLUS, then try it and see if it works.

Acknowledgements

S is an elegant, widely accepted, and enduring software system with outstanding conceptual integrity, thanks to the insight, taste, and effort of John Chambers. In 1998, the Association for Computing Machinery (ACM) presented him with its Software System Award, for 'the S system, which has forever altered the way people analyze, visualize, and manipulate data'. R was inspired by the S environment that was developed by John Chambers, and which had substantial input from Douglas Bates, Rick Becker, Bill Cleveland, Trevor Hastie, Daryl Pregibon and Allan Wilks.

R was initially written by Ross Ihaka and Robert Gentleman at the Department of Statistics of the University of Auckland in New Zealand. Subsequently, a large group of individuals contributed to R by sending code and bug reports. John Chambers graciously contributed advice and encouragement in the early days of R, and later became a member of the core team. The current R is the result of a collaborative effort with contributions from all over the world.

Since mid-1997 there has been a core group with write access to the R source, currently consisting of Douglas Bates, John Chambers, Peter Dalgaard, Seth Falcon, Robert Gentleman, Kurt Hornik, Stefano Iacus, Ross Ihaka, Friedrich Leisch, Uwe Ligges, Thomas Lumley, Martin Maechler, Guido Masarotto (up to June 2003), Duncan Murdoch, Paul Murrell, Martyn Plummer, Brian Ripley, Deepayan Sarkar, Heiner Schwarte (up to October 1999), Duncan Temple Lang, Luke Tierney and Simon Urbanek.

R would not be what it is today without the invaluable help of the following people, who contributed by donating code, bug fixes and documentation: Valerio Aimale, Thomas Baier, Roger Bivand, Ben Bolker, David Brahm, Göran Broström, Patrick Burns, Vince Carey, Saikat DebRoy, Brian D'Urso, Lyndon Drake, Dirk Eddelbuettel, John Fox, Paul Gilbert, Torsten Hothorn, Robert King, Kjetil Kjernsmo, Philippe Lambert, Jan de Leeuw, Jim Lindsey, Patrick Lindsey, Catherine Loader, Gordon Maclean, John Maindonald, David Meyer, Jens Oehlschlägel, Steve Oncley, Richard O'Keefe, Hubert Palme, José C. Pinheiro, Anthony Rossini, Jonathan Rougier, Günther Sawitzki, Bill Simpson, Gordon Smyth, Adrian Trapletti, Terry Therneau, Bill Venables, Gregory R. Warnes, Andreas Weingessel, Morten Welinder, Simon Wood, and Achim Zeileis.

If you use R you should cite it in your written work. To cite the base package, put:

R Development Core Team (2013) *R: A Language and Environment for Statistical Computing*. R Foundation for Statistical Computing, Vienna, Austria. ISBN 3-900051-07-0, URL http://www.R-project.org/.

You can see the most up-to-date citation by typing `citation()` at the prompt. To cite individual contributed packages, you may find the appropriate citation in the description of the package, but failing that you will need to construct the citation from the author's name, date, and title of the package from the reference manual for the package that is available on CRAN (see p. 3).

Special thanks are due to the generations of graduate students on the annual GLIM course at Silwood. It was their feedback that enabled me to understand those aspects of R that are most difficult for beginners, and highlighted the concepts that require the most detailed explanation. Please tell me about the errors and omissions you find, and send suggestions for changes and additions to m.crawley@imperial.ac.uk.

The data files used in this book can be downloaded from http://www.bio.ic.ac.uk/research/mjcraw/therbook/index.htm.

M.J. Crawley
Ascot
September 2012

1

Getting Started

1.1 How to use this book

Try to put yourself in one of the following categories, then go to the appropriate category heading within this section to find some suggestions about how you might get best value from this book:

- Beginner in both computing and statistics (Section 1.1.1);
- Student needing help with project work (1.1.2);
- Done some R and some statistics, but keen to learn more of both (1.1.3);
- Done regression and ANOVA, but want to learn more advanced statistical modelling (1.1.4);
- Experienced in statistics, but a beginner in R (1.1.5);
- Experienced in computing, but a beginner in R (1.1.6);
- Familiar with statistics and computing, but need a friendly reference manual (1.1.7).

1.1.1 Beginner in both computing and statistics

The book is structured principally with you in mind. There are six key things to learn: how to arrange your data, how to read the data into R, how to check the data once within R, how to select the appropriate statistical model and apply it correctly, how to interpret the output, and how to present the analysis for publication. It is essential that you understand the basics thoroughly before trying to do the more complicated things, so study Chapters 3–6 carefully to begin with. Do all of the exercises that are illustrated in the text on your own computer. Now you need to do the hard part, which is selecting the right statistics to use. Model choice is extremely important, and is the thing that will develop most with experience. Do not by shy to ask for expert help with this. Never do an analysis that is more complicated than it needs to be, so start by reading about the classical tests to see if one of these fits your purposes (Chapter 8). Finally, try to understand the distinction between regression (Chapter 10) where the explanatory variable is continuous, and analysis of variance (Chapter 11), where the explanatory variable is categorical. One of these two is likely to be the most complicated method you will need.

The R Book, Second Edition. Michael J. Crawley.
© 2013 John Wiley & Sons, Ltd. Published 2013 by John Wiley & Sons, Ltd.

1.1.2   Student needing help with project work

The first thing to ensure is that you know the difference between your response variable and your explanatory variable, and the distinction between a continuous variable and a categorical variable (Chapter 5). Once you have mastered this, then use the key at the beginning of Chapter 9 to see what kind of statistics you need to employ. It is most likely that if your response variable is a count, where you typically have lots of zeros, then you will want to use either the classical tests (Chapter 8) or count data in tables (Chapter 15). If your response variable is a continuous measure (e.g. a weight) then you will want to use either regression (Chapter 10) if your explanatory variable is continuous (e.g. an altitude) or analysis of variance (Chapter 11) if your explanatory variable is categorical (e.g. genotype). Do not forget to use the appropriate graphics (scatter plots for regressions, box and whisker plots for ANOVA).

1.1.3   Done some R and some statistics, but keen to learn more of both

The best plan is to skim quickly through the introductory material in case you have missed out on some of the basics. Certainly you should read all of the material in Chapter 2 on the fundamentals of the R language and Chapter 5 on graphics. Then, if you know what statistical models you want to use, go directly to the relevant chapter (e.g. regression in Chapter 10 and then non-linear regression in Chapter 20). Use the index for finding help on specific topics.

1.1.4   Done regression and ANOVA, but want to learn more advanced statistical modelling

If you learned regression and ANOVA in another language, the best plan is to go directly to Chapters 10–12 to see how the output from linear models is handled by R. Once you have familiarized yourself with data input (Chapter 3) and dataframes (Chapter 4), you should be able to go directly to the chapters on generalized linear models (Chapter 13), spatial statistics (Chapter 26), survival analysis (Chapter 27), non-linear models (Chapter 20) or mixed-effects models (Chapter 19) without any difficulty.

1.1.5   Experienced in statistics, but a beginner in R

The first thing is to get a thorough understanding of dataframes and data input to R, for which you should study Chapters 3 and 4. Then, if you know what statistics you want to do (e.g. mixed-effects models in R), you should be able to go straight to the appropriate material (Chapter 19 in this case). To understand the output from models in R, you will want to browse Chapter 9 on statistical modelling in R. Then you will want to present your data in the most effective way, by reading Chapter 5 on graphics and Chapter 29 on changing the look of graphics.

1.1.6   Experienced in computing, but a beginner in R

Well-written R code is highly intuitive and very readable. The most unfamiliar parts of R are likely to be the way it handles functions and the way it deals with environments. It is impossible to anticipate the order in which more advanced users are likely to encounter material and hence want to learn about specific features of the language, but vectorized calculations, subscripts on dataframes, function-writing and suchlike are bound to crop up early (Chapter 2). If you see a name in some code, and you want to find out about it, just type the name immediately after a question mark at the R prompt >. If, for example, you want know what `rnbinom` does, type:

```
?rnbinom
```

Recognizing mathematical functions is quite straightforward because of their names and the fact that their arguments are enclosed in round brackets (). Subscripts on objects have square brackets [ ]. Multi-line blocks of R code are enclosed within curly brackets { }. Again, you may not be familiar with lists, or with applying functions to lists; elements within lists have double square brackets [[ ]].

Look at the sections at the head of Chapter 2 as a starting point. The index is probably your most sensible entry point for queries about specifics.

1.1.7 Familiar with statistics and computing, but need a friendly reference manual

If it is a *topic* you want to understand, then use the chapter list on pages v–vi and the Detailed Contents on pp. vii–xxi to find the most appropriate section. For aspects of the R *language*, look at the sections mentioned at the start of Chapter 2 on p. 12. You are likely to want to spend time browsing the contents of general material such as Chapter 5 on graphics and Chapter 29 on changing the look of graphics. Your best bet, in general, is likely to be to use the Index.

Get used to R's help pages. If you know the name of the function for which you require help, just type a question mark followed directly by the function name at R's screen prompt >. To find out what all the graphics parameters mean, for instance, just type:

```
?par
```

1.2 Installing R

I assume that you have a PC or an Apple Mac, and that you want to install R on the hard disc. If you have access to the internet then this could hardly be simpler. First go to the site called CRAN (this stands for Comprehensive R Archive Network). You can type its full address,

```
http://cran.r-project.org/
```

or simply type CRAN into Google and be transported effortlessly to the site. Once there, you need to 'Download and Install R' by running the appropriate precompiled binary distributions. Click to choose between Linux, Mac OS and Windows, then follow the (slightly different) instructions. You want the 'base' package and you want to run the setup program which will have a name like R*.exe (on a PC) or R*.dmg (on a Mac). When asked, say you want to 'Run' the file (rather than 'Save' it). Then just sit back and watch. If you do not have access to the internet, then get a friend to download R and copy it onto a memory stick for you.

1.3 Running R

To run R, just click on the R icon. If there is no icon, go to Programs, then to R, then click on the R icon. The first thing you see is the version number of R and the date of your version. It is a good idea to visit the CRAN site regularly to make sure that you have got the most up-to-date version of R. If you have an old version, it is best to uninstall your current version before downloading the new one.

The header explains that there is no warranty for this free software, and allows you to see the list of current contributors. Perhaps the most important information in the header is found under

```
citation()
```

which shows how to cite the R software in your written work. The R Development Core Team has done a huge amount of work and we, the R user community, should give them due credit whenever we publish work that has used R.

Below the header you will see a blank line with a > symbol in the left-hand margin. This is called the **prompt** and is R's way of saying 'What now?'. This is where you type in your commands, as introduced on p. 13. When working, you will sometimes see + at the left-hand side of the screen instead of >. This means that the last command you typed is incomplete. The commonest cause of this is that you have forgotten one or more brackets. If you can see what is missing (e.g. a final right-hand bracket) then just type the missing character and press enter, at which point the command will execute. If you have made a mistake, then press the Esc key and the command line prompt > will reappear. Then use the Up arrow key to retrieve your last command, at which point you can correct the mistake, using the Left and Right arrow keys.

1.4   The Comprehensive R Archive Network

CRAN is your first port of call for everything to do with R. It is from here that you download and install R, find contributed packages to solve particular problems, find the answers to frequently asked questions, read about the latest developments, get programming tips and much more besides. These are the current headings on the main website:

CRAN
Mirrors
What's new?
Task Views
Search

About R
R Homepage
The R Journal

Software
R Sources
R Binaries
Packages
Other

Documentation
Manuals
FAQs
Contributed

It is well worth browsing through *The R Journal* (formerly *R News*). This is the refereed journal of the R project for statistical computing. It features short to medium-length articles covering topics that might be of interest to users or developers of R, including:

- Add-on packages – short introductions to or reviews of R extension packages.

- Programmer's Niche – hints for programming in R.

- Help Desk – hints for newcomers explaining aspects of R that might not be so obvious from reading the manuals and FAQs.

- Applications – demonstrating how a new or existing technique can be applied in an area of current interest using R, providing a fresh view of such analyses in R that is of benefit beyond the specific application.

1.4.1 Manuals

There are several manuals available on CRAN:

- *An Introduction to R* gives an introduction to the language and how to use R for doing statistical analysis and graphics.
- A draft of the *R Language Definition*, which documents the language *per se* – that is, the objects that it works on, and the details of the expression evaluation process, which are useful to know when programming R functions. This is perhaps the most important of all the manuals.
- *Writing R Extensions* covers how to create your own packages, write R help files, and use the foreign language (C, C++, Fortran, . . .) interfaces.
- *R Data Import/Export* describes the import and export facilities available either in R itself or via packages which are available from CRAN.
- *R Installation and Administration*, which is self-explanatory.
- *R: A Language and Environment for Statistical Computing* (referred to on the website as 'The R Reference Index') contains all the help files of the R standard and recommended packages in printable form.

These manuals are also available in R itself by choosing Help/Manuals (in PDF) from the menu bar. There are also answers to *Frequently Asked Questions* (FAQs) and *The R Journal*, as mentioned above. The most useful part of the site, however, is the Search facility which allows you to investigate the contents of most of the R documents, functions, and searchable mail archives.

1.4.2 Frequently asked questions

R has three collections of answers to FAQs:

- the R FAQ, which is the general collection and contains useful information for users on all platforms (Linux, Mac, Unix, Windows);
- the R Mac OS X FAQ for all users of Apple operating systems;
- the R Windows FAQ for all users of Microsoft operating systems.

You need to read the first of these, plus the appropriate one for your platform.

1.4.3 Contributed documentation

This contains a wide range of longer (more than 100 pages) and shorter manuals, tutorials, and exercises provided by users of R. You should browse these to find the ones most relevant to your needs.

1.5   Getting help in R

The simplest way to get help in R is to click on the Help button on the toolbar of the RGui window (this stands for R's Graphic User Interface). Alternatively, if you are connected to the internet, you can type CRAN into Google and search for the help you need at CRAN (see Section 1.4). However, if you know the name of the function you want help with, you just type a question mark ? at the command line prompt followed by the name of the function. So to get help on read.table, just type

```
?read.table
```

Sometimes you cannot remember the precise name of the function, but you know the subject on which you want help (e.g. data input in this case). Use the help.search function (without a question mark) with your query in double quotes like this:

```
help.search("data input")
```

and (with any luck) you will see the names of the R functions associated with this query. Then you can use ?read.table to get detailed help.

Other useful functions are find and apropos. The find function tells you what package something is in:

```
find("lowess")
```

```
[1] "package:stats"
```

while apropos returns a character vector giving the names of all objects in the search list that match your (potentially partial) enquiry:

```
apropos("lm")
```

```
[1]    ". __C__anova.glm"     ".__C__anova.glm.null"    ". __C__glm"
[4]    ". __C__glm.null"      ". __C__lm"               ". __C__mlm"
[7]    "anova.glm"            "anova.glmlist"           "anova.lm"
[10]   "anova.lmlist"         "anova.mlm"               "anovalist.lm"
[13]   "contr.helmert"        "glm"                     "glm.control"
[16]   "glm.fit"              "glm.fit.null"            "hatvalues.lm"
[19]   "KalmanForecast"       "KalmanLike"              "KalmanRun"
[22]   "KalmanSmooth"         "lm"                      "lm.fit"
[25]   "lm.fit.null"          "lm.influence"            "lm.wfit"
[28]   "lm.wfit.null"         "model.frame.glm"         "model.frame.lm"
[31]   "model.matrix.lm"      "nlm"                     "nlminb"
[34]   "plot.lm"              "plot.mlm"                "predict.glm"
[37]   "predict.lm"           "predict.mlm"             "print.glm"
[40]   "print.lm"             "residuals.glm"           "residuals.lm"
[43]   "rstandard.glm"        "rstandard.lm"            "rstudent.glm"
[46]   "rstudent.lm"          "summary.glm"             "summary.lm"
[49]   "summary.mlm"          "kappa.lm"
```

1.5.1   Worked examples of functions

To see a worked example just type the function name (e.g. linear models, lm)

```
example(lm)
```

and you will see the printed and graphical output produced by the lm function.

1.5.2 Demonstrations of R functions

These can be useful for seeing the range of things that R can do. Here are some for you to try:

```
demo(persp)
demo(graphics)
demo(Hershey)
demo(plotmath)
```

1.6 Packages in R

Finding your way around the contributed packages can be tricky, simply because there are so many of them, and the name of the package is not always as indicative of its function as you might hope. There is no comprehensive cross-referenced index, but there is a very helpful feature called 'Task Views' on CRAN, which explains the packages available under a limited number of usefully descriptive headings. Click on Packages on the CRAN home page, then inside Contributed Packages, you can click on CRAN Task Views, which allows you to browse bundles of packages assembled by topic. Currently, there are 29 Task Views on CRAN as follows:

Bayesian	Bayesian Inference
ChemPhys	Chemometrics and Computational Physics
ClinicalTrials	Clinical Trial Design, Monitoring, and Analysis
Cluster	Cluster Analysis & Finite Mixture Models
DifferentialEquations	Differential Equations
Distributions	Probability Distributions
Econometrics	Computational Econometrics
Environmetrics	Analysis of Ecological and Environmental Data
ExperimentalDesign	Design of Experiments (DoE) & Analysis of Experimental Data
Finance	Empirical Finance
Genetics	Statistical Genetics
Graphics	Graphic Displays & Dynamic Graphics & Graphic Devices & Visualization
HighPerformanceComputing	High-Performance and Parallel Computing with R
MachineLearning	Machine Learning & Statistical Learning
MedicalImaging	Medical Image Analysis
Multivariate	Multivariate Statistics
NaturalLanguageProcessing	Natural Language Processing
OfficialStatistics	Official Statistics & Survey Methodology
Optimization	Optimization and Mathematical Programming
Pharmacokinetics	Analysis of Pharmacokinetic Data
Phylogenetics	Phylogenetics, Especially Comparative Methods
Psychometrics	Psychometric Models and Methods
ReproducibleResearch	Reproducible Research
Robust	Robust Statistical Methods
SocialSciences	Statistics for the Social Sciences
Spatial	Analysis of Spatial Data
Survival	Survival Analysis
TimeSeries	Time Series Analysis
gR	gRaphical Models in R

Table 1.1.   Libraries used in this book that come supplied as part of the base package of R.

lattice	lattice graphics for panel plots or trellis graphs
MASS	package associated with Venables and Ripley's book entitled *Modern Applied Statistics using S-PLUS*
mgcv	generalized additive models
nlme	mixed-effects models (both linear and non-linear)
nnet	feed-forward neural networks and multinomial log-linear models
spatial	functions for kriging and point pattern analysis
survival	survival analysis, including penalised likelihood

Click on the Task View to get an annotated list of the packages available under any particular heading. With any luck you will find the package you are looking for.

To use one of the built-in libraries (listed in Table 1.1), simply type the `library` function with the name of the library in brackets. Thus, to load the spatial library type:

```
library(spatial)
```

1.6.1   Contents of packages

It is easy to use the `help` function to discover the contents of library packages. Here is how you find out about the contents of the `spatial` library:

```
library(help=spatial)
```

```
                    Information on package "spatial"
Package:            spatial
Description:        Functions for kriging and point pattern analysis.
```

followed by a list of all the functions and data sets. You can view the full list of the contents of a library using `objects` with `search()` like this. Here are the contents of the spatial library:

```
objects(grep("spatial",search()))
```

```
[1]     "anova.trls"      "anovalist.trls"    "correlogram"      "expcov"
[5]     "gaucov"          "Kaver"             "Kenvl"            "Kfn"
[9]     "plot.trls"       "ppgetregion"       "ppinit"           "pplik"
[13]    "ppregion"        "predict.trls"      "prmat"            "Psim"
[17]    "semat"           "sphercov"          "SSI"              "Strauss"
[21]    "surf.gls"        "surf.ls"           "trls.influence"   "trmat"
[25]    "variogram"
```

Then, to find out how to use, say, Ripley's *K* (`Kfn`), just type:

```
?Kfn
```

1.6.2   Installing packages

The base package does not contain some of the libraries referred to in this book, but downloading these is very simple. Before you start, you should check whether you need to "Run as administrator" before you can install packages (right click on the R icon to find this). Run the R program, then from the command line use the `install.packages` function to download the libraries you want. You will be asked to highlight the mirror nearest to you for fast downloading (e.g. London), then everything else is automatic. The packages used in this book are

```
install.packages("akima")
install.packages("boot")
install.packages("car")
install.packages("lme4")
install.packages("meta")
install.packages("mgcv")
install.packages("nlme")
install.packages("deSolve")
install.packages("R2jags")
install.packages("RColorBrewer")
install.packages("RODBC")
install.packages("rpart")
install.packages("spatstat")
install.packages("spdep")
install.packages("tree")
```

If you want other packages, then go to CRAN and browse the list called 'Packages' to select the ones you want to investigate.

1.7 Command line versus scripts

When writing functions and other multi-line sections of input you will find it useful to use a text editor rather than execute everything directly at the command line. Some people prefer to use R's own built-in editor. It is accessible from the RGui menu bar. Click on File then click on New script. At this point R will open a window entitled Untitled - R Editor. You can type and edit in this, then when you want to execute a line or group of lines, just highlight them and press Ctrl+R (the Control key and R together). The lines are automatically transferred to the command window and executed.

By pressing Ctrl+S you can save the contents of the R Editor window in a file that you will have to name. It will be given a .R file extension automatically. In a subsequent session you can click on File/Open script... when you will see all your saved .R files and can select the one you want to open.

Other people prefer to use an editor with more features. Tinn-R ("this is not notepad" for R) is very good, or you might like to try RStudio, which has the nice feature of allowing you to scroll back through all of the graphics produced in a session. These and others are free to download from the web.

1.8 Data editor

There is a data editor within R that can be accessed from the menu bar by selecting Edit/Data editor.... You provide the name of the matrix or dataframe containing the material you want to edit (this has to be a dataframe that is active in the current R session, rather than one which is stored on file), and a Data Editor window appears. Alternatively, you can do this from the command line using the `fix` function (e.g. `fix(data.frame.name)`). Suppose you want to edit the bacteria dataframe which is part of the MASS library:

```
library(MASS)
attach(bacteria)
fix(bacteria)
```

The window has the look of a spreadsheet, and you can change the contents of the cells, navigating with the cursor or with the arrow keys. My preference is to do all of my data preparation and data editing in a spreadsheet before even thinking about using R. Once checked and edited, I save the data from the spreadsheet to a tab-delimited text file (*.txt) that can be imported to R very simply using the function called `read.table` (p. 20). One of the most persistent frustrations for beginners is that they cannot get their data imported into R. Things that typically go wrong at the data input stage and the necessary remedial actions are described on p. 139.

1.9   Changing the look of the R screen

The default settings of the command window are inoffensive to most people, but you can change them if you do not like them. The Rgui Configuration Editor under Edit/GUI preferences . . . is used to change the look of the screen. You can change the colour of the input line (default is red), the output line (default navy) or the background (default white). The default numbers of rows (25) and columns (80) can be changed, and you have control over the font (default Courier New) and font size (default 10).

1.10   Good housekeeping

To see what variables you have created in the current session, type:

```
objects()
```

```
[1]    "colour.factor"    "colours"       "dates"      "index"
[5]    "last.warning"     "nbnumbers"     "nbtable"    "nums"
[9]    "wanted"           "x"             "xmat"       "xv"
```

To see which packages and dataframes are currently attached:

```
search()
```

```
[1]     ".GlobalEnv"            "nums"                "nums"
[4]     "package:methods"       "package:stats"       "package:graphics"
[7]     "package:grDevices"     "package:utils"       "package:data sets"
[10]    "Autoloads"             "package:base"
```

At the end of a session in R, it is good practice to remove (`rm`) any variables names you have created (using, say, `x <- 5.6`) and to `detach` any dataframes you have attached earlier in the session. That way, variables with the same names but different properties will not get in each other's way in subsequent work:

```
rm(x,y,z)
detach(worms)
```

The `detach` command does not make the dataframe called worms disappear; it just means that the variables within worms, such as Slope and Area, are no longer accessible directly by name. To get rid of everything, including all the dataframes, type

```
rm(list=ls())
```

but be absolutely sure that you really want to be as draconian as this before you execute the command.

1.11   Linking to other computer languages

Advanced users can employ the functions .C and .Fortran to provide a standard interface to compiled code that has been linked into R, either at build time or via dyn.load. They are primarily intended for compiled C and Fortran code respectively, but the .C function can be used with other languages which can generate C interfaces, for example C++. The .Internal and .Primitive interfaces are used to call C code compiled into R at build time. Functions .Call and .External provide interfaces which allow compiled code (primarily compiled C code) to manipulate R objects.

2

Essentials of the R Language

There is an enormous range of things that R can do, and one of the hardest parts of learning R is finding your way around. Likewise, there is no obvious order in which different people will want to learn the different components of the R language. I suggest that you quickly scan down the following bullet points, which represent the order in which I have chosen to present the introductory material, and if you are relatively experienced in statistical computing, you might want to skip directly to the relevant section. I strongly recommend that beginners work thorough the material in the order presented, because successive sections build upon knowledge gained from previous sections. This chapter is divided into the following sections:

- 2.1 Calculations
- 2.2 Logical operations
- 2.3 Sequences
- 2.4 Testing and coercion
- 2.5 Missing values and things that are not numbers
- 2.6 Vectors and subscripts
- 2.7 Vectorized functions
- 2.8 Matrices and arrays
- 2.9 Sampling
- 2.10 Loops and repeats
- 2.11 Lists
- 2.12 Text, character strings and pattern matching
- 2.13 Dates and times
- 2.14 Environments
- 2.15 Writing R functions
- 2.16 Writing to file from R

The R Book, Second Edition. Michael J. Crawley.
© 2013 John Wiley & Sons, Ltd. Published 2013 by John Wiley & Sons, Ltd.

Other essential material is elsewhere: beginners will want to master data input (Chapter 3), dataframes (Chapter 4) and graphics (Chapter 5).

2.1 Calculations

The screen prompt > is an invitation to put R to work. The convention in this book is that material that you need to type into the command line after the screen prompt is shown in red in Courier New font. Just press the Return key to see the answer. You can use the command line as a calculator, like this:

```
> log(42/7.3)

[1] 1.749795
```

Each line can have at most 8192 characters, but if you want to see a lengthy instruction or a complicated expression on the screen, you can continue it on one or more further lines simply by ending the line at a place where the line is obviously incomplete (e.g. with a trailing comma, operator, or with more left parentheses than right parentheses, implying that more right parentheses will follow). When continuation is expected, the prompt changes from > to +, as follows:

```
> 5+6+3+6+4+2+4+8+
+ 3+2+7

[1]   50
```

Note that the + continuation prompt does not carry out arithmetic plus. If you have made a mistake, and you want to get rid of the + prompt and return to the > prompt, then press the Esc key and use the Up arrow to edit the last (incomplete) line.

From here onwards and throughout the book, the prompt character > will be omitted. The output from R is shown in blue in Courier New font, which uses absolute rather than proportional spacing, so that columns of numbers remain neatly aligned on the page or on the screen.

Two or more expressions can be placed on a single line so long as they are separated by semi-colons:

```
2+3; 5*7; 3-7

[1] 5
[1] 35
[1] -4
```

For very big numbers or very small numbers R uses the following scheme (called exponents):

`1.2e3`	means 1200 because the e3 means 'move the decimal point 3 places to the right';
`1.2e-2`	means 0.012 because the e-2 means 'move the decimal point 2 places to the left';
`3.9+4.5i`	is a complex number with real (3.9) and imaginary (4.5) parts, and i is the square root of -1.

2.1.1 Complex numbers in R

Complex numbers consist of a real part and an imaginary part, which is identified by lower-case i like this:

```
z <- 3.5-8i
```

The elementary trigonometric, logarithmic, exponential, square root and hyperbolic functions are all implemented for complex values. The following are the special R functions that you can use with complex numbers. Determine the real part:

```
Re(z)
```

```
[1] 3.5
```

Determine the imaginary part:

```
Im(z)
```

```
[1] -8
```

Calculate the modulus (the distance from z to 0 in the complex plane by Pythagoras; if x is the real part and y is the imaginary part, then the modulus is $\sqrt{x^2 + y^2}$):

```
Mod(z)
```

```
[1] 8.732125
```

Calculate the argument (`Arg(x+ yi)= atan(y/x)`):

```
Arg(z)
```

```
[1] -1.158386
```

Work out the complex conjugate (change the sign of the imaginary part):

```
Conj(z)
```

```
[1] 3.5+8i
```

Membership and coercion are dealt with in the usual way (p. 30):

```
is.complex(z)
```

```
[1] TRUE
```

```
as.complex(3.8)
```

```
[1] 3.8+0i
```

2.1.2 Rounding

Various sorts of rounding (rounding up, rounding down, rounding to the nearest integer) can be done easily. Take the number 5.7 as an example. The 'greatest integer less than' function is `floor`:

```
floor(5.7)
```

```
[1] 5
```

The 'next integer' function is `ceiling`:

```
ceiling(5.7)
```

```
[1] 6
```

You can round to the nearest integer by adding 0.5 to the number, then using `floor`. There is a built-in function for this, but we can easily write one of our own to introduce the notion of function writing. Call it rounded, then define it as a function like this:

```
rounded <- function(x) floor(x+0.5)
```

Now we can use the new function:

```
rounded(5.7)
```

```
[1] 6
```

```
rounded(5.4)
```

```
[1] 5
```

The hard part is deciding how you want to round negative numbers, because the concept of up and down is more subtle (remember that −5 is a bigger number than −6). You need to think, instead, of whether you want to round towards zero or away from zero. For negative numbers, rounding up means rounding *towards zero* so do not be surprised when the value of the positive part is different:

```
ceiling(-5.7)
```

```
[1] -5
```

With floor, negative values are rounded *away* from zero:

```
floor(-5.7)
```

```
[1] -6
```

You can simply strip off the decimal part of the number using the function trunc, which returns the integers formed by truncating the values in x towards zero:

```
trunc(5.7)
```

```
[1] 5
```

```
trunc(-5.7)
```

```
[1] -5
```

There is an R function called round that you can use by specifying 0 decimal places in the second argument:

```
round(5.7,0)
```

```
[1] 6
```

```
round(5.5,0)
```

```
[1] 6
```

```
round(5.4,0)
```

```
[1] 5
```

```
round(-5.7,0)
```

```
[1] -6
```

The number of decimal places is not the same as the number of significant digits. You can control the number of significant digits in a number using the function signif. Take a big number like 12 345 678 (roughly

12.35 million). Here is what happens when we ask for 4, 5 or 6 significant digits:

```
signif(12345678,4)
[1] 12350000
signif(12345678,5)
[1] 12346000
signif(12345678,6)
[1] 12345700
```

and so on. Why you would want to do this would need to be explained.

2.1.3   Arithmetic

The screen prompt in R is a fully functional calculator. You can add and subtract using the obvious + and - symbols, while division is achieved with a forward slash / and multiplication is done by using an asterisk * like this:

```
7 + 3 - 5 * 2
[1] 0
```

Notice from this example that multiplication (5×2) is done *before* the additions and subtractions. Powers (like squared or cube root) use the caret symbol ^ and are done before multiplication or division, as you can see from this example:

```
3^2 / 2
[1] 4.5
```

All the mathematical functions you could ever want are here (see Table 2.1). The log function gives logs to the base e (e = 2.718 282), for which the antilog function is exp:

```
log(10)
[1] 2.302585
exp(1)
[1] 2.718282
```

If you are old fashioned, and want logs to the base 10, then there is a separate function, log10:

```
log10(6)
[1] 0.7781513
```

Logs to other bases are possible by providing the log function with a second argument which is the base of the logs you want to take. Suppose you want log to base 3 of 9:

```
log(9,3)
[1] 2
```

Table 2.1. Mathematical functions used in R.

Function	Meaning
`log(x)`	log to base e of x
`exp(x)`	antilog of x (e^x)
`log(x,n)`	log to base n of x
`log10(x)`	log to base 10 of x
`sqrt(x)`	square root of x
`factorial(x)`	$x! = x \times (x-1) \times (x-2) \times \cdots \times 3 \times 2$
`choose(n,x)`	binomial coefficients $n!/(x!\,(n-x)!)$
`gamma(x)`	$\Gamma(x)$, for real x ($x-1$)!, for integer x
`lgamma(x)`	natural log of $\Gamma(x)$
`floor(x)`	greatest integer less than x
`ceiling(x)`	smallest integer greater than x
`trunc(x)`	closest integer to x between x and 0, e.g. trunc(1.5) = 1, trunc(−1.5) = −1; trunc is like floor for positive values and like ceiling for negative values
`round(x, digits=0)`	round the value of x to an integer
`signif(x, digits=6)`	give x to 6 digits in scientific notation
`runif(n)`	generates n random numbers between 0 and 1 from a uniform distribution
`cos(x)`	cosine of x in radians
`sin(x)`	sine of x in radians
`tan(x)`	tangent of x in radians
`acos(x)`, `asin(x)`, `atan(x)`	inverse trigonometric transformations of real or complex numbers
`acosh(x)`, `asinh(x)`, `atanh(x)`	inverse hyperbolic trigonometric transformations of real or complex numbers
`abs(x)`	the absolute value of x, ignoring the minus sign if there is one

The trigonometric functions in R measure angles in radians. A circle is 2π radians, and this is $360°$, so a right angle ($90°$) is $\pi/2$ radians. R knows the value of π as `pi`:

```
pi
```

```
[1] 3.141593
```

```
sin(pi/2)
```

```
[1] 1
```

```
cos(pi/2)
```

```
[1] 6.123032e-017
```

Notice that the cosine of a right angle does not come out as exactly zero, even though the sine came out as exactly 1. The `e-017` means 'times 10^{-17}'. While this is a very small number, it is clearly not exactly zero (so you need to be careful when testing for exact equality of real numbers; see p. 23).

2.1.4 Modulo and integer quotients

Integer quotients and remainders are obtained using the notation `%/%` (percent, divide, percent) and `%%` (percent, percent) respectively. Suppose we want to know the integer part of a division: say, how many 13s are there in 119:

```
119 %/% 13
```

```
[1]  9
```

Now suppose we wanted to know the remainder (what is left over when 119 is divided by 13): in maths this is known as **modulo**:

```
119 %% 13
```

```
[1]  2
```

Modulo is very useful for testing whether numbers are odd or even: odd numbers have modulo 2 value 1 and even numbers have modulo 2 value 0:

```
9 %% 2
```

```
[1]  1
```

```
8 %% 2
```

```
[1]  0
```

Likewise, you use modulo to test if one number is an exact multiple of some other number. For instance, to find out whether 15 421 is a multiple of 7 (which it is), then ask:

```
15421 %% 7 == 0
```

```
[1]  TRUE
```

Note the use of 'double equals' to test for equality (this is explained in detail on p. 26).

2.1.5   Variable names and assignment

There are three important things to remember when selecting names for your variables in R:

- Variable names in R are case sensitive, so y is not the same as Y.
- Variable names should not begin with numbers (e.g. $1x$) or symbols (e.g. $%x$).
- Variable names should not contain blank spaces (use `back.pay` not `back pay`).

In terms of your work–life balance, make your variable names as short as possible, so that you do not spend most of your time typing, and the rest of your time correcting spelling mistakes in your ridiculously long variable names.

Objects obtain values in R by assignment ('x *gets* a value'). This is achieved by the **gets arrow** `<-` which is a composite symbol made up from 'less than' and 'minus' with no space between them. Thus, to create a scalar constant x with value 5 we type:

```
x <- 5
```

and not `x = 5`. Notice that there is a potential ambiguity if you get the spacing wrong. Compare our `x <- 5`, 'x gets 5', with `x < - 5` where there is a space between the 'less than' and 'minus' symbol. In R, this is actually a question, asking 'is x less than minus 5?' and, depending on the current value of x, would evaluate to the answer either `TRUE` or `FALSE`.

2.1.6   Operators

R uses the following operator tokens:

`+ - * / %/% %% ^`	arithmetic (plus, minus, times, divide, integer quotient, modulo, power)
`>= < <= == !=`	relational (greater than, greater than or equals, less than, less than or equals, equals, not equals)
`! & |`	logical (not, and, or)
`~`	model formulae ('is modelled as a function of')
`<- ->`	assignment (gets)
`$`	list indexing (the 'element name' operator)
`:`	create a sequence

Several of these operators have different meaning inside model formulae. Thus `*` indicates the main effects plus interaction (rather than multiplication), `:` indicates the interaction between two variables (rather than generate a sequence) and `^` means all interactions up to the indicated power (rather than raise to the power). You will learn more about these ideas in Chapter 9.

2.1.7   Integers

Integer vectors exist so that data can be passed to C or Fortran code which expects them, and so that small integer data can be represented exactly and compactly. The range of integers is from $-2\,000\,000\,000$ to $+2\,000\,000\,000$ (`-2*10^9` to `+2*10^9`, which R could portray as `-2e+09` to `2e+09`).

Be careful. Do not try to change the class of a vector by using the `integer` function. Here is a numeric vector of whole numbers that you want to convert into a vector of integers:

```
x <- c(5,3,7,8)
is.integer(x)
```

```
[1] FALSE
```

```
is.numeric(x)
```

```
[1] TRUE
```

Applying the integer function to it replaces all your numbers with zeros; definitely not what you intended.

```
x <- integer(x)
x
```

```
[1] 0 0 0 0 0
```

Make the numeric object first, then convert the object to integer using the `as.integer` function like this:

```
x <- c(5,3,7,8)
x <- as.integer(x)
is.integer(x)
```

```
[1] TRUE
```

The integer function works as `trunc` when applied to real numbers, and removes the imaginary part when applied to complex numbers:

```
as.integer(5.7)
```

```
[1] 5
```

```
as.integer(-5.7)
```

```
[1] -5
```

```
as.integer(5.7 -3i)
```

```
[1] 5
Warning message:
imaginary parts discarded in coercion
```

2.1.8   Factors

Factors are categorical variables that have a fixed number of levels. A simple example of a factor might be a variable called gender with two levels: 'female' and 'male'. If you had three females and two males, you could create the factor like this:

```
gender <- factor(c("female", "male", "female", "male", "female"))
class(gender)
```

```
[1] "factor"
```

```
mode(gender)
```

```
[1] "numeric"
```

More often, you will create a dataframe by reading your data from a file using `read.table`. When you do this, all variables containing one or more character strings will be converted automatically into factors. Here is an example:

```
data <- read.table("c:\\temp\\daphnia.txt",header=T)
attach(data)
head(data)
```

```
  Growth.rate Water Detergent Daphnia
1    2.919086  Tyne    BrandA  Clone1
2    2.492904  Tyne    BrandA  Clone1
3    3.021804  Tyne    BrandA  Clone1
4    2.350874  Tyne    BrandA  Clone2
5    3.148174  Tyne    BrandA  Clone2
6    4.423853  Tyne    BrandA  Clone2
```

This dataframe contains a continuous response variable (`Growth.rate`) and three categorical explanatory variables (`Water`, `Detergent` and `Daphnia`), all of which are factors. In statistical modelling, factors are associated with analysis of variance (all the explanatory variables are categorical) and analysis of covariance (some of the explanatory variables are categorical and some are continuous).

There are some important functions for dealing with factors. You will often want to check that a variable is a factor (especially if the factor levels are numbers rather than characters):

```
is.factor(Water)
```

```
[1]  TRUE
```

To discover the *names* of the factor levels, we use the `levels` function:

```
levels(Detergent)
```

```
[1]  "BrandA"  "BrandB"  "BrandC"  "BrandD"
```

To discover the *number* of levels of a factor, we use the `nlevels` function:

```
nlevels(Detergent)
```

```
[1]  4
```

The same result is achieved by applying the `length` function to the levels of a factor:

```
length(levels(Detergent))
```

```
[1]  4
```

By default, factor levels are treated in alphabetical order. If you want to change this (as you might, for instance, in ordering the bars of a bar chart) then this is straightforward: just type the factor levels in the order that you want them to be used, and provide this vector as the second argument to the `factor` function.

Suppose we have an experiment with three factor levels in a variable called `treatment`, and we want them to appear in this order: 'nothing', 'single' dose and 'double' dose. We shall need to override R's natural tendency to order them 'double', 'nothing', 'single':

```
frame <- read.table("c:\\temp\\trial.txt",header=T)
attach(frame)
tapply(response,treatment,mean)
```

```
 double nothing single
    25      60     34
```

This is achieved using the `factor` function like this:

```
treatment <- factor(treatment,levels=c("nothing","single","double"))
```

Now we get the order we want:

```
tapply(response,treatment,mean)
```

```
nothing single double
    60      34     25
```

Only `==` and `!=` can be used for factors. Note, also, that a factor can only be compared to another factor with an identical set of levels (not necessarily in the same ordering) or to a character vector. For example, you cannot ask quantitative questions about factor levels, like `>` or `<=`, even if these levels are numeric.

To turn factor levels into numbers (integers) use the `unclass` function like this:

```
as.vector(unclass(Daphnia))
```

```
 [1] 1 1 1 2 2 2 3 3 3 1 1 1 2 2 2 3 3 3 1 1 1 2 2 2 3 3 3 1 1 1 2 2 2 3 3 3 1 1
[39] 1 2 2 2 3 3 3 1 1 1 2 2 2 3 3 3 1 1 1 2 2 2 3 3 3 1 1 1 2 2 2 3 3 3
```

Table 2.2. Logical and relational operations.

Symbol	Meaning
!	logical NOT
&	logical AND
|	logical OR
<	less than
<=	less than or equal to
>	greater than
=	greater than or equal to
==	logical equals (double =)
!=	not equal
&&	AND with IF
||	OR with IF
xor(x,y)	exclusive OR
isTRUE(x)	an abbreviation of identical(TRUE,x)

2.2 Logical operations

A crucial part of computing involves asking questions about things. Is one thing bigger than other? Are two things the same size? Questions can be joined together using words like 'and' 'or', 'not'. Questions in R typically evaluate to TRUE or FALSE but there is the option of a 'maybe' (when the answer is not available, NA). In R, < means 'less than', > means 'greater than', and ! means 'not' (see Table 2.2).

2.2.1 TRUE and T with FALSE and F

You can use T for TRUE and F for FALSE, but you should be aware that T and F might have been allocated as variables. So this is obvious:

```
TRUE == FALSE
```

```
[1] FALSE
```

```
T == F
```

```
[1] FALSE
```

This, however, is not so obvious:

```
T <- 0
T == FALSE
```

```
[1] TRUE
```

```
F <- 1
TRUE == F
```

```
[1] TRUE
```

But now, of course, T is not equal to F:

```
T != F
```

```
[1] TRUE
```

To be sure, always write TRUE and FALSE in full, and never use T or F as variable names.

2.2.2   Testing for equality with real numbers

There are international standards for carrying out floating point arithmetic, but on your computer these standards are beyond the control of R. Roughly speaking, integer arithmetic will be exact between -10^{16} and 10^{16}, but for fractions and other real numbers we lose accuracy because of round-off error. This is only likely to become a real problem in practice if you have to **subtract** similarly sized but very large numbers. A dramatic loss in accuracy under these circumstances is called 'catastrophic cancellation error'. It occurs when an operation on two numbers increases *relative error* substantially more than it increases *absolute error*.

You need to be careful in programming when you want to test whether or not two computed numbers are equal. R will assume that you mean 'exactly equal', and what *that* means depends upon machine precision. Most numbers are rounded to an accuracy of 53 binary digits. Typically therefore, two floating point numbers will not reliably be equal unless they were computed by the same algorithm, and not always even then. You can see this by squaring the square root of 2: surely these values are the same?

```
x <- sqrt(2)
x * x == 2
```

```
[1] FALSE
```

In fact, they are not the same. We can see by how much the two values differ by subtraction:

```
x * x - 2
```

```
[1] 4.440892e-16
```

This is not a big number, but it is not zero either. So how do we test for equality of real numbers? The best advice is not to do it. Try instead to use the alternatives 'less than' with 'greater than or equal to', or conversely 'greater than' with 'less than or equal to'. Then you will not go wrong. Sometimes, however, you really do want to test for equality. In those circumstances, do not use double equals to test for equality, but employ the `all.equal` function instead.

2.2.3   Equality of floating point numbers using `all.equal`

The nature of floating point numbers used in computing is the cause of some initially perplexing features. You would imagine that since 0.3 minus 0.2 is 0.1, and the logic presented below would evaluate to TRUE. Not so:

```
x <- 0.3 - 0.2
y <- 0.1
x == y
```

```
[1] FALSE
```

The function called `identical` gives the same result.

```
identical(x,y)
```

```
[1] FALSE
```

The solution is to use the function called `all.equal` which allows for insignificant differences:

```
all.equal(x,y)
```

```
[1] TRUE
```

Do not use `all.equal` directly in `if` expressions. Either use `isTRUE(all.equal(....))` or `identical` as appropriate.

2.2.4 Summarizing differences between objects using `all.equal`

The function `all.equal` is very useful in programming for checking that objects are as you expect them to be. Where differences occur, `all.equal` does a useful job in describing all the differences it finds. Here, for instance, it reports on the difference between a which is a vector of characters and b which is a factor:

```
a <- c("cat","dog","goldfish")
b <- factor(a)
```

In the `all.equal` function, the object on the left (a) is called the 'target' and the object on the right (b) is 'current':

```
all.equal(a,b)

[1] "Modes: character, numeric"
[2] "Attributes: < target is NULL, current is list >"
[3] "target is character, current is factor"
```

Recall that factors are stored internally as integers, so they have `mode = numeric`.

```
class(b)

[1] "factor"

mode(b)

[1] "numeric"
```

The reason why 'current is list' in line [2] of the output is that factors have two attributes and these are stored as a list – namely, their levels and their class:

```
attributes(b)

$levels
[1] "cat" "dog" "goldfish"


$class
[1] "factor"
```

The `all.equal` function is also useful for obtaining feedback on differences in things like the lengths of vectors:

```
n1 <- c(1,2,3)
n2 <- c(1,2,3,4)
all.equal(n1,n2)

[1] "Numeric: lengths (3, 4) differ"
```

It works well, too, for multiple differences:

```
n2 <- as.character(n2)
all.equal(n1,n2)

[1] "Modes: numeric, character"
[2] "Lengths: 3, 4"
[3] "target is numeric, current is character"
```

Note that 'target' is the first argument to the function and 'current' is the second. If you supply more than two objects to be compared, the third and subsequent objects are simply ignored.

2.2.5 Evaluation of combinations of TRUE and FALSE

It is important to understand how combinations of logical variables evaluate, and to appreciate how logical operations (such as those in Table 2.2) work when there are missing values, NA. Here are all the possible outcomes expressed as a logical vector called *x*:

```
x <- c(NA, FALSE, TRUE)
names(x) <- as.character(x)
```

To see the logical combinations of & (logical AND) we can use the `outer` function with *x* to evaluate all nine combinations of NA, FALSE and TRUE like this:

```
outer(x, x, "&")
```

	<NA>	FALSE	TRUE
<NA>	NA	FALSE	NA
FALSE	FALSE	FALSE	FALSE
TRUE	NA	FALSE	TRUE

Only TRUE & TRUE evaluates to TRUE. Note the behaviour of NA & NA and NA & TRUE. Where one of the two components is NA, the result will be NA if the outcome is ambiguous. Thus, NA & TRUE evaluates to NA, but NA & FALSE evaluates to FALSE. To see the logical combinations of | (logical OR) write:

```
outer(x, x, "|")
```

	<NA>	FALSE	TRUE
<NA>	NA	NA	TRUE
FALSE	NA	FALSE	TRUE
TRUE	TRUE	TRUE	TRUE

Only FALSE | FALSE evaluates to FALSE. Note the behaviour of NA | NA and NA | FALSE.

2.2.6 Logical arithmetic

Arithmetic involving logical expressions is very useful in programming and in selection of variables. If logical arithmetic is unfamiliar to you, then persevere with it, because it will become clear how useful it is, once the penny has dropped. The key thing to understand is that logical expressions evaluate to either true or false (represented in R by TRUE or FALSE), and that R can coerce TRUE or FALSE into numerical values: 1 for TRUE and 0 for FALSE. Suppose that *x* is a sequence from 0 to 6 like this:

```
x <- 0:6
```

Now we can ask questions about the contents of the vector called *x*. Is *x* less than 4?

```
x < 4
```

```
[1]  TRUE  TRUE  TRUE  TRUE FALSE FALSE FALSE
```

The answer is yes for the first four values (0, 1, 2 and 3) and no for the last three (4, 5 and 6). Two important logical functions are `all` and `any`. They check an entire vector but return a single logical value: TRUE or FALSE. Are all the *x* values bigger than 0?

```
all(x>0)
```

```
[1] FALSE
```

No. The first *x* value is a zero. Are any of the *x* values negative?

```
any(x<0)
```

```
[1] FALSE
```

No. The smallest *x* value is a zero.

We can use the answers of logical functions in arithmetic. We can count the true values of (x<4), using sum:

```
sum(x<4)
```

```
[1] 4
```

We can multiply (x<4) by other vectors:

```
(x<4)*runif(7)
```

```
[1]  0.9433433  0.9382651  0.6248691  0.9786844  0.0000000  0.0000000
[7]  0.0000000
```

Logical arithmetic is particularly useful in generating simplified factor levels during statistical modelling. Suppose we want to reduce a five-level factor (a, b, c, d, e) called `treatment` to a three-level factor called `t2` by lumping together the levels a and e (new factor level 1) and c and d (new factor level 3) while leaving b distinct (with new factor level 2):

```
(treatment <- letters[1:5])
```

```
[1]  "a"  "b"  "c"  "d"  "e"
```

```
(t2 <- factor(1+(treatment=="b")+2*(treatment=="c")+2*(treatment=="d")))
```

```
[1]  1  2  3  3  1
Levels:  1  2  3
```

The new factor `t2` gets a value 1 as default for all the factors levels, and we want to leave this as it is for levels a and e. Thus, we do not add anything to the 1 if the old factor level is a or e. For old factor level b, however, we want the result that `t2=2` so we add 1 (`treatment=="b"`) to the original 1 to get the answer we require. This works because the logical expression evaluates to 1 (TRUE) for every case in which the old factor level is b and to 0 (FALSE) in all other cases. For old factor levels c and d we want the result that `t2=3` so we add 2 to the baseline value of 1 if the original factor level is either c (2*(`treatment=="c"`)) or d (2*(`treatment=="d"`)). You may need to read this several times before the penny drops. Note that 'logical equals' is a double equals sign without a space in between (==). You need to understand the distinction between:

      x <- y   *x* is assigned the value of *y* (*x gets* the values of *y*);
      x = y   in a function or a list *x* is set to *y* unless you specify otherwise;
      x == y   produces TRUE if *x* is exactly equal to *y* and FALSE otherwise.

2.3 Generating sequences

An important way of creating vectors is to generate a sequence of numbers. The simplest sequences are in steps of 1, and the colon operator is the simplest way of generating such sequences. All you do is specify the first and last values separated by a colon. Here is a sequence from 0 up to 10:

```
0:10
```

```
[1]  0  1  2  3  4  5  6  7  8  9  10
```

Here is a sequence from 15 down to 5:

```
15:5
```

```
[1]  15  14  13  12  11  10  9  8  7  6  5
```

To generate a sequence in steps other than 1, you use the seq function. There are various forms of this, of which the simplest has three arguments: from, to, by (the initial value, the final value and the increment). If the initial value is smaller than the final value, the increment should be positive, like this:

```
seq(0, 1.5, 0.1)
```

```
[1]  0.0 0.1 0.2 0.3 0.4 0.5 0.6 0.7 0.8 0.9 1.0 1.1 1.2 1.3 1.4 1.5
```

If the initial value is larger than the final value, the increment should be negative, like this:

```
seq(6,4,-0.2)
```

```
[1]  6.0 5.8 5.6 5.4 5.2 5.0 4.8 4.6 4.4 4.2 4.0
```

In many cases, you want to generate a sequence to match an existing vector in length. Rather than having to figure out the increment that will get from the initial to the final value and produce a vector of exactly the appropriate length, R provides the along and length options. Suppose you have a vector of population sizes:

```
N <- c(55,76,92,103,84,88,121,91,65,77,99)
```

You need to plot this against a sequence that starts at 0.04 in steps of 0.01:

```
seq(from=0.04,by=0.01,length=11)
```

```
[1]  0.04  0.05  0.06  0.07  0.08  0.09  0.10  0.11  0.12  0.13  0.14
```

But this requires you to figure out the length of N. A simpler method is to use the along argument and specify the vector, N, whose length has to be matched:

```
seq(0.04,by=0.01,along=N)
```

```
[1]  0.04  0.05  0.06  0.07  0.08  0.09  0.10  0.11  0.12  0.13  0.14
```

Alternatively, you can get R to work out the increment (0.01 in this example), by specifying the start and the end values (from and to), and the name of the vector (N) whose length has to be matched:

```
seq(from=0.04,to=0.14,along=N)
```

```
[1]  0.04  0.05  0.06  0.07  0.08  0.09  0.10  0.11  0.12  0.13  0.14
```

An important application of the last option is to get the x values for drawing smooth lines through a scatterplot of data using predicted values from a model (see p. 207).

Notice that when the increment does not match the final value, then the generated sequence stops short of the last value (rather than overstepping it):

```
seq(1.4,2.1,0.3)
```

```
[1] 1.4 1.7 2.0
```

If you want a vector made up of sequences of unequal lengths, then use the `sequence` function. Suppose that most of the five sequences you want to string together are from 1 to 4, but the second one is 1 to 3 and the last one is 1 to 5, then:

```
sequence(c(4,3,4,4,4,5))
```

```
 [1] 1 2 3 4 1 2 3 1 2 3 4 1 2 3 4 1 2 3 4 1 2 3 4 5
```

2.3.1 Generating repeats

You will often want to generate repeats of numbers or characters, for which the function is `rep`. The object that is named in the first argument is repeated a number of times as specified in the second argument. At its simplest, we would generate five 9s like this:

```
rep(9,5)
```

```
[1] 9 9 9 9 9
```

You can see the issues involved by a comparison of these three increasingly complicated uses of the `rep` function:

```
rep(1:4, 2)
```

```
[1]  1  2  3  4  1  2  3  4
```

```
rep(1:4, each = 2)
```

```
[1]  1  1  2  2  3  3  4  4
```

```
rep(1:4, each = 2, times = 3)
```

```
 [1]  1  1  2  2  3  3  4  4  1  1  2  2
[13]  3  3  4  4  1  1  2  2  3  3  4  4
```

In the simplest case, the *entire* first argument is repeated (i.e. the sequence 1 to 4 appears twice). You often want each *element* of the sequence to be repeated, and this is accomplished with the `each` argument. Finally, you might want each number repeated and the whole series repeated a certain number of times (here three times).

When each element of the series is to be repeated a different number of times, then the second argument must be a vector of the same length as the vector comprising the first argument (length 4 in this example). So if we want one 1, two 2s, three 3s and four 4s we would write:

```
rep(1:4,1:4)
```

```
[1]  1  2  2  3  3  3  4  4  4  4
```

In a more complicated case, there is a different but irregular repeat of each of the elements of the first argument. Suppose that we need four 1s, one 2, four 3s and two 4s. Then we use the concatenation function `c` to create a vector of length 4 `c(4,1,4,2)` which will act as the second argument to the `rep` function:

```
rep(1:4,c(4,1,4,2))

[1]  1  1  1  1  2  3  3  3  3  4  4
```

Here is the most complex case with character data rather than numbers: each element of the series is repeated an irregular number of times:

```
rep(c("cat","dog","gerbil","goldfish","rat"),c(2,3,2,1,3))

[1] "cat"     "cat"        "dog" "dog" "dog" "gerbil"
[7] "gerbil" "goldfish" "rat" "rat" "rat"
```

This is the most general, and also the most useful form of the rep function.

2.3.2 Generating factor levels

The function gl ('generate levels') is useful when you want to encode long vectors of factor levels. The syntax for the three arguments is: 'up to', 'with repeats of', 'to total length'. Here is the simplest case where we want factor levels up to 4 with repeats of 3 repeated only once (i.e. to total length 12):

```
gl(4,3)

[1]  1  1  1  2  2  2  3  3  3  4  4  4
Levels: 1  2  3  4
```

Here is the function when we want that whole pattern repeated twice:

```
gl(4,3,24)

[1]   1  1  1  2  2  2  3  3  3  4  4  4
[13]  1  1  1  2  2  2  3  3  3  4  4  4

Levels:   1  2  3  4
```

If you want text for the factor levels, rather than numbers, use labels like this:

```
Temp <- gl(2, 2, 24, labels = c("Low", "High"))
Soft <- gl(3, 8, 24, labels = c("Hard","Medium","Soft"))
M.user <- gl(2, 4, 24, labels = c("N", "Y"))
Brand <- gl(2, 1, 24, labels = c("X", "M"))

data.frame(Temp,Soft,M.user,Brand)

  Temp Soft   M.user Brand
1  Low    Hard    N X
2  Low    Hard    N M
3  High Hard    N X
4  High Hard    N M
5  Low    Hard    Y X
6  Low    Hard    Y M
7  High Hard    Y X
8  High Hard    Y M
9  Low    Medium N X
```

```
10    Low   Medium N M
11    High  Medium N X
12    High  Medium N M
13    Low   Medium Y X
14    Low   Medium Y M
15    High  Medium Y X
16    High  Medium Y M
17    Low   Soft   N X
18    Low   Soft   N M
19    High  Soft   N X
20    High  Soft   N M
21    Low   Soft   Y X
22    Low   Soft   Y M
23    High  Soft   Y X
24    High  Soft   Y M
```

2.4   Membership: Testing and coercing in R

The concepts of membership and coercion may be unfamiliar. Membership relates to the class of an object in R. Coercion changes the class of an object. For instance, a logical variable has class `logical` and mode `logical`. This is how we create the variable:

```
lv <- c(T,F,T)
```

We can assess its membership by asking if it is a logical variable using the `is.logical` function:

```
is.logical(lv)
```

```
[1] TRUE
```

It is not a factor, and so it does not have levels:

```
levels(lv)
NULL
```

But we can coerce it be a two-level factor like this:

```
(fv <- as.factor(lv))
```

```
[1] TRUE FALSE TRUE
Levels: FALSE TRUE
```

```
is.factor(fv)
```

```
[1] TRUE
```

We can coerce a logical variable to be numeric: TRUE evaluates to 1 and FALSE evaluates to zero, like this:

```
(nv <- as.numeric(lv))
```

```
[1] 1 0 1
```

This is particularly useful as a shortcut when creating new factors with reduced numbers of levels (as we do in model simplification).

Table 2.3. Functions for testing (`is`) the attributes of different categories of object (arrays, lists, etc.) and for coercing (`as`) the attributes of an object into a specified form. Neither operation changes the attributes of the object unless you overwrite its name.

Type	Testing	Coercing
Array	is.array	as.array
Character	is.character	as.character
Complex	is.complex	as.complex
Dataframe	is.data.frame	as.data.frame
Double	is.double	as.double
Factor	is.factor	as.factor
List	is.list	as.list
Logical	is.logical	as.logical
Matrix	is.matrix	as.matrix
Numeric	is.numeric	as.numeric
Raw	is.raw	as.raw
Time series (ts)	is.ts	as.ts
Vector	is.vector	as.vector

In general, the expression `as(object, value)` is the way to coerce an object to a particular class. Membership functions ask `is.something` and coercion functions say `as.something`.

Objects have a type, and you can test the type of an object using an `is.type` function (Table 2.3). For instance, mathematical functions expect numeric input and text-processing functions expect character input. Some types of objects can be coerced into other types. A familiar type of coercion occurs when we interpret the TRUE and FALSE of logical variables as numeric 1 and 0, respectively. Factor levels can be coerced to numbers. Numbers can be coerced into characters, but non-numeric characters cannot be coerced into numbers.

```
as.numeric(factor(c("a","b","c")))

[1]  1  2  3

as.numeric(c("a","b","c"))

[1]  NA  NA  NA

Warning message:
NAs introduced by coercion

as.numeric(c("a","4","c"))

[1]  NA  4  NA
Warning message:
NAs introduced by coercion
```

If you try to coerce complex numbers to numeric the imaginary part will be discarded. Note that `is.complex` and `is.numeric` are never both TRUE.

We often want to coerce tables into the form of vectors as a simple way of stripping off their `dimnames` (using `as.vector`), and to turn matrices into dataframes (`as.data.frame`). A lot of testing involves the NOT operator `!` in functions to return an error message if the wrong type is supplied. For instance, if you were writing a function to calculate geometric means you might want to test to ensure that the input was numeric using the `!is.numeric` function:

```
geometric <- function(x){
if(!is.numeric(x)) stop ("Input must be numeric")
exp(mean(log(x))) }
```

Here is what happens when you try to work out the geometric mean of character data:

```
geometric(c("a","b","c"))
```

```
Error in geometric(c("a", "b", "c")) : Input must be numeric
```

You might also want to check that there are no zeros or negative numbers in the input, because it would make no sense to try to calculate a geometric mean of such data:

```
geometric <- function(x){
if(!is.numeric(x)) stop ("Input must be numeric")
if(min(x)<=0) stop ("Input must be greater than zero")
exp(mean(log(x))) }
```

Testing this:

```
geometric(c(2,3,0,4))
```

```
Error in geometric(c(2, 3, 0, 4)) : Input must be greater than zero
```

But when the data are OK there will be no messages, just the numeric answer:

```
geometric(c(10,1000,10,1,1))
```

```
[1]   10
```

When vectors are created by calculation from other vectors, the new vector will be as long as the longest vector used in the calculation and the shorter variable will be recycled as necessary: here A is of length 10 and B is of length 3:

```
A <- 1:10
B <- c(2,4,8)
A * B
```

```
[1]  2  8  24  8  20  48  14  32  72  20
Warning message: longer object length is not a multiple of shorter
object length in: A * B
```

The vector B is recycled three times in full and a warning message in printed to indicate that the length of the longer vector (A) is not a multiple of the shorter vector (B).

2.5   Missing values, infinity and things that are not numbers

Calculations can lead to answers that are plus infinity, represented in R by `Inf`, or minus infinity, which is represented as `-Inf`:

```
3/0
```

```
[1]   Inf
```

```
-12/0
```

```
[1]   -Inf
```

Calculations involving infinity can be evaluated: for instance,

```
exp(-Inf)
```

```
[1]   0
```

```
0/Inf
```

```
[1]   0
```

```
(0:3)^Inf
```

```
[1]   0   1   Inf   Inf
```

Other calculations, however, lead to quantities that are not numbers. These are represented in R by NaN ('not a number'). Here are some of the classic cases:

```
0/0
```

```
[1]   NaN
```

```
Inf-Inf
```

```
[1]   NaN
```

```
Inf/Inf
```

```
[1]   NaN
```

You need to understand clearly the distinction between NaN and NA (this stands for 'not available' and is the missing-value symbol in R; see below). The function is.nan is provided to check specifically for NaN, and is.na also returns TRUE for NaN. Coercing NaN to logical or integer type gives an NA of the appropriate type. There are built-in tests to check whether a number is finite or infinite:

```
is.finite(10)
```

```
[1]   TRUE
```

```
is.infinite(10)
```

```
[1]   FALSE
```

```
is.infinite(Inf)
```

```
[1]   TRUE
```

2.5.1   Missing values: NA

Missing values in dataframes are a real source of irritation, because they affect the way that model-fitting functions operate and they can greatly reduce the power of the modelling that we would like to do.

You may want to discover which values in a vector are missing. Here is a simple case:

```
y <- c(4,NA,7)
```

The missing value question should evaluate to FALSE TRUE FALSE. There are two ways of looking for missing values that you might think should work, but do not. These involve treating NA as if it was a piece of text and using double equals (==) to test for it. So this does not work:

```
y == NA

[1] NA NA NA
```

because it turns *all* the values into NA (definitively not what you intended). This does not work either:

```
y == "NA"

[1] FALSE NA FALSE
```

It correctly reports that the numbers are not character strings, but it returns NA for the missing value itself, rather than TRUE as required. This is how you do it properly:

```
is.na(y)

[1] FALSE TRUE FALSE
```

To produce a vector with the NA stripped out, use subscripts with the not ! operator like this:

```
y[! is.na(y)]

[1] 4 7
```

This syntax is useful in editing out rows containing missing values from large dataframes. Here is a very simple example of a dataframe with four rows and four columns:

```
y1 <- c(1,2,3,NA)
y2 <- c(5,6,NA,8)
y3 <- c(9,NA,11,12)
y4 <- c(NA,14,15,16)
full.frame <- data.frame(y1,y2,y3,y4)

reduced.frame <- full.frame[!is.na(full.frame$y1),]
```

so the new reduced.frame will have fewer rows than full.frame when the variable in full.frame called full.frame$y1 contains one or more missing values.

```
reduced.frame
```

```
  y1 y2  y3  y4
1  1  5   9  NA
2  2  6  NA  14
3  3 NA  11  15
```

Some functions do not work with their default settings when there are missing values in the data, and mean is a classic example of this:

```
x <- c(1:8,NA)
mean(x)

[1]   NA
```

In order to calculate the mean of the non-missing values, you need to specify that the NA are to be removed, using the na.rm=TRUE argument:

```
mean(x,na.rm=T)
```

```
[1]   4.5
```

Here is an example where we want to find the locations (7 and 8) of missing values within a vector called vmv:

```
vmv <- c(1:6,NA,NA,9:12)
vmv
```

```
[1]   1   2   3   4   5   6   NA   NA   9   10   11   12
```

Making an index of the missing values in an array could use the seq function, like this:

```
seq(along=vmv)[is.na(vmv)]
```

```
[1]   7   8
```

However, the result is achieved more simply using the which function like this:

```
which(is.na(vmv))
```

```
[1]   7   8
```

If the missing values are genuine counts of zero, you might want to edit the NA to 0. Use the is.na function to generate subscripts for this:

```
vmv[is.na(vmv)] <- 0
vmv
```

```
[1]   1   2   3   4   5   6   0   0   9   10   11   12
```

Or use the ifelse function like this:

```
vmv <- c(1:6,NA,NA,9:12)
ifelse(is.na(vmv),0,vmv)
```

```
[1]   1   2   3   4   5   6   0   0   9   10   11   12
```

Be very careful when doing this, because most missing values are not genuine zeros.

2.6   Vectors and subscripts

A vector is a variable with one or more values of the same type. For instance, the numbers of peas in six pods were 4, 7, 6, 5, 6 and 7. The vector called peas is one object of length = 6. In this case, the class of the object is numeric. The easiest way to create a vector in R is to concatenate (link together) the six values using the concatenate function, c, like this:

```
peas <- c(4, 7, 6, 5, 6, 7)
```

We can ask all sorts of questions about the vector called peas. For instance, what type of vector is it?

```
class(peas)
```

```
[1]   "numeric"
```

How big is the vector?

```
length(peas)
```

```
[1] 6
```

The great advantage of a vector-based language is that it is very simple to ask quite involved questions that involve all of the values in the vector. These vector functions are often self-explanatory:

```
mean(peas)
```

```
[1] 5.833333
```

```
max(peas)
```

```
[1] 7
```

```
min(peas)
```

```
[1] 4
```

Others might be more opaque:

```
quantile(peas)
```

```
  0%   25%   50%   75% 100%
4.00  5.25  6.00  6.75  7.00
```

Another way to create a vector is to input data from the keyboard using the function called `scan`:

```
peas <- scan()
```

The prompt appears `1:` which means type in the first number of peas (4) then press the return key, then the prompt `2:` appears (you type in 7) and so on. When you have typed in all six values, and the prompt `7:` has appeared, you just press the return key to tell R that the vector is now complete. R replies by telling you how many items it has read:

```
1: 4
2: 7
3: 6
4: 5
5: 6
6: 7
7:
```

```
Read 6 items
```

For more realistic applications, the usual way of creating vectors is to read the data from a pre-prepared computer file (as described in Chapter 3).

2.6.1 Extracting elements of a vector using subscripts

You will often want to use some but not all of the contents of a vector. To do this, you need to master the use of subscripts (or indices as they are also known). In R, subscripts involve the use of square brackets []. Our vector called `peas` shows the numbers of peas in six pods:

```
peas
```

```
[1] 4 7 6 5 6 7
```

The first element of peas is 4, the second 7, and so on. The elements are indexed left to right, 1 to 6. It could not be more straightforward. If we want to extract the fourth element of peas (which you can see is a 5) then this is what we do:

```
peas[4]
```

```
[1]  5
```

If we want to extract several values (say the 2nd, 3rd and 6th) we use a vector to specify the pods we want as subscripts, either in two stages like this:

```
pods <- c(2,3,6)
peas[pods]
```

```
[1]  7 6 7
```

or in a single step, like this:

```
peas[c(2,3,6)]
```

```
[1]  7 6 7
```

You can drop values from a vector by using negative subscripts. Here are all but the first values of peas:

```
peas[-1]
```

```
[1]  7 6 5 6 7
```

Here are all but the last (note the use of the length function to decide what is last):

```
peas[-length(peas)]
```

```
[1]  4 7 6 5 6
```

We can use these ideas to write a function called trim to remove (say) the largest two and the smallest two values from a vector called x. First we have to sort the vector, then remove the smallest two values (these will have subscripts 1 and 2), then remove the largest two values (which will have subscripts length(x) and length(x)-1):

```
trim <- function(x) sort(x)[-c(1,2,length(x)-1,length(x))]
```

We can use trim on the vector called peas, expecting to get 6 and 6 as the result:

```
trim(peas)
```

```
[1]  6 6
```

Finally, we can use sequences of numbers to extract values from a vector. Here are the first three values of peas:

```
peas[1:3]
```

```
[1]  4 7 6
```

Here are the even-numbered values of peas:

```
peas[seq(2,length(peas),2)]
```

```
[1]  7 5 7
```

or alternatively:

```
peas[1:length(peas) %% 2 == 0]

[1] 7 5 7
```

using the modulo function `%%` on the sequence 1 to 6 to extract the even numbers 2, 4 and 6. Note that vectors in R could have length 0, and this could be useful in writing functions:

```
y <- 4.3
z <- y[-1]
length(z)

[1]   0
```

2.6.2   Classes of vector

The vector called `peas` contained numbers: in the jargon, it is of class `numeric`. R allows vectors of six types, so long as all of the elements in one vector belong to the same class. The classes are logical, integer, real, complex, string (or character) or raw. You will use numeric, logical and character variables all the time. Engineers and mathematicians will use complex numbers. But you could go a whole career without ever needing to use integer or raw.

2.6.3   Naming elements within vectors

It is often useful to have the values in a vector labelled in some way. For instance, if our data are counts of 0, 1, 2, ... occurrences in a vector called `counts`,

```
(counts <- c(25,12,7,4,6,2,1,0,2))

[1]   25  12  7  4  6  2  1  0  2
```

so that there were 25 zeros, 12 ones and so on, it would be useful to name each of the counts with the relevant number 0 to 8:

```
names(counts) <- 0:8
```

Now when we inspect the vector called counts we see both the names and the frequencies:

```
counts

0   1   2   3   4   5   6   7   8
25  12  7   4   6   2   1   0   2
```

If you have computed a table of counts, and you want to *remove* the names, then use the `as.vector` function like this:

```
(st <- table(rpois(2000,2.3)))

0    1    2    3    4    5    6    7    8   9
205  455  510  431  233  102  43   13   7   1

as.vector(st)

[1]   205  455  510  431  233  102  43  13  7  1
```

2.6.4 Working with logical subscripts

Take the example of a vector containing the 11 numbers 0 to 10:

```
x <- 0:10
```

There are two quite different kinds of things we might want to do with this. We might want to *add up* the values of the elements:

```
sum(x)
```

```
[1]   55
```

Alternatively, we might want to *count* the elements that passed some logical criterion. Suppose we wanted to know how many of the values were less than 5:

```
sum(x<5)
```

```
[1]   5
```

You see the distinction. We use the vector function `sum` in both cases. But `sum(x)` adds up the values of the *x*s and `sum(x<5)` counts up the number of cases that pass the logical condition '*x* is less than 5'. This works because of *coercion* (p. 30). Logical TRUE has been coerced to numeric 1 and logical FALSE has been coerced to numeric 0.

That is all well and good, but how do you add up the values of just some of the elements of *x*? We specify a logical condition, but we do not want to count the number of cases that pass the condition, we want to add up all the values of the cases that pass. This is the final piece of the jigsaw, and involves the use of *logical subscripts*. Note that when we counted the number of cases, the counting was applied to the entire vector, using `sum(x<5)`. To find the sum of the values of *x* that are less than 5, we write:

```
sum(x[x<5])
```

```
[1]   10
```

Let us look at this in more detail. The logical condition `x<5` is either true or false:

```
x<5
```

```
[1] TRUE TRUE TRUE TRUE TRUE FALSE FALSE FALSE FALSE FALSE FALSE
```

You can imagine false as being numeric 0 and true as being numeric 1. Then the vector of subscripts `[x<5]` is five 1s followed by six 0s:

```
1*(x<5)
```

```
[1]  1  1  1  1  1  0  0  0  0  0  0
```

Now imagine multiplying the values of *x* by the values of the logical vector

```
x*(x<5)
```

```
[1]  0  1  2  3  4  0  0  0  0  0  0
```

When the function `sum` is applied, it gives us the answer we want: the sum of the values of the numbers $0 + 1 + 2 + 3 + 4 = 10$.

```
sum(x*(x<5))
```

```
[1]   10
```

This produces the same answer as `sum(x[x<5])`, but is rather less elegant.

Suppose we want to work out the sum of the three largest values in a vector. There are two steps: first `sort` the vector into descending order; then add up the values of the first three elements of the reverse-sorted array. Let us do this in stages. First, the values of y:

```
y <- c(8,3,5,7,6,6,8,9,2,3,9,4,10,4,11)
```

Now if you apply `sort` to this, the numbers will be in ascending sequence, and this makes life slightly harder for the present problem:

```
sort(y)
```

```
[1]    2   3   3   4   4   5   6   6   7   8   8   9
[13]   9  10  11
```

We can use the reverse function, `rev` like this (use the Up arrow key to save typing):

```
rev(sort(y))
```

```
[1]   11  10   9   9   8   8   7   6   6   5   4   4
[13]   3   3   2
```

So the answer to our problem is $11 + 10 + 9 = 30$. But how to compute this? A range of subscripts is simply a series generated using the colon operator. We want the subscripts 1 to 3, so this is:

```
rev(sort(y))[1:3]
```

```
[1]   11  10   9
```

So the answer to the exercise is just:

```
sum(rev(sort(y))[1:3])
```

```
[1]   30
```

Note that we have not changed the vector y in any way, nor have we created any new space-consuming vectors during intermediate computational steps.

You will often want to find out which value in a vector is the maximum or the minimum. This is a question about indices, and the answer you want is an integer indicating which element of the vector contains the maximum (or minimum) out of all the values in that vector. Here is the vector:

```
x <- c(2,3,4,1,5,8,2,3,7,5,7)
```

So the answers we want are 6 (the maximum) and 4 (the minimum). The slow way to do it is like this:

```
which(x == max(x))
```

```
[1] 6
```

```
which(x == min(x))
```

```
[1] 4
```

Better, however, to use the much quicker built-in functions `which.max` or `which.min` like this:

```
which.max(x)
```

```
[1] 6
```

```
which.min(x)
```

```
[1] 4
```

2.7 Vector functions

One of R's great strengths is its ability to evaluate functions over entire vectors, thereby avoiding the need for loops and subscripts. The most important vector functions are listed in Table 2.4. Here is a numeric vector:

```
y <- c(8,3,5,7,6,6,8,9,2,3,9,4,10,4,11)
```

Some vector functions produce a single number:

```
mean(y)
```

```
[1] 6.333333
```

Table 2.4. Vector functions used in R.

Operation	Meaning
`max(x)`	maximum value in x
`min(x)`	minimum value in x
`sum(x)`	total of all the values in x
`mean(x)`	arithmetic average of the values in x
`median(x)`	median value in x
`range(x)`	vector of min(x) and max(x)
`var(x)`	sample variance of x
`cor(x,y)`	correlation between vectors x and y
`sort(x)`	a sorted version of x
`rank(x)`	vector of the ranks of the values in x
`order(x)`	an integer vector containing the permutation to sort x into ascending order
`quantile(x)`	vector containing the minimum, lower quartile, median, upper quartile, and maximum of x
`cumsum(x)`	vector containing the sum of all of the elements up to that point
`cumprod(x)`	vector containing the product of all of the elements up to that point
`cummax(x)`	vector of non-decreasing numbers which are the cumulative maxima of the values in x up to that point
`cummin(x)`	vector of non-increasing numbers which are the cumulative minima of the values in x up to that point
`pmax(x,y,z)`	vector, of length equal to the longest of x, y or z, containing the maximum of x, y or z for the ith position in each
`pmin(x,y,z)`	vector, of length equal to the longest of x, y or z, containing the minimum of x, y or z for the ith position in each
`colMeans(x)`	column means of dataframe or matrix x
`colSums(x)`	column totals of dataframe or matrix x
`rowMeans(x)`	row means of dataframe or matrix x
`rowSums(x)`	row totals of dataframe or matrix x

Others produce two numbers:

```
range(y)
```

```
[1]  2 11
```

here showing that the minimum was 2 and the maximum was 11. Other functions produce several numbers:

```
fivenum(y)
```

```
[1]  2.0 4.0 6.0 8.5 11.0
```

This is Tukey's famous five-number summary: the minimum, the lower hinge, the median, the upper hinge and the maximum (the hinges are explained on p. 346).

Perhaps the single most useful vector function in R is table. You need to see it in action to appreciate just how good it is. Here is a huge vector called counts containing 10 000 random integers from a negative binomial distribution (counts of fungal lesions on 10 000 individual leaves, for instance):

```
counts <- rnbinom(10000,mu=0.92,size=1.1)
```

Here is a look at the first 30 values in counts:

```
counts[1:30]
```

```
 [1]  3 1 0 0 1 0 0 0 0 1 1 0 0 2 0 1 3 1 0 1 0 1 1 0 0 2 1 4 0 1
```

The question is this: how many zeros are there in the whole vector of 10 000 numbers, how many 1s, and so on right up to the largest value within counts? A formidable task for you or me, but for R it is just:

```
table(counts)
```

```
counts
   0    1    2    3    4    5    6    7    8   9   10   11   13
5039 2574 1240 607 291 141 54 29 11 9   3    1    1
```

There were 5039 zeros, 2574 ones, and so on up the largest counts (there was one 11 and one 13 in this realization; you will have obtained different random numbers on your computer).

2.7.1 Obtaining tables of means using `tapply`

One of the most important functions in all of R is tapply. It does not sound like much from the name, but you will use it time and again for calculating means, variances, sample sizes, minima and maxima. With weather data, for instance, we might want the 12 monthly mean temperatures rather than the whole-year average. We have a response variable, temperature, and a categorical explanatory variable, month:

```
data<-read.table("c:\\temp\\temperatures.txt",header=T)
attach(data)
names(data)
```

```
[1] "temperature" "lower" "rain" "month" "yr"
```

The function that we want to apply is mean. All we do is invoke the tapply function with three arguments: the response variable, the categorical explanatory variable and the name of the function that we want to apply:

```
tapply(temperature,month,mean)
```

```
       1          2          3          4          5          6
```

```
7.930051 8.671136 11.200508 13.813708 17.880847 20.306151
       7          8          9         10         11         12
22.673854 23.104924 19.344211 15.125976 10.720702 8.299830
```

It is easy to apply other functions in the same way: here are the monthly variances

```
tapply(temperature,month,var)
```

and the monthly minima

```
tapply(temperature,month,min)
```

If R does not have a built in function to do what you want (Table 2.4), then you can easily write your own. Here, for instance, is a function to calculate the standard error of each mean (these are called anonymous functions in R, because they are unnamed):

```
tapply(temperature,month,function(x) sqrt(var(x)/length(x)))
```

```
        1         2         3         4         5         6
0.1401489 0.1414445 0.1358934 0.1476242 0.1673197 0.1596439
```

```
        7         8         9        10        11        12
0.1539661 0.1516091 0.1309294 0.1155612 0.1291703 0.1398438
```

The `tapply` function is very flexible. It can produce multi-dimensional tables simply by replacing the one categorical variable (`month`) by a `list` of categorical variables. Here are the monthly means calculated separately for each year, as specified by `list(yr,month)`. The variable you name first in the list (`yr`) will appear as the row of the results table and the second will appear as the columns (`month`):

```
tapply(temperature,list(yr,month),mean)[,1:6]
```

```
            1         2         3        4        5        6
1987  3.170968  6.871429  8.132258 14.92667 15.60645 17.73667
1988  8.048387  8.248276  9.959375 12.74483 17.31935 18.71667
1989  8.841935  9.482143 11.919355 11.09333 20.40323 21.23667
1990  9.445161 11.028571 12.487097 13.80000 20.16129 18.51667
1991  6.980645  4.817857 12.022581 13.14333 15.58065 16.88000
1992  6.964516  8.686207 11.477419 13.35000 20.45806 22.21667
1993 10.119355  6.985714 11.209677 14.17000 17.79355 21.10000
1994  8.825806  7.217857 11.806452 12.61667 16.23226 20.86000
1995  8.309677 10.439286 10.667742 14.79667 18.74063 19.94483
1996  7.019355  6.065517  8.487097 13.99667 14.38710 21.93667
1997  4.932258 10.178571 13.370968 15.00667 18.17419 19.93000
1998  8.759375 11.242857 11.719355 12.55333 19.43226 19.35000
1999  9.523333  8.485714 11.790323 14.65000 18.94839 20.00667
2000  8.229032 10.324138 11.900000 12.59000 18.22581 20.63333
2001  7.067742  9.121429  9.012903 12.65667 18.96452 20.52667
2002  9.067742 11.396429 12.319355 15.68667 16.81290 19.67667
2003  8.012903  8.171429 13.425806 15.69000 17.36452 22.80000
2004  8.261290  8.993103 10.354839 15.17000 17.98065 21.73667
2005  9.116129  7.032143 10.787097 13.78333 17.12258 22.00000
```

The subscripts `[,1:6]` simply restrict the output to the first six months. You can see at once that January (month 1) 1993 was exceptionally warm and January 1987 exceptionally cold.

There is just one thing about `tapply` that might confuse you. If you try to apply a function that has built-in protection against missing values, then `tapply` may not do what you want, producing NA instead of the numerical answer. This is most likely to happen with the mean function because its default is to produce NA when there are one or more missing values. The remedy is to provide an extra argument to `tapply`, specifying that you want to see the average of the non-missing values. Use `na.rm=TRUE` to remove the missing values like this:

```
tapply(temperature,yr,mean,na.rm=TRUE)
```

You might want to trim some of the extreme values before calculating the mean (the arithmetic mean is famously sensitive to outliers). The `trim` option allows you to specify the fraction of the data (between 0 and 0.5) that you want to be omitted from the left- and right-hand tails of the sorted vector of values before computing the mean of the central values:

```
tapply(temperature,yr,mean,trim=0.2)
```

```
    1987     1988     1989     1990     1991     1992     1993
13.46000 13.74500 14.99726 15.16301 13.92237 14.32091 14.28000
```

2.7.2 The aggregate function for grouped summary statistics

Suppose that we have two response variables (y and z) and two explanatory variables (x and w) that we might want to use to summarize functions like mean or variance of y and/or z. The `aggregate` function has a formula method which allows elegant summaries of four kinds:

one to one	`aggregate(y ~ x, mean)`
one to many	`aggregate(y ~ x + w, mean)`
many to one	`aggregate(cbind(y,z) ~ x, mean)`
many to many	`aggregate(cbind(y,z) ~ x + w, mean)`

This is very useful for removing pseudoreplication from dataframes. Here is an example using a dataframe with two continuous variables (`Growth.rate` and pH) and three categorical explanatory variables (`Water`, `Detergent` and `Daphnia`):

```
data<-read.table("c:\\temp\\pHDaphnia.txt",header=T)
names(data)
```

```
[1] "Growth.rate" "Water" "Detergent" "Daphnia" "pH"
```

Here is one-to-one use of `aggregate` to find mean growth rate in the two water samples:

```
aggregate(Growth.rate~Water,data,mean)
```

```
  Water Growth.rate
1  Tyne    3.685862
2  Wear    4.017948
```

Here is a one-to-many use to look at the interaction between `Water` and `Detergent`:

```
aggregate(Growth.rate~Water+Detergent,data,mean)
```

```
  Water Detergent Growth.rate
1  Tyne    BrandA     3.661807
2  Wear    BrandA     4.107857
3  Tyne    BrandB     3.911116
4  Wear    BrandB     4.108972
5  Tyne    BrandC     3.814321
6  Wear    BrandC     4.094704
7  Tyne    BrandD     3.356203
8  Wear    BrandD     3.760259
```

Finally, here is a many-to-many use to find mean pH as well as mean Growth.rate for the interaction between Water and Detergent:

```
aggregate(cbind(pH,Growth.rate)~Water+Detergent,data,mean)
```

```
  Water Detergent       pH  Growth.rate
1  Tyne    BrandA 4.883908     3.661807
2  Wear    BrandA 5.054835     4.107857
3  Tyne    BrandB 5.043797     3.911116
4  Wear    BrandB 4.892346     4.108972
5  Tyne    BrandC 4.847069     3.814321
6  Wear    BrandC 4.912128     4.094704
7  Tyne    BrandD 4.809144     3.356203
8  Wear    BrandD 5.097039     3.760259
```

2.7.3 Parallel minima and maxima: pmin and pmax

Here are three vectors of the same length, x, y and z. The parallel minimum function, pmin, finds the minimum from any one of the three variables for each subscript, and produces a *vector* as its result (of length equal to the longest of x, y, or z):

```
x
```

```
[1] 0.99822644 0.98204599 0.20206455 0.65995552 0.93456667 0.18836278
```

```
y
```

```
[1] 0.51827913 0.30125005 0.41676059 0.53641449 0.07878714 0.49959328
```

```
z
```

```
[1] 0.26591817 0.13271847 0.44062782 0.65120395 0.03183403 0.36938092
```

```
pmin(x,y,z)
```

```
[1] 0.26591817 0.13271847 0.20206455 0.53641449 0.03183403 0.18836278
```

Thus the first and second minima came from z, the third from x, the fourth from y, the fifth from z, and the sixth from x. The functions min and max produce *scalar* results, not vectors.

2.7.4 Summary information from vectors by groups

The vector function `tapply` is one of the most important and useful vector functions to master. The 't' stands for 'table' and the idea is to apply a function to produce a table from the values in the vector, based on one or more grouping variables (often the grouping is by factor levels). This sounds much more complicated than it really is:

```
data <- read.table("c:\\temp\\daphnia.txt",header=T)
attach(data)
names(data)

[1]  "Growth.rate"  "Water"  "Detergent"  "Daphnia"
```

The response variable is `Growth.rate` and the other three variables are factors (the analysis is on p. 528). Suppose we want the mean growth rate for each detergent:

```
tapply(Growth.rate,Detergent,mean)

BrandA    BrandB    BrandC    BrandD
3.88      4.01      3.95      3.56
```

This produces a table with four entries, one for each level of the factor called `Detergent`. To produce a two-dimensional table we put the two grouping variables in a list. Here we calculate the median growth rate for water type and daphnia clone:

```
tapply(Growth.rate,list(Water,Daphnia),median)

        Clone1    Clone2    Clone3
Tyne    2.87      3.91      4.62
Wear    2.59      5.53      4.30
```

The first variable in the list creates the rows of the table and the second the columns. More detail on the `tapply` function is given in Chapter 6 (p. 245).

2.7.5 Addresses within vectors

There is an important function called `which` for finding addresses within vectors. The vector y looks like this:

```
y <- c(8,3,5,7,6,6,8,9,2,3,9,4,10,4,11)
```

Suppose we wanted to know which elements of y contained values bigger than 5. We type:

```
which(y>5)

[1]  1   4   5   6   7   8   11   13   15
```

Notice that the answer to this enquiry is *a set of subscripts*. We do not use subscripts inside the `which` function itself. The function is applied to the whole array. To see the values of y that are larger than 5, we just type:

```
y[y>5]

[1]  8   7   6   6   8   9   9   10   11
```

Note that this is a shorter vector than y itself, because values of 5 or less have been left out:

```
length(y)
```

```
[1]    15
```

```
length(y[y>5])
```

```
[1]    9
```

2.7.6 Finding closest values

Finding the value in a vector that is closest to a specified value is straightforward using which. The vector xv contains 1000 random numbers from a normal distribution with mean $= 100$ and standard deviation $= 10$:

```
xv <- rnorm(1000,100,10)
```

Here, we want to find the value of xv that is closest to 108.0. The logic is to work out the difference between 108 and each of the 1000 random numbers, then find which of these differences is the smallest. This is what the R code looks like:

```
which(abs(xv-108)==min(abs(xv-108)))
```

```
[1]    332
```

The closest value to 108.0 is in location 332 within xv. But just how close to 108.0 is this 332nd value? We use 332 as a subscript on xv to find this out:

```
xv[332]
```

```
[1]    108.0076
```

Now we can write a function to return the closest value to a specified value (sv) in any vector (xv):

```
closest <- function(xv,sv){
xv[which(abs(xv-sv)==min(abs(xv-sv)))]  }
```

and run it like this:

```
closest(xv,108)
```

```
[1]    108.0076
```

2.7.7 Sorting, ranking and ordering

These three related concepts are important, and one of them (order) is difficult to understand on first acquaintance. Let us take a simple example:

```
houses <- read.table("c:\\temp\\houses.txt",header=T)
attach(houses)
names(houses)
```

```
[1]    "Location"    "Price"
```

We apply the three different functions to the vector called `Price`:

```
ranks <- rank(Price)
sorted <- sort(Price)
ordered <- order(Price)
```

Then we make a dataframe out of the four vectors like this:

```
view <- data.frame(Price,ranks,sorted,ordered)
view
```

	Price	ranks	sorted	ordered
1	325	12.0	95	9
2	201	10.0	101	6
3	157	5.0	117	10
4	162	6.0	121	12
5	164	7.0	157	3
6	101	2.0	162	4
7	211	11.0	164	5
8	188	8.5	188	8
9	95	1.0	188	11
10	117	3.0	201	2
11	188	8.5	211	7
12	121	4.0	325	1

Rank

The prices themselves are in no particular sequence. The `ranks` column contains the value that is the rank of the particular data point (value of `Price`), where 1 is assigned to the lowest data point and `length(Price)` – here 12 – is assigned to the highest data point. So the first element, a price of 325, happens to be the highest value in `Price`. You should check that there are 11 values smaller than 325 in the vector called `Price`. Fractional ranks indicate ties. There are two 188s in Price and their ranks are 8 and 9. Because they are tied, each gets the average of their two ranks $(8 + 9)/2 = 8.5$. The lowest price is 95, indicated by a rank of 1.

Sort

The sorted vector is very straightforward. It contains the values of `Price` sorted into ascending order. If you want to sort into descending order, use the reverse order function `rev` like this:

```
y <- rev(sort(x))
```

Note that `sort` *is potentially very dangerous*, because it uncouples values that might need to be in the same row of the dataframe (e.g. because they are the explanatory variables associated with a particular value of the response variable). It is bad practice, therefore, to write `x <- sort(x)`, not least because there is no 'unsort' function.

Order

This is the most important of the three functions, and much the hardest to understand on first acquaintance. The numbers in this column are subscripts between 1 and 12. The order function returns an integer vector *containing the permutation that will sort the input into ascending order.* You will need to think about this one. The lowest value of `Price` is 95. Look at the dataframe and ask yourself what is the subscript in the original vector called `Price` where 95 occurred. Scanning down the column, you find it in row number 9. This is the first value in ordered, `ordered[1]`. Where is the next smallest value (101) to be found within `Price`? It is in position 6, so this is `ordered[2]`. The third smallest value of `Price` (117) is in position 10, so this is `ordered[3]`. And so on.

This function is particularly useful in sorting dataframes, as explained on p. 166. Using `order` with subscripts is a much safer option than using `sort`, because with `sort` the values of the response variable and the explanatory variables could be uncoupled with potentially disastrous results if this is not realized at the time that modelling was carried out. The beauty of `order` is that we can use `order(Price)` as a subscript for `Location` to obtain the price-ranked list of locations:

```
Location[order(Price)]
```

```
[1]    Reading      Staines      Winkfield    Newbury
[5]    Bracknell    Camberley    Bagshot      Maidenhead
[9]    Warfield     Sunninghill  Windsor      Ascot
```

When you see it used like this, you can see exactly why the function is called *order*. If you want to reverse the order, just use the `rev` function like this:

```
Location[rev(order(Price))]
```

```
[1]    Ascot        Windsor      Sunninghill  Warfield
[5]    Maidenhead   Bagshot      Camberley    Bracknell
[9]    Newbury      Winkfield    Staines      Reading
```

Make sure you understand why some of the brackets are round and some are square.

2.7.8 Understanding the difference between `unique` and `duplicated`

The difference is best seen with a simple example. Here is a vector of names:

```
names <- c("Williams","Jones","Smith","Williams","Jones","Williams")
```

We can see how many times each name appears using `table`:

```
table(names)
```

```
names
 Jones  Smith  Williams
    2      1        3
```

It is clear that the vector contains just three different names. The function called `unique` extracts these three unique names, creating a vector of length 3, unsorted, in the order in which the names are encountered in the vector:

```
unique(names)
```

```
[1]  "Williams"  "Jones"  "Smith"
```

In contrast, the function called `duplicated` produces a vector, of the same length as the vector of names, containing the logical values either FALSE or TRUE, depending upon whether or not that name has appeared already (reading from the left). You need to see this in action to understand what is happening, and why it might be useful:

```
duplicated(names)
```

```
[1] FALSE FALSE FALSE TRUE TRUE TRUE
```

The first three names are not duplicated (FALSE), but the last three are all duplicated (TRUE). We can mimic the unique function by using this vector as subscripts like this:

```
names[!duplicated(names)]
```

```
[1] "Williams" "Jones" "Smith"
```

Note the use of the NOT operator (!) in front of the duplicated function. There you have it: if you want a shortened vector, containing only the unique values in names, then use unique, but if you want a vector of the same length as names then use duplicated. You might use this to extract values from a different vector (salaries, for instance) if you wanted the mean salary, ignoring the repeats:

```
salary <- c(42,42,48,42,42,42)
mean(salary)
```

```
[1] 43
```

```
salary[!duplicated(names)]
```

```
[1] 42 42 48
```

```
mean(salary[!duplicated(names)])
```

```
[1] 44
```

Note that this is not the same answer as would be obtained by omitting the duplicate salaries, because two of the people (Jones and Williams) had the same salary (42). Here is the wrong answer:

```
mean(salary[!duplicated(salary)])
```

```
[1] 45
```

2.7.9   Looking for runs of numbers within vectors

The function called rle, which stands for 'run length encoding' is most easily understood with an example. Here is a vector of 150 random numbers from a Poisson distribution with mean 0.7:

```
(poisson <- rpois(150,0.7))
```

```
[1]   1 1 0 0 2 1 0 1 0 0 1 1 1 0 1 1 2 1 1 0 1 0 2 1 1 2 0 2 0 1 0 0 0 2 0
[36]   1 0 4 0 0 1 0 1 0 1 0 2 1 1 1 0 1 0 1 0 0 0 0 0 0 0 2 0 0 0 0 1 0 0 0
[71]   2 1 1 1 1 0 1 0 1 0 0 1 1 0 1 0 2 1 1 2 0 1 0 1 0 0 0 1 1 0 1 2 2 0 1
[106]  0 0 0 0 0 0 1 0 0 2 1 2 0 2 0 2 2 1 1 0 2 0 1 1 2 2 2 1 1 1 1 0 0 0 1
[141]  0 2 1 4 0 0 2 1 0 1
```

We can do our own run length encoding on the vector by eye: there is a run of two 1s, then a run of two 0s, then a single 2, then a single 1, then a single 0, and so on. So the run lengths are 2, 2, 1, 1, 1, 1, ... The values associated with these runs were 1, 0, 2, 1, 0, 1, ... Here is the output from rle:

```
rle(poisson)
```

```
Run Length Encoding
```

```
lengths: int [1:93] 2 2 1 2 1 1 2 3 1 2 1 ...
values : num [1:93] 1 0 2 1 0 1 0 1 2 1 ...
```

The object produced by `rle` is a list of two vectors: the lengths of the runs and the values that did the running. To find the longest run, and the value associated with that longest run, we use the indexed lists like this:

```
max(rle(poisson)[[1]])
```

```
[1]   7
```

So the longest run in this vector of numbers was 7. But 7 of what? We use `which` to find the location of the 7 in lengths, then apply this index to values to find the answer:

```
which(rle(poisson)[[1]]==7)
```

```
[1]    55
```

```
rle(poisson)[[2]][55]
```

```
[1]   0
```

So, not surprisingly given that the mean was just 0.7, the longest run was of zeros.

Here is a function to return the length of the run and its value for any vector:

```
run.and.value <- function (x) {
       a <- max(rle(poisson)[[1]])
       b <- rle(poisson)[[2]][which(rle(poisson)[[1]] == a)]
cat("length = ",a," value = ",b, "\n")}
```

Testing the function on the vector of 150 Poisson data gives:

```
run.and.value(poisson)
```

```
length = 7   value = 0
```

It is sometimes of interest to know the number of runs in a given vector (for instance, the lower the number of runs, the more aggregated the numbers; and the greater the number of runs, the more regularly spaced out). We use the `length` function for this:

```
length(rle(poisson)[[2]])
```

```
[1]    93
```

indicating that the 150 values were arranged in 93 runs (this is an intermediate value, characteristic of a random pattern). The value 93 appears in square brackets [1:93] in the output of the run length encoding function.

In a different example, suppose we had n_1 values of 1 representing 'present' and n_2 values of 0 representing 'absent'; then the minimum number of runs would be 2 (a solid block of 1s then a sold block of 0s). The maximum number of runs would be $2n + 1$ if they alternated (until the smaller number $n = \text{min}(n1,n2)$ ran out). Here is a simple **runs test** based on 1000 randomizations of 25 ones and 30 zeros:

```
n1 <- 25
n2 <- 30
y <- c(rep(1,n1),rep(0,n2))
len <- numeric(10000)
for (i in 1:10000) len[i] <- length(rle(sample(y))[[2]])
quantile(len,c(0.025,0.975))
```

```
2.5%    97.5%
  21       35
```

Thus, for these data ($n_1 = 25$ and $n_2 = 30$) an aggregated pattern would score 21 or fewer runs, and a regular pattern would score 35 or more runs. Any scores between 21 and 35 fall within the realm of random patterns.

2.7.10 Sets: `union`, `intersect` and `setdiff`

There are three essential functions for manipulating sets. The principles are easy to see if we work with an example of two sets:

```
setA <- c("a", "b", "c", "d", "e")
setB <- c("d", "e", "f", "g")
```

Make a mental note of what the two sets have in common, and what is unique to each.

The **union** of two sets is everything in the two sets taken together, but counting elements only once that are common to both sets:

```
union(setA,setB)
```

```
[1]   "a"   "b"   "c"   "d"   "e"   "f"   "g"
```

The **intersection** of two sets is the material that they have in common:

```
intersect(setA,setB)
```

```
[1]   "d"   "e"
```

Note, however, that the **difference** between two sets is order-dependent. It is the material that *is* in the first named set, that *is not* in the second named set. Thus `setdiff(A,B)` gives a different answer than `setdiff(B,A)`. For our example:

```
setdiff(setA,setB)
```

```
[1]   "a"   "b"   "c"
```

```
setdiff(setB,setA)
```

```
[1]   "f"   "g"
```

Thus, it should be the case that `setdiff(setA,setB)` plus `intersect(setA,setB)` plus `setdiff(setB,setA)` is the same as the union of the two sets. Let us check:

```
all(c(setdiff(setA,setB),intersect(setA,setB),setdiff(setB,setA))==
       union(setA,setB))
```

```
[1]   TRUE
```

There is also a built-in function `setequal` for testing if two sets are equal:

```
setequal(c(setdiff(setA,setB),intersect(setA,setB),setdiff(setB,setA)),
       union(setA,setB))
```

```
[1]   TRUE
```

You can use %in% for comparing sets. The result is a logical vector whose length matches the vector on the left:

```
setA %in% setB
```

```
[1]   FALSE   FALSE   FALSE   TRUE   TRUE
```

```
setB %in% setA
```

```
[1]   TRUE   TRUE   FALSE   FALSE
```

Using these vectors of logical values as subscripts, we can demonstrate, for instance, that setA[setA %in% setB] is the same as intersect(setA,setB):

```
setA[setA %in% setB]
```

```
[1]   "d"   "e"
```

```
intersect(setA,setB)
```

```
[1]   "d"   "e"
```

2.8   Matrices and arrays

An array is a multi-dimensional object. The dimensions of an array are specified by its dim attribute, which gives the maximal indices in each dimension. So for a three-dimensional array consisting of 24 numbers in a sequence 1:24, with dimensions $2 \times 4 \times 3$, we write:

```
y <- 1:24
dim(y) <- c(2,4,3)
y

, , 1

     [,1]  [,2]  [,3]  [,4]
[1,]    1     3     5     7
[2,]    2     4     6     8

, , 2

     [,1]  [,2]  [,3]  [,4]
[1,]    9    11    13    15
[2,]   10    12    14    16

, , 3

     [,1]  [,2]  [,3]  [,4]
[1,]   17    19    21    23
[2,]   18    20    22    24
```

This produces three two-dimensional tables, because the third dimension is 3. This is what happens when you change the dimensions:

```
dim(y) <- c(3,2,4)
y
```

```
, , 1

     [,1] [,2]
[1,]    1    4
[2,]    2    5
[3,]    3    6

, , 2

     [,1] [,2]
[1,]    7   10
[2,]    8   11
[3,]    9   12

, , 3

     [,1] [,2]
[1,]   13   16
[2,]   14   17
[3,]   15   18

, , 4

     [,1] [,2]
[1,]   19   22
[2,]   20   23
[3,]   21   24
```

Now we have four two-dimensional tables, each of three rows and two columns. Keep looking at these two examples until you are sure that you understand exactly what has happened here.

A matrix is a two-dimensional array containing numbers. A dataframe is a two-dimensional list containing (potentially a mix of) numbers, text or logical variables in different columns. When there are two subscripts [5,3] to an object like a matrix or a dataframe, the first subscript refers to the row number (5 in this example; the rows are defined as **margin** number 1) and the second subscript refers to the column number (3 in this example; the columns are margin number 2). There is an important and powerful convention in R, such that *when a subscript appears as a blank it is understood to mean 'all of'*. Thus:

- [,4] means all rows in column 4 of an object;

- [2,] means all columns in row 2 of an object.

2.8.1 Matrices

There are several ways of making a matrix. You can create one directly like this:

```
X <- matrix(c(1,0,0,0,1,0,0,0,1),nrow=3)
X

     [,1]      [,2]      [,3]
[1,]    1         0         0
[2,]    0         1         0
[3,]    0         0         1
```

where, by default, the numbers are entered column-wise. The class and attributes of X indicate that it is a matrix of three rows and three columns (these are its dim attributes):

```
class(X)
```

```
[1]    "matrix"
```

```
attributes(X)
```

```
$dim
[1]   3   3
```

In the next example, the data in the vector appear row-wise, so we indicate this with byrow=T:

```
vector <- c(1,2,3,4,4,3,2,1)
V <- matrix(vector,byrow=T,nrow=2)
V
```

```
        [,1]       [,2]       [,3]       [,4]
[1,]       1          2          3          4
[2,]       4          3          2          1
```

Another way to convert a vector into a matrix is by providing the vector object with two dimensions (rows and columns) using the dim function like this:

```
dim(vector) <- c(4,2)
```

We can check that vector has now become a matrix:

```
is.matrix(vector)
```

```
[1]   TRUE
```

We need to be careful, however, because we have made no allowance at this stage for the fact that the data were entered row-wise into vector:

```
vector
```

```
        [,1]       [,2]
[1,]       1          4
[2,]       2          3
[3,]       3          2
[4,]       4          1
```

The matrix we want is the transpose, t, of this matrix:

```
(vector <- t(vector))
```

```
        [,1]       [,2]       [,3]       [,4]
[1,]       1          2          3          4
[2,]       4          3          2          1
```

2.8.2   Naming the rows and columns of matrices

At first, matrices have numbers naming their rows and columns (see above). Here is a 4×5 matrix of random integers from a Poisson distribution with mean 1.5:

```
X <- matrix(rpois(20,1.5),nrow=4)
X
```

	[,1]	[,2]	[,3]	[,4]	[,5]
[1,]	1	0	2	5	3
[2,]	1	1	3	1	3
[3,]	3	1	0	2	2
[4,]	1	0	2	1	0

Suppose that the rows refer to four different trials and we want to label the rows 'Trial.1' etc. We employ the function rownames to do this. We could use the paste function (see p. 87) but here we take advantage of the prefix option:

```
rownames(X) <- rownames(X,do.NULL=FALSE,prefix="Trial.")
X
```

	[,1]	[,2]	[,3]	[,4]	[,5]
Trial.1	1	0	2	5	3
Trial.2	1	1	3	1	3
Trial.3	3	1	0	2	2
Trial.4	1	0	2	1	0

For the columns we want to supply a vector of different names for the five drugs involved in the trial, and use this to specify the colnames(X):

```
drug.names <- c("aspirin", "paracetamol", "nurofen", "hedex", "placebo")
colnames(X) <- drug.names
X
```

	aspirin	Paracetamol	nurofen	hedex	placebo
Trial.1	1	0	2	5	3
Trial.2	1	1	3	1	3
Trial.3	3	1	0	2	2
Trial.4	1	0	2	1	0

Alternatively, you can use the dimnames function to give names to the rows and/or columns of a matrix. In this example we want the rows to be unlabelled (NULL) and the column names to be of the form 'drug.1', 'drug.2', etc. The argument to dimnames has to be a list (rows first, columns second, as usual) with the elements of the list of exactly the correct lengths (4 and 5 in this particular case):

```
dimnames(X) <- list(NULL,paste("drug.",1:5,sep=""))
X
```

	drug.1	drug.2	drug.3	drug.4	drug.5
[1,]	1	0	2	5	3
[2,]	1	1	3	1	3
[3,]	3	1	0	2	2
[4,]	1	0	2	1	0

2.8.3 Calculations on rows or columns of the matrix

We could use subscripts to select parts of the matrix, with a blank meaning 'all of the rows' or 'all of the columns'. Here is the mean of the rightmost column (number 5), calculated over all the rows (blank then comma),

```
mean(X[,5])
```

```
[1]    2
```

or the variance of the bottom row, calculated over all of the columns (a blank in the second position),

```
var(X[4,])
```

```
[1]    0.7
```

There are some special functions for calculating summary statistics on matrices:

```
rowSums(X)
```

```
[1]    11   9   8   4
```

```
colSums(X)
```

```
[1]   6   2   7   9   8
```

```
rowMeans(X)
```

```
[1]    2.2   1.8   1.6   0.8
```

```
colMeans(X)
```

```
[1]   1.50   0.50   1.75   2.25   2.00
```

These functions are built for speed, and blur some of the subtleties of dealing with NA or NaN. If such subtlety is an issue, then use `apply` instead (p. 61). Remember that columns are margin number 2 and rows are margin number 1:

```
apply(X,2,mean)
```

```
[1]   1.50   0.50   1.75   2.25   2.00
```

You might want to sum groups of rows within columns, and `rowsum` (singular and all lower case, in contrast to `rowSums`, above) is a very efficient function for this. In this example, we want to group together row 1 and row 4 (as group A) and row 2 and row 3 (group B). Note that the grouping vector has to have length equal to the number of rows:

```
group=c("A","B","B","A")
rowsum(X, group)
```

	[,1]	[,2]	[,3]	[,4]	[,5]
A	2	0	4	6	3
B	4	2	3	3	5

You could achieve the same ends (but more slowly) with `tapply` or `aggregate`:

```
tapply(X, list(group[row(X)], col(X)), sum)
```

	1	2	3	4	5
A	2	0	4	6	3
B	4	2	3	3	5

Note the use of `row(X)` and `col(X)`, with `row(X)` used as a subscript on `group`.

```
aggregate(X,list(group),sum)
```

	Group.1	V1	V2	V3	V4	V5
1	A	2	0	4	6	3
2	B	4	2	3	3	5

Suppose that we want to shuffle the elements of each column of a matrix independently. We apply the function sample to each column (margin number 2) like this:

```
apply(X,2,sample)
```

	[,1]	[,2]	[,3]	[,4]	[,5]
[1,]	1	1	2	1	3
[2,]	3	1	0	1	3
[3,]	1	0	3	2	0
[4,]	1	0	2	5	2

```
apply(X,2,sample)
```

	[,1]	[,2]	[,3]	[,4]	[,5]
[1,]	1	1	0	5	2
[2,]	1	1	2	1	3
[3,]	3	0	2	2	3
[4,]	1	0	3	1	0

and so on, for as many shuffled samples as you need.

2.8.4 Adding rows and columns to the matrix

In this particular case we have been asked to add a row at the bottom showing the column means, and a column at the right showing the row variances:

```
X <- rbind(X,apply(X,2,mean))
X <- cbind(X,apply(X,1,var))
X
```

	[,1]	[,2]	[,3]	[,4]	[,5]	[,6]
[1,]	1.0	0.0	2.00	5.00	3	3.70000
[2,]	1.0	1.0	3.00	1.00	3	1.20000
[3,]	3.0	1.0	0.00	2.00	2	1.30000
[4,]	1.0	0.0	2.00	1.00	0	0.70000
[5,]	1.5	0.5	1.75	2.25	2	0.45625

Note that the number of decimal places varies across columns, with one in columns 1 and 2, two in columns 3 and 4, none in column 5 (integers) and five in column 6. The default in R is to print the minimum number of decimal places consistent with the contents of the column as a whole.

Next, we need to label the sixth column as 'variance' and the fifth row as 'mean':

```
colnames(X) <- c(1:5,"variance")
rownames(X) <- c(1:4,"mean")
X
```

	1	2	3	4	5	variance
1	1.0	0.0	2.00	5.00	3	3.70000
2	1.0	1.0	3.00	1.00	3	1.20000

```
3        3.0     1.0     0.00    2.00    2    1.30000
4        1.0     0.0     2.00    1.00    0    0.70000
mean     1.5     0.5     1.75    2.25    2    0.45625
```

When a matrix with a single row or column is created by a subscripting operation, for example `row <- mat[2,]`, it is by default turned into a vector. In a similar way, if an array with dimension, say, $2 \times 3 \times 1 \times 4$ is created by subscripting it will be coerced into a $2 \times 3 \times 4$ array, losing the unnecessary dimension. After much discussion this has been determined to be a *feature* of R. To prevent this happening, add the option `drop = FALSE` to the subscripting. For example:

```
rowmatrix <- mat[2, , drop = FALSE]
colmatrix <- mat[, 2, drop = FALSE]
a <- b[1, 1, 1, drop = FALSE]
```

The `drop = FALSE` option should be used defensively when programming. For example, the statement

```
somerows <- mat[index,]
```

will return a vector rather than a matrix if `index` happens to have length 1, and this might cause errors later in the code. It should be written as:

```
somerows <- mat[index , , drop = FALSE]
```

2.8.5 The sweep function

The `sweep` function is used to 'sweep out' array summaries from vectors, matrices, arrays or dataframes. In this example we want to express a matrix in terms of the departures of each value from its column mean.

```
matdata <- read.table("c: \\temp \\sweepdata.txt")
```

First, you need to create a vector containing the parameters that you intend to sweep out of the matrix. In this case we want to compute the four column means:

```
(cols <- apply(matdata,2,mean))
```

```
V1       V2      V3       V4
4.60     13.30   0.44     151.60
```

Now it is straightforward to express all of the data in `matdata` as departures from the relevant column means:

```
sweep(matdata,2,cols)
```

```
      V1      V2      V3      V4
1    -1.6    -1.3    -0.04   -26.6
2     0.4    -1.3     0.26    14.4
3     2.4     1.7     0.36    22.4
4     2.4     0.7     0.26   -23.6
5     0.4     4.7    -0.14   -15.6
6     4.4    -0.3    -0.24     3.4
7     2.4     1.7     0.06   -36.6
8    -2.6    -0.3     0.06    17.4
9    -3.6    -3.3    -0.34    30.4
10   -4.6    -2.3    -0.24    14.4
```

Note the use of `margin = 2` as the second argument to indicate that we want the sweep to be carried out on the columns (rather than on the rows). A related function, `scale`, is used for centring and scaling data in terms of standard deviations (p. 254).

You can see what `sweep` has done by doing the calculation long-hand. The operation of this particular sweep is simply one of subtraction. The only issue is that the subtracted object has to have the same dimensions as the matrix to be swept (in this example, 10 rows of 4 columns). Thus, to sweep out the column means, the object to be subtracted from `matdata` must have the each column mean repeated in each of the 10 rows of 4 columns:

```
(col.means <- matrix(rep(cols,rep(10,4)),nrow=10))
```

```
        [,1]     [,2]     [,3]       [,4]
 [1,]   4.6     13.3     0.44      151.6
 [2,]   4.6     13.3     0.44      151.6
 [3,]   4.6     13.3     0.44      151.6
 [4,]   4.6     13.3     0.44      151.6
 [5,]   4.6     13.3     0.44      151.6
 [6,]   4.6     13.3     0.44      151.6
 [7,]   4.6     13.3     0.44      151.6
 [8,]   4.6     13.3     0.44      151.6
 [9,]   4.6     13.3     0.44      151.6
[10,]   4.6     13.3     0.44      151.6
```

Then the same result as we got from `sweep` is obtained simply by

```
matdata-col.means
```

Suppose that you want to obtain the subscripts for a column-wise or a row-wise sweep of the data. Here are the row subscripts repeated in each column:

```
apply(matdata,2,function (x) 1:10)
```

```
        V1    V2    V3    V4
 [1,]    1     1     1     1
 [2,]    2     2     2     2
 [3,]    3     3     3     3
 [4,]    4     4     4     4
 [5,]    5     5     5     5
 [6,]    6     6     6     6
 [7,]    7     7     7     7
 [8,]    8     8     8     8
 [9,]    9     9     9     9
[10,]   10    10    10    10
```

Here are the column subscripts repeated in each row:

```
t(apply(matdata,1,function (x) 1:4))
```

```
        [,1]     [,2]     [,3]       [,4]
1        1        2        3          4
2        1        2        3          4
3        1        2        3          4
4        1        2        3          4
5        1        2        3          4
```

6	1	2	3	4
7	1	2	3	4
8	1	2	3	4
9	1	2	3	4
10	1	2	3	4

Here is the same procedure using `sweep`:

```
sweep(matdata,1,1:10,function(a,b) b)
```

	[,1]	[,2]	[,3]	[,4]
[1,]	1	1	1	1
[2,]	2	2	2	2
[3,]	3	3	3	3
[4,]	4	4	4	4
[5,]	5	5	5	5
[6,]	6	6	6	6
[7,]	7	7	7	7
[8,]	8	8	8	8
[9,]	9	9	9	9
[10,]	10	10	10	10

```
sweep(matdata,2,1:4,function(a,b) b)
```

	[,1]	[,2]	[,3]	[,4]
[1,]	1	2	3	4
[2,]	1	2	3	4
[3,]	1	2	3	4
[4,]	1	2	3	4
[5,]	1	2	3	4
[6,]	1	2	3	4
[7,]	1	2	3	4
[8,]	1	2	3	4
[9,]	1	2	3	4
[10,]	1	2	3	4

2.8.6 Applying functions with `apply`, `sapply` and `lapply`

The `apply` function is used for applying functions to the rows or columns of matrices or dataframes. For example, here is a matrix with four rows and six columns:

```
(X <- matrix(1:24,nrow=4))
```

	[,1]	[,2]	[,3]	[,4]	[,5]	[,6]
[1,]	1	5	9	13	17	21
[2,]	2	6	10	14	18	22
[3,]	3	7	11	15	19	23
[4,]	4	8	12	16	20	24

Note that placing the expression to be evaluated in parentheses (as above) causes the value of the result to be printed on the screen. Often you want to apply a function across one of the margins of a matrix. Margin 1 refers to the rows and margin 2 to the columns. Here are the row totals (four of them):

```
apply(X,1,sum)
```

```
[1]   66   72   78   84
```

and here are the column totals (six of them):

```
apply(X,2,sum)
```

```
[1]   10   26   42   58   74   90
```

Note that in both cases, the answer produced by apply is a vector rather than a matrix. You can apply functions to the individual elements of the matrix rather than to the margins. The margin you specify influences only the shape of the resulting matrix.

```
apply(X,1,sqrt)
```

```
           [,1]        [,2]        [,3]        [,4]
[1,]   1.000000    1.414214    1.732051    2.000000
[2,]   2.236068    2.449490    2.645751    2.828427
[3,]   3.000000    3.162278    3.316625    3.464102
[4,]   3.605551    3.741657    3.872983    4.000000
[5,]   4.123106    4.242641    4.358899    4.472136
[6,]   4.582576    4.690416    4.795832    4.898979
```

```
apply(X,2,sqrt)
```

```
           [,1]        [,2]        [,3]        [,4]        [,5]        [,6]
[1,]   1.000000    2.236068    3.000000    3.605551    4.123106    4.582576
[2,]   1.414214    2.449490    3.162278    3.741657    4.242641    4.690416
[3,]   1.732051    2.645751    3.316625    3.872983    4.358899    4.795832
[4,]   2.000000    2.828427    3.464102    4.000000    4.472136    4.898979
```

Here are the shuffled numbers from each of the rows, using sample without replacement:

```
apply(X,1,sample)
```

```
        [,1]      [,2]      [,3]      [,4]
[1,]       5        14        19         8
[2,]      21        10         7        16
[3,]      17        18        15        24
[4,]       1        22        23         4
[5,]       9         2         3        12
[6,]      13         6        11        20
```

Note that the resulting matrix has six rows and four columns (i.e. it has been transposed).

You can supply your own function definition (here $x^2 + x$) within apply like this:

```
apply(X,1,function(x) x^ 2+x)
```

```
        [,1]      [,2]      [,3]      [,4]
[1,]       2         6        12        20
[2,]      30        42        56        72
[3,]      90       110       132       156
[4,]     182       210       240       272
```

```
[5,]       306       342       380       420
[6,]       462       506       552       600
```

This is an anonymous function because the function is not named.

If you want to `apply` a function to a vector (rather than to the margin of a matrix) then use `sapply`. Here is the code to generate a list of sequences from 1:3 up to 1:7 (see p. 30):

```
sapply(3:7, seq)
```

```
[[1]]
[1]   1     2     3
[[2]]
[1]   1     2     3     4
[[3]]
[1]   1     2     3     4     5
[[4]]
[1]   1     2     3     4     5     6
[[5]]
[1]   1     2     3     4     5     6     7
```

The function `sapply` is most useful with complicated iterative calculations. The following data show decay of radioactive emissions over a 50-day period, and we intend to use non-linear least squares (see p. 715) to estimate the decay rate a in $y = \exp(-ax)$:

```
sapdecay <- read.table("c:\\temp\\sapdecay.txt",header=T)
attach(sapdecay)
names(sapdecay)
```

```
[1]   "x"   "y"
```

We need to write a function to calculate the sum of the squares of the differences between the observed (`y`) and predicted (`yf`) values of `y`, when provided with a specific value of the parameter `a`:

```
sumsq <- function(a,xv=x,yv=y)
        { yf <- exp(-a*xv)
        sum((yv-yf)^2) }
```

We can get a rough idea of the decay constant, a, for these data by linear regression of log(y) against x, like this:

```
lm(log(y)~x)
```

```
Coefficients:
```

```
(Intercept)        x
0.04688    -0.05849
```

So our parameter a is somewhere close to 0.058. We generate a range of values for a spanning an interval on either side of 0.058:

```
a <- seq(0.01,0.2,.005)
```

Now we can use `sapply` to apply the sum of squares function for each of these values of a (without writing a loop), and plot the deviance against the parameter value for a:

```
plot(a,sapply(a,sumsq),type="l")
```

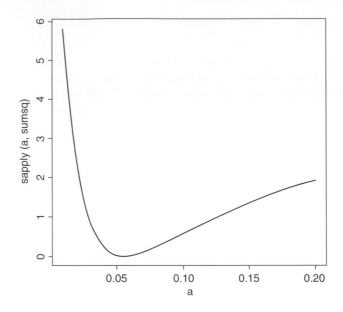

This shows that the least-squares estimate of a is indeed close to 0.06 (this is the value of a associated with the minimum deviance). To extract the minimum value of a we use min with subscripts (square brackets):

```
a[min(sapply(a,sumsq))==sapply(a,sumsq)]
```

```
[1]   0.055
```

Finally, we could use this value of a to generate a smooth exponential function to fit through our scatter of data points:

```
plot(x,y)
xv <- seq(0,50,0.1)
lines(xv,exp(-0.055*xv))
```

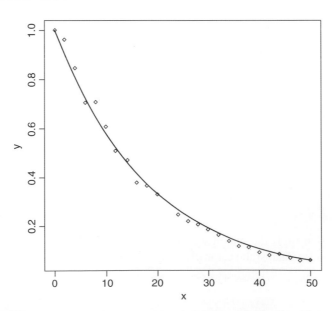

Here is the same procedure streamlined by using the `optimize` function. Write a function showing how the sum of squares depends on the value of the parameter a:

```
fa <- function(a) sum((y-exp(-a*x))^2)
```

Now use `optimize` with a specified range of values for a, here `c(0.01,0.1)`, to find the value of a that minimizes the sum of squares:

```
optimize(fa,c(0.01,0.1))

$minimum
[1]   0.05538411
$objective
[1]   0.01473559
```

The value of a is that minimizes the sum of squares is 0.055 38 and the minimum value of the sum of squares is 0.0147.

What if we had chosen a different way of assessing the fit of the model to the data? Instead of minimizing the sum of the squares of the residuals, we might want to minimize the sum of the absolute values of the residuals. We need to write a new function to calculate this quantity:

```
fb <- function(a) sum(abs(y-exp(-a*x)))
```

Then we use `optimize` as before:

```
optimize(fb,c(0.01,0.1))

$minimum
[1]   0.05596058
$objective
[1]   0.3939221
```

The results differ only in the fourth digit after the decimal point, and you could not choose between the two methods from a plot of the model. Sums of squares are not the only way of doing statistics, just the conventional way.

2.8.7 Using the `max.col` function

The task is to work out the number of plots on which a species is dominant in the Park Grass dataframe. This involves scanning each row of a matrix and reporting on the column number that contains the maximum value.

```
data <- read.table("c:\\temp\\pgfull.txt",header=T)
attach(data)
names(data)

 [1] "AC"       "AE"       "AM"  "AO"  "AP"  "AR"    "AS"
 [8] "AU"       "BH"       "BM"  "CC"  "CF"  "CM"    "CN"
[15] "CX"       "CY"       "DC"  "DG"  "ER"  "FM"    "FP"
[22] "FR"       "GV"       "HI"  "HL"  "HP"  "HS"    "HR"
[29] "KA"       "LA"       "LC"  "LH"  "LM"  "LO"    "LP"
[36] "OR"       "PL"       "PP"  "PS"  "PT"  "QR"    "RA"
[43] "RB"       "RC"       "SG"  "SM"  "SO"  "TF"    "TG"
[50] "TO"       "TP"       "TR"  "VC"  "VK"  "plot"  "lime"
[57] "species"  "hay"      "pH"
```

The species names are represented by two-letter codes (so, for example, 'AC' is *Agrostis capillaris*). We define the dominant as the species that has the maximum biomass on a given plot. The first task is to create a dataframe that contains only the species abundances (we do not want the plot numbers, or the treatments, or the values of any covariates). For the Park Grass data, the first 54 columns contain species abundance values, so we select all of the rows in the first 54 columns like this:

```
species <- data[,1:54]
```

Now we use the function `max.col` to go through all of the 89 rows, and for each row return the column number that contains the maximum biomass:

```
max.col(species)
```

```
 [1] 22 22 22  1 32 32 22  1 22 22 22  1 22 22  1  1 22 22 22  4  2  2 51  2  1
[26]  1 22 22  1  1  2  5  1  4  2  2  1  4 22 22 22  4  2  2 25 25  2  2  5 25
[51] 32  1 22 22  2  2  1  1 51  2  2 27  2  2  2  2 35 51 51  1  2  2  1  1 32
[76] 32  1  1  1  1  1  1 14  1  2  1  1  2  2
```

To get the identity of the dominant, we then extract the name of this column, using the index returned by `max.col` as a subscript to the object called `names(species)`:

```
names(species)[max.col(species)]
```

Finally, we use `table` to count up the total number of plots on which each species was dominant. The code looks like this:

```
table(names(species)[max.col(species)])
```

```
AC AE AO AP CN FR HL HS LH LP TP
26 23  4  2  1 19  3  1  5  1  4
```

So AC was dominant on more plots than any other species, with AE in second place and FR in third. The total number of species that were dominant on one or more plots is given by determining the length of this table:

```
length(table(names(species)[max.col(species)]))
```

```
[1] 11
```

So the number of species that were present in the system, but never attained dominance was $54 - 11 = 43$:

```
length(names(species))-length(table(names(species)[max.col(species)]))
```

```
[1] 43
```

There is no such function as 'min.col', but you can easily emulate it by using `max.col` with the negatives of your data. It makes no sense to do it with this example, because several species are absent from every plot, and the function would just pick one of the absent species at random. But, anyway,

```
max.col(-species)
```

picks out the identity (the column number) of one of the zeros from each row of the dataframe. In a case where there was a unique minimum in each row, then this would find it.

2.8.8 Restructuring a multi-dimensional array using `aperm`

There are circumstances where you may want to reorder the dimensions of an array. Here is an example of an array with three dimensions: two sexes, three ages and four income groups. For simplicity and ease of illustration the values in the array are just the numbers 1 to 24 in order ($2 \times 3 \times 4 = 24$):

```
data <- array(1:24, 2:4)
```

The second argument to the `array` function specifies the number of levels in dimensions 1, 2, and 3 using the sequence-generator 2 : 4 to produce the numbers 2, 3 and 4. This is what the array looks like:

```
data

, , 1

     [,1] [,2] [,3]
[1,]    1    3    5
[2,]    2    4    6

, , 2

     [,1] [,2] [,3]
[1,]    7    9   11
[2,]    8   10   12

, , 3

     [,1] [,2] [,3]
[1,]   13   15   17
[2,]   14   16   18

, , 4

     [,1] [,2] [,3]
[1,]   19   21   23
[2,]   20   22   24
```

There are four sub-tables, each with 2 rows and 3 columns. Now we give names to the factor levels in each of the three dimensions: these are called the `dimnames` attributes and are allocated as `lists` like this:

```
dimnames(data)[[1]] <- list("male","female")
dimnames(data)[[2]] <- list("young","mid","old")
dimnames(data)[[3]] <- list("A","B","C","D")
dimnames(data)

[[1]]
[1] "male" "female"

[[2]]
[1] "young" "mid" "old"

[[3]]
[1] "A" "B" "C" "D"
```

You can see the advantage of naming the dimensions by comparing the output of the array with (below) and without names (above):

```
data

, , A

       young  mid old
male       1    3   5
female     2    4   6

, , B

       young  mid old
male       7    9  11
female     8   10  12

, , C

       young  mid old
male      13   15  17
female    14   16  18

, , D

       young  mid old
male      19   21  23
female    20   22  24
```

Suppose, however, that we want the four income groups (A–D) to be the columns in each of the sub-tables, and the separate sub-tables to represent the two genders. This is a job for aperm. We need to specify the order 'age then income then gender' in terms of the order of their dimensions (row, column, sub-table, namely 2 then 3 then 1) like this:

```
new.data <- aperm(data,c(2,3,1))
new.data

, , male

       A  B  C  D
young  1  7 13 19
mid    3  9 15 21
old    5 11 17 23

, , female

       A  B  C  D
young  2  8 14 20
mid    4 10 16 22
old    6 12 18 24
```

This will be tricky to see at first, but you should persevere, because aperm is a very useful function.

2.9 Random numbers, sampling and shuffling

When debugging a program it is often useful to be able to get the same string of random numbers as you had last time. Use the set.seed function to control this:

```
set.seed(375)
runif(3)
```

```
[1] 0.9613669 0.6918535 0.7302684
```

```
runif(3)
```

```
[1] 0.9228566 0.1603804 0.9642799
```

```
runif(3)
```

```
[1] 0.52880907 0.08660864 0.29075809
```

If you reset the seed with the same value, you get the same random numbers as last time:

```
set.seed(375)
runif(3)
```

```
[1] 0.9613669 0.6918535 0.7302684
```

You might want to obtain part of the same series of random numbers, and we use .Random.seed like this:

```
current<-.Random.seed
runif(3)
```

```
[1] 0.9228566 0.1603804 0.9642799
```

```
runif(3)
```

```
[1] 0.52880907 0.08660864 0.29075809
```

```
runif(3)
```

```
[1] 0.02590182 0.85520652 0.31350305
```

Resetting .Random.seed recreates the same series of random numbers:

```
.Random.seed<-current
runif(3)
```

```
[1] 0.9228566 0.1603804 0.9642799
```

Randomization is central to a great many scientific and statistical procedures. Generating random numbers from a variety of probability distributions is explained in Chapter 7 (p. 272). Here we are concerned with randomizing (shuffling or sampling from) the elements of a vector, as we might use when planning a designed experiment (e.g. allocating treatments to individuals). There are two ways of sampling:

- sampling without replacement (where all of the values in the vector appear in the output, but in a randomized sequence; i.e. the values have been shuffled);

- sampling with replacement (where some values are omitted, and other values appear more than once in the output).

2.9.1   The sample function

The default `sample` function shuffles the contents of a vector into a random sequence while maintaining all the numerical values intact. It is extremely useful for randomization in experimental design, in simulation and in computationally intensive hypothesis testing. The vector `y` looks like this:

```
y <- c(8,3,5,7,6,6,8,9,2,3,9,4,10,4,11)
```

Here are two different shufflings of `y`:

```
sample(y)
```

```
[1]      8     8     9     9     2    10     6     7     3    11     5     4
[13]     6     3     4
```

```
sample(y)
```

```
[1]      9     3     9     8     8     6     5    11     4     6     4     7
[13]     3     2    10
```

The order of the values is different each time `sample` is invoked, but the same numbers are shuffled in every case, and all the numbers in the original vector appear once in the output (so if there are two 9s in the original data, there will be two 9s in the shuffled vector). This is called *sampling without replacement*. You can specify the size of the sample you want as an optional second argument. Suppose we want five random elements from `y`, in any one sample:

```
sample(y,5)
```

```
[1]      9     4    10     8    11
```

```
sample(y,5)
```

```
[1]      9     3     4     2     8
```

The option `replace=T` allows for *sampling with replacement*, which is the basis of bootstrapping (see p. 570). The vector produced by the `sample` function with `replace=T` is the same length as the vector sampled, but some values are left out at random and other values, again at random, appear two or more times. In this sample, 10 has been left out, and there are now three 9s:

```
sample(y,replace=T)
```

```
[1]      9     6    11     2     9     4     6     8     8     4     4     4
[13]     3     9     3
```

In this next case, there are two 10s and only one 9:

```
sample(y,replace=T)
```

```
[1]      3     7    10     6     8     2     5    11     4     6     3     9
[13]    10     7     4
```

More advanced options in `sample` include specifying different probabilities with which each element is to be sampled (`prob=`). For example, if we want to take four numbers at random from the sequence 1:10 without replacement where the probability of selection (p) is 5 times greater for the middle numbers (5 and 6) than for the first or last numbers, and we want to do this five times, we could write:

```
p <- c(1, 2, 3, 4, 5, 5, 4, 3, 2, 1)
x <- 1:10
```

```
sapply(1:5,function(i) sample(x,4,prob=p))
```

	[,1]	[,2]	[,3]	[,4]	[,5]
[1,]	8	7	4	10	8
[2,]	7	5	7	8	7
[3,]	4	4	3	4	5
[4,]	9	10	8	7	6

Thus, the four random numbers in the first trial were 8, 7, 4 and 9 (i.e. column 1). To learn more about sapply, see p. 63.

2.10   Loops and repeats

The classic, Fortran-like loop is available in R. The syntax is a little different, but the idea is identical; you request that an index, i, takes on a sequence of values, and that one or more lines of commands are executed as many times as there are different values of i. Here is a loop executed five times with the values of i from 1 to 5; we print the square of each value:

```
for (i in 1:5) print(i^2)
```

```
[1]    1
[1]    4
[1]    9
[1]    16
[1]    25
```

For multiple lines of code, you use curly brackets {} to enclose material over which the loop is to work. Note that the 'hard return' (the Enter key) at the end of each command line is an essential part of the structure (you can replace the hard returns by semicolons if you like, but clarity is improved if you put each command on a separate line):

```
j <- k <- 0
for (i in 1:5) {
j <- j+1
k <- k+i*j
print(i+j+k) }
```

```
[1]    3
[1]    9
[1]    20
[1]    38
[1]    65
```

Here we use a for loop to write a function to calculate factorial x (written $x!$) which is

$$x! = x \times (x-1) \times (x-2) \times (x-3) \ldots \times 2 \times 1$$

So $4! = 4 \times 3 \times 2 = 24$. Here is the function:

```
fac1 <- function(x) {
        f <- 1
```

```
if (x<2) return (1)
for (i in 2:x) {
f <- f*i}
f }
```

That seems rather complicated for such a simple task, but we can try it out for the numbers 0 to 5:

```
sapply(0:5,fac1)
```

```
[1]   1   1   2   6   24   120
```

There are two other looping functions in R: `repeat` and `while`. We demonstrate their use for the purpose of illustration, but we can do much better in terms of writing a compact function for finding factorials (see below). First, the `while` function:

```
fac2 <- function(x) {
f <- 1
t <- x
while(t>1) {
f <- f*t
t <- t-1 }
return(f) }
```

The key point is that if you want to use `while`, you need to set up an indicator variable (`t` in this case) and change its value *within* each iteration (`t <- t-1`). We test the function on the numbers 0 to 5:

```
sapply(0:5,fac2)
```

```
[1]   1   1   2   6   24   120
```

Finally, we demonstrate the use of the `repeat` function:

```
fac3 <- function(x) {
f <- 1
t <- x
repeat {
if (t<2) break
f <- f*t
t <- t-1 }
return(f) }
```

Because the `repeat` function contains no explicit limit, you need to be careful not to program an infinite loop. You must include a logical escape clause that leads to a `break` command:

```
sapply(0:5,fac3)
```

```
[1]   1   1   2   6   24   120
```

It is almost always better to use a built-in function that operates on the entire vector and hence removes the need for loops or repeats of any sort. In this case, we can make use of the cumulative product function, `cumprod`. Here it is in action:

```
cumprod(1:5)
```

```
[1]   1   2   6   24   120
```

This is already pretty close to what we need for our factorial function. It does not work for 0! of course, because the whole vector would end up full of zeros if the first element in the vector was zero (try 0:5 and see). The factorial of $x > 0$ is the maximum value from the vector produced by cumprod:

```
fac4 <- function(x) max(cumprod(1:x))
```

This definition has the desirable side effect that it also gets 0! correct, because when x is 0 the function finds the maximum of 1 and 0 which is 1.

```
max(cumprod(1:0))
```

```
[1]   1
```

```
sapply(0:5,fac4)
```

```
[1]   1   1   2   6   24   120
```

Alternatively, you could adapt an existing built-in function to do the job. $x!$ is the same as $\Gamma(x + 1)$, so

```
fac5 <- function(x) gamma(x+1)
sapply(0:5,fac5)
```

```
[1]   1   1   2   6   24   120
```

Until quite recently there was no built-in factorial function in R, but now there is:

```
sapply(0:5,factorial)
```

```
[1]   1   1   2   6   24   120
```

2.10.1   Creating the binary representation of a number

Here is a function that uses the while function in converting a specified number to its binary representation. The trick is that the smallest digit (0 for even or 1 for odd numbers) is always at the right-hand side of the answer (in location 32 in this case):

```
binary <- function(x) {
    i <- 0
    string <- numeric(32)
    while(x>0) {
        string[32-i]<  x %% 2
        x <- x%% 2
        i <- i+1 }
first <- match(1,string)
string[first:32] }
```

The leading zeros (1 to first − 1) within the string are not printed. We run the function to find the binary representation of the numbers 15 to 17:

```
sapply(15:17,binary)
```

```
[[1]]
[1]   1   1   1   1

[[2]]
[1]   1   0   0   0   0
```

```
[[3]]
[1]   1    0    0    0    1
```

The next function uses `while` to generate the Fibonacci series 1, 1, 2, 3, 5, 8, ... in which each term is the sum of its two predecessors. The key point about `while` loops is that the logical variable controlling their operation is altered inside the loop. In this example, we alter n, the number whose Fibonacci number we want, starting at n, reducing the value of n by 1 each time around the loop, and ending when n gets down to 0. Here is the code:

```
fibonacci <- function(n) {
    a <- 1
    b <- 0
    while(n>0)
            {swap <- a
            a <- a+b
            b <- swap
            n <- n-1 }
    b }
```

An important general point about computing involves the use of the `swap` variable above. When we replace a by a + b on line 6 we lose the original value of a. If we had not stored this value in `swap`, we could not set the new value of b to the old value of a on line 7. Now test the function by generating the Fibonacci numbers 1 to 10:

```
sapply(1:10,fibonacci)
```

```
[1]   1   1   2   3   5   8   13   21   34   55
```

2.10.2 Loop avoidance

It is good R programming practice to avoid using loops wherever possible. The use of vector functions (p. 41) makes this particularly straightforward in many cases. Suppose that you wanted to replace all of the negative values in an array by zeros. In the old days, you might have written something like this:

```
for (i in 1:length(y)) { if(y[i] < 0) y[i] <- 0 }
```

Now, however, you would use logical subscripts (p. 39) like this:

```
y[y<0] <- 0
```

The `ifelse` function

Sometimes you want to do one thing if a condition is true and a different thing if the condition is false (rather than do nothing, as in the last example). The `ifelse` function allows you to do this for entire vectors without using for loops. We might want to replace any negative values of y by -1 and any positive values and zero by $+1$:

```
z <- ifelse (y < 0, -1, 1)
```

Next we use `ifelse` to convert the continuous variable called Area into a new, two-level factor with values big and small defined by the median Area of the fields:

```
data <- read.table("c:\\temp\\worms.txt",header=T)
attach(data)
ifelse(Area>median(Area),"big","small")
```

```
[1] "big"     "big"    "small" "small" "big" "big"   "big"   "small"
[9] "small" "small" "small" "big"    "big" "small" "big" "big"
[17] "small" "big"    "small" "small"
```

You should use the much more powerful function called cut when you want to convert a continuous variable like Area into many levels (p. 838).

Another use of ifelse is to override R's natural inclinations. The log of zero in R is -Inf, as you see in these 20 random numbers from a Poisson process with a mean count of 1.5:

```
y <- log(rpois(20,1.5))
y
```

```
[1]        -Inf 0.6931472        -Inf 0.0000000        -Inf 0.0000000
[7]   0.0000000        -Inf 0.6931472 1.6094379 1.3862944 0.0000000
[13] 1.3862944        -Inf 0.0000000 0.0000000 0.6931472 0.6931472
[19] 0.0000000        -Inf
```

However, if we want the log of zero to be represented by NA in our particular application we can write:

```
ifelse(y<0,NA,y)
```

```
[1]          NA 0.6931472          NA 0.0000000          NA 0.0000000
[7]   0.0000000          NA 0.6931472 1.6094379 1.3862944 0.0000000
[13] 1.3862944          NA 0.0000000 0.0000000 0.6931472 0.6931472
[19] 0.0000000          NA
```

2.10.3   The slowness of loops

To see how slow loops can be, we compare two ways of finding the maximum number in a vector of 10 million random numbers from a uniform distribution:

```
x <- runif(10000000)
```

First, using the vector function max:

```
system.time(max(x))
```

```
user system elapsed
 0.03 0.00 0.03
```

As you see, this operation took just 0.03 seconds to solve using the vector function max to look at the 10 million numbers in x. Using a loop, however, took more than 9 seconds:

```
pc <- proc.time()
```

```
cmax <- x[1]
for (i in 2:10000000) {
if(x[i]>cmax) cmax <- x[i] }
```

```
proc.time()-pc
```

```
user system elapsed
 9.39 0.13 9.51
```

The functions `system.time` and `proc.time` produce a vector of three numbers, showing the user, system and total elapsed times for the currently running R process. It is the third number (elapsed time in seconds, 9.51 in this case) that is typically the most useful.

2.10.4   Do not 'grow' data sets by concatenation or recursive function calls

Here is an extreme example of what *not* to do. We want to create a vector containing 100 000 numbers in sequence from 1 to 100 000. First, the quickest way using the built-in sequence generator which is invoked by the colon symbol (`:`)

```
test1 <- function(){
y <- 1:100000
}
```

Now we obtain the same result using a loop, where we tell R in advance how long the final vector is going to be, using the `numeric` function. This is called preallocation.

```
test2 <- function(){
y <- numeric(100000)
for (i in 1:100000) y[i] <- i
}
```

Finally, the most inefficient way. Each time we go round the loop we concatenate the new value onto the right-hand end of the vector that has been created up to this point. We start with a `NULL` vector, then build it up, one step at a time, which looks like a neat idea, but is extremely inefficient, because changing the size of a vector takes roughly the same size as setting a vector up from scratch, and we change the length of our vector 100 000 times in this example. This ill-advised procedure is called re-dimensioning.

```
test3 <- function(){
y <- NULL
for (i in 1:100000) y <- c(y,i)
}
```

To compare the efficiency of the three methods, we shall work out how long each takes to complete the task. The function called `proc.time` determines how much real time and computer processing unit (CPU) time (in seconds) the currently running R process has already taken:

```
proc.time()

 user system elapsed
 53.15 5.14 2483.00
```

The user time is the CPU time charged for the execution of user instructions of the calling process, the system time is the CPU time charged for execution by the system on behalf of the calling process, and the elapsed time includes other stuff that the computer is doing, unrelated to your R session.

The function `system.time` calls the function `proc.time`, then evaluates your expression, and then calls `proc.time` once more, returning the difference between the two `proc.time` calls as its output. We can compare the efficiencies of our three different functions using `system.time` like this:

```
system.time(test1())

user system elapsed
   0      0       0
```

```
system.time(test2())
```

```
user system elapsed
0.16   0.02    0.17
```

```
system.time(test3())
```

```
user system elapsed
8.95   0.02    8.97
```

The first method is so lightening fast that it does not even register on the clock. The loop using a pre-determined vector length is also very fast (0.16 seconds). In contrast, the last method, where we grew the vector at each iteration, is staggeringly slow (8.95 seconds). The moral: *do not grow vectors by repeated concatenation.*

2.10.5   Loops for producing time series

Wherever we can, we use vectorized functions in R because this leads to compact, efficient and easily readable code. Sometimes, however, we need to resort to using loops. Suppose we are interested in the dynamics of a population which is governed by two parameters: the per capita reproductive rate (λ) and the maximum supportable population (N_{max}), which for convenience we shall set to 1.0). Next year's population $N(t+1)$ is given by this year's population, $N(t)$, multiplied by lambda, multiplied again by the fraction of N_{max} that is currently unrealized (i.e. $(N_{max} - N(t))/N_{max} = 1 - N(t)$ in the current case). Thus, we have a difference equation

$$N(t+1) = \lambda N(t)[1 - N(t)]$$

To simulate the dynamics of this population in R, we start by writing the difference equation as a function (call it next.year for instance):

```
next.year <- function(x) lambda * x * (1 - x)
```

So if we begin with a population of $N = 0.6$ and set $\lambda = 3.7$ we can predict next year's population like this:

```
lambda <- 3.7
next.year(0.6)
```

```
[1] 0.888
```

The population has increased by 48% (0.888 / 0.6 = 1.48). What happens in the second year?

```
next.year(0.888)
```

```
[1] 0.3679872
```

The population crashes to less than half its previous value (0.367 987 2/0.888 = 0.4144). We could go on repeating these calculations, modelling year after year, but this is an obvious case where using a loop would be the best solution. Let us assume that we want to model the population over 20 years. It is good practice in cases like this to define a vector to contain the 20 population sizes at the outset (*preallocation*) using numeric like this:

```
N <- numeric(20)
```

We set the initial population size (0.6) like this:

```
N[1] <- 0.6
```

Now if we run through a loop to simulate the years 2 through 20 using an index called t (for time), we can invoke the function called next.year repeatedly, employing t as a subscript like this:

```
for (t in 2:20) N[t] <- next.year(N[t-1])
```

Finally, we might want to plot a time series of the population dynamics over the course of 20 years.

```
plot(N,type="l")
```

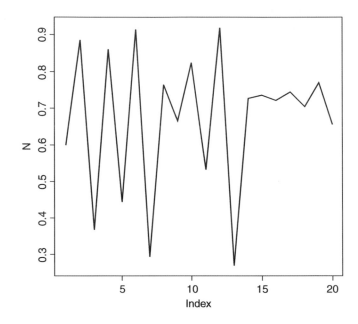

This famous difference equation is known as the quadratic map, and it played a central role in the development of chaos theory (May, 1974). For large values of λ (as we used in the example above), the function is capable of producing series of numbers that are, to all intents and purposes, random. This led to a definition of **chaos** as behaviour that exhibited *extreme sensitivity to initial conditions*: tiny differences in initial population size would lead to radically different time series in population dynamics.

2.11 Lists

Lists are extremely important objects in R. You will have heard of the problems of 'comparing apples and oranges' or how two things are 'as different as chalk and cheese'. You can think of lists as a way of getting around these problems. Here are four completely different objects: a numeric vector, a logical vector, a vector of character strings and a vector of complex numbers:

```
apples <- c(4,4.5,4.2,5.1,3.9)
oranges <- c(TRUE, TRUE, FALSE)
chalk <- c("limestone", "marl","oolite", "CaC03")
cheese <- c(3.2-4.5i,12.8+2.2i)
```

We cannot bundle them together into a dataframe, because the vectors are of different lengths:

```
data.frame(apples,oranges,chalk,cheese)
```

```
Error in data.frame(apples, oranges, chalk, cheese) :
arguments imply differing number of rows: 5, 3, 4, 2
```

Despite their differences, however, we can bundle them together in a single list called items:

```
items <- list(apples,oranges,chalk,cheese)
items
```

```
[[1]]
[1] 4.0 4.5 4.2 5.1 3.9

[[2]]
[1] TRUE TRUE FALSE

[[3]]
[1] "limestone" "marl" "oolite" "CaCO3"

[[4]]
[1] 3.2-4.5i 12.8+2.2i
```

Subscripts on vectors, matrices, arrays and dataframes have one set of square brackets [6], [3,4] or [2,3,2,1], but subscripts on lists have double square brackets [[2]] or [[i,j]]. If we want to extract the chalk from the list, we use subscript [[3]]:

```
items[[3]]
```

```
[1] "limestone" "marl" "oolite" "CaCO3"
```

If we want to extract the third element within chalk (oolite) then we use single subscripts *after* the double subscripts like this:

```
items[[3]][3]
```

```
[1] "oolite"
```

R is forgiving about failure to use double brackets on their own, but not when you try to access a component of an object within a list:

```
items[3]
```

```
[[1]]
[1] "limestone" "marl" "oolite" "CaCO3"
```

```
items[3][3]
```

```
[[1]]
NULL
```

There is another indexing convention in R which is used to extract named components from lists using the element names operator $. This is known as 'indexing tagged lists'. For this to work, the elements of the list must have names. At the moment our list called items has no names:

```
names(items)
```

```
NULL
```

You can give names to the elements of a list in the function that creates the list by using the equals sign like this:

```
items <- list(first=apples,second=oranges,third=chalk,fourth=cheese)
```

Now you can extract elements of the list by name

```
items$fourth
```

```
[1] 3.2-4.5i 12.8+2.2i
```

2.11.1   Lists and `lapply`

We can ask a variety of questions about our new list object:

```
class(items)
```

```
[1] "list"
```

```
mode(items)
```

```
[1] "list"
```

```
is.numeric(items)
```

```
[1] FALSE
```

```
is.list(items)
```

```
[1] TRUE
```

```
length(items)
```

```
[1] 4
```

Note that the length of a list is the number of items in the list, not the lengths of the individual vectors within the list.

The function `lapply` applies a specified function to each of the elements of a list in turn (without the need for specifying a loop, and not requiring us to know how many elements there are in the list). A useful function to apply to lists is the `length` function; this asks how many elements comprise each component of the list. Technically we want to know the length of each of the vectors making up the list:

```
lapply(items,length)
```

```
$first
[1] 5

$second
[1] 3

$third
[1] 4

$fourth
[1] 2
```

This shows that `items` consists of four vectors, and shows that there were 5 elements in the first vector, 3 in the second 4 in the third and 2 in the fourth. But 5 of what, and 3 of what? To find out, we apply the function class to the list:

```
lapply(items,class)

$first
[1] "numeric"

$second
[1] "logical"

$third
[1] "character"

$fourth
[1] "complex"
```

So the answer is there were 5 numbers in the first vector, 3 logical variables in the second, 4 character strings in the third vector and 2 complex numbers in the fourth.

Applying numeric functions to lists will only work for objects of class numeric or complex, or objects (like logical values) that can be coerced into numbers. Here is what happens when we use lapply to apply the function mean to items:

```
lapply(items,mean)

$first
[1] 4.34

$second
[1] 0.6666667

$third
[1] NA

$fourth
[1] 8-1.15i

Warning message:
In mean.default(X[[3L]], ...) :
 argument is not numeric or logical: returning NA
```

We get a warning message pointing out that the third vector cannot be coerced to a number (it is not numeric, complex or logical), so NA appears in the output. The second vector produces the answer 2/3 because logical false (FALSE) is coerced to numeric 0 and logical true (TRUE) is coerced to numeric 1.

The summary function works for lists:

```
summary(items)

       Length  Class    Mode
first  5       -none-   numeric
second 3       -none-   logical
third  4       -none-   character
fourth 2       -none-   complex
```

but the most useful overview of the contents of a list is obtained with `str`, the structure function:

```
str(items)
```

```
List of 4
 $ first : num [1:5] 4 4.5 4.2 5.1 3.9
 $ second: logi [1:3] TRUE TRUE FALSE
 $ third : chr [1:4] "limestone" "marl" "oolite" "CaCO3"
 $ fourth: cplx [1:2] 3.2-4.5i 12.8+2.2i
```

2.11.2   Manipulating and saving lists

Saving lists to files is tricky, because lists typically have different numbers of items in each row so we cannot use `write.table`. Here is a dataframe on species presence (1) or absence (0), with species' Latin binomials in the first column as the row names:

```
data<-read.csv("c:\\temp\\pa.csv",row.names=1)
data
```

	Carmel	Derry	Daneswall	Erith	Foggen	Highbury	Slatewell	Uppington	York
Bartsia alpina	0	0	1	0	0	0	1	0	0
Cleome serrulata	1	1	0	0	0	1	0	0	0
Conopodium majus	0	0	0	0	0	0	0	1	1
Corydalis sempervirens	1	0	0	1	0	1	0	0	0
Nitella flexilis	1	0	0	0	0	0	0	0	1
Ranunculus baudotii	1	0	1	1	0	0	0	1	0
Rhododendron luteum	1	1	1	1	1	0	1	1	1
Rodgersia podophylla	0	1	0	0	0	1	0	0	0
Tiarella wherryi	0	0	1	1	1	0	0	0	0
Veronica opaca	1	0	0	0	0	1	1	1	0

There are two kinds of operations you might want to do with a dataframe like this:

- produce lists of the sites at which each species is found;
- produce lists of the species found in any given site.

We shall do each of these tasks in turn.

The issue is that the numbers of place names differ from species to species, and the numbers of species differ from place to place. However, it is easy to create a list showing the column numbers that contain locations for each species:

```
sapply(1:10,function(i) which(data[i,]>0))
```

```
[[1]]
[1] 3 7

[[2]]
[1] 1 2 6

[[3]]
[1] 8 9
```

```
[[4]]
[1] 1 4 6

[[5]]
[1] 1 9

[[6]]
[1] 1 3 4 8

[[7]]
[1] 1 2 3 4 5 7 8 9

[[8]]
[1] 2 6

[[9]]
[1] 3 4 5

[[10]]
[1] 1 6 7 8
```

This indicates that *Bartsia alpina* (the first species) is found in locations 3 and 7 (Daneswall and Slatewell). If we save this list (calling it spp for instance), then we can extract the column names at which each species is present, using the elements of spp as subscripts on the column names of data, like this:

```
spp<-sapply(1:10,function(i) which(data[i,]>0))
sapply(1:10, function(i)names(data)[spp[[i]]] )

[[1]]
[1] "Daneswall" "Slatewell"

[[2]]
[1] "Carmel"    "Derry"      "Highbury"

[[3]]
[1] "Uppington" "York"

[[4]]
[1] "Carmel"    "Erith"      "Highbury"

[[5]]
[1] "Carmel"    "York"

[[6]]
[1] "Carmel"    "Daneswall" "Erith"       "Uppington"

[[7]]
[1] "Carmel"    "Derry"      "Daneswall" "Erith"   "Foggen" "Slatewell" "Uppington" "York"

[[8]]
[1] "Derry"     "Highbury"

[[9]]
[1] "Daneswall" "Erith"      "Foggen"
```

```
[[10]]
[1] "Carmel"      "Highbury"   "Slatewell"    "Uppington"
```

This completes the first task.

The second task is to get species lists for each location. We apply a similar method to extract the appropriate species (this time from `rownames(data)`) on the basis that the presence score for this site is `data[,j] > 0`:

```
sapply(1:9, function (j) rownames(data)[data[,j]>0] )
[[1]]
[1] "Cleome serrulata" "Corydalis sempervirens" "Nitella flexilis" "Ranunculus baudotii"
[5] "Rhododendron luteum" "Veronica opaca"
[[2]]
[1] "Cleome serrulata" "Rhododendron luteum" "Rodgersia podophylla"
[[3]]
[1] "Bartsia alpina" "Ranunculus baudotii" "Rhododendron luteum" "Tiarella wherryi"
[[4]]
[1] "Corydalis sempervirens" "Ranunculus baudotii" "Rhododendron luteum" "Tiarella wherryi"
[[5]]
[1] "Rhododendron luteum" "Tiarella wherryi"
[[6]]
[1] "Cleome serrulata" "Corydalis sempervirens" "Rodgersia podophylla" "Veronica opaca"
[[7]]
[1] "Bartsia alpina" "Rhododendron luteum" "Veronica opaca"
[[8]]
[1] "Conopodium majus" "Ranunculus baudotii" "Rhododendron luteum" "Veronica opaca"
[[9]]
[1] "Conopodium majus" "Nitella flexilis" "Rhododendron luteum"
```

Because the species lists for different sites are of different lengths, the simplest solution is to create a separate file for each species list. We need to create a set of nine file names incorporating the site name, then use `write.table` in a loop:

```
spplists<-sapply(1:9, function (j) rownames(data)[data[,j]>0] )

for (i in 1:9) {

slist<-data.frame(spplists[[i]])
names(slist)<-names(data)[i]

fn<-paste("c:\\temp\\",names(data)[i],".txt",sep="")
write.table(slist,fn)
}
```

We have produced nine separate files. Here, for instance, are the contents of the file called `c:\\temp\\Carmel.txt` as viewed in a text editor like Notepad:

```
"Carmel"
"1" "Cleome serrulata"
"2" "Corydalis sempervirens"
"3" "Nitella flexilis"
```

```
"4"  "Ranunculus baudotii"
"5"  "Rhododendron luteum"
"6"  "Veronica opaca"
```

Perhaps the simplest and best solution is to turn the whole presence/absence matrix into a dataframe. Then both tasks become very straightforward. Start by using `stack` to create a dataframe of place names and presence/absence information:

```
newframe<-stack(data)
head(newframe)
```

```
  values     ind
1      0 Carmel
2      1 Carmel
3      0 Carmel
4      1 Carmel
5      1 Carmel
6      1 Carmel
```

Now extract the species names from the row names, repeat the list of names nine times, and add the resulting vector species names to the dataframe:

```
newframe<-data.frame(newframe, rep(rownames(data),9))
```

Finally, give the three columns of the new dataframe sensible names:

```
names(newframe)<-c("present","location","species")
head(newframe)
```

```
  present location                 species
1       0   Carmel          Bartsia alpina
2       1   Carmel        Cleome serrulata
3       0   Carmel        Conopodium majus
4       1   Carmel Corydalis sempervirens
5       1   Carmel        Nitella flexilis
6       1   Carmel      Ranunculus baudotii
```

Unlike the lists, you can easily save this object to file:

```
write.table(newframe,"c:\\temp\\spplists.txt")
```

Now it is simple to do both our tasks. Here a location list for `species = Bartsia alpina`:

```
newframe[newframe$species=="Bartsia alpina" & newframe$present==1,2]
```

```
[1] Daneswall Slatewell
```

and here is a species list for `location = Carmel`:

```
newframe[newframe$location=="Carmel" & newframe$present==1,3]
[1] Cleome serrulata    Corydalis sempervirens  Nitella flexilis Ranunculus baudotii
[5] Rhododendron luteum Veronica  opaca
```

Lists are great, but dataframes are better. The cost of the dataframe is the potentially substantial redundancy in storage requirement. In practice, with relatively small dataframes, this seldom matters.

2.12   Text, character strings and pattern matching

In R, character strings are defined by double quotation marks:

```
a <- "abc"
b <- "123"
```

Numbers can be coerced to characters (as in b above), but non-numeric characters cannot be coerced to numbers:

```
as.numeric(a)
```

```
[1]   NA
```

```
Warning message:
```

```
NAs introduced by coercion
```

```
as.numeric(b)
```

```
[1]   123
```

One of the initially confusing things about character strings is the distinction between the `length` of a character object (a vector), and the numbers of characters (`nchar`) in the strings that comprise that object. An example should make the distinction clear:

```
pets <- c("cat","dog","gerbil","terrapin")
```

Here, `pets` is a vector comprising four character strings:

```
length(pets)
```

```
[1]   4
```

and the individual character strings have 3, 3, 6 and 7 characters, respectively:

```
nchar(pets)
```

```
[1]   3   3   6   7
```

When first defined, character strings are not factors:

```
class(pets)
```

```
[1]   "character"
```

```
is.factor(pets)
```

```
[1]   FALSE
```

However, if the vector of characters called `pets` was part of a dataframe (e.g. if it was input using `read.table`) then R would coerce all the character variables to act as factors:

```
df <- data.frame(pets)
```

```
is.factor(df$pets)
```

```
[1]   TRUE
```

There are built-in vectors in R that contain the 26 letters of the alphabet in lower case (letters) and in upper case (LETTERS):

```
letters
```

```
[1]   "a" "b" "c" "d" "e" "f" "g" "h" "i" "j" "k" "l" "m" "n" "o" "p"
[17]  "q" "r" "s" "t" "u" "v" "w" "x" "y" "z"
```

```
LETTERS
```

```
[1]   "A" "B" "C" "D" "E" "F" "G" "H" "I" "J" "K" "L" "M" "N" "O" "P"
[17]  "Q" "R" "S" "T" "U" "V" "W" "X" "Y" "Z"
```

To discover which number in the alphabet the letter n is, you can use the `which` function like this:

```
which(letters=="n")
```

```
[1]   14
```

For the purposes of printing you might want to suppress the quotes that appear around character strings by default. The function to do this is called `noquote`:

```
noquote(letters)
```

```
[1] a b c d e f g h i j k l m n o p q r s t u v w x y z
```

2.12.1   Pasting character strings together

You can amalgamate individual strings into vectors of character information:

```
c(a,b)
```

```
[1]   "abc"   "123"
```

This shows that the concatenation produces a vector of two strings. It does *not* convert two 3-character strings into one 6-character string. The R function to do that is `paste`:

```
paste(a,b,sep="")
```

```
[1]   "abc123"
```

The third argument, `sep=""`, means that the two character strings are to be pasted together without any separator between them: the default for `paste` is to insert a single blank space, like this:

```
paste(a,b)
```

```
[1]   "abc 123"
```

Notice that you do *not* lose blanks that are within character strings when you use the `sep=""` option in `paste`.

```
paste(a,b," a longer phrase containing blanks",sep="")
```

```
[1]   "abc123 a longer phrase containing blanks"
```

If one of the arguments to `paste` is a vector, each of the elements of the vector is pasted to the specified character string to produce an object of the same length as the vector:

```
d <- c(a,b,"new")
```

```
e <- paste(d,"a longer phrase containing blanks")
```

```
e
```

```
[1]   "abc a longer phrase containing blanks"
```

```
[2]   "123 a longer phrase containing blanks"
```

```
[3]   "new a longer phrase containing blanks"
```

You may need to think about why there are three lines of output.

In this next example, we have four fields of information and we want to paste them together to make a file path for reading data into R:

```
drive <- "c:"
folder <- "temp"
file <- "file"
extension <- ".txt"
```

Now use the function `paste` to put them together:

```
paste(drive, folder, file, extension)
```

```
[1] "c: temp file .txt"
```

This has the essence of what we want, but it is not quite there yet. We need to replace the blank spaces " " that are the default separator with "" no space, and to insert slashes "\\" between the drive and the directory, and the directory and file names:

```
paste(drive, "\\",folder, "\\",file, extension,sep="")
```

```
[1] "c:\\temp\\file.txt"
```

2.12.2   Extracting parts of strings

We being by defining a phrase:

```
phrase <- "the quick brown fox jumps over the lazy dog"
```

The function called `substr` is used to extract substrings of a specified number of characters from within a character string. Here is the code to extract the first, the first and second, the first, second and third, . . . , the first 20 characters from our phrase:

```
q <- character(20)
for (i in 1:20) q[i] <- substr(phrase,1,i)
q
  [1] "t"                    "th"                   "the"
  [4] "the "                 "the q"                "the qu"
```

```
 [7]  "the qui"              "the quic"            "the quick"
[10]  "the quick "           "the quick b"          "the quick br"
[13]  "the quick bro"        "the quick brow"       "the quick brown"
[16]  "the quick brown "     "the quick brown f"   "the quick brown fo"
[19]  "the quick brown fox"  "the quick brown fox "
```

The second argument in `substr` is the number of the character at which extraction is to begin (in this case always the first), and the third argument is the number of the character at which extraction is to end (in this case, the `i`th).

2.12.3   Counting things within strings

Counting the total number of characters in a string could not be simpler; just use the `nchar` function directly, like this:

```
nchar(phrase)
```

```
[1]  43
```

So there are 43 characters including the blanks between the words. To count the numbers of separate individual characters (including blanks) you need to split up the character string into individual characters (43 of them), using `strsplit` like this:

```
strsplit(phrase,split=character(0))
```

```
[[1]]
 [1]   "t"  "h"  "e"  " "  "q"  "u"  "i"  "c"  "k"  " "  "b"  "r"
[13]   "o"  "w"  "n"  " "  "f"  "o"  "x"  " "  "j"  "u"  "m"  "p"
[25]   "s"  "a "  "o"  "v"  "e"  "r"  " "  "t"  "h"  "e"  " "  "l"
[37]   "a"  "z"  "y"  " "  "d"  "o"  "g"
```

You could use `NULL` in place of `split=character(0)` (see below). The `table` function can then be used for counting the number of occurrences of each of the characters:

```
table(strsplit(phrase,split=character(0)))
```

```
  a b c d e f g h i j k l m n o p q r s t u v w x y z
8 1 1 1 1 3 1 1 2 1 1 1 1 1 4 1 1 2 1 2 2 1 1 1 1 1
```

This demonstrates that all of the letters of the alphabet were used at least once within our phrase, and that there were eight blanks within the string called `phrase`. This suggests a way of counting the number of words in a phrase, given that this will always be one more than the number of blanks (so long as there are no leading or trailing blanks in the string):

```
words <- 1+table(strsplit(phrase,split=character(0)))[1]
words
```

```
9
```

What about the lengths of the words within phrase? Here are the separate words:

```
strsplit(phrase, " ")
```

```
[[1]]
[1] "the" "quick" "brown" "fox" "jumps" "over" "the" "lazy" "dog"
```

We work out their lengths using `lapply` to apply the function `nchar` to each element of the list produced by `strsplit`. Then we use `table` to count how many words of each length are present:

```
table(lapply(strsplit(phrase, " "), nchar))

3 4 5
4 2 3
```

showing there were 4 three-letter words, 2 four-letter words and 3 five-letter words.

This is how you reverse a character string. The logic is that you need to break it up into individual characters, then reverse their order, then paste them all back together again. It seems long-winded until you think about what the alternative would be:

```
strsplit(phrase,NULL)

[[1]]
[1]    "t"    "h"    "e"    " "    "q"    "u"    "i"    "c"    "k"    " "    "b"    "r"
[13]   "o"    "w"    "n"    " "    "f"    "o"    "x"    " "    "j"    "u"    "m"    "p"
[25]   "s"    "a "   "o"    "v"    "e"    "r"    " "    "t"    "h"    "e"    " "    "l"
[37]   "a"    "z"    "y"    " "    "d"    "o"    "g"

lapply(strsplit(phrase,NULL),rev)

[[1]]
[1] "g" "o" "d" " " "y" "z" "a" "l" " " "e" "h" "t"
[13] " " "r" "e" "v" "o" " " "s" "p" "m" "u" "j" " "
[25] "x" "o" "f" " " "n" "w" "o" "r" "b" " " "k" "c"
[37] "i" "u" "q" " " "e" "h" "t"

sapply(lapply(strsplit(phrase, NULL), rev), paste, collapse="")

[1] "god yzal eht revo spmuj xof nworb kciuq eht"
```

The `collapse` argument is necessary to reduce the answer back to a single character string. Note that the word lengths are retained, so this would be a poor method of encryption.

When we specify a particular string to form the basis of the split, we end up with a list made up from the components of the string that *do not contain the specified string*. This is hard to understand without an example. Suppose we split our `phrase` using 'the':

```
strsplit(phrase,"the")

[[1]]

[1]   ""   " quick brown fox jumps over "   " lazy dog"
```

There are three elements in this list: the first one is the empty string "" because the first three characters within phrase were exactly 'the'; the second element contains the part of the phrase between the two occurrences of the string 'the'; and the third element is the end of the phrase, following the second 'the'. Suppose that we want to extract the characters between the first and second occurrences of 'the'. This is achieved very simply, using subscripts to extract the second element of the list:

```
strsplit(phrase,"the")[[1]] [2]

[1]   " quick brown fox jumps over "
```

Note that the first subscript in double square brackets refers to the number within the list (there *is* only one list in this case), and the second subscript refers to the second element within this list. So if we want to know how many characters there are between the first and second occurrences of the word 'the' within our phrase, we put:

```
nchar(strsplit(phrase,"the")[[1]] [2])
```

```
[1]   28
```

2.12.4   Upper- and lower-case text

It is easy to switch between upper and lower cases using the `toupper` and `tolower` functions:

```
toupper(phrase)
```

```
[1]   THE QUICK BROWN FOX JUMPS OVER THE LAZY DOG"
```

```
tolower(toupper(phrase))
```

```
[1]   "the quick brown fox jumps over the lazy dog"
```

2.12.5   The `match` function and relational databases

The `match` function answers the question: 'Where (if at all) do the values in the second vector appear in the first vector?' It is a really important function, but it is impossible to understand without an example:

```
first <- c(5,8,3,5,3,6,4,4,2,8,8,8,4,4,6)
second <- c(8,6,4,2)
match(first,second)
```

```
[1] NA 1 NA NA NA 2 3 3 4 1 1 1 3 3 2
```

The first thing to note is that `match` produces a vector of subscripts (index values) and that these are subscripts within the *second* vector. The length of the vector produced by `match` is the length of the *first* vector (15 in this example). If elements of the first vector do not occur anywhere in the second vector, then `match` produces NA. It works like this. Where does 5 (from the first position in the first vector) appear in the second vector? Answer: it does not (NA). Then, where does 8 (the second element of the first vector) appear in the second vector? Answer: in position number 1. And so on. Why would you ever want to use this? The answer turns out to be very general and extremely useful in data management.

Large and/or complicated databases are always best stored as relational databases (e.g. Oracle or Access). In these, data are stored in sets of two-dimensional spreadsheet-like objects called tables. Data are divided into small tables with strict rules as to what data they can contain. You then create relationships between the tables that allow the computer to look from one table to another in order to assemble the data you want for a particular application. The relationship between two tables is based on fields whose values (if not their variable names) are common to both tables. The rules for constructing effective relational databases were first proposed by Dr E.F. Codd of the IBM Research Laboratory at San Jose, California, in an extremely influential paper in 1970:

- All data are in tables.

- There is a separate table for each set of related variables.

- The order of the records within tables is irrelevant (so you can add records without reordering the existing records).

- The first column of each table is a unique ID number for every row in that table (a simple way to make sure that this works is to have the rows numbered sequentially from 1 at the top, so that when you add new rows you are sure that they get unique identifiers).

- There is no unnecessary repetition of data so the storage requirement is minimized, and when we need to edit a record, we only need to edit it once (the last point is very important).

- Each piece of data is 'granular' (meaning as small as possible); so you would split a customer's name into title (Dr), first name (Charles), middle name (Urban), surname (Forrester), and preferred form of address (Chuck), so that if they were promoted, for instance, we would only need to convert Dr to Prof. in the title field.

These are called the 'normalization rules' for creating bullet-proof databases. The use of Structured Query Language (SQL) in R to interrogate relational databases is discussed in Chapter 3 (p. 154). Here, the only point is to see how the match function relates information in one vector (or table) to information in another.

Take a medical example. You have a vector containing the anonymous identifiers of nine patients (subjects):

```
subjects <- c("A", "B", "G", "M", "N", "S", "T", "V", "Z")
```

Suppose you wanted to give a new drug to all the patients identified in the second vector called suitable.patients, and the conventional drug to all the others. Here are the suitable patients:

```
suitable.patients <- c("E", "G", "S", "U", "Z")
```

Notice that there are several suitable patients who are not part of this trial (E and U). This is what the match function does:

```
match(subjects, suitable.patients)
```

```
[1] NA NA 2 NA NA 3 NA NA 5
```

For each of the individuals in the first vector (subjects) it finds the subscript in the second vector (suitable patients), returning NA if that patient does not appear in the second vector. The key point to understand is that the vector produced by match is *the same length as the first vector* supplied to match, and that the numbers in the result are *subscripts within the second vector*. The last bit is what people find hard to understand at first.

Let us go through the output term by term and see what each means. Patient A is not in the suitable vector, so NA is returned. The same is true for patient B. Patient G is suitable, so we get a number in the third position. That number is a 2 because patient G is the second element of the vector called suitable.patients. Neither patient M nor N is in the second vector, so they both appear as NA. Patient S is suitable and so produces a number. The number is 3 because that is the position of S with the second vector.

To complete the job, we want to produce a vector of the drugs to be administered to each of the subjects. We create a vector containing the two treatment names:

```
drug <- c("new", "conventional")
```

Then we use the result of the match to give the right drug to the right patient:

```
drug[ifelse(is.na(match(subjects, suitable.patients)),2,1)]
```

```
[1] "conventional" "conventional" "new" "conventional"
```

```
[5] "conventional" "new" "conventional" "conventional"
[9] "new"
```

Note the use of `ifelse` with `is.na` to produce a subscript 2 (to use with drug) for the unsuitable patients, and a 1 when the result of the match is not NA (i.e. for the suitable patients). You may need to work through this example several times (but it is well worth mastering it).

2.12.6 Pattern matching

We need a dataframe with a serious amount of text in it to make these exercises relevant:

```
wf <- read.table("c:\\temp\\worldfloras.txt",header=T)
attach(wf)
names(wf)

[1]    "Country"     "Latitude"     "Area"      "Population"     "Flora"
[6]    "Endemism"    "Continent"

Country
```

As you can see, there are 161 countries in this dataframe (strictly, 161 places, since some of the entries, such as Sicily and Balearic Islands, are not countries). The idea is that we want to be able to select subsets of countries on the basis of specified patterns within the character strings that make up the country names (factor levels). The function to do this is `grep`. This searches for matches to a pattern (specified in its first argument) within the character vector which forms the second argument. It returns a vector of indices (subscripts) within the vector appearing as the second argument, where the pattern was found in whole or in part. The topic of pattern matching is very easy to master once the penny drops, but it hard to grasp without simple, concrete examples. Perhaps the simplest task is to select all the countries containing a particular letter – for instance, upper-case R:

```
as.vector(Country[grep("R",as.character(Country))])

[1]       "Central African Republic"      "Costa Rica"
[3]       "Dominican Republic"            "Puerto Rico"
[5]       "Reunion"                       "Romania"
[7]       "Rwanda"                        "USSR"
```

To restrict the search to countries whose *first* name begins with R use the ^ character like this:

```
as.vector(Country[grep("^ R",as.character(Country))])

[1]    "Reunion"    "Romania"    "Rwanda"
```

To select those countries with multiple names with upper-case R as the first letter of their second or subsequent names, we specify the character string as "blank R" like this:

```
as.vector(Country[grep(" R",as.character(Country))])

[1]       "Central African Republic"      "Costa Rica"
[3]       "Dominican Republic"            "Puerto Rico"
```

To find all the countries with two or more names, just search for a blank " ":

```
as.vector(Country[grep(" ",as.character(Country))])
```

[1]	"Balearic Islands"	"Burkina Faso"
[3]	"Central African Republic"	"Costa Rica"
[5]	"Dominican Republic"	"El Salvador"
[7]	"French Guiana"	"Germany East"
[9]	"Germany West"	"Hong Kong"
[11]	"Ivory Coast"	"New Caledonia"
[13]	"New Zealand"	"Papua New Guinea"
[15]	"Puerto Rico"	"Saudi Arabia"
[17]	"Sierra Leone"	"Solomon Islands"
[19]	"South Africa"	"Sri Lanka"
[21]	"Trinidad & Tobago"	"Tristan da Cunha"
[23]	"United Kingdom"	"Viet Nam"
[25]	"Yemen North"	"Yemen South"

To find countries with names ending in 'y' use the $ symbol like this:

```
as.vector(Country[grep("y$",as.character(Country))])
```

```
[1]    "Hungary"   "Italy"   "Norway"   "Paraguay"   "Sicily"   "Turkey"
[7]    "Uruguay"
```

To recap: the start of the character string is denoted by ^ and the end of the character string is denoted by $. For conditions that can be expressed as groups (say, series of numbers or alphabetically grouped lists of letters), use square brackets inside the quotes to indicate the range of values that is to be selected. For instance, to select countries with names containing upper-case letters from C to E inclusive, write:

```
as.vector(Country[grep("[C-E]",as.character(Country))])
```

[1]	"Cameroon"	"Canada"
[3]	"Central African Republic"	"Chad"
[5]	"Chile"	"China"
[7]	"Colombia"	"Congo"
[9]	"Corsica"	"Costa Rica"
[11]	"Crete"	"Cuba"
[13]	"Cyprus"	"Czechoslovakia"
[15]	"Denmark"	"Dominican Republic"
[17]	"Ecuador"	"Egypt"
[19]	"El Salvador"	"Ethiopia"
[21]	"Germany East"	"Ivory Coast"
[23]	"New Caledonia"	"Tristan da Cunha"

Notice that this formulation picks out countries like Ivory Coast and Tristan da Cunha that contain upper-case Cs in places other than as their first letters. To restrict the choice to first letters use the ^ operator before the list of capital letters:

```
as.vector(Country[grep("^[C-E]",as.character(Country))])
```

[1]	"Cameroon"	"Canada"
[3]	"Central African Republic"	"Chad"

```
[5]       "Chile"                    "China"
[7]       "Colombia"                 "Congo"
[9]       "Corsica"                  "Costa Rica"
[11]      "Crete"                    "Cuba"
[13]      "Cyprus"                   "Czechoslovakia"
[15]      "Denmark"                  "Dominican Republic"
[17]      "Ecuador"                  "Egypt"
[19]      "El Salvador"              "Ethiopia"
```

How about selecting the counties *not* ending with a specified patterns? The answer is simply to *use negative subscripts* to drop the selected items from the vector. Here are the countries that do not end with a letter between 'a' and 't':

```
as.vector(Country[-grep("[a-t]$",as.character(Country))])
```

```
[1]   "Hungary" "Italy"   "Norway" "Paraguay" "Peru"     "Sicily"
[7]   "Turkey"  "Uruguay" "USA"    "USSR"     "Vanuatu"
```

You see that USA and USSR are included in the list because we specified lower-case letters as the endings to omit. To omit these other countries, put ranges for both upper- and lower-case letters inside the square brackets, separated by a space:

```
as.vector(Country[-grep("[A-T a-t]$",as.character(Country))])
```

```
[1]   "Hungary"  "Italy"    "Norway"   "Paraguay"  "Peru"   "Sicily"
[7]   "Turkey"   "Uruguay"  "Vanuatu"
```

2.12.7   Dot . as the 'anything' character

Countries with 'y' as their second letter are specified by ^.y. The ^ shows 'starting', then a single dot means one character of any kind, so y is the specified second character:

```
as.vector(Country[grep("^.y",as.character(Country))])
```

```
[1]   "Cyprus"  "Syria"
```

To search for countries with 'y' as third letter:

```
as.vector(Country[grep("^..y",as.character(Country))])
```

```
[1]   "Egypt"  "Guyana"  "Seychelles"
```

If we want countries with 'y' as their sixth letter:

```
as.vector(Country[grep("^.{5}y",as.character(Country))])
```

```
[1]   "Norway"  "Sicily"  "Turkey"
```

(Five 'anythings' is shown by '.' then curly brackets {5} then *y*.) Which are the countries with four or fewer letters in their names?

```
as.vector(Country[grep("^.{,4}$",as.character(Country))])
```

```
[1]   "Chad"   "Cuba"   "Iran"   "Iraq"   "Laos"   "Mali"   "Oman"
[8]   "Peru"   "Togo"   "USA"    "USSR"
```

The '.' means 'anything' while the {,4} means 'repeat up to four' anythings (dots) before $ (the end of the string). So to find all the countries with 15 or more characters in their name:

```
as.vector(Country[grep("^.{15,}$",as.character(Country))])
```

```
[1]    "Balearic Islands"      "Central African Republic"
[3]    "Dominican Republic"    "Papua New Guinea"
[5]    "Solomon Islands"       "Trinidad & Tobago"
[7]    "Tristan da Cunha"
```

2.12.8 Substituting text within character strings

Search-and-replace operations are carried out in R using the functions sub and gsub. The two substitution functions differ only in that sub replaces only the first occurrence of a pattern within a character string, whereas gsub replaces all occurrences. An example should make this clear. Here is a vector comprising seven character strings, called text:

```
text <- c("arm", "leg", "head", "foot", "hand", "hindleg" "elbow")
```

We want to replace all lower-case 'h' with upper-case 'H':

```
gsub("h","H",text)
```

```
[1] "arm" "leg" "Head" "foot" "Hand" "Hindleg" "elbow"
```

Now suppose we want to convert the first occurrence of a lower-case 'o' into an upper-case 'O'. We use sub for this (not gsub):

```
sub("o","O",text)
```

```
[1] "arm" "leg" "head" "fOot" "hand" "hindleg" "elbOw"
```

You can see the difference between sub and gsub in the following, where both instances of 'o' in foot are converted to upper case by gsub but not by sub:

```
gsub("o","O",text)
```

```
[1] "arm" "leg" "head" "fOOt" "hand" "hindleg" "elbOw"
```

More general patterns can be specified in the same way as we learned for grep (above). For instance, to replace the first character of every string with upper-case 'O' we use the dot notation (. stands for 'anything') coupled with ^ (the 'start of string' marker):

```
gsub("^.","O",text)
```

```
[1] "Orm" "Oeg" "Oead" "Ooot" "Oand" "Oindleg" "Olbow"
```

It is useful to be able to manipulate the cases of character strings. Here, we capitalize the first character in each string:

```
gsub("(\\w*)(\\w*)", "\\U\\1\\L\\2",text, perl=TRUE)
```

```
[1] "Arm" "Leg" "Head" "Foot" "Hand" "Hindleg" "Elbow"
```

Here we convert all the characters to upper case:

```
gsub("(\\w*)", "\\U\\1",text, perl=TRUE)
```

```
[1] "ARM" "LEG" "HEAD" "FOOT" "HAND" "HINDLEG" "ELBOW"
```

2.12.9 Locations of a pattern within a vector using `regexpr`

Instead of substituting the pattern, we might want to know *if* it occurs in a string and, if so, *where* it occurs within each string. The result of `regexpr`, therefore, is a numeric vector (as with `grep`, above), but now indicating the position of the first instance of the pattern within the string (rather than just *whether* the pattern was there). If the pattern does not appear within the string, the default value returned by `regexpr` is –1. An example is essential to get the point of this:

```
text
```

```
[1] "arm" "leg" "head" "foot" "hand" "hindleg" "elbow"
```

```
regexpr("o",text)
```

```
[1] -1   -1    -1    2    -1    -1    4
```

```
attr(,"match.length")
```

```
[1] -1   -1    -1    1    -1    -1    1
```

This indicates that there were lower-case 'o's in two of the elements of text, and that they occurred in positions 2 and 4, respectively. Remember that if we wanted just the subscripts showing which elements of text contained an 'o' we would use `grep` like this:

```
grep("o",text)
```

```
[1]   4 7
```

and we would extract the character strings like this:

```
text[grep("o",text)]
```

```
[1]   "foot"   "elbow"
```

Counting how many 'o's there are in each string is a different problem again, and this involves the use of `gregexpr`:

```
freq <- as.vector(unlist (lapply(gregexpr("o",text),length)))
```

```
present <- ifelse(regexpr("o",text)<0,0,1)
```

```
freq*present
```

```
[1]       0      0      0      2      0      0      1
```

indicating that there are no 'o's in the first three character strings, two in the fourth and one in the last string. You will need lots of practice with these functions to appreciate all of the issues involved.

The function `charmatch` is for matching characters. If there are multiple matches (two or more) then the function returns the value 0 (e.g. when all the elements contain 'm'):

```
charmatch("m", c("mean", "median", "mode"))
```

```
[1]    0
```

If there is a unique match the function returns the index of the match within the vector of character strings (here in location number 2):

```
charmatch("med", c("mean", "median", "mode"))
```

```
[1]    2
```

2.12.10   Using %in% and which

You want to know all of the matches between one character vector and another:

```
stock <- c("car","van")
requests <- c("truck","suv","van","sports","car","waggon","car")
```

Use which to find the locations in the first-named vector of any and all of the entries in the second-named vector:

```
which(requests %in% stock)
```

```
[1]  3  5  7
```

If you want to know *what* the matches are as well as *where* they are:

```
requests [which(requests %in% stock)]
```

```
[1]  "van"  "car"  "car"
```

You could use the match function to obtain the same result (p. 91):

```
stock[match(requests,stock)] [!is.na(match(requests,stock))]
```

```
[1]  "van"  "car"  "car"
```

but this is more clumsy. A slightly more complicated way of doing it involves sapply:

```
which(sapply(requests,  "%in%",  stock))
```

```
van   car   car
3     5     7
```

Note the use of quotes around the %in% function. Note also that the match must be perfect for this to work ('car' with 'car' is not the same as 'car' with 'cars').

2.12.11   More on pattern matching

For the purposes of specifying these patterns, certain characters are called **metacharacters**, specifically \ | ( ) [{ ^ $ } * + ? Any metacharacter with special meaning in your string may be quoted by preceding it with a backslash: \\, \{, \$ or *, for instance. You might be used to specifying one or more 'wildcards' by * in DOS-like applications. In R, however, the regular expressions used are those specified by POSIX (Portable Operating System Interface) 1003.2, either extended or basic, depending on the value of the extended argument, unless perl = TRUE when they are those of PCRE (see ?grep for details).

Note that the square brackets in these class names [ ] are part of the symbolic names, and must be included in addition to the brackets delimiting the bracket list. For example, [[:alnum:]] means [0-9A-Za-z], except the latter depends upon the locale and the character encoding, whereas the former is independent of locale and character set. The interpretation below is that of the POSIX locale:

[:alnum:]	Alphanumeric characters: [:alpha:] and [:digit:].
[:alpha:]	Alphabetic characters: [:lower:] and [:upper:].
[:blank:]	Blank characters: space and tab.
[:cntrl:]	Control characters in ASCII, octal codes 000 through 037, and 177 (DEL).
[:digit:]	Digits: 0 1 2 3 4 5 6 7 8 9.

[:graph:]   Graphical characters: [:alnum:] and [:punct:].
[:lower:]   Lower-case letters in the current locale.
[:print:]   Printable characters: [:alnum:], [:punct:]} and space.
[:punct:]   Punctuation characters:
             ! " # $ % & () * +$, - ./: ; <=>? @ [\] ^ _ ' { | } ~.
[:space:]   Space characters: tab, newline, vertical tab, form feed, carriage return, space.
[:upper:]   Upper-case letters in the current locale.
[:xdigit:]   Hexadecimal digits: 0 1 2 3 4 5 6 7 8 9 A B C D E F a b c d e f.

Most metacharacters lose their special meaning inside lists. Thus, to include a literal], place it first in the list. Similarly, to include a literal ^, place it anywhere but first. Finally, to include a literal -, place it first or last. Only these and \ remain special inside character classes. To recap:

- dot . matches any single character.

- caret ^ matches the empty string at the beginning of a line.

- dollar sign $ matches the empty string at the end of a line.

- symbols \< and \> respectively match the empty string at the beginning and end of a word.

- the symbol \b matches the empty string at the edge of a word, and \B matches the empty string provided it is not at the edge of a word.

A regular expression may be *followed* by one of several repetition quantifiers:

?	the preceding item is optional and will be matched at most once.
*	the preceding item will be matched zero or more times.
+	the preceding item will be matched one or more times.
{n}	the preceding item is matched exactly *n* times.
{n, }	the preceding item is matched *n* or more times.
{,m}	the preceding item is matched up to *m* times.
{n,m}	the preceding item is matched at least *n* times, but not more than *m* times.

You can use the OR operator | so that "abba|cde" matches either the string "abba" or the string "cde".
   Here are some simple examples to illustrate the issues involved:

```
text <- c("arm","leg","head", "foot","hand", "hindleg", "elbow")
```

The following lines demonstrate the 'consecutive characters' {n} in operation:

```
grep("o{1}",text,value=T)
```

```
[1]  "foot"  "elbow"
```

```
grep("o{2}",text,value=T)
```

```
[1]  "foot"
```

```
grep("o{3}",text,value=T)
```

```
character(0)
```

The following lines demonstrate the use of {n, } '*n* or more' character counting in words:

```
grep("[[:alnum:]]{4, }",text,value=T)

[1]  "head"  "foot"  "hand"  "hindleg"  "elbow"

grep("[[:alnum:]]{5, }",text,value=T)

[1]  "hindleg"  "elbow"

grep("[[:alnum:]]{6, }",text,value=T)

[1]  "hindleg"

grep("[[:alnum:]]{7, }",text,value=T)

[1] "hindleg"
```

2.12.12 `Perl` regular expressions

The `perl = TRUE` argument switches to the PCRE library that implements regular expression pattern matching using the same syntax and semantics as Perl 5.6 or later (with just a few differences). For details (and there are many) see `?regexp`.

2.12.13 Stripping patterned text out of complex strings

Suppose that we want to tease apart the information in these complicated strings:

```
(entries <- c ("Trial 1 58 cervicornis (52 match)", "Trial 2 60
terrestris (51 matched)", "Trial 8 109 flavicollis (101 matches)"))

[1]  "Trial 1 58 cervicornis (52 match)"
[2]  "Trial 2 60 terrestris (51 matched)"
[3]  "Trial 8 109 flavicollis (101 matches)"
```

The first task is to remove the material on numbers of matches including the brackets:

```
gsub(" *$", "", gsub("\\(.*\\)$", "", entries))

[1]  "Trial 1 58 cervicornis"    "Trial 2 60 terrestris"
[3]  "Trial 8 109 flavicollis"
```

The first argument " *$", "", removes the 'trailing blanks', while the second deletes everything .* between the left \\ (and right \\) hand brackets "\\(.*\\)$", substituting this with nothing "". The next job is to strip out the material in brackets and to extract that material, ignoring the brackets themselves:

```
pos <- regexpr("\\(.*\\)$", entries)
substring(entries, first=pos+1, last=pos+attr(pos,"match.length")-2)

[1]   "52 match"  "51 matched"   "101 matches"
```

To see how this has worked it is useful to inspect the values of `pos` that have emerged from the `regexpr` function:

```
pos

[1]   25   23   25
```

```
attr(,"match.length")
```

```
[1]    10    12    13
```

The left-hand bracket appears in position 25 in the first and third elements (note that there are two blanks before 'cervicornis') but in position 23 in the second element. Now the lengths of the strings matching the pattern `\\(.*\\)$` can be checked; it is the number of 'anything' characters between the two brackets, plus one for each bracket: 10, 12 and 13.

Thus, to extract the material in brackets, but to ignore the brackets themselves, we need to locate the first character to be extracted (pos+1) and the last character to be extracted `pos+attr(pos,"match.length")-2`, then use the `substring` function to do the extracting. Note that first and last are vectors of length 3 (= `length(entries)`).

2.13 Dates and times in R

The measurement of time is highly idiosyncratic. Successive years start on different days of the week. There are months with different numbers of days. Leap years have an extra day in February. Americans and Britons put the day and the month in different places: 3/4/2006 is March 4 for the former and April 3 for the latter. Occasional years have an additional 'leap second' added to them because friction from the tides is slowing down the rotation of the earth from when the standard time was set on the basis of the tropical year in 1900. The cumulative effect of having set the atomic clock too slow accounts for the continual need to insert leap seconds (32 of them since 1958). There is currently a debate about abandoning leap seconds and introducing a 'leap minute' every century or so instead. Calculations involving times are complicated by the operation of time zones and daylight saving schemes in different countries. All these things mean that working with dates and times is excruciatingly complicated. Fortunately, R has a robust system for dealing with this complexity. To see how R handles dates and times, have a look at `Sys.time()`:

```
Sys.time()
```

```
[1] "2014-01-24 16:24:54 GMT"
```

This description of date and time is strictly hierarchical from left to right: the longest time scale (years) comes first, then month, then day, separated by hyphens, then there is a blank space, followed by the time, with hours first (using the 24-hour clock), then minutes, then seconds, separated by colons. Finally, there is a character string explaining the time zone (GMT stands for Greenwich Mean Time). This representation of the date and time as a character string is user-friendly and familiar, but it is no good for calculations. For that, we need a single numeric representation of the combined date and time. The convention in R is to base this on seconds (the smallest time scale that is accommodated in `Sys.time`). You can always aggregate upwards to days or year, but you cannot do the reverse. The baseline for expressing today's date and time in seconds is 1 January 1970:

```
as.numeric(Sys.time())
```

```
[1] 1390580694
```

This is fine for plotting time series graphs, but it is not much good for computing monthly means (e.g. is the mean for June significantly different from the July mean?) or daily means (e.g. is the Monday mean significantly different from the Friday mean?). To answer questions like these we have to be able to access a broad set of categorical variables associated with the date: the year, the month, the day of the week, and so forth. To accommodate this, R uses the POSIX system for representing times and dates:

```
class(Sys.time())
```

```
[1] "POSIXct" "POSIXt"
```

You can think of the class POSIXct, with suffix 'ct', as *continuous time* (i.e. a number of seconds), and POSIXlt, with suffix 'lt', as *list time* (i.e. a list of all the various categorical descriptions of the time, including day of the week and so forth). It is hard to remember these acronyms, but it is well worth making the effort. Naturally, you can easily convert to one representation to the other:

```
time.list <- as.POSIXlt(Sys.time())
```

```
unlist(time.list)
```

```
sec min hour mday mon year wday yday isdst
 54  24   16   24   0  114    5   23     0
```

Here you see the nine components of the list. The time is represented by the number of seconds (sec), minutes (min) and hours (on the 24-hour clock). Next comes the day of the month (mday, starting from 1), then the month of the year (mon, starting at January = 0), then the year (starting at 0 = 1900). The day of the week (wday) is coded from Sunday = 0 to Saturday = 6. The day within the year (yday) is coded from 0 = January 1. Finally, there is a logical variable isdst which asks whether daylight saving time is in operation (0 = FALSE in this case). The ones you are most likely to use include year (to get yearly mean values), mon (to get monthly means) and wday (to get means for the different days of the week).

2.13.1 Reading time data from files

It is most likely that your data files contain dates in Excel format, for example 03/09/2014 (a character string showing day/month/year separated by slashes).

```
data <- read.table("c:\\temp\\dates.txt",header=T)
attach(data)
head(data)
```

```
  x       date
1 3 15/06/2014
2 1 16/06/2014
3 6 17/06/2014
4 7 18/06/2014
5 8 19/06/2014
6 9 20/06/2014
```

When you read character data into R using read.table, the default option is to convert the character variables into factors. Factors are of mode numeric and class factor:

```
mode(date)
```

```
[1] "numeric"
```

```
class(date)
```

```
[1] "factor"
```

For our present purposes, the point is that the data are not recognized by R as being dates. To convert a factor or a character string into a POSIXlt object, we employ an important function called 'strip time', written strptime.

2.13.2 The `strptime` function

To convert a factor or a character string into dates using the `strptime` function, we provide a format statement enclosed in double quotes to tell R exactly what to expect, in what order, and separated by what kind of symbol. For our present example we have day (as two digits), then slash, then month (as two digits), then slash, then year (with the century, making four digits).

```
Rdate <- strptime(as.character(date),"%d/%m/%Y")
```

```
class(Rdate)
```

```
[1] "POSIXlt" "POSIXt"
```

It is always a good idea at this stage to add the R-formatted date to your dataframe:

```
data <- data.frame(data,Rdate)
```

```
head(data)
```

```
  x       date       Rdate
1 3 15/06/2014 2014-06-15
2 1 16/06/2014 2014-06-16
3 6 17/06/2014 2014-06-17
4 7 18/06/2014 2014-06-18
5 8 19/06/2014 2014-06-19
6 9 20/06/2014 2014-06-20
```

Now, at last, we can do things with the date information. We might want the mean value of x for each day of the week. The name of this object is `Rdate$wday`:

```
tapply(x,Rdate$wday,mean)
```

```
    0     1     2     3     4     5     6
5.660 2.892 5.092 7.692 8.692 9.692 8.892
```

The lowest mean is on Mondays (wday = 1) and the highest on Fridays (wday = 5).

It is hard to remember all the format codes for strip time, but they are roughly mnemonic and they are always preceded by a percent symbol. Here is the full list of format components:

%a	Abbreviated weekday name
%A	Full weekday name
%b	Abbreviated month name
%B	Full month name
%c	Date and time, locale-specific
%d	Day of the month as decimal number (01–31)
%H	Hours as decimal number (00–23) on the 24-hour clock
%I	Hours as decimal number (01–12) on the 12-hour clock
%j	Day of year as decimal number (0–366)
%m	Month as decimal number (0–11)
%M	Minute as decimal number (00–59)
%p	AM/PM indicator in the locale
%S	Second as decimal number (00–61, allowing for two 'leap seconds')
%U	Week of the year (00–53) using the first Sunday as day 1 of week 1

%w	Weekday as decimal number (0–6, Sunday is 0)
%W	Week of the year (00–53) using the first Monday as day 1 of week 1
%x	Date, locale-specific
%X	Time, locale-specific
%Y	Year with century
%y	Year without century
%Z	Time zone as a character string (output only)

Note the difference between the upper case for year %Y (this is the unambiguous year including the century, 2014), and the potentially ambiguous lower case %y (it is not clear whether 14 means 1914 or 2014).

There is a useful function called `weekdays` (note the plural) for turning the day number into the appropriate name:

```
y <- strptime("01/02/2014",format="%d/%m/%Y")
weekdays(y)
```

```
[1] "Saturday"
```

which is converted from

```
y$wday
```

```
[1] 6
```

because the days of the week are numbered from Sunday = 0.

Here is another kind of date, with years in two-digit form (%y), and the months as abbreviated names (%b) using no separators:

```
other.dates <- c("1jan99", "2jan05", "31mar04", "30jul05")
strptime(other.dates, "%d%b%y")
```

```
[1]  "1999-01-01"  "2005-01-02"  "2004-03-31"  "2005-07-30"
```

Here is yet another possibility with year, then month in full, then week of the year, then day of the week abbreviated, all separated by a single blank space:

```
yet.another.date <- c("2016 January 2 Mon","2017 February 6 Fri","2018
  March 10 Tue")
strptime(yet.another.date,"%Y %B %W %a")
```

```
[1] "2016-01-11" "2017-02-10" "2018-03-06"
```

The system is clever in that it knows the date of the Monday in week number 2 of January in 2016, and of the Tuesday in week 10 of 2018 (the information on month is redundant in this case):

```
yet.more.dates <- c("2016 2 Mon","2017 6 Fri","2018 10 Tue")
strptime(yet.more.dates,"%Y %W %a")
```

```
[1] "2016-01-11" "2017-02-10" "2018-03-06"
```

2.13.3 The `difftime` function

The function `difftime` calculates a difference of two date-time objects and returns an object of class `difftime` with an attribute indicating the units. You can use various arithmetic operations on a `difftime` object including `round`, `signif`, `floor`, `ceiling`, `trunc`, `abs`, `sign` and certain logical operations. You can create a `difftime` object like this:

```
as.difftime(yet.more.dates,"%Y %W %a")

Time differences in days
[1]  1434 1830 2219
attr(,"tzone")
[1]  ""
```

or like this:

```
difftime("2014-02-06","2014-07-06")

Time difference of -149.9583 days

round(difftime("2014-02-06","2014-07-06"),0)

Time difference of -150 days
```

2.13.4 Calculations with dates and times

You can do the following calculations with dates and times:

- time + number

- time − number

- time1 − time2

- time1 'logical operation' time2

where the logical operations are one of ==, ! =, <, <=, > or >=. You can add or subtract a number of seconds or a `difftime` object from a date-time object, but you cannot add two date-time objects. Subtraction of two date-time objects is equivalent to using `difftime`. Unless a time zone has been specified, `POSIXlt` objects are interpreted as being in the current time zone in calculations.

The thing you need to grasp is that you should convert your dates and times into POSIXlt objects *before* starting to do any calculations. Once they are POSIXlt objects, it is straightforward to calculate means, differences and so on. Here we want to calculate the number of days between two dates, 22 October 2015 and 22 October 2018:

```
y2 <- as.POSIXlt("2015-10-22")
y1 <- as.POSIXlt("2018-10-22")
```

Now you can do calculations with the two dates:

```
y1-y2

Time difference of 1096 days
```

2.13.5 The `difftime` and `as.difftime` functions

Working out the time difference between two dates and times involves the `difftime` function, which takes two date-time objects as its arguments. The function returns an object of class `difftime` with an attribute indicating the units. For instance, how many days elapsed between 15 August 2013 and 21 October 2015?

```
difftime("2015-10-21","2013-8-15")
```

```
Time difference of 797 days
```

If you want only the number of days to use in calculation, then write

```
as.numeric(difftime("2015-10-21","2013-8-15"))
```

```
[1] 797
```

If you have times but no dates, then you can use `as.difftime` to create appropriate objects for calculations:

```
t1 <- as.difftime("6:14:21")
```

```
t2 <- as.difftime("5:12:32")
```

```
t1-t2
```

```
Time difference of 1.030278 hours
```

You will often want to create POSIXlt objects from components stored in different vectors within a dataframe. For instance, here is a dataframe with the hours, minutes and seconds from an experiment with two factor levels in four separate columns:

```
times <- read.table("c:\\temp\\times.txt",header=T)
attach(times)
head(times)
```

```
  hrs min sec experiment
1   2  23   6          A
2   3  16  17          A
3   3   2  56          A
4   2  45   0          A
5   3   4  42          A
6   2  56  25          A
```

Because the times are not in POSIXlt format, you need to paste together the hours, minutes and seconds into a character string, using colons as the separator:

```
paste(hrs,min,sec,sep=":")
```

```
 [1] "2:23:6"  "3:16:17" "3:2:56" "2:45:0"  "3:4:42"  "2:56:25" "3:12:28"
 [8] "1:57:12" "2:22:22" "1:42:7" "2:31:17" "3:15:16" "2:28:4"  "1:55:34"
[15] "2:17:7"  "1:48:48"
```

Now save this object as a difftime vector called `duration`:

```
duration <- as.difftime (paste(hrs,min,sec,sep=":"))
```

Then you can carry out calculations like mean and variance using the `tapply` function:

```
tapply(duration,experiment,mean)
```

```
       A          B
2.829375    2.292882
```

which gives the answer in decimal hours.

2.13.6 Generating sequences of dates

You may want to generate sequences of dates by years, months, weeks, days of the month or days of the week. Here are four sequences of dates, all starting on 4 November 2015, the first going in increments of one day:

```
seq(as.POSIXlt("2015-11-04"), as.POSIXlt("2015-11-15"), "1 day")
```

```
 [1] "2015-11-04 GMT" "2015-11-05 GMT" "2015-11-06 GMT" "2015-11-07 GMT"
 [5] "2015-11-08 GMT" "2015-11-09 GMT" "2015-11-10 GMT" "2015-11-11 GMT"
 [9] "2015-11-12 GMT" "2015-11-13 GMT" "2015-11-14 GMT" "2015-11-15 GMT"
```

the second with increments of 2 weeks:

```
seq(as.POSIXlt("2015-11-04"), as.POSIXlt("2016-04-05"), "2 weeks")
```

```
 [1] "2015-11-04 GMT" "2015-11-18 GMT" "2015-12-02 GMT" "2015-12-16 GMT"
 [5] "2015-12-30 GMT" "2016-01-13 GMT" "2016-01-27 GMT" "2016-02-10 GMT"
 [9] "2016-02-24 GMT" "2016-03-09 GMT" "2016-03-23 GMT"
```

the third with increments of 3 months:

```
seq(as.POSIXlt("2015-11-04"), as.POSIXlt("2018-10-04"), "3 months")
```

```
 [1] "2015-11-04 GMT" "2016-02-04 GMT" "2016-05-04 BST" "2016-08-04 BST"
 [5] "2016-11-04 GMT" "2017-02-04 GMT" "2017-05-04 BST" "2017-08-04 BST"
 [9] "2017-11-04 GMT" "2018-02-04 GMT" "2018-05-04 BST" "2018-08-04 BST"
```

the fourth with increments of years:

```
seq(as.POSIXlt("2015-11-04"), as.POSIXlt("2026-02-04"), "year")
```

```
 [1] "2015-11-04 GMT" "2016-11-04 GMT" "2017-11-04 GMT" "2018-11-04 GMT"
 [5] "2019-11-04 GMT" "2020-11-04 GMT" "2021-11-04 GMT" "2022-11-04 GMT"
 [9] "2023-11-04 GMT" "2024-11-04 GMT" "2025-11-04 GMT" "2026-11-04 GMT"
```

If you specify a number, rather than a recognized character string, in the by part of the sequence function, then the number is assumed to be a number of seconds, so this generates the time as well as the date:

```
seq(as.POSIXlt("2015-11-04"), as.POSIXlt("2015-11-05"), 8955)
```

```
 [1] "2015-11-04 00:00:00 GMT" "2015-11-04 02:29:15 GMT"
 [3] "2015-11-04 04:58:30 GMT" "2015-11-04 07:27:45 GMT"
 [5] "2015-11-04 09:57:00 GMT" "2015-11-04 12:26:15 GMT"
 [7] "2015-11-04 14:55:30 GMT" "2015-11-04 17:24:45 GMT"
 [9] "2015-11-04 19:54:00 GMT" "2015-11-04 22:23:15 GMT"
```

As with other forms of seq, you can specify the length of the vector to be generated, instead of specifying the final date:

```
seq(as.POSIXlt("2015-11-04"), by="month", length=10)
```

```
[1] "2015-11-04 GMT" "2015-12-04 GMT" "2016-01-04 GMT" "2016-02-04 GMT"
[5] "2016-03-04 GMT" "2016-04-04 BST" "2016-05-04 BST" "2016-06-04 BST"
[9] "2016-07-04 BST" "2016-08-04 BST"
```

or you can generate a vector of dates to match the length of an existing vector, using `along=` instead of `length=`:

```
results <- runif(16)
seq(as.POSIXlt("2015-11-04"), by="month", along=results )
```

```
[1]  "2015-11-04 GMT" "2015-12-04 GMT" "2016-01-04 GMT" "2016-02-04 GMT"
[5]  "2016-03-04 GMT" "2016-04-04 BST" "2016-05-04 BST" "2016-06-04 BST"
[9]  "2016-07-04 BST" "2016-08-04 BST" "2016-09-04 BST" "2016-10-04 BST"
[13] "2016-11-04 GMT" "2016-12-04 GMT" "2017-01-04 GMT" "2017-02-04 GMT"
```

You can use the `weekdays` function to extract the days of the week from a series of dates:

```
weekdays(seq(as.POSIXlt("2015-11-04"), by="month", along=results ))
```

```
[1]  "Wednesday" "Friday"    "Monday"    "Thursday" "Friday" "Monday"
[7]  "Wednesday" "Saturday"  "Monday"    "Thursday" "Sunday" "Tuesday"
[13] "Friday"    "Sunday"    "Wednesday" "Saturday"
```

Suppose that you want to find the dates of all the Mondays in a sequence of dates. This involves the use of logical subscripts (see p. 39). The subscripts evaluating to TRUE will be selected, so the logical statement you need to make is `wday == 1`. (because Sunday is `wday == 0`). You create an object called `y` containing the first 100 days in 2016 (note that the start date is 31 December 2015), then convert this vector of dates into a POSIXlt object, a list called `x`, like this:

```
y <- as.Date(1:100,origin="2015-12-31")
x <- as.POSIXlt(y)
```

Now, because `x` is a list, you can use the `$` operator to access information on weekday, and you find, of course, that they are all 7 days apart, starting from the 4 January 2016:

```
x[x$wday==1]
```

```
[1]  "2016-01-04 UTC" "2016-01-11 UTC" "2016-01-18 UTC" "2016-01-25 UTC"
[5]  "2016-02-01 UTC" "2016-02-08 UTC" "2016-02-15 UTC" "2016-02-22 UTC"
[9]  "2016-02-29 UTC" "2016-03-07 UTC" "2016-03-14 UTC" "2016-03-21 UTC"
[13] "2016-03-28 UTC" "2016-04-04 UTC"
```

Suppose you want to list the dates of the first Monday in each month. This is the date with `wday == 1` (as above) but only on its first occurrence in each month of the year. This is slightly more tricky, because several months will contain five Mondays, so you cannot use `seq` with `by = "28 days"` to solve the problem (this would generate 13 dates, not the 12 required). Here are the dates of all the Mondays in the year of 2016:

```
y <- as.POSIXlt(as.Date(1:365,origin="2015-12-31"))
```

Here is what we know so far:

```
data.frame(monday=y[y$wday==1],month=y$mo[y$wday==1])[1:12,]
```

```
     monday month
1 2016-01-04     0
2 2016-01-11     0
3 2016-01-18     0
4 2016-01-25     0
```

```
5   2016-02-01      1
6   2016-02-08      1
7   2016-02-15      1
8   2016-02-22      1
9   2016-02-29      1
10  2016-03-07      2
11  2016-03-14      2
12  2016-03-21      2
```

You want a vector to mark the 12 Mondays you require: these are those where month is not duplicated (i.e. you want to take the first row from each month). For this example, the first Monday in January is in row 1 (obviously), the first in February in row 5, the first in March in row 10, and so on. You can use the not duplicated function !duplicated to tag these rows

```
wanted <- !duplicated(y$mo[y$wday==1])
```

Finally, select the 12 dates of the first Mondays using wanted as a subscript like this:

```
y[y$wday==1][wanted]
```

```
[1] "2016-01-04 UTC" "2016-02-01 UTC" "2016-03-07 UTC" "2016-04-04 UTC"
[5] "2016-05-02 UTC" "2016-06-06 UTC" "2016-07-04 UTC" "2016-08-01 UTC"
[9] "2016-09-05 UTC" "2016-10-03 UTC" "2016-11-07 UTC" "2016-12-05 UTC"
```

Note that every month is represented, and none of the dates is later than the 7th of the month as required.

2.13.7 Calculating time differences between the rows of a dataframe

A common action with time data is to compute the time difference between successive rows of a dataframe. The vector called duration created above is of class difftime and contains 16 times measured in decimal hours:

```
class(duration)
```

```
[1] "difftime"
```

```
duration
```

```
Time differences in hours
```

```
 [1] 2.385000 3.271389 3.048889 2.750000 3.078333 2.940278 3.207778
 [8] 1.953333 2.372778 1.701944 2.521389 3.254444 2.467778 1.926111
[15] 2.285278 1.813333
```

```
attr(,"tzone")
[1] ""
```

You can compute the differences between successive rows using subscripts, like this:

```
duration[1:15]-duration[2:16]
```

```
Time differences in hours
```

```
 [1] -0.8863889  0.2225000  0.2988889 -0.3283333  0.1380556
 [6] -0.2675000  1.2544444 -0.4194444  0.6708333 -0.8194444
[11] -0.7330556  0.7866667  0.5416667 -0.3591667  0.4719444
```

You might want to make the differences between successive rows into part of the dataframe (for instance, to relate change in time to one of the explanatory variables in the dataframe). Before doing this, you need to decide on the row in which you want to put the first of the differences. You should be guided by whether the change in time between rows 1 and 2 is related to the explanatory variables in row 1 or row 2. Suppose it is row 1 that we want to contain the first time difference (−0.8864). Because we are working with differences (see p. 785) the vector of differences is shorter by one than the vector from which it was calculated:

```
length(duration[1:15]-duration[2:16])
```

```
[1]   15
```

```
length(duration)
```

```
[1]   16
```

so you need to add one NA to the bottom of the vector (in row 16):

```
diffs <- c(duration[1:15]-duration[2:16],NA)
diffs
```

```
 [1] -0.8863889  0.2225000  0.2988889 -0.3283333  0.1380556 -0.2675000
 [7]  1.2544444 -0.4194444  0.6708333 -0.8194444 -0.7330556  0.7866667
[13]  0.5416667 -0.3591667  0.4719444  NA
```

Now you can make this new vector part of the dataframe called `times`:

```
times$diffs <- diffs
times
```

```
     hrs   min  sec experiment       diffs
1      2    23    6          A  -0.8863889
2      3    16   17          A   0.2225000
3      3     2   56          A   0.2988889
4      2    45    0          A  -0.3283333
5      3     4   42          A   0.1380556
6      2    56   25          A  -0.2675000
7      3    12   28          A   1.2544444
8      1    57   12          A  -0.4194444
9      2    22   22          B   0.6708333
10     1    42    7          B  -0.8194444
11     2    31   17          B  -0.7330556
12     3    15   16          B   0.7866667
13     2    28    4          B   0.5416667
14     1    55   34          B  -0.3591667
15     2    17    7          B   0.4719444
16     1    48   48          B          NA
```

You need to take care when doing things with differences. For instance, is it really appropriate that the difference in row 8 is between the last measurement on treatment A and the first measurement on treatment B? Perhaps what you really want are the time differences within the treatments, so you need to insert another NA in row number 8? If so, then:

```
times$diffs[8] <- NA
```

2.13.8 Regression using dates and times

Here is an example where the number of individual insects was monitored each month over the course of 13 months:

```
data <- read.table("c:\\temp\\timereg.txt",header=T)
attach(data)
head(data)
```

```
  survivors          date
1        100 01/01/2011
2         52 01/02/2011
3         28 01/03/2011
4         12 01/04/2011
5          6 01/05/2011
6          5 01/06/2011
```

The first job, as usual, is to use `strptime` to convert the character string `"01/01/2011"` into a date-time object:

```
dl <- strptime(date,"%d/%m/%Y")
```

You can see that the object called `dl` is of class `POSIXlt` and mode `list`:

```
class(dl)
```

```
[1] "POSIXlt" "POSIXt"
```

```
mode(dl)
```

```
[1] "list"
```

We start by looking at the data using `plot` with the date `dl` on the *x* axis:

```
windows(7,4)
par(mfrow=c(1,2))
plot(dl,survivors,pch=16,xlab ="month")
plot(dl,log(survivors),pch=16,xlab ="month")
```

 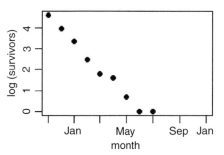

Inspection of the relationship suggests an exponential decay in numbers surviving, so we shall analyse a model in which `log(survivors)` is modelled as a function of time. There are lots of zeros at the end of the time series (once the last of the individuals was dead), so we shall use `subset` to leave out all of the zeros from the model. Let us try to do the regression analysis of `log(survivors)` against date:

```
model <- lm(log(survivors)~dl,subset=(survivors>0))
```

```
Error in model.frame.default(formula = log(survivors) ~ dl,
subset = (survivors > :
 invalid type (list) for variable 'dl'
```

Oops. Why did that not work? The answer is that you cannot have a list as an explanatory variable in a linear model, and as we have just seen, dl is a list. We need to convert from a list (class = POSIXlt) to a continuous numeric variable (class = POSIXct):

```
dc <- as.POSIXct(dl)
```

Now the regression works perfectly when we use the new continuous explanatory variable dc:

```
model <- lm(log(survivors)~dc,subset=(survivors>0))
```

You would get the same effect by using as.numeric(dl) in the model formula. We can use the output from this model to add a regression line to the plot of log(survivors) against time using:

```
abline(model)
```

You need to take care in reporting the values of slopes in regressions involving date-time objects, because the slopes are rates of change of the response variable *per second*. Here is the summary:

```
summary(model)
```

```
Call:
lm(formula = log(survivors) ~ dc, subset = (survivors > 0))
```

```
Residuals:
     Min       1Q   Median       3Q      Max
-0.27606 -0.18306  0.04492  0.13760  0.39277
```

```
Coefficients:
              Estimate Std. Error t value Pr(>|t|)
(Intercept)  3.040e+02  1.531e+01   19.86 2.05e-07 ***
dc          -2.315e-07  1.174e-08  -19.72 2.15e-07 ***
```

```
Residual standard error: 0.2383 on 7 degrees of freedom
Multiple R-squared:      0.9823, Adjusted R-squared: 0.9798
F-statistic:             389 on 1 and 7 DF, p-value: 2.152e-07
```

The slope is -2.315×10^{-7}; the change in log(survivors) *per second*. It might be more useful to express this as the monthly rate. So, with 60 seconds per minute, 60 minutes per hour, 24 hours per day, and (say) 30 days per month, the appropriate rate is

```
 -2.315E-07 * 60 * 60 * 24 * 30
```

```
[1] -0.600048
```

We can check this out by calculating how many survivors we would expect from 100 starters after two months:

```
100*exp(-0.600048 * 2)
```

```
[1] 30.11653
```

which compares well with our observed count of 28 (see above).

2.13.9   Summary of dates and times in R

The key thing to understand is the difference between the two representations of dates and times in R. They have unfortunately non-memorable names.

- POSIXlt gives a **list** containing separate vectors for the year, month, day of the week, day within the year, and suchlike. It is very useful as a categorical explanatory variable (e.g. to get monthly means from data gathered over many years using `date$mon`).

- POSIXct gives a **vector** containing the date and time expressed as a continuous variable that you can use in regression models (it is the number of seconds since the beginning of 1970).

You can use other functions like `date`, but I do not recommend them. If you stick with POSIX you are less likely to get confused.

2.14   Environments

R is built around a highly sophisticated system of naming and locating objects. When you start a session in R, the variables you create are in the global environment `.GlobalEnv`, which is known more familiarly as the user's workspace. This is the first place in which R looks for things. Technically, it is the first item on the search path. It can also be accessed by `globalenv()`.

An environment consists of a **frame**, which is collection of named objects, and a **pointer** to an enclosing environment. The most common example is the frame of variables that is local to a function call; its **enclosure** is the environment where the function was defined. The enclosing environment is distinguished from the parent frame, which is the environment of the **caller** of a function.

There is a strict hierarchy in which R looks for things: it starts by looking in the frame, then in the enclosing frame, and so on.

2.14.1   Using `with` **rather than** `attach`

When you `attach` a dataframe you can refer to the variables within that dataframe by name.

Advanced R users do not routinely employ `attach` in their work, because it can lead to unexpected problems in resolving names (e.g. you can end up with multiple copies of the same variable name, each of a different length and each meaning something completely different). Most modelling functions like `lm` or `glm` have a `data=` argument so `attach` is unnecessary in those cases. Even when there is no `data=` argument it is preferable to wrap the call using `with` like this:

```
with(dataframe, function(...))
```

The `with` function evaluates an R expression in an environment constructed from data. You will often use the `with` function with other functions like `tapply` or `plot` which have no built-in data argument. If your dataframe is part of the built-in package called `datasets` (like `OrchardSprays`) you can refer to the dataframe directly by name:

```
with(OrchardSprays,boxplot(decrease~treatment))
```

Here we calculate the number of 'no' (not infected) cases in the `bacteria` dataframe which is part of the MASS library:

```
library(MASS)

with(bacteria,tapply((y=="n"),trt,sum))

placebo    drug     drug+
12         18       13
```

Here we plot brain weight against body weight for `mammals` on log–log axes:

```
with(mammals,plot(body,brain,log="xy"))
```

without attaching either dataframe. Here is an unattached dataframe called `reg.data`:

```
reg.data <- read.table("c:\\temp\\regression.txt",header=T)
```

with which we carry out a linear regression and print a summary:

```
with (reg.data, {
        model <- lm(growth~tannin)
        summary(model) })
```

The linear model fitting function `lm` knows to look in `reg.data` to find the variables called `growth` and `tannin` because the `with` function has used `reg.data` for constructing the environment from which `lm` is called. Groups of statements (different lines of code) to which the `with` function applies are contained within curly brackets. An alternative is to define the data environment as an argument in the call to lm like this:

```
summary(lm(growth~tannin,data=reg.data))
```

You should compare these outputs with the same example using `attach` on p. 450. Note that whatever form you choose, you still need to get the dataframe into your current environment by using `read.table` (if, as here, it is to be read from an external file), or from a library (like MASS to get `bacteria` and `mammals`, as above). To see the names of the dataframes in the built-in package called `datasets`, type:

```
data()
```

To see all available data sets (including those in the installed packages), type:

```
data(package = .packages(all.available = TRUE))
```

2.14.2 Using `attach` in this book

I use `attach` throughout this book because experience has shown that it makes the code easier to understand for beginners. In particular, using `attach` provides simplicity and brevity, so that we can:

- refer to variables by name, so x rather than `dataframe$x`

- write shorter models, so `lm(y~x)` rather than `lm(y~x,data=dataframe)`

- go straight to the intended action, so `plot(y~x)` not `with(dataframe,plot(y~x))`

Nevertheless, readers are encouraged to use `with` or `data=` for their own work, and to avoid using `attach` wherever possible.

2.15   Writing R functions

You typically write functions in R to carry out operations that require two or more lines of code to execute, and that you do not want to type lots of times. We might want to write simple functions to calculate measures of central tendency (p. 116), work out factorials (p. 71) and such-like.

Functions in R are objects that carry out operations on *arguments* that are supplied to them and return one or more values. The syntax for writing a function is

```
function (argument list) { body }
```

The first component of the function declaration is the keyword `function`, which indicates to R that you want to create a function. An argument list is a comma-separated list of formal arguments. A formal argument can be a symbol (i.e. a variable name such as *x* or *y*), a statement of the form `symbol = expression` (e.g. `pch=16`) or the special formal argument `...` (triple dot). The body can be any valid R expression or set of R expressions over one or more lines. Generally, the body is a group of expressions contained in curly brackets `{ }`, with each expression on a separate line (if the body fits on a single line, no curly brackets are necessary). Functions are typically assigned to symbols, but they need not be. This will only begin to mean anything after you have seen several examples in operation.

2.15.1   Arithmetic mean of a single sample

The mean is the sum of the numbers $\sum y$ divided by the number of numbers $n = \sum 1$ (summing over the number of numbers in the vector called *y*). The R function for *n* is `length(y)` and for $\sum y$ is `sum(y)`, so a function to compute arithmetic means is

```
arithmetic.mean <- function(x)    sum(x)/length(x)
```

We should test the function with some data where we know the right answer:

```
y <- c(3,3,4,5,5)
```

```
arithmetic.mean(y)
```

```
[1]    4
```

Needless to say, there is a built-in function for arithmetic means called `mean`:

```
mean(y)
```

```
[1]    4
```

2.15.2   Median of a single sample

The median (or 50th percentile) is the middle value of the sorted values of a vector of numbers:

```
sort(y)[ceiling(length(y)/2)]
```

There is slight hitch here, of course, because if the vector contains an even number of numbers, then there *is* no middle value. The logic here is that we need to work out the arithmetic average of the two values of *y* on either side of the middle. The question now arises as to how we know, in general, whether the vector *y* contains an odd or an even number of numbers, so that we can decide which of the two methods to use. The trick here is to use modulo 2 (p. 18). Now we have all the tools we need to write a general function to calculate medians. Let us call the function `med` and define it like this:

```
med <- function(x) {
odd.even <- length(x)%%2
if (odd.even == 0) (sort(x)[length(x)/2]+sort(x)[1+ length(x)/2])/2
else sort(x)[ceiling(length(x)/2)]
}
```

Notice that when the `if` statement is true (i.e. we have an even number of numbers) then the expression immediately following the `if` function is evaluated (this is the code for calculating the median with an even number of numbers). When the `if` statement is false (i.e. we have an odd number of numbers, and `odd.even == 1`) then the expression following the `else` function is evaluated (this is the code for calculating the median with an odd number of numbers). Let us try it out, first with the odd-numbered vector y, then with the even-numbered vector y[-1], after the first element of y (y[1] = 3) has been dropped (using the negative subscript):

```
med(y)
```

```
[1]   4
```

```
med(y[-1])
```

```
[1]   4.5
```

You could write the same function in a single (long) line by using `ifelse` instead of `if`. You need to remember that the second argument in `ifelse` is the action to be performed when the condition is true, and the third argument is what to do when the condition is false:

```
med <- function(x) ifelse(length(x)%%2==1, sort(x)[ceiling(length(x)/2)],
        (sort(x)[length(x)/2]+sort(x)[1+ length(x)/2])/2 )
```

Again, you will not be surprised that there is a built-in function for calculating medians, and helpfully it is called `median`.

2.15.3   Geometric mean

For processes that change multiplicatively rather than additively, neither the arithmetic mean nor the median is an ideal measure of central tendency. Under these conditions, the appropriate measure is the geometric mean. The formal definition of this is somewhat abstract: the geometric mean is the nth root of the product of the data. If we use capital Greek pi (Π) to represent multiplication, and $\hat{y}$ (pronounced y-hat) to represent the geometric mean, then

$$\hat{y} = \sqrt[n]{\Pi y}.$$

Let us take a simple example we can work out by hand: the numbers of insects on 5 plants were as follows: 10, 1, 1000, 1, 10. Multiplying the numbers together gives 100 000. There are five numbers, so we want the fifth root of this. Roots are hard to do in your head, so we will use R as a calculator. Remember that roots are fractional powers, so the fifth root is a number raised to the power $1/5 = 0.2$. In R, powers are denoted by the ^ symbol:

```
100000^0.2
```

```
[1]   10
```

So the geometric mean of these insect numbers is 10 insects per stem. Note that two of the data were exactly like this, so it seems a reasonable estimate of central tendency. The arithmetic mean, on the other hand, is a hopeless measure of central tendency in this case, because the large value (1000) is so influential: it is given by $(10 + 1 + 1000 + 1 + 10)/5 = 204.4$, and none of the data is close to it.

```
insects <- c(1,10,1000,10,1)
```

```
mean(insects)
```

```
[1]   204.4
```

Another way to calculate geometric mean involves the use of logarithms. Recall that to multiply numbers together we add up their logarithms. And to take roots, we divide the logarithm by the root. So we should be able to calculate a geometric mean by finding the antilog (exp) of the average of the logarithms (log) of the data:

```
exp(mean(log(insects)))
```

```
[1]   10
```

So here is a function to calculate geometric mean of a vector of numbers x:

```
geometric <- function (x) exp(mean(log(x)))
```

We can test it with the insect data:

```
geometric(insects)
```

```
[1]   10
```

The use of geometric means draws attention to a general scientific issue. Look at the figure below, which shows numbers varying through time in two populations. Now ask yourself which population is the more variable. Chances are, you will pick the upper line:

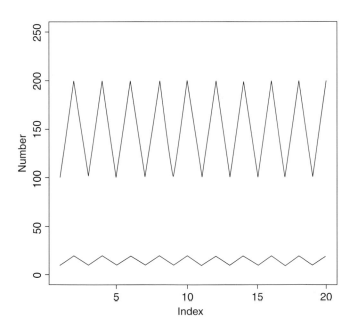

But now look at the scale on the *y* axis. The upper population is fluctuating 100, 200, 100, 200 and so on. In other words, it is doubling and halving, doubling and halving. The lower curve is fluctuating 10, 20, 10, 20, 10, 20 and so on. It, too, is doubling and halving, doubling and halving. So the answer to the question is that they are equally variable. It is just that one population has a higher mean value than the other (150 vs. 15 in this case). In order not to fall into the trap of saying that the upper curve is more variable than the lower curve, it is good practice to graph the logarithms rather than the raw values of things like population sizes that change multiplicatively, as below.

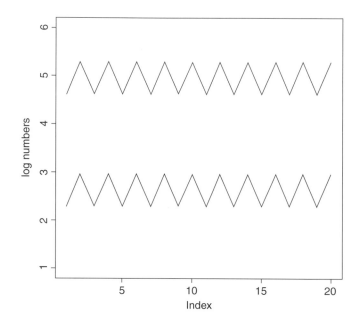

Now it is clear that both populations are equally variable. Note the change of scale, as specified using the `ylim=c(1,6)` option within the `plot` function (p. 193).

2.15.4 Harmonic mean

Consider the following problem. An elephant has a territory which is a square of side 2 km. Each morning, the elephant walks the boundary of this territory. He begins the day at a sedate pace, walking the first side of the territory at a speed of 1 km/hr. On the second side, he has sped up to 2 km/hr. By the third side he has accelerated to an impressive 4 km/hr, but this so wears him out, that he has to return on the final side at a sluggish 1 km/hr. So what is his average speed over the ground? You might say he travelled at 1, 2, 4 and 1 km/hr so the average speed is $(1 + 2 + 4 + 1)/4 = 8/4 = 2$ km/hr. But that is wrong. Can you see how to work out the right answer? Recall that velocity is defined as distance travelled divided by time taken. The distance travelled is easy: it is just $4 \times 2 = 8$ km. The time taken is a bit harder. The first edge was 2 km long, and travelling at 1 km/hr this must have taken 2 hr. The second edge was 2 km long, and travelling at 2 km/hr this must have taken 1 hr. The third edge was 2 km long and travelling at 4 km/hr this must have taken 0.5 hr. The final edge was 2 km long and travelling at 1 km/hr this must have taken 2 hr. So the total time taken was $2 + 1 + 0.5 + 2 = 5.5$ hr. So the average speed is not 2 km/hr but $8/5.5 = 1.4545$ km/hr. The way to solve this problem is to use the **harmonic mean**.

The harmonic mean is the reciprocal of the average of the reciprocals. The average of our reciprocals is:

$$\frac{1}{1} + \frac{1}{2} + \frac{1}{4} + \frac{1}{1} = \frac{2.75}{4} = 0.6875.$$

The reciprocal of this average is the harmonic mean

$$\frac{4}{2.75} = \frac{1}{0.6875} = 1.4545.$$

In symbols, therefore, the harmonic mean, $\tilde{y}$ (y-curl), is given by

$$\tilde{y} = \frac{1}{\left(\sum(1/y)\right)/n} = \frac{n}{\sum(1/y)}.$$

An R function for calculating harmonic means, therefore, could be

```
harmonic <- function (x) 1/mean(1/x)
```

and testing it on our elephant data gives

```
harmonic(c(1,2,4,1))
```

```
[1]  1.454545
```

2.15.5   Variance

A measure of variability is perhaps the most important quantity in statistical analysis. The greater the variability in the data, the greater will be our uncertainty in the values of parameters estimated from the data, and the less will be our ability to distinguish between competing hypotheses about the data.

The variance of a sample is measured as a function of 'the sum of the squares of the difference between the data and the arithmetic mean'. This important quantity is called the 'sum of squares':

$$SS = \sum (y - \bar{y})^2.$$

Naturally, this quantity gets bigger with every new data point you add to the sample. An obvious way to compensate for this is to measure variability as the average of the squared departures from the mean (the 'mean square deviation'.). There is a slight problem, however. Look at the formula for the sum of squares, SS, above and ask yourself what you need to know before you can calculate it. You have the data, y, but the only way you can know the sample mean, $\bar{y}$, is to calculate it from the data (you will never know $\bar{y}$ in advance).

2.15.6   Degrees of freedom

To complete our calculation of the variance we need the **degrees of freedom** (d.f.). This important concept in statistics is defined as follows:

$$\text{d.f.} = n - k,$$

which is the sample size, n, minus the number of parameters, k, estimated from the data. For the variance, we have estimated one parameter from the data, $\bar{y}$, and so there are $n - 1$ degrees of freedom. In a linear regression, we estimate two parameters from the data, the slope and the intercept, and so there are $n - 2$ degrees of freedom in a regression analysis.

Variance is denoted by the lower-case Latin letter s squared: s^2. The square root of variance, s, is called the standard deviation. We always calculate variance as

$$\text{variance} = s^2 = \frac{\text{sum of squares}}{\text{degrees of freedom}}.$$

Consider the following data:

```
y <- c(13,7,5,12,9,15,6,11,9,7,12)
```

We need to write a function to calculate the sample variance: we call it variance and define it like this:

```
variance <- function(x) sum((x - mean(x))^2)/(length(x)-1)
```

and use it like this:

```
variance(y)
```

```
[1]   10.25455
```

Our measure of variability in these data, the variance, is thus 10.254 55. It is said to be an unbiased estimator because we divide the sum of squares by the degrees of freedom $(n - 1)$ rather than by the sample size, n, to compensate for the fact that we have estimated one parameter from the data. So the variance is *close* to the average squared difference between the data and the mean, especially for large samples, but it is not exactly equal to the mean squared deviation. Needless to say, R has a built-in function to calculate variance called var:

```
var(y)
```

```
[1]   10.25455
```

2.15.7 Variance ratio test

How do we know if two variances are significantly different from one another? One of several sensible ways to do this is to carry out Fisher's F test, which is simply the ratio of the two variances (see p. 287). Here is a function to print the p value (p. 347) associated with a comparison of the larger and smaller variances:

```
variance.ratio <- function(x,y)  {

       v1 <- var(x)
       v2 <- var(y)
   if (var(x)  > var(y)){
       vr <- var(x)/var(y)
       df1 <- length(x)-1
       df2 <- length(y)-1}
   else {
        vr <- var(y)/var(x)
        df1 <- length(y)-1
        df2 <- length(x)-1}
2*(1-pf(vr,df1,df2))  }
```

The last line of our function works out the probability of getting an F ratio as big as `vr` or bigger by chance alone if the two variances were really the same, using the cumulative probability of the F distribution, which is an R function called `pf`. We need to supply `pf` with three *arguments*: the size of the variance ratio (`vr`), the number of degrees of freedom in the numerator (`df1 = 9`) and the number of degrees of freedom in the denominator (`df2 = 9`).

Here are some data to test our function. They are normally distributed random numbers but the first set has a variance of 4 and the second a variance of 16 (i.e. standard deviations of 2 and 4, respectively):

```
a <- rnorm(10,15,2)
```

```
b <- rnorm(10,15,4)
```

Here is our function in action:

```
variance.ratio(a,b)
```

```
[1]    0.01593334
```

We can compare our p with the p value given by the built-in function called `var.test`:

```
var.test(a,b)
```

```
F test to compare two variances

data: a and b
F = 0.1748, num df = 9, denom df = 9, p-value = 0.01593
alternative hypothesis: true ratio of variances is not equal to 1
95 percent confidence interval:
0.04340939 0.70360673
sample estimates:
ratio of variances
       0.1747660
```

2.15.8 Using variance

Variance is used in two main ways: for establishing measures of unreliability (e.g. confidence intervals) and for testing hypotheses (e.g. Student's t test). Here we will concentrate on the former; the latter is discussed in Chapter 8.

Consider the properties that you would like a measure of unreliability to possess. As the variance of the data increases, what would happen to the unreliability of estimated parameters? Would it go up or down? Unreliability would go up as variance increased, so we would want to have the variance on the top (the numerator) of any divisions in our formula for unreliability:

$$\text{unreliability} \propto s^2.$$

What about sample size? Would you want your estimate of unreliability to go up or down as sample size, n, increased? You would want unreliability to go down as sample size went up, so you would put sample size on the bottom of the formula for unreliability (i.e. in the denominator):

$$\text{unreliability} \propto \frac{s^2}{n}.$$

Finally, consider the units in which unreliability is measured. What are the units in which our current measure is expressed? Sample size is dimensionless, but variance is based on the sum of squared differences, so it has dimensions of mean squared. So if the mean was a length in cm, the variance would be an area in cm^2. This is an unfortunate state of affairs. It would make good sense to have the dimensions of the unreliability measure and of the parameter whose unreliability it is measuring the same. That is why all unreliability measures are enclosed inside a big square root term. Unreliability measures are called *standard errors*. What we have just worked out is the *standard error of the mean*,

$$se_{\bar{y}} = \sqrt{\frac{s^2}{n}},$$

where s^2 is the variance and n is the sample size. There is no built-in R function to calculate the standard error of a mean, but it is easy to write one:

```
se <- function(x) sqrt(var(x)/length(x))
```

You can refer to functions from within other functions. Recall that a confidence interval (CI) is 't from tables times the standard error':

$$\text{CI} = t_{\alpha/2,\text{d.f.}} \times se.$$

The R function qt gives the value of Student's t with $1 - \alpha/2 = 0.975$ and degrees of freedom d.f. $=$ length(x)-1. Here is a function called ci95 which uses our function se to compute 95% confidence intervals for a mean:

```
ci95 <- function(x) {
        t.value <- qt(0.975,length(x)-1)
        standard.error <- se(x)
        ci <- t.value*standard.error
cat("95 Confidence Interval = ", mean(x) -ci, "to ", mean(x) +ci,"\n") }
```

We can test the function with 150 normally distributed random numbers with mean 25 and standard deviation 3:

```
x <- rnorm(150,25,3)
```

```
ci95(x)
```

```
95% Confidence Interval = 24.76245 to 25.74469
```

If we were to repeat the experiment, we could be 95% certain that the mean of the new sample would lie between 24.76 and 25.74.

We can use the se function to investigate how the standard error of the mean changes with the sample size. First we generate one set of data from which we shall take progressively larger samples:

```
xv <- rnorm(30)
```

Now in a loop take samples of size 2, 3, 4, ..., 30:

```
sem <- numeric(30)
sem[1] <- NA
```

```
for(i in 2:30) sem[i] <- se(xv[1:i])
```

```
plot(1:30,sem,ylim=c(0,0.8),
    ylab="standard error of mean",xlab="sample size n",pch=16)
```

You can see clearly that as the sample size falls below about $n = 15$, so the standard error of the mean increases rapidly. The blips in the line are caused by outlying values being included in the calculations of the standard error with increases in sample size. The smooth curve is easy to compute: since the values in xv came from a standard normal distribution with mean 0 and standard deviation 1, so the average curve would be $1/\sqrt{n}$, which we can add to our graph using `lines`:

```
lines(2:30,1/sqrt(2:30))
```

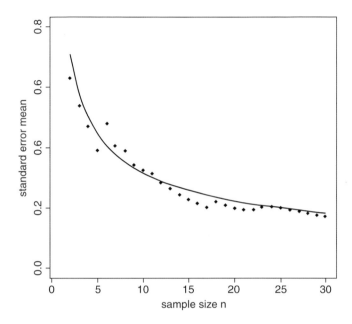

You can see that our single simulation captured the essence of the shape but was wrong in detail, especially for the samples with the lowest replication. However, our single sample was reasonably good for $n > 24$.

2.15.9 Deparsing: A graphics function for error bars

There is no function in the base package of R for drawing error bars on bar charts, although several contributed packages use the `arrows` function for this purpose (p. 204). Here is a simple, stripped-down function that is supplied with three arguments: the heights of the bars (yv), the lengths (up and down) of the error bars (z) and the labels for the bars on the x axis (nn).

The process of deparsing turns an unevaluated expression into a character string. One of the important uses of deparsing is in functions that produce output that you want to label with the particular names of the variables that were passed to the function. For instance, if the function is written in terms of a continuous response variable y and a categorical explanatory variable x, you might want to label the axes of a plot produced by the function with, say, clipping and biomass in place of x and y. For instance, if the function is written in terms of a continuous response variable yv, you might want to label the axes of a plot produced by the function with, say, biomass in place of yv. Inside the error.bars function, the barplot function uses the deparse function to create the appropriate text for ylab.

```
error.bars <- function(yv,z,nn){
xv <-
barplot(yv,ylim=c(0,(max(yv)+max(z))),names=nn,ylab=deparse(substitute(yv)
))
g=(max(xv)-min(xv))/50
for (i in 1:length(xv)) {
     lines(c(xv[i],xv[i]),c(yv[i]+z[i],yv[i]-z[i]))
     lines(c(xv[i]-g,xv[i]+g),c(yv[i]+z[i], yv[i]+z[i]))
     lines(c(xv[i]-g,xv[i]+g),c(yv[i]-z[i], yv[i]-z[i]))
}}
```

Here is the `error.bars` function in action with the plant competition data (p. 426):

```
comp <- read.table("c:\\temp\\competition.txt",header=T)
attach(comp)
names(comp)
```

```
[1]  "biomass"  "clipping"
```

```
se <- rep(28.75,5)
labels <- as.character(levels(clipping))
ybar <- as.vector(tapply(biomass,clipping,mean))
```

Now invoke the function with the means, standard errors and bar labels:

```
error.bars(ybar,se,labels)
```

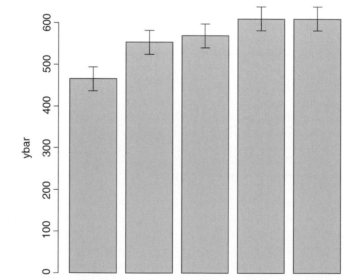

Here is a function to plot error bars on a scatterplot in both the *x* and *y* directions:

```
xy.error.bars <- function (x,y,xbar,ybar){
  plot(x, y, pch=16, ylim=c(min(y-ybar),max(y+ybar)),
    xlim=c(min(x-xbar),max(x+xbar)))
```

```
arrows(x, y-ybar, x, y+ybar, code=3, angle=90, length=0.1)
arrows(x-xbar, y, x+xbar, y, code=3, angle=90, length=0.1) }
```

We test it with these data:

```
x <- rnorm(10,25,5)
y <- rnorm(10,100,20)
xb <- runif(10)*5
yb <- runif(10)*20

xy.error.bars(x,y,xb,yb)
```

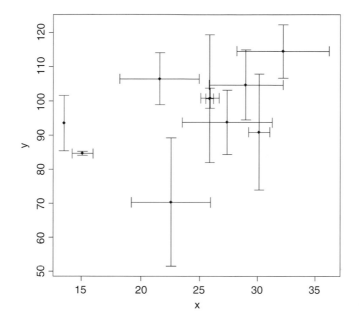

2.15.10 The switch function

When you want a function to do different things in different circumstances, then the switch function can be useful. Here we write a function that can calculate any one of four different measures of central tendency: arithmetic mean, geometric mean, harmonic mean or median (see pp. 115–119 for explanations of the separate functions). The character variable called measure should take one value of Mean, Geometric, Harmonic or Median; any other text will lead to the error message Measure not included. Alternatively, you can specify the number of the switch (e.g. 1 for Mean, 4 for Median).

```
central <- function(y, measure) {
    switch(measure,
        Mean = mean(y),
        Geometric = exp(mean(log(y))),
        Harmonic = 1/mean(1/y),
        Median = median(y),
    stop("Measure not included")) }
```

Note that you have to include the character strings in quotes as arguments to the function, but they must not be in quotes within the `switch` function itself.

```
central(rnorm(100,10,2),"Harmonic")
```

```
[1]  9.554712
```

```
central(rnorm(100,10,2),4)
```

```
[1]   10.46240
```

2.15.11   The evaluation environment of a function

When a function is called or invoked a new *evaluation frame* is created. In this frame the formal arguments are matched with the supplied arguments according to the rules of **argument matching** (below). The statements in the body of the function are evaluated sequentially in this environment frame.

The first thing that occurs in a function evaluation is the matching of the formal to the actual or supplied arguments. This is done by a three-pass process:

- **Exact matching on tags**. For each named supplied argument the list of formal arguments is searched for an item whose name matches exactly.

- **Partial matching on tags**. Each named supplied argument is compared to the remaining formal arguments using partial matching. If the name of the supplied argument matches exactly with the first part of a formal argument then the two arguments are considered to be matched.

- **Positional matching**. Any unmatched formal arguments are bound to unnamed supplied arguments, in order. If there is a ... argument, it will take up the remaining arguments, tagged or not.

- If any arguments remain unmatched an error is declared.

Supplied arguments and default arguments are treated differently. The supplied arguments to a function are evaluated in the evaluation frame of the calling function. The default arguments to a function are evaluated in the evaluation frame of the function. In general, supplied arguments behave as if they are local variables initialized with the value supplied and the name of the corresponding formal argument. Changing the value of a supplied argument within a function will not affect the value of the variable in the calling frame.

2.15.12   Scope

The **scoping rules** are the set of rules used by the evaluator to find a value for a symbol. A symbol can be either **bound** or **unbound**. All of the formal arguments to a function provide bound symbols in the body of the function. Any other symbols in the body of the function are either local variables or unbound variables. A local variable is one that is defined within the function, typically by having it on the left-hand side of an assignment. During the evaluation process if an unbound symbol is detected then R attempts to find a value for it: the environment of the function is searched first, then its enclosure and so on until the global environment is reached. The value of the first match is then used.

2.15.13   Optional arguments

Here is a function called `charplot` that produces a scatterplot of x and y using solid red circles as the plotting symbols: there are two essential arguments (x and y) and two optional (pc and co) to control selection of the plotting symbol and its colour:

```
charplot <- function(x,y,pc=16,co="red"){
plot(y~x,pch=pc,col=co)}
```

The optional arguments are given their default values using = in the argument list. To execute the function you need only provide the vectors of x and y:

```
charplot(1:10,1:10)
```

to get solid red circles. You can get a different plotting symbol simply by adding a third argument

```
charplot(1:10,1:10,17)
```

which produces red solid triangles (pch=17). If you want to change only the colour (the fourth argument) then you have to specify the variable name because the optional arguments would not then be presented in sequence. So, for navy blue solid circles, you put:

```
charplot(1:10,1:10,co="navy")
```

To change both the plotting symbol and the colour you do not need to specify the variable names, so long as the plotting symbol is the third argument and the colour is the fourth:

```
charplot(1:10,1:10,15,"green")
```

This produces solid green squares. Reversing the optional arguments does not work:

```
charplot(1:10,1:10,"green",15)
```

(this uses the letter g as the plotting symbol and colour no. 15). If you specify both variable names, then the order does not matter:

```
charplot(1:10,1:10,co="green",pc=15)
```

This produces solid green squares despite the arguments being out of sequence.

2.15.14 Variable numbers of arguments (...)

Some applications are much more straightforward if the number of arguments does not need to be specified in advance. There is a special formal name ... (triple dot) which is used in the argument list to specify that an arbitrary number of arguments are to be passed to the function. Here is a function that takes any number of vectors and calculates their means and variances:

```
many.means <- function (...) {
     data <- list(...)
     n <- length(data)
     means <- numeric(n)
     vars <- numeric(n)
     for (i in 1:n) {
          means[i] <- mean(data[[i]])
          vars[i] <- var(data[[i]])
     }
     print(means)
     print(vars)
     invisible(NULL)
}
```

The main features to note are these. The function definition has ... as its only argument. The 'triple dot' argument ... allows the function to accept additional arguments of unspecified name and number, and this introduces tremendous flexibility into the structure and behaviour of functions. The first thing done inside the function is to create an object called `data` out of the list of vectors that are actually supplied in any particular case. The length of this list is the number of vectors, not the lengths of the vectors themselves (these could differ from one vector to another, as in the example below). Then the two output variables (`means` and `vars`) are defined to have as many elements as there are vectors in the parameter list. The loop goes from 1 to the number of vectors, and for each vector uses the built-in functions `mean` and `var` to compute the answers we require. It is important to note that because `data` is a list, we use double `[[ ]]` subscripts in addressing its elements.

Now try it out. To make things difficult we shall give it three vectors of different lengths. All come from the standard normal distribution (with mean 0 and variance 1) but x is 100 in length, y is 200 and z is 300 numbers long:

```
x <- rnorm(100)
y <- rnorm(200)
z <- rnorm(300)
```

Now we invoke the function:

```
many.means(x,y,z)
```

```
[1]    -0.039181830    0.003613744    0.050997841
[1]        1.146587       0.989700       0.999505
```

As expected, all three means (top row) are close to 0, and all three variances are close to 1.

You can use ... to absorb some arguments into an intermediate function which can then be extracted by functions called subsequently. R has a form of *lazy evaluation* of function arguments in which arguments are not evaluated until they are needed (in some cases the argument will never be evaluated).

2.15.15   Returning values from a function

Often you want a function to return a single value (like a mean or a maximum), in which case you simply leave the last line of the function unassigned (i.e. there is no 'gets arrow' on the last line). Here is a function to return the median value of the parallel maxima (built-in function `pmax`) of two vectors supplied as arguments:

```
parmax <- function (a,b) {
c <- pmax(a,b)
median(c) }
```

Here is the function in action: the unassigned last line `median(c)` returns the answer

```
x <- c(1,9,2,8,3,7)
y <- c(9,2,8,3,7,2)
parmax(x,y)
```

```
[1]   8
```

If you want to return two or more variables from a function you should use `return` with a list containing the variables to be returned. Suppose we wanted the median value of both the parallel maxima and the parallel minima to be returned:

```
parboth <- function (a,b) {
c <- pmax(a,b)
d <- pmin(a,b)
answer <- list(median(c),median(d))
names(answer)[[1]] <- "median of the parallel maxima"
names(answer)[[2]] <- "median of the parallel minima"
return(answer) }
```

Here it is in action with the same x and y data as above:

```
parboth(x,y)

$"median of the parallel maxima"

[1]   8

$"median of the parallel minima"

[1]   2
```

The point is that you make the multiple returns into a list, then return the list. The provision of multi-argument returns (e.g. `return(median(c),median(d))` in the example above) has been deprecated in R and a warning is given, as multi-argument returns were never documented in S, and whether or not the list was named differs from one version of S to another.

2.15.16   Anonymous functions

Here is an example of an anonymous function. It generates a vector of values but the function is not allocated a name (although the answer could be).

```
(function(x,y){ z <- 2* x^2 + y^2; x+y+z }) (0:7, 1)

[1] 2 5 12 23 38 57 80 107
```

The function first uses the supplied values of x and y to calculate z, then returns the value of x + y + z evaluated for eight values of x (from 0 to 7) and one value of y (1). Anonymous functions are used most frequently with `apply`, `tapply`, `sapply` and `lapply` (p. 63).

2.15.17   Flexible handling of arguments to functions

Because of the **lazy evaluation** practised by R, it is very simple to deal with missing arguments in function calls, giving the user the opportunity to specify the absolute minimum number of arguments, but to override the default arguments if they want to. As a simple example, take a function `plotx2` that we want to work when provided with either one or two arguments. In the one-argument case (only an integer $x > 1$ provided), we want it to plot z^2 against z for $z = 1$ to x in steps of 1. In the second case, when y is supplied, we want it to plot y against z for $z = 1$ to x:

```
plotx2 <- function (x, y = z^2) {
      z <- 1:x
      plot(z,y,type="l") }
```

In many other languages, the first line would fail because z is not defined at this point. But R does not evaluate an expression until the body of the function actually calls for it to be evaluated (i.e. never, in the case where y is supplied as a second argument). Thus for the one-argument case we get a graph of z^2 against z and in the two-argument case we get a graph of y against z (in this example, the straight line 1:12 vs. 1:12). We rescale the `windows` (width then height in inches) so that the graphs come out looking roughly square rather than elongated:

```
windows(7,4)
par(mfrow=c(1,2))
plotx2(12)
plotx2(12,1:12)
```

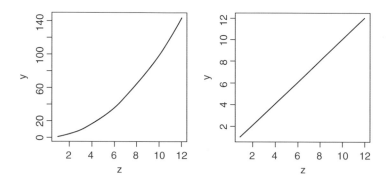

It is possible to access the actual (not default) expressions used as arguments inside the function. The mechanism is implemented via promises. You can find an explanation of promises by typing `?promise` at the command prompt.

2.15.18 Structure of an object: `str`

Here is one of the simplest objects in R – a vector of length 7 containing real numbers:

```
(y <- seq(0.9,0.3,-0.1))
```

```
[1] 0.9 0.8 0.7 0.6 0.5 0.4 0.3
```

We can ask R about the structure of the object called `y` using `str`:

```
str(y)
```

```
num [1:7] 0.9 0.8 0.7 0.6 0.5 0.4 0.3
```

We discover that it is `numeric` (in both class and mode), a vector of length 7 `[1:7]`, and (because the vector is short) we see all of the values listed. For longer vectors we would see the first few values, depending on what would fit on a single printed line (as affected by the number of decimal places displayed).

What about a slightly more complicated object? Here is a dataframe with two columns:

```
data <- read.table("c:\\temp\\spino.txt",header=T)
str(data)
```

```
'data.frame': 109 obs. of 2 variables:
 $ condition: Factor w/ 5 levels "better","much.better",..: 4 1 1 4 4 4 1 5 4 1 ...
 $ treatment: Factor w/ 3 levels "drug.A","drug.B",..: 1 2 2 3 2 2 1 1 2 2 ...
```

We learn that `data` is a dataframe with 109 rows and 2 columns, then we get detailed information on each of the columns in turn. The first is a variable called `condition` which is a factor with five levels (the first two levels of which (in alphabetical order) are `better` and `much.better`). The second variable is called `treatment` and is a factor with three levels. The numbers are the integer representations of the factor levels in the first 10 rows of the dataframe. Because we can see only factor levels 1 and 2, we would need to do more work to discover what factor level 4 of condition, or level 3 of treatment, actually represented:

```
levels(data$condition);levels(data$treatment)

[1] "better"  "much.better"  "much.worse"  "no.change"  "worse"
[1] "drug.A"  "drug.B"       "placebo"
```

We often want to know about the structure of model objects. Here is the simplest case, with a linear regression model (see p. 450 for details):

```
reg <- read.table("c:\\temp\\tannin.txt",header=T)
reg.model <- lm(growth~tannin,data=reg)
str(reg.model)
List of 12
 $ coefficients : Named num [1:2] 11.76 -1.22
 ..- attr(*, "names")= chr [1:2]  "(Intercept)" "tannin"
 $ residuals : Named num [1:9] 0.244 -0.539 -1.322 2.894 -0.889 ...
 ..- attr(*, "names")= chr [1:9] "1" "2" "3" "4" ...
 $ effects : Named num [1:9] -20.67 -9.42 -1.32 2.83 -1.01 ...
 ..- attr(*, "names")= chr [1:9]  "(Intercept)" "tannin" "" "" ...
 $ rank : int 2
 $ fitted.values: Named num [1:9] 11.76 10.54 9.32 8.11 6.89 ...
 ..- attr(*, "names")= chr [1:9] "1" "2" "3" "4" ...
 $ assign : int [1:2] 0 1
 $ qr :List of 5
 ..$ qr : num [1:9, 1:2] -3 0.333 0.333 0.333 0.333 ...
 .. ..- attr(*, "dimnames")=List of 2
 .. .. ..$ : chr [1:9] "1" "2" "3" "4" ...
 .. .. ..$ : chr [1:2] "(Intercept)" "tannin"
 .. ..- attr(*, "assign")= int [1:2] 0 1
 ..$ qraux: num [1:2] 1.33 1.26
 ..$ pivot: int [1:2] 1 2
 ..$ tol : num 1e-07
 ..$ rank : int 2
 ..- attr(*, "class")= chr "qr"
 $ df.residual : int 7
 $ xlevels : Named list()
 $ call : language lm(formula = growth ~ tannin, data = reg)
 $ terms :Classes 'terms', 'formula' length 3 growth ~ tannin
 .. ..- attr(*, "variables")= language list(growth, tannin)
 .. ..- attr(*, "factors")= int [1:2, 1] 0 1
 .. .. ..- attr(*, "dimnames")=List of 2
 .. .. .. ..$ : chr [1:2] "growth" "tannin"
```

```
.. .. .. ..$ : chr "tannin"
.. ..- attr(*, "term.labels")= chr "tannin"
.. ..- attr(*, "order")= int 1
.. ..- attr(*, "intercept")= int 1
.. ..- attr(*, "response")= int 1
.. ..- attr(*, ".Environment")=<environment: R_GlobalEnv>
.. ..- attr(*, "predvars")= language list(growth, tannin)
.. ..- attr(*, "dataClasses")= Named chr [1:2] "numeric" "numeric"
.. .. ..- attr(*, "names")= chr [1:2] "growth" "tannin"
$ model :'data.frame': 9 obs. of 2 variables:
..$ growth: int [1:9] 12 10 8 11 6 7 2 3 3
..$ tannin: int [1:9] 0 1 2 3 4 5 6 7 8
..- attr(*, "terms")=Classes 'terms', 'formula' length 3 growth ~
    tannin
.. .. ..- attr(*, "variables")= language list(growth, tannin)
.. .. ..- attr(*, "factors")= int [1:2, 1] 0 1
.. .. .. ..- attr(*, "dimnames")=List of 2
.. .. .. .. ..$ : chr [1:2] "growth" "tannin"
.. .. .. .. ..$ : chr "tannin"
.. .. ..- attr(*, "term.labels")= chr "tannin"
.. .. ..- attr(*, "order")= int 1
.. .. ..- attr(*, "intercept")= int 1
.. .. ..- attr(*, "response")= int 1
.. .. ..- attr(*, ".Environment")=<environment: R_GlobalEnv>
.. .. ..- attr(*, "predvars")= language list(growth, tannin)
.. .. ..- attr(*, "dataClasses")= Named chr [1:2] "numeric" "numeric"
.. .. .. ..- attr(*, "names")= chr [1:2] "growth" "tannin"
- attr(*, "class")= chr "lm"
```

There are 12 elements in the list representing the structure of this linear model object: coefficients, residuals, effects, rank, fitted values, assign, qr, residual degrees of freedom, xlevels, call, terms and model. Each of these, in turn, is broken down into components; for instance, the two coefficients are numbers (11.76 and −1.22), and their names are (Intercept) and tannin. You should work down the list and see if you can figure out why each row is an important part of the model.

For more complicated models, the structure is even more involved. Here is the structure of a generalized linear model with a binary response and binomial errors:

```
data <- read.table("c:\\temp\\spino.txt",header=T)
attach(data)
y <- factor(1+(condition=="better")+(condition=="much.better"))
model <- glm(y~treatment,binomial)
summary(model)

Call:
glm(formula = y ~ treatment, family = binomial)

Deviance Residuals:
    Min       1Q   Median       3Q      Max
-0.9741  -0.9741  -0.7747   1.3953   1.6431
```

```
Coefficients:
                  Estimate Std. Error z value Pr(>|z|)
(Intercept)        -0.6131     0.3444  -1.780    0.075 .
treatmentdrug.B     0.1141     0.4617   0.247    0.805
treatmentplacebo   -0.4367     0.5581  -0.783    0.434
---
Signif. codes: 0 '***' 0.001 '**' 0.01 '*' 0.05 '.' 0.1 ' ' 1

(Dispersion parameter for binomial family taken to be 1)

Null deviance:     139.67 on 108 degrees of freedom
Residual deviance: 138.54 on 106 degrees of freedom
AIC:               144.54
```

We have carried out a one-way analysis of deviance with a two-level response (improved or not) and a three-level factor as explanatory variable (`treatment`). There was no significant difference between `drug.B` and the placebo, nor between either of these and `drug.A` (the intercept). Here is the structure of the object called `model`:

```
str(model)
```

As you will see, this is a very large object, comprising a list with 30 components covering all aspects of the model: the coefficients, fitted values, effects plus all the details of the family and the model formula. I recommend you work your way slowly down the whole list and try to understand why each of the rows represents an essential piece of information about the model.

2.16 Writing from R to file

You often want to save an object that you have created in R.

2.16.1 Saving your work

To save your current R session, so that you can load it again later and continue your work where you left off, use `save` like this:

```
save(list = ls(all=TRUE), file= "c:\\temp\\session")
```

Then, on another occasion, when you want to restore the data, use `load` like this:

```
load(file= "c:\\temp\\session")
```

2.16.2 Saving history

It is very useful to be able to see all of the lines of R code that one has typed during a particular session. You may want to copy the lines into a text editor to make minor alterations, or you may simply want to paste multiple lines back into R to repeat certain operations. To see all of your lines of input code just type:

```
history(Inf)
```

This opens a window called R History through which you can scroll, highlight and copy using Ctrl + C. You could then open a new Untitled R Editor window (File > New Script) and paste the selected lines of code using Ctrl + V. Alternatively, you might want to save the entire history to file, for use on a subsequent occasion:

```
savehistory(file = "c:\\temp\\session18.txt")
```

To retrieve the history for use on another occasion use:

```
loadhistory(file = "c:\\temp\\session18.txt")
```

Then you can access it by `history(Inf)` in the new session.

2.16.3   Saving graphics

For speed and simplicity, you can click on a graph (the bar on top of the R Graphics Device goes darker blue) then press Ctrl + C (to copy the graph), then switch to a word processor and paste using Ctrl + V. For publication-quality graphics, however, you will want to save each figure in a separate file as a PDF or PostScript file. There are a great many options (see `?pdf` and `?postscript` for details) but the basics are very simple. Here we set the graphics device to produce a PDF:

```
pdf("c:\\temp\\fig1.pdf")
```

Now, any plot directives are sent to this file. To switch off writing graphics to file, type:

```
dev.off()
```

2.16.4   Saving data produced within R to disc

It is often convenient to generate numbers within R and then to use them somewhere else (in a spreadsheet, say). Here are 1000 random integers from a negative binomial distribution with mean mu=1.2 and clumping parameter or aggregation parameter (k) size = 1.0, that I want to save as a single column of 1000 rows in a file called nbnumbers.txt in the temp directory on the c: drive:

```
nbnumbers <- rnbinom(1000, size=1, mu=1.2)
```

There is general point to note here about the number and order of arguments provided to built-in functions like rnbinom. This function can have two of three optional arguments: size, mean (mu) and probability (prob) (see ?rnbinom). R knows that the unlabelled number 1000 refers to the number of numbers required because of its position, first in the list of arguments. If you are prepared to specify the names of the arguments, then the order in which they appear is irrelevant: rnbinom(1000, size=1, mu=1.2) and rnbinom(1000, mu=1.2, size=1) would give the same output. But if optional arguments are not labelled, then their order is crucial: so rnbinom(1000, 0.9, 0.6) is different from rnbinom(1000, 0.6, 0.9) because if there are no labels, then the second argument *must* be size and the third argument *must* be prob.

To export the numbers I use write like this, specifying that the numbers are to be output in a single column (i.e. with third argument 1 because the default is 5 columns):

```
write(nbnumbers,"c:\\temp\\nbnumbers.txt",1)
```

Sometimes you will want to save a table or a matrix of numbers to file. There is an issue here, in that the write function transposes rows and columns. It is much simpler to use the write.table function which does *not* transpose the rows and columns. Here is a matrix of 1000 rows and 100 columns made up of random integers from a Poisson distribution with mean 0.75:

```
xmat <- matrix(rpois(100000,0.75),nrow=1000)
write.table(xmat,"c:\\temp\\table.txt",col.names=F,row.names=F)
```

The function adds made-up row names and column names unless (as here) you specify otherwise. You have saved 1000 rows each of 100 Poisson random numbers with $\lambda = 0.75$.

Suppose that you have counted the number of different entries in the vector of negative binomial numbers (above):

```
nbtable <- table(nbnumbers)
nbtable
```

0	1	2	3	4	5	6	7	8	9	11	15
445	248	146	62	41	33	13	4	1	5	1	1

and you want write this output to a file. If you want to save both the counts and their frequencies in adjacent columns, use

```
write.table(nbtable,"c:\\temp\\table.txt",col.names=F,row.names=F)
```

but if you only want to export a single column of frequencies (445, 248, ...) use

```
write.table(unclass(nbtable),"c:\\temp\\table.txt",col.names=F,row.names=F)
```

2.16.5 Pasting into an Excel spreadsheet

Writing a vector from R to the Windows Clipboard uses the function `writeClipboard(x)` where x is a single vector of **characters**, so you need to build up a spreadsheet in Excel by pasting (Ctrl + V) one column at a time. Remember that character strings in dataframes are converted to factors on input unless you protect them by `as.is(name)` on input. For example:

```
writeClipboard(as.character(factor.name))
```

Go into Excel and press Ctrl + V, and then back into R and type:

```
writeClipboard(as.character(numeric.variable))
```

Then go into Excel and press Ctrl + V in the second column, and so on.

2.16.6 Writing an Excel readable file from R

Suppose you want to transfer an entire dataframe called `data` to Excel (rather than one column, as above):

```
data <- read.table("c:\\temp\\worms.txt",header=T)
write.table(data,"clipboard",sep="\t",col.names=NA)
```

Then, in Excel, just press Ctrl + V or click on the Paste icon (the Clipboard). Your variable names will appear in the first row of the spreadsheet, with (unheaded) row numbers in the leftmost column.

2.17 Programming tips

- Know exactly what you are trying to achieve.

- Keep it simple.

- Clever is good, but clear is better.

- Test each line as you go along, to make sure it does what you want it to do.

- Put plenty of comments in the code, using # for documentation.

- Use variable names and function names that are self-explanatory.

- Do not use `attach` in programs.

- Use `with`, or refer to variables within named dataframes.

- Try different ways of doing the same thing, and select the fastest method.

- Use indents (tabs) to improve clarity of loops and `if` statements.

- Build up the program from small, independently tested functions.

- Stop tinkering once it works effectively.

3

Data Input

You can get numbers into R through the keyboard, from the Clipboard or from an external file. For a single variable of up to 10 numbers or so, it is probably quickest to type the numbers at the command line, using the *concatenate* function c like this:

```
y <- c (6,7,3,4,8,5,6,2)
```

For intermediate sized variables, you might want to enter data from the keyboard using the scan function. For larger data sets, and certainly for sets with several variables, you should make a dataframe externally (e.g. in a spreadsheet) and read it into R using read.table (p. 139).

3.1 Data input from the keyboard

The scan function is useful if you want to type (or paste) a few numbers into a vector called x from the keyboard:

```
x <-scan()
```

```
1:
```

At the 1: prompt type your first number, then press the Enter key. When the 2: prompt appears, type in your second number and press Enter, and so on. When you have put in all the numbers you need (suppose there are eight of them) then simply press the Enter key at the 9: prompt.

```
1:  6
2:  7
3:  3
4:  4
5:  8
6:  5
7:  6
8:  2
9:
```

```
Read 8 items
```

The R Book, Second Edition Michael J. Crawley.
© 2013 John Wiley & Sons, Ltd. Published 2013 by John Wiley & Sons, Ltd.

You can also use `scan` to paste in columns of numbers from the Clipboard. In the spreadsheet, highlight the column of numbers you want, then type Ctrl + C (the accelerator keys for Copy). Now go back into R. At the `1:` prompt just type Ctrl + V (the accelerator keys for Paste) and the numbers will be scanned into the named variable (x in this example). You can then paste in another set of numbers, or press Return to complete data entry. If you try to read in a group of numbers from a *row* of cells in Excel, the characters will be pasted into a single multi-digit number (definitely *not* what is likely to have been intended). So, if you are going to paste numbers from Excel, make sure the numbers are in columns, not in rows, in the spreadsheet. If necessary, use Edit/Paste Special/Transpose in Excel to turn a row into a column before copying the numbers.

3.2   Data input from files

The easiest way to get data into R is to make your data into the shape of a dataframe before trying to read it into R. As explained in detail in Chapter 4, you should put all of the values of each variable into a single column, and put the name of the variable in row 1 (called the 'header' row). You will sometimes see the rows and the columns of the dataframe referred to as *cases* and *fields* respectively. In our terminology, the fields are the *variables* and the cases are *rows*.

Where you have text strings containing blanks (e.g. place names like 'New Brighton') then `read.table` is no good because it will think that 'New' is the value of one variable and 'Brighton' is another, causing the input to fail because the number of data items does not match the number of columns. In such cases, use a comma to separate the fields, and read the data from file using the function `read.csv` (the suffix `.csv` stands for 'comma separated values'). The function `read.csv2` is the variant for countries where a comma is used as the decimal point: in this case, a semicolon is the field separator.

3.2.1   The working directory

It is useful to have a working directory, because this makes accessing and writing files so much easier, since you do not need to type the (potentially long) path name each time. For instance, we can simplify data entry by using setwd to set the working directory to `"c:\\temp"`:

```
setwd("c:\\temp")
```

To find out the name of the current working directory, use `getwd()` like this:

```
getwd()
```

```
[1] "c:/temp"
```

If you have not changed the working directory, then you will see something like this:

```
getwd()
```

```
[1] "C:/Documents and Settings/mjcraw/My Documents"
```

which is the working directory assumed by R when you start a new session. If you plan to switch back to the default working directory during a session, then it is sensible to save the long default path before you change it, like this:

```
mine<-getwd()
setwd("c:\\temp")
...
setwd(mine)
```

If you want to view file names from R, then use the `dir` function like this:

```
dir("c:\\temp")
```

To pick a file from a directory using the browser, invoke the `file.choose` function. This is most used in association with `read.table` if you have forgotten the name or the location of the file you want to read into a dataframe:

```
data<-read.table(file.choose(),header=T)
```

Just click on the file you want to read, and the contents will be assigned to the frame called data.

3.2.2   Data input using `read.table`

Here is an example of the standard means of data entry, creating a dataframe within R using `read.table`:

```
setwd("c:\\temp\\")
data <- read.table(yield.txt",header=T)
head(data)
```

```
  year wheat barley oats rye  corn
1 1980   5.9   4.4  4.1 3.8   4.4
2 1981   5.8   4.4  4.3 3.7   4.1
3 1982   6.2   4.9  4.4 4.1   4.0
4 1983   6.4   4.7  4.3 3.7   4.1
5 1984   7.7   5.6  4.9 4.7   4.7
6 1985   6.3   5.0  4.6 4.6   4.3
```

You can save time by using `read.delim`, because then you can omit `header=T`:

```
data <- read.delim("yield.txt")
```

Or you could go the whole way down the labour-saving route and write your own text-minimizing function, which uses the `paste` function to add the suffix `.txt` as well as shortening the function name from `read.table` to `rt` like this:

```
rt <- function(x) read.table(paste("c:\\temp\\",x,".txt",sep=""),
  header=TRUE)
```

Then simply type:

```
data <- rt("yields")
```

3.2.3   Common errors when using `read.table`

It is important to note that `read.table` would fail if there were any spaces in any of the variable names in row 1 of the dataframe (the header row, see p. 161), such as Field Name, Soil pH or Worm Density, or between any of the words within the same factor level (as in many of the field names). You should replace all these spaces by dots '.' before saving the dataframe in Excel (use Edit/Replace with " " replaced by " . "). Now the dataframe can be read into R. There are three things to remember:

- The whole path and file name needs to be enclosed in double quotes: `"c:\\abc.txt"`.

- `header=T` says that the first row contains the variable names ('T' stands for TRUE).

- Always use double `\\` rather than `\` in the file path definition.

The most common cause of failure is that the number of variable names (characters strings in row 1) does not match the number of columns of information. In turn, the commonest cause of this is that you have left blank spaces in your variable names:

```
state name    population    home ownership    cars    insurance
```

This is wrong because R expects seven columns of numbers when there are only five. Replace the spaces within the names by dots and it will work fine:

```
state.name    population    home.ownership    cars    insurance
```

The next most common cause of failure is that the data file contains blank spaces where there are missing values. Replace these blanks with NA in your spreadsheet before reading the file into R.

Finally, there can be problems when you are trying to read variables that consist of character strings containing blank spaces (as in files containing place names). You can use `read.table` so long as you export the file from the spreadsheet using commas to separate the fields, and you tell `read.table` that the separators are commas using `sep=","` (to override the default blanks or tabs (`\t`):

```
map <- read.table("c:\\temp\\bowens.csv",header=T,sep=",")
```

However, it is quicker and easier to use `read.csv` in this case (see below).

3.2.4   Separators and decimal points

The default field separator character in `read.table` is `sep=" "`. This separator is white space, which is produced by one or more spaces, one or more tabs `\t`, one or more newlines `\n`, or one or more carriage returns. If you do have a different separator between the variables sharing the same line (i.e. other than a tab within a `.txt` file) then there may well be a special `read` function for your case. Note that all the alternatives to `read.table` have the sensible default that `header=TRUE` (the first row contains the variable names):

- for comma-separated fields use `read.csv("c:\\temp\\file.csv")`;
- for semicolon-separated fields `read.csv2("c:\\temp\\file.csv")`;
- for tab-delimited fields with decimal points as a commas, use `read.delim2("c:\\temp\\file.txt")`.

You would use comma or semicolon separators if you had character variables that might contain one or more blanks (e.g. country names like 'United Kingdom' or 'United States of America').

If you want to specify `row.names` then one of the columns of the dataframe must be a vector of unique row names. This can be a single number giving the column of the table which contains the row names, or character string giving the variable name of the table column containing the row names (see p. 176). Otherwise, if `row.names` is missing, rows numbers are generated automatically on the left of the dataframe.

The default behaviour of `read.table` is to convert character variables into factors. If you do *not* want this to happen (you want to keep a variable as a character vector) then use `as.is` to specify the columns that should not be converted to factors:

```
murder <- read.table("c:\\temp\\murders.txt",header=T,as.is="region")
```

3.2.5   Data input directly from the web

You will typically use `read.table` to read data from a file, but the function also works for complete URLs. In computing, URL stands for 'universal resource locator', and is a specific character string that constitutes a reference to an Internet resource, combining domain names with file path syntax, where forward slashes are used to separate folder and file names:

```
data2 <- read.table
("http://www.bio.ic.ac.uk/research/mjcraw/therbook/data/cancer.txt",
  header=T)
head(data2)
```

```
  death treatment status
1     4     DrugA      1
2    26     DrugA      1
3     2     DrugA      1
4    25     DrugA      1
5     7     DrugA      1
6     6     DrugA      0
```

3.3   Input from files using scan

For dataframes, read.table is superb. But look what happens when you try to use read.table with a more complicated file structure:

```
read.table("c:\\temp\\rt.txt")
```

```
Error in scan(file, what, nmax, sep, dec, quote, skip, nlines, na.strings
    line 1 did not have 4 elements
```

It simply cannot cope with lines having different numbers of fields. However, scan and readLines come into their own with these complicated, non-standard files.

The scan function reads data into a list when it is used to read from a file. It is much less friendly for reading dataframes than read.table, but it is substantially more flexible for tricky or non-standard files.

By default, scan assumes that you are inputting double precision numbers. If not, then you need to use the what argument to explain what exactly you are inputting (e.g. 'character', 'logical', 'integer', 'complex' or 'list'). The file name is interpreted relative to the current working directory (given by getwd()), unless you specify an absolute path. If what is itself a list, then scan assumes that the *lines* of the data file are records, each containing one or more items and the number of *fields* on that line is given by length(what).

By default, scan expects to read space-delimited or tab-delimited input fields. If your file has separators other than blank spaces or tab markers (\t), then you can specify the separator option (e.g. sep=",") to specify the character which delimits fields. A field is always delimited by an end-of-line marker unless it is quoted. If sep is the default (" "), the character \ in a quoted string escapes the following character, so quotes may be included in the string by escaping them ("she.said \"What!\".to.him"). If sep is non-default, the fields may be quoted in the style of .csv files where separators inside quotes (' ' or " ") are ignored and quotes may be put inside strings by doubling them. However, if sep = "\n" it is assumed by default that one wants to read entire lines verbatim.

With scan, you will often want to skip the header row (because this contains variable names rather than data). The option for this is skip = 1 (you can specify any number of lines to be skipped before beginning to read data values). If a single record occupies more than one line of the input file then use the option multi.line =TRUE.

3.3.1   Reading a dataframe with scan

To illustrate the issues involved with scan, and to demonstrate why read.table is the preferred way of reading data into dataframes, we use scan to input the worms dataframe (for details, see Chapter 4).

We want to skip the first row because that is a header containing the variable names, so we specify skip = 1. There are seven columns of data, so we specify seven fields of character variables "" in the list supplied to what:

```
scan("t:\\data\\worms.txt",skip=1,what=as.list(rep("",7)))
```

```
Read 20 records
```

```
[[1]]
 [1] "Nashs.Field"        "Silwood.Bottom"      "Nursery.Field"
 [4] "Rush.Meadow"        "Gunness.Thicket"     "Oak.Mead"
 [7] "Church.Field"       "Ashurst"             "The.Orchard"
[10] "Rookery.Slope"      "Garden.Wood"         "North.Gravel"
[13] "South.Gravel"       "Observatory.Ridge"   "Pond.Field"
[16] "Water.Meadow"       "Cheapside"           "Pound.Hill"
[19] "Gravel.Pit"         "Farm.Wood"
```

```
[[2]]
 [1] "3.6" "5.1" "2.8" "2.4" "3.8" "3.1" "3.5" "2.1" "1.9" "1.5" "2.9" "3.3"
[13] "3.7" "1.8" "4.1" "3.9" "2.2" "4.4" "2.9" "0.8"
```

```
[[3]]
 [1] "11" "2"  "3"  "5"  "0"  "2"  "3"  "0"  "0"  "4"  "10" "1"  "2"  "6"  "0"
[16] "0"  "8"  "2"  "1"  "10"
```

```
[[4]]
 [1] "Grassland" "Arable"    "Grassland" "Meadow"    "Scrub"     "Grassland"
 [7] "Grassland" "Arable"    "Orchard"   "Grassland" "Scrub"     "Grassland"
[13] "Grassland" "Grassland" "Meadow"    "Meadow"    "Scrub"     "Arable"
[19] "Grassland" "Scrub"
```

```
[[5]]
 [1] "4.1" "5.2" "4.3" "4.9" "4.2" "3.9" "4.2" "4.8" "5.7" "5"   "5.2" "4.1"
[13] "4"   "3.8" "5"   "4.9" "4.7" "4.5" "3.5" "5.1"
```

```
[[6]]
 [1] "FALSE" "FALSE" "FALSE" "TRUE"  "FALSE" "FALSE" "FALSE" "FALSE" "FALSE"
[10] "TRUE"  "FALSE" "FALSE" "FALSE" "FALSE" "TRUE"  "TRUE"  "TRUE"  "FALSE"
[19] "FALSE" "TRUE"
```

```
[[7]]
 [1] "4" "7" "2" "5" "6" "2" "3" "4" "9" "7" "8" "1" "2" "0" "6" "8" "4" "5" "1"
[20] "3"
```

As you can see, scan has created a list of seven vectors of character string information. To convert this list into a dataframe, we use the as.data.frame function which turns the lists into columns in the dataframe (so long as the columns are all the same length):

```
data <-
as.data.frame(scan("t:\\data\\worms.txt",skip=1,what=as.list(rep("",7))))
```

In its present form, the variable names manufactured by scan are ridiculously long, so we need to replace them with the original variable names that are in the first row of the file. For this we can use scan again, but specify that we want to read only the first line, by specifying nlines=1 and removing the skip option:

```
header <- unlist(scan("t:\\data\\worms.txt",nlines=1,what=as.list
  (rep("",7))))
```

Now, we replace the manufactured names by the correct variable names in `data`:

```
names(data)<-header
head(data)
```

```
        Field.Name Area Slope Vegetation Soil.pH Damp Worm.density
1      Nashs.Field  3.6    11  Grassland     4.1    F            4
2   Silwood.Bottom  5.1     2     Arable     5.2    F            7
3    Nursery.Field  2.8     3  Grassland     4.3    F            2
4      Rush.Meadow  2.4     5     Meadow     4.9    T            5
5 Gunness.Thicket  3.8     0      Scrub     4.2    F            6
6         Oak.Mead  3.1     2  Grassland     3.9    F            2
```

This has produced the right result, but you can see why you would want to use `read.table` rather than `scan` for reading dataframes.

3.3.2   Input from more complex file structures using `scan`

Here is an image of a file containing information on the identities of the neighbours of five individuals from a population: the first individual has one neighbour (number 138), the second individual has two neighbours (27 and 44), the third individual has four neighbours, and so on.

```
138
27      44
19      20      345     48
115     2366
59
```

See if you can figure out why the following three different readings of the same file `rt.txt` produce such different objects; the first has 10 items, the second 5 items and the third 20 items:

```
scan("c:\\temp\\rt.txt")
```

```
Read 10 items
 [1]  138   27   44   19   20  345   48  115 2366   59
```

```
scan("c:\\temp\\rt.txt",sep="\n")
```

```
Read 5 items
[1]         138      2744 192034548     1152366             59
```

```
scan("c:\\temp\\rt.txt",sep="\t")
```

```
Read 20 items
 [1]  138   NA   NA   NA   27   44   NA   NA   19   20  345   48  115
[14] 2366   NA   NA   59   NA   NA   NA
```

The only difference between the three calls to `scan` is in the specification of the separator. The first uses the default which is blanks or tabs (the 10 items are the 10 numbers that we are interested in, but information about their grouping has been lost). The second uses the new line `"\n"` control character as the separator (the contents of each of the five lines have been stripped out and trimmed to create meaningless numbers, except for 138 and 59 which were the only numbers on their respective lines). The third uses tabs `"\t"` as separators (we have no information on lines, but at least the numbers have retained their integrity, and missing values (NA) have been used to pad out each line to the same length, 4. To get the result we want we need to use the information on the number of lines from method 2 and the information on the contents of each line from method 3. The first step is easy:

```
length(scan("c:\\temp\\rt.txt",sep="\n"))
```

```
Read 5 items
[1] 5
```

So we have five lines of information in this file. To find the number of items per line were divide the total number of items

```
length(scan("c:\\temp\\rt.txt",sep="\t"))
```

```
Read 20 items
[1] 20
```

by the number of lines: 20 / 5 = 4. To extract the information on each line, we want to take a line at a time, and extract the missing values (i.e. remove the NAs). So, for line 1 this would be

```
scan("c:\\temp\\rt.txt",sep="\t")[1:4]
```

```
Read 20 items
[1] 138  NA  NA  NA
```

then, to remove the NA we use `na.omit`, to remove the `Read 20 items` we use `quiet=T` and to leave only the numerical value we use `as.numeric`:

```
as.numeric(na.omit(scan("c:\\temp\\rt.txt",sep="\t",quiet=T)[1:4]))
```

```
[1] 138
```

To complete the job, we need to apply this logic to each of the five lines in turn, to produce a list of vectors of variable lengths (1, 2, 4, 2 and 1):

```
sapply(1:5, function(i)
            as.numeric(na.omit(
            scan("c:\\temp\\rt.txt",sep="\t",quiet=T)[(4*i-3):
            (4*i)]))))
```

```
[[1]]
[1]   138

[[2]]
[1]    27   44

[[3]]
[1]    19   20  345   48

[[4]]
[1]   115 2366

[[5]]
[1]    59
```

That was about as complicated a procedure as you are likely to encounter in reading information from a file. In hindsight, we might have created the data as a dataframe with missing values explicitly added to the rows that had less than four numbers. Then a single `read.table` statement would have been enough.

3.4   Reading data from a file using `readLines`

An alternative is to use `readLines` instead of `scan`. Here is a repeat of the example of reading the worms dataframe (above).

3.4.1   Input a dataframe using `readLines`

```
line<-readLines("c:\\temp\\worms.txt")
line
```

```
 [1]  "Field.Name\tArea\tSlope\tVegetation\tSoil.pH\tDamp\tWorm.density"
 [2]  "Nashs.Field\t3.6\t11\tGrassland\t4.1\tFALSE\t4"
 [3]  "Silwood.Bottom\t5.1\t2\tArable\t5.2\tFALSE\t7"
 [4]  "Nursery.Field\t2.8\t3\tGrassland\t4.3\tFALSE\t2"
 [5]  "Rush.Meadow\t2.4\t5\tMeadow\t4.9\tTRUE\t5"
 [6]  "Gunness.Thicket\t3.8\t0\tScrub\t4.2\tFALSE\t6"
 [7]  "Oak.Mead\t3.1\t2\tGrassland\t3.9\tFALSE\t2"
 [8]  "Church.Field\t3.5\t3\tGrassland\t4.2\tFALSE\t3"
 [9]  "Ashurst\t2.1\t0\tArable\t4.8\tFALSE\t4"
[10]  "The.Orchard\t1.9\t0\tOrchard\t5.7\tFALSE\t9"
[11]  "Rookery.Slope\t1.5\t4\tGrassland\t5\tTRUE\t7"
[12]  "Garden.Wood\t2.9\t10\tScrub\t5.2\tFALSE\t8"
[13]  "North.Gravel\t3.3\t1\tGrassland\t4.1\tFALSE\t1"
[14]  "South.Gravel\t3.7\t2\tGrassland\t4\tFALSE\t2"
[15]  "Observatory.Ridge\t1.8\t6\tGrassland\t3.8\tFALSE\t0"
[16]  "Pond.Field\t4.1\t0\tMeadow\t5\tTRUE\t6"
[17]  "Water.Meadow\t3.9\t0\tMeadow\t4.9\tTRUE\t8"
[18]  "Cheapside\t2.2\t8\tScrub\t4.7\tTRUE\t4"
[19]  "Pound.Hill\t4.4\t2\tArable\t4.5\tFALSE\t5"
[20]  "Gravel.Pit\t2.9\t1\tGrassland\t3.5\tFALSE\t1"
[21]  "Farm.Wood\t0.8\t10\tScrub\t5.1\tTRUE\t3"
```

Each line has become a single character string. As you can see, we shall need to strip out all of the tab marks (\t), thereby separating the data entries and creating seven columns of information. The function for this is `strsplit`:

```
db<-strsplit(line,"\t")
db
```

```
[[1]]
[1] "Field.Name"       "Area" "Slope" "Vegetation" "Soil.pH" "Damp" "Worm.density"

[[2]]
[1] "Nashs.Field"      "3.6"   "11"   "Grassland"  "4.1"     "F"    "4"

[[3]]
[1] "Silwood.Bottom"   "5.1"   "2"    "Arable"     "5.2"     "F"    "7"

[[4]]
[1] "Nursery.Field"    "2.8"   "3"    "Grassland"  "4.3"     "F"    "2"
```

and so on for 21 elements of the list. The new challenge is to get this into a form that we can turn into a dataframe. The first issue is to get rid of the list structure using the `unlist` function:

```
bb<-unlist(db)
```

```
 [1] "Field.Name"                 "Area"              "Slope"      "Vegetation"    "Soil.pH"     "Damp"

 [7] "Worm.density" "Nashs.Field"                       "3.6"                "11" "Grassland"         "4.1"

[13] "F"                          "4" "Silwood.Bottom"              "5.1"              "2" "Arable"

[19] "5.2"                        "F"               "7" "Nursery.Field"             "2"        "3"

[25] "Grassland"                 "4.3"                "F" "Rush.Meadow"          "2.4"
```

and so on up to 147 items. We need to give this vector dimensions: the seven variable names come first, so the appropriate dimensionality is (7,21):

```
dim(bb)<-c(7,21)
bb
```

```
      [,1]           [,2]           [,3]               [,4]             [,5]           [,6]               [,7]
[1,] "Field.Name"   "Nashs.Field"  "Silwood.Bottom"   "Nursery.Field"  "Rush.Meadow"  "Gunness.Thicket"  "Oak.Mead"
[2,] "Area"         "3.6"          "5.1"              "2.8"            "2.4"          "3.8"              "3.1"
[3,] "Slope"        "11"           "2"                "3"              "5"            "0"                "2"
[4,] "Vegetation"   "Grassland"    "Arable"           "Grassland"      "Meadow"       "Scrub"            "Grassland"
[5,] "Soil.pH"      "4.1"          "5.2"              "4.3"            "4.9"          "4.2"              "3.9"
[6,] "Damp"         "F"            "F"                "F"              "T"            "F"                "F"
[7,] "Worm.density" "4"            "7"                "2"              "5"            "6"                "2"
```

and so on for 21 rows. This is closer, but the rows and columns are the wrong way around. We need to transpose this object and drop the first row (because this is the header row containing the variable names):

```
t(bb)[-1,]
```

```
        [,1]                   [,2]   [,3]   [,4]          [,5]   [,6]   [,7]
 [1,]   "Nashs.Field"          "3.6"  "11"   "Grassland"   "4.1"  "F"    "4"
 [2,]   "Silwood.Bottom"       "5.1"  "2"    "Arable"      "5.2"  "F"    "7"
 [3,]   "Nursery.Field"        "2.8"  "3"    "Grassland"   "4.3"  "F"    "2"
 [4,]   "Rush.Meadow"          "2.4"  "5"    "Meadow"      "4.9"  "T"    "5"
 [5,]   "Gunness.Thicket"      "3.8"  "0"    "Scrub"       "4.2"  "F"    "6"
 [6,]   "Oak.Mead"             "3.1"  "2"    "Grassland"   "3.9"  "F"    "2"
 [7,]   "Church.Field"         "3.5"  "3"    "Grassland"   "4.2"  "F"    "3"
 [8,]   "Ashurst"              "2.1"  "0"    "Arable"      "4.8"  "F"    "4"
 [9,]   "The.Orchard"          "1.9"  "0"    "Orchard"     "5.7"  "F"    "9"
[10,]   "Rookery.Slope"        "1.5"  "4"    "Grassland"   "5"    "T"    "7"
[11,]   "Garden.Wood"          "2.9"  "10"   "Scrub"       "5.2"  "F"    "8"
[12,]   "North.Gravel"         "3.3"  "1"    "Grassland"   "4.1"  "F"    "1"
[13,]   "South.Gravel"         "3.7"  "2"    "Grassland"   "4"    "F"    "2"
[14,]   "Observatory.Ridge"    "1.8"  "6"    "Grassland"   "3.8"  "F"    "0"
[15,]   "Pond.Field"           "4.1"  "0"    "Meadow"      "5"    "T"    "6"
[16,]   "Water.Meadow"         "3.9"  "0"    "Meadow"      "4.9"  "T"    "8"
[17,]   "Cheapside"            "2.2"  "8"    "Scrub"       "4.7"  "T"    "4"
[18,]   "Pound.Hill"           "4.4"  "2"    "Arable"      "4.5"  "F"    "5"
[19,]   "Gravel.Pit"           "2.9"  "1"    "Grassland"   "3.5"  "F"    "1"
[20,]   "Farm.Wood"            "0.8"  "10"   "Scrub"       "5.1"  "T"    "3"
```

That's more like it. Now the function as.data.frame should work:

```
frame<-as.data.frame(t(bb)[-1,])
head(frame)
```

	V1	V2	V3	V4	V5	V6	V7
1	Nashs.Field	3.6	11	Grassland	4.1	F	4
2	Silwood.Bottom	5.1	2	Arable	5.2	F	7
3	Nursery.Field	2.8	3	Grassland	4.3	F	2
4	Rush.Meadow	2.4	5	Meadow	4.9	T	5
5	Gunness.Thicket	3.8	0	Scrub	4.2	F	6
6	Oak.Mead	3.1	2	Grassland	3.9	F	2

All we need to do now is add in the variable names: these are in the first row of the transpose of the object called bb (above):

```
names(frame)<-t(bb)[1,]
head(frame)
```

	Field.Name	Area	Slope	Vegetation	Soil.pH	Damp	Worm.density
1	Nashs.Field	3.6	11	Grassland	4.1	F	4
2	Silwood.Bottom	5.1	2	Arable	5.2	F	7
3	Nursery.Field	2.8	3	Grassland	4.3	F	2
4	Rush.Meadow	2.4	5	Meadow	4.9	T	5
5	Gunness.Thicket	3.8	0	Scrub	4.2	F	6

The complexity of this procedure makes you appreciate just how useful the function read.table really is.

3.4.2 Reading non-standard files using readLines

Here is a repeat of the example of the neighbours file that we analysed using scan on p. 141:

```
readLines("c:\\temp\\rt.txt")
```

```
[1] "138\t\t\t"        "27\t44\t\t"      "19\t20\t345\t48" "115\t2366\t\t"
[5] "59\t\t\t"
```

The readLines function had produced a vector of length 5 (one element for each row in the file), with the contents of each row reduced to a single character string (including the literal tab markers \t). We need to spit this up within the lines using strsplit: we split first on the tabs, then on the new lines in order to see the distinction:

```
strsplit(readLines("c:\\temp\\rt.txt"),"\t")
```

```
[[1]]
[1] "138" ""     ""

[[2]]
[1]  "27" "44"   ""

[[3]]
[1]  "19" "20"   "345"  "48"
```

```
[[4]]
[1] "115" "2366"        ""

[[5]]
[1]   "59"        ""        ""
```

```
strsplit(readLines("c:\\temp\\rt.txt"),"\n")
```

```
[[1]]
[1] "138\t\t\t"

[[2]]
[1] "27\t44\t\t"

[[3]]
[1] "19\t20\t345\t48"

[[4]]
[1] "115\t2366\t\t"

[[5]]
[1] "59\t\t\t"
```

The split by tab markers is closest to what we want to achieve, so we shall work on that. First, turn the character strings into numbers:

```
rows<-lapply(strsplit(readLines("c:\\temp\\rt.txt"),"\t"),as.numeric)
rows
```

```
[[1]]
[1]   138    NA   NA

[[2]]
[1]    27    44   NA

[[3]]
[1]    19    20  345   48

[[4]]
[1]   115  2366   NA

[[5]]
[1]    59    NA   NA
```

Now all that we need to do is to remove the NAs from each of the vectors:

```
sapply(1:5, function(i) as.numeric(na.omit(rows[[i]])))
```

```
[[1]]
[1]   138

[[2]]
[1]    27    44

[[3]]
[1]    19    20  345   48
```

```
[[4]]
[1]   115 2366
[[5]]
[1]    59
```

Both scan and readLines were fiddly, but they got what we were looking for in the end. You will need a lot of practice before you appreciate when to use scan and when to use readLines. Writing lists to files is tricky (see p. 82 for an explanation of the options available).

3.5   Warnings when you attach the dataframe

Suppose that you read a file from data, then attach it:

```
murder <- read.table("c:\\temp\\murders.txt",header=T,as.is="region")
```

The following warning will be produced if your attach function causes a duplication of one or more names:

```
attach(murder)
```

```
The following object(s) are masked _by_ .GlobalEnv:
      murder
```

The reason in the present case is that we have created a dataframe called murder and attached a variable from this dataframe which is also called murder. This shows the cause of the problem: the dataframe name and the third variable name are identical:

```
head(murder)
```

```
        state population murder region
1      Alabama       3615   15.1  South
2       Alaska        365   11.3   West
3      Arizona       2212    7.8   West
4     Arkansas       2110   10.1  South
5   California      21198   10.3   West
6     Colorado       2541    6.8   West
```

This ambiguity might cause difficulties later. A much better plan is to give the dataframe a unique name (like murders, for instance). We use the attach function so that the variables inside a dataframe can be accessed directly by name. Technically, this means that the dataframe is attached to the R search path, so that the dataframe is searched by R when evaluating a variable. We could tabulate the numbers of murders by region, for instance:

```
table(region)
```

```
region
North.Central       Northeast         South          West
           12               9            16            13
```

If we had not attached the dataframe, then we would have had to specify the name of the dataframe first like this:

```
table(murder$region)
```

3.6   Masking

You may have attached the same dataframe twice. Alternatively, you may have two dataframes attached that have one or more variable names in common. The commonest cause of masking occurs with simple variable names like x and y. It is very easy to end up with multiple variables of the same name within a single session that mean totally different things. The warning after using `attach` should alert you to the possibility of such problems. If the vectors sharing the same name are of different lengths, then R is likely to stop you before you do anything too silly, but if the vectors are of the same length then you run the serious risk of fitting the wrong explanatory variable (e.g. fitting the wrong one from two vectors both called x) or having the wrong response variable (e.g. from two vectors both called y). The moral is:

- use longer, more self-explanatory variable names;
- do not calculate variables with the same name as a variable inside a dataframe;
- always `detach` dataframes once you have finished using them;
- remove calculated variables once you are finished with them (`rm`; see p. 10).

The best practice, however, is not to use `attach` in the first place, but to use functions like `with` instead (see p. 113). If you get into a real tangle, it is often easiest to quit R and start another R session.

The opposite problem occurs when you assign values to an existing variable name (perhaps by accident); the original contents of the name are lost.

```
z <- 10
. . .
. . .
z <- 2.5
```

Now, z is 2.5 and there is no way to retrieve the original value of 10.

3.7   Input and output formats

Formatting is controlled using **escape sequences**, typically within double quotes:

```
\n   newline
\r   carriage return
\t   tab character
\b   backspace
\a   bell
\f   form feed
\v   vertical tab
```

Here is an example of the `cat` function. The default produces the output on the computer screen, but you can save the output to file using a `file=file.name` argument:

```
data<-read.table("c:\\temp\\catdata.txt",header=T)
attach(data)
names(data)

[1] "y"     "soil"

model<-lm(y~soil)
```

Suppose that you wanted to produce a slightly different layout for the ANOVA table than that produced by `summary.aov(model)`:

```
summary.aov(model)
```

```
           Df   Sum Sq   Mean Sq   F value   Pr(>F)
ind         2    99.2     49.60     4.245     0.025 *
Residuals  27   315.5     11.69
```

with the sum of squares column before the degrees of freedom column, plus a row for the total sum of squares, different row labels, and no *p* value, like this:

```
ANOVA table
Source              SS         d.f.     MS              F
Treatment           99.2       2        49.6            4.244691
Error               315.5      27       11.68519
Total               414.7      29
```

First, extract the necessary numbers from the `summary.aov` object:

```
df1<-unlist(summary.aov(model)[[1]]  [1])[1]
df2<-unlist(summary.aov(model)[[1]]  [1])[2]
ss1<-unlist(summary.aov(model)[[1]]  [2])[1]
ss2<-unlist(summary.aov(model)[[1]]  [2])[2]
```

Here is the R code to produce the ANOVA table, using `cat` with multiple tabs (`"\t\t"`) and single new-line markers (`"\n"`) at the end of each line:

```
{cat("ANOVA table","\n")
cat("Source","\t\t","SS","\t","d.f.","\t","MS","\t\t","F","\n")
cat("Treatment","\t",ss1,"\t",df1,"\t",ss1/df1,"\t\t",
                               (ss1/df1)/(ss2/df2),"\n")
cat("Error","\t\t",ss2,"\t",df2,"\t",ss2/df2,"\n")
cat("Total","\t\t",ss1+ss2,"\t",df1+df2,"\n")}
```

Note the use of curly brackets to group the five `cat` functions into a single print object.

3.8 Checking files from the command line

It can be useful to check whether a given filename exists in the path where you think it should be. The function is `file.exists` and is used like this:

```
file.exists("c:\\temp\\Decay.txt")
```

```
[1]   TRUE
```

For more on file handling, see `?files`.

3.9 Reading dates and times from files

You need to be very careful when dealing with dates and times in any sort of computing. R has a robust system for working with dates and times, but it takes some getting used to. Typically, you will read dates and times into R as character strings, then convert them into dates and times using the `strptime` function to explain exactly what the elements of the character string mean (e.g. which are the days, which are the months, what are the separators, and so on; see p. 103 for an explanation of the formats supported).

3.10 Built-in data files

There are many built-in data sets within the `datasets` package of R. You can see their names by typing:

```
data()
```

To see the data sets in extra installed packages as well, type:

```
data(package = .packages(all.available = TRUE))
```

You can read the documentation for a particular data set with the usual query:

```
?lynx
```

Many of the contributed packages contain data sets, and you can view their names using the `try` function. This evaluates an expression and traps any errors that occur during the evaluation. The `try` function establishes a handler for errors that uses the default error handling protocol:

```
try(data(package="spatstat"));Sys.sleep(3)
try(data(package="spdep"));Sys.sleep(3)
try(data(package="MASS"))
```

where `try` is a wrapper to run an expression that might fail and allow the user's code to handle error recovery, so this would work even if one of the packages was missing.

Built-in data files can be attached in the normal way; then the variables within them accessed by their names:

```
attach(OrchardSprays)
decrease
```

3.11 File paths

There are several useful R functions for manipulating file paths. A file path is a character string that looks something like this:

```
c:\\temp\\thesis\\chapter1\\data\\problemA
```

and you would not want to type all of that every time you wanted to read data or save material to file. You can set the default file path for a session using the current working directory:

```
setwd("c:\\temp\\thesis\\chapter1\\data")
```

The `basename` function removes all of the path up to and including the last path separator (if any):

```
basename("c:\\temp\\thesis\\chapter1\\data\\problemA")
```

```
[1] "problemA"
```

The `dirname` function returns the part of the path up to but excluding the last path separator, or "." if there is no path separator:

```
dirname("c:\\temp\\thesis\\chapter1\\data\\problemA")
```

```
[1] "c:/temp/thesis/chapter1/data"
```

Note that this function returns forward slashes as the separator, replacing the double backslashes.

Suppose that you want to construct the path to a file from components in a platform-independent way. The `file.path` function does this very simply:

```
A <- "c:"
B <- "temp"
C <- "thesis"
D <- "chapter1"
E <- "data"
F <  "problemA"
```

```
file.path(A,B,C,D,E,F)
```

```
[1] "c:/temp/thesis/chapter1/data/problemA"
```

The default separator is platform-dependent (/ in the example above, not \, but you can specify the separator (`fsep`) like this:

```
file.path(A,B,C,D,E,F,fsep="\\")
```

```
[1] "c:\\temp\\thesis\\chapter1\\data\\problemA"
```

3.12 Connections

Connections are ways of getting information into and out of R, such as your keyboard and your screen. The three standard connections are known as `stdin()` for input, `stdout()` for output, and `stderr()` for reporting errors. They are text-mode connections of class `terminal` which cannot be opened or closed, and are read-only, write-only and write-only respectively. When R is reading a script from a file, the *file* is the 'console': this is traditional usage to allow in-line data.

Functions to create, open and close connections include `file`, `url`, `gzfile`, `bzfile`, `xzfile`, `pipe` and `socketConnction`. The intention is that `file` and `gzfile` can be used generally for text input (from files and URLs) and binary input, respectively. The functions `file`, `pipe`, `fifo`, `url`, `gzfile`, `bzfile`, `xzfile`, `unz` and `socketConnection` return a connection object which inherits from class `connection`: `isOpen` returns a logical value, indicating whether the connection is currently open; `isIncomplete` returns a logical value, indicating whether the last read attempt was blocked, or for an output text connection whether there is unflushed output. The functions are used like this:

```
file(description = "", open = "", blocking = TRUE,
     encoding = getOption("encoding"), raw = FALSE)
```

For `file` the description is a path to the file to be opened or a complete URL, or "" (the default) or "clipboard". You can use `file` with `description` = "clipboard" in modes "r" and "w" only. There is a 32Kb limit on the text to be written to the Clipboard. This can be raised by using, for example, `file("clipboard-128")` to give 128Kb. For `gzfile` the description is the path to a file compressed by `gzip`: it can also open for reading uncompressed files and those compressed by `bzip2`, `xz` or `lzma`. For `bzfile` the description is the path to a file compressed by `bzip2`. For `xzfile` the description is the path to a file compressed by `xz` or `lzma`. A maximum of 128 connections can be allocated (not necessarily open) at any one time.

Not all modes are applicable to all connections: for example, URLs can only be opened for reading. Only file and socket connections can be opened for both reading and writing. For compressed files, the type of compression involves trade-offs: `gzip`, `bzip2` and `xz` are successively less widely supported, need more resources for both compression and decompression, and achieve more compression (although individual files may buck the general trend). Typical experience is that `bzip2` compression is 15% better on text files than `gzip` compression, and `xz` with maximal compression 30% better. With current computers decompression times even with `compress = 9` are typically modest, and reading compressed files is usually faster than uncompressed ones because of the reduction in disc activity.

3.13 Reading data from an external database

Open Data Base Connectivity (ODBC) provides a standard software interface for accessing database management systems (DBMS) that is independent of programming languages, database systems and operating systems. Thus, any application can use ODBC to query data from a database, regardless of the platform it is on or the DBMS it uses. ODBC accomplishes this by using a driver as a translation layer between the application and the DBMS. The application thus only needs to know ODBC syntax, and the driver can then pass the query to the DBMS in its native format, returning the data in a format that the application can understand. Communication with the database uses SQL ('Structured Query Language').

The example we shall use is called Northwind. This is a relational database that is downloaded for Access from Microsoft Office, and which is used in introductory texts on SQL (e.g. Kauffman et al., 2001). The database refers to a fictional company that operates as a food wholesaler. You should download the database from `http://office.microsoft.com`. There are many related tables in the database, each with its unique row identifier (ID):

- **Categories**: 18 rows describing the categories of goods sold (from baked goods to oils) including Category name as well as Category.ID

- **Products**: 45 rows with details of the various products sold by Northwind including CategoryID (as above) and Product.ID

- **Suppliers**: 10 rows with details of the firms that supply goods to Northwind

- **Shippers**: 3 rows with details of the companies that ship goods to customers

- **Employees**: 9 rows with details of the people who work for Northwind

- **Customers**: 27 rows with details of the firms that buy goods from Northwind

- **Orders**: empty rows with orders from various customers tagged by OrderID

- **Order Details**: 2155 rows containing one to many rows for each order (typically 2 or 3 rows per order), each row containing the product, number required, unit price and discount, as well as the (repeated) OrderID

- **Order Status**: four categories – new, invoiced, shipped or completed

- **Order Details:** empty rows

- **Order Details Status:** six categories – none, no stock, back-ordered, allocated, invoiced and shipped

- **Inventory**: details of the 45 products held on hand and reordered.

The tables are related to one another like this:

- Orders and Order Details are related by a unique OrderID;
- Categories and Products are related by a unique CategoryID;
- Employees and Orders are linked by a unique EmployeeID;
- Products and Order Details are linked by a unique ProductID;
- and so on.

3.13.1   Creating the DSN for your computer

To analyse data from these tables in R we need first to set up a **channel** through which to access the information. Outside R, you need to give your computer a data source name (DSN) to associate with the Access database. This is a bit fiddly, but here is how to do it in Windows 7, assuming you have Microsoft Office 2010 and have already downloaded the Northwind database (you will need to figure out the equivalent directions for Mac or Linux, or if you are not using Office 2010):

- Go to Computer, click on Drive C: and then scroll down and click on the Windows folder.
- Scroll down and click on SysWOW64.
- Scroll way down and double-click on odbcad32.exe.
- The ODBC Data Source Administrator window should appear.
- Click on the Add button.
- Scroll down to Microsoft Access Driver (*.mdb,*.accdb) and click on Finish.
- The ODBC Microsoft Access Setup window should appear.
- In the top box write "northwind" (with no quote marks).
- In the second box write "example" (with no quote marks).
- Then, under Database, click on Select.
- Browse through the directory to the place where you put Northwind.accdb.
- Accept and you should see "northwind" has been added to the ODBC Data Source list.
- Click on OK at the bottom of the window to complete the task.

Fortunately, you need only do this once. Now you can access Northwind from R.

3.13.2   Setting up R to read from the database

If you have not done so already, you will need to obtain the package called RODBC that provides access to databases from within R. So start up R (run as administrator) then type:

```
install.packages("RODBC")
```

Open the library and tell R the name of the channel through which it can access the database (as set up in the previous section) using the function odbcConnect:

```
library(RODBC)
channel  <-  odbcConnect("northwind")
```

You communicate with the database from R using SQL. The syntax is very simple. You create a dataframe in R from a table (or more typically from several related tables) in the database using the function sqlQuery like this:

```
new.datatframe <- sqlQuery(channel, query)
```

The channel is defined using the odbcConnect function, as shown above. The skill is in creating the (often complicated) character string called query. The key components of an SQL query are:

SELECT	a list of the variables required (or * for all variables)
FROM	the name of the table containing these variables
WHERE	specification of which rows of the table(s) are required
JOIN	the tables to be joined and the variables on which to join them
GROUP BY	columns with factors to act as grouping levels
HAVING	conditions applied after grouping
ORDER BY	sorted on which variables
LIMIT	offsets or counts

The simplest cases require only SELECT and FROM.

Let us start by creating a dataframe in R called containing all of the variables (*) and rows from the table called Categories in Northwind:

```
query <-  "SELECT * FROM Categories"
stock <- sqlQuery(channel, query)
```

This is what the R dataframe looks like – there are just two fields ID and Category:

```
stock
```

```
     ID                    Category
1    1      Baked Goods & Mixes
2    2                 Beverages
3    3                     Candy
4    4  Canned Fruit & Vegetables
5    5               Canned Meat
6    6                    Cereal
7    7                     Chips
8    8                Condiments
9    9            Dairy Products
10  10         Dried Fruit & Nuts
11  11                    Grains
12  12           Jams & Preserves
13  14                     Pasta
14  15                    Sauces
15  16                    Snacks
16  17                     Soups
17  18                      Oils
```

More usually, you will want to select only some of the variables from a table. From the table called `Inventory`, for example, we might want to extract only `ProductID` and `OnHand` to make a dataframe called `quantities`:

```
query <-  "SELECT ProductID,OnHand FROM Inventory"
quantities <- sqlQuery(channel, query)
head(quantities,10)
```

```
   ProductID OnHand
1          1    200
2          2    300
3          3    400
4          4    200
5          5    150
6          6    100
7          7    100
8          8     50
9          9     50
10        10    100
```

This would be a lot more informative if we could see the product names rather than their ID numbers. The names are in a different table called `Products`. We need to **relate** the two tables `Inventory` and `Produces`. The variable relating the two tables is `ProductID` from `Inventory` and `ID` from `Products`. The convention is to expand the variable names by adding the table name in front of the variable name, and separating the two names by a full stop, like this:

```
query <- "SELECT Products.ProductName, Inventory.OnHand FROM
Inventory INNER JOIN Products ON Inventory.ProductID = Products.ID "

quantities <- sqlQuery(channel, query)
head(quantities,10)
```

```
                              ProductName OnHand
1                  Northwind Traders Chai    200
2                 Northwind Traders Syrup    300
3         Northwind Traders Cajun Seasoning 400
4              Northwind Traders Olive Oil    200
5       Northwind Traders Boysenberry Spread 150
6             Northwind Traders Dried Pears   100
7             Northwind Traders Curry Sauce   100
8               Northwind Traders Walnuts     50
9          Northwind Traders Fruit Cocktail   50
10 Northwind Traders Chocolate Biscuits Mix  100
```

So that is how we relate two tables.

The next task is data selection. This involves the WHERE command in SQL. Here are the cases with more than 150 items on hand:

```
query <- "SELECT Products.ProductName, Inventory.OnHand FROM
Inventory INNER JOIN Products ON Inventory.ProductID = Products.ID
WHERE Inventory.OnHand > 150 "
```

```
supply <- sqlQuery(channel, query)
head(supply,10)
```

```
                    ProductName OnHand
1              Northwind Traders Chai    200
2             Northwind Traders Syrup    300
3 Northwind Traders Cajun Seasoning    400
4          Northwind Traders Olive Oil    200
5          Northwind Traders Crab Meat    200
6      Northwind Traders Chicken Soup    500
```

Using WHERE with text is more challenging because the text needs to be enclosed in quotes and quotes are always tricky in character strings. The solution is to mix single and double quotes to paste together the query you want:

```
name <- "NWTDFN-14"
```

```
query <- paste("SELECT ProductName FROM Products WHERE
ProductCode='",name,"'",sep="")
```

```
code <- sqlQuery(channel, query)
head(code,10)
```

```
             ProductName
1 Northwind Traders Walnuts
```

This is what the query looks like as a single character string:

```
query
```

```
[1] "SELECT ProductName FROM Products WHERE ProductCode='NWTDFN-14'"
```

You will need to ponder this query really hard in order to see why we had to do what we did to select the ProductCode on the basis of its being "NWTDFN-14". The problem is that quotes must be part of the character string that forms the query, but quotes generally mark the end of a character string. There is another example of selecting records on the basis of character strings on p. 197 where we use data from a large relational database to produce species distribution maps.

While you have the Northwind example up and running, you should practise joining together three tables, and use some of the other options for selecting records, using LIKE with wildcards * (e.g. WHERE Variable.name LIKE "Product*").

4

Dataframes

Learning how to handle your data, how to enter them into the computer, and how to read them into R are among the most important topics you will need to master. R handles data in objects known as dataframes. A **dataframe** is an object with rows and columns (a bit like a matrix). The rows contain different observations from your study, or measurements from your experiment (these are sometimes called *cases*). The columns contain the values of different variables (these are often called *fields*). The values in the body of a matrix can only be numbers, but the values in the body of a dataframe can be numbers, but they could also be text (e.g. the names of factor levels for categorical variables, like male or female in a variable called gender), they could be calendar dates (e.g. 23/5/04), or they could be logical variables (TRUE or FALSE). Here is a spreadsheet in the form of a dataframe with seven variables, the leftmost of which comprises the row names, and other variables are numeric (Area, Slope, Soil pH and Worm Density), categorical (Field Name and Vegetation) or logical (Damp is either true = T or false = F).

Field Name	Area	Slope	Vegetation	Soil pH	Damp	Worm Density
Nash's Field	3.6	11	Grassland	4.1	F	4
Silwood Bottom	5.1	2	Arable	5.2	F	7
Nursery Field	2.8	3	Grassland	4.3	F	2
Rush Meadow	2.4	5	Meadow	4.9	T	5
Gunness' Thicket	3.8	0	Scrub	4.2	F	6
Oak Mead	3.1	2	Grassland	3.9	F	2
Church Field	3.5	3	Grassland	4.2	F	3
Ashurst	2.1	0	Arable	4.8	F	4
The Orchard	1.9	0	Orchard	5.7	F	9
Rookery Slope	1.5	4	Grassland	5	T	7
Garden Wood	2.9	10	Scrub	5.2	F	8
North Gravel	3.3	1	Grassland	4.1	F	1
South Gravel	3.7	2	Grassland	4	F	2
Observatory Ridge	1.8	6	Grassland	3.8	F	0
Pond Field	4.1	0	Meadow	5	T	6
Water Meadow	3.9	0	Meadow	4.9	T	8
Cheapside	2.2	8	Scrub	4.7	T	4

(Continued)

The R Book, Second Edition. Michael J. Crawley.
© 2013 John Wiley & Sons, Ltd. Published 2013 by John Wiley & Sons, Ltd.

(Continued)

Field Name	Area	Slope	Vegetation	Soil pH	Damp	Worm Density
Pound Hill	4.4	2	Arable	4.5	F	5
Gravel Pit	2.9	1	Grassland	3.5	F	1
Farm Wood	0.8	10	Scrub	5.1	T	3

Perhaps the most important thing about analysing your own data properly is getting your dataframe absolutely right. The expectation is that you will have used a spreadsheet such as Excel to enter and edit the data, and that you will have used plots to check for errors. The thing that takes some practice is learning exactly how to put your numbers into the spreadsheet. There are countless ways of doing it wrong, but only one way of doing it right. And this way is *not* the way that most people find intuitively to be the most obvious.

The key thing is this: all the values of the same variable must go in the same column. It does not sound like much, but this is what people tend to get wrong. If you had an experiment with three treatments (control, preheated and prechilled), and four measurements per treatment, it might seem like a good idea to create the spreadsheet like this:

control	preheated	prechilled			
6.1	6.3	7.1			
5.9	6.2	8.2			
5.8	5.8	7.3			
5.4	6.3	6.9			

However, this is not a dataframe, because values of the response variable appear in three different columns, rather than all in the same column. The correct way to enter these data is to have two columns: one for the response variable and one for the levels of the experimental factor called Treatment (control, preheated and prechilled). Here are the same data, entered correctly as a dataframe:

Response	Treatment			
6.1	Control			
5.9	Control			
5.8	Control			
5.4	Control			
6.3	Preheated			
6.2	Preheated			
5.8	Preheated			
6.3	Preheated			
7.1	Prechilled			
8.2	Prechilled			
7.3	Prechilled			
6.9	Prechilled			

A good way to practice this layout is to use the Excel function called PivotTable (found under Data on the main menu bar) on your own data: it requires your spreadsheet to be in the form of a dataframe, with each of the variables in its own column.

Once you have made your dataframe in a spreadsheet and corrected all the inevitable data entry and spelling errors, then you need to save the dataframe in a file format that can be read by R. Much the simplest way is to save all your dataframes from the spreadsheet as tab-delimited text files. In Excel, for instance, you click on File/Save As.../ then from the 'Save as type' options choose 'Text (Tab delimited)'. There is no need to add a suffix, because Excel will automatically add '.txt' to your file name. This file can then be read into R directly as a dataframe, using the `read.table` function as explained in Chapter 3.

It is important to note that `read.table` would fail if there were any spaces in any of the variable names in row 1 of the dataframe (the header row), such as Field Name, Soil pH or Worm Density (above), or between any of the words within the same factor level (as in many of the field names). These should be replaced by dots '.' before the dataframe is saved from the spreadsheet. Also, it is good idea to remove any apostrophes, as these can sometimes cause problems because there is more than one ASCII code for quotation marks. Now the dataframe can be read into R. Think of a name for the dataframe (say, 'worms' in this case) and then allocate the data from the file to the dataframe name using the gets arrow <- like this:

```
worms <- read.table("c:\\temp\\worms.txt",header=T)
```

Once the file has been imported to R we often want to do four things:

- use `attach` to make the variables accessible by name within the R session;
- use `names` to get a list of the variable names;
- use `head` to look at the first few rows of the data;
- use `tail` to look at the last few rows of the data.

Typically, the commands are issued in sequence, whenever a new dataframe is imported from file (but see p. 113 for superior alternatives to `attach`):

```
attach(worms)
names(worms)
```

```
[1]  "Field.Name"  "Area"   "Slope"        "Vegetation"
[5]  "Soil.pH"     "Damp"   "Worm.density"
```

```
head(worms)
```

	Field.Name	Area	Slope	Vegetation	Soil.pH	Damp	Worm.density
1	Nashs.Field	3.6	11	Grassland	4.1	FALSE	4
2	Silwood.Bottom	5.1	2	Arable	5.2	FALSE	7
3	Nursery.Field	2.8	3	Grassland	4.3	FALSE	2
4	Rush.Meadow	2.4	5	Meadow	4.9	TRUE	5
5	Gunness.Thicket	3.8	0	Scrub	4.2	FALSE	6
6	Oak.Mead	3.1	2	Grassland	3.9	FALSE	2

```
tail(worms)
```

	Field.Name	Area	Slope	Vegetation	Soil.pH	Damp	Worm.density
15	Pond.Field	4.1	0	Meadow	5.0	TRUE	6
16	Water.Meadow	3.9	0	Meadow	4.9	TRUE	8
17	Cheapside	2.2	8	Scrub	4.7	TRUE	4
18	Pound.Hill	4.4	2	Arable	4.5	FALSE	5
19	Gravel.Pit	2.9	1	Grassland	3.5	FALSE	1
20	Farm.Wood	0.8	10	Scrub	5.1	TRUE	3

To see the contents of the whole dataframe, just type its name:

```
worms
```

	Field.Name	Area	Slope	Vegetation	Soil.pH	Damp	Worm.density
1	Nashs.Field	3.6	11	Grassland	4.1	FALSE	4
2	Silwood.Bottom	5.1	2	Arable	5.2	FALSE	7
3	Nursery.Field	2.8	3	Grassland	4.3	FALSE	2
4	Rush.Meadow	2.4	5	Meadow	4.9	TRUE	5
5	Gunness.Thicket	3.8	0	Scrub	4.2	FALSE	6
6	Oak.Mead	3.1	2	Grassland	3.9	FALSE	2
7	Church.Field	3.5	3	Grassland	4.2	FALSE	3
8	Ashurst	2.1	0	Arable	4.8	FALSE	4
9	The.Orchard	1.9	0	Orchard	5.7	FALSE	9
10	Rookery.Slope	1.5	4	Grassland	5.0	TRUE	7
11	Garden.Wood	2.9	10	Scrub	5.2	FALSE	8
12	North.Gravel	3.3	1	Grassland	4.1	FALSE	1
13	South.Gravel	3.7	2	Grassland	4.0	FALSE	2
14	Observatory.Ridge	1.8	6	Grassland	3.8	FALSE	0
15	Pond.Field	4.1	0	Meadow	5.0	TRUE	6
16	Water.Meadow	3.9	0	Meadow	4.9	TRUE	8
17	Cheapside	2.2	8	Scrub	4.7	TRUE	4
18	Pound.Hill	4.4	2	Arable	4.5	FALSE	5
19	Gravel.Pit	2.9	1	Grassland	3.5	FALSE	1
20	Farm.Wood	0.8	10	Scrub	5.1	TRUE	3

Notice that R has expanded our abbreviated T and F into TRUE and FALSE. The object called worms now has all the attributes of a dataframe. For example, you can summarize it, using summary:

```
summary(worms)
```

```
     Field.Name        Area              Slope            Vegetation
Ashurst     : 1   Min.    :0.800   Min.    : 0.00   Arable   :3
Cheapside   : 1   1st Qu.:2.175    1st Qu.: 0.75    Grassland:9
Church.Field: 1   Median :3.000    Median : 2.00    Meadow   :3
Farm.Wood   : 1   Mean    :2.990   Mean    : 3.50   Orchard  :1
Garden.Wood : 1   3rd Qu.:3.725    3rd Qu.: 5.25    Scrub    :4
Gravel.Pit  : 1   Max.    :5.100   Max.    :11.00
(Other)     :14
```

```
   Soil.pH          Damp          Worm.density
Min.    :3.500   Mode :logical   Min.    :0.00
1st Qu.:4.100    FALSE:14        1st Qu.:2.00
Median :4.600    TRUE :6         Median :4.00
Mean    :4.555   NA's :0         Mean    :4.35
3rd Qu.:5.000                    3rd Qu.:6.25
Max.    :5.700                   Max.    :9.00
```

Values of continuous variables are summarized under six headings: one parametric (the arithmetic mean) and five non-parametric (maximum, minimum, median, 25th percentile or first quartile, and 75th percentile or third quartile). Tukey's famous five-number function (fivenum; see p. 42) is slightly different, with hinges rather than first and third quartiles. Levels of categorical variables are counted. Note that the field names are not listed in full because they are unique to each row; six of them are named, then R says 'plus 14 others' (Other) :14.

The two functions by and aggregate allow summary of the dataframe on the basis of factor levels. For instance, it might be interesting to know the means of the numeric variables for each vegetation type. The function for this is by:

```
by(worms,Vegetation,mean)
```

```
Vegetation: Arable
Field.Name          Area         Slope     Vegetation        Soil.pH          Damp Worm.density
       NA       3.866667      1.333333             NA       4.833333      0.000000      5.333333
------------------------------------------------------------------------------------------------
Vegetation: Grassland
Field.Name          Area         Slope     Vegetation        Soil.pH          Damp Worm.density
       NA      2.9111111     3.6666667             NA      4.1000000     0.1111111     2.4444444
------------------------------------------------------------------------------------------------
Vegetation: Meadow
Field.Name          Area         Slope     Vegetation        Soil.pH          Damp Worm.density
       NA       3.466667      1.666667             NA       4.933333      1.000000      6.333333
------------------------------------------------------------------------------------------------
Vegetation: Orchard
Field.Name          Area         Slope     Vegetation        Soil.pH          Damp Worm.density
       NA            1.9           0.0             NA            5.7           0.0           9.0
------------------------------------------------------------------------------------------------
Vegetation: Scrub
Field.Name          Area         Slope     Vegetation        Soil.pH          Damp Worm.density
       NA          2.425         7.000             NA          4.800         0.500         5.250
```

Notice that the logical variable Damp has been coerced to numeric ($TRUE = 1$, $FALSE = 0$) and then averaged. Warning messages are printed for the non-numeric variables to which the function mean is not applicable (e.g. the factor levels for Field.name and Vegetation), but this is a useful and quick overview of the effects of the five types of vegetation.

You can also fit models using by: here is worm density as a function of soil pH for each vegetation type:

```
by(worms, Vegetation, function(x) lm(Worm.density ~ Soil.pH, data=x))
```

```
Vegetation: Arable
Call:
lm(formula = Worm.density ~ Soil.pH, data = x)

Coefficients:
(Intercept)        Soil.pH
     -9.689          3.108

------------------------------------------------------------------------------
---------------
etc. for each level of vegetation in alphabetical order
------------------------------------------------------------------------------
---------------
Vegetation: Scrub

Call:
lm(formula = Worm.density ~ Soil.pH, data = x)
```

```
Coefficients:
(Intercept)      Soil.pH
    4.4758        0.1613
```

4.1   Subscripts and indices

The key thing about working effectively with dataframes is to become completely at ease with using subscripts (or indices, as some people call them). In R, subscripts appear in square brackets [ ]. A dataframe is a two-dimensional object, comprising rows and columns. The rows are referred to by the first (left-hand) subscript, the columns by the second (right-hand) subscript. Thus

```
worms[3,5]
```

```
[1]   4.3
```

is the value in row 3 of `Soil.pH` (the variable in column 5). To extract a range of values (say the 14th to 19th rows) from worm density (the variable in the seventh column) we use the colon operator : to generate a series of subscripts (14, 15, 16, 17, 18 and 19):

```
worms[14:19,7]
```

```
[1]   0   6   8   4   5   1
```

To extract a group of rows and a group of columns, you need to generate a series of subscripts for both the row and column subscripts. Suppose we want `Area` and `Slope` (columns 2 and 3) from rows 1 to 5:

```
worms[1:5,2:3]
```

```
     Area      Slope
1     3.6         11
2     5.1          2
3     2.8          3
4     2.4          5
5     3.8          0
```

This next point is very important, and is hard to grasp without practice. To select *all* the entries in a *row* the syntax is 'number comma blank'. Similarly, to select all the entries in a *column* the syntax is 'blank comma number'. Thus, to select all the columns in row 3 we type

```
worms[3,]
```

```
      Field.Name Area Slope Vegetation Soil.pH  Damp Worm.density
3 Nursery.Field  2.8     3  Grassland      4.3 FALSE            2
```

whereas to select all the rows in column 3 we need

```
worms[,3]
```

```
 [1] 11   2   3   5   0   2   3   0   0   4  10   1   2   6   0   0   8   2   1  10
```

This is a key feature of the R language, and one that causes problems for beginners. Note that these two apparently similar commands create *objects of different classes*:

```
class(worms[3,])
```

```
[1]  "data.frame"
```

```
class(worms[,3])

[1]  "integer"
```

You can create sets of rows or columns. For instance, to extract all the rows for `Field.Name` and `Soil.pH` (columns 1 and 5) use the concatenate function, `c`, to make a vector of the required column numbers `c(1,5)`:

```
worms[,c(1,5)]
```

```
          Field.Name Soil.pH
1        Nashs.Field     4.1
2     Silwood.Bottom     5.2
3      Nursery.Field     4.3
4        Rush.Meadow     4.9
5    Gunness.Thicket     4.2
6           Oak.Mead     3.9
7       Church.Field     4.2
8            Ashurst     4.8
9        The.Orchard     5.7
10      Rookery.Slope    5.0
11       Garden.Wood     5.2
12       North.Gravel    4.1
13       South.Gravel    4.0
14  Observatory.Ridge    3.8
15         Pond.Field    5.0
16       Water.Meadow    4.9
17          Cheapside    4.7
18         Pound.Hill    4.5
19         Gravel.Pit    3.5
20          Farm.Wood    5.1
```

The commands for selecting rows and columns from the dataframe are summarized in Table 4.1.

4.2 Selecting rows from the dataframe at random

In bootstrapping or cross-validation we might want to select certain rows from the dataframe at random. We use the `sample` function to do this: the default `replace = FALSE` performs shuffling (each row is selected once and only once), while the option `replace = TRUE` (sampling with replacement) allows for multiple copies of certain rows and the omission of others. Here we use the default `replace = F` to select a unique 8 of the 20 rows at random:

```
worms[sample(1:20,8),]
```

```
        Field.Name Area Slope Vegetation Soil.pH  Damp Worm.density
7     Church.Field  3.5     3  Grassland     4.2 FALSE            3
3    Nursery.Field  2.8     3  Grassland     4.3 FALSE            2
19      Gravel.Pit  2.9     1  Grassland     3.5 FALSE            1
4      Rush.Meadow  2.4     5     Meadow     4.9  TRUE            5
18      Pound.Hill  4.4     2     Arable     4.5 FALSE            5
```

12	North.Gravel	3.3	1	Grassland	4.1	FALSE	1
1	Nashs.Field	3.6	11	Grassland	4.1	FALSE	4
15	Pond.Field	4.1	0	Meadow	5.0	TRUE	6

Table 4.1. Selecting parts of a dataframe called `data`. Suppose that n is one of the row numbers in your dataframe that you want to select or remove, and m is one of the columns. Note that the syntax `[n,]` selects all of the columns, while `[,m]` selects all of the rows.

command	meaning
`data[n,]`	select all of the columns from row n of the dataframe
`data[-n,]`	drop the whole of row n from the dataframe
`data[1:n,]`	select all of the columns from rows 1 to n of the dataframe
`data[-(1:n),]`	drop all of the columns from rows 1 to n of the dataframe
`data[c(i,j,k),]`	select all of the columns from rows i, j, and k of the dataframe
`data[x > y,]`	use a logical test ($x > y$) to select all columns from certain rows
`data[,m]`	select all of the rows from column m of the dataframe
`data[,-m]`	drop the whole of column m from the dataframe
`data[,1:m]`	select all of the rows from columns 1 to m of the dataframe
`data[,-(1:m)]`	drop all of the rows from columns 1 to m of the dataframe
`data[,c(i,j,k)]`	select all of the rows from columns i, j, and k of the dataframe
`data[,x > y]`	use a logical test ($x > y$) to select all rows from certain columns
`data[,c(1:m,i,j,k)]`	add duplicate copies of columns i, j, and k to the dataframe
`data[x > y,a != b]`	extract certain rows ($x > y$) and certain columns ($a \mathrel{!=} b$)
`data[c(1:n,i,j,k),]`	add duplicate copies of rows i, j, and k to the dataframe

Note that the row numbers are in random sequence (not sorted), so that if you want a sorted random sample you will need to order the dataframe after the randomization.

4.3   Sorting dataframes

It is common to want to sort a dataframe by rows, but rare to want to sort by columns. Because we are sorting by rows (the first subscript) we specify the order of the row subscripts *before* the comma. Thus, to sort the dataframe on the basis of values in one of the columns (say, `Slope`), we write

```
worms[order(Slope),]
```

	Field.Name	Area	Slope	Vegetation	Soil.pH	Damp	Worm.density
5	Gunness.Thicket	3.8	0	Scrub	4.2	FALSE	6
8	Ashurst	2.1	0	Arable	4.8	FALSE	4
9	The.Orchard	1.9	0	Orchard	5.7	FALSE	9
15	Pond.Field	4.1	0	Meadow	5.0	TRUE	6
16	Water.Meadow	3.9	0	Meadow	4.9	TRUE	8
12	North.Gravel	3.3	1	Grassland	4.1	FALSE	1
19	Gravel.Pit	2.9	1	Grassland	3.5	FALSE	1
2	Silwood.Bottom	5.1	2	Arable	5.2	FALSE	7
6	Oak.Mead	3.1	2	Grassland	3.9	FALSE	2
13	South.Gravel	3.7	2	Grassland	4.0	FALSE	2
18	Pound.Hill	4.4	2	Arable	4.5	FALSE	5
3	Nursery.Field	2.8	3	Grassland	4.3	FALSE	2
7	Church.Field	3.5	3	Grassland	4.2	FALSE	3

10	Rookery.Slope	1.5	4	Grassland	5.0	TRUE	7
4	Rush.Meadow	2.4	5	Meadow	4.9	TRUE	5
14	Observatory.Ridge	1.8	6	Grassland	3.8	FALSE	0
17	Cheapside	2.2	8	Scrub	4.7	TRUE	4
11	Garden.Wood	2.9	10	Scrub	5.2	FALSE	8
20	Farm.Wood	0.8	10	Scrub	5.1	TRUE	3
1	Nashs.Field	3.6	11	Grassland	4.1	FALSE	4

There are some points to notice here. Because we wanted the sorting to apply to all the columns, the column subscript (after the comma) is blank: `[order(Slope),]`. The original row numbers are retained in the leftmost column. Where there are ties for the sorting variable (e.g. there are five ties for Slope = 0) then the rows are in their original order. If you want the dataframe in reverse order (ascending order) then use the rev function outside the order function like this:

```
worms[rev(order(Slope)),]
```

	Field.Name	Area	Slope	Vegetation	Soil.pH	Damp	Worm.density
1	Nashs.Field	3.6	11	Grassland	4.1	FALSE	4
20	Farm.Wood	0.8	10	Scrub	5.1	TRUE	3
11	Garden.Wood	2.9	10	Scrub	5.2	FALSE	8
17	Cheapside	2.2	8	Scrub	4.7	TRUE	4
14	Observatory.Ridge	1.8	6	Grassland	3.8	FALSE	0
4	Rush.Meadow	2.4	5	Meadow	4.9	TRUE	5
10	Rookery.Slope	1.5	4	Grassland	5.0	TRUE	7
7	Church.Field	3.5	3	Grassland	4.2	FALSE	3
3	Nursery.Field	2.8	3	Grassland	4.3	FALSE	2
18	Pound.Hill	4.4	2	Arable	4.5	FALSE	5
13	South.Gravel	3.7	2	Grassland	4.0	FALSE	2
6	Oak.Mead	3.1	2	Grassland	3.9	FALSE	2
2	Silwood.Bottom	5.1	2	Arable	5.2	FALSE	7
19	Gravel.Pit	2.9	1	Grassland	3.5	FALSE	1
12	North.Gravel	3.3	1	Grassland	4.1	FALSE	1
16	Water.Meadow	3.9	0	Meadow	4.9	TRUE	8
15	Pond.Field	4.1	0	Meadow	5.0	TRUE	6
9	The.Orchard	1.9	0	Orchard	5.7	FALSE	9
8	Ashurst	2.1	0	Arable	4.8	FALSE	4
5	Gunness.Thicket	3.8	0	Scrub	4.2	FALSE	6

Notice now that when there are ties (e.g. Slope = 0), the original rows are also in reverse order.

More complicated sorting operations might involve two or more variables. This is achieved very simply by separating a series of variable names by commas within the order function. R will sort on the basis of the left-hand variable, with ties being broken by the second variable, and so on. Suppose that we want to order the rows of the database on worm density within each vegetation type:

```
worms[order(Vegetation,Worm.density),]
```

	Field.Name	Area	Slope	Vegetation	Soil.pH	Damp	Worm.density
8	Ashurst	2.1	0	Arable	4.8	FALSE	4
18	Pound.Hill	4.4	2	Arable	4.5	FALSE	5
2	Silwood.Bottom	5.1	2	Arable	5.2	FALSE	7
14	Observatory.Ridge	1.8	6	Grassland	3.8	FALSE	0

12	North.Gravel	3.3	1	Grassland	4.1	FALSE	1
19	Gravel.Pit	2.9	1	Grassland	3.5	FALSE	1
3	Nursery.Field	2.8	3	Grassland	4.3	FALSE	2
6	Oak.Mead	3.1	2	Grassland	3.9	FALSE	2
13	South.Gravel	3.7	2	Grassland	4.0	FALSE	2
7	Church.Field	3.5	3	Grassland	4.2	FALSE	3
1	Nashs.Field	3.6	11	Grassland	4.1	FALSE	4
10	Rookery.Slope	1.5	4	Grassland	5.0	TRUE	7
4	Rush.Meadow	2.4	5	Meadow	4.9	TRUE	5
15	Pond.Field	4.1	0	Meadow	5.0	TRUE	6
16	Water.Meadow	3.9	0	Meadow	4.9	TRUE	8
9	The.Orchard	1.9	0	Orchard	5.7	FALSE	9
20	Farm.Wood	0.8	10	Scrub	5.1	TRUE	3
17	Cheapside	2.2	8	Scrub	4.7	TRUE	4
5	Gunness.Thicket	3.8	0	Scrub	4.2	FALSE	6
11	Garden.Wood	2.9	10	Scrub	5.2	FALSE	8

Notice that as with single-condition sorts, when there are ties (as in grassland with worm density = 2), the rows are in their original sequence (here, 3, 6, 13). We might want to override this by specifying a third sorting condition (e.g. soil pH):

```
worms[order(Vegetation,Worm.density,Soil.pH),]
```

	Field.Name	Area	Slope	Vegetation	Soil.pH	Damp	Worm.density
8	Ashurst	2.1	0	Arable	4.8	FALSE	4
18	Pound.Hill	4.4	2	Arable	4.5	FALSE	5
2	Silwood.Bottom	5.1	2	Arable	5.2	FALSE	7
14	Observatory.Ridge	1.8	6	Grassland	3.8	FALSE	0
19	Gravel.Pit	2.9	1	Grassland	3.5	FALSE	1
12	North.Gravel	3.3	1	Grassland	4.1	FALSE	1
6	Oak.Mead	3.1	2	Grassland	3.9	FALSE	2
13	South.Gravel	3.7	2	Grassland	4.0	FALSE	2
3	Nursery.Field	2.8	3	Grassland	4.3	FALSE	2
7	Church.Field	3.5	3	Grassland	4.2	FALSE	3
1	Nashs.Field	3.6	11	Grassland	4.1	FALSE	4
10	Rookery.Slope	1.5	4	Grassland	5.0	TRUE	7
4	Rush.Meadow	2.4	5	Meadow	4.9	TRUE	5
15	Pond.Field	4.1	0	Meadow	5.0	TRUE	6
16	Water.Meadow	3.9	0	Meadow	4.9	TRUE	8
9	The.Orchard	1.9	0	Orchard	5.7	FALSE	9
20	Farm.Wood	0.8	10	Scrub	5.1	TRUE	3
17	Cheapside	2.2	8	Scrub	4.7	TRUE	4
5	Gunness.Thicket	3.8	0	Scrub	4.2	FALSE	6
11	Garden.Wood	2.9	10	Scrub	5.2	FALSE	8

The rule is this: if in doubt, sort using more variables than you think you need. That way you can be absolutely certain that the rows are in the order you expect them to be in. This is exceptionally important when you begin to make assumptions about the variables associated with a particular value of the response variable on the basis of its row number.

Perhaps you want only certain columns in the sorted dataframe? Suppose we want vegetation, worm.density, soil pH and slope, and we want them in that order from left to right. We specify the column numbers in the sequence we want them to appear as a vector: `c(4,7,5,3)`:

```
worms[order(Vegetation,Worm.density),c(4,7,5,3)]
```

```
    Vegetation Worm.density Soil.pH Slope
8      Arable            4     4.8     0
18     Arable            5     4.5     2
2      Arable            7     5.2     2
14  Grassland            0     3.8     6
12  Grassland            1     4.1     1
19  Grassland            1     3.5     1
3   Grassland            2     4.3     3
6   Grassland            2     3.9     2
13  Grassland            2     4.0     2
7   Grassland            3     4.2     3
1   Grassland            4     4.1    11
10  Grassland            7     5.0     4
4      Meadow            5     4.9     5
15     Meadow            6     5.0     0
16     Meadow            8     4.9     0
9     Orchard            9     5.7     0
20      Scrub            3     5.1    10
17      Scrub            4     4.7     8
5       Scrub            6     4.2     0
11      Scrub            8     5.2    10
```

You can select the columns on the basis of their variables names, but this is more fiddly to type, because you need to put the variable names in quotes like this:

```
worms[order(Vegetation,Worm.density),
            c("Vegetation", "Worm.density", "Soil.pH", "Slope")]
```

4.4 Using logical conditions to select rows from the dataframe

A very common operation is selecting certain rows from the dataframe on the basis of values in one or more of the variables (the columns of the dataframe). Suppose we want to restrict the data to cases from damp fields. We want all the columns, so the syntax for the subscripts is ['which rows', blank]:

```
worms[Damp == T,]
```

```
        Field.Name Area Slope Vegetation Soil.pH Damp Worm.density
4      Rush.Meadow  2.4     5     Meadow     4.9 TRUE            5
10   Rookery.Slope  1.5     4  Grassland     5.0 TRUE            7
15      Pond.Field  4.1     0     Meadow     5.0 TRUE            6
16     Water.Meadow 3.9     0     Meadow     4.9 TRUE            8
17       Cheapside  2.2     8      Scrub     4.7 TRUE            4
20       Farm.Wood  0.8    10      Scrub     5.1 TRUE            3
```

Note that because `Damp` is a logical variable (with just two potential values, TRUE or FALSE) we can refer to true or false in abbreviated form, T or F. Also notice that the T in this case is not enclosed in quotes: the T means true, not the character string "T". The other important point is that the symbol for the logical condition is == (two successive equals signs with no gap between them; see p. 26).

The logic for the selection of rows can refer to values (and functions of values) in more than one column. Suppose that we wanted the data from the fields where worm density was higher than the median (>median(Worm.density)) and soil pH was less than 5.2. In R, the logical operator for AND is the & ('ampersand') symbol:

```
worms[Worm.density > median(Worm.density) & Soil.pH < 5.2,]
```

```
          Field.Name Area Slope Vegetation Soil.pH  Damp Worm.density
4        Rush.Meadow  2.4     5     Meadow     4.9  TRUE            5
5    Gunness.Thicket  3.8     0      Scrub     4.2 FALSE            6
10     Rookery.Slope  1.5     4  Grassland     5.0  TRUE            7
15        Pond.Field  4.1     0     Meadow     5.0  TRUE            6
16      Water.Meadow  3.9     0     Meadow     4.9  TRUE            8
18        Pound.Hill  4.4     2     Arable     4.5 FALSE            5
```

Suppose that we want to extract all the columns that contain numbers (rather than characters or logical variables) from the dataframe. The function `is.numeric` can be applied across all the columns of worms using `sapply` to create the appropriate subscripts like this:

```
worms[,sapply(worms,is.numeric)]
```

```
   Area Slope Soil.pH Worm.density
1   3.6    11     4.1            4
2   5.1     2     5.2            7
3   2.8     3     4.3            2
4   2.4     5     4.9            5
5   3.8     0     4.2            6
6   3.1     2     3.9            2
7   3.5     3     4.2            3
8   2.1     0     4.8            4
9   1.9     0     5.7            9
10  1.5     4     5.0            7
11  2.9    10     5.2            8
12  3.3     1     4.1            1
13  3.7     2     4.0            2
14  1.8     6     3.8            0
15  4.1     0     5.0            6
16  3.9     0     4.9            8
17  2.2     8     4.7            4
18  4.4     2     4.5            5
19  2.9     1     3.5            1
20  0.8    10     5.1            3
```

We might want to extract the columns that were factors:

```
worms[,sapply(worms,is.factor)]
```

```
         Field.Name Vegetation
1       Nashs.Field  Grassland
2    Silwood.Bottom     Arable
3     Nursery.Field  Grassland
4       Rush.Meadow     Meadow
5   Gunness.Thicket      Scrub
6          Oak.Mead  Grassland
7      Church.Field  Grassland
8           Ashurst     Arable
9       The.Orchard    Orchard
10    Rookery.Slope  Grassland
11      Garden.Wood      Scrub
12     North.Gravel  Grassland
13     South.Gravel  Grassland
14 Observatory.Ridge Grassland
15       Pond.Field     Meadow
16     Water.Meadow     Meadow
17        Cheapside      Scrub
18       Pound.Hill     Arable
19       Gravel.Pit  Grassland
20        Farm.Wood      Scrub
```

Because worms is a dataframe, the characters have all been coerced to factors, so worms[,sapply(worms,is.character)] produces the answer NULL.

To drop a row or rows from the dataframe, use **negative subscripts**. Thus to drop the middle 10 rows (i.e. row numbers 6 to 15 inclusive) do this:

```
worms[-(6:15),]
```

```
         Field.Name Area Slope Vegetation Soil.pH  Damp Worm.density
1       Nashs.Field  3.6    11  Grassland     4.1 FALSE            4
2    Silwood.Bottom  5.1     2     Arable     5.2 FALSE            7
3     Nursery.Field  2.8     3  Grassland     4.3 FALSE            2
4       Rush.Meadow  2.4     5     Meadow     4.9  TRUE            5
5   Gunness.Thicket  3.8     0      Scrub     4.2 FALSE            6
16     Water.Meadow  3.9     0     Meadow     4.9  TRUE            8
17        Cheapside  2.2     8      Scrub     4.7  TRUE            4
18       Pound.Hill  4.4     2     Arable     4.5 FALSE            5
19       Gravel.Pit  2.9     1  Grassland     3.5 FALSE            1
20        Farm.Wood  0.8    10      Scrub     5.1  TRUE            3
```

Here are all the rows that are not grasslands (recall that the logical symbol ! means NOT):

```
worms[!(Vegetation=="Grassland"),]
```

```
         Field.Name Area Slope Vegetation Soil.pH  Damp Worm.density
2    Silwood.Bottom  5.1     2     Arable     5.2 FALSE            7
4       Rush.Meadow  2.4     5     Meadow     4.9  TRUE            5
5   Gunness.Thicket  3.8     0      Scrub     4.2 FALSE            6
8           Ashurst  2.1     0     Arable     4.8 FALSE            4
9       The.Orchard  1.9     0    Orchard     5.7 FALSE            9
11      Garden.Wood  2.9    10      Scrub     5.2 FALSE            8
```

15	Pond.Field	4.1	0	Meadow	5.0	TRUE	6
16	Water.Meadow	3.9	0	Meadow	4.9	TRUE	8
17	Cheapside	2.2	8	Scrub	4.7	TRUE	4
18	Pound.Hill	4.4	2	Arable	4.5	FALSE	5
20	Farm.Wood	0.8	10	Scrub	5.1	TRUE	3

If you want to use minus signs rather than logical NOT to drop rows from the dataframe, the expression you use must evaluate to numbers. The which function is useful for this. Let us use this technique to drop the non-damp fields:

```
worms[-which(Damp==F),]
```

	Field.Name	Area	Slope	Vegetation	Soil.pH	Damp	Worm.density
4	Rush.Meadow	2.4	5	Meadow	4.9	TRUE	5
10	Rookery.Slope	1.5	4	Grassland	5.0	TRUE	7
15	Pond.Field	4.1	0	Meadow	5.0	TRUE	6
16	Water.Meadow	3.9	0	Meadow	4.9	TRUE	8
17	Cheapside	2.2	8	Scrub	4.7	TRUE	4
20	Farm.Wood	0.8	10	Scrub	5.1	TRUE	3

which achieves the same end as the more elegant

```
worms[!Damp==F,]
```

or, even simpler,

```
worms[Damp==T,]
```

4.5   Omitting rows containing missing values, NA

In statistical modelling it is often useful to have a dataframe that contains no missing values in the response or explanatory variables. You can create a shorter dataframe using the na.omit function. Here is a sister dataframe of worms in which certain values are NA:

```
data <- read.table("c:\\temp\\worms.missing.txt",header=T)
data
```

	Field.Name	Area	Slope	Vegetation	Soil.pH	Damp	Worm.density
1	Nashs.Field	3.6	11	Grassland	4.1	FALSE	4
2	Silwood.Bottom	5.1	NA	Arable	5.2	FALSE	7
3	Nursery.Field	2.8	3	Grassland	4.3	FALSE	2
4	Rush.Meadow	2.4	5	Meadow	4.9	TRUE	5
5	Gunness.Thicket	3.8	0	Scrub	4.2	FALSE	6
6	Oak.Mead	3.1	2	Grassland	3.9	FALSE	2
7	Church.Field	3.5	3	Grassland	NA	NA	NA
8	Ashurst	2.1	0	Arable	4.8	FALSE	4
9	The.Orchard	1.9	0	Orchard	5.7	FALSE	9
10	Rookery.Slope	1.5	4	Grassland	5.0	TRUE	7
11	Garden.Wood	2.9	10	Scrub	5.2	FALSE	8
12	North.Gravel	3.3	1	Grassland	4.1	FALSE	1
13	South.Gravel	3.7	2	Grassland	4.0	FALSE	2

14	Observatory.Ridge	1.8	6	Grassland	3.8	FALSE	0
15	Pond.Field	4.1	0	Meadow	5.0	TRUE	6
16	Water.Meadow	3.9	0	Meadow	4.9	TRUE	8
17	Cheapside	2.2	8	Scrub	4.7	TRUE	4
18	Pound.Hill	4.4	2	Arable	4.5	FALSE	5
19	Gravel.Pit	NA	1	Grassland	3.5	FALSE	1
20	Farm.Wood	0.8	10	Scrub	5.1	TRUE	3

By inspection we can see that we should like to leave out row 2 (one missing value), row 7 (three missing values) and row 19 (one missing value). This could not be simpler:

```
na.omit(data)
```

	Field.Name	Area	Slope	Vegetation	Soil.pH	Damp	Worm.density
1	Nashs.Field	3.6	11	Grassland	4.1	FALSE	4
3	Nursery.Field	2.8	3	Grassland	4.3	FALSE	2
4	Rush.Meadow	2.4	5	Meadow	4.9	TRUE	5
5	Gunness.Thicket	3.8	0	Scrub	4.2	FALSE	6
6	Oak.Mead	3.1	2	Grassland	3.9	FALSE	2
8	Ashurst	2.1	0	Arable	4.8	FALSE	4
9	The.Orchard	1.9	0	Orchard	5.7	FALSE	9
10	Rookery.Slope	1.5	4	Grassland	5.0	TRUE	7
11	Garden.Wood	2.9	10	Scrub	5.2	FALSE	8
12	North.Gravel	3.3	1	Grassland	4.1	FALSE	1
13	South.Gravel	3.7	2	Grassland	4.0	FALSE	2
14	Observatory.Ridge	1.8	6	Grassland	3.8	FALSE	0
15	Pond.Field	4.1	0	Meadow	5.0	TRUE	6
16	Water.Meadow	3.9	0	Meadow	4.9	TRUE	8
17	Cheapside	2.2	8	Scrub	4.7	TRUE	4
18	Pound.Hill	4.4	2	Arable	4.5	FALSE	5
20	Farm.Wood	0.8	10	Scrub	5.1	TRUE	3

and you see that rows 2, 7 and 19 have been omitted in creating the new dataframe. Alternatively, you can use the na.exclude function. This differs from na.omit only in the class of the na.action attribute of the result, which gives different behaviour in functions making use of naresid and napredict: when na.exclude is used the residuals and predictions are padded to the correct length by inserting NAs for cases omitted by na.exclude (in this example they would be of length 20, whereas na.omit would give residuals and predictions of length 17).

```
new.frame <- na.exclude(data)
```

The function to test for the presence of missing values across a dataframe is complete.cases:

```
complete.cases(data)
```

```
[1]    TRUE FALSE  TRUE  TRUE  TRUE  TRUE FALSE  TRUE  TRUE  TRUE  TRUE
[12]   TRUE  TRUE  TRUE  TRUE  TRUE  TRUE  TRUE FALSE  TRUE
```

You could use this as a less efficient analogue of na.omit(data), but why would you?

```
data[complete.cases(data),]
```

It is well worth checking the individual variables separately, because it is possible that one or more variables contribute most of the missing values, and it may be preferable to remove these variables from the modelling rather than lose the valuable information about the other explanatory variables associated with these cases. Use `summary` to count the missing values for each variable in the dataframe, or use `apply` with the function `is.na` to sum the missing values in each variable:

```
apply(apply(data,2,is.na),2,sum)

Field.Name   Area   Slope   Vegetation   Soil.pH   Damp   Worm.density
         0      1       1            0         1      1              1
```

You can see that in this case no single variable contributed more missing values than any other.

4.5.1   Replacing NAs with zeros

You would need to think carefully before doing this, but there might be circumstances when you wanted to replace the missing values NA by zero (or by some other missing-value indicator). Continuing the missing-worms example, above, where the dataframe called `data` contained five missing values, this is how to replace all the NAs by zeros:

```
data[is.na(data)]<-0
```

4.6   Using `order` and `!duplicated` to eliminate pseudoreplication

In this rather more complicated example, you are asked to extract a single record for each vegetation type, and that record is to be the case within each vegetation type that has the greatest worm density. There are two steps to this: first order all of the rows in a new dataframe using `rev(order(Worm.density))`, then select the subset of these rows which is not duplicated (`!duplicated`) within each vegetation type in the new dataframe (using `new$Vegetation`):

```
new <- worms[rev(order(Worm.density)),]
new[!duplicated(new$Vegetation),]

       Field.Name Area Slope Vegetation Soil.pH  Damp Worm.density
9      The.Orchard  1.9     0    Orchard     5.7 FALSE            9
16    Water.Meadow  3.9     0     Meadow     4.9  TRUE            8
11     Garden.Wood  2.9    10      Scrub     5.2 FALSE            8
10   Rookery.Slope  1.5     4  Grassland     5.0  TRUE            7
2  Silwood.Bottom  5.1     2     Arable     5.2 FALSE            7
```

4.7   Complex ordering with mixed directions

Sometimes there are multiple sorting variables, but the variables have to be sorted in opposing directions. In this example, the task is to order the database first by vegetation type in alphabetical order (the default) and then within each vegetation type to sort by worm density in decreasing order (highest densities first). The trick here is to use `order` (rather than `rev(order)`) but to put a minus sign in front of `Worm.density` like this:

```
worms[order(Vegetation,-Worm.density),]
```

	Field.Name	Area	Slope	Vegetation	Soil.pH	Damp	Worm.density
2	Silwood.Bottom	5.1	2	Arable	5.2	FALSE	7
18	Pound.Hill	4.4	2	Arable	4.5	FALSE	5
8	Ashurst	2.1	0	Arable	4.8	FALSE	4
10	Rookery.Slope	1.5	4	Grassland	5.0	TRUE	7
1	Nashs.Field	3.6	11	Grassland	4.1	FALSE	4
7	Church.Field	3.5	3	Grassland	4.2	FALSE	3
3	Nursery.Field	2.8	3	Grassland	4.3	FALSE	2
6	Oak.Mead	3.1	2	Grassland	3.9	FALSE	2
13	South.Gravel	3.7	2	Grassland	4.0	FALSE	2
12	North.Gravel	3.3	1	Grassland	4.1	FALSE	1
19	Gravel.Pit	2.9	1	Grassland	3.5	FALSE	1
14	Observatory.Ridge	1.8	6	Grassland	3.8	FALSE	0
16	Water.Meadow	3.9	0	Meadow	4.9	TRUE	8
15	Pond.Field	4.1	0	Meadow	5.0	TRUE	6
4	Rush.Meadow	2.4	5	Meadow	4.9	TRUE	5
9	The.Orchard	1.9	0	Orchard	5.7	FALSE	9
11	Garden.Wood	2.9	10	Scrub	5.2	FALSE	8
5	Gunness.Thicket	3.8	0	Scrub	4.2	FALSE	6
17	Cheapside	2.2	8	Scrub	4.7	TRUE	4
20	Farm.Wood	0.8	10	Scrub	5.1	TRUE	3

Using the minus sign only works when sorting numerical variables. For factor levels you can use the rank function to make the levels numeric like this:

```
worms[order(-rank(Vegetation),-Worm.density),]
```

	Field.Name	Area	Slope	Vegetation	Soil.pH	Damp	Worm.density
11	Garden.Wood	2.9	10	Scrub	5.2	FALSE	8
5	Gunness.Thicket	3.8	0	Scrub	4.2	FALSE	6
17	Cheapside	2.2	8	Scrub	4.7	TRUE	4
20	Farm.Wood	0.8	10	Scrub	5.1	TRUE	3
9	The.Orchard	1.9	0	Orchard	5.7	FALSE	9
16	Water.Meadow	3.9	0	Meadow	4.9	TRUE	8
15	Pond.Field	4.1	0	Meadow	5.0	TRUE	6
4	Rush.Meadow	2.4	5	Meadow	4.9	TRUE	5
10	Rookery.Slope	1.5	4	Grassland	5.0	TRUE	7
1	Nashs.Field	3.6	11	Grassland	4.1	FALSE	4
7	Church.Field	3.5	3	Grassland	4.2	FALSE	3
3	Nursery.Field	2.8	3	Grassland	4.3	FALSE	2
6	Oak.Mead	3.1	2	Grassland	3.9	FALSE	2
13	South.Gravel	3.7	2	Grassland	4.0	FALSE	2
12	North.Gravel	3.3	1	Grassland	4.1	FALSE	1
19	Gravel.Pit	2.9	1	Grassland	3.5	FALSE	1
14	Observatory.Ridge	1.8	6	Grassland	3.8	FALSE	0
2	Silwood.Bottom	5.1	2	Arable	5.2	FALSE	7
18	Pound.Hill	4.4	2	Arable	4.5	FALSE	5
8	Ashurst	2.1	0	Arable	4.8	FALSE	4

It is less likely that you will want to select *columns* on the basis of logical operations, but it is perfectly possible. Suppose that for some reason you want to select the columns that contain the character 'S' (upper-case S). In R the function for this is `grep`, which returns the subscript (a number or set of numbers) indicating which character strings within a vector of character strings contained an upper-case S. The names of the variables within a dataframe are obtained by the `names` function:

```
names(worms)
```

```
[1]    "Field.Name"    "Area"    "Slope"         "Vegetation"
[5]    "Soil.pH"       "Damp"    "Worm.density"
```

so we want our function `grep` to pick out variables numbers 3 and 5 because they are the only ones containing upper-case S:

```
grep("S",names(worms))
```

```
[1]  3  5
```

Finally, we can use these numbers as subscripts `[,c(3,5)]` to select columns 3 and 5:

```
worms[,grep("S",names(worms))]
```

```
     Slope  Soil.pH
1      11     4.1
2       2     5.2
3       3     4.3
4       5     4.9
5       0     4.2
6       2     3.9
7       3     4.2
8       0     4.8
9       0     5.7
10      4     5.0
11     10     5.2
12      1     4.1
13      2     4.0
14      6     3.8
15      0     5.0
16      0     4.9
17      8     4.7
18      2     4.5
19      1     3.5
20     10     5.1
```

4.8   A dataframe with row names instead of row numbers

You can suppress the creation of row numbers and allocate your own unique names to each row by altering the syntax of the `read.table` function. The first column of the worms database contains the names of the fields in which the other variables were measured. Up to now, we have read this column as if it was the first variable (p. 161).

```
worms2 <- read.table("c:\\temp\\worms.txt",header=T,row.names=1)
worms2
```

	Area	Slope	Vegetation	Soil.pH	Damp	Worm.density
Nashs.Field	3.6	11	Grassland	4.1	FALSE	4
Silwood.Bottom	5.1	2	Arable	5.2	FALSE	7
Nursery.Field	2.8	3	Grassland	4.3	FALSE	2
Rush.Meadow	2.4	5	Meadow	4.9	TRUE	5
Gunness.Thicket	3.8	0	Scrub	4.2	FALSE	6
Oak.Mead	3.1	2	Grassland	3.9	FALSE	2
Church.Field	3.5	3	Grassland	4.2	FALSE	3
Ashurst	2.1	0	Arable	4.8	FALSE	4
The.Orchard	1.9	0	Orchard	5.7	FALSE	9
Rookery.Slope	1.5	4	Grassland	5.0	TRUE	7
Garden.Wood	2.9	10	Scrub	5.2	FALSE	8
North.Gravel	3.3	1	Grassland	4.1	FALSE	1
South.Gravel	3.7	2	Grassland	4.0	FALSE	2
Observatory.Ridge	1.8	6	Grassland	3.8	FALSE	0
Pond.Field	4.1	0	Meadow	5.0	TRUE	6
Water.Meadow	3.9	0	Meadow	4.9	TRUE	8
Cheapside	2.2	8	Scrub	4.7	TRUE	4
Pound.Hill	4.4	2	Arable	4.5	FALSE	5
Gravel.Pit	2.9	1	Grassland	3.5	FALSE	1
Farm.Wood	0.8	10	Scrub	5.1	TRUE	3

Notice that the field names column is not now headed by a variable name, and that the row numbers, as intended, have been suppressed.

4.9 Creating a dataframe from another kind of object

We have seen that the simplest way to create a dataframe in R is to read a table of data from an external file using the read.table function. Alternatively, you can create a dataframe by using the data.frame function to bind together a number of vectors. Here are three vectors of the same length:

```
x <- runif(10)
y <- letters[1:10]
z <- sample(c(rep(T,5),rep(F,5)))
```

To make them into a dataframe called new, just type:

```
new <- data.frame(y,z,x)
new
```

	y	z	x
1	a	TRUE	0.72675982
2	b	FALSE	0.83847227

3	c	FALSE	0.61765685
4	d	TRUE	0.78541650
5	e	FALSE	0.51168828
6	f	TRUE	0.53526324
7	g	TRUE	0.05552335
8	h	TRUE	0.78486234
9	i	FALSE	0.68385443
10	j	FALSE	0.89367837

Note that the order of the columns is controlled simply by the sequence of the vector names (left to right) specified within the `data.frame` function.

In this next example, we create a table of counts of random integers from a Poisson distribution, then convert the table into a dataframe. First, we make a table object:

```
y <- rpois(1500,1.5)
table(y)
```

```
y
   0     1     2     3     4     5     6     7
 344   502   374   199    63    11     5     2
```

Now it is simple to convert this table object into a dataframe with two variables, the count and the frequency, using the `as.data.frame` function:

```
short<-as.data.frame(table(y))
short
```

```
    y   Freq
1   0    344
2   1    502
3   2    374
4   3    199
5   4     63
6   5     11
7   6      5
8   7      2
```

In some cases you might want to expand a dataframe like the one above such that it had a separate row for every distinct count (i.e. 344 rows with $y = 0$, 502 rows with $y = 1$, 374 rows with $y = 2$, and so on). This is very straightforward using subscripts. We need to create a vector of indices containing 344 repeats of 1, 502 repeats of 2 and so on. Note that these repeats are of the row numbers (1, 2, 3, ..., 8), not repeats of the values of y (0, 1, 2, ..., 7).

```
index<-rep(1:8,short$Freq)
```

This simple command has produced a vector with the right number of repeats of each of the row numbers

```
length(index)
```

```
[1] 1500
```

```
hist(index,-0.5:8.5)
```

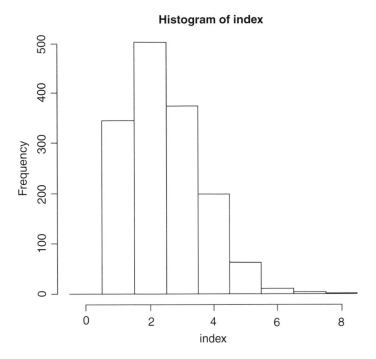

To get the long version of the dataframe, we just use index as the row specifier [index,]:

```
long<-short[index,]
```

Here is a look at the bottom of this long dataframe:

```
tail(long)
```

```
      y Freq
7.1 6      5
7.2 6      5
7.3 6      5
7.4 6      5
8   7      2
8.1 7      2
```

Note the way that R has handled the duplicate row numbers, creating a nested series to indicate the repeats of each of the original row numbers.

A longer-winded alternative might use `lapply` with `rep` to do the same thing:

```
long2 <- as.data.frame(lapply(short,
               function(x) rep(x, short$Freq)))
tail(long2)
```

```
       y Freq
1495 6      5
1496 6      5
1497 6      5
1498 6      5
```

```
1499 7    2
1500 7    2
```

Note the use of the anonymous function in `lapply` to generate the repeats of each row by the value specified in `Freq`. Before you did anything useful with this longer dataframe, you would probably want to get rid of the redundant column called `Freq`.

4.10   Eliminating duplicate rows from a dataframe

Sometimes a dataframe will contain duplicate rows where all the variables have exactly the same values in two or more rows. Here is a simple example:

```
dups <- read.table("c:\\temp\\dups.txt",header=T)
dups
```

```
   var1  var2  var3  Var4
1    1     2     3     1
2    1     2     2     1
3    3     2     1     1
4    4     4     2     1
5    3     2     1     1
6    6     1     2     5
7    1     2     3     2
```

Note that row number 5 is an exact duplicate of row number 3. To create a dataframe with all the duplicate rows stripped out, use the `unique` function like this:

```
unique(dups)
```

```
   var1  var2  var3  var4
1    1     2     3     1
2    1     2     2     1
3    3     2     1     1
4    4     4     2     1
6    6     1     2     5
7    1     2     3     2
```

Notice that the row names in the new dataframe are the same as in the original, so that you can spot that row number 5 was removed by the operation of the function `unique`.

To view the rows that are duplicates in a dataframe (if any) use the `duplicated` function to create a vector of TRUE and FALSE to act as the filter:

```
dups[duplicated(dups),]
```

```
   var1  var2  var3  var4
5    3     2     1     1
```

4.11   Dates in dataframes

There is an introduction to the complexities of using dates and times on pp. 101–113. Here we illustrate a simple example:

```
nums <- read.table("c:\\temp\\sortdata.txt",header=T)
attach(nums)
head(nums)
```

```
       name       date    response treatment
1    albert 25/08/2003 0.05963704         A
2       ann 21/05/2003 1.46555993         A
3      john 12/10/2003 1.59406539         B
4       ian 02/12/2003 2.09505949         A
5   michael 18/10/2003 2.38330748         B
6       ann 02/07/2003 2.86983693         B
```

The idea is to order the rows by date. The ordering is to be applied to all four columns of the dataframe. Note that ordering on the basis of our variable called date does not work in the way we want it to:

```
nums[order(date),]
```

```
        name       date    response treatment
53    rachel 01/08/2003 32.98792196         B
65    albert 02/06/2003 38.41979568         A
6        ann 02/07/2003  2.86983693         B
10    cecily 02/11/2003  6.81467570         A
4        ian 02/12/2003  2.09505949         A
29   michael 03/05/2003 15.59890900         B
67   william 03/09/2003 38.95014474         A
```

This is because of the format used for depicting the date is a character string in which the first characters are the day, then the month, then the year, so the dataframe has been sorted into alphabetical order, rather than date order as required. In order to sort by date we need first to convert our variable into date-time format using the strptime function (see p. 103 for details):

```
dates <- strptime(date,format="%d/%m/%Y")
dates
```

```
[1]    "2003-08-25"  "2003-05-21"  "2003-10-12"  "2003-12-02"  "2003-10-18"
[6]    "2003-07-02"  "2003-09-27"  "2003-06-05"  "2003-06-11"  "2003-11-02"
```

Note how strptime has produced a date object with year first, then a hyphen, then month, then a hyphen, then day, and this will sort into the desired sequence. We bind the new variable to the dataframe like this:

```
nums <- cbind(nums,dates)
```

Now that the new variable is in the correct format, the dates can be sorted correctly:

```
nums[order(dates),]
```

```
        name       date    response treatment       dates
49    albert 21/04/2003 30.66632632         A 2003-04-21
63     james 24/04/2003 37.04140266         A 2003-04-24
24      john 27/04/2003 12.70257306         A 2003-04-27
33   william 30/04/2003 18.05707279         B 2003-04-30
29   michael 03/05/2003 15.59890900         B 2003-05-03
71       ian 06/05/2003 39.97237868         A 2003-05-06
```

```
50     rachel 09/05/2003 30.81807436         B 2003-05-09
69 elizabeth 12/05/2003 39.39536726         B 2003-05-12
```

4.12   Using the `match` function in dataframes

The `worms` dataframe above contains fields of five different vegetation types:

```
unique(worms$Vegetation)
```

```
[1] Grassland Arable    Meadow    Scrub     Orchard
Levels: Arable Grassland Meadow Orchard Scrub
```

and we want to know the appropriate herbicides to use in each of the 20 fields. The herbicides are in a separate dataframe that contains the recommended herbicides for a much larger set of plant community types:

```
herbicides <- read.table("c:\\temp\\herbicides.txt",header=T)
herbicides
```

```
         Type    Herbicide
1    Woodland    Fusilade
2     Conifer    Weedwipe
3      Arable    Twinspan
4        Hill    Weedwipe
5     Bracken    Fusilade
6       Scrub    Weedwipe
7   Grassland    Allclear
8       Chalk    Vanquish
9      Meadow    Propinol
10       Lawn    Vanquish
11    Orchard    Fusilade
12      Verge    Allclear
```

The task is to create a vector of length 20 (one for every field in `worms`) containing the name of the appropriate herbicide. The first value needs to be Allclear because Nash's Field is grassland, and the second needs to be Twinspan because Silwood Bottom is arable, and so on. The first argument in match is `worms$Vegetation` and the second argument in match is `herbicides$Type`. The result of this match is used as a vector of subscripts to extract the relevant herbicides from `herbicides$Herbicide` like this:

```
herbicides$Herbicide[match(worms$Vegetation,herbicides$Type)]
```

```
[1]  Allclear Twinspan Allclear Propinol Weedwipe Allclear Allclear
[8]  Twinspan Fusilade Allclear Weedwipe Allclear Allclear Allclear
[15] Propinol Propinol Weedwipe Twinspan Allclear Weedwipe
```

```
Levels: Allclear Fusilade Propinol Twinspan Vanquish Weedwipe
```

You could add this information as a new column in the `worms` dataframe:

```
worms$hb <- herbicides$Herbicide[match(worms$Vegetation,herbicides$Type)]
```

or create a new dataframe called `recs` containing the herbicide recommendations:

```
recs <- data.frame(
     worms,hb=herbicides$Herbicide[match(worms$Vegetation,herbicides$Type)])
recs
```

	Field.Name	Area	Slope	Vegetation	Soil.pH	Damp	Worm.density	hb
1	Nashs.Field	3.6	11	Grassland	4.1	FALSE	4	Allclear
2	Silwood.Bottom	5.1	2	Arable	5.2	FALSE	7	Twinspan
3	Nursery.Field	2.8	3	Grassland	4.3	FALSE	2	Allclear
4	Rush.Meadow	2.4	5	Meadow	4.9	TRUE	5	Propinol
5	Gunness.Thicket	3.8	0	Scrub	4.2	FALSE	6	Weedwipe
6	Oak.Mead	3.1	2	Grassland	3.9	FALSE	2	Allclear
7	Church.Field	3.5	3	Grassland	4.2	FALSE	3	Allclear
8	Ashurst	2.1	0	Arable	4.8	FALSE	4	Twinspan
9	The.Orchard	1.9	0	Orchard	5.7	FALSE	9	Fusilade
10	Rookery.Slope	1.5	4	Grassland	5.0	TRUE	7	Allclear
11	Garden.Wood	2.9	10	Scrub	5.2	FALSE	8	Weedwipe
12	North.Gravel	3.3	1	Grassland	4.1	FALSE	1	Allclear
13	South.Gravel	3.7	2	Grassland	4.0	FALSE	2	Allclear
14	Observatory.Ridge	1.8	6	Grassland	3.8	FALSE	0	Allclear
15	Pond.Field	4.1	0	Meadow	5.0	TRUE	6	Propinol
16	Water.Meadow	3.9	0	Meadow	4.9	TRUE	8	Propinol
17	Cheapside	2.2	8	Scrub	4.7	TRUE	4	Weedwipe
18	Pound.Hill	4.4	2	Arable	4.5	FALSE	5	Twinspan
19	Gravel.Pit	2.9	1	Grassland	3.5	FALSE	1	Allclear
20	Farm.Wood	0.8	10	Scrub	5.1	TRUE	3	Weedwipe

4.13   Merging two dataframes

Suppose we have two dataframes, the first containing information on plant life forms and the second containing information of time of flowering. We want to produce a single dataframe showing information on both life form and flowering time. Both dataframes contain variables for genus name and species name:

```
(lifeforms <- read.table("c:\\temp\\lifeforms.txt",header=T))
```

	Genus	species	lifeform
1	Acer	platanoides	tree
2	Acer	palmatum	tree
3	Ajuga	reptans	herb
4	Conyza	sumatrensis	annual
5	Lamium	album	herb

```
(flowering <- read.table("c:\\temp\\fltimes.txt",header=T))
```

	Genus	species	flowering
1	Acer	platanoides	May
2	Ajuga	reptans	June
3	Brassica	napus	April
4	Chamerion	angustifolium	July
5	Conyza	bilbaoana	August
6	Lamium	album	January

Because at least one of the variable names is identical in the two dataframes (in this case, two variables are identical, namely `Genus` and `species`) we can use the simplest of all merge commands:

```
merge(flowering,lifeforms)
```

```
    Genus      species flowering lifeform
1    Acer platanoides       May      tree
2   Ajuga     reptans      June      herb
3  Lamium       album   January      herb
```

The important point to note is that the merged dataframe contains only those rows which had *complete* entries in both dataframes. Two rows from the `lifeforms` database were excluded because there were no flowering time data for them (*Acer platanoides* and *Conyza sumatrensis*), and three rows from the `flowering` database were excluded because there were no life-form data for them (*Chamerion angustifolium*, *Conyza bilbaoana* and *Brassica napus*).

If you want to include all the species, with missing values (NA) inserted when flowering times or life forms are not known, then use the `all=T` option:

```
(both <- merge(flowering,lifeforms,all=T))
```

```
      Genus        species flowering lifeform
1       Acer    platanoides       May      tree
2       Acer       palmatum      <NA>      tree
3      Ajuga        reptans      June      herb
4   Brassica         napus     April      <NA>
5  Chamerion angustifolium      July      <NA>
6     Conyza      bilbaoana    August      <NA>
7     Conyza    sumatrensis      <NA>    annual
8     Lamium          album   January      herb
```

One complexity that often arises is that the same variable has *different names* in the two dataframes that need to be merged. The simplest solution is often to edit the variable names in your spreadsheet before reading them into R, but failing this, you need to specify the names in the first dataframe (known conventionally as the *x* dataframe) and the second dataframe (known conventionally as the *y* dataframe) using the `by.x` and `by.y` options in merge. We have a third dataframe containing information on the seed weights of all eight species, but the variable `Genus` is called `name1` and the variable `species` is called `name2`.

```
(seeds <- read.table("c:\\temp\\seedwts.txt",header=T))
```

```
      name1          name2 seed
1      Acer    platanoides 32.0
2    Lamium          album 12.0
3     Ajuga        reptans  4.0
4 Chamerion angustifolium  1.5
5    Conyza      bilbaoana  0.5
6  Brassica         napus  7.0
7      Acer       palmatum 21.0
8    Conyza    sumatrensis  0.6
```

Just using `merge(both,seeds)` fails miserably: you should try it, to see what happens. We need to inform the merge function that `Genus` and `name1` are synonyms (different names for the same variable), as are `species` and `name2`.

```
merge(both,seeds,by.x=c("Genus","species"),by.y=c("name1","name2"))
```

	Genus	species	flowering	lifeform	seed
1	Acer	palmatum	<NA>	tree	21.0
2	Acer	platanoides	May	tree	32.0
3	Ajuga	reptans	June	herb	4.0
4	Brassica	napus	April	<NA>	7.0
5	Chamerion	angustifolium	July	<NA>	1.5
6	Conyza	bilbaoana	August	<NA>	0.5
7	Conyza	sumatrensis	<NA>	annual	0.6
8	Lamium	album	January	herb	12.0

Note that the variable names used in the merged dataframe are the names used in the *x* dataframe.

4.14 Adding margins to a dataframe

Suppose we have a dataframe showing sales by season and by person:

```
frame <- read.table("c:\\temp\\sales.txt",header=T)
frame
```

	name	spring	summer	autumn	winter
1	Jane.Smith	14	18	11	12
2	Robert.Jones	17	18	10	13
3	Dick.Rogers	12	16	9	14
4	William.Edwards	15	14	11	10
5	Janet.Jones	11	17	11	16

We want to add margins to this dataframe showing departures of the seasonal means from the overall mean (as an extra row at the bottom) and departures of the people's means (as an extra column on the right). Finally, we want the sales in the body of the dataframe to be represented by departures from the overall mean.

```
people <- rowMeans(frame[,2:5])
people <- people-mean(people)
people
```

```
[1]  0.30  1.05 -0.70 -0.95  0.30
```

It is very straightforward to add a new column to the dataframe using cbind:

```
(new.frame <- cbind(frame,people))
```

	name	spring	summer	autumn	winter	people
1	Jane.Smith	14	18	11	12	0.30
2	Robert.Jones	17	18	10	13	1.05
3	Dick.Rogers	12	16	9	14	-0.70
4	William.Edwards	15	14	11	10	-0.95
5	Janet.Jones	11	17	11	16	0.30

Robert Jones is the most effective sales person ($+1.05$) and William Edwards is the least effective (-0.95). The column means are calculated in a similar way:

```
seasons <- colMeans(frame[,2:5])
seasons <- seasons-mean(seasons)
seasons
```

```
spring summer autumn winter
0.35    3.15   -3.05   -0.45
```

Sales are highest in summer ($+3.15$) and lowest in autumn (-3.05).

Now there is a hitch, however, because there are only four column means but there are six columns in new.frame, so we cannot use rbind directly. The simplest way to deal with this is to make a copy of one of the rows of the new dataframe

```
new.row <- new.frame[1,]
```

and then edit this to include the values we want: a label in the first column to say 'seasonal means' then the four column means, and then a zero for the grand mean of the effects:

```
new.row[1] <- "seasonal effects"
new.row[2:5] <- seasons
new.row[6] <- 0
new.row
```

```
                name spring summer autumn winter people
1 seasonal effects    0.35   3.15  -3.05  -0.45      0
```

Now we can use rbind to add our new row to the bottom of the extended dataframe:

```
(new.frame <- rbind(new.frame,new.row))
```

```
                name spring summer autumn winter people
1         Jane.Smith  14.00  18.00  11.00  12.00   0.30
2       Robert.Jones  17.00  18.00  10.00  13.00   1.05
3        Dick.Rogers  12.00  16.00   9.00  14.00  -0.70
4    William.Edwards  15.00  14.00  11.00  10.00  -0.95
5        Janet.Jones  11.00  17.00  11.00  16.00   0.30
6   seasonal effects   0.35   3.15  -3.05  -0.45   0.00
```

The last task is to replace the counts of sales in the dataframe new.frame[1:5,2:5] by departures from the overall mean sale per person per season (the grand mean, gm = 13.45). We need to use unlist to stop R from estimating a separate mean for each column, then create a vector of length 4 containing repeated values of the grand mean (one for each column of sales). Finally, we use sweep to subtract the grand mean from each value:

```
gm <- mean(unlist(new.frame[1:5,2:5]))
gm <- rep(gm,4)
new.frame[1:5,2:5] <- sweep(new.frame[1:5,2:5],2,gm)
new.frame
```

```
                name spring summer autumn winter people
1         Jane.Smith   0.55   4.55  -2.45  -1.45   0.30
2       Robert.Jones   3.55   4.55  -3.45  -0.45   1.05
3        Dick.Rogers  -1.45   2.55  -4.45   0.55  -0.70
4    William.Edwards   1.55   0.55  -2.45  -3.45  -0.95
```

```
5        Janet.Jones  -2.45   3.55  -2.45   2.55   0.30
6 seasonal effects     0.35   3.15  -3.05  -0.45   0.00
```

To complete the table we want to put the grand mean in the bottom right-hand corner:

```
new.frame[6,6] <- gm[1]
new.frame
```

```
                name spring summer autumn winter people
1        Jane.Smith   0.55   4.55  -2.45  -1.45   0.30
2      Robert.Jones   3.55   4.55  -3.45  -0.45   1.05
3       Dick.Rogers  -1.45   2.55  -4.45   0.55  -0.70
4   William.Edwards   1.55   0.55  -2.45  -3.45  -0.95
5        Janet.Jones  -2.45   3.55  -2.45   2.55   0.30
6 seasonal effects     0.35   3.15  -3.05  -0.45  13.45
```

The best per-season performance was shared by Jane Smith and Robert Jones who each sold 4.55 units more than the overall average in summer.

4.15   Summarizing the contents of dataframes

The usual function to obtain cross-classified summary functions like the mean or median for a single vector is `tapply` (p. 245), but there are three useful functions for summarizing whole dataframes:

- `summary`   summarize all the contents of all the variables;

- `aggregate`   create a table after the fashion of `tapply`;

- `by`   perform functions for each level of specified factors.

Use of `summary` and `by` with the worms database was described on p. 163. The `aggregate` function is used like `tapply` to apply a function (mean in this case) to the levels of a specified categorical variable (Vegetation in this case) for a specified range of variables (Area, Slope, Soil.pH and Worm.density) which are specified using their subscripts as a column index, `worms[,c(2,3,5,7)]`:

```
aggregate(worms[,c(2,3,5,7)],by=list(veg=Vegetation),mean)
```

```
        veg      Area     Slope  Soil.pH Worm.density
1    Arable 3.866667 1.333333 4.833333     5.333333
2 Grassland 2.911111 3.666667 4.100000     2.444444
3    Meadow 3.466667 1.666667 4.933333     6.333333
4   Orchard 1.900000 0.000000 5.700000     9.000000
5     Scrub 2.425000 7.000000 4.800000     5.250000
```

The by argument needs to be a `list` even if, as here, we have only one classifying factor. Here are the aggregated summaries cross-classified by Vegetation and Damp:

```
aggregate(worms[,c(2,3,5,7)],by=list(veg=Vegetation,d=Damp),mean)
```

```
        veg     d      Area     Slope  Soil.pH Worm.density
1    Arable FALSE 3.866667 1.333333 4.833333     5.333333
2 Grassland FALSE 3.087500 3.625000 3.987500     1.875000
```

```
3    Orchard FALSE 1.900000 0.000000 5.700000      9.000000
4      Scrub FALSE 3.350000 5.000000 4.700000      7.000000
5 Grassland  TRUE 1.500000 4.000000 5.000000      7.000000
6     Meadow  TRUE 3.466667 1.666667 4.933333      6.333333
7      Scrub  TRUE 1.500000 9.000000 4.900000      3.500000
```

Note that this summary is unbalanced because there were no damp arable or orchard sites and no dry meadows.

Graphics

Producing high-quality graphics is one of the main reasons for doing statistical computing. The particular plot function you need will depend on the number of variables you want to plot and the pattern you wish to highlight. The plotting functions in this chapter are dealt with under four headings:

- plots with two variables;
- plots for a single sample;
- multivariate plots;
- special plots for particular purposes.

Changes to the detailed look of the graphs are dealt with in Chapter 29.

5.1 Plots with two variables

With two variables (typically the *response variable* on the y axis and the *explanatory variable* on the x axis), the kind of plot you should produce depends upon the nature of your explanatory variable. When the explanatory variable is a continuous variable, such as length or weight or altitude, then the appropriate plot is a **scatterplot**. In cases where the explanatory variable is categorical, such as genotype or colour or gender, then the appropriate plot is either a **box-and-whisker plot** (when you want to show the scatter in the raw data) or a **barplot** (when you want to emphasize the effect sizes).

The most frequently used plotting functions for two variables in R are the following:

- `plot(x,y)`   scatterplot of y against x;
- `plot(factor, y)`   box-and-whisker plot of y at each factor level;
- `barplot(y)`   heights from a vector of y values (one bar per factor level).

The R Book, Second Edition. Michael J. Crawley.
© 2013 John Wiley & Sons, Ltd. Published 2013 by John Wiley & Sons, Ltd.

5.2   Plotting with two continuous explanatory variables: Scatterplots

The `plot` function draws axes and adds a scatterplot of points. Two extra functions, `points` and `lines`, add extra points or lines to an existing plot. There are two ways of specifying `plot`, `points` and `lines` and you should choose whichever you prefer:

- Cartesian `plot(x,y)`

- formula `plot(y~x)`

The advantage of the formula-based plot is that the plot function and the model fit look and feel the same (response variable, tilde, explanatory variable). If you use Cartesian plots (eastings first, then northings, like the grid reference on a map) then the plot has 'x then y' while the model has 'y then x'.

At its most basic, the `plot` function needs only two arguments: first the name of the explanatory variable (x in this case), and second the name of the response variable (y in this case): `plot(x,y)`. The data we want to plot are read into R from a file:

```
data1 <- read.table("c:\\temp\\scatter1.txt",header=T)
attach(data1)
names(data1)
```

```
[1] "x1" "y1"
```

Producing the scatterplot could not be simpler: just type

```
plot(x1,y1,col="red")
```

with the vector of x values first, then the vector of y values (changing the colour of the points is optional).

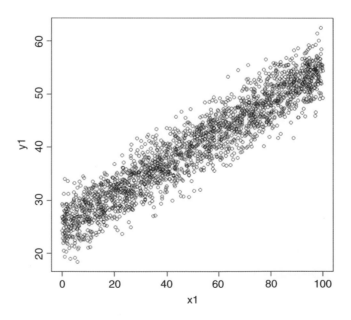

Notice that the axes are labelled with the variable names, unless you chose to override these with `xlab` and `ylab`. It is often a good idea to have longer, more explicit labels for the axes than are provided by the variable names that are used as default options (x1 and y1 in this case). Suppose we want to change the label x1 into the longer label 'Explanatory variable' and the label on the *y* axis from y1 to 'Response variable'. Then we use xlab and ylab like this:

```
plot(x1,y1,col="red",xlab="Explanatory variable",ylab="Response
variable")
```

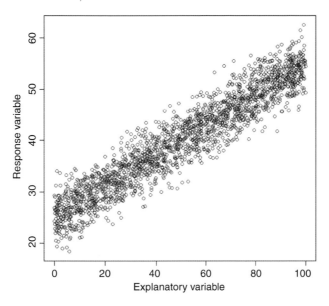

The great thing about graphics in R is that it is extremely straightforward to add things to your plots. In the present case, we might want to add a regression line through the cloud of data points. The function for this is abline which can take as its argument the linear model object lm(y1~x1) as explained on p. 491:

```
abline(lm(y1~x1))
```

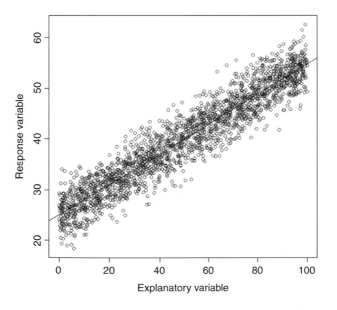

Just as it is easy to add lines to the plot, so it is straightforward to add more points. The extra points are in another file:

```
data2 <- read.table("c:\\temp\\scatter2.txt",header=T)
attach(data2)
names(data2)
```

```
[1] "x2" "y2"
```

The new points (x2,y2) are added using the points function like this:

```
points(x2,y2,col="blue",pch=16)
```

and we can finish by adding a regression line through the extra points:

```
abline(lm(y2~x2))
```

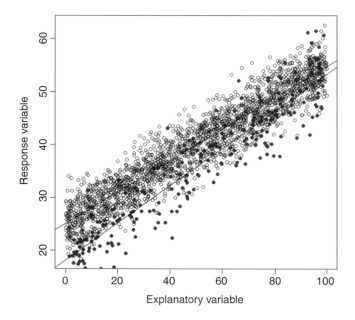

This example shows a very important feature of the plot function. Notice that several of the lower values from the second (blue) data set have *not* appeared on the graph. This is because (unless we say otherwise at the outset) R chooses 'pretty' scaling for the axes based on the data range in the *first* set of points to be drawn. If, as here, the range of subsequent data sets lies outside the scale of the *x* and *y* axes, then points are simply left off without any warning message.

One way to cure this problem is to plot *all* the data with type="n" so that the axes are scaled to encompass all the points from all the data sets (using the concatenation function, c), then to use points and lines to add both sets of data to the blank axes, like this:

```
plot(c(x1,x2),c(y1,y2),xlab="Explanatory variable",
                        ylab="Response variable",type="n")
points(x1,y1,col="red")
points(x2,y2,col="blue",pch=16)
```

```
abline(lm(y1~x1))
abline(lm(y2~x2))
```

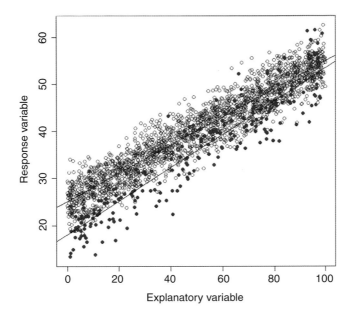

Now all of the points from both data sets appear on the scatterplot. You might want to take control of the selection of the limits for the *x* and *y* axes, rather than accept the 'pretty' default values. In the last plot, for instance, the minimum on the *y* axis was about 13 (but it is not exactly obvious). You might want to specify that the minimum on the *y* axis was zero. This is achieved with the ylim argument, which is a vector of length 2, specifying the minimum and maximum values for the *y* axis: ylim=c(0,70). You will want to control the scaling of the axes when you want two comparable graphs side by side, or when you want to overlay several lines or sets of points on the same axes.

A good way to find out the axis values is to use the range function applied to the data sets in aggregate:

```
range(c(x1,x2))
```

```
[1]   0.02849861  99.93262000
```

```
range(c(y1,y2))
```

```
[1]  13.41794  62.59482
```

Here the *x* axis needs to go from 0.02 up to 99.93 (0 to 100 would be pretty) and the *y* axis needs to go from 13.4 up to 62.6 (0 to 70 would be pretty). This is how the axes are drawn; the points and lines are added exactly as before:

```
plot(c(x1,x2),c(y1,y2),xlim=c(0,100),ylim=c(0,70),
        xlab="Explanatory variable",ylab="Response variable",type="n")
points(x1,y1,col="red")
points(x2,y2,col="blue",pch=16)
abline(lm(y1~x1))
abline(lm(y2~x2))
```

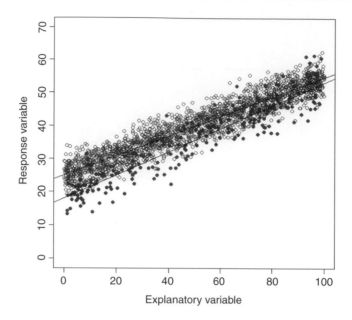

Adding a legend to the plot to explain the difference between the two colours of points would be useful. The thing to understand about the legend function is that the number of lines of text inside the legend box is determined by the length of the vector containing the labels (2 in this case: c("treatment","control"). The other two vectors must be of the same length as this: one for the plotting symbols pch=c(1,16) and one for the colours col=c("red","blue"). The legend function can be used with locator(1) to allow you to select exactly where on the plot surface the legend box should be placed. Click the mouse button when the cursor is where you want the *top left* of the box around the legend to be.

```
legend(locator(1),c("treatment","control"),pch=c(1,16), col=c("red","blue"))
```

This is about as complicated as you would want to make any figure. Adding more information would begin to detract from the message.

5.2.1 Plotting symbols: `pch`

There are 256 different plotting symbols available in Windows (0 to 255). Here is a graphic showing all of them in sequence, from bottom left to top right:

```
plot(0:10,0:10,xlim=c(0,32),ylim=c(0,40),type="n",xaxt="n",yaxt="n",
xlab="",ylab="")
x <- seq(1,31,2)
s <- -16
f <- -1
for (y in seq(2,40,2.5)) {
s <- s+16
f <- f+16
y2 <- rep(y,16)
points(x,y2,pch=s:f,cex=0.7)
text(x,y-1,as.character(s:f),cex=0.6)   }
```

δ	ñ	ò	ó	ô	õ	ö	÷	ø	ù	ú	û	ü	ý	þ	ÿ
240	241	242	243	244	245	246	247	248	249	250	251	252	253	254	255
à	á	â	ã	ä	å	æ	ç	è	é	ê	ë	ì	í	î	ï
224	225	226	227	228	229	230	231	232	233	234	235	236	237	238	239
Ð	Ñ	Ò	Ó	Ô	Õ	Ö	×	Ø	Ù	Ú	Û	Ü	Ý	Þ	ß
208	209	210	211	212	213	214	215	216	217	218	219	220	221	222	223
À	Á	Â	Ã	Ä	Å	Æ	Ç	È	É	Ê	Ë	Ì	Í	Î	Ï
192	193	194	195	196	197	198	199	200	201	202	203	204	205	206	207
º	±	²	³	´	µ	¶	·	¸	¹	º	»	¼	½	¾	¿
176	177	178	179	180	181	182	183	184	185	186	187	188	189	190	191
	¡	¢	£	¤	¥	¦	§	¨	©	ª	«	¬		®	¯
160	161	162	163	164	165	166	167	168	169	170	171	172	173	174	175
	‘	’	“	”	•	–	—	~	™	Š	›	œ		ž	Ÿ
144	145	146	147	148	149	150	151	152	153	154	155	156	157	158	159
€		‚	ƒ	„	…	†	‡	ˆ	‰	Š	‹	Œ		Ž	
128	129	130	131	132	133	134	135	136	137	138	139	140	141	142	143
p	q	r	s	t	u	v	w	x	y	z	{	|	}	~	
112	113	114	115	116	117	118	119	120	121	122	123	124	125	126	127
`	a	b	c	d	e	f	g	h	i	j	k	l	m	n	o
96	97	98	99	100	101	102	103	104	105	106	107	108	109	110	111
P	Q	R	S	T	U	V	W	X	Y	Z	[	\	]	^	_
80	81	82	83	84	85	86	87	88	89	90	91	92	93	94	95
@	A	B	C	D	E	F	G	H	I	J	K	L	M	N	O
64	65	66	67	68	69	70	71	72	73	74	75	76	77	78	79
0	1	2	3	4	5	6	7	8	9	:	;	<	=	>	?
48	49	50	51	52	53	54	55	56	57	58	59	60	61	62	63
	!	"	#	$	%	&	'	(	)	±	+	'	-	.	/
32	33	34	35	36	37	38	39	40	41	42	43	44	45	46	47
●	▲	✦	●	+	○	□	◇	△	▽						
16	17	18	19	20	21	22	23	24	25	26	27	28	29	30	31
□	○	△	+	✕	◇	▽	▣	✳	✢	⊕	✖	⊞	⊠	◪	■
0	1	2	3	4	5	6	7	8	9	10	11	12	13	14	15

The basic plotting symbols (pch) are shown in the bottom two rows, with their pch number immediately beneath. The default value, pch=1, is a small open circle in black. Note that values between 26 and 32 are not implemented at present and are ignored (blanks are plotted). Values for pch between 33 and 127 represent the ASCII character set, while values between 128 and 255 are the symbols from the Windows character set. The symbols for pch=19 and pch=20 are solid circles of different sizes (pch = 20 is the 'bullet point' and is two-thirds the size of pch=19). The difference between pch=16 and pch=19 is that the latter uses a border, and so it is larger when line width lwd is large relative to character expansion cex. The symbol for pch=46 is the 'dot' and is treated specially (it is a rectangle of side 0.01 inch, scaled by cex, and if cex = 1 (the default), each side is at least one pixel, which is 1/72 inch on the pdf, postscript and xfig devices, so that the dot does not disappear on scaling down the image.

5.2.2 Colour for symbols in plots

The plotting symbols (pch) numbered 21 to 25 allow you to specify the background (fill) colour and the border colour separately. In the illustration below, the background colours (bg) 1 to 8 are shown in the columns and numbered on the *x* axis. The border colours (col) 1 to 8 are shown in the rows and numbered on the *y* axis.

```
plot(0:9,0:9,pch=16,type="n",xaxt="n",yaxt="n",ylab="col",xlab="bg")
axis(1,at=1:8)
axis(2,at=1:8)
for (i in 1:8)  points(1:8,rep(i,8),pch=c(21,22,24),bg=1:8,col=i)
```

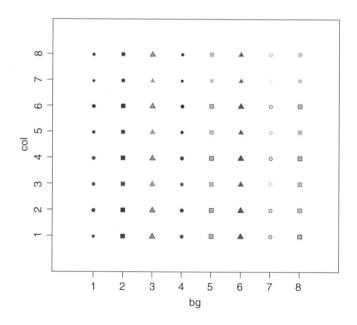

Some combinations are visually more effective than others. Black borders in row 1 (col = 1) are effective with all the shapes and fill colours, but red borders in row 2 (col = 2) work well only with green, pale blue, yellow and grey fill colours. If you specify only col, then open (white) symbols bordered with the specified colour are produced. If you specify only bg, then solid symbols filled with the specified colour, but bordered in black, are produced. If you specify both col and bg with the same colours, then solid, apparently unbordered symbols are produced. If you specify both col and bg using different colours, then solid symbols with contrasting borders are produced. Note that for bordered plotting symbols you need to use the argument pt.bg in the legend function in place of the usual bg for the interior colour of the symbol (because bg controls the background colour of the whole legend box).

5.2.3   Adding text to scatterplots

It is very easy to add text to graphics. Suppose you wanted to add the text '(b)' to a plot at the location $x = 80$ and $y = 65$; just type text(80,65,"(b)").

In this example we want to produce a map of place names, and the place names are in a file called map.places.csv, but their coordinates are in another, much longer file called bowens.csv, containing many more place names than we want to plot. If you have factor level names with spaces in them (e.g. multiple words), then the best format for reading files is comma-delimited (.csv) rather than the standard tab-delimited (.txt). You read them into a dataframe in R using read.csv in place of read.table:

```
map.places <- read.csv("c:\\temp\\map.places.csv",header=T)
attach(map.places)
names(map.places)

[1]  "wanted"

map.data <- read.csv("c:\\temp\\bowens.csv",header=T)
attach(map.data)
names(map.data)

[1]  "place"  "east"  "north"
```

There is a slight complication to do with the coordinates. The northernmost places are in a different 100 km square so, for instance, a northing of 3 needs to be altered to 103. It is convenient that all of the values that need to be changed have northings < 60 in the dataframe:

```
nn <- ifelse(north<60,north+100,north)
```

This says change all of the northings for which north < 60 is TRUE to nn <- north+100, and leave unaltered all the others (FALSE) as nn <- north.

The default graphics window in R is a square measuring 7 inches by 7 inches (quaintly old-fashioned, isn't it?). But our map is rectangular, roughly 80 km wide and 50 km high (a rectangle in landscape format). Allowing for the margins, we want to make the plotting region 9 units wide and 7 units high. We achieve this via the windows function:

```
windows(9,7)
```

We begin by plotting a blank space (type="n") of the right size (eastings from 20 to 100 and northings from 60 to 110) with blank axis labels and no tick marks or numbers:

```
plot(c(20,100),c(60,110),type="n",xlab="",ylab="",xaxt="n", yaxt="n")
```

The trick is to select the appropriate places in the vector called place and use text to plot each name in the correct position (east[ii],nn[ii]). For each place name in wanted we find the correct subscript for that name within place using the which function to find ii:

```
for (i in 1:length(wanted)){
ii <- which(place == as.character(wanted[i]))
text(east[ii], nn[ii], as.character(place[ii]), cex = 0.6) }
```

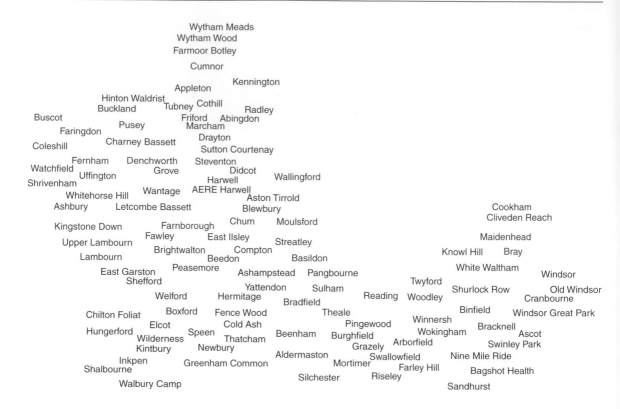

5.2.4   Identifying individuals in scatterplots

The best way to identify multiple individuals in scatterplots is to use a combination of colours and symbols. A useful trick is to use as.numeric to convert a grouping factor (the variable acting as the subject identifier) into a colour and/or a symbol. Here is an example where reaction time is plotted against duration of sleep deprivation for 18 subjects:

```
data <- read.table("c:\\temp\\sleep.txt",header=T)
attach(data)
plot(Days,Reaction)
```

I think you will agree that the raw scatterplot is uninformative; the individuals need to stand out more clearly from one another. The main purpose of the graphic is to show the relationship between sleep deprivation (measured in days) and reaction time. Another aim is to draw attention to the differences between the 18 subjects in their mean reaction times, and to differences in the rate of increase of reaction time with the duration of sleep deprivation. Because there are so many subjects, the graph is potentially very confusing. One improvement is to join together the time series for the individual subjects, using a non-intrusive line colour. Let us do that first. We need to create a vector s to contain the numeric values (1 to 18) of the Subject identity numbers (which range, with gaps, between 308 and 372):

```
s <- as.numeric(factor(Subject))
```

This vector will be used in subscripts to select the *x* and *y* coordinates of each subject's time series in turn. Next, the subjects, k, are taken one at a time in a loop, and lines with type="b" (both points and lines) are drawn in a non-intrusive colour (gray is useful for this):

```
plot(Days,Reaction,type="n")
for (k in 1:max(s)){
x <- Days[s==k]
y <- Reaction[s==k]
lines(x,y,type="b",col="gray")
}
```

Next, we need to select plotting symbols and colours for each subject. The colour-filled symbols pch=21, pch=22 and pch=24 are very useful here. Let us use the non-black colours (bcol from 2 to 8) for each of the first two plotting symbols (sym), then use colours 2 to 5 for the third plotting symbol for the remaining subjects:

```
sym <- rep(c(21,22,24),c(7,7,4))
bcol <- c(2:8,2:8,2:5)
```

Finally, we can take each subject in turn and use points to add the coloured symbols (each with black edges, col=1) to the graph:

```
for (k in 1:max(s)){
points(Days[s==k],Reaction[s==k],pch=sym[k],bg=bcol[k],col=1)
}
```

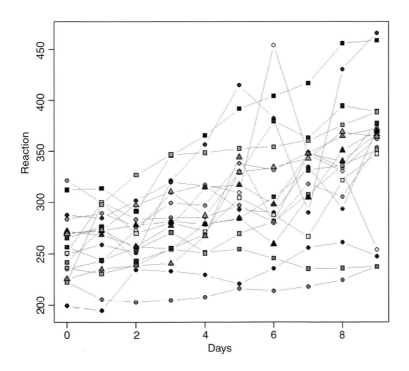

I think that there is insufficient room on the plotting surface to insert a legend with 18 labels in it. For a plot as complicated as this, it is best to put the explanations of the plotting symbols in the text. Perhaps the clearest pattern to emerge from the graphic is that subject 331 (the yellow-filled circle) clearly had a hangover on day 6, because he/she was the third fastest reactor after 9 days of deprivation.

5.2.5   Using a third variable to label a scatterplot

The following example concerns the response of a grass species *Festuca rubra* as measured by its biomass in small samples (FR) to two explanatory variables, soil pH and total hay yield (the mass of all plant species combined). A scatterplot of pH against hay shows the locations of the various samples. The idea is to use the text function to label each of the points on the scatterplot with the dry mass of *F. rubra* in that particular sample, to see whether there is systematic variation in the mass of *Festuca* with changes in hay yield and soil pH.

```
data <- read.table("c:\\temp\\pgr.txt",header=T)
attach(data)
names(data)

[1]  "FR"  "hay"  "pH"

plot(hay,pH)
text(hay, pH, labels=round(FR, 2), pos=1, offset=0.5,cex=0.7)
```

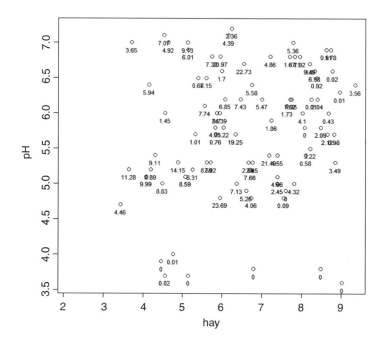

The labels are *centred* on the *x* value of the point (pos=1) and are *offset half a character below* the point (offset=0.5). They show the value of FR rounded to two significant digits (labels=round(FR, 2)) at 70% character expansion (cex=0.7). There is an obvious problem with this method when there is lots of overlap between the labels (as in the top right), but the technique works well for more widely spaced points. The plot shows that high values of *Festuca* biomass are concentrated at intermediate values of both soil pH and hay yield.

    You can also use a third variable to choose the colour of the points in your scatterplot. Here the points with FR above median are shown in red, the others in black:

```
plot(hay,pH,pch=16,col=ifelse(FR>median(FR),"red","black"))
legend(locator(1),c("FR>median","FR<=median"),pch=16,col=c("red","black"))
```

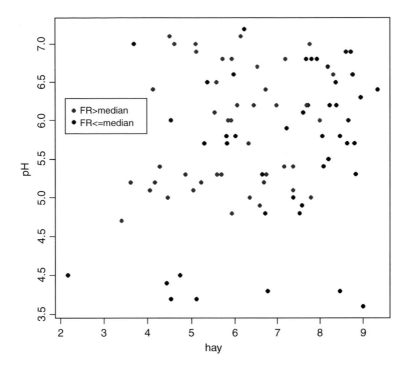

For three-dimensional plots see `image`, `contour` and `wireframe` on p. 931.

5.2.6   Joining the dots

Sometimes you want to join the points on a scatterplot by lines. The trick is to ensure that the points on the x axis are ordered: if they are not ordered, the result is a mess, as you will see below.

```
smooth <- read.table("c:\\temp\\smoothing.txt",header=T)
attach(smooth)
names(smooth)
```

```
[1]   "x"   "y"
```

Begin by producing a vector of subscripts representing the ordered values of the explanatory variable. Then draw lines with this vector as subscripts to both the x and y variables:

```
plot(x,y,pch=16)
sequence <- order(x)
lines(x[sequence],y[sequence])
```

If you do not order the x values, and just use the `lines` function, this is what happens:

```
plot(x,y,pch=16)
lines(x,y)
```

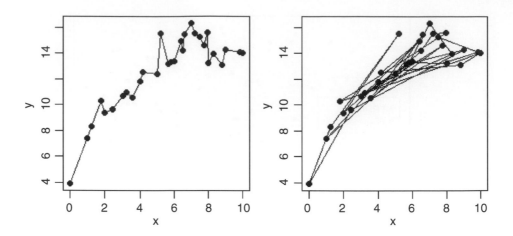

There is a plot option `type="b"` (this stands for 'both' points and lines) which draws the points and joins them together with lines. You can choose the plotting symbol (`pch`) and the line type (`lty`) to be used.

5.2.7   Plotting stepped lines

When plotting square edges between two points, you need to decide whether to go across and then up, or up and then across. The issue should become clear with an example. We have two vectors from 0 to 10:

```
x<-0:10
y<-0:10
plot(x,y)
```

There are three ways we can join the dots: with a straight line

```
lines(x,y,col="red")
```

with a stepped line going across first then up, using lower-case 's'

```
lines(x,y,col="blue",type="s")
```

or with a stepped green line going up first, then across using upper-case 'S' ('upper case, up first' is the way to remember it):

```
lines(x,y,col="green",type="S")
```

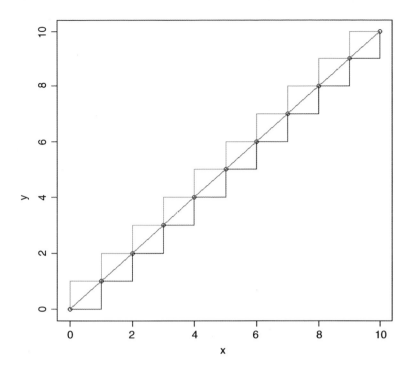

5.3 Adding other shapes to a plot

Once you have produced a set of axes using `plot` it is straightforward to locate and insert other kinds of things. Here are two unlabelled axes, without tick marks (`xaxt="n"`), both scaled from 0 to 10 but without any of the 11 points drawn on the axes (`type="n"`):

```
plot(0:10,0:10,xlab="",ylab="",xaxt="n",yaxt="n",type="n")
```

You can easily add extra graphical objects to plots:

- `rect`   rectangles
- `arrows`   arrows and headed bars
- `polygon`   more complicated filled shapes, including objects with curved sides

For the purposes of demonstration we shall add a single-headed arrow, a double-headed arrow, a rectangle and a six-sided polygon to this space.

We want to put a solid square object in the top right-hand corner, and we know the precise coordinates to use. The syntax for the `rect` function is to provide four numbers:

```
rect(xleft, ybottom, xright, ytop)
```

Thus, to plot the square from (6,6) to (9,9) involves:

```
rect(6,6,9,9)
```

You can fill the shape with solid colour (`col`) or with shading lines (`density`, `angle`) as described on p. 920.

5.3.1 Placing items on a plot with the cursor, using the `locator` function

You might want to point with the cursor and get R to tell you the coordinates of the corners of the rectangle. You can use the `locator()` function for this. The `rect` function does not accept `locator` as its argument, but you can easily write a function (here called `corners`) to do this:

```
corners <- function(){
      coos <- c(unlist(locator(1)),unlist(locator(1)))
      rect(coos[1],coos[2],coos[3],coos[4])
}
```

Run the function like this:

```
corners()
```

Then click in the bottom left-hand corner and again in the top right-hand corner, and a rectangle will be drawn from your screen-supplied pointers.

Drawing arrows is straightforward. The syntax for the `arrows` function is to draw a line from the point (`x0`, `y0`) to the point (`x1`, `y1`) with the arrowhead, by default, at the 'second' end (`x1`, `y1`):

```
arrows(x0, y0, x1, y1)
```

Thus, to draw an arrow from (1,1) to (3,8) with the head at (3,8) type:

```
arrows(1,1,3,8)
```

A horizontal double-headed arrow from (1,9) to (5,9) is produced by adding `code=3` like this:

```
arrows(1,9,5,9,code=3)
```

A vertical bar with two square ends (e.g. like an error bar) uses `angle = 90` instead of the default `angle = 30`):

```
arrows(4,1,4,6,code=3,angle=90)
```

Here is a function that draws an arrow from the cursor position of your first click to the position of your second click:

```
click.arrows <- function(){
      coos <- c(unlist(locator(1)),unlist(locator(1)))
      arrows(coos[1],coos[2],coos[3],coos[4])
}
```

To run this, type

```
click.arrows()
```

then click the cursor on the two ends.

We now wish to draw a polygon. To do this, it is often useful to save the values of a series of locations. Here we intend to save the coordinates of six points in a vector called `locations` to define a polygon for plotting:

```
locations <- locator(6)
```

After you have clicked over the sixth location, control returns to the screen. What kind of object has locator produced?

```
class(locations)
```

```
[1]  "list"
```

It has produced a list, and we can extract the vectors of *x* and *y* values from the list using $ to name the elements of the list (R has created the helpful names x and y):

```
locations
```

```
$x
[1]  5.484375  7.027344  9.019531  8.589844  6.792969  5.230469
```

```
$y
[1]  3.9928797  4.1894975  2.5510155  0.7377620  0.6940691  2.1796262
```

Now we draw a lavender-coloured polygon like this:

```
polygon(locations,col="lavender")
```

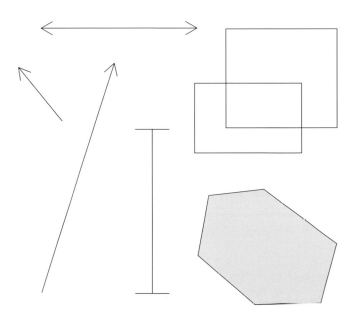

Note that the polygon function has automatically closed the shape, drawing a line from the last point to the first.

5.3.2   Drawing more complex shapes with polygon

The polygon function can be used to draw more complicated shapes, including curved ones. In this example we are asked to shade the area beneath a standard normal curve for values of z that are less than or equal to −1. First draw the probability density (dnorm) line for the standard normal (mean = 0 and standard deviation = 1):

```
z <- seq(-3,3,0.01)
pd <- dnorm(z)
plot(z,pd,type="l")
```

Then fill the area to the left of z ≤ −1 in red:

```
polygon(c(z[z<=-1],-1),c(pd[z<=-1],pd[z==-3]),col="red")
```

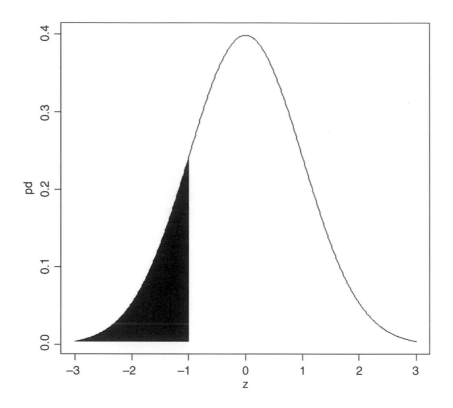

Note the insertion of the point (-1, pd[z == -3]) to create the right-angled corner to the polygon on the z axis at z = −1 and pd set to the same value as when z is −3 to make sure that the bottom line is horizontal.

5.4   Drawing mathematical functions

The curve function is convenient for this. Here is a plot of $x^3 - 3x$ between $x = -2$ and $x = 2$:

```
curve(x^3-3*x, -2, 2)
```

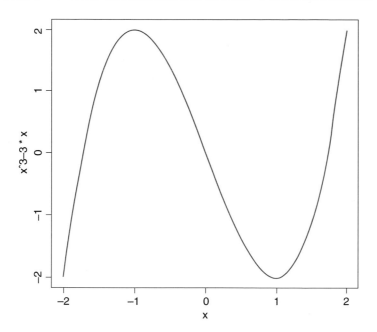

Here is the more cumbersome code to do the same thing using `plot`:

```
x <- seq(-2,2,0.01)
y <- x^3-3*x
plot(x,y,type="l")
```

With `plot`, you need to decide how many segments you want to generate to create the curve (using `seq` with steps of 0.01 in this example), then calculate the matching y values, then use `plot` with `type="l"`. This stands for 'type = line' (rather than the default `points`) and can cause problems if you misread the symbol as a number 'one' rather than a lower-case letter 'L'.

5.4.1 Adding smooth parametric curves to a scatterplot

Up to this point our response variable was shown as a scatter of data points. In many cases, however, we want to show the response as a smooth curve. The important tip is that to produce reasonably smooth looking curves in R you should draw about 100 straight-line sections between the minimum and maximum values of your x axis.

The Ricker curve is named after the famous Canadian fish biologist who introduced this two-parameter hump-shaped model for describing recruitment to a fishery y as a function of the density of the parental stock, x. We wish to compare two Ricker curves with the following parameter values:

$$y_A = 482x\,e^{-0.045x}, \qquad y_B = 518x\,e^{-0.055x}.$$

The first decision to be made is the range of x values for the plot. In our case this is easy because we know from the literature that the minimum value of x is 0 and the maximum value of x is 100. Next we need to generate about 100 values of x at which to calculate and plot the smoothed values of y:

```
xv <- 0:100
```

Next, calculate vectors containing the values of y_A and y_B at each of these x values:

```
yA <- 482*xv*exp(-0.045*xv)
yB <- 518*xv*exp(-0.055*xv)
```

We are now ready to draw the two curves, but we do not know how to scale the y axis. We could find the maximum and minimum values of y_A and y_B then use `ylim` to specify the extremes of the y axis, but it is more convenient to use the option `type="n"` to draw the axes without any data, then use `lines` to add the two smooth functions later. The blank axes are produced like this:

```
plot(c(xv,xv),c(yA,yB),xlab="stock",ylab="recruits",type="n")
```

We want to draw the smooth curve for y_A as a dashed blue line (`lty = 2, col = "blue"`),

```
lines(xv,yA,lty=2,col="blue")
```

and the curve for y_B as a solid red line (`lty = 1, col = "red"`),

```
lines(xv,yB,lty=1,col="red")
```

Next, we want to see which (if either) of these lines best describes our field data, by overlaying a scatter of points (as black solid circles, `pch = 16`) on the smooth curves:

```
info <- read.table("c:\\temp\\plotfit.txt",header=T)
attach(info)
names(info)
```

```
[1] "x" "y"
```

```
points(x,y,pch=16)
```

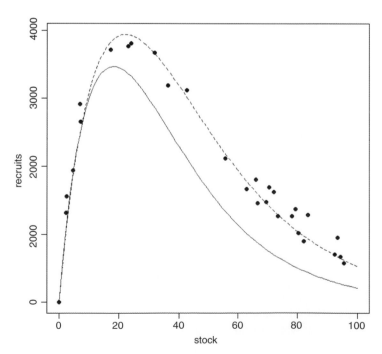

You can see that the blue dotted line is a much better description of our data than is the solid red line. Estimating the parameters of non-linear functions like the Ricker curve from data is explained in Chapter 20.

5.4.2   Fitting non-parametric curves through a scatterplot

It is common to want to fit a non-parametric smoothed curve through data, especially when there is no obvious candidate for a parametric function. R offers a range of options:

- `lowess` (a non-parametric curve fitter);

- `loess` (a modelling tool);

- `gam` (fits generalized additive models; p. 666);

- `lm` for polynomial regression (fit a linear model involving powers of *x*).

We will illustrate each of these options using the jaws data. First, we load the data:

```
data <- read.table("c:\\temp\\jaws.txt",header=T)
attach(data)
names(data)
```

```
[1]  "age"  "bone"
```

Before we fit our various curves to the data, we need to consider how best to display the results together. Without doubt, the graphical parameter you will change most often just happens to be the least intuitive to use. This is the number of graphs per screen, called somewhat unhelpfully, `mfrow`. This stands for 'multiple frames by rows'. The idea is simple, but the syntax is hard to remember. You need to specify the number of rows of plots you want, and number of plots per row, in a vector of two numbers. The first number is the number of rows and the second number is the number of graphs per row. The vector is made using concatenate `c` in the normal way. The default single-plot screen is `par(mfrow=c(1,1))`. Two plots side by side is `par(mfrow=c(1,2))` and a panel of four plots in a 2 × 2 square is `par(mfrow=c(2,2))`.

To move from one plot to the next, you need to execute a new `plot` function. Control stays within the same plot frame while you execute functions like `points`, `lines` or `text`. Remember to return to the default single plot when you have finished your multiple plot by executing `par(mfrow=c(1,1))`. If you have more than two graphs per row or per column, the character expansion `cex` is set to 0.5 and you get half-size characters and labels.

```
par(mfrow=c(2,2))
```

Let us now plot our four graphs with different smooth functions fitted through the jaws data. First, the simple non-parametric smoother called `lowess`. You provide the `lowess` function with arguments for the explanatory variable and the response variable, then provide this object as an argument to the `lines` function like this:

```
plot(age,bone,pch=16,main="lowess")
lines(lowess(age,bone),col="red")
```

It is a reasonable fit overall, but a poor descriptor of the jaw size for the lowest five ages. Let us try `loess`, which is a model-fitting function. We use the fitted model to `predict` the jaw sizes:

```
plot(age,bone,pch=16,main="loess")
model <- loess(bone~age)
xv <- 0:50
yv <- predict(model,data.frame(age=xv))
lines(xv,yv,col="red")
```

This is much better at describing the jaw size of the youngest animals, but shows a slight decrease for the oldest animals which might not be realistic. Next, we use a generalized additive model (gam, from the library mgcv) to fit bone as s(age), a smooth function of age:

```
library(mgcv)
plot(age,bone,pch=16,main="gam")
model <- gam(bone~s(age))
xv <- 0:50
yv <- predict(model,list(age=xv))
lines(xv,yv,col="red")
```

The line is almost indistinguishable from the line produced by loess. Finally, a polynomial:

```
plot(age,bone,pch=16,main="cubic polynomial"))
model <- lm(bone~age+I(age^2)+I(age^3))
xv <- 0:50
yv <- predict(model,list(age=xv))
lines(xv,yv,col="red")
```

As so often with polynomials, the line is more curvaceous than we really want. Note the use of capital I (the 'as is' function) in front of the quadratic and cubic terms. The fit is good for young animals, but is rather wavy where we might expect to see an asymptote. It tips up at the end, whereas the last two smoothers tipped down.

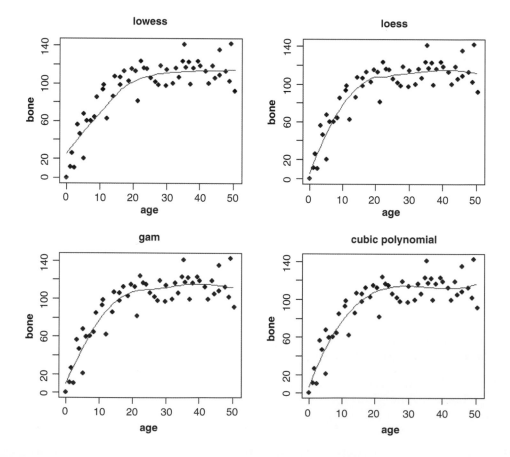

Because it is a built-in function and does not require any external packages to be loaded, my recommendation is for loess (top right); it is a reasonable fit, and is not over-curvaceous. You fit a model, then use predict with a specified vector of values for the explanatory variable, then draw the curve using lines.

5.5 Shape and size of the graphics window

The default graphics window is a square, measuring 7 inches by 7 inches (I know it should be metric, but it is not). This is fine for most purposes, but it needs to be changed if you want to put two graphs side by side, using par(mfrow=c(1,2)).

```
data <- read.table("c:\\temp\\pollute.txt",header=T)
attach(data)
```

If you use the default window the graphs will come out looking far too narrow:

```
par(mfrow=c(1,2))
plot(Population,Pollution)
plot(Temp,Pollution)
```

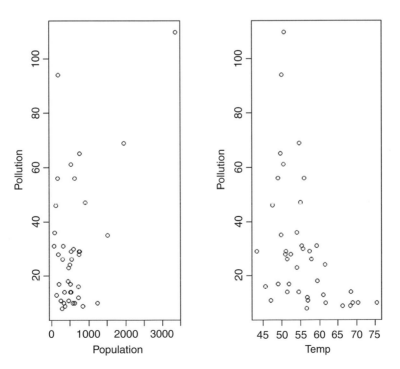

The simplest solution is to use the mouse to drag up the base of the graphics window until you obtain a more pleasing shape. Alternatively, you can invoke the windows function, specifying first the width then the height in inches. The best choice for this case is (7,4):

```
windows(7,4)
par(mfrow=c(1,2))
plot(Population,Pollution)
plot(Temp,Pollution)
```

5.6 Plotting with a categorical explanatory variable

When the explanatory variable is categorical rather than continuous, we cannot produce a scatterplot. Instead, we choose between a **barplot** and a **boxplot**. I prefer box-and-whisker plots because they convey so much more information, and this is the default plot in R with a categorical explanatory variable.

Categorical variables are **factors** with two or more **levels** (see p. 20). Our first example uses the factor called month (with levels 1 to 12) to investigate weather patterns at Silwood Park:

```
weather <- read.table("c:\\temp\\SilwoodWeather.txt",header=T)
attach(weather)
names(weather)
```

```
[1]  "upper"  "lower"  "rain"  "month"  "yr"
```

There is one bit of housekeeping we need to do before we can plot the data. We need to declare month to be a factor. At the moment, R thinks it is just a number:

```
month <- factor(month)
```

Now we can plot using a categorical explanatory variable (month) and, because the first variable is a factor, we get a boxplot rather than a scatterplot:

```
plot(month,upper)
```

Note that there are no axis labels in the default box-and-whisker plot, and to get informative labels we should need to type:

```
plot(month,upper,ylab="daily maximum temperature",xlab="month")
```

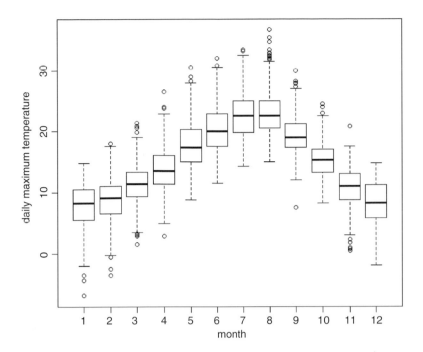

The boxplot summarizes a great deal of information very clearly. The horizontal line shows the **median** upper daily temperature for each month. The bottom and top of the box show the 25th and 75th **percentiles**, respectively (i.e. the location of the middle 50% of the data, also called the first and third **quartiles**). The vertical dashed lines are called the 'whiskers'. For the upper whisker, we see one of two things: either the maximum value or, when there are outliers present, the largest data point that is less than 1.5 times the **interquartile range** above the 75th percentile. The quantity '1.5 times the interquartile range of the data' is roughly 2 standard deviations, and the interquartile range is the difference in the response variable between its first and third quartiles. Points more than 1.5 times the interquartile range *above the third quartile* and points more than 1.5 times the interquartile range *below the first quartile* are defined as **outliers** and plotted individually. Thus, when there are no outliers the whiskers simply show the maximum and minimum values (as shown here only in month 12). Boxplots not only show the location and spread of data but also indicate skewness (which shows up as asymmetry in the sizes of the upper and lower parts of the box). For example, in February the range of lower temperatures was much greater than the range of higher temperatures. Boxplots are also excellent for spotting errors in the data when the errors are represented by extreme outliers. Note that the box-and-whisker plot is based entirely on the data points themselves; there are no estimated parameters like means or standard deviations. The whiskers always end at data points, so the upper and lower whiskers are typically asymmetric, even when there are outliers both above and below (e.g. in November).

5.6.1 Boxplots with notches to indicate significant differences

Boxplots are very good at showing the distribution of the data points around the median, but they are not so good at indicating whether or not the median values are significantly different from one another. Tukey invented **notches** to get the best of both worlds. The notches are drawn as a 'waist' on either side of the median and are intended to give a rough impression of the significance of the differences between two medians. Boxes in which *the notches do not overlap* are likely to prove to have significantly different medians under an appropriate test. Boxes with overlapping notches probably do not have significantly different medians. The size of the notch increases with the magnitude of the interquartile range and declines with the square root of the replication, like this:

$$\text{notch} = \pm 1.58 \frac{\text{IQR}}{\sqrt{n}},$$

where IQR is the interquartile range and n is the replication per sample. Notches are based on assumptions of asymptotic normality of the median and roughly equal sample sizes for the two medians being compared, and are said to be rather insensitive to the underlying distributions of the samples. The idea is to give roughly a 95% confidence interval for the difference in two medians, but the theory behind this is somewhat vague.

Here are the Silwood Weather data (above) with the option `notches=TRUE`:

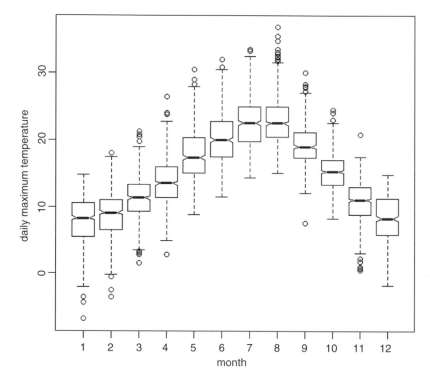

There is no significant difference in daily maximum temperature between July and August (the notches for months 7 and 8 overlap completely), but maxima in September are significantly lower than in August. If the boxes do not overlap (e.g. months 9 and 10) then the difference in their medians will be highly significant under the appropriate test.

When the sample sizes are small and/or the within-sample variance is high, the notches are not drawn as you might expect them (i.e. as a waist within the box). Instead, the notches are extended *above* the 75th percentile and/or *below* the 25th percentile. This looks odd, but it is an intentional feature, supposed to act as a warning of the likely invalidity of the test (see p. 217).

5.6.2   Barplots with error bars

Rather than use `plot` to produce a boxplot, an alternative is to use a `barplot` to show the heights of the mean values from the different treatments. We need to begin by calculating the heights of the bars, typically by using the function `tapply` to work out the mean values for each level of the categorical explanatory variable. Data for this example come from an experiment on plant competition, with five factor levels in a single categorical variable called `clipping`: a control (unclipped), two root clipping treatments (`r5` and `r10`) and two shoot clipping treatments (`n25` and `n50`) in which the leaves of neighbouring plants were reduced by 25% and 50%. The response variable is yield at maturity (a dry weight) called `biomass`.

```
trial <- read.table("c:\\temp\\compexpt.txt",header=T)
attach(trial)
names(trial)
```

```
[1]  "biomass"  "clipping"
```

First, calculate the heights of the bars using `tapply` to compute the five mean values:

```
means <- tapply(biomass,clipping,mean)
```

Then the barplot is produced very simply:

```
barplot(means,xlab="treatment",ylab="mean yield",col="green")
```

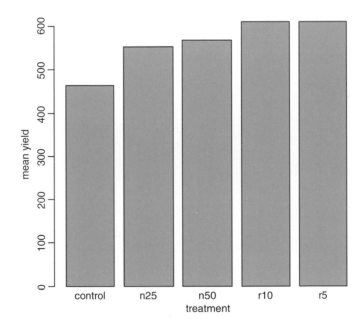

Unless we add error bars to such a barplot, the graphic gives no indication of the extent of the uncertainty associated with each of the estimated treatment means, and hence is unsuitable for publication. There is no built-in function for drawing error bars on barplots, but it easy to write a function to do this. One obvious issue is that the y axis as drawn by the previous call to `barplot` is likely to be too short to accommodate the error bar extending from the top of the tallest bar. Another issue is that it is not obvious where to centre each of the error bars (i.e. the x coordinates of the middles of the bars).

The next decision to make is what kind of bar to draw. Many journals prefer plus or minus one standard error of the mean. An old fashioned approach is to use plus or minus the 95% confidence interval of the mean. Perhaps the most informative error bar is plus or minus one half of the least significant difference between two means (because then non-overlapping bars indicates significant difference, and overlapping bars indicates non-significance; see p. 515). On the assumption that you want to publish your work in *Science* or *Nature*, we shall use plus or minus one standard error of the mean, because this is their error bar of choice. First, work out the error variance from the ANOVA table of `lm(y~x)` where x is categorical. Now calculate the replication per factor level, and use this to compute `sem`, the standard error of the mean. Work out the mean values that will be represented by the heights of the bars using `tapply`. To scale the top of the y axis, add the standard error to the largest of the means. Determine the labels for the bars from the factor levels of the explanatory variable using `nn <- as.character(levels(x))`. Find the locations of the centres of the bars along the x axis using `xs <- barplot`. Here is the function in full:

```
seBars <- function(x,y){
    model <- lm(y~factor(x))
    reps <- length(y)/length(levels(x))
    sem <- summary(model)$sigma/sqrt(reps)
    m <- as.vector(tapply(y,x,mean))
    upper <- max(m)+sem
    nn <- as.character(levels(x))
    xs <- barplot(m,ylim=c(0,upper),names=nn,
        ylab=deparse(substitute(y)),xlab=deparse(substitute(x)))
for (i in 1:length(xs)) {
    arrows(xs[i],m[i]+sem,xs[i],m[i]-sem,angle=90,code=3,length=0.1) }
}
```

You run it like this, specifying the categorical variable first, then the continuous response variable:

```
seBars(clipping,biomass)
```

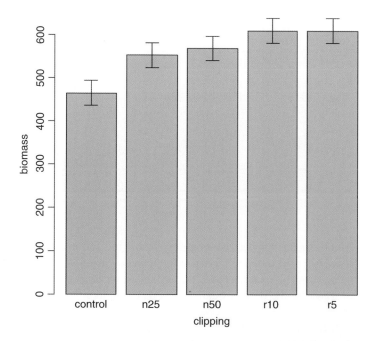

For comparison, here are the box-and-whisker plots for the same data, without and with notches:

```
windows(7,4)
par(mfrow=c(1,2))
plot(clipping,biomass)
plot(clipping,biomass,notch=T)
```

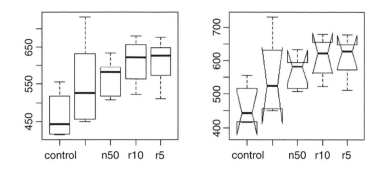

illustrating the curious behaviour of the notches when the sample sizes are small.

5.6.3   Plots for multiple comparisons

When there are many levels of a categorical explanatory variable, we need to be cautious about the statistical issues involved with multiple comparisons (see p. 531). Here we contrast two graphical techniques for displaying multiple comparisons: boxplots with notches, and Tukey's 'honest significant difference'.

The data show the response of yield to a categorical variable (fact) with eight levels representing eight different genotypes of seed (cultivars) used in the trial:

```
data <- read.table("c:\\temp\\box.txt",header=T)
attach(data)
names(data)

[1]  "fact"  "response"

plot(response~factor(fact),notch=TRUE)
```

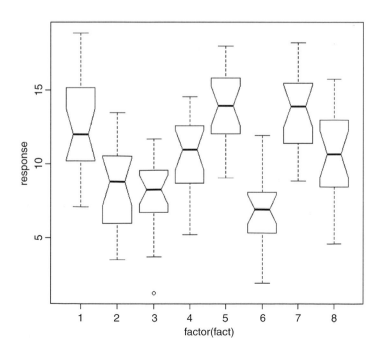

Because the genotypes (factor levels) are unordered, it is hard to judge from the plot which levels might be significantly different from which others. We start, therefore, by calculating an index which will rank the mean values of response across the different factor levels:

```
index <- order(tapply(response,fact,mean))
ordered <- factor(rep(index,rep(20,8)))
boxplot(response~ordered,notch=T,names=as.character(index),
  xlab="ranked treatments",ylab="response")
```

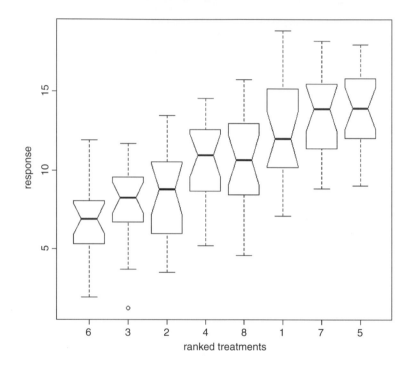

There are several points to clarify here. We plot the response as a function of the factor called `ordered` (rather than `fact`) so that the boxes are ranked from lowest mean yield on the left (cultivar 6) to greatest mean on the right (cultivar 5). We change the names of the boxes to reflect the values of `index` (i.e. the original values of `fact`: otherwise they would read 1 to 8). Note that the vector called `index` is of length 8 (the number of boxes on the plot), but `ordered` is of length 160 (the number of values of response). Looking at the notches, no two adjacent pairs of medians appear to be significantly different, but the median of treatment 4 appears to be significantly greater than the median of treatment 6, and the median of treatment 5 appears to be significantly greater than the median of treatment 8 (but only just).

The statistical analysis of these data might involve user-specified contrasts (p. 434), once it is established that there are significant differences to be explained. This we assess with a one-way analysis of variance to test the hypothesis that at least one of the means is significantly different from the others (see p. 501):

```
model <- aov(response~factor(fact))
summary(model)
```

```
              Df  Sum Sq  Mean Sq  F value  Pr(>F)
factor(fact)   7   925.7   132.24    17.48  <2e-16  ***
Residuals    152  1150.1     7.57
```

Indeed, there is compelling evidence ($p<0.0001$) for accepting that there are significant differences between the mean yields of the eight different crop cultivars.

Alternatively, if you want to do multiple comparisons, then because there is no *a priori* way of specifying contrasts between the eight treatments, you might use Tukey's honest significant difference (see p. 531):

```
plot(TukeyHSD(model))
```

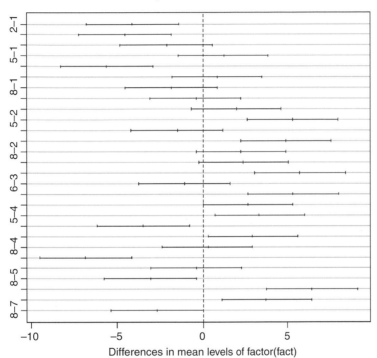

Comparisons having intervals that do not overlap the vertical dashed line are significantly different. The vertical dashed line indicates no difference between the mean values for the factor-level comparisons indicated on the y axis. Thus, we can say that the contrast between cultivars 8 and 7 (8–7) falls just short of significance (despite the fact that their notches do not overlap; see above), but the comparisons 7–6 and 8–6 are both significant (their *boxes* do not overlap, let alone their notches). The missing comparison labels on the y axis of the HSD plot have to be inferred from a knowledge of the number of factor levels (8 in this example). So, since 8 vs. 7 is labelled, the next one up must be 8–6 and the one above that is 7–6, then we find 8–5 labelled, so it must be 7–5 above that and 6–5 above that, then 8–4 labelled, and so on.

5.6.4 Using colour palettes with categorical explanatory variables

You can create a vector of colours from a palette, then refer to the colours by their subscripts within the palette. The key is to create the right number of colours for your needs. Here, we use the built-in heat.colors to shade the temperature bars in Silwood Weather. We want the colours to grade from cold to hot then back to cold again from January to December:

```
data <- read.table("c:\\temp\\silwoodweather.txt",header=T)
attach(data)
month <- factor(month)

season <- heat.colors(12)
temp <- c(11,10,8,5,3,1,2,3,5,8,10,11)
plot(month,upper,col=season[temp])
```

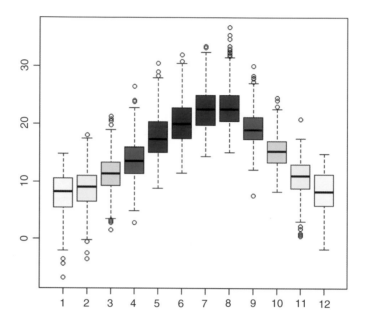

Colouring the other parts of the box-and-whisker plot is explained on p. 918.

5.7 Plots for single samples

When we have a just one variable, the choice of plots is more restricted:

- `hist(y)` histograms to show a frequency distribution
- `plot(y)` index plots to show the values of *y* in sequence
- `plot.ts(y)` time series plots
- `pie(x)` compositional plots like pie diagrams

5.7.1 Histograms and bar charts

A common mistake among beginners is to confuse histograms and bar charts. Histograms have the response variable on the *x* axis, and the *y* axis shows the frequency (or the probability density) of different values of the response. In contrast, a bar chart has the response variable on the *y* axis and a categorical explanatory variable on the *x* axis.

Let us look at an example: the response variable is the growth rate of daphnia in different water qualities; there are four different detergents and three different clones of daphnia.

```
data<-read.table("c:\\temp\\daphnia.txt",header=T)
attach(data)
names(data)
```

```
[1] "Growth.rate" "Water"        "Detergent"     "Daphnia"
```

The histogram shows the frequency with which each growth rate was observed over the experiment as a whole. There are many different bar charts we could draw: here are the mean growth rates cross-classified by clone and detergent:

```
par(mfrow=c(1,2))
hist(Growth.rate,seq(0,8,0.5),col="green",main="")
y <- as.vector(tapply(Growth.rate,list(Daphnia,Detergent),mean))
barplot(y,col="green",ylab="Growth rate",xlab="Treatment")
```

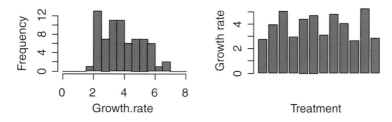

There is a superficial similarity between the two plots in that both have numerous green vertical bars. But there the similarity ends. The histogram on the left has Growth.rate on the *x* axis, but the bar plot on the right has Growth.rate on the *y* axis. The *y* axis on the histogram shows the count (frequency) of the number of times that values from a given interval of growth rates were observed in the whole experiment. The *y* axis on the bar plot shows the arithmetic mean growth rate for that particular experimental treatment. There is no need to labour the point, but you must be absolutely sure that you understand the difference between a histogram and a bar plot, and try not to refer to a bar chart as a histogram or vice versa.

5.7.2   Histograms

The divisions of the *x* axis into which the values of the response variable are distributed and then counted are called **bins**. Histograms are profoundly tricky, because what you see depends on the subjective judgements of where exactly to put the bin margins. Wide bins produce one picture, narrow bins produce a different picture, unequal bins produce confusion.

```
par(mfrow=c(2,2))
hist(Growth.rate,seq(0,8,0.25),col="green",main="")
hist(Growth.rate,seq(0,8,0.5),col="green",main="")
hist(Growth.rate,seq(0,8,2),col="green",main="")
hist(Growth.rate,c(0,3,4,8),col="green",main="")
```

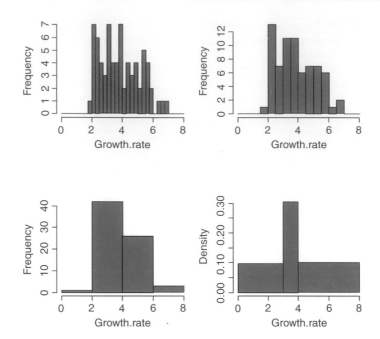

The bins are 0.25 units wide in the lop left-hand histrogram, 0.5 wide in the top right, 2.0 wide in the bottom left, and there are three different widths (3, 1, then 4) in the bottom right. The narrower the bins, the lower the peak frequencies (note that the y scale changes: 7, 12, 40). Small bins produce multimodality (top left), broad bins unimodality (bottom right). When there are different bin widths (bottom right), the default in R is for `hist` to convert the counts (frequencies) into densities (so that the total green area is 1.0).

The convention adopted in R for showing bin boundaries is to employ square and round brackets, so that [a,b) means 'greater than or equal to a but less than b' [square then round), and (a,b] means 'greater than a but less than or equal to b' (round then square]. The point is that it must be unequivocal which bin gets a given number when that number falls exactly on a boundary between two bins. You need to take care that the bins can accommodate both your minimum and maximum values.

The function `cut` takes a continuous vector and cuts it up into bins which you can then use for counting. To show how it works, we shall use `cut` with the daphnia data to produce the density distribution shown above in the bottom right. First, we create a vector of bin edges. To do this, we need to know the range of the growth rates:

```
range(Growth.rate)
```

```
[1] 1.761603 6.918344
```

So a lower bound of 0 and an upper bound of 8 will encompass all of the data. We want edges at 3 and 4, so the vector of bin edges is:

```
edges <- c(0,3,4,8)
```

The next bit is what can seem confusing at first. We create a new vector called `bin` which contains the names of the bins (the factor levels) into which each value of growth rate will be placed. Obviously, this new vector is the same length as `Growth.rate`. It is a factor with as many levels as there are bins (three in this case). The names of the factor levels indicate the bin margins and the edge convention, indicated by round and square brackets (0,3] in this default case:

```
bin <- cut(Growth.rate,edges)
bin
```

```
 [1] (0,3] (0,3] (3,4] (0,3] (3,4] (4,8] (4,8] (3,4] (4,8] (0,3] (3,4]
[12] (0,3] (3,4] (4,8] (4,8] (4,8] (4,8] (4,8] (0,3] (3,4] (3,4] (3,4]
[23] (3,4] (3,4] (3,4] (4,8] (4,8] (0,3] (0,3] (3,4] (3,4] (4,8] (4,8]
[34] (0,3] (3,4] (4,8] (0,3] (0,3] (3,4] (3,4] (3,4] (4,8] (4,8] (4,8]
[45] (4,8] (3,4] (0,3] (3,4] (4,8] (4,8] (4,8] (3,4] (4,8] (4,8] (0,3]
[56] (3,4] (0,3] (4,8] (4,8] (4,8] (0,3] (3,4] (4,8] (0,3] (0,3] (0,3]
[67] (4,8] (4,8] (4,8] (0,3] (0,3] (0,3]
Levels: (0,3] (3,4] (4,8]
```

```
is.factor(bin)
```

```
[1] TRUE
```

As you can see, the default of the `cut` function is to produce bins with the round bracket on the left and the square bracket on the right: (0,3] (3,4] and (4,8]. This is the option `right = TRUE` (the right-hand value will be *included* in the bin (square bracket), and the left-hand value will appear in the next bin to the left, if one exists). If you want to include the left-hand value in the bin and exclude the right-hand value (as you might with a mapping study), then you need to specify the option `right = FALSE` in the `cut` function (see the example on p. 842). Counting the number of cases in each bin could not be simpler:

```
table(bin)
```

```
bin
(0,3] (3,4] (4,8]
   21    22    29
```

To get the heights of the bars for the density plot we need to allow for the areas of the rectangles. First, the total of the counts,

```
sum(table(bin))
```

```
[1] 72
```

and the relative widths of the bins,

```
diff(edges)
```

```
[1] 3 1 4
```

```
(table(bin)/sum(table(bin)))/diff(edges)
```

```
bin
      (0,3]        (3,4]        (4,8]
0.09722222 0.30555556 0.10069444
```

These are the heights of the three bars in the density plot (bottom right, above). They do not add to 1 because the bars are of different widths. It is the total area of the three bars that is 1 under this convention.

5.7.3 Histograms of integers

Histograms are excellent for showing the mode, the spread and the symmetry (skew) of a set of data, but the R function `hist` is deceptively simple. Here is a histogram of 1000 random integers drawn from a Poisson distribution with a mean of 1.7. With the default 'pretty' scaling to produce eight bars, the histogram produces a graphic that does not clearly distinguish between the zeros and the ones:

```
values <- rpois(1000,1.70)
hist(values,main="",xlab="random numbers from a Poisson with mean 1.7")
```

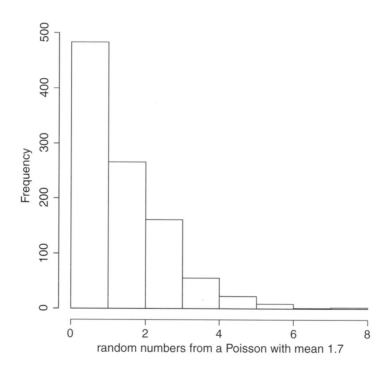

With low-value integer data like this, it is much better to specify the bins explicitly, using the `breaks` argument. The most sensible breaks for count data are –0.5 to + 0.5 to capture the zeros, 0.5 to 1.5 to capture the 1s, and so on; `breaks=(-0.5:8.5)` generates such a sequence automatically. Now the histogram makes clear that 1s are roughly twice as frequent as zeros:

```
hist(values,breaks=(-0.5:8.5),main="",
           xlab="random numbers from a Poisson with mean 1.7")
```

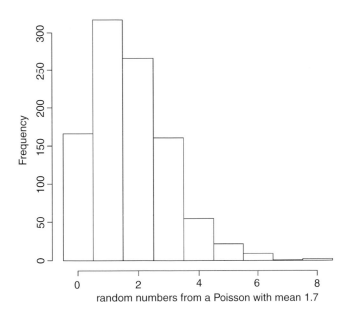

That's more like it. Now we can see that the mode is 1 (not 0), and that 2s are substantially more frequent than 0s. The distribution is said to be 'skewed to the right' (or 'positively skewed') because the long tail is on the right-hand side of the histogram.

5.7.4 Overlaying histograms with smooth density functions

If it is in any way important, then you should always specify the break points yourself. Unless you do this, the hist function may not take your advice about the number of bars or the width of bars. For small-integer data (less than 20, say), the best plan is to have one bin for each value. You create the breaks by starting at −0.5 to accommodate the zeros and going up to max(y) + 0.5 to accommodate the biggest count. Here are 158 random integers from a negative binomial distribution with $\mu = 1.5$ and $k = 1.0$:

```
y <- rnbinom(158,mu=1.5,size=1)
bks <-   -0.5:(max(y)+0.5)
hist(y,bks,main="")
```

To get the best fit of a density function for this histogram we should estimate the parameters of our particular sample of negative binomially distributed counts:

```
mean(y)
```

```
[1]   1.772152
```

```
var(y)
```

```
[1]   4.228009
```

```
mean(y)^2/(var(y)-mean(y))
```

```
[1]   1.278789
```

In R, the parameter k of the negative binomial distribution is known as size and the mean is known as mu. We want to generate the probability density for each count between 0 and 11, for which the R function is dnbinom:

```
xs <- 0:11
ys <- dnbinom(xs,size=1.2788,mu=1.772)
lines(xs,ys*158)
```

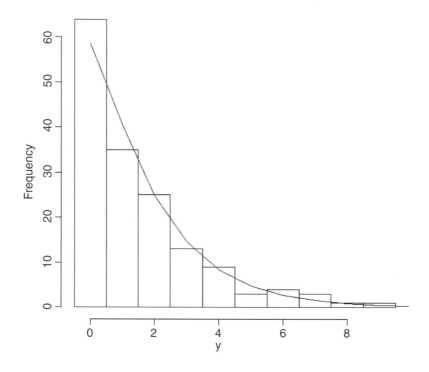

Not surprisingly, since we generated the data, the negative binomial distribution is a very good description of the frequency distribution. The frequency of 1s is a bit low and of 0s is a bit high, but the other frequencies are very well described.

5.7.5 Density estimation for continuous variables

The problems associated with drawing histograms of continuous variables are much more challenging. The subject of density estimation is an important issue for statisticians, and whole books have been written about it (Silverman, 1986; Scott, 1992). You can get a feel for what is involved by browsing the ?density help window. The algorithm used in density.default disperses the mass of the empirical distribution function over a regular grid of at least 512 points, uses the fast Fourier transform to convolve this approximation with a discretized version of the kernel, and then uses linear approximation to evaluate the density at the specified points. The choice of bandwidth is a compromise between smoothing enough to rub out insignificant bumps, and smoothing too much so that real peaks are eliminated. The rule of thumb for bandwidth is

$$b = \frac{\max(x) - \min(x)}{2(1 + \log_2 n)}$$

(where *n* is the number of data points; for details see Venables and Ripley, 2002). We can compare `hist` with Venables and Ripley's `truehist` for the Old Faithful eruptions data:

```
library(MASS)
attach(faithful)
```

The rule of thumb for bandwidth gives:

```
(max(eruptions)-min(eruptions))/(2*(1+log(length(eruptions),base=2)))
```

```
[1]  0.192573
```

but this produces much too bumpy a fit. A bandwidth of 0.6 looks much better:

```
windows(7,4)
par(mfrow=c(1,2))
hist(eruptions,15,freq=FALSE,main="",col=27)
lines(density(eruptions,width=0.6,n=200))
truehist(eruptions,nbins=15,col=27)
lines(density(eruptions,n=200))
```

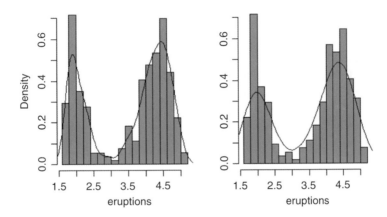

Note that although we asked for 15 bins, we actually got 18. Note also, that although both histograms have 18 bins, they differ substantially in the heights of several of the bars. The left `hist` has two peaks above density = 0.5 while `truehist` on the right has three. There is a sub-peak in the trough of `hist` at about 3.5 but not of `truehist`. And so on. Such are the problems with histograms. Note, also, that the default probability density curve (on the right) picks out the heights of the peaks and troughs much less well than our bandwidth of 0.6 (on the left).

5.7.6 Index plots

The other plot that is useful for single samples is the index plot. Here, `plot` takes a single argument which is a continuous variable and plots the values on the *y* axis, with the *x* coordinate determined by the position of the number in the vector (its 'index', which is 1 for the first number, 2 for the second, and so on up to `length(y)` for the last value). This kind of plot is especially useful for error checking. Here is a data set that has not yet been quality checked, with an index plot of `response$y`:

```
response <- read.table("c:\\temp\\das.txt",header=T)
plot(response$y)
```

The error stands out like a sore thumb. We should check whether this might have been a data entry error, such as a decimal point in the wrong place. But which value is it, precisely, that is wrong? What is clear is that it is the only point for which $y > 15$, so we can use the `which` function to find out its index (the subscript within y):

```
which(response$y > 15)
```

```
[1]   50
```

We can then use this value as the subscript to see the precise value of the erroneous y:

```
response$y[50]
```

```
[1]   21.79386
```

Having checked in the lab notebook, it is obvious that this number should be 2.179 rather than 21.79, so we replace the 50th value of y with the correct value:

```
response$y[50]  <-  2.179386
```

Now we can repeat the index plot to see if there are any other obvious mistakes

```
plot(response$y)
```

That's more like it.

5.7.7 Time series plots

When a time series is complete, the time series plot is straightforward, because it just amounts to joining the dots in an ordered set of y values. The issues arise when there are missing values in the time series, particularly groups of missing values for which periods we typically know nothing about the behaviour of the time series.

There are two functions in R for plotting time series data: `ts.plot` and `plot.ts`. Here is `ts.plot` in action, producing three time series on the same axes using different line types:

```
data(UKLungDeaths)
ts.plot(ldeaths, mdeaths, fdeaths, xlab="year", ylab="deaths", lty=c(1:3))
```

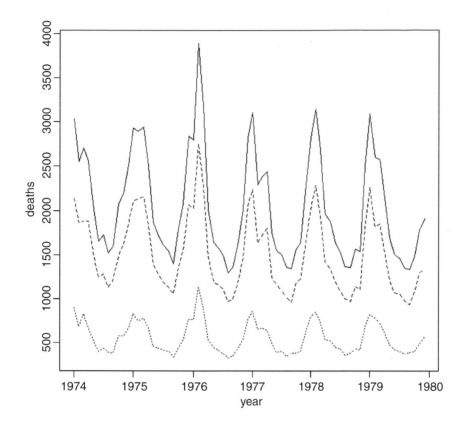

The upper, solid line shows total deaths, the heavier dashed line shows male deaths and the faint dotted line shows female deaths. The difference between the sexes is clear, as is the pronounced seasonality, with deaths peaking in midwinter.

The alternative function `plot.ts` works for plotting objects inheriting from `class=ts` (rather than simple vectors of numbers in the case of `ts.plot`).

```
data(sunspots)
plot(sunspots)
```

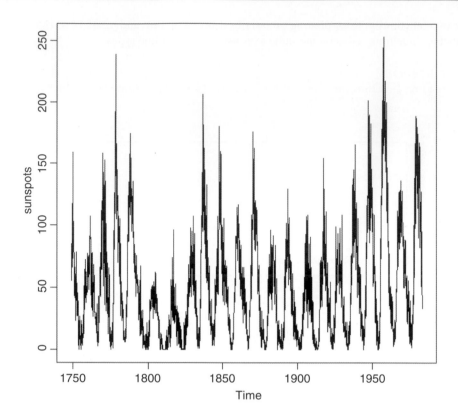

The simple statement plot(sunspots) works because sunspots inherits from the time series class, and has the dates for plotting on the *x* axis built into the object:

```
class(sunspots)
```

```
[1]  "ts"
```

```
is.ts(sunspots)
```

```
 [1]   TRUE
```

```
str(sunspots)
```

```
 Time-Series [1:2820] from 1749 to 1984: 58 62.6 70 55.7 85 83.5 94.8   ...
```

5.7.8   Pie charts

Statisticians do not like pie charts because they think that people should know what 50% looks like. Pie charts, however, can sometimes be useful to illustrate the proportional make-up of a sample in presentations. The function pie takes a vector of numbers, turns them into proportions, and divides up the circle on the basis of those proportions. It is essential to use a label to indicate which pie segment is which. The label is provided as a vector of character strings, here called data$names. Because there are blank spaces in some of the names ('oil shales' and 'methyl clathrates') we cannot use read.table with a tab-delimited text file to enter the data. Instead, we save the file called piedata as a comma-delimited file, with a '.csv' extention, and input the data to R using `read.csv` in place of `read.table`, like this:

```
data <- read.csv("c:\\temp\\piedata.csv")
data
```

```
          names amounts
1          coal       4
2           oil       2
3           gas       1
4     oil shales       3
5 methyl clathrates    6
```

The pie chart is created like this:

```
pie(data$amounts,labels=as.character(data$names))
```

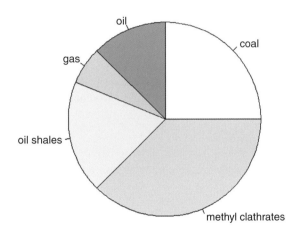

You can change the colours of the segments if you want to (p. 910).

5.7.9 The `stripchart` function

For sample sizes that are too small to use box-and-whisker plots, an alternative plotting method is to use the `stripchart` function. The point of using `stripchart` is to look carefully at the location of individual values within the small sample, and to compare values across cases. The `stripchart` plot can be specified by a model formula y~factor and the strips can be specified to run vertically rather than horizontally. Here is an example from the built-in `OrchardSprays` data set where the response variable is called `decrease` and there is a single categorical variable called `treatment` (with eight levels A–H). Note the use of `with` instead of `attach`:

```
data(OrchardSprays)
with(OrchardSprays,
     stripchart(decrease ~ treatment,
     ylab = "decrease", vertical = TRUE, log = "y"))
```

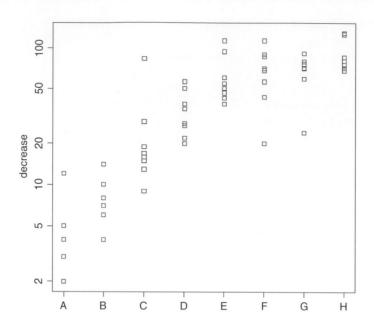

This has the general layout of the box-and-whisker plot, but shows all the raw data values. Note the logarithmic *y* axis, `log = "y"`, and the vertical alignment of the eight strip charts.

5.7.10 A plot to test for normality

Here is a simple function that plots a data set and compares it to a plot of normally distributed data with the same mean and standard deviation:

```
normal.plot <- function(y) {
s <- sd(y)
plot(c(0,3),c(min(0,mean(y)-s * 4*
      qnorm(0.75)),max(y)),xaxt="n",xlab="",type="n",ylab="")
#    for your data's boxes and whiskers, centred at x = 1
top <- quantile(y,0.75)
bottom <- quantile(y,0.25)
w1u <- quantile(y,0.91)
w2u <- quantile(y,0.98)
w1d <- quantile(y,0.09)
w2d <- quantile(y,0.02)
rect(0.8,bottom,1.2,top)
lines(c(0.8,1.2),c(mean(y),mean(y)),lty=3)
lines(c(0.8,1.2),c(median(y),median(y)))
lines(c(1,1),c(top,w1u))
lines(c(0.9,1.1),c(w1u,w1u))
lines(c(1,1),c(w2u,w1u),lty=3)
lines(c(0.9,1.1),c(w2u,w2u),lty=3)
```

```
nou <- length(y[y>w2u])
points(rep(1,nou),jitter(y[y>w2u]))
lines(c(1,1),c(bottom,w1d))
lines(c(0.9,1.1),c(w1d,w1d))
lines(c(1,1),c(w2d,w1d),lty=3)
lines(c(0.9,1.1),c(w2d,w2d),lty=3)
nod <- length(y[y<w2d])
points(rep(1,nod),jitter(y[y<w2d]))

#for the normal box and whiskers, centred at x = 2

n75 <- mean(y)+ s * qnorm(0.75)
n25 <- mean(y)- s * qnorm(0.75)
n91 <- mean(y)+ s * 2* qnorm(0.75)
n98 <- mean(y)+ s * 3* qnorm(0.75)
n9 <- mean(y)- s * 2* qnorm(0.75)
n2 <- mean(y)- s * 3* qnorm(0.75)

rect(1.8,n25,2.2,n75)
lines(c(1.8,2.2),c(mean(y),mean(y)),lty=3)
lines(c(2,2),c(n75,n91))
lines(c(1.9,2.1),c(n91,n91))
lines(c(2,2),c(n98,n91),lty=3)
lines(c(1.9,2.1),c(n98,n98),lty=3)
lines(c(2,2),c(n25,n9))
lines(c(1.9,2.1),c(n9,n9))
lines(c(2,2),c(n9,n2),lty=3)
lines(c(1.9,2.1),c(n2,n2),lty=3)
lines(c(1.2,1.8),c(top,n75),lty=3,col="gray")
lines(c(1.1,1.9),c(w1u,n91),lty=3,col="gray")
lines(c(1.1,1.9),c(w2u,n98),lty=3,col="gray")
lines(c(1.2,1.8),c(bottom,n25),lty=3,col="gray")
lines(c(1.1,1.9),c(w1d,n9),lty=3,col="gray")
lines(c(1.1,1.9),c(w2d,n2),lty=3,col="gray")

# label the two boxes

axis(1,c(1,2),c("data","normal"))    }
```

The plots are like extended box-and-whisker plots, in that they show the median inside a box defined by the 25th and 75th percentiles, with solid whiskers running from the 9th percentile to the 91st percentile, and dotted whiskers running to the 2nd and 98th percentiles. Outliers are defined as values outside the 2nd to the 98th percentiles and are plotted as open circles.

Here are our strongly non-normal test data:

```
y <- rnbinom(100,1,0.2)
```

Here is the test:

```
normal.plot(y)
```

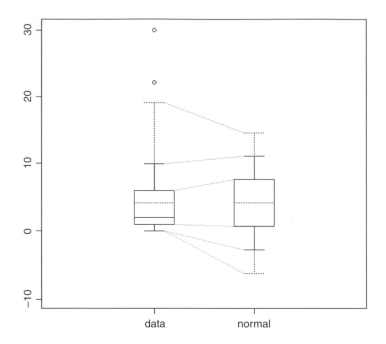

Our data (on the left) are non-normal in several obvious ways: the median is lower than the mean (the solid line is below the horizontal dotted line inside the box), the 75th percentile is rather low (the top of the normal box on the right is higher), and our data have two serious outliers (the open circles). Most obviously, however, our data have no negative values, which normally distributed data with a mean and standard deviation as specified would certainly be expected to have (the 9th and 2nd percentiles on the right-hand box are both well below zero, but our minimum value was 0).

5.8   Plots with multiple variables

Initial data inspection using plots is even more important when there are many variables, any one of which might contain mistakes or omissions. The principal plot functions when there are multiple variables are:

- `pairs` for a matrix of scatterplots of every variable against every other;
- `coplot` for conditioning plots where y is plotted against x for different values of z;
- `xyplot` where a set of panel plots is produced.

We illustrate these functions with the ozone data.

5.8.1   The `pairs` function

With two or more continuous explanatory variables (i.e. in a multiple regression; see p. 395) it is valuable to be able to check for subtle dependencies between the explanatory variables. The `pairs` function plots every variable in the dataframe on the y axis against every other variable on the x axis: you will see at once what this means from the following example:

```
ozonedata <- read.table("c:\\temp\\ozone.data.txt",header=T)
attach(ozonedata)
names(ozonedata)

[1]  "rad"  "temp"  "wind"  "ozone"
```

The `pairs` function needs only the name of the whole dataframe as its first argument. We exercise the option to add a non-parametric smoother to the scatterplots:

```
pairs(ozonedata,panel=panel.smooth)
```

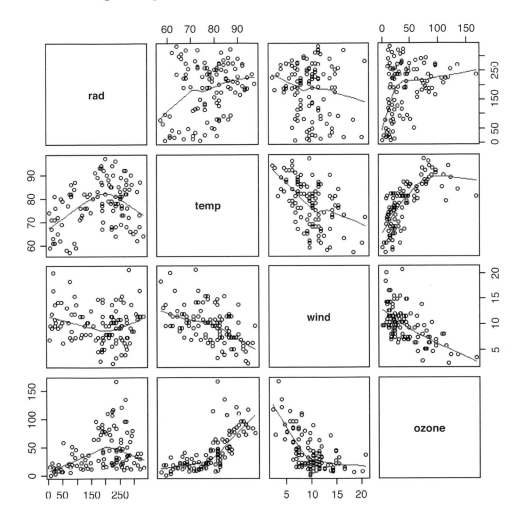

The response variables are named in the rows and the explanatory variables are named in the columns. Thus, in the upper row, labelled `rad`, the response variable (on the *y* axis) is solar radiation. In the bottom row the response variable, `ozone`, is on the *y* axis of all three panels. There appears to be a strong negative non-linear relationship between `ozone` and `wind` speed, a positive non-linear relationship between air temperature and ozone (middle panel in the bottom row) and an indistinct, perhaps humped, relationship between ozone and solar radiation (left-most panel in the bottom row). As to the explanatory variables, there appears to be a negative correlation between wind speed and temperature.

5.8.2 The `coplot` function

A real difficulty with multivariate data is that the relationship between two variables may be obscured by the effects of other processes. When you draw a two-dimensional plot of *y* against *x*, then all of the effects of the other explanatory variables are squashed flat onto the plane of the paper. In the simplest case, we have one response variable (ozone) and just two explanatory variables (wind speed and air temperature). The function is written like this:

```
coplot(ozone~wind|temp,panel = panel.smooth)
```

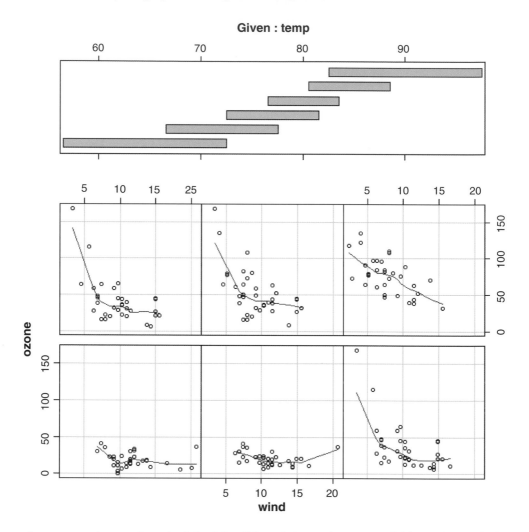

We have the response (`ozone`) on the left of the tilde and the explanatory variable on the *x* axis (`wind`) on the right, with the conditioning variable after the conditioning operator | (here read as 'given `temp`'). An option employed here is to fit a non-parametric smoother through the scatterplot to emphasize the contrasting trends in each of the panels.

The `coplot` panels are ordered from lower left to upper right, associated with the values of the conditioning variable in the upper panel (`temp`) from left to right. Thus, the lower left-hand plot is for the lowest temperatures (56–72°F) and the upper right plot is for the highest temperatures (82–96°F). This `coplot` highlights an interesting interaction. At the two lowest levels of the conditioning variable, `temp`, there is little or no relationship between ozone concentration and wind speed, but in the four remaining panels (at higher temperatures) there is a distinct negative relationship between wind speed and ozone. The hard thing to understand about `coplot` involves the 'shingles' that are shown in the upper margin (given `temp` in this case). The overlap between the shingles is intended to show how much overlap there is between one panel and the next in terms of the data points they have in common. In this default configuration, half of the data in a panel is shared with the panel to the left, and half of the data is shared with the panel to the right (`overlap = 0.5`). You can alter the shingle as far as the other extreme, when all the data points in a panel are unique to that panel (there is no overlap between adjacent shingles; `overlap = -0.05`).

5.8.3 Interaction plots

These are useful when the response to one factor depends upon the level of another factor. They are a particularly effective graphical means of interpreting the results of factorial experiments (p. 516). Here is an experiment with grain yields in response to irrigation and fertilizer application:

```
yields <- read.table("c:\\temp\\splityield.txt",header=T)
attach(yields)
names(yields)

[1] "yield" "block" "irrigation" "density" "fertilizer"
```

The interaction plot has a rather curious syntax, because the response variable (`yield`) comes *last* in the list of arguments. The factor listed first forms the *x* axis of the plot (three levels of `fertilizer`), and the factor listed second produces the family of lines (two levels of `irrigation`). The lines join the mean values of the response for each combination of factor levels:

```
interaction.plot(fertilizer,irrigation,yield)
```

The interaction plot shows that the mean response to fertilizer depends upon the level of irrigation, as evidenced by the fact that the lines are not parallel.

5.9 Special plots

5.9.1 Design plots

An effective way of visualizing effect sizes in designed experiments is the `plot.design` function which is used just like a model formula:

```
plot.design(Growth.rate~Water*Detergent*Daphnia)
```

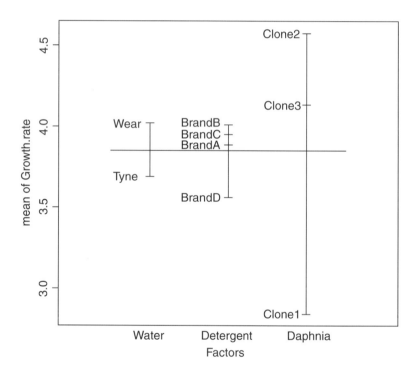

This shows the main effects of the three factors, drawing attention to the major differences between the daphnia clones and the small differences between the detergent brands A, B and C. The default (as here) is to plot means, but other functions can be specified such as `median`, `var` or `sd`. Alternatively, you can supply your own anonymous function. Here, for instance, are the standard errors for the different factor levels:

```
plot.design(Growth.rate~Water*Detergent*Daphnia,
                        fun=function(x)  sqrt(var(x)/3)  )
```

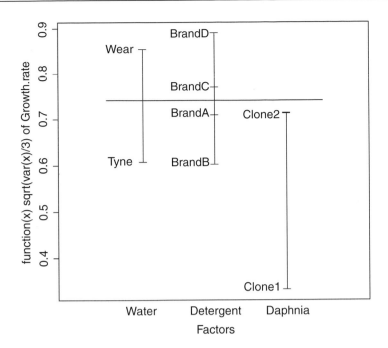

5.9.2 Bubble plots

The bubble plot is useful for illustrating variation in a third variable across different locations in the *x–y* plane. Here is a simple function for drawing bubble plots (see also p. 940):

```
bubble.plot <- function(xv,yv,rv,bs=0.1){
        r <- rv/max(rv)
        yscale <- max(yv)-min(yv)
        xscale <- max(xv)-min(xv)
        plot(xv,yv,type="n", xlab=deparse(substitute(xv)),
            ylab=deparse(substitute(yv)))
        for (i in 1:length(xv)) bubble(xv[i],yv[i],r[i],bs,xscale,yscale)  }
        bubble <- function (x,y,r,bubble.size,xscale,yscale) {
                theta <- seq(0,2*pi,pi/200)
                yv <- r*sin(theta)*bubble.size*yscale
                xv <- r*cos(theta)* bubble.size*xscale
                lines(x+xv,y+yv)  }
```

The example data are on grass yields at different combinations of biomass and soil pH:

```
ddd <- read.table("c:\\temp\\pgr.txt",header=T)
attach(ddd)
names(ddd)
[1]  "FR"  "hay"  "pH"
bubble.plot(hay,pH,FR)
```

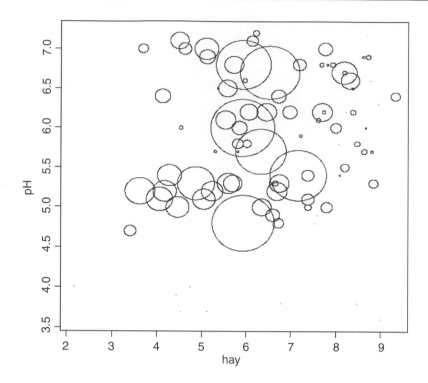

In the vicinity of hay = 6 and pH = 6 *Festuca rubra* shows one very high value, four intermediate values, two low values and one very low value. Evidently, hay crop and soil pH are not the only factors determining the abundance of *F. rubra* in this experiment.

5.9.3 Plots with many identical values

Sometimes, especially with count data, it happens that two or more points fall in exactly the same location in a scatterplot. In such a case, the repeated values of *y* are hidden, one buried beneath the other, and you might want to indicate the number of cases represented at each point on the scatterplot.

```
numbers <- read.table("c:\\temp\\longdata.txt",header=T)
attach(numbers)
names(numbers)
[1]  "xlong"  "ylong"
```

The first option is to 'jitter' the points within the plot function. This means to increase or decrease their *x* and/or *y* coordinates by a small random amount until each data point shows separately:

```
plot(jitter(xlong,amount=1),jitter(ylong,amount=1),xlab="input",ylab="count")
```

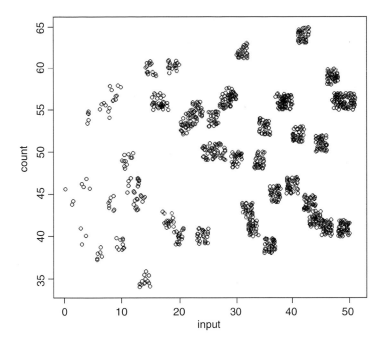

You need to experiment with the amount argument to get the degree of scatter you require (this specifies the limit on the *x* or *y* axis of the amount of jitter on either side of the actual value).

An alternative function is called sunflowerplot, so called because it produces one 'petal' of a flower for each value of *y* (if there is more than one) that is located at that particular point. Here it is in action:

```
sunflowerplot(xlong,ylong)
```

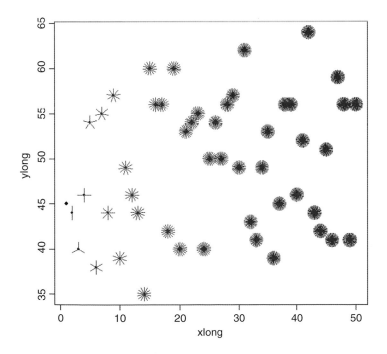

As you can see, the replication at each point increases as x increases from 1 on the left to 50 on the right. The petals stop being particularly informative once there are more than about 20 of them (about half way along the x axis). Single values (as on the extreme left) are shown without any petals, while two points in the same place have two petals. As an option, you can specify two vectors containing the unique values of x and y with a third vector containing the frequency of each combination (the number of repeats of each value).

5.10 Saving graphics to file

For publication-quality graphics, you are likely to want to save each of your plots as a PDF or PostScript file. You do this simply by specifying the 'device' before you start plotting, then turning the device off once you have finished. The default device is your computer screen, and you can obtain a rough and ready copy of the graph (press Ctrl + C) which you can then paste into a document outside R (press Ctrl + V).

```
data <- read.table("c:\\temp\\pollute.txt",header=T)
attach(data)
```

You are most likely to want to save to a PDF file. Here is how you do so:

```
pdf("c:\\temp\\pollution.pdf",width=7,height=4)
par(mfrow=c(1,2))
plot(Population,Pollution)
plot(Temp,Pollution)
dev.off()
```

Here is how you save to a PostScript file:

```
postscript("c:\\temp\\pollution.ps",width=7,height=4)
par(mfrow=c(1,2))
plot(Population,Pollution)
plot(Temp,Pollution)
dev.off()
```

There are numerous options for the `pdf` and `postscript` functions, but width and height are the ones you are likely to want to change most often. The sizes are in inches. You can specify any non-default arguments that you want to change (`width, height, onefile, family, title, fonts, paper, encoding, pointsize, bg, fg, pagecentre, useDingbats, colormodel, fillOddEven` and `compress`) using the functions `pdf.options(..., reset = FALSE)` and `ps.options(..., reset = FALSE)` before you invoke either `pdf` or `postscript`. The logical option `reset = TRUE` resets all the options to their default, 'factory-fresh' values. Don't forget to set `dev.off()` once you have finished.

5.11 Summary

It is worth restating the really important things about plotting.

- **Plots**: `plot(x,y)` gives a scatterplot if x is continuous, and a box-and-whisker plot if x is a factor. Some people prefer the alternative syntax `plot(y~x)` using 'tilde' as in a model formula; one advantage is that this has a `subset` option.

- **Type** of plot: Options include lines `type="l"` or null (axes only) `type="n"`.
- **Lines**: `lines(x,y)` plots a smooth function of `y` against `x` using the `x` and `y` values provided. You might prefer `lines(y~x)`.
- **Line types**: Useful dotted or dashed lines; `lty=2` (an option in `plot` or `lines`).
- **Points**: `points(x,y)` adds another set of data points to a plot. You might prefer `points(y~x)`.
- **Plotting characters** for different data sets: `pch=16` or `pch="*"` (an option in `points` or `plot`).
- **Axes**: setting non-default limits to the *x* or *y* axis scales uses `xlim=c(0,25)` and/or `ylim=c(0,1)` as an option in `plot`.
- **Labels**: use `xlab` and `ylab` to label the *x* and *y* axes.
- **Scales**: use `ylim` and `xlim` to control the top and bottom values on your axes.

6

Tables

The alternative to using graphics is to summarize your data in tabular form. Broadly speaking, if you want to convey *detail* use a table, and if you want to show *effects* then use graphics. You are more likely to want to use a table to summarize data when your explanatory variables are categorical (such as people's names, or different commodities) than when they are continuous (in which case a scatterplot is likely to be more informative; see p. 189).

There are two very important functions that you need to distinguish:

- `table` for counting things;

- `tapply` for averaging things, and applying other functions across factor levels.

6.1 Tables of counts

The `table` function is perhaps the most useful of all the simple vector functions, because it does so much work behind the scenes. We have a vector of objects (they could be numbers or character strings) and we want to know how many of each is present in the vector. Here are 1000 integers from a Poisson distribution with mean 0.6:

```
counts<-rpois(1000,0.6)
```

We want to count up all of the zeros, ones, twos, and so on. A big task, but here is the `table` function in action:

```
table(counts)
counts
  0   1   2   3   4   5
539 325 110  24   1   1
```

There were 539 zeros, 325 ones, 110 twos, 24 threes, 1 four, 1 five and nothing larger than 5. That is a lot of work (imagine tallying them for yourself). The function works for characters as well as for numbers, and for

The R Book, Second Edition. Michael J. Crawley.
© 2013 John Wiley & Sons, Ltd. Published 2013 by John Wiley & Sons, Ltd.

multiple classifying variables:

```
infections<-read.table("c:\\temp\\disease.txt",header=T)
attach(infections)
head(infections)
```

```
  status gender
1  clear   male
2  clear   male
3  clear   male
4  clear   male
5  clear   male
6  clear   male
```

and so on for 1000 rows. You want to know how many males and females were infected and how many were clear of infection:

```
table(status,gender)
```

```
          gender
status    females male
clear         284  515
infected       53   68
```

If you want the genders as the rows rather than the columns, then put gender first in the argument list to table:

```
table(gender,status)
```

```
          status
gender    clear infected
females     284       53
male        515       68
```

The `table` function is likely to be one of the R functions you use most often in your own work.

6.2 Summary tables

The most important function in R for generating summary tables is the somewhat obscurely named `tapply` function. It is called `tapply` because it applies a named function (such as mean or variance) across specified margins (factor levels) to create a table. If you have used the PivotTable function in Excel you will be familiar with the concept.

Here is `tapply` in action:

```
data<-read.table("c:\\temp\\Daphnia.txt",header=T)
attach(data)
names(data)
```

```
[1] "Growth.rate" "Water" "Detergent" "Daphnia"
```

The response variable is growth rate of the animals, and there are three categorical explanatory variables: the river from which the water was sampled, the kind of detergent experimentally added, and the clone of daphnia employed in the experiment. In the simplest case we might want to tabulate the mean growth rates for the four brands of detergent tested,

```
tapply(Growth.rate,Detergent,mean)
```

```
  BrandA    BrandB    BrandC    BrandD
3.884832  4.010044  3.954512  3.558231
```

or for the two rivers,

```
tapply(Growth.rate,Water,mean)
```

```
    Tyne      Wear
3.685862  4.017948
```

or for the three daphnia clones,

```
tapply(Growth.rate,Daphnia,mean)
```

```
  Clone1    Clone2    Clone3
2.839875  4.577121  4.138719
```

Two-dimensional summary tables are created by replacing the single explanatory variable (the second argument in the function call) by a `list` indicating which variable is to be used for the rows of the summary table and which variable is to be used for creating the columns of the summary table. To get the daphnia clones as the rows and detergents as the columns, we write `list(Daphnia,Detergent)` – rows first then columns – and use `tapply` to create the summary table as follows:

```
tapply(Growth.rate,list(Daphnia,Detergent),mean)
```

```
          BrandA    BrandB    BrandC    BrandD
Clone1  2.732227  2.929140  3.071335  2.626797
Clone2  3.919002  4.402931  4.772805  5.213745
Clone3  5.003268  4.698062  4.019397  2.834151
```

If we wanted the median values (rather than the means), then we would just alter the third argument of the `tapply` function like this:

```
tapply(Growth.rate,list(Daphnia,Detergent),median)
```

```
          BrandA    BrandB    BrandC    BrandD
Clone1  2.705995  3.012495  3.073964  2.503468
Clone2  3.924411  4.282181  4.612801  5.416785
Clone3  5.057594  4.627812  4.040108  2.573003
```

To obtain a table of the standard errors of the means (where each mean is based on six numbers: two replicates and three rivers) the function we want to apply is $\sqrt{s^2/n}$. There is no built-in function for the standard error of a mean, so we create what is known as an **anonymous function** inside the `tapply` function with `function(x) sqrt(var(x)/length(x))` like this:

```
tapply(Growth.rate,list(Daphnia,Detergent), function(x) sqrt(var(x)/length(x)))
```

```
          BrandA    BrandB    BrandC    BrandD
Clone1  0.2163448  0.2319320  0.3055929  0.1905771
Clone2  0.4702855  0.3639819  0.5773096  0.5520220
Clone3  0.2688604  0.2683660  0.5395750  0.4260212
```

When `tapply` is asked to produce a three-dimensional table, it produces a stack of two-dimensional tables, the number of stacked tables being determined by the number of levels of the categorical variable that comes *third* in the list (`Water` in this case):

```
tapply(Growth.rate,list(Daphnia,Detergent,Water),mean)
```

```
, , Tyne

        BrandA    BrandB    BrandC    BrandD
Clone1 2.811265 2.775903 3.287529 2.597192
Clone2 3.307634 4.191188 3.620532 4.105651
Clone3 4.866524 4.766258 4.534902 3.365766

, , Wear

        BrandA    BrandB    BrandC    BrandD
Clone1 2.653189 3.082377 2.855142 2.656403
Clone2 4.530371 4.614673 5.925078 6.321838
Clone3 5.140011 4.629867 3.503892 2.302537
```

In cases like this, the function `ftable` (which stands for 'flat table') often produces more pleasing output:

```
ftable(tapply(Growth.rate,list(Daphnia,Detergent,Water),mean))
```

```
                  Tyne      Wear

Clone1 BrandA   2.811265  2.653189
       BrandB   2.775903  3.082377
       BrandC   3.287529  2.855142
       BrandD   2.597192  2.656403
Clone2 BrandA   3.307634  4.530371
       BrandB   4.191188  4.614673
       BrandC   3.620532  5.925078
       BrandD   4.105651  6.321838
Clone3 BrandA   4.866524  5.140011
       BrandB   4.766258  4.629867
       BrandC   4.534902  3.503892
       BrandD   3.365766  2.302537
```

Notice that the order of the rows, columns or tables is determined by the alphabetical sequence of the factor levels (e.g. Tyne comes before Wear in the alphabet). If you want to override this, you must specify that the factor levels are ordered in a non-standard way:

```
water<-factor(Water,levels=c("Wear","Tyne"))
```

Now the summary statistics for the Wear appear in the left-hand column of output:

```
ftable(tapply(Growth.rate,list(Daphnia,Detergent,water),mean))
```

```
                  Wear      Tyne

Clone1 BrandA   2.653189  2.811265
       BrandB   3.082377  2.775903
       BrandC   2.855142  3.287529
       BrandD   2.656403  2.597192
```

```
Clone2 BrandA   4.530371 3.307634
       BrandB   4.614673 4.191188
       BrandC   5.925078 3.620532
       BrandD   6.321838 4.105651
Clone3 BrandA   5.140011 4.866524
       BrandB   4.629867 4.766258
       BrandC   3.503892 4.534902
       BrandD   2.302537 3.365766
```

The function to be applied in generating the table can be supplied with extra arguments:

```
tapply(Growth.rate,Detergent,mean,trim=0.1)

   BrandA    BrandB    BrandC    BrandD
 3.874869  4.019206  3.890448  3.482322
```

The `trim` argument is part of the `mean` function, specifying the fraction (between 0 and 0.5) of the observations to be trimmed from each end of the sorted vector of values before the mean is computed. Values of `trim` outside that range are taken as the nearest endpoint.

An extra argument is essential if you want means when there are missing values:

```
tapply(Growth.rate,Detergent,mean,na.rm=T)
```

Without the argument specifying that you want to average over the non-missing values (`na.rm=T` means 'it is true that I want to remove the missing values') , the `mean` function will simply fail, producing `NA` as the answer.

You can use `tapply` to create new, abbreviated dataframes comprising summary parameters estimated from larger dataframe. Here, for instance, we want a dataframe of mean growth rate classified by detergent and daphina clone (i.e. averaged over river water and replicates). The trick is to convert the factors to numbers before using `tapply`, then use these numbers to extract the relevant levels from the original factors:

```
dets <- as.vector(tapply(as.numeric(Detergent),list(Detergent,Daphnia),mean))
levels(Detergent)[dets]
```

```
[1] "BrandA" "BrandB" "BrandC" "BrandD" "BrandA" "BrandB" "BrandC" "BrandD"
[9] "BrandA" "BrandB" "BrandC" "BrandD"
```

```
clones<-as.vector(tapply(as.numeric(Daphnia),list(Detergent,Daphnia),mean))
levels(Daphnia)[clones]
```

```
[1] "Clone1" "Clone1" "Clone1" "Clone1" "Clone2" "Clone2" "Clone2" "Clone2"
[9] "Clone3" "Clone3" "Clone3" "Clone3"
```

You will see that these vectors of factor levels are the correct length for the new reduced dataframe (12, rather than the original length 72). The 12 mean values that will form our response variable in the new, reduced dataframe are given by:

```
tapply(Growth.rate,list(Detergent,Daphnia),mean)

          Clone1   Clone2   Clone3
BrandA 2.732227 3.919002 5.003268
BrandB 2.929140 4.402931 4.698062
BrandC 3.071335 4.772805 4.019397
BrandD 2.626797 5.213745 2.834151
```

These can now be converted into a vector called means, and the three new vectors combined into a dataframe:

```
means <- as.vector(tapply(Growth.rate,list(Detergent,Daphnia),mean))
detergent <- levels(Detergent)[dets]
daphnia <- levels(Daphnia)[clones]
data.frame(means,detergent,daphnia)
```

```
      means detergent daphnia
1  2.732227    BrandA  Clone1
2  2.929140    BrandB  Clone1
3  3.071335    BrandC  Clone1
4  2.626797    BrandD  Clone1
5  3.919002    BrandA  Clone2
6  4.402931    BrandB  Clone2
7  4.772805    BrandC  Clone2
8  5.213745    BrandD  Clone2
9  5.003268    BrandA  Clone3
10 4.698062    BrandB  Clone3
11 4.019397    BrandC  Clone3
12 2.834151    BrandD  Clone3
```

The same result can be obtained using the as.data.frame.table function:

```
as.data.frame.table(tapply(Growth.rate,list(Detergent,Daphnia),mean))
```

```
     Var1   Var2      Freq
1  BrandA Clone1 2.732227
2  BrandB Clone1 2.929140
3  BrandC Clone1 3.071335
4  BrandD Clone1 2.626797
5  BrandA Clone2 3.919002
6  BrandB Clone2 4.402931
7  BrandC Clone2 4.772805
8  BrandD Clone2 5.213745
9  BrandA Clone3 5.003268
10 BrandB Clone3 4.698062
11 BrandC Clone3 4.019397
12 BrandD Clone3 2.834151
```

but you need to edit the variable names like this:

```
new<-as.data.frame.table(tapply(Growth.rate,list(Detergent,Daphnia),mean))
names(new)<-c("detergents","daphina","means")
head(new)
```

```
  detergents daphina     means
1     BrandA  Clone1 2.732227
2     BrandB  Clone1 2.929140
3     BrandC  Clone1 3.071335
4     BrandD  Clone1 2.626797
5     BrandA  Clone2 3.919002
6     BrandB  Clone2 4.402931
```

6.3   Expanding a table into a dataframe

For the purposes of model-fitting, we often want to expand a table of explanatory variables to create a dataframe with as many repeated rows as specified by a count. Here are the data:

```
count.table<-read.table("c:\\temp\\tabledata.txt",header=T)
attach(count.table)
head(count.table)
```

```
  count     sex    age   condition
1    12    male  young     healthy
2     7    male    old     healthy
3     9  female  young     healthy
4     8  female    old     healthy
5     6    male  young  parasitized
6     7    male    old  parasitized
```

The idea is to create a new dataframe with a separate row for each case. That is to say, we want 12 copies of the first row (for healthy young males), seven copies of the second row (for healthy old males), and so on. The trick is to use `lapply` to apply the repeat function `rep` to each variable in `count.table` such that each row is repeated by the number of times specified in the vector called `count`:

```
lapply(count.table,function(x)rep(x, count.table$count))
```

```
$count
 [1]    12 12 12 12 12 12 12 12 12 12 12 12  7   7   7   7   7   7   7   9
[21]  9  9  9  9  9  9  9  9  8  8  8  8  8  8  8  8  6  6  6  6
[41]  6  6  7  7  7  7  7  7  7  8  8  8  8  8  8  8  5  5  5
[61]  5  5
```

```
$sex
 [1]    male    male    male    male    male    male    male    male    male      male
[11]    male    male    male    male    male    male    male    male    male    female
[21]  female  female  female  female  female  female  female  female  female    female
[31]  female  female  female  female  female  female    male    male    male      male
[41]    male    male    male    male    male    male    male    male    male    female
[51]  female  female  female  female  female  female  female  female  female    female
[61]  female  female
```

```
Levels: female male
```

```
$age
 [1]   young  young  young  young  young  young  young  young  young     young
[11]   young  young    old    old    old    old    old    old    old     young
[21]   young  young  young  young  young  young  young  young    old       old
[31]     old    old    old    old    old    old  young  young  young     young
[41]   young  young    old    old    old    old    old    old    old     young
[51]   young  young  young  young  young  young  young    old    old       old
[61]     old    old
```

```
Levels: old young
```

```
$condition
 [1]  healthy        healthy healthy healthy     healthy      healthy      healthy
 [8]  healthy        healthy healthy healthy     healthy      healthy      healthy
[15]  healthy        healthy healthy healthy     healthy      healthy
[21]  healthy        healthy healthy healthy     healthy      healthy      healthy
[28]  healthy        healthy healthy healthy     healthy      healthy      healthy
[35]  healthy        healthy     parasitized parasitized parasitized parasitized
[41]  parasitized parasitized parasitized parasitized parasitized parasitized
[47]  parasitized parasitized parasitized parasitized parasitized parasitized
[53]  parasitized parasitized parasitized parasitized parasitized parasitized
[59]  parasitized parasitized parasitized parasitized

Levels:  healthy parasitized
```

Then we convert this object from a list to a data.frame using as.data.frame like this:

```
dbtable<-as.data.frame(lapply(count.table,
            function(x) rep(x, count.table$count)))
head(dbtable)
```

```
  count  sex    age condition
1    12 male young   healthy
2    12 male young   healthy
3    12 male young   healthy
4    12 male young   healthy
5    12 male young   healthy
6    12 male young   healthy
```

To tidy up, we probably want to remove the redundant vector of counts:

```
dbtable<-dbtable[,-1]
head(dbtable)
```

```
   sex    age condition
1 male young   healthy
2 male young   healthy
3 male young   healthy
4 male young   healthy
5 male young   healthy
6 male young   healthy
```

```
tail(dbtable)
```

```
      sex    age   condition
57 female young parasitized
58 female    old parasitized
59 female    old parasitized
60 female    old parasitized
61 female    old parasitized
62 female    old parasitized
```

Now we can use the contents of dbtable as explanatory variables in modelling other responses of each of the 62 cases (e.g. the animals' body weights). The alternative is to produce a long vector of row numbers and use this as a subscript on the rows of the short dataframe to turn it into a long dataframe with the same column structure (this is illustrated on p. 255).

6.4   Converting from a dataframe to a table

The reverse procedure of creating a table from a dataframe is much more straightforward, and involves nothing more than the `table` function:

```
table(dbtable)

, ,condition = healthy
        Age
sex            old       young
  female        8           9
   male         7          12
, ,condition = parasitized
        Age
sex            old       young
  female        5           8
   male         7           6
```

You might want this tabulated object itself to be another dataframe, in which case use:

```
as.data.frame(table(dbtable))

    sex      age     condition       Freq
1   female   old     healthy         8
2   male     old     healthy         7
3   female   young   healthy         9
4   male     young   healthy         12
5   female   old     parasitized     5
6   male     old     parasitized     7
7   female   young   parasitized     8
8   male     young   parasitized     6
```

You will see that R has invented the variable name `Freq` for the counts of the various contingencies. To change this to 'count' use `names` with the appropriate subscript [4]:

```
frame<-as.data.frame(table(dbtable))
names(frame)[4]<-"count"
frame

    sex      age     condition       count
1   female   old     healthy         8
2   male     old     healthy         7
3   female   young   healthy         9
4   male     young   healthy         12
5   female   old     parasitized     5
6   male     old     parasitized     7
7   female   young   parasitized     8
8   male     young   parasitized     6
```

6.5 Calculating tables of proportions with `prop.table`

The **margins** of a table (the row totals or the column totals) are often useful for calculating proportions instead of counts. Here is a data matrix called `counts`:

```
counts<-matrix(c(2,2,4,3,1,4,2,0,1,5,3,3),nrow=4)
counts
```

```
     [,1] [,2] [,3]
[1,]    2    1    1
[2,]    2    4    5
[3,]    4    2    3
[4,]    3    0    3
```

The proportions will be different when they are expressed as a fraction of the row totals or of the column totals. To find the proportions we use `prop.table(counts,margin)`. You need to remember that the row subscripts come first, which is why `margin=1` refers to the row totals:

```
prop.table(counts,1)
```

```
          [,1]       [,2]       [,3]
[1,] 0.5000000 0.2500000 0.2500000
[2,] 0.1818182 0.3636364 0.4545455
[3,] 0.4444444 0.2222222 0.3333333
[4,] 0.5000000 0.0000000 0.5000000
```

Use `margin=2` to express the counts as proportions of the relevant column total:

```
prop.table(counts,2)
```

```
          [,1]       [,2]        [,3]
[1,] 0.1818182 0.1428571 0.08333333
[2,] 0.1818182 0.5714286 0.41666667
[3,] 0.3636364 0.2857143 0.25000000
[4,] 0.2727273 0.0000000 0.25000000
```

To check that the column proportions sum to 1, use `colSums` like this:

```
colSums(prop.table(counts,2))
```

```
[1] 1 1 1
```

If you want the proportions expressed as a fraction of the grand total `sum(counts)`, then simply omit the margin number:

```
prop.table(counts)
```

```
           [,1]        [,2]        [,3]
[1,] 0.06666667 0.03333333 0.03333333
[2,] 0.06666667 0.13333333 0.16666667
[3,] 0.13333333 0.06666667 0.10000000
[4,] 0.10000000 0.00000000 0.10000000
```

```
sum(prop.table(counts))
```

```
[1] 1
```

In any particular case, you need to think carefully whether it makes sense to express your counts as proportions of the row totals, the column totals or the grand total.

6.6 The `scale` function

For a numeric matrix, you might want to scale the values within a column so that they have a mean of 0. You might also want to know the standard deviation of the values within each column. These two actions are carried out simultaneously with the `scale` function:

```
scale(counts)
```

```
           [,1]       [,2]       [,3]
[1,] -0.7833495 -0.439155 -1.224745
[2,] -0.7833495  1.317465  1.224745
[3,]  1.3055824  0.146385  0.000000
[4,]  0.2611165 -1.024695  0.000000
```

```
attr(,"scaled:center")
[1] 2.75 1.75 3.00
attr(,"scaled:scale")
[1] 0.9574271 1.7078251 1.6329932
```

The values in the table are the counts minus the column means of the counts. The means of the columns `attr(,"scaled:center")` are 2.75, 1.75 and 3.0, while the standard deviations of the columns `attr(,"scaled:scale")` are 0.96, 1.71 and 1.63. To check that the scales are the standard deviations (`sd`) of the counts within a column, you could use `apply` to the columns (margin = 2) like this:

```
apply(counts,2,sd)
```

```
[1] 0.9574271 1.7078251 1.6329932
```

6.7 The `expand.grid` function

This is a useful function for generating tables from factorial combinations of factor levels. Suppose we have three variables: `height` with five levels between 60 and 80 in steps of 5, `weight` with five levels between 100 and 300 in steps of 50, and two sexes. Then:

```
expand.grid(height = seq(60, 80, 5), weight = seq(100, 300, 50),
                               sex = c("Male","Female"))
```

```
   height weight    sex
1      60    100   Male
2      65    100   Male
3      70    100   Male
4      75    100   Male
5      80    100   Male

...
48     70    300 Female
49     75    300 Female
50     80    300 Female
```

6.8 The `model.matrix` **function**

Creating tables of dummy variables for use in statistical modelling is extremely easy with the `model.matrix` function. You will see what the function does with a simple example. Suppose that our dataframe contains a factor called `parasite` indicating the identity of a gut parasite; this variable has five levels: *vulgaris, kochii, splendens, viridis* and *knowlesii*. Note that there was no header row in the data file, so the variable name parasite had to be added subsequently, using `names`:

```
data<-read.table("c:\\temp\\parasites.txt")
names(data)<-"parasite"
attach(data)
head(data)

   parasite
1   vulgaris
2 splendens
3 knowlesii
4   vulgaris
5 knowlesii
6    viridis

levels(parasite)

[1] "knowlesii" "kochii"    "splendens" "viridis"    "vulgaris"
```

In our modelling we want to create a two-level dummy variable (present or absent) for each parasite species (in five extra columns), so that we can ask questions such as whether the mean value of the response variable is significantly different in cases where each parasite was present and when it was absent. So for the first row of the dataframe, we want `vulgaris` = TRUE, `knowlesii=FALSE, kochii=FALSE, splendens=FALSE` and `viridis=FALSE`.

The long-winded way of doing this is to create a new factor for each species separately:

```
vulgaris<-factor(1*(parasite=="vulgaris"))
kochii<-factor(1*(parasite=="kochii"))
table(vulgaris)

vulgaris
 0  1
99 52

table(kochii)

kochii
  0   1
134  17
```

and so on, with 1 for TRUE (meaning present) and 0 for FALSE (meaning absent). This is how easy it is to do with `model.matrix`:

```
model.matrix(~parasite-1)
```

	parasiteknowlesii	parasitekochii	parasitesplendens	parasiteviridis	parasitevulgaris
1	0	0	0	0	1
2	0	0	1	0	0

3	1	0	0	0	0
4	0	0	0	0	1
5	1	0	0	0	0
6	0	0	0	1	0

. . . .

etc. down to

. . . .

147	1	0	0	0	0
148	0	0	0	1	0
149	0	0	0	0	1
150	0	1	0	0	0
151	0	0	1	0	0

```
attr(,"assign")
[1] 1 1 1 1 1
attr(,"contrasts")
attr(,"contrasts")$parasite
[1] "contr.treatment"
```

The -1 in the model formula ensures that we create a dummy variable for each of the five parasite species (technically, it suppresses the creation of an intercept). Now we can join these five columns of dummy variables to the dataframe containing the response variable and the other explanatory variables. Suppose we had an original.frame. We just join the new columns to it,

```
new.frame<-data.frame(original.frame, model.matrix(~parasite-1))
attach(new.frame)
```

after which we can use variable names like parasiteknowlesii in statistical modelling.

6.9 Comparing table and tabulate

You will often want to count how many times different values are represented in a vector. This simple example illustrates the difference between the two functions. Here is table in action:

```
table(c(2,2,2,7,7,11))
```

```
2   7 11
3   2   1
```

It produces names for each element in the vector (2, 7, 11), and counts only those elements that are present (e.g. there are no zeros or ones in the output vector). The tabulate function counts all of the integers (turning real numbers into the nearest integer if necessary), starting at 1 and ending at the maximum (11 in this case), putting a zero in the resulting vector for every missing integer, like this:

```
tabulate(c(2,2,2,7,7,11))
```

```
[1] 0 3 0 0 0 0 2 0 0 0 1
```

Because there are no 1s in our example, a count of zero is returned for the first element. There are three 2s but then a long gap to two 7s, then another gap to the maximum 11. It is important that you understand that tabulate will ignore negative numbers and zeros without warning:

```
tabulate(c(2,0,-3,2,2,7,-1, 0,0,7,11))
```

```
[1] 0 3 0 0 0 0 2 0 0 0 1
```

For most applications, `table` is much more useful than `tabulate`, but there are occasions when you want the zero counts to be retained. The commonest case is where you are generating a set of vectors, and you want all the vectors to be the same length (e.g. so that you can bind them to a dataframe).

Suppose, for instance, that you want to make a dataframe containing three different realizations of a negative binomial distribution of counts, where the rows contain the frequencies of 0, 1, 2, 3, ... successes. Let us take an example where the negative binomial parameters are `size = 1` and `prob = 0.2` and generate 100 random numbers from it, repeated three times:

```
table(rnbinom(100,1,0.2))
```

0	1	2	3	4	5	6	7	8	9	10	11	12	13	14
24	10	12	6	13	4	3	5	2	5	4	3	5	2	2

```
table(rnbinom(100,1,0.2))
```

0	1	2	3	4	5	6	7	8	10	11	12	13	15	16
23	17	10	8	8	9	1	4	7	2	7	1	1	1	1

```
table(rnbinom(100,1,0.2))
```

0	1	2	3	4	5	6	7	8	9	10	11	12	15	16	19	20	23	24	25
15	15	13	12	6	9	6	3	3	3	4	2	2	1	1	1	1	1	1	1

The three realizations produce vectors of different lengths (15, 15 again (but with a 15 and a 16 but no 9 or 14), and 20 respectively). With `tabulate`, we can specify the length of the output vector (the number of bins, `nbins`), but we need to make sure it is long enough, because overruns will be ignored without warning. We also need to remember to add 1 to the random integers generated, so that the zeros are counted rather than ignored. From what we have seen (above), it looks as if 30 bins should work well enough, so here are six realizations with `nbins=30`:

```
tabulate(rnbinom(100,1,0.2)+1,30)
```

```
[1] 21 14 10  8 10  6 11  4  3  3  1  4  1  2  1  0  0  0  0  0  1  0  0  0  0  0  0  0  0  0
[1] 21 14 13  7  6 10 10  4  2  0  4  3  2  0  1  0  0  0  1  0  1  0  0  0  0  0  0  0  1  0
[1] 25 13 16  8  5  7  4  2  5  3  4  0  1  2  0  1  0  1  0  0  0  2  0  0  0  0  0  1  0  0
[1] 22 21 12 13  8  5  3  3  2  4  2  2  1  1  0  1  0  0  0  0  0  0  0  0  0  0  0  0  0  0
[1] 18 10 13 11 12  6  5  4  4  5  2  1  0  3  1  1  0  0  0  1  0  0  0  0  0  0  0  0  0  1
[1] 26 15  9 12  4  7  7  5  3  2  1  3  3  0  0  2  0  0  0  0  0  0  0  0  0  0  1  0  0
```

This looks fine, until you check the fifth row. There are only 98 numbers here, so two unknown values of counts greater than 29 have been generated but ignored without warning. To see how often this might happen, we can run the test 1000 times and tally the number of numbers presented by `tabulate`:

```
totals<-numeric(1000)
for (i in 1:1000) totals[i] <- sum(tabulate(rnbinom(100,1,0.2)+1,30))
table(totals)
```

```
totals
 98   99  100
  5  114  881
```

As you see, we lost one or more numbers on 119 occasions out of 1000. So take care with `tabulate`, and remember that 1 was added to all the counts to accommodate the zeros.

Mathematics

You can do a lot of maths in R. Here we concentrate on the kinds of mathematics that find most frequent application in scientific work and statistical modelling:

- functions;

- continuous distributions;

- discrete distributions;

- matrix algebra;

- calculus;

- differential equations.

7.1 Mathematical functions

For the kinds of functions you will meet in statistical computing there are only three mathematical rules that you need to learn: these are concerned with powers, exponents and logarithms. In the expression x^b the explanatory variable is raised to the **power** b. In e^x the explanatory variable appears as a power – in this special case, of $e = 2.718\,28$, of which x is the **exponent**. The inverse of e^x is the **logarithm** of x, denoted by $\log(x)$ – note that all our logs are to the base e and that, for us, writing $\log(x)$ is the same as $\ln(x)$.

It is also useful to remember a handful of mathematical facts that are useful for working out **behaviour at the limits**. We would like to know what happens to y when x gets very large (e.g. $x \to \infty$) and what happens to y when x goes to 0 (i.e. what the intercept is, if there is one). These are the most important rules:

- Anything to the power zero is 1: $x^0 = 1.$

- One raised to any power is still 1: $1^x = 1.$

- Infinity plus 1 is infinity: $\infty + 1 = \infty.$

- One over infinity (the reciprocal of infinity, ∞^{-1}) is zero: $\frac{1}{\infty} = 0.$

- A number > 1 raised to the power infinity is infinity: $1.2^\infty = \infty.$

- A fraction (e.g. 0.99) raised to the power infinity is zero: $0.99^\infty = 0.$

The R Book, Second Edition. Michael J. Crawley.
© 2013 John Wiley & Sons, Ltd. Published 2013 by John Wiley & Sons, Ltd.

- Negative powers are reciprocals: $x^{-b} = \frac{1}{x^b}$.

- Fractional powers are roots: $x^{1/3} = \sqrt[3]{x}$.

- The base of natural logarithms, e, is 2.718 28, so $e^{\infty} = \infty$.

- Last, but perhaps most usefully: $e^{-\infty} = \frac{1}{e^{\infty}} = \frac{1}{\infty} = 0$.

There are built-in functions in R for logarithmic, probability and trigonometric functions (p. 17).

7.1.1 Logarithmic functions

The logarithmic function is given by

$$y = a \ln(bx).$$

Here the logarithm is to base e. The exponential function, in which the response y is the antilogarithm of the continuous explanatory variable x, is given by

$$y = ae^{bx}.$$

Both these functions are smooth functions, and to draw smooth functions in R you need to generate a series of 100 or more regularly spaced x values between `min(x)` and `max(x)`:

```
x <- seq(0,10,0.1)
```

In R the exponential function is `exp` and the natural log function (ln) is `log`. Let $a = b = 1$. To plot the exponential and logarithmic functions with these values together in a row, write

```
windows(7,4)
par(mfrow=c(1,2))
y <- exp(x)
plot(y~x,type="l",main="Exponential")
y <- log(x)
plot(y~x,type="l",main="Logarithmic")
```

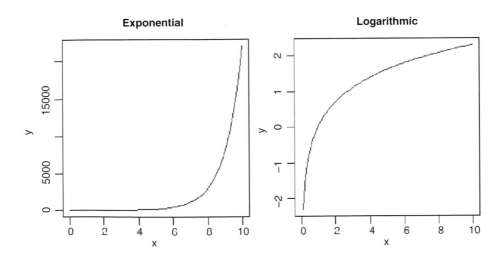

Note that the `plot` function can be used in an alternative way, specifying the Cartesian coordinates of the line using `plot(x,y)` rather than the formula `plot(y~x)` (see p. 190).

These functions are most useful in modelling process of exponential growth and decay.

7.1.2 Trigonometric functions

Here are the cosine (base/hypotenuse), sine (perpendicular/hypotenuse) and tangent (perpendicular/base) functions of x (measured in radians) over the range 0 to 2π. Recall that the full circle is 2π radians, so 1 radian $= 360/2\pi = 57.295\,78$ degrees.

```
windows(7,7)
par(mfrow=c(2,2))

x <- seq(0,2*pi,2*pi/100)
y1 <- cos(x)
y2 <- sin(x)
y3 <- tan(x)

plot(y1~x,type="l",main="cosine")
plot(y2~x,type="l",main="sine")
plot(y3~x,type="l",ylim=c(-3,3),main="tangent")
```

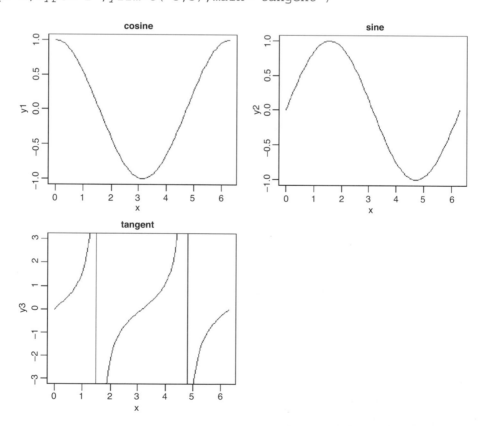

The tangent of x has discontinuities, shooting off to positive infinity at $x = \pi/2$ and again at $x = 3\pi/2$. Restricting the range of values plotted on the y axis (here from -3 to $+3$) therefore gives a better picture of the shape of the tan function. Note that R joins the plus infinity and minus infinity 'points' with a straight line at $x = \pi/2$ and at $x = 3\pi/2$ within the frame of the graph defined by `ylim`.

7.1.3 Power laws

There is an important family of two-parameter mathematical functions of the form

$$y = ax^b,$$

known as **power laws**. Depending on the value of the power, b, the relationship can take one of five forms. In the trivial case of $b = 0$ the function is $y = a$ (a horizontal straight line). The four more interesting shapes are as follows:

```
x <- seq(0,1,0.01)
y <- x^0.5
plot(x,y,type="l",main="0<b<1")
y <- x
plot(x,y,type="l",main="b=1")
y <- x^2
plot(x,y,type="l",main="b>1")
y <- 1/x
plot(x,y,type="l",main="b<0")
```

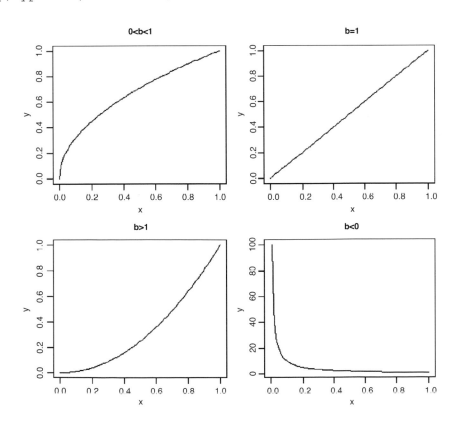

These functions are useful in a wide range of disciplines. The parameters a and b are easy to estimate from data because the function is linearized by a log–log transformation,

$$\log(y) = \log(ax^b) = \log(a) + b\log(x),$$

so that on log–log axes the intercept is $\log(a)$ and the slope is b. These are often called **allometric** relationships because when $b \neq 1$ the proportion of x that becomes y varies with x.

An important empirical relationship from ecological entomology that has applications in a wide range of statistical analysis is known as **Taylor's power law**. It has to do with the relationship between the variance and the mean of a sample. In elementary statistical models, the variance is assumed to be constant (i.e. the variance does not depend upon the mean). In field data, however, Taylor found that variance increased with the mean according to a power law, such that on log–log axes the data from most systems fell above a line through the origin with slope $= 1$ (the pattern shown by data that are Poisson distributed, where the variance is equal to the mean) and below a line through the origin with a slope of 2. Taylor's power law states that, for a particular system:

- log(variance) is a linear function of log(mean);

- the scatter about this straight line is small;

- the slope of the regression of log(variance) against log(mean) is greater than 1 and less than 2;

- the parameter values of the log–log regression are fundamental characteristics of the system.

7.1.4   Polynomial functions

Polynomial functions are functions in which x appears several times, each time raised to a different power. They are useful for describing curves with humps, inflections or local maxima like these:

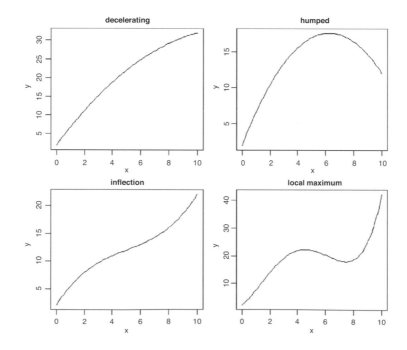

The top left-hand panel shows a decelerating positive function, modelled by the quadratic

```
x <- seq(0,10,0.1)
y1 <- 2+5*x-0.2*x^2
```

Making the negative coefficient of the x^2 term larger produces a curve with a hump as in the top right-hand panel:

```
y2 <- 2+5*x-0.4*x^2
```

Cubic polynomials can show points of inflection, as in the lower left-hand panel:

```
y3 <- 2+4*x-0.6*x^2+0.04*x^3
```

Finally, polynomials containing powers of 4 are capable of producing curves with local maxima, as in the lower right-hand panel:

```
y4 <- 2+4*x+2*x^2-0.6*x^3+0.04*x^4

par(mfrow=c(2,2))
plot(x,y1,type="l",ylab="y",main="decelerating")
plot(x,y2,type="l",ylab="y",main="humped")
plot(x,y3,type="l",ylab="y",main="inflection")
plot(x,y4,type="l",ylab="y",main="local maximum")
```

Inverse polynomials are an important class of functions which are suitable for setting up generalized linear models with gamma errors and inverse link functions:

$$\frac{1}{y} = a + bx + cx^2 + dx^3 + \ldots + zx^n.$$

Various shapes of function are produced, depending on the order of the polynomial (the maximum power) and the signs of the parameters:

```
par(mfrow=c(2,2))

y1 <- x/(2+5*x)
y2 <- 1/(x-2+4/x)
y3 <- 1/(x^2-2+4/x)

plot(x,y1,type="l",ylab="y",main="Michaelis-Menten")
plot(x,y2,type="l",ylab="y",main="shallow hump")
plot(x,y3,type="l",ylab="y",main="steep hump")
```

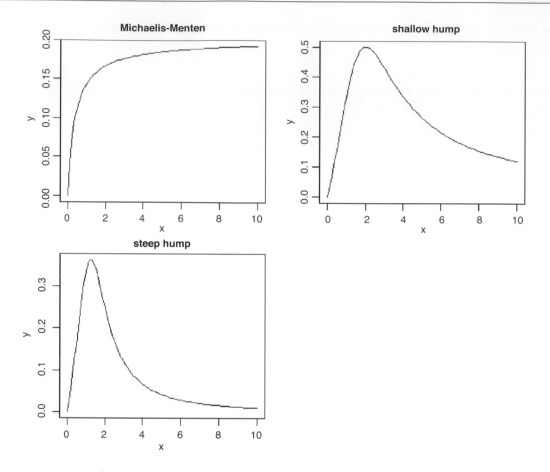

There are two ways of parameterizing the Michaelis–Menten equation:

$$y = \frac{ax}{1 + bx} \quad \text{and} \quad y = \frac{x}{c + dx}.$$

In the first case, the asymptotic value of y is a/b and in the second it is $1/d$.

7.1.5 Gamma function

The gamma function $\Gamma(t)$ is an extension of the factorial function, $t!$, to positive real numbers:

$$\Gamma(t) = \int_0^\infty x^{t-1} e^{-x} dx.$$

It looks like this:

```
par(mfrow=c(1,1))
t <- seq(0.2,4,0.01)
plot(t,gamma(t),type="l")
abline(h=1,lty=2)
```

Note that $\Gamma(t)$ is equal to 1 at both $t = 1$ and $t = 2$. For integer values of t, $\Gamma(t + 1) = t!$

7.1.6 Asymptotic functions

Much the most commonly used asymptotic function is

$$y = \frac{ax}{1 + bx},$$

which has a different name in almost every scientific discipline. For example, in biochemistry it is called Michaelis–Menten, and shows reaction rate as a function of enzyme concentration; in ecology it is called Holling's disc equation and shows predator feeding rate as a function of prey density. The graph passes through the origin and rises with diminishing returns to an asymptotic value at which increasing the value of x does not lead to any further increase in y.

The other common function is the asymptotic exponential

$$y = a(1 - e^{-bx}).$$

This, too, is a two-parameter model, and in many cases the two functions would describe data equally well (see p. 719 for an example of this comparison). Let us work out the behaviour at the limits of our two asymptotic functions, starting with the asymptotic exponential. For $x - 0$ we have

$$y = a(1 - e^{-b \times 0}) = a(1 - e^0) = a(1 - 1) = a \times 0 = 0,$$

so the graph goes through the origin. At the other extreme, for $x = \infty$, we have

$$y = a(1 - e^{-b \times \infty}) = a(1 - e^{-\infty}) = a(1 - 0) = a(1) = a,$$

which demonstrates that the relationship is asymptotic, and that the asymptotic value of y is a.

Turning to the Michaelis–Menten equation, or $x = 0$ the limit is easy:

$$y = \frac{a \times 0}{1 + b \times 0} = \frac{0}{1 + 0} = \frac{0}{1} = 0.$$

However, determining the behaviour at the limit $x = \infty$ is somewhat more difficult, because we end up with $y = \infty/(1 + \infty) = \infty/\infty$, which you might imagine is always going to be 1 no matter what the values of a and b. In fact, there is a special mathematical rule for this case, called l'Hospital's rule: when you get a ratio of infinity to infinity, you work out the ratio of the derivatives to obtain the behaviour at the limit. The numerator is ax so its derivative with respect to x is a. The denominator is $1 + bx$ so its derivative with respect to x is $0 + b = b$. So the ratio of the derivatives is a/b, and this is the asymptotic value of the Michaelis–Menten equation.

7.1.7   Parameter estimation in asymptotic functions

There is no way of linearizing the asymptotic exponential model, so we must resort to non-linear least squares (nls) to estimate parameter values for it (p. 715). One of the advantages of the Michaelis–Menten function is that it is easy to linearize. We use the **reciprocal transformation**

$$\frac{1}{y} = \frac{1 + bx}{ax}.$$

At first glance, this is no great help. But we can separate the terms on the right because they have a common denominator. Then we can cancel the xs, like this:

$$\frac{1}{y} = \frac{1}{ax} + \frac{bx}{ax} = \frac{1}{ax} + \frac{b}{a}$$

so if we put $Y = 1/y$, $X = 1/x$, $A = 1/a$, and $C = b/a$, we see that

$$Y = AX + C,$$

which is linear: C is the intercept and A is the slope. So to estimate the values of a and b from data, we would transform both x and y to reciprocals, plot a graph of $1/y$ against $1/x$, carry out a linear regression, then back-transform, to get:

$$a = \frac{1}{A},$$

$$b = aC.$$

Suppose that we knew that the graph passed through the two points (0.2, 44.44) and (0.6, 70.59). How do we work out the values of the parameters a and b? First, we calculate the four reciprocals. The slope of the linearized function, A, is the change in $1/y$ divided by the change in $1/x$:

```
(1/44.44 - 1/70.59)/(1/0.2 - 1/0.6)
```

```
[1]  0.002500781
```

so $a = 1/A = 1/0.0025 = 400$. Now we rearrange the equation and use one of the points (say $x = 0.2$, $y = 44.44$) to get the value of b:

$$b = \frac{1}{x}\left(\frac{ax}{y} - 1\right) = \frac{1}{0.2}\left(\frac{400 \times 0.2}{44.44} - 1\right) = 4.$$

7.1.8 Sigmoid (S-shaped) functions

The simplest S-shaped function is the **two-parameter logistic** where, for $0 \le y \le 1$,

$$y = \frac{e^{a+bx}}{1 + e^{a+bx}}$$

which is central to the fitting of generalized linear models for proportion data (Chapter 16).

The **three-parameter logistic** function allows y to vary on any scale:

$$y = \frac{a}{1 + be^{-cx}}.$$

The intercept is $a/(1 + b)$, the asymptotic value is a and the initial slope is measured by c. Here is the curve with parameters 100, 90 and 1.0:

```
par(mfrow=c(2,2))

x <- seq(0,10,0.1)
y <- 100/(1+90*exp(-1*x))

plot(x,y,type="l",main="three-parameter logistic")
```

The **four-parameter logistic** function has asymptotes at the left- (a) and right-hand (b) ends of the x axis and scales (c) the response to x about the midpoint (d) where the curve has its inflexion:

$$y = a + \frac{b - a}{1 + e^{c(d-x)}}.$$

Letting $a = 20$, $b = 120$, $c = 0.8$ and $d = 3$, the function

$$y = 20 + \frac{100}{1 + e^{0.8 \times (3-x)}}$$

looks like this:

```
y <- 20+100/(1+exp(0.8*(3-x)))

plot(x,y,ylim=c(0,140),type="l",main="four-parameter logistic")
```

Negative sigmoid curves have the parameter $c < 0$, as for the function

$$y = 20 + \frac{100}{1 + e^{-0.8 \times (3-x)}}.$$

An asymmetric S-shaped curve much used in demography and life insurance work is the **Gompertz growth model**,

$$y = ae^{be^{cx}}.$$

The shape of the function depends on the signs of the parameters b and c. For a negative sigmoid, b is negative (here -1) and c is positive (here $+0.02$):

```
x <-   -200:100

y <- 100*exp(-exp(0.02*x))

plot(x,y,type="l",main="negative Gompertz")
```

For a positive sigmoid both parameters are negative:

```
x <-   0:100
y <-   50*exp(-5*exp(-0.08*x))

plot(x,y,type="l",main="positive Gompertz")
```

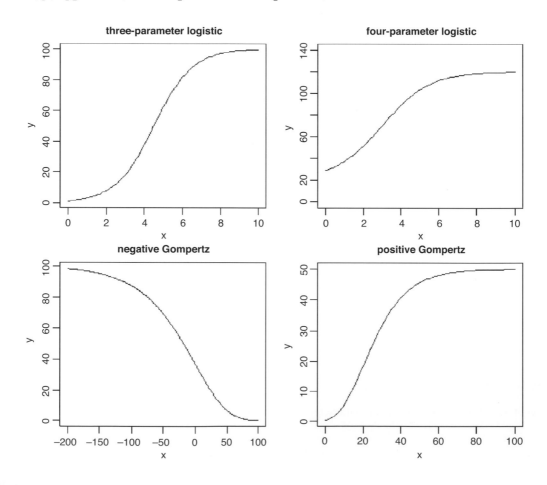

7.1.9   Biexponential model

This is a useful four-parameter non-linear function, which is the sum of two exponential functions of x:

$$y = ae^{bx} + ce^{dx}.$$

Various shapes depend upon the signs of the parameters b, c and d (a is assumed to be positive): the upper left-hand panel shows c positive, b and d negative (it is the sum of two exponential decay curves, so the fast decomposing material disappears first, then the slow, to produce two different phases); the upper right-hand panel shows c and d positive, b negative (this produces an asymmetric U-shaped curve); the lower left-hand panel shows c negative, b and d positive (this can, but does not always, produce a curve with a hump); and the lower right panel shows b and c positive, d negative. When b, c and d are all negative (not illustrated), the function is known as the **first-order compartment model** in which a drug administered at time 0 passes through the system with its dynamics affected by three physiological processes: elimination, absorption and clearance.

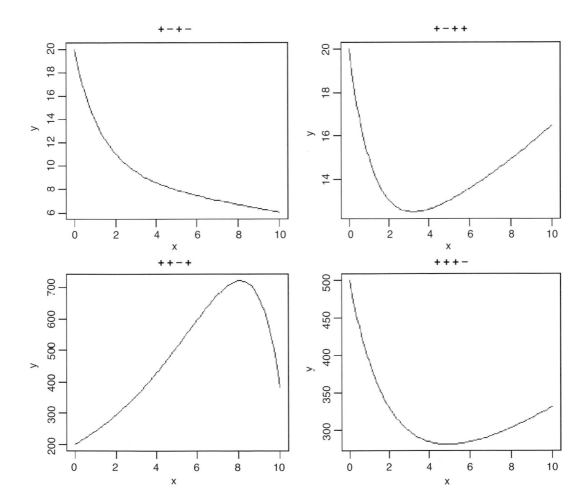

```
#1
a <- 10
b <-  -0.8
c <- 10
d <- -0.05
y <- a*exp(b*x)+c*exp(d*x)
plot(x,y,main="+ - + -",type="l")
#2
a <- 10
b <-  -0.8
c <- 10
d  <-  0.05
y <- a*exp(b*x)+c*exp(d*x)
plot(x,y,main="+ - + +",type="l")
#3
a <- 200
b <- 0.2
c <-  -1
d <- 0.7
y <- a*exp(b*x)+c*exp(d*x)
plot(x,y,main="+ + - +",type="l")
#4
a <- 200
b <- 0.05
c <- 300
d <-  -0.5
y <- a*exp(b*x)+c*exp(d*x)
plot(x,y,main="+ + + -",type="l")
```

7.1.10 Transformations of the response and explanatory variables

We have seen the use of transformation to linearize the relationship between the response and the explanatory variables:

- $\log(y)$ against x for exponential relationships;
- $\log(y)$ against $\log(x)$ for power functions;
- $\exp(y)$ against x for logarithmic relationships;
- $1/y$ against $1/x$ for asymptotic relationships;
- $\log(p/(1 - p))$ against x for proportion data.

Other transformations are useful for variance stabilization:

- $\sqrt{y}$ to stabilize the variance for count data;
- $\arcsin(y)$ to stabilize the variance of percentage data.

7.2   Probability functions

There are many specific probability distributions in R (normal, Poisson, binomial, etc.), and these are discussed in detail later. Here we look at the base mathematical functions that deal with elementary probability. The **factorial** function gives the number of permutations of n items. How many ways can four items be arranged? The first position could have any one of the 4 items in it, but by the time we get to choosing the second item we shall already have specified the first item so there are just $4 - 1 = 3$ ways of choosing the second item. There are only $4 - 2 = 2$ ways of choosing the third item, and by the time we get to the last item we have no degrees of freedom at all: the last number must be the one item out of four that we have not used in positions 1, 2 or 3. So with 4 items the answer is $4 \times (4-1) \times (4-2) \times (4-3)$ which is $4 \times 3 \times 2 \times 1 = 24$. In general, factorial($n$) is given by

$$n! = n(n - 1)(n - 2) \ldots \times 3 \times 2.$$

The R function is `factorial` and we can plot it for values of x from 0 to 10 using the step option `type="s"`, in plot with a logarithmic scale on the y axis `log="y"`,

```
par(mfrow=c(1,1))
x <- 0:6

plot(x,factorial(x),type="s",main="factorial x",log="y")
```

The other important base function for probability calculations in R is the `choose` function which calculates **binomial coefficients**. These show the number of ways there are of selecting x items out of n items when the item can be one of just two types (e.g. either male or female, black or white, solvent or insolvent). Suppose we have 8 individuals and we want to know how many ways there are that 3 of them could be males (and hence 5 of them females). The answer is given by

$$\binom{n}{x} = \frac{n!}{x!(n-x)!},$$

so with $n = 8$ and $x = 3$ we get

$$\binom{n}{x} = \frac{8!}{3!(8-3)!} = \frac{8 \times 7 \times 6}{3 \times 2} = 56,$$

and in R

```
choose(8,3)
```

```
[1] 56
```

Obviously there is only one way that all 8 individuals could be male or female, so there is only one way of getting 0 or 8 'successes'. One male could be the first individual you select, or the second, or the third, and so on. So there are 8 ways of selecting 1 out of 8. By the same reasoning, there must be 8 ways of selecting 7 males out of 8 individuals (the lone female could be in any one of the 8 positions). The following is a graph of the number of ways of selecting from 0 to 8 males out of 8 individuals:

```
plot(0:8,choose(8,0:8),type="s",main="binomial coefficients")
```

7.3 Continuous probability distributions

R has a wide range of built-in probability distributions, for each of which four functions are available: the probability density function (which has a d prefix); the cumulative probability (p); the quantiles of the distribution (q); and random numbers generated from the distribution (r). Each letter can be prefixed to the R function names in Table 7.1 (e.g. dbeta).

Table 7.1. The probability distributions supported by R. The meanings of the parameters are explained in the text.

R function	Distribution	Parameters
beta	beta	shape1, shape2
binom	binomial	sample size, probability
cauchy	Cauchy	location, scale
exp	exponential	rate (optional)
chisq	chi-squared	degrees of freedom
F	Fisher's F	df1, df2
gamma	gamma	shape
geom	geometric	probability
hyper	hypergeometric	m, n, k
lnorm	lognormal	mean, standard deviation
logis	logistic	location, scale
nbinom	negative binomial	size, probability
norm	normal	mean, standard deviation
pois	Poisson	mean
signrank	Wilcoxon signed rank statistic	sample size n
t	Student's t	degrees of freedom
unif	uniform	minimum, maximum (opt.)
weibull	Weibull	shape
wilcox	Wilcoxon rank sum	m, n

The **cumulative probability** function is a straightforward notion: it is an S-shaped curve showing, for any value of x, the probability of obtaining a sample value that is less than or equal to x. Here is what it looks like for the normal distribution:

```
curve(pnorm(x),-3,3)

arrows(-1,0,-1,pnorm(-1),col="red")

arrows(-1,pnorm(-1),-3,pnorm(-1),col="green")
```

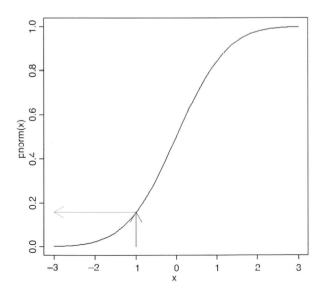

The value of x (−1) leads up to the cumulative probability (red arrow) and the probability associated with obtaining a value of this size (−1) or smaller is on the y axis (green arrow). The value on the y axis is 0.1586553:

```
pnorm(-1)
```

```
[1] 0.1586553
```

The **probability density** is the slope of this curve (its derivative). You can see at once that the slope is never negative. The slope starts out very shallow up to about $x = -2$, increases up to a peak (at $x = 0$ in this example) then gets shallower, and becomes very small indeed above about $x = 2$. Here is what the density function of the normal (dnorm) looks like:

```
curve(dnorm(x),-3,3)
```

For a discrete random variable, like the Poisson or the binomial, the probability density function is straightforward: it is simply a histogram with the y axis scaled as probabilities rather than counts, and the discrete values of x (0, 1, 2, 3, ...) on the horizontal axis. But for a continuous random variable, the definition of the probability density function is more subtle: it does not have probabilities on the y axis, but rather the derivative (the slope) of the cumulative probability function at a given value of x.

7.3.1 Normal distribution

This distribution is central to the theory of parametric statistics. Consider the following simple exponential function:

$$y = \exp(-|x|^m).$$

As the power (m) in the exponent increases, the function becomes more and more like a step function. The following panels show the relationship between y and x for $m = 1, 2, 3$ and 8, respectively:

```
par(mfrow=c(2,2))
x <- seq(-3,3,0.01)
y <- exp(-abs(x))
plot(x,y,type="l",main= "x")
y <- exp(-abs(x)^2)
plot(x,y,type="l",main= "x^2")
y <- exp(-abs(x)^3)
plot(x,y,type="l",main= "x^3")
y <- exp(-abs(x)^8)
plot(x,y,type="l",main= "x^8")
```

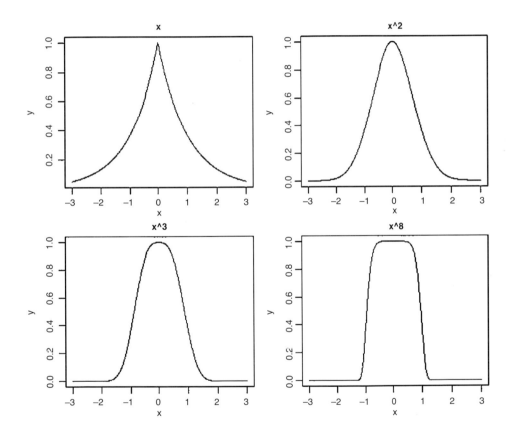

The second of these panels (top right), where $y = \exp(-x^2)$, is the basis of an extremely important and famous probability density function. Once it has been scaled, so that the integral (the area under the curve from $-\infty$ to $+\infty$) is unity, this is the normal distribution. Unfortunately, the scaling constants are rather cumbersome. When the distribution has mean 0 and standard deviation 1 (the **standard normal** distribution) the equation becomes:

$$f(z) = \frac{1}{\sqrt{2\pi}}e^{-z^2/2}.$$

Suppose we have measured the heights of 100 people. The mean height was 170 cm and the standard deviation was 8 cm. We can ask three sorts of questions about data like these: what is the probability that a randomly selected individual will be:

- shorter than a particular height?
- taller than a particular height?
- between one specified height and another?

The area under the whole curve is exactly 1; everybody has a height between minus infinity and plus infinity. True, but not particularly helpful. Suppose we want to know the probability that one of our people, selected at random from the group, will be less than 160 cm tall. We need to convert this height into a value of z; that is to say, we need to convert 160 cm into *a number of standard deviations from the mean*. What do we know about the standard normal distribution? It has a mean of 0 and a standard deviation of 1. So we can convert any value y, from a distribution with mean $\bar{y}$ and standard deviation s very simply by calculating

$$z = \frac{y - \bar{y}}{s}.$$

So we convert 160 cm into a number of standard deviations. It is less than the mean height (170 cm) so its value will be negative:

$$z = \frac{160 - 170}{8} = -1.25.$$

Now we need to find the probability of a value of the standard normal taking a value of −1.25 or smaller. This is *the area under the left-hand tail* (the integral) of the density function. The function we need for this is pnorm: we provide it with a value of z (or, more generally, with a quantile) and it provides us with the probability we want:

```
pnorm(-1.25)
```

```
[1] 0.1056498
```

So the answer to our first question (the shaded area, top left) is just over 10%.

Next, what is the probability of selecting one of our people and finding that they are taller than 185 cm (top right)? The first two parts of the exercise are exactly the same as before. First we convert our value of 185 cm into a number of standard deviations:

$$z = \frac{185 - 170}{8} = 1.875.$$

Then we ask what probability is associated with this, using pnorm:

```
pnorm(1.875)
```

```
[1] 0.9696036
```

But this is the answer to a different question. This is the probability that someone will be *less* than or equal to 185 cm tall (that is what the function pnorm has been written to provide). All we need to do is to work out the complement of this:

```
1-pnorm(1.875)
```

```
[1] 0.03039636
```

So the answer to the second question is about 3%.

Finally, we might want to know the probability of selecting a person between 165 cm and 180 cm. We have a bit more work to do here, because we need to calculate two z values:

$$z_1 = \frac{165 - 170}{8} = -0.625 \quad \text{and} \quad z_2 = \frac{180 - 170}{8} = 1.25.$$

The important point to grasp is this: we want the probability of selecting a person between these two z values, so we *subtract the smaller probability from the larger probability*:

```
pnorm(1.25)-pnorm(-0.625)
```

```
[1] 0.6283647
```

Thus we have a 63% chance of selecting a medium-sized person (taller than 165 cm and shorter than 180 cm) from this sample with a mean height of 170 cm and a standard deviation of 8 cm (bottom left).

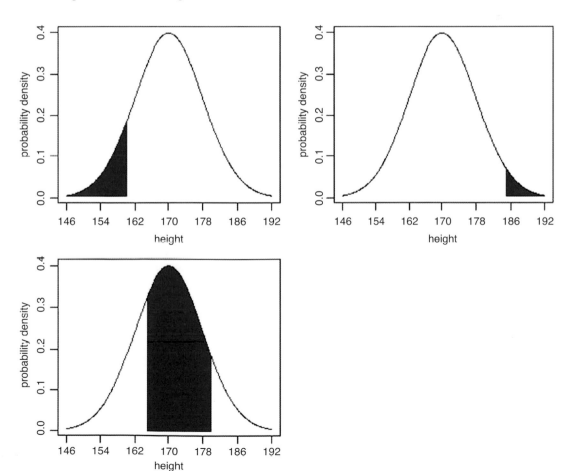

The trick with curved polygons like these is to finish off their closure properly. In the bottom left-hand panel, for instance, we want to return to the *x* axis at 180 cm then draw straight along the *x* axis to 165 cm. We do this by concatenating two extra points on the end of the vectors of z and p coordinates:

```
x <- seq(-3,3,0.01)
z <- seq(-3,-1.25,0.01)
p <- dnorm(z)
z <- c(z,-1.25,-3)
p <- c(p,min(p),min(p))
plot(x,dnorm(x),type="l",xaxt="n",ylab="probability density",xlab="height")
axis(1,at=-3:3,labels=c("146","154","162","170","178","186","192"))
polygon(z,p,col="red")

z <- seq(1.875,3,0.01)
p <- dnorm(z)
z <- c(z,3,1.875)
p <- c(p,min(p),min(p))
plot(x,dnorm(x),type="l",xaxt="n",ylab="probability density",xlab="height")
axis(1,at=-3:3,labels=c("146","154","162","170","178","186","192"))
polygon(z,p,col="red")

z <- seq(-0.635,1.25,0.01)
p <- dnorm(z)
z <- c(z,1.25,-0.635)
p <- c(p,0,0)
plot(x,dnorm(x),type="l",xaxt="n",ylab="probability density",xlab="height")
axis(1,at=-3:3,labels=c("146","154","162","170","178","186","192"))
polygon(z,p,col="red")
```

7.3.2   The central limit theorem

If you take repeated samples from a population with finite variance and calculate their averages, then the averages will be normally distributed. This is called the **central limit theorem**. Let us demonstrate it for ourselves. We can take five uniformly distributed random numbers between 0 and 10 and work out the average. The average will be low when we get, say, 2,3,1,2,1 and high when we get 9,8,9,6,8. Typically, of course, the average will be close to 5. Let us do this 10 000 times and look at the distribution of the 10 000 means. The data are rectangularly (uniformly) distributed on the interval 0 to 10, so the distribution of the raw data should be flat-topped:

```
par(mfrow=c(1,1))

hist(runif(10000)*10,main="")
```

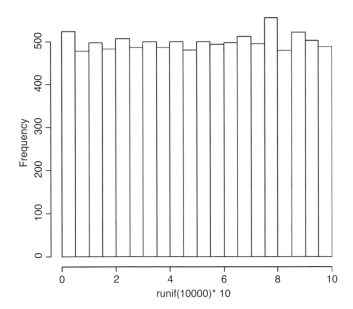

What about the distribution of sample means, based on taking just five uniformly distributed random numbers?

```
means <- numeric(10000)
for (i in 1:10000){
means[i] <- mean(runif(5)*10)
}
hist(means,ylim=c(0,1600),main="")
```

Nice, but how close is this to a normal distribution? One test is to draw a normal distribution with the same parameters on top of the histogram. But what are these parameters? The normal is a two-parameter distribution that is characterized by its mean and its standard deviation. We can estimate these two parameters from our sample of 10 000 means (your values will be slightly different because of the randomization):

```
mean(means)
```

```
[1]  4.998581
```

```
sd(means)
```

```
[1]  1.289960
```

Now we use these two parameters in the probability density function of the normal distribution (dnorm) to create a normal curve with our particular mean and standard deviation. To draw the smooth line of the normal curve, we need to generate a series of values for the x axis; inspection of the histograms suggest that sensible limits would be from 0 to 10 (the limits we chose for our uniformly distributed random numbers). A good rule of thumb is that for a smooth curve you need at least 100 values, so let us try this:

```
xv <- seq(0,10,0.1)
```

There is just one thing left to do. The probability density function has an integral of 1.0 (that is the area beneath the normal curve), but we had 10 000 samples. To scale the normal probability density function to our particular case, however, depends on the height of the highest bar (about 1500 in this case). The height, in turn, depends on the chosen bin widths; if we doubled with width of the bin there would be roughly twice as many numbers in the bin and the bar would be twice as high on the y axis. To get the height of the bars on our frequency scale, therefore, we multiply the total frequency, 10 000 by the bin width, 0.5 to get 5000. We multiply 5000 by the probability density to get the height of the curve. Finally, we use lines to overlay the smooth curve on our histogram:

```
yv <- dnorm(xv,mean=4.998581,sd=1.28996)*5000
lines(xv,yv)
```

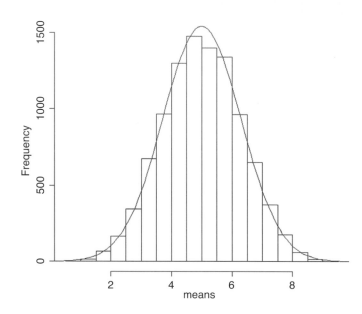

The fit is excellent. The central limit theorem really works. Almost any distribution, even a 'badly behaved' one like the uniform distribution we worked with here, will produce a normal distribution of sample means taken from it.

A simple example of the operation of the central limit theorem involves the use of dice. Throw one die lots of times and each of the six numbers should come up equally often: this is an example of a **uniform** distribution:

```
par(mfrow=c(2,2))

hist(sample(1:6,replace=T,10000),breaks=0.5:6.5,main="",xlab="one die")
```

Now throw two dice and add the scores together: this is the ancient game of craps. There are 11 possible scores from a minimum of 2 to a maximum of 12. The most likely score is 7 because there are six ways that this could come about:

$$1,6 \quad 6,1 \quad 2,5 \quad 5,2 \quad 3,4 \quad 4,3$$

For many throws of craps we get a **triangular** distribution of scores, centred on 7:

```
a <- sample(1:6,replace=T,10000)

b <- sample(1:6,replace=T,10000)

hist(a+b,breaks=1.5:12.5,main="", xlab="two dice")
```

There is already a clear indication of central tendency and spread. For three dice we get

```
c <- sample(1:6,replace=T,10000)
hist(a+b+c,breaks=2.5:18.5,main="", xlab="three dice")
```

and the bell shape of the normal distribution is starting to emerge. By the time we get to five dice, the **binomial** distribution is virtually indistinguishable from the normal:

```
d <- sample(1:6,replace=T,10000)
e <- sample(1:6,replace=T,10000)
hist(a+b+c+d+e,breaks=4.5:30.5,main="", xlab="five dice")
```

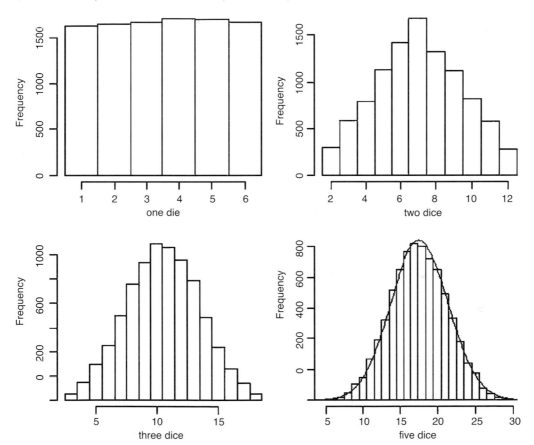

The smooth curve is given by a normal distribution with the same mean and standard deviation:

```
mean(a+b+c+d+e)
```

```
[1] 17.5937
```

```
sd(a+b+c+d+e)
```

```
[1] 3.837668
```

```
lines(seq(1,30,0.1),dnorm(seq(1,30,0.1),17.5937,3.837668)*10000)
```

7.3.3   Maximum likelihood with the normal distribution

The probability density of the normal is

$$f(y|\mu, \sigma) = \frac{1}{\sigma\sqrt{2\pi}} \exp\left[-\frac{(y - \mu)^2}{2\sigma^2}\right],$$

which is read as saying the probability density for a data value y, given (|) a mean of μ and a variance of σ^2, is calculated from this rather complicated-looking two-parameter exponential function. For any given combination of μ and σ^2, it gives a value between 0 and 1. Recall that likelihood is the product of the probability densities, for each of the values of the response variable, y. So if we have n values of y in our experiment, the likelihood function is

$$L(\mu, \sigma) = \prod_{i=1}^{n}\left(\frac{1}{\sigma\sqrt{2\pi}} \exp\left[-\frac{(y_i - \mu)^2}{2\sigma^2}\right]\right),$$

where the only change is that y has been replaced by y_i and we multiply together the probabilities for each of the n data points. There is a little bit of algebra we can do to simplify this: we can get rid of the product operator, $\prod$, in two steps. First, the constant term, multiplied by itself n times, can just be written as $1/(\sigma\sqrt{2\pi})^n$. Second, remember that the product of a set of antilogs (exp) can be written as the antilog of a sum of the values of x_i like this: $\prod \exp(x_i) = \exp\left(\sum x_i\right)$. This means that the product of the right-hand part of the expression can be written as

$$\exp\left[-\frac{\sum_{i=1}^{n}(y_i - \mu)^2}{2\sigma^2}\right],$$

so we can rewrite the likelihood of the normal distribution as

$$L(\mu, \sigma) = \frac{1}{\left(\sigma\sqrt{2\pi}\right)^n} \exp\left[-\frac{1}{2\sigma^2}\sum_{i=1}^{n}(y_i - \mu)^2\right].$$

The two parameters μ and σ are unknown, and the purpose of the exercise is to use statistical modelling to determine their maximum likelihood values from the data (the n different values of y). So how do we find the values of μ and σ that maximize this likelihood? The answer involves calculus: first we find the derivative of the function with respect to the parameters, then set it to zero, and solve.

It turns out that because of the exp function in the equation, it is easier to work out the log of the likelihood,

$$l(\mu, \sigma) = -\frac{n}{2}\log(2\pi) - n\log(\sigma) - \sum(y_i - \mu)^2/2\sigma^2,$$

and maximize this instead. Obviously, the values of the parameters that maximize the log-likelihood $l(\mu, \sigma) = \log(L(\mu, \sigma))$ will be the same as those that maximize the likelihood. From now on, we shall assume that summation is over the index i from 1 to n.

Now for the calculus. We start with the mean, μ. The derivative of the log-likelihood with respect to μ is

$$\frac{dl}{d\mu} = \sum(y_i - \mu)/\sigma^2.$$

Set the derivative to zero and solve for μ:

$$\sum (y_i - \mu)/\sigma^2 = 0 \quad \text{so} \quad \sum (y_i - \mu) = 0.$$

Taking the summation through the bracket, and noting that $\sum \mu = n\mu$,

$$\sum y_i - n\mu = 0 \quad \text{so} \quad \sum y_i = n\mu \quad \text{and} \quad \mu = \frac{\sum y_i}{n}.$$

The maximum likelihood estimate of μ is the arithmetic mean.

Next we find the derivative of the log-likelihood with respect to σ:

$$\frac{dl}{d\sigma} = -\frac{n}{\sigma} + \frac{\sum (y_i - \mu)^2}{\sigma^3},$$

recalling that the derivative of $\log(x)$ is $1/x$ and the derivative of $-1/x^2$ is $2/x^3$. Solving, we get

$$-\frac{n}{\sigma} + \frac{\sum (y_i - \mu)^2}{\sigma^3} = 0 \quad \text{so} \quad \sum (y_i - \mu)^2 = \sigma^3 \left(\frac{n}{\sigma}\right) = \sigma^2 n$$

$$\sigma^2 = \frac{\sum (y_i - \mu)^2}{n}.$$

The maximum likelihood estimate of the variance σ^2 is the mean squared deviation of the y values from the mean. This is a biased estimate of the variance, however, because it does not take account of the fact that we estimated the value of μ from the data. To unbias the estimate, we need to lose 1 degree of freedom to reflect this fact, and divide the sum of squares by $n - 1$ rather than by n (see p. 119 and restricted maximum likelihood estimators in Chapter 19).

Here, we illustrate R's built-in probability functions in the context of the normal distribution. The density function dnorm has a value of z (a quantile) as its argument. Optional arguments specify the mean and standard deviation (the default is the standard normal with mean 0 and standard deviation 1). Values of z outside the range -3.5 to $+3.5$ are very unlikely.

```
par(mfrow=c(2,2))
```

```
curve(dnorm,-3,3,xlab="z",ylab="Probability density",main="Density")
```

The probability function pnorm also has a value of z (a quantile) as its argument. Optional arguments specify the mean and standard deviation (default is the standard normal with mean 0 and standard deviation 1). It shows the cumulative probability of a value of z less than or equal to the value specified, and is an S-shaped curve:

```
curve(pnorm,-3,3,xlab="z",ylab="Probability",main="Probability")
```

Quantiles of the normal distribution qnorm have a cumulative probability as their argument. They perform the opposite function of pnorm, returning a value of z when provided with a probability.

```
curve(qnorm,0,1,xlab="p",ylab="Quantile (z)",main="Quantiles")
```

The normal distribution random number generator rnorm produces random real numbers from a distribution with specified mean and standard deviation. The first argument is the number of numbers that you want to be generated: here are 1000 random numbers with mean 0 and standard deviation 1:

```
y <- rnorm(1000)

hist(y,xlab="z",ylab="frequency",main="Random numbers")
```

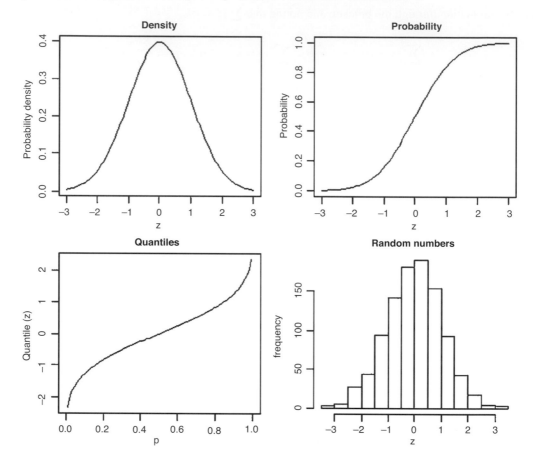

The four functions (d, p, q and r) work in similar ways with all the other probability distributions.

7.3.4 Generating random numbers with exact mean and standard deviation

If you use a random number generator like rnorm then, naturally, the sample you generate will not have exactly the mean and standard deviation that you specify, and two runs will produce vectors with different means and standard deviations. Suppose we want 100 normal random numbers with a mean of exactly 24 and a standard deviation of precisely 4:

```
yvals <- rnorm(100,24,4)

mean(yvals)

[1] 24.2958

sd(yvals)

[1] 3.5725
```

Close, but not spot on. If you want to generate random numbers with an exact mean and standard deviation, then do the following:

```
ydevs <- rnorm(100,0,1)
```

Now compensate for the fact that the mean is not exactly 0 and the standard deviation is not exactly 1 by expressing all the values as departures from the sample mean scaled in units of the sample's standard deviations:

```
ydevs <- (ydevs-mean(ydevs))/sd(ydevs)
```

Check that the mean is 0 and the standard deviation is exactly 1:

```
mean(ydevs)
```

```
[1] -2.449430e-17
```

```
sd(ydevs)
```

```
[1] 1
```

The mean is as close to 0 as makes no difference, and the standard deviation is 1. Now multiply this vector by your desired standard deviation and add to your desired mean value to get a sample with exactly the means and standard deviation required:

```
yvals <- 24 + ydevs*4
```

```
mean(yvals)
```

```
[1] 24
```

```
sd(yvals)
```

```
[1]  4
```

7.3.5 Comparing data with a normal distribution

Various tests for normality are described on p. 346. Here we are concerned with the task of comparing a histogram of real data with a smooth normal distribution with the same mean and standard deviation, in order to look for evidence of non-normality (e.g. skew or kurtosis).

```
par(mfrow=c(1,1))
fishes <- read.table("c:\\temp\\fishes.txt",header=T)
attach(fishes)
names(fishes)
```

```
[1]  "mass"
```

```
mean(mass)
```

```
[1]  4.194275
```

```
max(mass)
```

```
[1]  15.53216
```

Now the histogram of the mass of the fish is produced, specifying integer bins that are 1 gram in width, up to a maximum of 16.5 g:

```
hist(mass,breaks=-0.5:16.5,col="green",main="")
```

For the purposes of demonstration, we generate everything we need *inside* the `lines` function: the sequence of *x* values for plotting (0 to 16), and the height of the density function (the number of fish (`length(mass)`) times the probability density for each member of this sequence, for a normal distribution with `mean(mass)` and standard deviation `sqrt(var(mass))` as its parameters, like this:

```
lines(seq(0,16,0.1),length(mass)*dnorm(seq(0,16,0.1),mean(mass),sqrt(var(mass))))
```

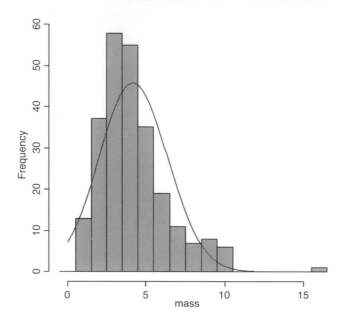

The distribution of fish sizes is clearly *not* normal. There are far too many fishes of 3 and 4 grams, too few of 6 or 7 grams, and too many really big fish (more than 8 grams). This kind of skewed distribution is probably better described by a gamma distribution (see Section 7.3.10) than a normal distribution.

7.3.6   Other distributions used in hypothesis testing

The main distributions used in hypothesis testing are: **chi-squared**, for testing hypotheses involving count data; **Fisher's *F***, in analysis of variance (ANOVA) for comparing two variances; and **Student's *t***, in small-sample work for comparing two parameter estimates. These distributions tell us the size of the test statistic that could be expected by chance alone when nothing was happening (i.e. when the null hypothesis was true). Given the rule that a big value of the test statistic tells us that something *is* happening, and hence that the null hypothesis is false, these distributions define what constitutes a big value of the test statistic (its **critical value**).

   For instance, if we are doing a chi-squared test, and our test statistic is 14.3 on 9 degrees of freedom (d.f.), we need to know whether this is a large value (meaning the null hypothesis is probably false) or a small value (meaning that the null hypothesis cannot be rejected). In the old days we would have looked up the value in chi-squared tables. We would have looked in the row labelled 9 (the degrees of freedom row) and the column headed by $\alpha = 0.05$. This is the conventional value for the acceptable probability of committing a Type I error: that is to say, we allow a 1 in 20 chance of rejecting the null hypothesis when it is actually true (see p. 358). Nowadays, we just type:

```
1-pchisq(14.3,9)
```

```
[1]  0.1120467
```

This indicates that 14.3 is actually a relatively small number when we have 9 d.f. We would conclude that nothing is happening, because a value of chi-squared as large as 14.3 has a greater than an 11% probability of arising by chance alone when the null hypothesis is true. We would want the probability to be less than 5% before we rejected the null hypothesis. So how large would the test statistic need to be, before we would reject the null hypothesis (i.e. what is the critical value of chi-squared)? We use qchisq to answer this. Its two arguments are $1 - \alpha$ and the number of degrees of freedom:

```
qchisq(0.95,9)
```

```
[1]  16.91898
```

So the test statistic would need to be larger than 16.92 in order for us to reject the null hypothesis when there were 9 d.f.

We could use pf and qf in an exactly analogous manner for Fisher's F. Thus, the probability of getting a variance ratio of 2.85 by chance alone when the null hypothesis is true, given that we have 8 d.f. in the numerator and 12 d.f. in the denominator, is just under 5% (i.e. the value is just large enough to allow us to reject the null hypothesis):

```
1-pf(2.85,8,12)
```

```
[1]  0.04992133
```

Note that with pf, degrees of freedom in the numerator (8) come first in the list of arguments, followed by d.f. in the denominator (12).

Similarly, with Student's t statistic and pt and qt. For instance, the value of t in tables for a two-tailed test at $\alpha/2 = 0.025$ with 10 d.f. is

```
qt(0.975,10)
```

```
[1]  2.228139
```

7.3.7 The chi-squared distribution

This is perhaps the second-best known of all the statistical distributions, introduced to generations of school children in their geography lessons and comprehensively misunderstood thereafter. It is a special case of the gamma distribution (p. 293) characterized by a single parameter, the number of degrees of freedom. The mean is equal to the degrees of freedom v ('nu', pronounced 'new'), and the variance is equal to $2v$. The density function looks like this:

$$f(x) = \frac{1}{2^{v/2}\Gamma(v/2)} x^{v/2-1} e^{-x/2},$$

where Γ is the gamma function (see p. 17). The chi-squared distribution is important because many quadratic forms follow it under the assumption that the data follow the normal distribution. In particular, the sample variance is a scaled chi-squared variable. Likelihood ratio statistics are also approximately distributed as a chi-squared (see the F distribution, below).

When the cumulative probability is used, an optional third argument can be provided to describe non-centrality. If the non-central chi-squared is the sum of v independent normal random variables, then the non-centrality parameter is equal to the sum of the squared means of the normal variables. Here are the cumulative probability plots for a non-centrality parameter (ncp) based on three normal means (of 1, 1.5 and 2) and another with 4 means and ncp = 10:

```
windows(7,4)
par(mfrow=c(1,2))
x <- seq(0,30,.25)
plot(x,pchisq(x,3,7.25),type="l",ylab="p(x)",xlab="x")
plot(x,pchisq(x,5,10),type="l",ylab="p(x)",xlab="x")
```

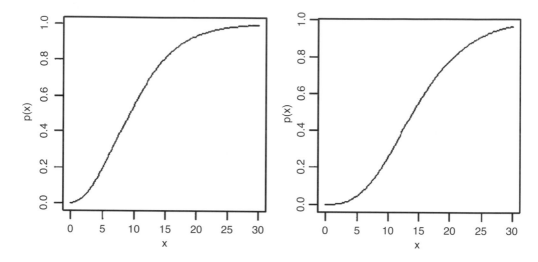

The cumulative probability on the left has 3 d.f. and non-centrality parameter $1^2 + 1.5^2 + 2^2 = 7.25$, while the distribution on the right has 4 d.f. and non-centrality parameter 10 (note the longer left-hand tail at low probabilities).

Chi-squared is also used to establish confidence intervals for sample variances. The quantity

$$\frac{(n-1)s^2}{\sigma^2}$$

is the degrees of freedom $(n-1)$ multiplied by the ratio of the sample variance s^2 to the unknown population variance σ^2. This follows a chi-squared distribution, so we can establish a 95% confidence interval for σ^2 as follows:

$$\frac{(n-1)s^2}{\chi^2_{1-\alpha/2}} \leq \sigma^2 \leq \frac{(n-1)s^2}{\chi^2_{\alpha/2}}$$

Suppose the sample variance $s^2 = 10.2$ on 8 d.f. Then the interval on σ^2 is given by

```
8*10.2/qchisq(.975,8)
```

```
[1] 4.65367
```

```
8*10.2/qchisq(.025,8)
```

```
[1] 37.43582
```

which means that we can be 95% confident that the population variance lies in the range $4.65 \leq \sigma^2 \leq 37.44$.

7.3.8 Fisher's *F* distribution

This is the famous variance ratio test that occupies the penultimate column of every ANOVA table. The ratio of treatment variance to error variance follows the *F* distribution, and you will often want to use the quantile qf to look up critical values of *F*. You specify, in order, the probability of your one-tailed test (this will usually be 0.95), then the two degrees of freedom – numerator first, then denominator. So the 95% value of *F* with 2 and 18 d.f. is

```
qf(.95,2,18)
```

```
[1] 3.554557
```

This is what the density function of *F* looks like for 2 and 18 d.f. (left) and 6 and 18 d.f. (right):

```
x <- seq(0.05,4,0.05)
plot(x,df(x,2,18),type="l",ylab="f(x)",xlab="x")
plot(x,df(x,6,18),type="l",ylab="f(x)",xlab="x")
```

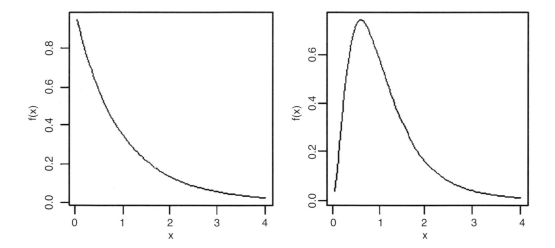

The *F* distribution is a two-parameter distribution defined by the density function

$$f(x) = \frac{r\Gamma(1/2(r+s))}{s\Gamma(1/2r)\Gamma(1/2s)} \frac{(rx/s)^{(r-1)/2}}{[1+(rx/s)]^{(r+s)/2}},$$

where *r* is the degrees of freedom in the numerator and *s* is the degrees of freedom in the denominator. The distribution is named after R.A. Fisher, the father of analysis of variance, and principal developer of quantitative genetics. It is central to hypothesis testing, because of its use in *assessing the significance of the differences between two variances*. The test statistic is calculated by dividing the larger variance by the smaller variance. The two variances are significantly different when this ratio is larger than the critical value of Fisher's *F*. The degrees of freedom in the numerator and in the denominator allow the calculation of the critical value of the test statistic. When there is a single degree of freedom in the numerator, the distribution is equal to the square of Student's *t*: $F = t^2$. Thus, while the rule of thumb for the critical value of *t* is 2, so the rule of thumb for $F = t^2 = 4$. To see how well the rule of thumb works, we can plot critical F against d.f. in the numerator:

```
windows(7,7)
par(mfrow=c(1,1))
df <- seq(1,30,.1)
plot(df,qf(.95,df,30),type="l",ylab="Critical F")
lines(df,qf(.95,df,10),lty=2)
```

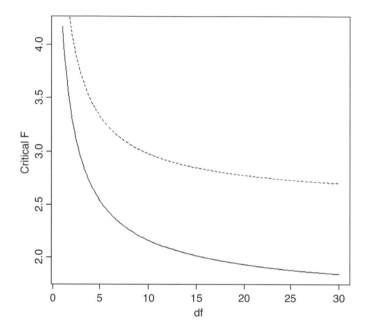

You see that the rule of thumb (critical $F = 4$) quickly becomes much too large once the d.f. in the numerator (on the x axis) is larger than 2. The lower (solid) line shows the critical values of F when the denominator has 30 d.f. and the upper (dashed) line shows the case in which the denominator has 10 d.f.

The shape of the density function of the F distribution depends on the degrees of freedom in the numerator.

```
x <- seq(0.01,3,0.01)
plot(x,df(x,1,10),type="l",ylim=c(0,1),ylab="f(x)")
lines(x,df(x,2,10),lty=6,col="red")
lines(x,df(x,5,10),lty=2,col="green")
lines(x,df(x,30,10),lty=3,col="blue")
legend(2,0.9,c("1","2","5","30"),col=(1:4),lty=c(1,6,2,3),
                           title="numerator d.f.")
```

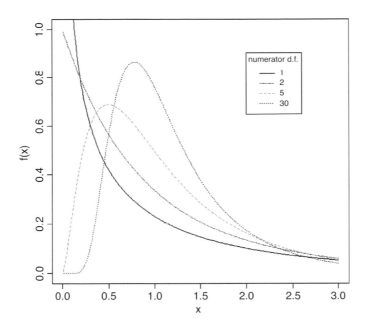

The probability density $f(x)$ declines monotonically when the numerator has 1 or 2 d.f., but rises to a maximum for 3 d.f. or more (5 and 30 are shown here): all the graphs have 10 d.f. in the denominator.

7.3.9 Student's *t* distribution

This famous distribution was first published by W.S. Gossett in 1908 under the pseudonym of 'Student' because his then employer, the Guinness brewing company in Dublin, would not permit employees to publish under their own names. It is a model with one parameter, r, with density function

$$f(x) = \frac{\Gamma\left(1/2(r+1)\right)}{(\pi r)^{1/2}\Gamma\left(1/2r\right)} \left(1 + \frac{x^2}{r}\right)^{-(r+1)/2},$$

where $-\infty < x < +\infty$. This looks very complicated, but if all the constants are stripped away, you can see just how simple the underlying structure really is:

$$f(x) = \left(1 + x^2\right)^{-1/2}.$$

We can plot this for values of x from -3 to $+3$ as follows:

```
curve( (1+x^2)^(-0.5), -3, 3,ylab="t(x)",col="red")
```

The main thing to notice is how fat the tails of the distribution are, compared with the normal distribution. The plethora of constants is necessary to scale the density function so that its integral is 1. If we define U as

$$U = \frac{n-1}{\sigma^2}s^2,$$

then this is chi-squared distributed on $n - 1$ d.f. (see above). Now define V as

$$V = \frac{n^{1/2}}{\sigma}(\bar{y} - \mu)$$

and note that this is normally distributed with mean 0 and standard deviation 1 (the standard normal distribution), so

$$\frac{V}{(U/(n-1))^{1/2}}$$

is the ratio of a normal distribution and a chi-squared distribution. You might like to compare this with the F distribution (above), which is the ratio of two chi-squared distributed random variables.

At what point does the rule of thumb for Student's $t = 2$ break down so seriously that it is actually misleading? To find this out, we need to plot the value of Student's t against sample size (actually against degrees of freedom) for small samples. We use qt (quantile of t) and fix the probability at the two-tailed value of 0.975:

```
plot(1:30,qt(0.975,1:30), ylim=c(0,12),type="l",

                ylab="Students t value",xlab="d.f.",col="red")
abline(h=2,lty=2,col="green")
```

As you see, the rule of thumb only becomes really hopeless for degrees of freedom less than about 5 or so. For most practical purposes $t \approx 2$ really is a good working rule of thumb. So what does the t distribution look like, compared to a normal? Let us redraw the standard normal as a dotted line (lty=2):

```
xvs <- seq(-4,4,0.01)

plot(xvs,dnorm(xvs),type="l",lty=2,

                ylab="Probability density",xlab="Deviates")
```

Now we can overlay Student's t with d.f. = 5 as a solid line to see the difference:

```
lines(xvs,dt(xvs,df=5),col="red")
```

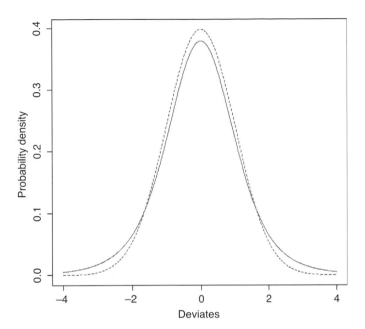

The difference between the normal (blue dashed line) and Student's *t* distributions (solid red line) is that the *t* distribution has 'fatter tails'. This means that extreme values are more likely with a *t* distribution than with a normal, and the confidence intervals are correspondingly broader. So instead of a 95% interval of $\pm$ 1.96 with a normal distribution we should have a 95% interval of $\pm$ 2.57 for a Student's *t* distribution with 5 degrees of freedom:

```
qt(0.975,5)
```

```
[1] 2.570582
```

In hypothesis testing we generally use two-tailed tests because typically we do not know the direction of the response in advance. This means that we put 0.025 in each of two tails, rather than 0.05 in one tail.

7.3.10 The gamma distribution

The gamma distribution is useful for describing a wide range of processes where the data are positively skew (i.e. non-normal, with a long tail on the right). It is a two-parameter distribution, where the parameters are traditionally known as shape and rate. Its density function is:

$$f(x) = \frac{1}{\beta^\alpha \Gamma(\alpha)} x^{\alpha-1} e^{-x/\beta},$$

where α is the shape parameter and β^{-1} is the rate parameter (alternatively, β is known as the scale parameter). Special cases of the gamma distribution are the **exponential** ($\alpha = 1$) and **chi-squared** ($\alpha = \nu/2, \beta = 2$).

   To see the effect of the shape parameter on the probability density, we can plot the gamma distribution for different values of shape and rate over the range 0.01 to 4:

```
x <- seq(0.01,4,.01)
par(mfrow=c(2,2))
```

```
y <- dgamma(x,.5,.5)
plot(x,y,type="l",col="red",main="alpha = 0.5")
y <- dgamma(x,.8,.8)
plot(x,y,type="l",col="red", main="alpha = 0.8")
y <- dgamma(x,2,2)
plot(x,y,type="l",col="red", main="alpha = 2")
y <- dgamma(x,10,10)
plot(x,y,type="l",col="red", main="alpha = 10")
```

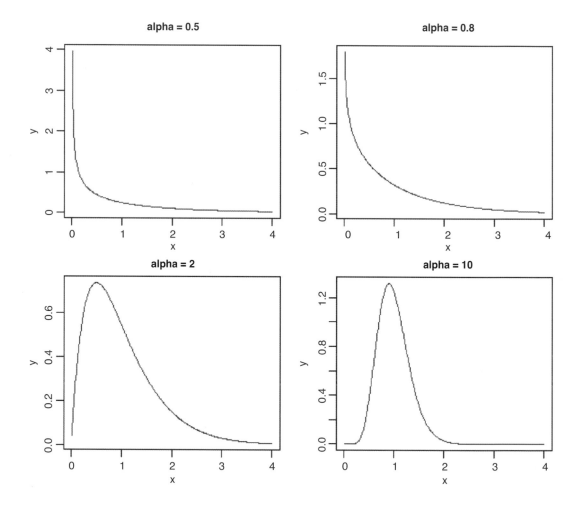

The graphs from top left to bottom right show different values of α: 0.5, 0.8, 2 and 10. Note how $\alpha < 1$ produces monotonic declining functions and $\alpha > 1$ produces humped curves that pass through the origin, with the degree of skew declining as α increases.

The mean of the distribution is $\alpha\beta$, the variance is $\alpha\beta^2$, the skewness is $2/\sqrt{\alpha}$ and the kurtosis is $6/\alpha$. Thus, for the exponential distribution we have a mean of β, a variance of β^2, a skewness of 2 and a kurtosis of 6, while for the chi-squared distribution we have a mean of v, a variance of $2v$ a skewness of $2\sqrt{2/v}$ and a kurtosis of $12/v$. Observe also that

$$\frac{1}{\beta} = \frac{mean}{variance},$$

$$shape = \frac{1}{\beta} \times mean.$$

We can now answer questions like this: what is the value of the 95% quantile expected from a gamma distribution with mean $= 2$ and variance $= 3$? This implies that rate is 2/3 and shape is 4/3 so:

```
qgamma(0.95,2/3,4/3)
```

```
[1] 1.732096
```

An important use of the gamma distribution is in describing continuous measurement data that are *not* normally distributed. Here is an example where body mass data for 200 fishes are plotted as a histogram and a gamma distribution with the same mean and variance is overlaid as a smooth curve:

```
fishes <- read.table("c:\\temp\\fishes.txt",header=T)
attach(fishes)
names(fishes)
```

```
[1] "mass"
```

First, we calculate the two parameter values for the gamma distribution:

```
rate <- mean(mass)/var(mass)
shape <- rate*mean(mass)
rate
```

```
[1]  0.8775119
```

```
shape
```

```
[1]  3.680526
```

We need to know the largest value of mass, in order to make the bins for the histogram:

```
max(mass)
```

```
[1]  15.53216
```

Now we can plot the histogram, using break points at 0.5 to get integer-centred bars up to a maximum of 16.5 to accommodate our biggest fish:

```
par(mfrow=c(1,1))
```

```
hist(mass,breaks=-0.5:16.5,col="green",main="")
```

The density function of the gamma distribution is overlaid using lines like this:

```
lines(seq(0.01,15,0.01),length(mass)*dgamma(seq(0.01,15,0.01),shape,rate))
```

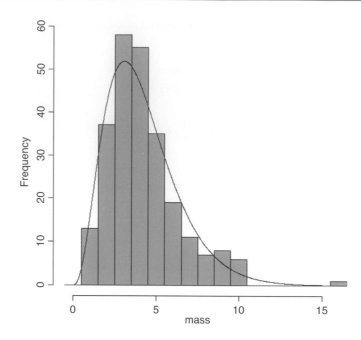

The fit is much better than when we tried to fit a normal distribution to these same data earlier (see p. 286).

7.3.11 The exponential distribution

This is a one-parameter distribution that is a special case of the gamma distribution. Much used in survival analysis, its density function is given on p. 874 and its use in survival analysis is explained on p. 884. The random number generator of the exponential is useful for Monte Carlo simulations of time to death when the hazard (the instantaneous risk of death) is constant with age. You specify the hazard, which is the reciprocal of the mean age at death:

```
rexp(15,0.1)
```

```
[1]    8.4679954 19.4649828 16.3599100 31.6182943   1.9592625   6.3877954
[7]  26.4725498 18.7831597 34.9983158 18.0820563   2.1303369   0.1319956
[13] 35.3649667   3.5672353   4.8672067
```

These are 15 random lifetimes with an expected value of $1/0.1 = 10$ years; they give a sample mean of 9.66 years.

7.3.12 The beta distribution

This has two positive constants, a and b, and x is bounded in the range $0 \leq x \leq 1$:

$$f(x) = \frac{\Gamma(a+b)}{\Gamma(a)\Gamma(b)} x^{a-1}(1-x)^{b-1}.$$

In R we generate a family of density functions like this:

```
par(mfrow=c(2,2))
```

```
x <- seq(0,1,0.01)

fx <- dbeta(x,2,3)
plot(x,fx,type="l",main="a=2 b=3",col="red")

fx <- dbeta(x,0.5,2)
plot(x,fx,type="l",main="a=0.5 b=2",col="red")

fx <- dbeta(x,2,0.5)
plot(x,fx,type="l",main="a=2 b=0.5",col="red")

fx <- dbeta(x,0.5,0.5)
plot(x,fx,type="l",main="a=0.5 b=0.5",col="red")
```

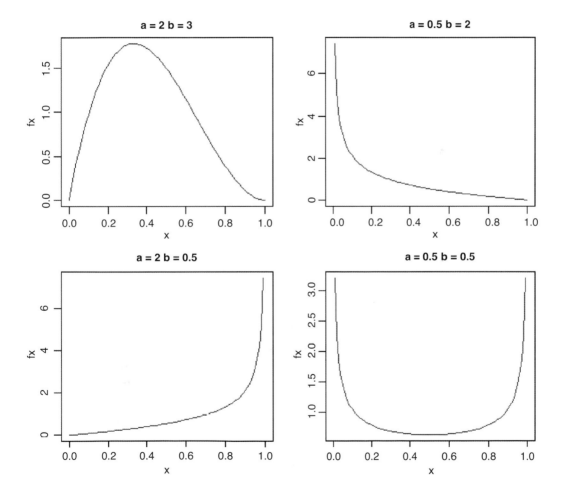

The important point is whether the parameters are greater or less than 1. When both are greater than 1 we get an *n*-shaped curve which becomes more skew as $b > a$ (top left). If $0 < a < 1$ and $b > 1$ then the slope of the density is negative (top right), while for $a > 1$ and $0 < b < 1$ the slope of the density is positive (bottom left). The function is U-shaped when both *a* and *b* are positive fractions. If $a = b = 1$, then we obtain the uniform distribution on [0,1].

Here are 10 random numbers from the beta distribution with shape parameters 2 and 3:

```
rbeta(10,2,3)
```

```
[1]  0.2908066  0.1115131  0.5217944  0.1691430  0.4456099
[6]  0.3917639  0.6534021  0.3633334  0.2342860  0.6927753
```

7.3.13   The Cauchy distribution

This is a long-tailed two-parameter distribution, characterized by a location parameter a and a scale parameter b. It is real-valued, symmetric about a (which is also its median), and is a curiosity in that it has long enough tails that the expectation does not exist – indeed, it has no moments at all (it often appears in counter-examples in maths books). The harmonic mean of a variable with positive density at 0 is typically distributed as Cauchy, and the Cauchy distribution also appears in the theory of Brownian motion (e.g. random walks). The general form of the distribution is

$$f(x) = \frac{1}{\pi b(1 + ((x - a)/b)^2)},$$

for $-\infty < x < \infty$. There is also a one-parameter version, with $a = 0$ and $b = 1$, which is known as the standard Cauchy distribution and is the same as Student's t distribution with one degree of freedom:

$$f(x) = \frac{1}{\pi(1 + x^2)},$$

for $-\infty < x < \infty$.

```
windows(7,4)
par(mfrow=c(1,2))
plot(-200:200,dcauchy(-200:200,0,10),type="l",ylab="p(x)",xlab="x",
col="red")
plot(-200:200,dcauchy(-200:200,0,50),type="l",ylab="p(x)",xlab="x",
col="red")
```

 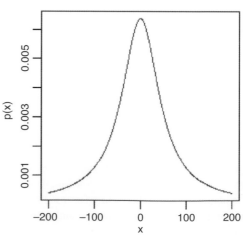

Note the very long, fat tail of the Cauchy distribution. The left-hand density function has scale $= 10$ and the right hand plot has scale $= 50$; both have location $= 0$.

7.3.14   The lognormal distribution

The lognormal distribution takes values on the positive real line. If the logarithm of a lognormal deviate is taken, the result is a normal deviate, hence the name. Applications for the lognormal include the distribution of particle sizes in aggregates, flood flows, concentrations of air contaminants, and failure times. The hazard function of the lognormal is increasing for small values and then decreasing. A mixture of heterogeneous items that individually have monotone hazards can create such a hazard function.

   Density, cumulative probability, quantiles and random generation for the lognormal distribution employ the functions `dlnorm`, `plnorm`, `qlnorm` and `rlnorm`. Here, for instance, is the density function: `dlnorm(x, meanlog=0, sdlog=1)`. The mean and standard deviation are optional, with default `meanlog = 0` and `sdlog = 1`. Note that these are *not* the mean and standard deviation; the lognormal distribution has mean $e^{\mu + \sigma^2/2}$, variance $(e^{\sigma^2} - 1)e^{2\mu + \sigma^2}$, skewness $(e^{\sigma^2} + 2)\sqrt{e^{\sigma^2} - 1}$ and kurtosis $e^{4\sigma^2} + 2e^{3\sigma^2} + 3e^{2\sigma^2} - 6$.

```
windows(7,7)

plot(seq(0,10,0.05),dlnorm(seq(0,10,0.05)),

                  type="l",xlab="x",ylab="LogNormal f(x)",col="x")
```

The extremely long tail and exaggerated positive skew are characteristic of the lognormal distribution. Logarithmic transformation followed by analysis with normal errors is often appropriate for data such as these.

7.3.15   The logistic distribution

The logistic is the canonical link function in generalized linear models with binomial errors and is described in detail in Chapter 16 on the analysis of proportion data. The cumulative probability is a symmetrical S-shaped distribution that is bounded above by 1 and below by 0. There are two ways of writing the cumulative probability equation:

$$p(x) = \frac{e^{a+bx}}{1 + e^{a+bx}}$$

and

$$p(x) = \frac{1}{1 + \beta e^{-\alpha x}}$$

The great advantage of the first form is that it linearizes under the log-odds transformation (see p. 630) so that

$$\ln\left(\frac{p}{q}\right) = a + bx,$$

where p is the probability of success and $q = 1 - p$ is the probability of failure. The logistic is a unimodal, symmetric distribution on the real line with tails that are longer than the normal distribution. It is often used to model growth curves, but has also been used in bioassay studies and other applications. A motivation for using the logistic with growth curves is that the logistic distribution function $f(x)$ has the property that the derivative of $f(x)$ with respect to x is proportional to $[f(x) - A][B - f(x)]$ with $A < B$. The interpretation is that the rate of growth is proportional to the amount already grown, times the amount of growth that is still expected.

```
windows(7,4)
par(mfrow=c(1,2))
plot(seq(-5,5,0.02),dlogis(seq(-5,5,.02)),
type="l",main="Logistic",col="red",xlab="x",ylab="p(x)")
plot(seq(-5,5,0.02),dnorm(seq(-5,5,.02)),
type="l",main="Normal",col="red",xlab="x",ylab="p(x)")
```

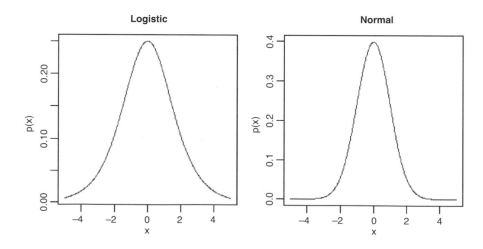

Here, the logistic density function `dlogis` (left) is compared with an equivalent normal density function `dnorm` (right) using the default mean 0 and standard deviation 1 in both cases. Note the much fatter tails of the logistic (there is still substantial probability at ± 4 standard deviations). Note also the difference in the scales of the two y axes (0.25 for the logistic, 0.4 for the normal).

7.3.16   The log-logistic distribution

The log-logistic is a very flexible four-parameter model for describing growth or decay processes:

$$y = a + b\left[\frac{\exp(c(\log(x) - d))}{1 + \exp(c(\log(x) - d))}\right].$$

Here are two cases. The first is a negative sigmoid with $c = -1.59$ and $a = -1.4$:

```
windows(7,4)
par(mfrow=c(1,2))
x <- seq(0.1,1,0.01)
y <-   -1.4+2.1*(exp(-1.59*log(x)-1.53)/(1+exp(-1.59*log(x)-1.53)))
plot(log(x),y,type="l", main="c = -1.59", col="red")
```

For the second we have $c = 1.59$ and $a = 0.1$:

```
y <- 0.1+2.1*(exp(1.59*log(x)-1.53)/(1+exp(1.59*log(x)-1.53)))
plot(log(x),y,type="l",main="c = 1.59",col="red")
```

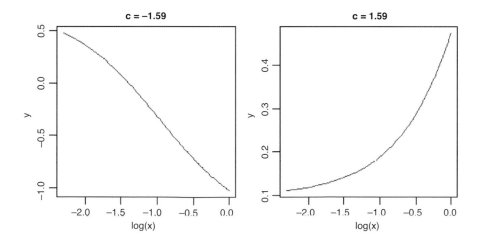

7.3.17   The Weibull distribution

The origin of the Weibull distribution is in *weakest link analysis*. If there are r links in a chain, and the strengths of each link Z_i are independently distributed on $(0, \infty)$, then the distribution of weakest link $V = \min(Z_j)$ approaches the Weibull distribution as the number of links increases.

   The Weibull is a two-paramter model that has the exponential distribution as a special case. Its value in demographic studies and survival analysis is that it allows for the death rate to increase or to decrease with age, so that all three types of survivorship curve can be analysed (as explained on p. 872). The density, survival and hazard functions with $\lambda = \mu^{-\alpha}$ are:

$$f(t) = \alpha \lambda t^{\alpha-1} e^{-\lambda t^\alpha},$$

$$S(t) = e^{-\lambda t^\alpha},$$

$$h(t) = \frac{f(t)}{S(t)} = \alpha \lambda t^{\alpha-1}.$$

The mean of the Weibull distribution is $\Gamma(1 + \alpha^{-1})\mu$ and the variance is $\mu^2(\Gamma(1 + 2/\alpha) - (\Gamma(1 + 1/\alpha))^2)$, and the parameter α describes the shape of the hazard function (the background to determining the likelihood equations is given by Aitkin *et al.*, 2009). For $\alpha = 1$ (the exponential distribution) the hazard is constant, while for $\alpha > 1$ the hazard increases with age and for $\alpha < 1$ the hazard decreases with age.

Because the Weibull, lognormal and log-logistic all have positive skewness, it is difficult to discriminate between them with small samples. This is an important problem, because each distribution has differently shaped hazard functions, and it will be hard, therefore, to discriminate between different assumptions about the age-specificity of death rates. In survival studies, parsimony requires that we fit the exponential rather than the Weibull unless the shape parameter α is significantly different from 1.

Here is a family of three Weibull distributions with $\alpha = 1$, 2 and 3 (red, green and blue lines, respectively):

```
windows(7,7)
a <- 3
l <- 1
t <- seq(0,1.8,.05)
ft <- a*l*t^(a-1)*exp(-l*t^a)
plot(t,ft,type="l",col="blue",ylab="f(t) ")
a <- 1
ft <- a*l*t^(a-1)*exp(-l*t^a)
lines(t,ft,type="l",col="red")
a <- 2
ft <- a*l*t^(a-1)*exp(-l*t^a)
lines(t,ft,type="l",col="green")
legend(1.4,1.1,c("1","2","3"),title="alpha",lty=c(1,1,1),col=c(2,3,4))
```

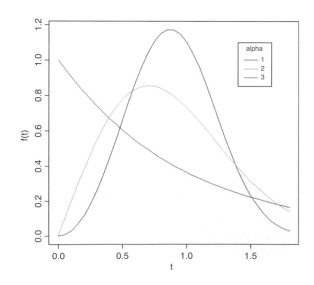

Note that for large values of α the distribution becomes symmetrical, while for $\alpha \leq 1$ the distribution has its mode at $t = 0$.

7.3.18   Multivariate normal distribution

If you want to generate two (or more) vectors of normally distributed random numbers that are correlated with one another to a specified degree, then you need the `mvrnorm` function from the `MASS` library:

```
library(MASS)
```

Suppose we want two vectors of 1000 random numbers each. The first vector has a mean of 50 and the second has a mean of 60. The difference from `rnorm` is that we need to specify their covariance as well as the standard deviations of each separate variable. This is achieved with a positive-definite symmetric matrix specifying the covariance matrix of the variables.

```
xy <- mvrnorm(1000,mu=c(50,60),matrix(c(4,3.7,3.7,9),2))
```

We can check how close the variances are to our specified values:

```
var(xy)
```

```
          [,1]       [,2]
[1,]  3.849190  3.611124
[2,]  3.611124  8.730798
```

Not bad: we said the covariance should be 3.70 and the simulated data are 3.611 124. We extract the two separate vectors x and y and plot them to look at the correlation:

```
x <- xy[,1]
y <- xy[,2]
plot(x,y,pch=16,ylab="y",xlab="x",col="blue")
```

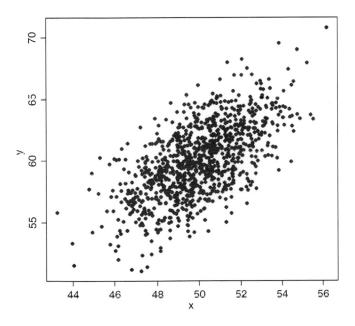

It is worth looking at the variances of *x* and *y* in more detail:

```
var(x)
```

```
[1] 3.84919
```

```
var(y)
```

```
[1] 8.730798
```

If the two samples were *independent*, then the variance of the sum of the two variables would be equal to the sum of the two variances. Is this the case here?

```
var(x+y)
```

```
[1] 19.80224
```

```
var(x)+var(y)
```

```
[1] 12.57999
```

No it is not. The variance of the sum (19.80) is much greater than the sum of the variances (12.58). This is because *x* and *y* are positively correlated; big values of *x* tend to be associated with big values of *y* and vice versa. This being so, we would expect the variance of the difference between *x* and *y* to be less than the sum of the two variances:

```
var(x-y)
```

```
[1] 5.357741
```

As predicted, the variance of the difference (5.36) is much less than the sum of the variances (12.58). We conclude that *the variance of a sum of two variables is only equal to the variance of the difference of two variables when the two variables are independent*. What about the covariance of *x* and *y*? We found this already by applying the `var` function to the matrix *xy* (above). We specified that the covariance should be 3.70 in calling the multivariate normal distribution, and the difference between 3.70 and 3.611 124 is simply due to the random selection of points. The covariance is related to the separate variances through the correlation coefficient ρ as follows (see p. 373):

$$\text{cov}(x, y) = \rho\sqrt{s_x^2 s_y^2}.$$

For our example, this checks out as follows, where the sample value of ρ is `cor(x,y)`:

```
cor(x,y)*sqrt(var(x)*var(y))
```

```
[1] 3.611124
```

which is our observed covariance between *x* and *y* with $\rho = 0.622\ 917\ 8$.

7.3.19 The uniform distribution

This is the distribution that the random number generator in your calculator hopes to emulate. The idea is to generate numbers between 0 and 1 where every possible real number on this interval has exactly the same probability of being produced. If you have thought about this, it will have occurred to you that there is something wrong here. Computers produce numbers by following recipes. If you are following a recipe then the outcome is predictable. If the outcome is predictable, then how can it be random? As John von Neumann once said: 'Anyone who uses arithmetic methods to produce random numbers is in a state of sin.' This raises the question as to what, exactly, a computer-generated random number is. The answer turns out to be scientifically very interesting and very important to the study of encryption (for instance, any pseudorandom number sequence generated by a linear recursion is insecure, since, from a sufficiently long subsequence of the outputs, one can predict the rest of the outputs). If you are interested, look up the **Mersenne twister** online. Here we are only concerned with how well the modern pseudorandom number generator performs. Here is the outcome of the R function `runif` simulating the throwing of a six-sided die 10 000 times: the histogram ought to be flat:

```
x <- ceiling(runif(10000)*6)
table(x)
```

```
x
   1    2    3    4    5    6
1680 1668 1654 1622 1644 1732
```

```
hist(x,breaks=0.5:6.5,main="")
```

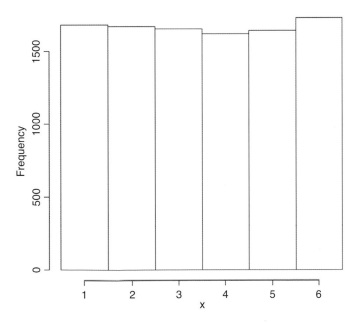

This is remarkably close to theoretical expectation, reflecting the very high efficiency of R's random-number generator. Try mapping 1 000 000 points to look for gaps:

```
x <- runif(1000000)
y <- runif(1000000)
plot(x,y,pch=".",col="blue")
```

The scatter of unfilled space (white dots amongst the sea produced by 1 000 000 blue dots pch=".") shows no evidence of clustering. For a more thorough check we can count the frequency of combinations of numbers: with 36 cells, the expected frequency is 1 000 000/36 = 27 777.78 numbers per cell. We use the cut function to produce 36 bins:

```
table(cut(x,6),cut(y,6))
```

	(-0.001,0.166]	(0.166,0.333]	(0.333,0.5]	(0.5,0.667]	(0.667,0.834]	(0.834,1]
(-0.000997,0.166]	27667	28224	27814	27601	27592	27659
(0.166,0.333]	27604	27790	27922	27687	27990	27701
(0.333,0.5]	27951	27668	27683	27773	27999	27959
(0.5,0.667]	27550	27767	27951	27912	27619	27577
(0.667,0.834]	27527	28106	27868	28262	27804	27460
(0.834,1]	27617	27662	27863	27867	27727	27577

As you can see the observed frequencies are remarkably close to expectation:

```
range(table(cut(x,6),cut(y,6)))
```

```
[1] 27460 28262
```

None of the cells contained fewer than 27 460 random points, and none more than 28 262.

7.3.20 Plotting empirical cumulative distribution functions

The function ecdf is used to compute or plot an empirical cumulative distribution function. Here it is in action for the fishes data (p. 286 and 296):

```
fishes <- read.table("c:\\temp\\fishes.txt",header=T)
attach(fishes)
names(fishes)
```

```
[1]  "mass"
```

```
plot(ecdf(mass))
```

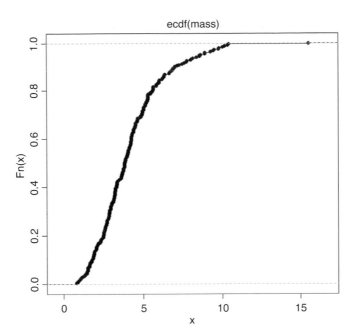

The pronounced positive skew in the data is evident from the fact that the left-hand side of the cumulative distribution is much steeper than the right-hand side (and see p. 350).

7.4 Discrete probability distributions

7.4.1 The Bernoulli distribution

This is the distribution underlying tests with a binary response variable. The response takes one of only two values: it is 1 with probability p (a 'success') and is 0 with probability $1 - p$ (a 'failure'). The density function is given by:

$$p(X) - p^x(1 - p)^{1-x}$$

The statistician's definition of **variance** is the expectation of x^2 minus the square of the expectation of x: $\sigma^2 = E(X^2) - [E(X)]^2$. We can see how this works with a simple distribution like the Bernoulli. There are just two outcomes in $f(x)$: a success, where $x = 1$ with probability p and a failure, where $x = 0$ with probability $1 - p$. Thus, the expectation of x is

$$E(X) = \sum xf(x) = 0 \times (1 - p) + 1 \times p = 0 + p = p$$

and the expectation of x^2 is

$$E(X^2) = \sum x^2 f(x) = 0^2 \times (1 - p) + 1^2 \times p = 0 + p = p,$$

so the variance of the Bernoulli distribution is

$$\mathrm{var}(X) = \mathrm{E}(X^2) - [\mathrm{E}(X)]^2 = p - p^2 = p(1 - p) = pq.$$

7.4.2 The binomial distribution

This is a one-parameter distribution in which p describes the probability of success in a binary trial. The probability of x successes out of n attempts is given by multiplying together the probability of obtaining one specific realization and the number of ways of getting that realization.

We need a way of generalizing the number of ways of getting x items out of n items. The answer is the combinatorial formula

$$\binom{n}{x} = \frac{n!}{x!(n - x)!},$$

where the 'exclamation mark' means 'factorial'. For instance, $5! = 5 \times 4 \times 3 \times 2 = 120$. This formula has immense practical utility. It shows you at once, for example, how unlikely you are to win the National Lottery in which you are invited to select six numbers between 1 and 49. We can use the built-in `factorial` function for this,

```
factorial(49)/(factorial(6)*factorial(49-6))
```

```
[1] 13983816
```

which is roughly a 1 in 14 million chance of winning the jackpot. You are more likely to die between buying your ticket and hearing the outcome of the draw. As we have seen (p. 17), there is a built-in R function for the combinatorial function,

```
choose(49,6)
```

```
[1] 13983816
```

and we use the `choose` function from here on.

The general form of the binomial distribution is given by

$$p(x) = \binom{n}{x} p^x (1 - p)^{n-x},$$

using the combinatorial formula above. The mean of the binomial distribution is np and the variance is $np(1 - p)$.

Since $1 - p$ is less than 1 it is obvious that **the variance is less than the mean** for the binomial distribution (except, of course, in the trivial case when $p = 0$ and the variance is 0). It is easy to visualize the distribution for particular values of n and p.

```
p <- 0.1
n <- 4
x <- 0:n
px <- choose(n,x)*p^x*(1-p)^(n-x)
barplot(px,names=x,xlab="outcome",ylab="probability",col="green")
```

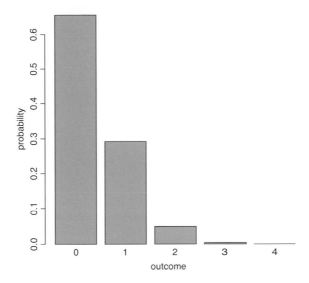

The four distribution functions available for the binomial in R (density, cumulative probability, quantiles and random generation) are used like this. The density function `dbinom(x, size, prob)` shows the probability for the specified count x (e.g. the number of parasitized fish) out of a sample of $n =$ `size`, with probability of success = `prob`. So if we catch four fish when 10% are parasitized in the parent population, we have `size = 4` and `prob = 0.1`, (as illustrated above). Much the most likely number of parasitized fish in our sample is 0.

The cumulative probability shows the sum of the probability densities up to and including $p(x)$, plotting cumulative probability against the number of successes, for a sample of $n =$ `size` and probability = `prob`. Our fishy plot looks like this:

```
barplot(pbinom(0:4,4,0.1),names=0:4,xlab="parasitized fish",

                          ylab="probability",col="red")
```

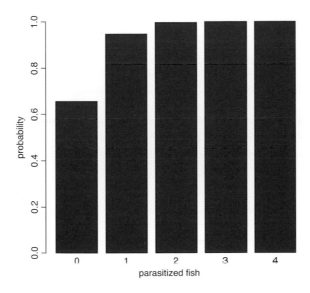

This shows that the probability of getting 2 or fewer parasitized fish out of a sample of 4 is very close to 1. Note that you can generate the series inside the density function $(0 : 4)$.

To obtain a confidence interval for the expected number of fish to be caught in a sample of $n =$ size and a probability $=$ prob, we need qbinom, the quantile function for the binomial. The lower and upper limits of the 95% confidence interval are

```
qbinom(.025,4,0.1)
```

```
[1]  0
```

```
qbinom(.975,4,0.1)
```

```
[1]  2
```

This means that with 95% certainty we shall catch between 0 and 2 parasitized fish out of 4 if we repeat the sampling exercise. We are very unlikely to get 3 or more parasitized fish out of a sample of 4 if the proportion parasitized really is 0.1.

This kind of calculation is very important in power calculations in which we are interested in determining whether or not our chosen sample size ($n = 4$ in this case) is capable of doing the job we ask of it. Suppose that the fundamental question of our survey is whether or not the parasite is present in a given lake. If we find one or more parasitized fish then the answer is clearly 'yes'. But how likely are we to miss out on catching any parasitized fish and hence of concluding, wrongly, that the parasites are not present in the lake? With our sample size of $n = 4$ and $p = 0.1$ we have a probability of missing the parasite of 0.9 for each fish caught and hence a probability of $0.9^4 = 0.6561$ of missing out altogether on finding the parasite. This is obviously unsatisfactory. We need to think again about the sample size. What is the smallest sample, n, that makes the probability of missing the parasite altogether less than 0.05?

We need to solve

$$0.05 = 0.9^n.$$

Taking logs,

$$\log(0.05) = n \, \log(0.9),$$

so

$$n = \frac{\log(0.05)}{\log(0.9)} = 28.433 \, 16$$

which means that to make our journey worthwhile we should keep fishing until we have found more than 28 unparasitized fishes, before we reject the hypothesis that parasitism is present at a rate of 10%. Of course, it would take a much bigger sample to reject a hypothesis of presence at a substantially lower rate.

Random numbers are generated from the binomial distribution like this. The first argument is the number of random numbers we want. The second argument is the sample size ($n = 4$) and the third is the probability of success ($p = 0.1$).

```
rbinom(10,4,0.1)
```

```
[1]  0  0  0  0  0  1  0  1  0  1
```

Here we repeated the sampling of 4 fish ten times. We got 1 parasitized fish out of 4 on three occasions, and 0 parasitized fish on the remaining seven occasions. We never caught 2 or more parasitized fish in any of these samples of 4.

7.4.3 The geometric distribution

Suppose that a series of independent Bernoulli trials with probability p are carried out at times 1, 2, 3, ... Now let W be the waiting time until the first success occurs. So

$$P(W > x) = (1 - p)^x,$$

which means that

$$P(W = x) = P(W > x - 1) - P(W > x).$$

The density function, therefore, is

$$f(x) = p(1 - p)^{x-1}.$$

```
fx <- dgeom(0:20,0.2)
barplot(fx,names=0:20,xlab="outcome",ylab="probability",col="cyan")
```

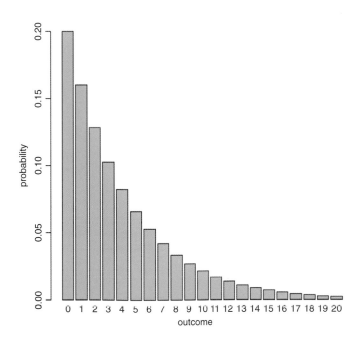

For the geometric distribution,

- the mean is $\frac{1-p}{p}$,

- the variance is $\frac{1-p}{p^2}$.

The geometric has a very long tail. Here are 100 random numbers from a geometric distribution with $p = 0.1$: the modes are 0 and 1, but outlying values as large as 33 and 44 have been generated:

```
table(rgeom(100,0.1))
```

```
 0  1  2  3  4  5  6  7  8  9 10 11 12 13 14 15 17 18 21 22 24 28 29 31 33 44
14 14  8  5  1 13  3  5  3  5  2  5  3  2  3  1  1  2  1  1  2  2  1  1  1  1
```

7.4.4 The hypergeometric distribution

'Balls in urns' are the classic sort of problem solved by this distribution. The density function of the hypergeometric is

$$f(x) = \frac{\binom{b}{x}\binom{N-b}{n-x}}{\binom{N}{n}}.$$

Suppose that there are N coloured balls in the statistician's famous urn: b of them are blue and $r = N - b$ of them are red. Now a sample of n balls is removed from the urn; this is sampling *without replacement*. Now $f(x)$ gives the probability that x of these n balls are blue.

The built-in functions for the hypergeometric are used like this: dhyper(q,b,r,n) and rhyper(m,b,r,n). Here

- q is a vector of values of a random variable representing the number of blue balls out of a sample of size n drawn from an urn containing b blue balls and r red ones.

- b is the number of blue balls in the urn. This could be a vector with non-negative integer elements.

- r is the number of red balls in the urn = $N - b$. This could also be a vector with non-negative integer elements.

- n number of balls drawn from an urn with b blue and r red balls. This can be a vector like b and r.

- p vector of probabilities with values between 0 and 1.

- m the number of hypergeometrically distributed random numbers to be generated.

Let the urn contain $N = 20$ balls, of which 6 are blue and 14 are red. We take a sample of $n = 5$ balls so x could be 0, 1, 2, 3, 4 or 5 of them blue, but since the proportion blue is only 6/20 the higher frequencies are most unlikely. Our example is evaluated like this:

```
ph <- dhyper(0:5,6,14,5)
```

```
barplot(ph,names=(0:5),col="red",xlab="outcome",ylab="probability")
```

We are very unlikely to get more than 3 red balls out of 5. The most likely outcome is that we get 1 or 2 red balls out of 5. We can simulate a set of Monte Carlo trials of size 5. Here are the numbers of red balls obtained in 20 realizations of our example:

```
rhyper(20,6,14,5)
```

```
[1] 1 1 1 2 1 2 0 1 3 2 3 0 2 0 1 1 2 1 1 2
```

The binomial distribution is a limiting case of the hypergeometric which arises as N, b and r approach infinity in such a way that b/N approaches p, and r/N approaches $1 - p$ (see p. 308). This is because as the numbers get large, the fact that we are sampling without replacement becomes irrelevant. The binomial distribution assumes sampling with replacement from a finite population, or sampling without replacement from an infinite population.

7.4.5 The multinomial distribution

Suppose that there are t possible outcomes from an experimental trial, and the outcome i has probability p_i. Now allow n independent trials where $n = n_1 + n_2 + \ldots + n_t$ and ask what is the probability of obtaining the vector of N_i occurrences of the ith outcome:

$$P(N_i = n_i) = \frac{n!}{n_1!n_2!n_3!\ldots n_t!} p_1^{n_1} p_2^{n_2} p_3^{n_3} \cdots p_t^{n_t},$$

where i goes from 1 to t. Take an example with three outcomes, (say black, red and blue, so $t = 3$), where the first outcome is twice as likely as the other two ($p_1 = 0.5, p_2 = 0.25, p_3 = 0.25$), noting that the probabilities sum to 1. It is sensible to start by writing a function called `multi` to carry out the calculations for any numbers of successes a, b and c (black, red and blue, respectively) given our three probabilities (above):

```
multi <- function(a,b,c) {
factorial(a+b+c)/(factorial(a)*factorial(b)*factorial(c))*0.5^a*0.25^b*0.25^c}
```

We illustrate just one case, in which the third outcome (blue) is fixed at four successes out of 24 trials. This means that the first and second outcomes must add to $24 - 4 = 20$. We plot the probability of obtaining different numbers of blacks from 0 to 20:

```
barplot(sapply(0:20,function (i) multi(i,20-i,4)),names=0:20,cex.names=0.7,
                    xlab="outcome",ylab="probability",col="yellow")
```

The most likely outcome for this example is that we would get 13 or 14 successes of type 1 (black) in a trial of size $= 24$ with probabilities 0.5, 0.25 and 0.25 for the three types of outcome, when the number of successes of the third case was 4 out of 24. Note the use of cex.names=0.7 to make the labels sufficiently small that all of the bars are given outcome names.

7.4.6 The Poisson distribution

This is one of the most useful and important of the discrete probability distributions for describing count data. We know how many times something happened (e.g. kicks from cavalry horses, lightening strikes, bomb hits), but we have no way of knowing how many times it did not happen. The Poisson is a one-parameter distribution with the interesting property that its variance is equal to its mean. A great many processes show variance increasing with the mean, often faster than linearly (see the negative binomial distribution below). The density function of the Poisson shows the probability of obtaining a count of x when the mean count per unit is λ:

$$p(x) = \frac{e^{-\lambda}\lambda^x}{x!}.$$

The zero term of the Poisson (the probability of obtaining a count of zero) is obtained by setting $x = 0$:

$$p(0) = e^{-\lambda},$$

which is simply the antilog of minus the mean. Given $p(0)$, it is clear that $p(1)$ is just

$$p(1) = p(0)\lambda = \lambda e^{-\lambda},$$

and any subsequent probability is readily obtained by multiplying the previous probability by the mean and dividing by the count,

$$p(x) = p(x-1)\frac{\lambda}{x}.$$

Functions for the density, cumulative distribution, quantiles and random number generation of the Poisson distribution are obtained by dpois(x, lambda), ppois(q, lambda), qpois(p, lambda) and rpois(n, lambda), where lambda is the mean count per sample.

The Poisson distribution holds a central position in three quite separate areas of statistics:

- in the description of random spatial point patterns (see p. 838);

- as the frequency distribution of counts of rare but independent events (see p. 314);

- as the error distribution in GLMs for count data (see p. 579).

If we wanted 600 simulated counts from a Poisson distribution with a mean of, say, 0.90 blood cells per slide, we just type:

```
count <- rpois(600,0.9)
```

We can use table to see the frequencies of each count generated:

```
table(count)
```

```
count
  0    1    2    3    4    5
244  212  104   33    6    1
```

or hist to see a histogram of the counts:

```
hist(count,breaks = - 0.5:6.5,main="")
```

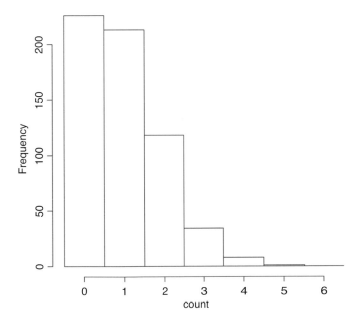

Note the use of the vector of break points on integer increments from −0.5 to create integer bins for the histogram bars.

7.4.7   The negative binomial distribution

This discrete, two-parameter distribution is useful for describing the distribution of count data, where the variance is often much greater than the mean. The two parameters are the mean μ and the clumping parameter k, given by

$$k = \frac{\mu^2}{\sigma^2 - \mu}$$

The smaller the value of k, the greater the degree of clumping. The density function is

$$p(x) = \left(1 + \frac{\mu}{k}\right)^{-k} \frac{(k+x-1)!}{x!(k-1)!} \left(\frac{\mu}{\mu+k}\right)^{x}.$$

The zero term is found by setting $x = 0$ and simplifying:

$$p(0) = \left(1 + \frac{\mu}{k}\right)^{-k}.$$

Successive terms in the distribution can then be computed iteratively from

$$p(x) = p(x-1)\left(\frac{k+x-1}{x}\right)\left(\frac{\mu}{\mu+k}\right).$$

An initial estimate of the value of k can be obtained from the sample mean and variance,

$$k \approx \frac{\bar{x}^2}{s^2 - \bar{x}}.$$

Since k cannot be negative, it is clear that the negative binomial distribution should not be fitted to data where the variance is less than the mean.

The maximum likelihood estimate of k is found numerically, by iterating progressively more fine-tuned values of k until the left- and right-hand sides of the following equation are equal:

$$n \ln\left(1 + \frac{\mu}{k}\right) = \sum_{x=0}^{max}\left(\frac{A(x)}{k+x}\right)$$

where the vector $A(x)$ contains the total frequency of values *greater* than x. You could write a function to work out the probability densities like this:

```
negbin <- function(x,u,k)
```

```
(1+u/k)^(-k)*(u/(u+k))^x*gamma(k+x)/(factorial(x)*gamma(k))
```

then use the function to produce a barplot of probability densities for a range of x values (say 0 to 10), for a distribution with specified mean and aggregation parameter (say $\mu = 0.8$, $k = 0.2$) like this:

```
xf <- sapply(0:10, function(i) negbin(i,0.8,0.2))
```

```
barplot(xf,names=0:10,xlab="count",ylab="probability density",col="green")
```

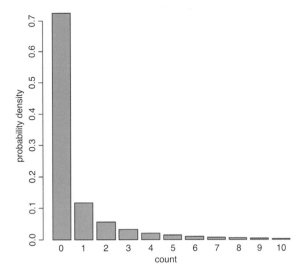

There is another, quite different way of looking at the negative binomial distribution. Here, the response variable is the waiting time W_r for the rth success:

$$f(x) = \binom{x-1}{r-1} p^r (1-p)^{x-r}.$$

It is important to realize that x starts at r and increases from there (obviously, the rth success cannot occur before the rth attempt). The density function `dnbinom(x, size, prob)` represents the number of failures x (e.g. tails in coin tossing) before `size` successes (or heads in coin tossing) are achieved, when the probability of a success (a head) is `prob`.

Suppose we are interested in the distribution of waiting times until the fifth success occurs in a negative binomial process with $p = 0.1$. We start the sequence of x values at 5:

```
plot(5:100,dnbinom(5:100,5,0.1),type="s",xlab="x",ylab="f(x)")
```

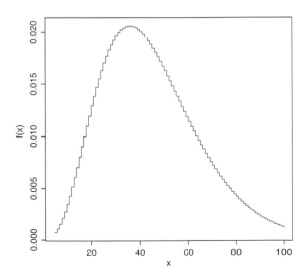

This shows that the most likely waiting time for the 5th success, when the probability of a success is 1/10, is about 31 trials after the fifth trial. Note that the negative binomial distribution is quite strongly skew to the right.

It is easy to generate negative binomial data using the random number generator: `rnbinom(n, size, prob)`. The number of random numbers required is n. When the second parameter, `size`, is set to 1 the distribution becomes the geometric (see above). The final parameter, `prob`, is the probability of success per trial, p. Here we generate 100 counts with a mean of 0.6:

```
count <- rnbinom(100,1,0.6)
```

We can use `table` to see the frequency of the different counts:

```
table(count)
```

```
0    1    2   3   5   6
65   18   13  2   1   1
```

It is sensible to check that the mean really is 0.6 (or very close to it):

```
mean(count)
```

```
[1] 0.61
```

The variance will be substantially greater than the mean:

```
var(count)
```

```
[1] 1.129192
```

This gives an estimate of k of

$$\frac{0.61^2}{1.129 - 0.61} = 0.717.$$

The following data show the number of spores counted on 238 buried glass slides. We are interested in whether these data are well described by a negative binomial distribution. If they are we would like to find the maximum likelihood estimate of the aggregation parameter k.

```
x <- 0:12
freq <- c(131,55,21,14,6,6,2,0,0,0,0,2,1)
barplot(freq,names=x,ylab="frequency",xlab="spores",col="purple")
```

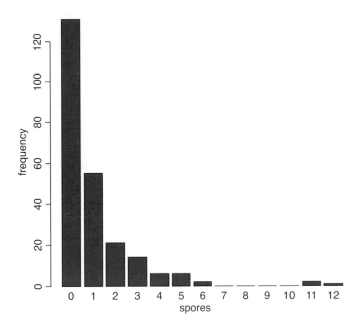

We start by looking at the variance–mean ratio of the counts. We cannot use mean and variance directly, because our data are *frequencies* of counts, rather than counts themselves. This is easy to rectify: we use rep to create a vector of counts y in which each count (x) is repeated the relevant number of times (freq). Now we can use mean and var directly:

```
y <- rep(x,freq)
mean(y)
```

```
[1]  1.004202
```

```
var(y)
```

```
[1]  3.075932
```

This shows that the data are highly aggregated (the variance–mean ratio is roughly 3, recalling that it would be 1 if the data were Poisson distributed). Our rough estimate of k is therefore

```
mean(y)^2/(var(y)-mean(y))
[1]  0.4867531
```

Here is a function that takes a vector of frequencies of counts x (between 0 and length(x) $-$ 1) and computes the maximum likelihood estimate of k, the aggregation parameter:

```
kfit  <- function(x)  {
     lhs <- numeric()
     rhs <- numeric()
     y  <- 0:(length(x) - 1)
     j <- 0:(length(x)-2)
     m  <- sum(x * y)/(sum(x))
     s2  <- (sum(x * y^2) - sum(x * y)^2/sum(x))/(sum(x)- 1)
     k1  <- m^2/(s2 - m)
     a <- numeric(length(x)-1)
```

```
        for(i in 1:(length(x) - 1)) a[i]   <- sum(x [- c(1:i)])
        i <- 0
        for (k in seq(k1/1.2,2*k1,0.001)) {
        i <- i+1
        lhs[i]   <- sum(x) * log(1 + m/k)
        rhs[i]   <- sum(a/(k + j))
          }
        k <- seq(k1/1.2,2*k1,0.001)
        plot(k, abs(lhs-rhs),xlab="k",ylab="Difference",type="l",col="red")
        d <- min(abs(lhs-rhs))
        sdd <- which(abs(lhs-rhs)==d)
        k[sdd]
}
```

We can try it out with our spore count data.

```
kfit(freq)
```

```
[1]  0.5826276
```

The minimum difference is close to zero and occurs at about $k = 0.58$. The printout shows that the maximum likelihood estimate of k is 0.582 (to the 3 decimal places we simulated; the last 4 decimals (6276) are meaningless and would not be printed in a more polished function).

How would a negative binomial distribution with a mean of 1.0042 and a k value of 0.582 describe our count data? The expected frequencies are obtained by multiplying the probability density (above) by the total sample size (238 slides in this case).

```
nb <- 238*(1+1.0042/0.582)^(-0.582)*factorial(.582+(0:12)-1)/
(factorial(0:12)*factorial(0.582-1))*(1.0042/(1.0042+0.582))^(0:12)
```

We shall compare the observed and expected frequencies using barplot. We intend to alternate the observed and expected frequencies. There are three steps to the procedure:

- Concatenate the observed and expected frequencies in an alternating sequence.

- Create list of labels to name the bars (alternating blanks and counts).

- Produce a legend to describe the different bar colours.

The concatenated list of frequencies (called both) is made like this, putting the 13 observed counts (freq) in the odd-numbered bars and the 13 expected counts (nb) in the even-numbered bars (note the use of modulo %% to do this):

```
both <- numeric(26)

both[1:26 %% 2 != 0] <- freq

both[1:26 %% 2 == 0] <- nb
```

Because adjacent blue and green bars refer to the same count (the observed and expected frequencies) we do not want to use barplot's built-in names argument for labelling the bars (it would want to write a label on every bar, 26 labels in all). Instead, we want to write the count just once for each pair of bars, located beneath the observed and (green) bars, using as.character(0:12). The trick is to produce a vector of length 26 containing the repeated bar labels, then replace the even-numbered entries with blanks like this (using modulo to pick out the even numbers):

```
labs <- as.character(rep(0:12,each=2))

labs[1:26%%2==0] <- ""
```

Now we can draw the combined barplot specifying cex.names=0.8 to ensure that all the bar labels are small enough to be printed:

```
barplot(both,col=rep(c(3,4),13),ylab="frequency",names=labs,cex.names=0.8)
```

The legend function creates a legend to show which bars represent the observed frequencies (black in this case) and which represent the expected, negative binomial frequencies (open bars). Just click when the cursor is in the position where you want the *top left-hand corner* of the legend box to be:

```
legend(locator(1),c("observed","expected"),fill=c(3,4))
```

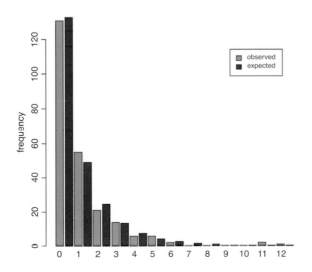

The fit is very close, so we can be reasonably confident in describing the observed counts as negative binomially distributed. The tail of the observed distribution is rather fatter than the expected negative binomial tail, so we might want to measure the lack of fit between observed and expected distributions. A simple way to do this is to use Pearson's chi-squared, taking care to use only those cases where the expected frequency nb is greater than 5:

```
sum(((freq-nb)^2/nb)[nb > 5])
```

```
[1] 1.634975
```

This is based on five legitimate comparisons,

```
sum(nb>5)
```

```
[1] 5
```

and hence on $5 - p - 1 = 2$ d.f. because we have estimated $p = 2$ parameters from the data in estimating the expected distribution (the mean and k of the negative binomial) and lost one degree of freedom for contingency (the total number of counts must add up to 238). Our calculated value of chi-squared $= 1.63$ is much less than the value in tables:

```
qchisq(0.95,2)
```

```
[1] 5.991465
```

so we accept the hypothesis that our data are not significantly different from a negative binomial with mean $= 1.0042$ and $k = 0.582$.

7.4.8   The Wilcoxon rank-sum statistic

This function calculates the distribution of the Wilcoxon rank-sum statistic (also known as Mann–Whitney), and returns values for the exact probability at discrete values of q: dwilcox(q, m, n). Here q is a vector of quantiles, m is the number of observations in sample x (a positive integer not greater than 50), and n is the number of observations in sample y (also a positive integer not greater than 50). The Wilcoxon rank-sum statistic is the sum of the ranks of x in the combined sample c(x,y). The Wilcoxon rank-sum statistic takes on values W between the limits

$$\frac{m(m + 1)}{2} \leq W \leq \frac{m(m + 2n + 1)}{2}.$$

This statistic can be used for a non-parametric test of location shift between the parent populations x and y.

7.5   Matrix algebra

There is a comprehensive set of functions for handling matrices in R. We begin with a matrix called a that has three rows and two columns. Data are typically entered into matrices columnwise, so the first three numbers (1, 0, 4) go in column 1 and the second three numbers (2, –1, 1) go in column 2:

```
a <- matrix(c(1,0,4,2,-1,1),nrow=3)
a
```

```
     [,1] [,2]
[1,]    1    2
```

```
[2,]      0    -1
[3,]      4     1
```

Our second matrix, called *b*, has the same number of columns as *A* has rows (i.e. three in this case). Entered columnwise, the first two numbers (1, –1) go in column 1, the second two numbers (2, 1) go in column 2, and the last two numbers (1, 0) go in column 3:

```
b <- matrix(c(1,-1,2,1,1,0),nrow=2)
b
```

```
        [,1]  [,2]  [,3]
[1,]      1     2     1
[2,]     -1     1     0
```

7.5.1 Matrix multiplication

To multiply one matrix by another matrix you take the rows of the first matrix and the columns of the second matrix. Put the first row of *a* side by side with the first column of *b*:

```
a[1,]
```

```
[1]  1 2
```

```
b[,1]
```

```
[1]  1 -1
```

and work out the point products:

```
a[1,]*b[,1]
```

```
[1]  1 -2
```

then add up the point products

```
sum(a[1,]*b[,1])
```

```
[1]  -1
```

The sum of the point products is –1 and this is the first element of the product matrix. Next, put the first row of *a* with the second column of *b*:

```
a[1,]
```

```
[1]  1 2
```

```
b[,2]
```

```
[1]  2 1
```

```
a[1,]*b[,2]
```

```
[1]  2 2
```

```
sum(a[1,]*b[,2])
```

```
[1]  4
```

so the point products are 2, 2 and the sum of the point products is $2 + 2 = 4$. So 4 goes in row 1 and column 2 of the answer. Then take the last column of b and match it against the first row of a:

```
a[1,]*b[,3]
```

```
[1] 1  0
```

```
sum(a[1,]*b[,3])
```

```
[1] 1
```

so the sum of the point products is $1 + 0 = 1$. This goes in row 1, column 3 of the answer. And so on. We repeat these steps for row 2 of matrix a $(0, -1)$ and then again for row 3 of matrix a $(4, 1)$ to obtain the complete matrix of the answer. In R, the symbol for matrix multiplication is %*%. Here is the full answer:

```
a %*% b
```

```
     [,1] [,2] [,3]
[1,]   -1    4    1
[2,]    1   -1    0
[3,]    3    9    4
```

where you see the values we calculated by hand $(-1, 4, 1)$ in the first row.

It is important to understand that with matrices a times b is not the same as b times a. The matrix resulting from a multiplication has the number of rows of the matrix on the left (a has 3 rows in the case above). But b has just two rows, so multiplication

```
b %*% a
```

```
     [,1] [,2]
[1,]    5    1
[2,]   -1   -3
```

produces a matrix with 2 rows. The value 5 in row 1 column 1 of the answer is the sum of the point products $(1 \times 1) + (2 \times 0) + (1 \times 4) = 1 + 0 + 4 = 5$.

7.5.2 Diagonals of matrices

To create a diagonal matrix of 3 rows and 3 columns, with 1s on the diagonal use the diag function like this:

```
(ym <- diag(1,3,3))
```

```
     [,1] [,2] [,3]
[1,]    1    0    0
[2,]    0    1    0
[3,]    0    0    1
```

You can alter the values of the diagonal elements of a matrix like this:

```
diag(ym) <- 1:3
ym
```

```
     [,1] [,2] [,3]
[1,]    1    0    0
[2,]    0    2    0
[3,]    0    0    3
```

or extract a vector containing the diagonal elements of a matrix like this:

```
diag(ym)
```

```
[1] 1 2 3
```

You might want to extract the diagonal of a variance–covariance matrix:

```
M   <- cbind(X=1:5, Y=rnorm(5))
var(M)
```

```
          X          Y
X   2.50000000 0.04346324
Y   0.04346324 0.88056034
```

```
diag(var(M))
```

```
       X          Y
2.5000000  0.8805603
```

7.5.3 Determinant

The determinant of the square (2×2) array

$$\begin{bmatrix} a & b \\ c & d \end{bmatrix}$$

is defined for any numbers a, b, c and d as

$$\begin{vmatrix} a & b \\ c & d \end{vmatrix} \equiv ad - bc.$$

Suppose that A is a square matrix of order (3×3):

$$A = \begin{bmatrix} a_{11} & a_{12} & a_{13} \\ a_{21} & a_{22} & a_{23} \\ a_{31} & a_{32} & a_{23} \end{bmatrix}.$$

Then the third-order determinant of A is defined to be the number

$$\det A = a_{11} \begin{vmatrix} a_{22} & a_{23} \\ a_{32} & a_{33} \end{vmatrix} - a_{12} \begin{vmatrix} a_{21} & a_{23} \\ a_{32} & a_{33} \end{vmatrix} + a_{13} \begin{vmatrix} a_{21} & a_{22} \\ a_{31} & a_{32} \end{vmatrix}.$$

Applying the rule $\begin{vmatrix} a & b \\ c & d \end{vmatrix} \equiv ad - bc$ to this equation gives

$$\det A = a_{11}a_{22}a_{33} - a_{11}a_{23}a_{32} + a_{12}a_{23}a_{31} - a_{12}a_{21}a_{33} + a_{13}a_{21}a_{32} - a_{13}a_{22}a_{31}.$$

Take a numerical example:

$$A = \begin{bmatrix} 1 & 2 & 3 \\ 2 & 1 & 1 \\ 4 & 1 & 2 \end{bmatrix}.$$

This has determinant

$$\det A = (1 \times 1 \times 2) - (1 \times 1 \times 1) + (2 \times 1 \times 4) - (2 \times 2 \times 2) + (3 \times 2 \times 1) - (3 \times 1 \times 4)$$
$$= 2 - 1 + 8 - 8 + 6 - 12 = -5.$$

Here is the example in R using the determinant function `det`:

```
A <- matrix(c(1,2,4,2,1,1,3,1,2),nrow=3)
A
```

```
     [,1] [,2] [,3]
[1,]    1    2    3
[2,]    2    1    1
[3,]    4    1    2
```

```
det(A)
```

```
[1] -5
```

The great thing about determinants is that if any row or column of a determinant is multiplied by a scalar λ, then the value of the determinant is multiplied by λ (since a factor λ will appear in each of the products). Also, if all the elements of a row or a column are zero then the determinant $|A| = 0$. Again, if all the corresponding elements of two rows or columns of $|A|$ are equal then $|A| = 0$.

For instance, here is the bottom row of A multiplied by 3:

```
B <- A
B[3,] <- 3*B[3,]
B
```

```
     [,1] [,2] [,3]
[1,]    1    2    3
[2,]    2    1    1
[3,]   12    3    6
```

and here is the determinant:

```
det(B)
```

```
[1] -15
```

Here is an example when all the elements of column 2 are zero, so det $C = 0$:

```
C <- A
C[,2] <- 0
C
```

```
     [,1] [,2] [,3]
[1,]    1    0    3
```

```
[2,]      2      0      1
[3,]      4      0      2

det(C)

[1]  0
```

If det $A \neq 0$ then the rows and columns of A must be linearly independent. This important concept is expanded in terms of contrast coefficients on p. 434.

7.5.4 Inverse of a matrix

The operation of division is not defined for matrices. However, for a square matrix that has $|A| \neq 0$ a multiplicative inverse matrix denoted by A^{-1} can be defined. This multiplicative inverse A^{-1} is unique and has the property that

$$A^{-1}A = AA^{-1} = I,$$

where I is the unit matrix. So if A is a square matrix for which $|A| \neq 0$ the matrix inverse is defined by the relationship

$$A^{-1} = \frac{\text{adj } A}{|A|},$$

where the adjoint matrix of A (adj A) is the matrix of cofactors of A. The cofactors of A are computed as $A_{ij} = (-1)^{i+j} M_{ij}$, where M_{ij} are the 'minors' of the elements a_{ij} (these are the determinants of the matrices of A from which row i and column j have been deleted). The properties of the inverse matrix can be laid out for two non-singular square matrices, A and B, of the same order as follows:

$$AA^{-1} = A^{-1}A = I,$$
$$(AB)^{-1} = B^{-1}A^{-1},$$
$$(A^{-1})' = (A')^{-1},$$
$$(A^{-1})^{-1} = A,$$
$$|A| = \frac{1}{|A^{-1}|}.$$

Here is R's version of the inverse of the 3 × 3 matrix A (above) using the `ginv` function from the MASS library:

```
library(MASS)
ginv(A)

                 [,1]   [,2]   [,3]
[1,]  -2.000000e-01    0.2    0.2
[2,]  -2.224918e-16    2.0   -1.0
[3,]   4.000000e-01   -1.4    0.6
```

where the number in row 2 column 1 is zero (except for rounding error). Here is the penultimate rule, $(A^{-1})^{-1} = A$, evaluated by R:

```
ginv(ginv(A))
```

```
     [,1] [,2] [,3]
[1,]    1    2    3
[2,]    2    1    1
[3,]    4    1    2
```

Here is the last rule, $|A| = 1/|A^{-1}|$:

```
1/det(ginv(A))
```

```
[1] -5
```

7.5.5 Eigenvalues and eigenvectors

We have a square matrix A and two column vectors X and K, where

$$AX = K,$$

and we want to discover the scalar multiplier λ such that

$$AX = \lambda X.$$

This is equivalent to $(A - \lambda I)X = 0$, where I is the unit matrix. This can only have one non-trivial solution when the determinant associated with the coefficient matrix A vanishes, so we must have

$$|A - \lambda I| = 0.$$

When expanded, this determinant gives rise to an algebraic equation of degree n in λ called the **characteristic equation**. It has n roots $\lambda_1, \lambda_2, \ldots, \lambda_n$, each of which is called an **eigenvalue**. The corresponding solution vector X_i is called an **eigenvector** of A corresponding to λ_i.

Here is an example from population ecology. The matrix A shows the demography of different age classes: the top row shows fecundity (the number of females born per female of each age) and the sub-diagonals show survival rates (the fraction of one age class that survives to the next age class). When these numbers are constants the matrix is known as the **Leslie matrix**. In the absence of density dependence the constant parameter values in A will lead either to exponential increase in total population size (if $\lambda_1 > 1$) or exponential decline (if $\lambda_1 < 1$) once the initial transients in age structure have damped away. Once exponential growth has been achieved, then the age structure, as reflected by the proportion of individuals in each age class, will be a constant. This is known as the first eigenvector.

Consider the Leslie matrix, L, which is to be multiplied by a column matrix of age-structured population sizes, n:

```
L <- c(0,0.7,0,0,6,0,0.5,0,3,0,0,0.3,1,0,0,0)
L <- matrix(L,nrow=4)
```

Note that the elements of the matrix are entered in columnwise, not row-wise sequence. We make sure that the Leslie matrix is properly conformed:

```
L
```

```
     [,1] [,2] [,3] [,4]
[1,]  0.0  6.0  3.0    1
```

```
[2,]   0.7   0.0   0.0    0
[3,]   0.0   0.5   0.0    0
[4,]   0.0   0.0   0.3    0
```

The top row contains the age-specific fecundities (e.g. 2-year-olds produce six female offspring per year), and the sub-diagonal contains the survivorships (70% of 1-year-olds become 2-year-olds, etc.). Now the population sizes at each age go in a column vector, *n*:

```
n <- c(45,20,17,3)
n <- matrix(n,ncol=1)
n
```

```
       [,1]
[1,]    45
[2,]    20
[3,]    17
[4,]     3
```

Population sizes next year in each of the four age classes are obtained by matrix multiplication:

```
L  %*%  n
```

```
        [,1]
[1,]  174.0
[2,]   31.5
[3,]   10.0
[4,]    5.1
```

We can check this the long way. The number of juveniles next year (the first element of *n*) is the sum of all the babies born last year:

```
45*0+20*6+17*3+3*1
```

```
[1]  174
```

We write a function to carry out the matrix multiplication, giving next year's population vector as a function of this year's:

```
fun <- function(x) L %*% x
```

Now we can simulate the population dynamics over a period long enough (say, 40 generations) for the age structure to approach stability. So long as the population growth rate $\lambda > 1$ the population will increase exponentially, once the age structure has stabilized:

```
n <- c(45,20,17,3)
n <- matrix(n,ncol=1)
structure <- numeric(160)
dim(structure) <- c(40,4)

for (i in 1:40) {
n <- fun(n)
structure[i,] <- n
}
matplot(1:40,log(structure),type="l")
```

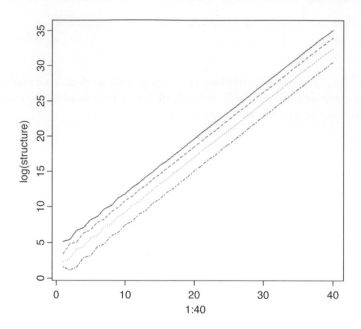

You can see that after some initial transient fluctuations, the age structure has more or less stabilized by year 20 (the lines for log population size of juveniles (top line), 1-, 2- and 3-year-olds are parallel). By year 40 the population is growing exponentially in size, multiplying by a constant of λ each year.

The population growth rate (the per-year multiplication rate, λ) is approximated by the ratio of total population sizes in the 40th and 39th years:

```
sum(structure[40,])/sum(structure[39,])
```

```
[1] 2.164035
```

and the approximate stable age structure is obtained from the 40th value of n:

```
structure[40,]/sum(structure[40,])
```

```
[1] 0.709769309 0.230139847 0.052750539 0.007340305
```

The exact values of the population growth rate and the stable age distribution are obtained by matrix algebra: they are the dominant eigenvalue and dominant eigenvector, respectively. Use the function `eigen` applied to the Leslie matrix, L, like this:

```
eigen(L)
```

```
$values
[1]   2.1694041+0.0000000i  -1.9186627+0.0000000i  -0.1253707+0.0975105i
[4]  -0.1253707-0.0975105i

$vectors
                [,1]             [,2]                    [,3]                    [,4]
[1,]  0.949264118+0i  -0.93561508+0i  -0.01336028-0.03054433i  -0.01336028+0.03054433i
[2,]  0.306298338+0i   0.34134741+0i  -0.03616819+0.14241169i  -0.03616819-0.14241169i
[3,]  0.070595039+0i  -0.08895451+0i   0.36511901-0.28398118i   0.36511901+0.28398118i
[4,]  0.009762363+0i   0.01390883+0i  -0.87369452+0.00000000i  -0.87369452+0.00000000i
```

The dominant eigenvalue is 2.1694 (compared with our empirical approximation of 2.1640 after 40 years). The stable age distribution is given by the first eigenvector (column 1, above), which we need to turn into proportions:

```
eigen(L)$vectors[,1]/sum(eigen(L)$vectors[,1])
```

```
[1]  0.710569659+0i  0.229278977+0i  0.052843768+0i  0.007307597+0i
```

This compares with our approximation (above) in which the proportion in the first age class was 0.709 77 after 40 years (rather than 0.710 57).

7.5.6   Matrices in statistical models

Perhaps the main use of matrices in R is in statistical calculations, in generalizing the calculation of sums of squares and sums of products (see p. 450 for background). Here are the data used in Chapter 10 to introduce the calculation of sums of squares in linear regression:

```
numbers <- read.table("c:\\temp\\tannin.txt",header=T)
attach(numbers)
names(numbers)
```

```
[1]  "growth"  "tannin"
```

The response variable is growth (y) and the explanatory variable is tannin concentration (x) in the diet of a group of insect larvae. We need the famous five (see p. 453): the sum of the y values,

```
growth
```

```
[1]  12 10 8 11 6 7 2 3 3
```

```
sum(growth)
```

```
[1]  62
```

the sum of the squares of the y values,

```
growth^2
```

```
[1]  144 100 64 121 36 49 4 9 9
```

```
sum(growth^2)
```

```
[1]  536
```

the sum of the x values,

```
tannin
```

```
[1]  0 1 2 3 4 5 6 7 8
```

```
sum(tannin)
```

```
[1]  36
```

the sum of the squares of the x values,

```
tannin^2
```

```
[1]  0 1 4 9 16 25 36 49 64
```

```
sum(tannin^2)
```

```
[1]  204
```

and finally, to measure the covariation between *x* and *y*, we need the sum of the products,

```
growth*tannin
```

```
[1]  0 10 16 33 24 35 12 21 24
```

```
sum(growth*tannin)
```

```
[1]  175
```

You can see at once that for more complicated models (such as multiple regression) it is essential to be able to generalize and streamline this procedure. This is where matrices come in. Matrix multiplication involves the calculation of sums of products where a row vector is multiplied by a column vector of the same length to obtain a single value. Thus, we should be able to obtain the required sum of products, 175, by using matrix multiplication symbol %*% in place of the regular multiplication symbol:

```
growth %*% tannin
```

```
        [,1]
[1,]   175
```

That works fine. But what about sums of squares? Surely if we use matrix multiplication on the same vector we will get an object with many rows (nine in this case). Not so.

```
growth %*% growth
```

```
        [,1]
[1,]   536
```

R has coerced the left-hand vector of growth into a row vector in order to obtain the desired result. You can override this, if for some reason you wanted the answer to have nine rows, by specifying the transpose t() of the right-hand growth vector,

```
growth %*% t(growth)
```

```
        [,1] [,2] [,3] [,4] [,5] [,6] [,7] [,8] [,9]
[1,]   144  120   96  132   72   84   24   36   36
[2,]   120  100   80  110   60   70   20   30   30
[3,]    96   80   64   88   48   56   16   24   24
[4,]   132  110   88  121   66   77   22   33   33
[5,]    72   60   48   66   36   42   12   18   18
[6,]    84   70   56   77   42   49   14   21   21
[7,]    24   20   16   22   12   14    4    6    6
[8,]    36   30   24   33   18   21    6    9    9
[9,]    36   30   24   33   18   21    6    9    9
```

but, of course, that is not what we want. R's default is what we need. So this should also work in obtaining the sum of squares of the explanatory variable:

```
tannin %*% tannin
```

```
        [,1]
[1,]   204
```

So far, so good. But how do we obtain the sums using matrix multiplication? The trick here is to matrix multiply the vector by a vector of 1s: here are the sum of the *y* values:

```
growth %*% rep(1,9)
        [,1]
[1,]    62
```

and the sum of the *x* values,

```
tannin %*% rep(1,9)
        [,1]
[1,]    36
```

Finally, can we use matrix multiplication to arrive at the sample size, *n*? We do this by matrix multiplying a row vector of 1s by a column vector of 1s. This rather curious operation produces the right result, by adding up the nine 1s that result from the nine repeats of the calculation 1 × 1:

```
rep(1,9 )%*% rep(1,9)
        [,1]
[1,]     9
```

But how do we get all of the famous five in a single matrix? The thing to understand is the dimensionality of such a matrix. It needs to contain sums as well as sums of products. We have two variables (growth and tannin) and their matrix multiplication produces a single scalar value (see above). In order to get to the sums of squares as well as the sums of products we use cbind to create a 9 × 2 matrix like this:

```
a <- cbind(growth,tannin)
a

     growth tannin
[1,]     12      0
[2,]     10      1
[3,]      8      2
[4,]     11      3
[5,]      6      4
[6,]      7      5
[7,]      2      6
[8,]      3      7
[9,]      3      8
```

To obtain a results table with 2 rows rather than 9 rows we need to multiply the transpose of matrix a by matrix a:

```
t(a) %*% a
        growth tannin
growth     536    175
tannin     175    204
```

That's OK as far as it goes, but it has only given us the sums of squares (536 and 204) and the sum of products (175). How do we get the sums as well? The trick is to bind a column of 1s onto the left of matrix *a*:

```
b <- cbind(1,growth,tannin)
b
```

```
        growth tannin
[1,] 1      12      0
[2,] 1      10      1
[3,] 1       8      2
[4,] 1      11      3
[5,] 1       6      4
[6,] 1       7      5
[7,] 1       2      6
[8,] 1       3      7
[9,] 1       3      8
```

It would look better if the first column had a variable name: let's call it `sample`:

```
dimnames(b)[[2]] [1] <- "sample"
```

Now to get a summary table of sums as well as sums of products, we matrix multiply b by itself. We want the answer to have three rows (rather than nine) so we matrix multiply the transpose of b (which has three rows) by *b* (which has nine rows):

```
t(b) %*% b
```

```
        sample growth tannin
sample       9     62     36
growth      62    536    175
tannin      36    175    204
```

So there you have it. All of the famous five, plus the sample size, in a single matrix multiplication.

7.5.7 Statistical models in matrix notation

We continue this example to show how matrix algebra is used to generalize the procedures used in linear modelling (such as regression or analysis of variance) based on the values of the famous five. We want to be able to determine the parameter estimates (such as the intercept and slope of a linear regression) and to apportion the total sum of squares between variation explained by the model (SSR) and unexplained variation (SSE). Expressed in matrix terms, the linear regression model is

$$Y = Xb + e,$$

and we want to determine the least-squares estimate of *b*, given by

$$b = (X'X)^{-1}X'Y,$$

and then carry out the analysis of variance

$$b'X'Y'.$$

We look at each of these in turn.

The response variable Y, 1 and the errors e are simple $n \times 1$ column vectors, X is an $n \times 2$ matrix and β is a 2×1 vector of coefficients, as follows:

$$Y = \begin{bmatrix} 12 \\ 10 \\ 8 \\ 11 \\ 6 \\ 7 \\ 2 \\ 3 \\ 3 \end{bmatrix}, \quad X = \begin{bmatrix} 1 & 0 \\ 1 & 1 \\ 1 & 2 \\ 1 & 3 \\ 1 & 4 \\ 1 & 5 \\ 1 & 6 \\ 1 & 7 \\ 1 & 8 \end{bmatrix}, \quad e = \begin{bmatrix} e_1 \\ e_2 \\ e_3 \\ e_4 \\ e_5 \\ e_6 \\ e_7 \\ e_8 \\ e_9 \end{bmatrix}, \quad 1 = \begin{bmatrix} 1 \\ 1 \\ 1 \\ 1 \\ 1 \\ 1 \\ 1 \\ 1 \\ 1 \end{bmatrix}, \quad \beta = \begin{bmatrix} \beta_0 \\ \beta_1 \end{bmatrix}.$$

The y vector and the 1 vector are created like this:

```
Y <- growth
```

```
one <- rep(1,9)
```

The sample size is given by $1'1$ (transpose of vector 1 times vector 1):

```
t(one) %*% one
```

```
        [,1]
[1,]      9
```

The vector of explanatory variable(s) X is created by binding a column of ones to the left:

```
X <- cbind(1,tannin)
X
```

```
        tannin
[1,] 1       0
[2,] 1       1
[3,] 1       2
[4,] 1       3
[5,] 1       4
[6,] 1       5
[7,] 1       6
[8,] 1       7
[9,] 1       8
```

In this notation

$$\sum y^2 = y_1^2 + y_2^2 + \ldots + y_n^2 = Y'Y,$$

```
t(Y) %*% Y
```

```
        [,1]
[1,]    536
```

$$\sum y = n\bar{y} = y_1 + y_2 + \ldots + y_n = 1'Y,$$

```
t(one) %*% Y

       [,1]
[1,]    62
```

$$\left(\sum y\right)^2 = Y'11'Y.$$

```
t(Y) %*% one %*% t(one) %*% Y

       [,1]
[1,] 3844
```

For the matrix of explanatory variables, we see that $X'X$ gives a 2×2 matrix containing n, $\sum x$ and $\sum x^2$. The numerical values are easy to find using matrix multiplication:

```
t(X) %*% X

            tannin
         9      36
tannin  36     204
```

Note that $X'X$ (a 2×2 matrix) is completely different from XX' (a 9×9 matrix). The matrix $X'Y$ gives a 2×1 matrix containing $\sum y$ and the sum of products $\sum xy$:

```
t(X) %*% Y

          [,1]
           62
tannin    175
```

Now, using the beautiful symmetry of the normal equations,

$$b_0 n + b_1 \sum x = \sum y,$$

$$b_0 \sum x + b_1 \sum x^2 = \sum xy,$$

we can write the regression directly in matrix form as

$$X'Xb = X'Y$$

because we already have the necessary matrices to form the left- and right-hand sides. To find the least-squares parameter values b we need to divide both sides by $X'X$. This involves calculating the inverse of the $X'X$ matrix. The inverse exists only when the matrix is square and when its determinant is non-singular. The inverse contains $-\bar{x}$ and $\sum x^2$ as its terms, with $SSX = \sum (x - \bar{x})^2$, the sum of the squared differences between the x values and mean x, or $n.SSX$ as the denominator:

$$(X'X)^{-1} = \begin{bmatrix} \dfrac{\sum x^2}{n\sum(x-\bar{x})^2} & \dfrac{-\bar{x}}{\sum(x-\bar{x})^2} \\[2ex] \dfrac{-\bar{x}}{\sum(x-\bar{x})^2} & \dfrac{1}{\sum(x-\bar{x})^2} \end{bmatrix}.$$

When every element of a matrix has a common factor, it can be taken outside the matrix. Here, the term $1/(n.SSX)$ can be taken outside to give

$$(X'X)^{-1} = \dfrac{1}{n\sum(x-\bar{x})^2} \begin{bmatrix} \sum x^2 & -\sum x \\[2ex] -\sum x & n \end{bmatrix}.$$

Computing the numerical value of this is easy using the matrix function `ginv`:

```
library(MASS)
ginv(t(X) %*% X)
```

```
             [,1]           [,2]
[1,]   0.37777778  -0.06666667
[2,]  -0.06666667   0.01666667
```

Now we can solve the normal equations

$$(X'X)^{-1}(X'X)b = (X'X)^{-1}X'Y,$$

using the fact that $(X'X)^{-1}(X'X) = I$ to obtain the important general result:

$$b = (X'X)^{-1}X'Y.$$

```
ginv(t(X) %*% X) %*% t(X) %*% Y
```

```
             [,1]
[1,]  11.755556
[2,]  -1.216667
```

which you will recognize from our hand calculations as the intercept and slope respectively (see p. 455). The ANOVA computations are as follows. The correction factor is

$$CF = Y'11'Y/n.$$

```
CF <- t(Y) %*% one %*% t(one) %*% Y/9
CF
```

```
           [,1]
[1,]  427.1111
```

The total sum of squares, *SSY*, is $Y'Y - CF$:

```
t(Y) %*% Y - CF
```

```
          [,1]
[1,] 108.8889
```

The regression sum of squares, *SSR*, is $b'X'Y - CF$:

```
b <- ginv(t(X) %*% X) %*% t(X) %*% Y
t(b) %*% t(X) %*% Y - CF
```

```
         [,1]
[1,] 88.81667
```

The error sum of squares, *SSE*, is $Y'Y - b'X'Y$:

```
t(Y) %*% Y - t(b) %*% t(X) %*% Y
```

```
         [,1]
[1,] 20.07222
```

You should check these figures against the hand calculations on p. 457. Obviously, this is not a sensible way to carry out a single linear regression, but it demonstrates how to generalize the calculations for cases that have two or more continuous explanatory variables.

7.6   Solving systems of linear equations using matrices

Suppose we have two equations containing two unknown variables:

$$3x + 4y = 12,$$
$$x + 2y = 8.$$

We can use the function `solve` to find the values of the variables if we provide it with two matrices:

- a square matrix *A* containing the *coefficients* (3, 1, 4 and 2, columnwise);
- a column vector *kv* containing the *known values* (12 and 8).

We set the two matrices up like this (columnwise, as usual):

```
A <- matrix(c(3,1,4,2),nrow=2)
A
```

```
     [,1] [,2]
[1,]    3    4
[2,]    1    2
```

```
kv <- matrix(c(12,8),nrow=2)
kv
```

```
     [,1]
[1,]   12
[2,]    8
```

Now we can solve the simultaneous equations, using the `solve` function like this:

```
solve(A,kv)

       [,1]
[1,]    -4
[2,]     6
```

to give $x = -4$ and $y = 6$ (which you can easily verify by hand). The function is most useful when there are many simultaneous equations to be solved.

7.7 Calculus

The rules of differentiation and integration are known to R. You will use them in modelling (e.g. in calculating starting values in non-linear regression) and for numeric minimization using `optim`. Read the help files `?D` and `?integrate` to understand the limitations of these functions.

7.7.1 Derivatives

The R function for symbolic and algorithmic derivatives of simple expressions is D. Here are some simple examples to give you the idea. See also `?deriv`.

```
D(expression(2*x^3),"x")

2 * (3 * x^2)

D(expression(log(x)),"x")

1/x

D(expression(a*exp(-b * x)),"x")

-(a * (exp(-b * x) * b))

D(expression(a/(1+b*exp(-c * x))),"x")

a * (b * (exp(-c * x) * c))/(1 + b * exp(-c * x))^2

trig.exp <- expression(sin(cos(x + y^2)))
D(trig.exp, "x")

-(cos(cos(x + y^2)) * sin(x + y^2))
```

7.7.2 Integrals

The R function is `integrate`. Here are some simple examples to give you the idea:

```
integrate(dnorm,0,Inf)

0.5 with absolute error < 4.7e-05

integrate(dnorm,-Inf,Inf)

1 with absolute error < 9.4e-05

integrate(function(x) rep(2, length(x)), 0, 1)
```

```
2 with absolute error < 2.2e-14

integrand   <- function(x)  {1/((x+1)*sqrt(x))}
integrate(integrand, lower = 0, upper = Inf)

3.141593 with absolute error < 2.7e-05

xv <- seq(0,10,0.1)
plot(xv,integrand(xv),type="l")
```

The area under the curve is $\pi = 3.141\ 593$.

7.7.3   Differential equations

We need to solve a system of ordinary differential equations (ODEs) using classical Runge–Kutta integration from the deSolve package (Soetaert et al., 2012):

```
install.packages("deSolve")
```

```
library(deSolve)
```

The example involves a simple resource-limited plant–herbivore interaction where $V =$ vegetation and $N =$ herbivore population. We need to specify two differential equations: one for the vegetation (dV/dt) and one for the herbivore population (dN/dt):

$$\frac{dV}{dt} = rV\left(\frac{K-V}{K}\right) - bVN,$$

$$\frac{dN}{dt} = cVN - dN.$$

The steps involved in solving these ODEs in R are as follows:

- Define a function (called `phmodel` in this case) containing the equations.

- Write the vegetation equation as dv using `with`.

- Write the herbivore equation as dn using `with`.

- Combine these vectors into a list called `result`.

- Generate a time series over which to solve the equations in `times`.

- Set the parameter values in `parameters`.

- Set the starting values for V and N in `initial`.

- Use `ode` to create a matrix with the time series of V and N in `output`.

None of this is at all complicated, but there are lots of steps, so it looks a bit daunting.

First we write a function called `phmodel` (for plant–herbivore model) which tells R the structure of the two equations, showing how change in each population is related to the functional and numerical responses, and then puts the results into a list:

```
phmodel <- function(t,state,parameters){
with(as.list(c(state,parameters)),{
    dv <- r*v*(K-v)/K-b*v*n
    dn <- c*v*n-d*n
    result <- c(dv,dn)
    list(result)
})}
```

The rightmost curly bracket ends the function, the plain right bracket closes the `with` function and the leftmost curly bracket ends the definition of the equations.

To run the model we need to create a vector of times over which to calculate the population dynamics,

```
times <- seq(0,500,length=501)
```

then define the numeric values of the five parameters (these values will determine the behaviour of the two populations)

```
parameters <- c(r=0.4,K=1000,b=0.02,c=0.01,d=0.3)
```

and set the initial conditions (plant = 50 and herbivores = 10 at the start):

```
initial <- c(v=50,n=10)
```

That is the end of the preliminaries.

Solving the equations could not be easier. The important function is ode (ordinary differential equation solver). The function takes four arguments: the starting values, the vector of times, the function containing the equations, and the list containing the parameter values:

```
output <- ode(y=initial,time=times,func=phmodel,parms=parameters)
```

The output object is a matrix with three columns: time, plant abundance (v) and herbivore abundance (n):

```
head(output)
```

```
      time          v          n
[1,]     0 50.00000 10.00000
[2,]     1 58.29220 12.75106
[3,]     2 62.99695 17.40172
[4,]     3 60.70065 24.09264
[5,]     4 50.79407 31.32860
[6,]     5 37.68312 36.12636
```

Plotting the two time series is done like this:

```
plot(output[,1],output[,2],
  ylim=c(0,60),type="n",ylab="abundance",xlab="time")
lines(output[,1],output[,2],col="green")
lines(output[,1],output[,3],col="red")
```

The graph shows plant abundance as a green line against time and herbivore abundance as a red line:

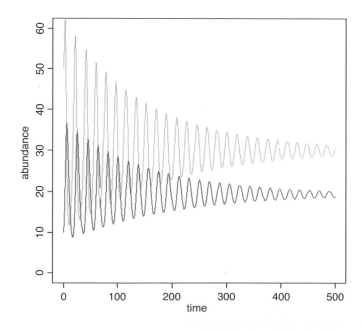

The system exhibits damped oscillations to a stable point equilibrium at which dV/dt and dN/dt are both equal to zero, so equilibrium plant abundance $V^* = d/c = 0.3/0.01 = 30$ and equilibrium herbivore abundance $N^* = r(K - V^*)/bK = 19.4$.

An alternative is to plot the output as a phase plane, with herbivore abundance on the *x* axis and plant abundance on the *y* axis:

```
plot(output[,3],output[,2],
    ylim=c(0,70),xlim=c(0,70),type="n",ylab="plant",xlab="herbivore")
```

```
lines(output[,2],output[,3],col="red")
```

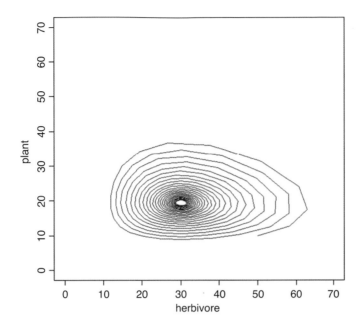

8

Classical Tests

There is absolutely no point in carrying out an analysis that is more complicated than it needs to be. Occam's razor applies to the choice of statistical model just as strongly as to anything else: simplest is best. The so-called classical tests deal with some of the most frequently used kinds of analysis for single-sample and two-sample problems.

8.1 Single samples

Suppose we have a single sample. The questions we might want to answer are these:

- What is the mean value?

- Is the mean value significantly different from current expectation or theory?

- What is the level of uncertainty associated with our estimate of the mean value?

In order to be reasonably confident that our inferences are correct, we need to establish some facts about the distribution of the data:

- Are the values normally distributed or not?

- Are there outliers in the data?

- If data were collected over a period of time, is there evidence for serial correlation?

Non-normality, outliers and serial correlation can all invalidate inferences made by standard parametric tests like Student's t test. It is much better in cases with non-normality and/or outliers to use a non-parametric technique such as Wilcoxon's signed-rank test. If there is serial correlation in the data, then you need to use time series analysis or mixed-effects models.

The R Book, Second Edition. Michael J. Crawley.
© 2013 John Wiley & Sons, Ltd. Published 2013 by John Wiley & Sons, Ltd.

8.1.1 Data summary

To see what is involved in summarizing a single sample, read the data called *y* from the file called `classic.txt`:

```
data <- read.table("c:\\temp\\classic.txt",header=T)
names(data)

[1]  "y"

attach(data)
```

As usual, we begin with a set of single sample plots: an index plot (scatterplot with a single argument, in which data are plotted in the order in which they appear in the dataframe), a box-and-whisker plot (see p. 212) and a frequency plot (a histogram with bin widths chosen by R):

```
par(mfrow=c(2,2))
plot(y)
boxplot(y)
hist(y,main="")
y2 <- y
y2[52] <- 21.75
plot(y2)
```

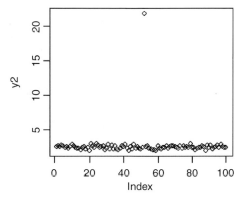

The index plot (bottom right) is particularly valuable for drawing attention to mistakes in the dataframe. Suppose that the 52nd value had been entered as 21.75 instead of 2.175: the mistake stands out like a sore thumb in the plot.

Summarizing the data could not be simpler. We use the built-in function called summary like this:

```
summary(y)
```

```
  Min. 1st Qu.  Median    Mean 3rd Qu.    Max.
 1.904   2.241   2.414   2.419   2.568   2.984
```

This gives us six pieces of information about the vector y. The smallest value is 1.904 (labelled Min. for minimum) and the largest value is 2.984 (labelled Max. for maximum). There are two measures of central tendency: the median is 2.414 and the arithmetic mean in 2.419. The other two figures (labelled 1st Qu. and 3rd Qu.) are the first and third quartiles (the 25th and 75th percentiles; see p. 115).

An alternative is Tukey's 'five-number summary' which comprises minimum, lower hinge, median, upper hinge and maximum for the input data. Hinges are close to the first and third quartiles (compare with summary, above), but different for small samples (see below):

```
fivenum(y)
```

```
[1] 1.903978 2.240931 2.414137 2.569583 2.984053
```

Notice that in this case Tukey's 'hinges' are not exactly the same as the 25th and 75th percentiles produced by summary. In our sample of 100 numbers the hinges are the average of the 25th and 26th sorted numbers and the average of the 75th and 76th sorted numbers. This is how the fivenum summary is produced: x takes the sorted values of y, and n is the length of y. Then five numbers, d, are calculated to use as subscripts to extract five averaged values from x like this:

```
x <- sort(y)
n <- length(y)
d <- c(1, 0.5 * floor(0.5 * (n + 3)), 0.5 * (n + 1), n + 1 - 0.5 *
                    floor(0.5 * (n + 3)), n)
0.5 * (x[floor(d)] + x[ceiling(d)])
```

```
[1] 1.903978 2.240931 2.414137 2.569583 2.984053
```

where the d values are

```
[1]    1.0   25.5   50.5   75.5 100.0
```

with floor and ceiling providing the lower and upper subscripts for averaging (25 with 26 and 75 with 76 for the lower and upper hinges, respectively).

8.1.2   Plots for testing normality

The simplest test of normality (and in many ways the best) is the 'quantile–quantile plot'. This plots the ranked samples from our distribution against a similar number of ranked quantiles taken from a normal distribution. If our sample is normally distributed then the line will be straight. Departures from normality show up as various sorts of non-linearity (e.g. S-shapes or banana shapes). The functions you need are qqnorm and qqline (quantile–quantile plot against a normal distribution):

```
par(mfrow=c(1,1))
qqnorm(y)
qqline(y,lty=2)
```

This shows a slight S-shape, but there is no compelling evidence of non-normality (our distribution is somewhat skew to the left; see the histogram, above). A novel plot for illustrating non-normality is shown on p. 232.

8.1.3 Testing for normality

We might use `shapiro.test` for testing whether the data in a vector come from a normal distribution. The null hypothesis is that the sample data *are* normally distributed. Let us generate some data that are log-normally distributed, in the hope that they will *fail* the normality test:

```
x <- exp(rnorm(30))
shapiro.test(x)
```

```
        Shapiro-Wilk normality test
data: x
W = 0.5701, p-value = 3.215e-08
```

They certainly do fail: $p < 0.000\,001$. A p value is *not* the probability that the null hypothesis is true (this is a common misunderstanding). On the contrary, the p value is based on the assumption that the null hypothesis *is* true. A p value is an estimate of the probability that a particular result ($W = 0.5701$ in this case), or a result more extreme than the result observed, could have occurred by chance, *if the null hypothesis were true*. In short, the p value is a measure of the credibility of the null hypothesis. A large p value (say, $p = 0.23$) means that there is no compelling evidence on which to reject the null hypothesis. Of course, saying 'we do not reject the null hypothesis' and 'the null hypothesis is true' are two quite different things. For instance, we may have failed to reject a false null hypothesis because our sample size was too low, or because our measurement error was too large. Thus, p values are interesting, but they do not tell the whole story: effect sizes and sample sizes are equally important in drawing conclusions.

8.1.4   An example of single-sample data

We can investigate the issues involved by examining the data from Michelson's famous experiment in 1879 to measure the speed of light (see Michelson, 1880). The dataframe called `light` contains his results (km s^{-1}), but with 299 000 subtracted.

```
light <- read.table("t:\\data\\light.txt",header=T)
attach(light)
hist(speed,main="",col="green")
```

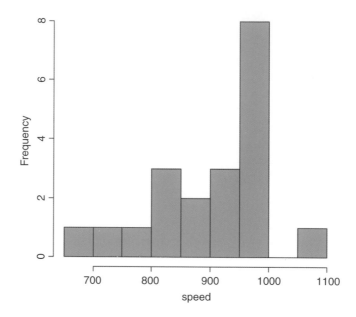

We get a summary of the non-parametric descriptors of the sample like this:

```
summary(speed)
```

```
  Min. 1st Qu.  Median    Mean 3rd Qu.    Max.
   650     850     940     909     980    1070
```

From this, you see at once that the median (940) is substantially bigger than the mean (909), as a consequence of the strong negative skew in the data seen in the histogram. The **interquartile range** is the difference between the first and third quartiles: $980 - 850 = 130$. This is useful in the detection of outliers: a good rule of thumb is that an **outlier** is a value that is more than 1.5 times the interquartile range above the third quartile or below the first quartile ($130 \times 1.5 = 195$). In this case, therefore, outliers would be measurements of speed that were less than $850 - 195 = 655$ or greater than $980 + 195 = 1175$. You will see that there are no large outliers in this data set, but one or more small outliers (the minimum is 650).

We want to test the hypothesis that Michelson's estimate of the speed of light is significantly different from the value of 299 990 thought to prevail at the time. Since the data have all had 299 000 subtracted from them, the test value is 990. Because of the non-normality, the use of Student's t test in this case is ill advised. The correct test is Wilcoxon's signed-rank test.

```
wilcox.test(speed,mu=990)
```

```
        Wilcoxon signed rank test with continuity correction
```

```
data:  speed
V = 22.5, p-value = 0.00213
alternative hypothesis: true location is not equal to 990

Warning message:
In wilcox.test.default(speed, mu = 990) :
  cannot compute exact p-value with ties
```

We reject the null hypothesis and accept the alternative hypothesis because $p = 0.002\ 13$ (i.e. much less than 0.05). The speed of light is significantly less than 299 990.

8.2 Bootstrap in hypothesis testing

You have probably heard the old phrase about 'pulling yourself up by your own bootlaces'. That is where the term 'bootstrap' comes from. It is used in the sense of getting 'something for nothing'. The idea is very simple. You have a single sample of n measurements, but you can sample from this in very many ways, so long as you allow some values to appear more than once, and other samples to be left out (i.e. *sampling with replacement*). All you do is calculate the sample mean lots of times, once for each sampling from your data, then obtain the confidence interval by looking at the extreme highs and lows of the estimated means using a quantile function to extract the interval you want (e.g. a 95% interval is specified using `c(0.0275, 0.975)` to locate the lower and upper bounds).

   Our sample mean value of y is 909 (see the previous example). The question we have been asked to address is this: how likely is it that the population mean that we are trying to estimate with our random sample of 100 values is as big as 990? We take 10 000 random samples with replacement using $n = 100$ from the 100 values of light and calculate 10 000 values of the mean. Then we ask: what is the probability of obtaining a mean as large as 990 by inspecting the right-hand tail of the cumulative probability distribution of our 10 000 bootstrapped mean values? This is not as hard as it sounds:

```
a <- numeric(10000)
for(i in 1:10000) a[i] <- mean(sample(speed,replace=T))
hist(a,main="",col="blue")
```

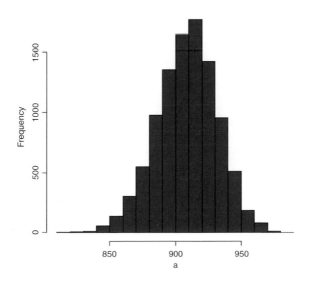

The test value of 990 is way off the scale to the right, so a mean of 990 is clearly most unlikely, given the data with `max(a) = 979`. In our 10 000 samples of the data, we never obtained a mean value greater than 979, so the probability that the mean is 990 is clearly $p < 0.0001$.

8.3 Skew and kurtosis

So far, and without saying so explicitly, we have encountered the first two moments of a sample distribution. The quantity $\sum y$ was used in the context of defining the arithmetic mean of a single sample: this is the first moment $\bar{y} = \sum y/n$. The quantity $\sum(y - \bar{y})^2$, the sum of squares, was used in calculating sample variance, and this is the second moment of the distribution $s^2 = \sum(y - \bar{y})^2/(n - 1)$. Higher-order moments involve powers of the difference greater than 2 such as $\sum(y - \bar{y})^3$ and $\sum(y - \bar{y})^4$.

8.3.1 Skew

Skew (or skewness) is the dimensionless version of the third moment about the mean,

$$m_3 = \frac{\sum (y - \bar{y})^3}{n},$$

which is rendered dimensionless by dividing by the cube of the standard deviation of y (because this is also measured in units of y^3),

$$s_3 = \mathrm{sd}(y)^3 = (\sqrt{s^2})^3.$$

The skew is then given by

$$\mathrm{skew} = \gamma_1 = \frac{m_3}{s_3}.$$

It measures the extent to which a distribution has long, drawn-out *tails* on one side or the other. A normal distribution is symmetrical and has $\gamma_1 = 0$. Negative values of γ_1 mean skew to the left (negative skew) and positive values mean skew to the right.

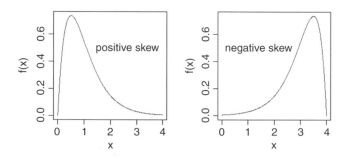

```
windows(7,4)
par(mfrow=c(1,2))
x <- seq(0,4,0.01)
plot(x,dgamma(x,2,2),type="l",ylab="f(x)",xlab="x",col="red")
```

```
text(2.7,0.5,"positive skew")
plot(4-x,dgamma(x,2,2),type="l",ylab="f(x)",xlab="x",col="red")
text(1.3,0.5,"negative skew")
```

To test whether a particular value of skew is significantly different from 0 (and hence the distribution from which it was calculated is significantly non-normal) we divide the estimate of skew by its approximate standard error:

$$se_{\gamma_1} = \sqrt{\frac{6}{n}}.$$

It is straightforward to write an R function to calculate the degree of skew for any vector of numbers, x, like this:

```
skew <- function(x){
     m3 <- sum((x-mean(x))^3)/length(x)
     s3 <- sqrt(var(x))^3
     m3/s3
}
```

Note the use of the `length(x)` function to work out the sample size, n, whatever the size of the vector x. The last expression inside the function is not assigned a variable name, and is returned as the value of `skew(x)` when this is executed from the command line.

Let us test the following data set:

```
data <- read.table("c:\\temp\\skewdata.txt",header=T)
attach(data)
names(data)

[1]   "values"

hist(values)
```

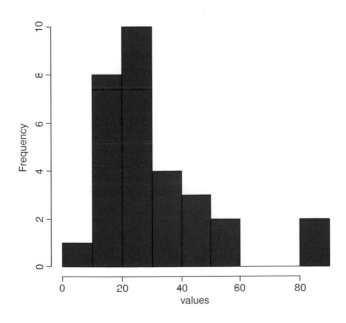

The data appear to be positively skew (i.e. to have a longer tail on the right than on the left). We use the new function skew to quantify the degree of skewness:

```
skew(values)
```

```
[1] 1.318905
```

Now we need to know whether a skew of 1.319 is significantly different from zero. We do a *t* test, dividing the observed value of skew by its standard error $\sqrt{6/n}$:

```
skew(values)/sqrt(6/length(values))
```

```
[1]  2.949161
```

Finally, we ask what is the probability of getting a *t* value of 2.949 by chance alone, when the skew value really is zero:

```
1-pt(2.949,28)
```

```
[1]  0.003185136
```

We conclude that these data show significant non-normality ($p < 0.0032$).

The next step might be to look for a transformation that normalizes the data by reducing the skewness. One way of drawing in the larger values is to take square roots, so let us try this to begin with:

```
skew(sqrt(values))/sqrt(6/length(values))
```

```
[1]  1.474851
```

This is not significantly skew. Alternatively, we might take the logs of the values:

```
skew(log(values))/sqrt(6/length(values))
```

```
[1]  -0.6600605
```

This is now slightly skew to the left (negative skew), but the value of Student's *t* is smaller than with a square root transformation, so we might prefer a log transformation in this case.

8.3.2 Kurtosis

This is a measure of non-normality that has to do with the peakyness, or flat-toppedness, of a distribution. The normal distribution is bell-shaped, whereas a kurtotic distribution is other than bell-shaped. In particular, a more flat-topped distribution is said to be **platykurtic**, and a more pointy distribution is said to be **leptokurtic**.

Kurtosis is the dimensionless version of the fourth moment about the mean,

$$m_4 = \frac{\sum (y - \bar{y})^4}{n},$$

which is rendered dimensionless by dividing by the square of the variance of y (because this is also measured in units of y^4),

$$s_4 = (\text{var}(y))^2 = (s^2)^2.$$

Kurtosis is then given by

$$\text{kurtosis} = \gamma_2 = \frac{m_4}{s_4} - 3.$$

The -3 is included because a normal distribution has $m_4/s_4 = 3$. This formulation therefore has the desirable property of giving zero kurtosis for a normal distribution, while a flat-topped (platykurtic) distribution has a negative value of kurtosis, and a pointy (leptokurtic) distribution has a positive value of kurtosis. The approximate standard error of kurtosis is

$$se_{\gamma_2} = \sqrt{\frac{24}{n}}.$$

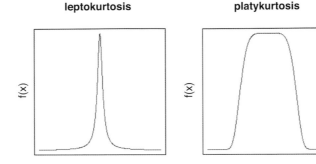

```
plot(-200:200,dcauchy(-200:200,0,10),type="l",ylab="f(x)",xlab="",yaxt="n",
            xaxt="n",main="leptokurtosis",col="red")
xv <- seq(-2,2,0.01)
plot(xv,exp(-abs(xv)^6),type="l",ylab="f(x)",xlab="",yaxt="n",
            xaxt="n",main="platykurtosis",col="red")
```

An R function to calculate kurtosis might look like this:

```
kurtosis <- function(x) {
m4 <- sum((x-mean(x))^4)/length(x)
s4 <- var(x)^2
m4/s4 - 3 }
```

For our present data, we find that kurtosis is not significantly different from normal:

```
kurtosis(values)
```

```
[1]  1.297751
```

```
kurtosis(values)/sqrt(24/length(values))
```

```
[1]  1.45093
```

8.4 Two samples

The classical tests for two samples include:

- comparing two variances (Fisher's F test, `var.test`);
- comparing two sample means with normal errors (Student's t test, `t.test`);
- comparing two means with non-normal errors (Wilcoxon's rank test, `wilcox.test`);
- comparing two proportions (the binomial test, `prop.test`);

- correlating two variables (Pearson's or Spearman's rank correlation, `cor.test`);

- testing for independence of two variables in a contingency table (chi-squared, `chisq.test`, or Fisher's exact test, `fisher.test`).

8.4.1 Comparing two variances

Before we can carry out a test to compare two sample means (see below), we need to test whether the sample variances are significantly different (see p. 356). The test could not be simpler. It is called Fisher's F test after the famous statistician and geneticist R.A. Fisher, who worked at Rothamsted in south-east England. To compare two variances, all you do is divide the larger variance by the smaller variance. Obviously, if the variances are the same, the ratio will be 1. In order to be significantly different, the ratio will need to be significantly bigger than 1 (because the larger variance goes on top, in the numerator). How will we know a significant value of the variance ratio from a non-significant one? The answer, as always, is to look up the *critical value* of the variance ratio. In this case, we want critical values of Fisher's F. The R function for this is `qf`, which stands for 'quantiles of the F distribution'.

For our example of ozone levels in market gardens (see p. 354) there were 10 replicates in each garden, so there were $10 - 1 = 9$ degrees of freedom for each garden. In comparing two gardens, therefore, we have 9 d.f. in the numerator and 9 d.f. in the denominator. Although F tests in analysis of variance are typically one-tailed (the treatment variance is expected to be larger than the error variance if the means are significantly different; see p. 501), in this case we have no expectation as to which garden was likely to have the higher variance, so we carry out a two-tailed test ($p = 1 - \alpha/2$). Suppose we work at the traditional $\alpha = 0.05$, then we find the critical value of F like this:

```
qf(0.975,9,9)
```

```
4.025994
```

This means that a calculated variance ratio will need to be greater than or equal to 4.02 in order for us to conclude that the two variances are significantly different at $\alpha = 0.05$.

To see the test in action, we can compare the variances in ozone concentration for market gardens B and C:

```
f.test.data <- read.table("c:\\temp\\f.test.data.txt",header = T)
attach(f.test.data)
names(f.test.data)
```

```
[1]  "gardenB"  "gardenC"
```

First, we compute the two variances:

```
var(gardenB)
```

```
[1]   1.333333
```

```
var(gardenC)
```

```
[1]   14.22222
```

The larger variance is clearly in garden C, so we compute the F ratio like this:

```
F.ratio <- var(gardenC)/var(gardenB)
F.ratio
```

```
[1]   10.66667
```

The variance in garden C is more than 10 times as big as the variance in garden B. The critical value of F for this test (with 9 d.f. in both the numerator and the denominator) is 4.026 (see qf, above), so, since the calculated value is larger than the critical value we reject the null hypothesis. The null hypothesis was that the two variances were not significantly different, so we accept the alternative hypothesis that the two variances are significantly different. In fact, it is better practice to present the p value associated with the calculated F ratio rather than just to reject the null hypothesis; to do this we use pf rather than qf. We double the resulting probability to allow for the two-tailed nature of the test:

```
2*(1-pf(F.ratio,9,9))
```

```
[1]  0.001624199
```

so the probability that the variances are the same is $p < 0.002$. Because the variances are significantly different, it would be wrong to compare the two sample means using Student's t test.

There is a built-in function called var.test for speeding up the procedure. All we provide are the names of the two variables containing the raw data whose variances are to be compared (we do not need to work out the variances first):

```
var.test(gardenB,gardenC)

        F test to compare two variances
data:  gardenB  and  gardenC
F = 0.0938,  num  df = 9,  denom  df = 9,  p-value = 0.001624
alternative hypothesis: true ratio of variances is not equal to 1
95 percent  confidence  interval:
0.02328617  0.37743695
sample estimates:
ratio  of  variances
         0.09375
```

Note that the variance ratio, F, is given as roughly 1/10 rather than roughly 10 because var.test put the variable name that came first in the alphabet (gardenB) on top (i.e. in the numerator) instead of the bigger of the two variances. But the p value of 0.0016 is the same as we calculated by hand (above), and we reject the null hypothesis. These two variances are highly significantly different. This test is highly sensitive to outliers, so use it with care.

It is important to know whether variance differs significantly from sample to sample. Constancy of variance (**homoscedasticity**) is the most important assumption underlying regression and analysis of variance (p. 490). For comparing the variances of two samples, Fisher's F test is appropriate (p. 354). For multiple samples you can choose between the Bartlett test and the Fligner–Killeen test. Here are both tests in action:

```
refs <- read.table("c:\\temp\\refuge.txt",header=T)
attach(refs)
names(refs)
```

```
[1]  "B"  "T"
```

where T is an ordered factor with nine levels. Each level produces 30 estimates of yields except for level 9 which is a single zero. We begin by looking at the variances:

```
tapply(B,T,var)
```

1	2	3	4	5	6	7	8	9
1354.024	2025.431	3125.292	1077.030	2542.599	2221.982	1445.490	1459.955	NA

When it comes to the variance tests we shall have to leave out level 9 of T because the tests require at least two replicates at each factor level. We need to know which data point refers to treatment $T = 9$:

```
which(T==9)
```

```
[1]  31
```

So we shall omit the 31st data point using negative subscripts. First Bartlett:

```
bartlett.test(B[-31],T[-31])
```

```
        Bartlett test of homogeneity of variances
data:  B[-31] and T[-31]
Bartlett's K-squared = 13.1986, df = 7, p-value = 0.06741
```

So there is no significant difference between the eight variances ($p = 0.067$). Now Fligner:

```
fligner.test(B[-31],T[-31])
```

```
        Fligner-Killeen test of homogeneity of variances
data:  B[-31] and T[-31]
Fligner-Killeen:med chi-squared = 14.3863, df = 7, p-value = 0.04472
```

Hmm. This test says that there *are* significant differences between the variances ($p < 0.05$). What you do next depends on your outlook. There are obviously some close-to-significant differences between these eight variances, but if you simply look at a plot of the data, `plot(T,B)`, the variances appear to be very well behaved. A linear model shows some slight pattern in the residuals and some evidence of non-normality:

```
model <- lm(B~T)
plot(model)
```

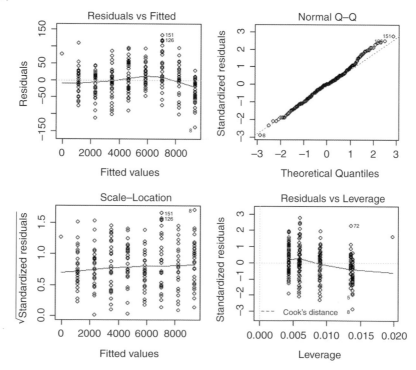

The various tests can give wildly different interpretations. Here are the ozone data from three market gardens:

```
ozone <- read.table("c:\\temp\\gardens.txt",header=T)
attach(ozone)
names(ozone)

[1]  "gardenA"  "gardenB"  "gardenC"

y <- c(gardenA,gardenB,gardenC)
garden <- factor(rep(c("A","B","C"),c(10,10,10)))
```

The question is whether the variance in ozone concentration differs from garden to garden or not. Fisher's F test comparing gardens B and C says that variance is significantly greater in garden C:

```
var.test(gardenB,gardenC)

        F test to compare two variances
data: gardenB and gardenC
F = 0.0938, num df = 9, denom df = 9, p-value = 0.001624
alternative hypothesis: true ratio of variances is not equal to 1
95 percent confidence interval:
0.02328617  0.37743695
sample estimates:
ratio of variances
        0.09375
```

Bartlett's test, likewise, says there is a highly significant difference in variance across gardens:

```
bartlett.test(y~garden)
Bartlett test of homogeneity of variances
data: y  by garden
Bartlett's K-squared = 16.7581, df = 2, p-value = 0.0002296
```

In contrast, the Fligner–Killeen test (preferred over Bartlett's test by many statisticians) says there is no compelling evidence for non-constancy of variance (**heteroscedasticity**) in these data:

```
fligner.test(y~garden)

        Fligner-Killeen test of homogeneity of variances
data: y  by garden
Fligner-Killeen: med chi-squared = 1.8061, df = 2, p-value = 0.4053
```

The reason for the difference is that Fisher and Bartlett are sensitive to outliers, whereas Fligner–Killeen is not (it is a non-parametric test which uses the ranks of the absolute values of the centred samples, and weights $a(i) = qnorm((1 + i/(n+1))/2)$. Of the many tests for homogeneity of variances, this is the most robust against departures from normality (Conover *et al.*, 1981). In this particular case, I think the Flinger test is too forgiving: gardens B and C both had a mean of 5 parts per hundred milllion (pphm; well below the damage threshold of 8 pphm), but garden B never suffered damaging levels of ozone whereas garden C experienced damaging ozone levels on 30% of days. That difference is scientifically important, and deserves to be statistically significant.

8.4.2   Comparing two means

Given what we know about the variation from replicate to replicate within each sample (the within-sample variance), how likely is it that our two sample means were drawn from populations with the same average? If it is highly likely, then we shall say that our two sample means are not significantly different. If it is rather unlikely, then we shall say that our sample means are significantly different. But perhaps a better way to proceed is to work out the probability that the two samples were indeed drawn from populations with the same mean. If this probability is very low (say, less than 5% or less than 1%) then we can be reasonably certain (95% or 99% in these two examples) than the means really are different from one another. Note, however, that we can never be 100% certain; the apparent difference might just be due to random sampling – we just happened to get a lot of low values in one sample, and a lot of high values in the other.

There are two classical tests for comparing two sample means:

- Student's *t* test when the samples are independent, the variances constant, and the errors are normally distributed;

- Wilcoxon's rank-sum test when the samples are independent but the errors are *not* normally distributed (e.g. they are ranks or scores or some sort).

What to do when these assumptions are violated (e.g. when the variances are different) is discussed later on.

8.4.3   Student's *t* test

Student was the pseudonym of W.S. Gossett who published his influential paper in *Biometrika* in 1908. The archaic employment laws in place at the time allowed his employer, the Guinness Brewing Company, to prevent him publishing independent work under his own name. Student's *t* distribution, later perfected by R.A. Fisher, revolutionized the study of small-sample statistics where inferences need to be made on the basis of the sample variance s^2 with the population variance σ^2 unknown (indeed, usually unknowable). The test statistic is the number of standard errors of the difference by which the two sample means are separated:

$$t = \frac{\text{difference between the two means}}{\text{standard error of the difference}} = \frac{\bar{y}_A - \bar{y}_B}{se_{\text{diff}}}.$$

We know the standard error of the mean (see p. 43) but we have not yet met the standard error of the difference between two means. For two independent (i.e. non-correlated) variables, *the variance of a difference is the sum of the separate variances*. This important result allows us to write down the formula for the standard error of the difference between two sample means:

$$se_{\text{diff}} = \sqrt{\frac{s_A^2}{n_A} + \frac{s_B^2}{n_B}}$$

We now have everything we need to carry out Student's *t* test. Our null hypothesis is that the two population means are the same, and we shall accept this unless the value of Student's *t* is so large that it is unlikely that such a difference could have arisen by chance alone. Everything varies, so in real studies our two sample means will *never* be exactly the same, no matter what the parent population means. For the ozone example introduced on p. 354, each sample has 9 degrees of freedom, so we have 18 d.f. in total. Another way of thinking of this is to reason that the complete sample size as 20, and we have estimated two parameters from the data, $\bar{y}_A$ and $\bar{y}_B$, so we have $20 - 2 = 18$ d.f. We typically use 5% as the chance of rejecting the null hypothesis when it is true (this is the Type I error rate). Because we did not know in advance which of the two gardens was going to have the higher mean ozone concentration (and we usually do not), this is a two-tailed test, so the *critical value* of Student's *t* is:

```
qt(0.975,18)
```

```
[1]  2.100922
```

This means that our test statistic needs to be bigger than 2.1 in order to reject the null hypothesis, and hence to conclude that the two means are significantly different at $\alpha = 0.05$.

The dataframe is attached like this:

```
t.test.data <- read.table("c:\\temp\\t.test.data.txt",header=T)
attach(t.test.data)
par(mfrow=c(1,1))
names(t.test.data)
```

```
[1]  "gardenA"  "gardenB"
```

A useful graphical test for two samples employs the `notch` option of `boxplot`:

```
ozone <- c(gardenA,gardenB)
label <- factor(c(rep("A",10),rep("B",10)))
boxplot(ozone~label,notch=T,xlab="Garden",ylab="Ozone")
```

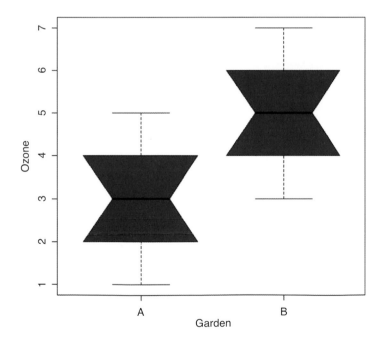

Because the notches of two plots do not overlap, we conclude that the medians are significantly different at the 5% level. Note that the variability is similar in both gardens, both in terms of the range (the whiskers) and the interquartile range (the boxes).

To carry out a *t* test long-hand, we begin by calculating the variances of the two samples:

```
s2A <- var(gardenA)
s2B <- var(gardenB)
```

The value of the test statistic for Student's t is the difference divided by the standard error of the difference. The numerator is the difference between the two means, and the denominator is the square root of the sum of the two variances divided by their sample sizes:

```
(mean(gardenA)-mean(gardenB))/sqrt(s2A/10+s2B/10)
```

which gives the value of Student's t as

```
[1]   -3.872983
```

With t-tests you can ignore the minus sign; it is only the absolute value of the difference between the two sample means that concerns us. So the calculated value of the test statistic is 3.87 and the critical value is 2.10 (qt(0.975,18), above). Since the calculated value of the test statistic is larger than the critical value, we reject the null hypothesis. Notice that the wording is exactly the same as it was for the F test (above). Indeed, the wording is always the same for all kinds of tests, and you should try to memorize it. The abbreviated form is easier to remember: 'larger reject, smaller accept'. The null hypothesis was that the two population means are not significantly different, so we reject this and accept the alternative hypothesis that the two means are significantly different. Again, rather than merely rejecting the null hypothesis, it is better to state the probability that data as extreme as this (or more extreme) would be observed if the population mean values really were the same. For this we use pt rather than qt, and in this instance 2*pt because we are doing a two-tailed test:

```
2*pt(-3.872983,18)
```

```
[1]   0.001114540
```

We conclude that $p < 0.005$.

You will not be surprised to learn that there is a built-in function to do all the work for us. It is called, helpfully, t.test and is used simply by providing the names of the two vectors containing the samples on which the test is to be carried out (gardenA and gardenB in our case).

```
t.test(gardenA,gardenB)
```

There is rather a lot of output. You often find this: the simpler the statistical test, the more voluminous the output.

```
      Welch Two Sample t-test
data: gardenA  and gardenB
t = -3.873, df = 18, p-value = 0.001115
alternative hypothesis: true difference in means is not equal to 0
95 percent confidence interval:
-3.0849115  -0.9150885
sample estimates:
mean of x mean of y
      3       5
```

The result is exactly the same as we obtained the long way. The value of t is -3.873 and, since the sign is irrelevant in a t test, we reject the null hypothesis because the test statistic is larger than the critical value of 2.1. The mean ozone concentration is significantly higher in garden B than in garden A. The computer output also gives a p value and a confidence interval. Note that, because the means are significantly different, *the confidence interval on the difference does not include zero* (in fact, it goes from -3.085 up to -0.915). You might present the result like this: 'Ozone concentration was significantly higher in garden B (mean = 5.0 pphm) than in garden A (mean = 3.0 pphm; $t = 3.873$, $p = 0.0011$ (2-tailed),   d.f. = 18).'

There is a formula-based version of `t.test` that you can use when your explanatory variable consists of a two-level factor (see `?t.test`).

8.4.4 Wilcoxon rank-sum test

This is a non-parametric alternative to Student's *t* test, which we could use if the errors were non-normal. The Wilcoxon rank-sum test statistic, *W*, is calculated as follows. Both samples are put into a single array with their sample names clearly attached (*A* and *B* in this case, as explained below). Then the aggregate list is sorted, taking care to keep the sample labels with their respective values. A rank is assigned to each value, with ties getting the appropriate average rank (two-way ties get (rank i + (rank i + 1))/2, three-way ties get (rank i + (rank i + 1) + (rank i + 2))/3, and so on). Finally the ranks are added up for each of the two samples, and significance is assessed on the size of the smaller sum of ranks.

First we make a combined vector of the samples:

```
ozone <- c(gardenA,gardenB)
ozone
```

```
[1]   3   4   4   3   2   3   1   3   5   2   5   5   6   7   4   4   3   5   6   5
```

Then we make a list of the sample names:

```
label <- c(rep("A",10),rep("B",10))
label
```

```
[1] "A" "A" "A" "A" "A" "A" "A" "A" "A" "A" "B" "B" "B" "B" "B" "B" "B" "B" "B" "B"
```

Now use the built-in function `rank` to get a vector containing the ranks, smallest to largest, within the combined vector:

```
combined.ranks <- rank(ozone)
combined.ranks
```

```
[1]    6.0 10.5 10.5  6.0  2.5  6.0  1.0  6.0 15.0  2.5 15.0 15.0
[13] 18.5 20.0 10.5 10.5  6.0 15.0 18.5 15.0
```

Notice that the ties have been dealt with by averaging the appropriate ranks. Now all we need to do is calculate the sum of the ranks for each garden. We use `tapply` with `sum` as the required operation

```
tapply(combined.ranks,label,sum)
```

```
  A    B
 66  144
```

Finally, we compare the smaller of the two values (66) with values in tables of Wilcoxon rank sums (e.g. Snedecor and Cochran, 1980, p. 555), and reject the null hypothesis if our value of 66 is *smaller* than the value in tables. For samples of size 10 and 10 like ours, the 5% value in tables is 78. Our value of 66 is smaller than this, so we reject the null hypothesis. The two sample means are significantly different (in agreement with our earlier *t* test).

We can carry out the whole procedure automatically, and avoid the need to use tables of critical values of Wilcoxon rank sums, by using the built-in function `wilcox.test`:

```
wilcox.test(gardenA,gardenB)
```

```
        Wilcoxon rank sum test with continuity correction
data:  gardenA and gardenB
```

```
W = 11, p-value = 0.002988
alternative hypothesis: true location shift is not equal to 0
```

```
Warning message:
In wilcox.test.default(gardenA, gardenB) :
  cannot compute exact p-value with ties
```

The function uses a normal approximation algorithm to work out a z value, and from this a p value to assess the hypothesis that the two means are the same. This p value of 0.002 988 is much less than 0.05, so we reject the null hypothesis, and conclude that the mean ozone concentrations in gardens A and B are significantly different. The warning message at the end draws attention to the fact that there are ties in the data (repeats of the same ozone measurement), and this means that the p value cannot be calculated exactly (this is seldom a real worry).

It is interesting to compare the p values of the t test and the Wilcoxon test with the same data: $p = 0.001\ 115$ and 0.002 988, respectively. The non-parametric test is much more appropriate than the t test when the errors are not normal, and the non-parametric test is about 95% as powerful with normal errors, and can be *more* powerful than the t test if the distribution is strongly skewed by the presence of outliers. Typically, as here, the t test will give the lower p value, so the Wilcoxon test is said to be conservative: if a difference is significant under a Wilcoxon test it would be even more significant under a t test.

8.5   Tests on paired samples

Sometimes, two-sample data come from paired observations. In this case, we might expect a correlation between the two measurements, because they were either made on the same individual, or taken from the same location. You might recall that the variance of a difference is the average of

$$(y_A - \mu_A)^2 + (y_B - \mu_B)^2 - 2(y_A - \mu_A)(y_B - \mu_B),$$

which is the variance of sample A, plus the variance of sample B, minus twice the covariance of A and B. When the covariance of A and B is *positive*, this is a great help because it reduces the variance of the difference, which makes it easier to detect significant differences between the means. Pairing is not always effective, because the correlation between y_A and y_B may be weak.

The following data are a composite biodiversity score based on a kick sample of aquatic invertebrates:

```
streams <- read.table("c:\\temp\\streams.txt",header=T)
attach(streams)
names(streams)
```

```
[1]  "down"  "up"
```

The elements are paired because the two samples were taken on the same river, one upstream and one downstream from the same sewage outfall.

If we ignore the fact that the samples are paired, it appears that the sewage outfall has no impact on biodiversity score ($p = 0.6856$):

```
t.test(down,up)

        Welch Two Sample t-test

data:  down and up
t = -0.4088, df = 29.755, p-value = 0.6856
```

```
alternative hypothesis: true difference in means is not equal to 0
95 percent confidence interval:
 -5.248256   3.498256
sample estimates:
mean of x mean of y
    12.500      13.375
```

However, if we allow that the samples are paired (simply by specifying the option `paired=T`), the picture is completely different:

```
t.test(down,up,paired=TRUE)

        Paired t-test

data:   down and up
t = -3.0502, df = 15, p-value = 0.0081
alternative hypothesis: true difference in means is not equal to 0
95 percent confidence interval:
 -1.4864388 -0.2635612
sample estimates:
mean of the differences
              -0.875
```

This is a good example of the benefit of writing TRUE rather than T. Because we have a variable called T (p. 22) the test would fail if we typed `t.test(down,up,paired=T)`. Now the difference between the means is highly significant ($p = 0.0081$). The moral is clear. If you can do a paired t test, then you should always do the paired test. It can never do any harm, and sometimes (as here) it can do a huge amount of good. In general, if you have information on *blocking* or *spatial correlation* (in this case, the fact that the two samples came from the same river), then you should always use it in the analysis.

Here is the same paired test carried out as a one-sample t test based on the differences between the pairs (upstream diversity minus downstream diversity):

```
difference <- up - down
t.test(difference)

        One Sample t-test

data:   difference
t = 3.0502, df = 15, p-value = 0.0081
alternative hypothesis: true mean is not equal to 0
95 percent confidence interval:
 0.2635612 1.4864388
sample estimates:
mean of x
    0.875
```

As you see, the result is identical to the two-sample t test with `paired=TRUE` ($p = 0.0081$). The upstream values of the biodiversity score were greater by 0.875 on average, and this difference is highly significant. Working with the differences has halved the number of degrees of freedom (from 30 to 15), but it has more than compensated for this by reducing the error variance, because there is such a strong positive correlation between y_A and y_B.

8.6   The sign test

This is one of the simplest of all statistical tests. Suppose that you cannot *measure* a difference, but you can *see* it (e.g. in judging a diving contest). For example, nine springboard divers were scored as better or worse, having trained under a new regime and under the conventional regime (the regimes were allocated in a randomized sequence to each athlete: new then conventional, or conventional then new). Divers were judged twice: one diver was worse on the new regime, and 8 were better. What is the evidence that the new regime produces significantly better scores in competition? The answer comes from a two-tailed binomial test. How likely is a response of 1/9 (or 8/9 or more extreme than this, i.e. 0/9 or 9/9) if the populations are actually the same (i.e. $p = 0.5$)? We use a binomial test for this, specifying the number of 'failures' (1) and the total sample size (9):

```
binom.test(1,9)
```

```
            Exact binomial test

data:   1 and 9
number of successes = 1, number of trials = 9, p-value = 0.03906
alternative hypothesis: true probability of success is not equal to 0.5
95 percent confidence interval:
 0.002809137 0.482496515
sample estimates:
probability of success
              0.1111111
```

We would conclude that the new training regime is significantly better than the traditional method, because $p < 0.05$.

It is easy to write a function to carry out a sign test to compare two samples, x and y:

```
sign.test <- function(x, y)
{
if(length(x) != length(y)) stop("The two variables must be the same length")
      d <- x - y
      binom.test(sum(d > 0), length(d))
}
```

The function starts by checking that the two vectors are the same length, then works out the vector of the differences, d. The binomial test is then applied to the number of positive differences (sum(d > 0)) and the total number of numbers (length(d)). If there was no difference between the samples, then on average, the sum would be about half of length(d). Here is the sign test used to compare the ozone levels in gardens A and B (see above):

```
sign.test(gardenA,gardenB)
```

```
            Exact binomial test

data:   sum(d > 0) and length(d)
number of successes = 0, number of trials = 10, p-value = 0.001953
alternative hypothesis: true probability of success is not equal to 0.5
95 percent confidence interval:
 0.0000000 0.3084971
sample estimates:
probability of success
              0
```

Note that the p value (0.002) from the sign test is larger than in the equivalent t test ($p = 0.0011$) that we carried out earlier. This will generally be the case: other things being equal, the parametric test will be more powerful than the non-parametric equivalent.

8.7 Binomial test to compare two proportions

Suppose that only four females were promoted, compared to 196 men. Is this an example of blatant sexism, as it might appear at first glance? Before we can judge, of course, we need to know the number of male and female candidates. It turns out that 196 men were promoted out of 3270 candidates, compared with 4 promotions out of only 40 candidates for the women. Now, if anything, it looks like the females did better than males in the promotion round (10% success for women versus 6% success for men).

The question then arises as to whether the apparent positive discrimination in favour of women is statistically significant, or whether this sort of difference could arise through chance alone. This is easy in R using the built-in binomial proportions test `prop.test` in which we specify two vectors, the first containing the number of successes for females and males `c(4,196)` and second containing the total number of female and male candidates `c(40,3270)`:

```
prop.test(c(4,196),c(40,3270))

    2-sample test for equality of proportions with continuity correction
data:   c(4, 196) out of c(40, 3270)
X-squared = 0.5229, df = 1, p-value = 0.4696
alternative hypothesis: two.sided
95 percent confidence interval:
 -0.06591631  0.14603864
sample estimates:
    prop 1      prop 2
0.10000000 0.05993884

Warning message:
In prop.test(c(4, 196), c(40, 3270)) :
  Chi-squared approximation may be incorrect
```

There is no evidence in favour of positive discrimination ($p = 0.4696$). A result like this will occur more than 45% of the time by chance alone. Just think what would have happened if one of the successful female candidates had not applied. Then the same promotion system would have produced a female success rate of 3/39 instead of 4/40 (7.7% instead of 10%). In small samples, small changes have big effects.

8.8 Chi-squared contingency tables

A great deal of statistical information comes in the form of *counts* (whole numbers or integers): the number of animals that died, the number of branches on a tree, the number of days of frost, the number of companies that failed, the number of patients who died. With count data, the number 0 is often the value of a response variable (consider, for example, what a 0 would mean in the context of the examples just listed). The analysis of count data is discussed in more detail in Chapters 14 and 15.

The dictionary definition of contingency is: 'A possible or uncertain event on which other things depend or are conditional' (OED, 2012). In statistics, however, the contingencies are *all the events that could possibly happen*. A contingency table shows the counts of how many times each of the contingencies actually happened in a particular sample. Consider the following example that has to do with the relationship between hair colour and eye colour in white people. For simplicity, we just chose two contingencies for hair colour: 'fair' and 'dark'. Likewise we just chose two contingencies for eye colour: 'blue' and 'brown'. Each of these two categorical variables, eye colour and hair colour, has two levels ('blue' and 'brown', and 'fair' and 'dark', respectively). Between them, they define four possible outcomes (the contingencies): fair hair and blue eyes, fair hair and brown eyes, dark hair and blue eyes, and dark hair and brown eyes. We take a random sample of white people and count how many of them fall into each of these four categories. Then we fill in the 2 × 2 contingency table like this:

	Blue eyes	Brown eyes
Fair hair	38	11
Dark hair	14	51

These are our observed frequencies (or counts). The next step is very important. In order to make any progress in the analysis of these data we need a *model* which predicts the expected frequencies. What would be a sensible model in a case like this? There are all sorts of complicated models that you might select, but the simplest model (Occam's razor, or the principle of parsimony) is that hair colour and eye colour are *independent*. We may not believe that this is actually true, but the hypothesis has the great virtue of being falsifiable. It is also a very sensible model to choose because it makes it possible to predict the expected frequencies based on the assumption that the model is true. We need to do some simple probability work. What is the probability of getting a random individual from this sample whose hair was fair? A total of 49 people (38 + 11) had fair hair out of a total sample of 114 people. So the probability of fair hair is 49/114 and the probability of dark hair is 65/114. Notice that because we have only two levels of hair colour, these two probabilities add up to 1 ((49 + 65)/114). What about eye colour? What is the probability of selecting someone at random from this sample with blue eyes? A total of 52 people had blue eyes (38 + 14) out of the sample of 114, so the probability of blue eyes is 52/114 and the probability of brown eyes is 62/114. As before, these sum to 1 ((52 + 62)/114). It helps to add the subtotals to the margins of the contingency table like this:

	Blue eyes	Brown eyes	Row totals
Fair hair	38	11	49
Dark hair	14	51	65
Column totals	52	62	114

Now comes the important bit. We want to know the expected frequency of people with fair hair *and* blue eyes, to compare with our observed frequency of 38. Our model says that the two are independent. This is essential information, because it allows us to calculate the expected probability of fair hair and blue eyes. If, and only if, the two traits are independent, then the probability of having fair hair and blue eyes is the product of the two probabilities. So, following our earlier calculations, the probability of fair hair and blue eyes is 49/114 × 52/114. We can do exactly equivalent things for the other three cells of the contingency table:

	Blue eyes	Brown eyes	Total count in each row
Fair hair	49/114 × 52/114	49/114 × 62/114	49
Dark hair	65/114 × 52/114	65/114 × 62/114	65
Total count in each column	52	62	114

Now we need to know how to calculate the expected frequency. It couldn't be simpler. It is just the probability multiplied by the total sample ($n = 114$). So the expected frequency of blue eyes and fair hair is

49/114 $\times$ 52/114 $\times$ 114 = 22.35, which is much less than our observed frequency of 38. It is beginning to look as if our hypothesis of independence of hair and eye colour is false.

You might have noticed something useful in the last calculation: two of the sample sizes cancel out. Therefore, the expected frequency in each cell is just the row total (R) times the column total (C) divided by the grand total (G) like this:

$$E = \frac{R \times C}{G}.$$

We can now work out the four expected frequencies:

	Blue eyes	Brown eyes	Row totals
Fair hair	22.35	26.65	49
Dark hair	29.65	35.35	65
Column totals	52	62	114

Notice that the row and column totals (the so-called 'marginal totals') are retained under the model. It is clear that the observed frequencies and the expected frequencies are different. But in sampling, everything always varies, so this is no surprise. The important question is whether the expected frequencies are *significantly* different from the observed frequencies.

We can assess the significance of the differences between observed and expected frequencies in a variety of ways:

- Pearson's chi-squared;

- G test;

- Fisher's exact test.

8.8.1 Pearson's chi-squared

We begin with Pearson's chi-squared test. The test statistic χ^2 is

$$\chi^2 = \sum \frac{(O - E)^2}{E},$$

where O is the observed frequency and E is the expected frequency. It makes the calculations easier if we write the observed and expected frequencies in parallel columns, so that we can work out the corrected squared differences more easily.

	O	E	$(O - E)^2$	$\frac{(O-E)^2}{E}$
Fair hair and blue eyes	38	22.35	244.92	10.96
Fair hair and brown eyes	11	26.65	244.92	9.19
Dark hair and blue eyes	14	29.65	244.92	8.26
Dark hair and brown eyes	51	35.35	244.92	6.93

All we need to do now is to add up the four components of chi-squared to get $\chi^2 = 35.33$.

The question now arises: is this a big value of chi-squared or not? This is important, because if it *is* a bigger value of chi-squared than we would expect by chance, then we should reject the null hypothesis. If, on the other hand, it is within the range of values that we would expect by chance alone, then we should accept the null hypothesis.

We always proceed in the same way at this stage. We have a calculated value of the test statistic: $\chi^2 = 35.33$. We compare this value of the test statistic with the relevant critical value. To work out the critical value of chi-squared we need two things:

- the number of degrees of freedom, and

- the degree of certainty with which to work.

In general, a contingency table has a number of rows (r) and a number of columns (c), and the degrees of freedom is given by

$$\text{d.f.} = (r - 1) \times (c - 1).$$

So we have $(2 - 1) \times (2 - 1) = 1$ degree of freedom for a 2×2 contingency table. You can see why there is only one degree of freedom by working through our example. Take the 'fair hair, brown eyes' box (the top right in the table) and ask how many values this could possibly take. The first thing to note is that the count could not be more than 49, otherwise the row total would be wrong. But in principle, the number in this box is free to take any value between 0 and 49. We have one degree of freedom for this box. But when we have fixed this box to be, say, 11,

	Blue eyes	Brown eyes	Row totals
Fair hair		11	49
Dark hair			65
Column totals	52	62	114

you will see that we have no freedom at all for any of the other three boxes. The top left box has to be $49 - 11 = 38$ because the row total is fixed at 49. Once the top left box is defined as 38 then the bottom left box has to be $52 - 38 = 14$ because the column total is fixed (the total number of people with blue eyes was 52). This means that the bottom right box has to be $65 - 14 = 51$. Thus, *because the marginal totals are constrained*, a 2×2 contingency table has just one degree of freedom.

The next thing we need to do is say how certain we want to be about the falseness of the null hypothesis. The more certain we want to be, the larger the value of chi-squared we would need to reject the null hypothesis. It is conventional to work at the 95% level. That is our certainty level, so our uncertainty level is $100 - 95 = 5\%$. Expressed as a decimal, this is called alpha ($\alpha = 0.05$). Technically, alpha is the probability of *rejecting* the null hypothesis when it is *true*. This is called a Type I error. A Type II error is *accepting* the null hypothesis when it is *false*.

Critical values in R are obtained by use of *quantiles* (q) of the appropriate statistical distribution. For the chi-squared distribution, this function is called qchisq. The function has two arguments: the certainty level ($p = 0.95$), and the degrees of freedom (d.f. $= 1$):

```
qchisq(0.95,1)
```

```
[1]  3.841459
```

The critical value of chi-squared is 3.841. Since the calculated value of the test statistic is *greater* than the critical value we *reject* the null hypothesis.

What have we learned so far? We have rejected the null hypothesis that eye colour and hair colour are independent. But that is not the end of the story, because we have not established the *way* in which they are related (e.g. is the correlation between them positive or negative?). To do this we need to look carefully at the data, and compare the observed and expected frequencies. If fair hair and blue eyes were positively correlated, would the observed frequency be greater or less than the expected frequency? A moment's thought should convince you that the observed frequency will be greater than the expected frequency when the traits are positively correlated (and less when they are negatively correlated). In our case we expected only 22.35 but we observed 38 people (nearly twice as many) to have both fair hair and blue eyes. So it is clear that fair hair and blue eyes are *positively* associated.

In R the procedure is very straightforward. We start by defining the counts as a 2×2 matrix like this:

```
count <- matrix(c(38,14,11,51),nrow=2)
count
```

```
      [,1] [,2]
[1,]    38   11
[2,]    14   51
```

Notice that you enter the data *columnwise* (not row-wise) into the matrix. Then the test uses the `chisq.test` function, with the matrix of counts as its only argument:

```
chisq.test(count)
```

```
        Pearson's Chi-squared test with Yates' continuity correction
data:   count
X-squared = 33.112, df = 1, p-value = 8.7e-09
```

The calculated value of chi-squared is slightly different from ours, because Yates' correction has been applied as the default (see Sokal and Rohlf, 1995, p. 736). If you switch the correction off (`correct=F`), you get the value we calculated by hand:

```
chisq.test(count,correct=F)
```

```
        Pearson's Chi-squared test
data:   count
X-squared = 35.3338, df = 1, p-value = 2.778e-09
```

It makes no difference at all to the interpretation: there is a highly significant positive association between fair hair and blue eyes for this group of people. If you need to extract the frequencies expected under the null hypothesis of independence then use:

```
chisq.test(count,correct=F)$expected
```

```
          [,1]      [,2]
[1,] 22.35088 26.64912
[2,] 29.64912 35.35088
```

8.8.2 *G test of contingency*

The idea is exactly the same. We are looking for evidence of non-independence of hair colour and eye colour. Even the distribution of the critical value is the same: chi-squared. The difference is in the test statistic. Instead of computing Pearson's chi-squared $\sum (O - E)^2/E$, we compute the deviance from a log-linear model (see p. 562):

$$G = 2 \sum O \ln \left(\frac{O}{E} \right).$$

Here are the calculations:

	O	E	$\ln\left(\frac{O}{E}\right)$	$O\ln\left(\frac{O}{E}\right)$
Fair hair and blue eyes	38	22.35	0.5307598	20.168874
Fair hair and brown eyes	11	26.65	-0.8848939	-9.733833
Dark hair and blue eyes	14	29.65	-0.7504048	-10.505667
Dark hair and brown eyes	51	35.35	0.3665272	18.692889

The test statistic G is twice the sum of the right-hand column: $2 \times 18.622\,26 = 37.244\,53$. This value is compared with chi-squared in tables with 1 d.f. as before. The calculated value of the test statistic is much greater than the critical value (3.841) so we reject the null hypothesis of independence. Hair colour and eye colour are correlated in this group of people. We need to look at the data to see which way the correlation goes. We see far more people with fair hair and blue eyes (38) than expected under the null hypothesis of independence (22.35) so the correlation is *positive*. Pearson's chi-squared was $\chi^2 = 35.33$ (above) so the test statistic values are slightly different ($\chi^2 = 37.24$ in the G test) but the interpretation is identical.

8.8.3 Unequal probabilities in the null hypothesis

So far we have assumed equal probabilities, but `chisq.test` can deal with cases with unequal probabilities. This example has 21 individuals distributed over four categories:

```
chisq.test(c(10,3,2,6))

        Chi-squared test for given probabilities

data:   c(10, 3, 2, 6)
X-squared = 7.381, df = 3, p-value = 0.0607
```

The four counts are not significantly different if the probability of appearing in each of the four cells is 0.25 (the calculated p-value is greater than 0.05). However, if the null hypothesis was that the third and fourth cells had 1.5 times the probability of the first two cells, then these counts *are* highly significant.

```
chisq.test(c(10,3,2,6),p=c(0.2,0.2,0.3,0.3))

        Chi-squared test for given probabilities

data:   c(10, 3, 2, 6)
X-squared = 11.3016, df = 3, p-value = 0.0102

Warning message:
In chisq.test(c(10, 3, 2, 6), p = c(0.2, 0.2, 0.3, 0.3)) :
  Chi-squared approximation may be incorrect
```

Note the warning message associated with the low expected frequencies in cells 1 and 2.

8.8.4 Chi-squared tests on table objects

You can use the `chisq.test` function with `table` objects as well as vectors. To test the random number generator as a simulator of the throws of a six-sided die we could simulate 100 throws like this, then use table to count the number of times each number appeared:

```
die <- ceiling(runif(100,0,6))
table(die)
```

```
die
 1  2  3  4  5  6
23 15 20 14 12 16
```

So we observed only 12 fives in this trail and 23 ones. But is this a significant departure from fairness of the die? chisq.test will answer this:

```
chisq.test(table(die))
```

```
        Chi-squared test for given probabilities
```

```
data:    table(die)
X-squared = 5, df = 5, p-value = 0.4159
```

No. This is a fair die ($p = 0.4159$). Note that the syntax is chisq.test(table(die)) not chisq.test(die) and that there are 5 degrees of freedom in this case.

8.8.5   Contingency tables with small expected frequencies: Fisher's exact test

When one or more of the expected frequencies is less than 4 (or 5 depending on the rule of thumb you follow) then it is wrong to use Pearson's chi-squared or log-linear models (G tests) for your contingency table. This is because small expected values inflate the value of the test statistic, and it no longer can be assumed to follow the chi-squared distribution. The individual counts are a, b, c and d like this:

	Column 1	Column 2	Row totals
Row 1	a	b	$a + b$
Row 2	c	d	$c + d$
Column totals	$a + c$	$b + d$	n

The probability of any one particular outcome is given by

$$p = \frac{(a + b)!(c + d)!(a + c)!(b + d)!}{a!b!c!d!n!}$$

where n is the grand total.

Our data concern the distribution of eight ants' nests over 10 trees of each of two species of tree (A and B). There are two categorical explanatory variables (ants and trees), and four contingencies, ants (present or absent) and trees (A or B). The response variable is the vector of four counts c(6,4,2,8) entered columnwise:

	Tree A	Tree B	Row totals
With ants	6	2	8
Without ants	4	8	12
Column totals	10	10	20

We can calculate the probability for this particular outcome:

```
factorial(8)*factorial(12)*factorial(10)*factorial(10)/
     (factorial(6)*factorial(2)*factorial(4)*factorial(8)*factorial(20))
```

```
[1]  0.07501786
```

But this is only part of the story. We need to compute the probability of outcomes that are *more extreme* than this. There are two of them. Suppose only 1 ant colony had been found on tree B. Then the table values would be 7, 1, 3, 9 but the row and column totals would be exactly the same (*the marginal totals are constrained*). The numerator always stays the same, so this case has probability

```
factorial(8)*factorial(12)*factorial(10)*factorial(10)/
        (factorial(7)*factorial(3)*factorial(1)*factorial(9)*factorial(20))
```

```
[1]   0.009526078
```

There is an even more extreme case if no ant colonies at all were found on tree B. Now the table elements become 8, 0, 2, 10 with probability

```
factorial(8)*factorial(12)*factorial(10)*factorial(10)/
        (factorial(8)*factorial(2)*factorial(0)*factorial(10)*factorial(20))
```

```
[1]   0.0003572279
```

and we need to add these three probabilities together:

```
0.07501786 + 0.009526078 + 0.000352279
```

```
[1]   0.08489622
```

But there was no *a priori* reason for expecting that the result would be in this direction. It might have been tree A that happened to have relatively few ant colonies. We need to allow for extreme counts in the opposite direction by doubling this probability (all Fisher's exact tests are two-tailed):

```
2*(0.07501786 + 0.009526078 + 0.000352279)
```

```
[1]   0.1697924
```

This shows that there is no evidence of any correlation between tree and ant colonies. The observed pattern, or a more extreme one, could have arisen by chance alone with probability $p = 0.17$.

There is a built-in function called `fisher.test`, which saves us all this tedious computation. It takes as its argument a 2×2 matrix containing the counts of the four contingencies. We make the matrix like this (compare with the alternative method of making a matrix, above):

```
x <- as.matrix(c(6,4,2,8))
dim(x) <- c(2,2)
x
```

```
      [,1] [,2]
[1,]    6    2
[2,]    4    8
```

We then run the test like this:

```
fisher.test(x)
```

```
        Fisher's Exact Test for Count Data

data:  x
p-value = 0.1698
alternative hypothesis: true odds ratio is not equal to 1
95 percent confidence interval:
 0.6026805 79.8309210
```

```
sample estimates:
odds ratio
   5.430473
```

You see the same non-significant *p*-value that we calculated by hand. Another way of using the function is to provide it with two vectors containing factor levels, instead of a two-dimensional matrix of counts. This saves you the trouble of counting up how many combinations of each factor level there are:

```
table <- read.table("c:\\temp\\fisher.txt",header=TRUE)
head(table)
```

```
  tree nests
1    A  ants
2    B  ants
3    A  none
4    A  ants
5    B  none
6    A  none
```

```
attach(table)
fisher.test(tree,nests)
```

```
        Fisher's Exact Test for Count Data

data:   tree and nests
p-value = 0.1698
alternative hypothesis: true odds ratio is not equal to 1
95 percent confidence interval:
 0.6026805 79.8309210
sample estimates:
odds ratio
   5.430473
```

The `fisher.test` procedure can be used with matrices much bigger than 2 × 2.

8.9   Correlation and covariance

With two continuous variables, *x* and *y*, the question naturally arises as to whether their values are correlated with each other. Correlation is defined in terms of the variance of *x*, the variance of *y*, and the covariance of *x* and *y* (the way the two vary together, which is to say the way they covary) on the assumption that both variables are normally distributed. We have symbols already for the two variances: s_x^2 and s_y^2. We denote the covariance of *x* and *y* by cov(*x*, *y*), so the correlation coefficient *r* is defined as

$$r = \frac{\mathrm{cov}(x, y)}{\sqrt{s_x^2 s_y^2}}.$$

We know how to calculate variances, so it remains only to work out the value of the covariance of *x* and *y*. Covariance is defined as *the expectation of the vector product xy*. The covariance of *x* and *y* is the expectation of the product minus the product of the two expectations. Note that when *x* and *y* are independent (i.e. they are not correlated) then the covariance between *x* and *y* is 0, so $E[xy] = E[x].E[y]$ (i.e. the product of their mean values).

Let us work through a numerical example:

```
data <- read.table("c:\\temp\\twosample.txt",header=T)
attach(data)
plot(x,y,pch=21,col="red",bg="orange")
```

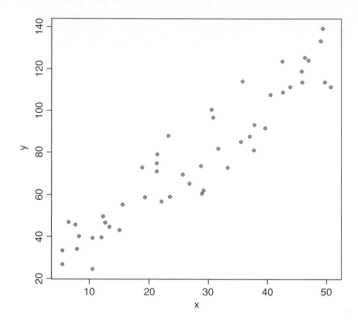

There is clearly a strong positive correlation between the two variables. First, we need the variance of x and the variance of y:

```
var(x)
```

```
[1]   199.9837
```

```
var(y)
```

```
[1]   977.0153
```

The covariance of x and y, $\text{cov}(x, y)$, is given by the var function when we supply it with two vectors like this:

```
var(x,y)
```

```
[1]   414.9603
```

Thus, the correlation coefficient should be $414.96/\sqrt{199.98 \times 977.02}$:

```
var(x,y)/sqrt(var(x)*var(y))
```

```
[1]   0.9387684
```

Let us see if this checks out:

```
cor(x,y)
```

```
[1]   0.9387684
```

So now you know the definition of the correlation coefficient: it is the covariance divided by the geometric mean of the two variances.

8.9.1 Data dredging

The R function `cor` returns the correlation matrix of a data matrix, or a single value showing the correlation between one vector and another (as above):

```
pollute <- read.table("c:\\temp\\Pollute.txt",header=T)
attach(pollute)
cor(pollute)
```

```
             Pollution        Temp    Industry  Population        Wind        Rain    Wet.days
Pollution   1.00000000 -0.43360020  0.64516550  0.49377958  0.09509921  0.05428389  0.36956363
Temp       -0.43360020  1.00000000 -0.18788200 -0.06267813 -0.35112340  0.38628047 -0.43024212
Industry    0.64516550 -0.18788200  1.00000000  0.95545769  0.23650590 -0.03121727  0.13073780
Population  0.49377958 -0.06267813  0.95545769  1.00000000  0.21177156 -0.02606884  0.04208319
Wind        0.09509921 -0.35112340  0.23650590  0.21177156  1.00000000 -0.01246601  0.16694974
Rain        0.05428389  0.38628047 -0.03121727 -0.02606884 -0.01246601  1.00000000  0.49605834
Wet.days    0.36956363 -0.43024212  0.13073780  0.04208319  0.16694974  0.49605834  1.00000000
```

The phrase 'data dredging' is used disparagingly to describe the act of trawling through a table like this, desperately looking for big values which might suggest relationships that you can publish. This behaviour is not to be encouraged. The raw correlation suggests that there is a very strong positive relationship between `Industry` and `Population` ($r = 0.9555$). The correct approach is model simplification (see p. 391), which indicates that people live in places with less, not more, polluted air. Note that the correlations are identical in opposite halves of the matrix (in contrast to regression, where regression of y on x would give different parameter values and standard errors than regression of x on y). The correlation between two vectors produces a single value:

```
cor(Pollution,Wet.days)
```

```
[1]  0.3695636
```

Correlations with single explanatory variables can be highly misleading if (as is typical) there is substantial correlation amongst the explanatory variables (collinearity; see p. 490).

8.9.2 Partial correlation

With more than two variables, you often want to know the correlation between x and y when a third variable, say, z, is held constant. The *partial correlation coefficient* measures this. It enables correlation due to a shared common cause to be distinguished from direct correlation. It is given by

$$r_{xy.z} = \frac{r_{xy} - r_{xz}r_{yz}}{\sqrt{(1 - r_{xz}^2)(1 - r_{yz}^2)}}.$$

Suppose we had four variables and we wanted to look at the correlation between x and y holding the other two, z and w, constant. Then

$$r_{xy.zw} = \frac{r_{xy.z} - r_{xw.z}r_{yw.z}}{\sqrt{(1 - r_{xw.z}^2)(1 - r_{yw.z}^2)}}.$$

You will need partial correlation coefficients if you want to do path analysis. R has a package called `sem` for carrying out structural equation modelling (including the production of `path.diagram`) and another called `corpcor` for converting correlations into partial correlations using the `cor2pcor` function (or vice versa with `pcor2cor`).

8.9.3   Correlation and the variance of differences between variables

Samples often exhibit positive correlations that result from pairing, as in the upstream and downstream invertebrate biodiversity data that we investigated earlier. There is an important general question about the effect of correlation on the variance of differences between variables. In the extreme, when two variables are so perfectly correlated that they are identical, then the difference between one variable and the other is zero. So it is clear that the variance of a difference will decline as the strength of positive correlation increases.

The following data show the depth of the water table (in centimetres below the surface) in winter and summer at 10 locations:

```
data <- read.table("c:\\temp\\wtable.txt",header=T)
attach(data)
names(data)
```

```
[1] "summer" "winter"
```

We begin by asking whether there is a correlation between summer and winter water table depths across locations:

```
cor(summer, winter)
```

```
[1] 0.6596923
```

There is a reasonably strong positive correlation ($p = 0.037\,95$, which is marginally significant; see below). Not surprisingly, places where the water table is high in summer tend to have a high water table in winter as well. If you want to determine the significance of a correlation (i.e. the p value associated with the calculated value of r) then use `cor.test` rather than `cor`. This test has non-parametric options for Kendall's *tau* or Spearman's rank, depending on the method you specify (`method="k"` or `method="s"`), but the default method is Pearson's product-moment correlation (`method="p"`):

```
cor.test(summer, winter)

        Pearson's product-moment correlation

data:   summer and winter
t = 2.4828, df = 8, p-value = 0.03795
alternative hypothesis: true correlation is not equal to 0
95 percent confidence interval:
 0.05142655 0.91094772
sample estimates:
     cor
0.6596923
```

Now, let us investigate the relationship between the correlation coefficient and the three variances: the summer variance, the winter variance, and the *variance of the differences* (winter minus summer water table depth):

```
varS <- var(summer)
varW <- var(winter)
varD <- var(winter-summer)
varS;varW;varD
```

```
[1] 15.13203
[1] 7.541641
[1] 8.579066
```

The correlation coefficient ρ is related to these three variances by:

$$\rho = \frac{\sigma_y^2 + \sigma_z^2 - \sigma_{y-z}^2}{2\sigma_y\sigma_z}$$

So, using the values we have just calculated, we get the correlation coefficient to be

```
(varS+varW-varD)/(2*sqrt(varS)*sqrt(varW))
```

```
[1] 0.6596923
```

which checks out. We can also see whether the variance of the difference is equal to the sum of the component variances (see p. 362):

```
varD
```

```
[1] 8.579066
```

```
varS+varW
```

```
[1] 22.67367
```

No, it is not. They would be equal only if the two samples were *independent*. In fact, we know that the two variables are positively correlated, so the variance of the difference should be *less* than the sum of the variances by an amount equal to $2 \times r \times s_1 \times s_2$:

```
varS+varW-varD
```

```
[1] 14.09461
```

```
2 * cor(summer,winter) * sqrt(varS) * sqrt(varW)
```

```
[1] 14.09461
```

That's more like it.

8.9.4 Scale-dependent correlations

Another major difficulty with correlations is that scatterplots can give a highly misleading impression of what is going on. The moral of this exercise is very important: things are not always as they seem. The following data show the number of species of mammals (y) in forests of differing productivity (x):

```
productivity <- read.table("c:\\temp\\productivity.txt",header=T)
attach(productivity)
head(productivity)
```

```
   x y f
1  1 3 a
2  2 4 a
3  3 2 a
4  4 1 a
5  5 3 a
6  6 1 a
```

```
plot(x,y,pch=21,col="blue",bg="green",
                    xlab="Productivity",ylab="Mammal species")
```

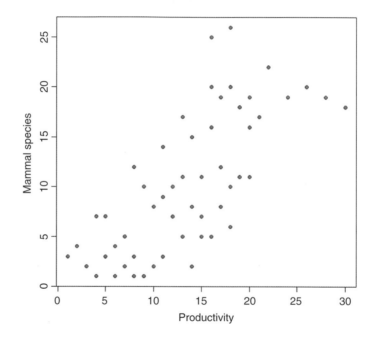

```
cor.test(x,y)
```

```
        Pearson's product-moment correlation
```

```
data:  x and y
t = 7.5229, df = 52, p-value = 7.268e-10
alternative hypothesis: true correlation is not equal to 0
95 percent confidence interval:
 0.5629686 0.8293555
sample estimates:
     cor
0.7219081
```

There is evidently a significant *positive* correlation ($p < 0.000\ 001$) between mammal species and productivity: increasing productivity is associated with increasing species richness. However, when we look at the relationship for each region (f) separately using coplot, we see exactly the opposite relationship:

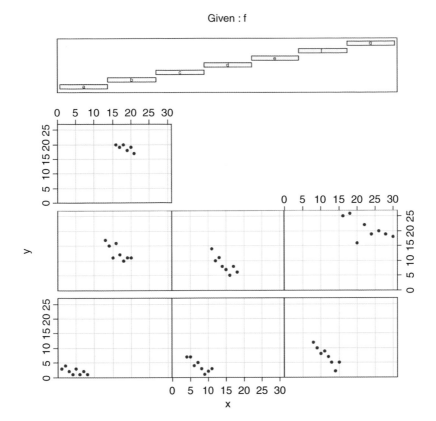

The pattern is obvious. In every single case, increasing productivity is associated with *reduced* mammal species richness within each region (regions are labelled a–g from bottom left). The lesson is clear: you need to be extremely careful when looking at *correlations across different scales*. Things that are positively correlated over short time scales may turn out to be negatively correlated in the long term. Things that appear to be positively correlated at large spatial scales may turn out (as in this example) to be negatively correlated at small scales.

8.10 Kolmogorov–Smirnov test

People know this test for its wonderful name, rather than for what it actually does. It is an extremely simple test for asking one of two different questions:

- Are two sample distributions the same, or are they significantly different from one another in one or more (unspecified) ways?

- Does a particular sample distribution arise from a particular hypothesized distribution?

The two-sample problem is the one most often used. The apparently simple question is actually very broad. It is obvious that two distributions could be different because their means were different. But two distributions with exactly the same mean could be significantly different if they differed in variance, or in skew or kurtosis (see p. 350).

The Kolmogorov–Smirnov test works on **cumulative distribution functions**. These give the probability that a randomly selected value of X is less than or equal to x:

$$F(x) = P[X \leq x].$$

This sounds somewhat abstract. Suppose we had insect wing sizes (y) for two geographically separated populations (A and B) and we wanted to test whether the distribution of wing lengths was the same in the two places:

```
data <- read.table("c:\\temp\\ksdata.txt",header=T)
attach(data)
names(data)
```

```
[1] "y"     "site"
```

We start by extracting the data for the two populations, and describing the samples:

```
table(site)
```

```
site
 A  B
10 12
```

There are 10 samples from site A and 12 from site B.

```
tapply(y,site,mean)
```

```
       A         B
4.355266 11.665089
```

```
tapply(y,site,var)
```

```
       A        B
27.32573 90.30233
```

Their means are quite different, but the size of the difference in their variances precludes using a t test. We start by plotting the cumulative probabilities for the two samples on the same axes, bearing in mind that there are 10 values if A and 12 values of B:

```
plot(seq(0,1,length=12),cumsum(sort(B)/sum(B)),type="l",
                ylab="Cumulative probability",xlab="Index",col="red")
lines(seq(0,1,length=10),cumsum(sort(A)/sum(A)),col="blue")
```

It certainly looks as if population A (the blue line) is different. We test the significance of the difference between the two distributions with ks.test like this:

```
ks.test(y[site=="A"],y[site=="B"])
```

```
        Two-sample Kolmogorov-Smirnov test
```

```
data:   y[site == "A"] and y[site == "B"]
D = 0.55, p-value = 0.04889
alternative hypothesis: two-sided
```

The test works despite the difference in length of the two vectors, and shows a marginally significant difference between the two sites ($p = 0.049$).

The other test involves comparing one sample with the probability function of a named distribution. Let us test whether the larger sample from site B is normally distributed, using pnorm as the probability function, with specified mean and standard deviation:

```
ks.test(y[site=="B"],"pnorm",mean(y[site=="B"]),sd(y[site=="B"]))
```

```
        One-sample Kolmogorov-Smirnov test
```

```
data:   y[site == "B"]
D = 0.1844, p-value = 0.7446
alternative hypothesis: two-sided
```

There is no evidence that the samples from site B depart significantly from normality. Note, however, that the Shapiro–Wilk test

```
shapiro.test(y[site=="B"])
```

```
        Shapiro-Wilk normality test
```

```
data:   y[site == "B"]
W = 0.876, p-value = 0.0779
```

comes much closer to suggesting significant non-normality (above), while

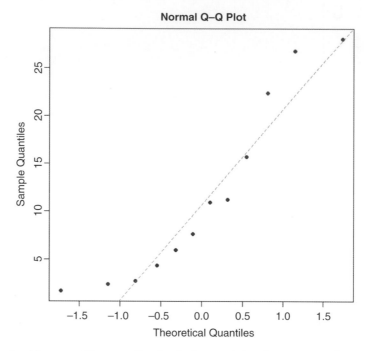

the standard model-checking quantile–quantile plot looks suspiciously non-normal:

```
qqnorm(y[site=="B"],pch=16,col="blue")
qqline(y[site=="B"],col="green",lty=2)
```

8.11   Power analysis

The power of a test is the probability of rejecting the null hypothesis when it is false. It has to do with Type II errors: β is the probability of accepting the null hypothesis when it is false. In an ideal world, we would obviously make β as small as possible. But there is a snag. The smaller we make the probability of committing a Type II error, the greater we make the probability of committing a Type I error, and rejecting the null hypothesis when, in fact, it is correct. This is a classic trade-off. A compromise is called for. Most statisticians work with $\alpha = 0.05$ and $\beta = 0.2$. The power of a test is defined as $1 - \beta = 0.8$ under the standard assumptions.

The issues involved are your choices of alpha and beta (the trade-off between Type I and Type II errors), the size of the effect you want to detect as being significant, the variance of the samples, and the sample size. If we are doing a two-sample t test, the value of the test statistic is the difference between the two means, d, divided by the standard error of the difference between two means (assuming equal variances, s^2, and equal sample sizes, n):

$$t = \frac{d}{\sqrt{2\frac{s^2}{n}}}$$

Let us rearrange this expression to find the sample size as a function of the other variables:

$$\sqrt{2\frac{s^2}{n}} = \frac{d}{t} \quad \Rightarrow \quad 2\frac{s^2}{n} = \frac{d^2}{t^2}$$

so

$$n = \frac{2s^2t^2}{d^2}.$$

The value of t depends on our choice of power $(1 - \beta = 0.8)$ and significance level $(\alpha = 0.025)$. Roughly speaking, the quantile associated with the 0.025 tail of a normal distribution is 1.96, and the quantile associated with 0.8 is 0.84. We add these quantiles to estimate $t = 2.8$, so t^2 is roughly 7.8. To get our rule of thumb, we round this up to 8. Now the formula for n is $2 \times 8 \times$ variance/the square of the difference:

$$n = \frac{16s^2}{d^2}.$$

The smaller the effect size that we want to be able to detect as being significant, the larger the sample size will need to be.

Suppose that the control value of our response variable is known from the literature to have a mean of 20 and a standard deviation of 2 (so the variance is 4). The rule of thumb would give the following relationship:

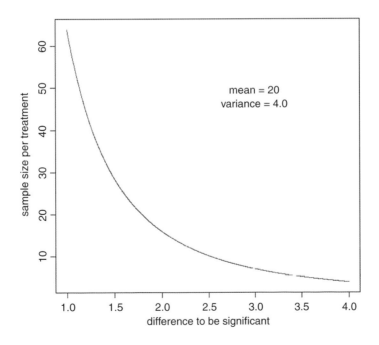

So if you want to be able to detect an effect size of 1.0 you will need at least 60 samples per treatment. The standard idea of a 'big-enough' sample $(n = 30)$ would enable you to detect an effect size of about 1.5 in this example. If you could only afford 10 replicates per treatment, you should not expect to be able to detect effects smaller than about 2.5.

There are built-in functions in R for carrying out power analyses for ANOVA, proportion data and *t* tests:

`power.t.test`	power calculations for one- and two-sample *t* tests;
`power.prop.test`	power calculations two-sample test for proportions;
`power.anova.test`	power calculations for balanced one-way ANOVA tests.

The arguments to the `power.t.test` function are n (the number of observations per group), delta (the difference in means we want to be able to detect; you will need to think hard about this value), sd (the standard deviation of the sample), sig.level (the significance level, i.e. Type I error probability, where you will often accept the default value of 5%), power (the power you want the test to have, where you will often accept the default value of 80%), type (the type of *t* test you want to carry out: two-sample, one-sample or paired) and alternative (whether you want to do a one- or a two-tailed test, where you will typically want to do the default, two-tailed test). One of the parameters n, delta, power, sd and sig.level must be passed as NULL, and that parameter will be calculated from the others. This sounds like a lot of work, but you will typically use all of the defaults so you only need to specify the difference, delta, and the standard deviation, sd, to work out the sample size n that will give you the power you want.

So how many replicates do we need in each of two samples to detect a difference of 10% with power = 80% when the mean is 20 (i.e. delta = 2.0) and the standard deviation is about 3.5?

```
power.t.test(delta=2,sd=3.5,power=0.8)

     Two-sample t test power calculation

              n = 49.05349
          delta = 2
             sd = 3.5
      sig.level = 0.05
          power = 0.8
    alternative = two.sided

  NOTE: n is number in *each* group
```

The (perhaps rather shocking) answer is that you need at least 50 replicates from each sample (100 data points in all).

If you had been working with a rule of thumb like '30 is a big enough sample' then you would be severely disappointed in this case. You simply could not have detected a difference of 10% with this experimental design. You need 50 replicates in each sample (100 replicates in all) to achieve a power of 80%. You can work out what size of difference your sample of 30 would allow you to detect, by specifying *n* (15 in each treatment) and omitting *delta*:

```
power.t.test(n=15,sd=3.5,power=0.8)

     Two-sample t test power calculation
              n = 15
          delta = 3.709303
             sd = 3.5
      sig.level = 0.05
          power = 0.8
    alternative = two.sided
```

This shows that you could have detected an 18.5% change (100 × 3.709/20), which is roughly double the effect size you hoped to be able to detect (10% = 2.0). The work you need to do before carrying out a power analysis before designing your experiment is to find values for the standard deviation (from the literature or by carrying out a pilot experiment) and the size of the difference your want to detect (from discussions with your sponsor and your colleagues). Experiments in ecology are often planned to be able to detect 50% effects. Aspiring to estimate effects as small as 10% would lead to impossibly large sample sizes (see the discussion in Perry et al., 2003).

8.12 Bootstrap

We want to use bootstrapping to obtain a 95% confidence interval for the mean of a vector of numbers called values:

```
data <- read.table("c:\\temp\\skewdata.txt",header=T)
attach(data)
names(data)
```

```
[1]  "values"
```

We shall sample with replacement from values using `sample(values,replace=T)`, then work out the mean, repeating this operation 10 000 times, and storing the 10 000 different mean values in a vector called ms:

```
ms <- numeric(10000)
for (i in 1:10000){
ms[i] <- mean(sample(values,replace=T)) }
```

The answer to our problem is provided by the `quantile` function applied to ms: we want to know the values of ms associated with its 0.025 and 0.975 tails:

```
quantile(ms,c(0.025,0.975))
```

```
2.5%           97.5%
24.97918       37.62932
```

Thus the intervals below and above the mean are

```
mean(values)-quantile(ms,c(0.025,0.975))
```

```
2.5%           97.5%
5.989472       -6.660659
```

How does this compare with the parametric confidence interval, $CI = 1.96 \times \sqrt{s^2/n}$?

```
1.96*sqrt(var(values)/length(values))
```

```
[1]  6.569802
```

Close, but not identical. Our bootstrapped intervals are skew because the data are skewed, but the parametric interval, of course, is symmetric.

Now let us see how to do the same thing using the `boot` function from the library called `boot`:

```
install.packages("boot")
library(boot)
```

The syntax of boot is very simple:

```
                  boot(data, statistic, R)
```

The trick to using `boot` lies in understanding how to write the statistic function. `R` is the number of resamplings you want to do (R=10000 in this example), and `data` is the name of the data object to be resampled (`values` in this case). The attribute we want to estimate repeatedly is the mean value of `values`. Thus, the first argument to our function must be `values`. The second argument is an index (a vector of subscripts) that is used within `boot` to select random assortments of `values`. Our `statistic` function can use the built-in function `mean` to calculate the mean value of the sample of `values`.

```
mymean <- function(values,i) mean(values[i])
```

The key point is that we write `mean(values[i])` not `mean(values)`. Now we can run the bootstrap for 10 000 iterations:

```
myboot <- boot(values,mymean,R=10000)
myboot
```

```
ORDINARY NONPARAMETRIC BOOTSTRAP
Call:
boot(data = values, statistic = mymean, R = 10000)
Bootstrap Statistics :
        original            Bias      std. error
   t1* 30.96866     -0.08155796       3.266455
```

The output is interpreted as follows. The `original` is the mean of the whole sample:

```
mean(values)
```

```
[1]  30.96866
```

while `bias` is the difference between the arithmetic mean and the mean of the bootstrapped samples which are in the variable called `myboot$t`:

```
mean(myboot$t)-mean(values)
```

```
[1]  -0.08155796
```

and `std. error` is the standard deviation of the simulated values in `myboot$t`:

```
sqrt(var(myboot$t))
```

```
           [,1]
[1,]   3.266455
```

The components of `myboot` can be used to do other things. For instance, we can compare our homemade vector (`ms` above) with a histogram of `myboot$t`:

```
windows(7,4)
par(mfrow=c(2,1))
hist(ms)
hist(myboot$t)
```

They differ in detail because they were generated with different series of random numbers. Here are the 95% intervals for comparison with ours, calculated from the quantiles of `myboot$t`:

```
mean(values)-quantile(myboot$t,c(0.025,0.975))
```

```
2.5%            97.5%
6.126120     -6.599232
```

There is a function `boot.ci` for calculating confidence intervals from the `boot` object:

```
boot.ci(myboot)

BOOTSTRAP CONFIDENCE INTERVAL CALCULATIONS
Based on 10000 bootstrap replicates
CALL :
boot.ci(boot.out = myboot)
Intervals :
    Level        Normal                  Basic
     95%      (24.65, 37.45)         (24.37, 37.10)
    Level       Percentile               BCa
     95%      (24.84, 37.57)         (25.63, 38.91)
Calculations and Intervals on Original Scale
Warning message:
bootstrap variances needed for studentized intervals in:
boot.ci(myboot)
```

`Normal` is the parametric CI based on the standard error of the mean and the sample size (p. 514). The `Percentile` interval is the quantile from the bootstrapped estimates:

```
quantile(myboot$t,c(0.025,0.975))

    2.5%         97.5%
 24.84254      37.56789
```

which, as we saw earlier, was close to our home-made values (above). The BCa interval is the bias-corrected accelerated percentile. It is not greatly different in this case, but it is the interval preferred by statisticians. A more complex example of the use of bootstrapping involving a generalized linear model is explained on p. 570. For other examples see `?boot`, and for more depth read the Davison and Hinkley (1997) book from which the `boot` package was developed (as programmed by A.J. Canty).

9

Statistical Modelling

The hardest part of any statistical work is getting started. And one of the hardest things about getting started is choosing the right kind of statistical analysis. The choice depends on the nature of your data and on the particular question you are trying to answer. The key is to understand what kind of *response* variable you have, and to know the nature of your *explanatory* variables. The response variable is the thing you are working on: it is the variable whose variation you are attempting to understand. This is the variable that goes on the y axis of the graph. The explanatory variable goes on the x axis of the graph; you are interested in the extent to which variation in the response variable is associated with variation in the explanatory variable. You also need to consider the *way* that the variables in your analysis measure what they purport to measure. A continuous measurement is a variable such as height or weight that can take any real numbered value. A categorical variable is a factor with two or more levels: sex is a factor with two levels (male and female), and colour might be a factor with seven levels (red, orange, yellow, green, blue, indigo, violet).

It is essential, therefore, that you can answer the following questions:

- Which of your variables is the response variable?

- Which are the explanatory variables?

- Are the explanatory variables continuous or categorical, or a mixture of both?

- What kind of response variable do you have: is it a continuous measurement, a count, a proportion, a time at death, or a category?

These simple keys will lead you to the appropriate statistical method:

The explanatory variables
(a)	All explanatory variables continuous	**Regression**
(b)	All explanatory variables categorical	**Analysis of variance (ANOVA)**
(c)	Explanatory variables both continuous and categorical	**Analysis of covariance (ANCOVA)**

The R Book, Second Edition. Michael J. Crawley.
© 2013 John Wiley & Sons, Ltd. Published 2013 by John Wiley & Sons, Ltd.

The response variable

(a)	Continuous	**Normal regression, ANOVA or ANCOVA**
(b)	Proportion	**Logistic regression**
(c)	Count	**Log-linear models**
(d)	Binary	**Binary logistic analysis**
(e)	Time at death	**Survival analysis**

The object is to determine the values of the parameters in a specific model that lead to *the best fit of the model to the data*. The data are sacrosanct, and they tell us what actually happened under a given set of circumstances. It is a common mistake to say 'the data were fitted to the model' as if the data were something flexible, and we had a clear picture of the structure of the model. On the contrary, what we are looking for is the minimal adequate model to describe the data. The model is fitted to data, not the other way around. The best model is the model that produces the least unexplained variation (the *minimal residual deviance*), subject to the constraint that all the parameters in the model should be statistically significant.

You have to specify the model. It embodies your mechanistic understanding of the explanatory variables involved, and of the way that they are related to the response variable. You want the model to be *minimal* because of the principle of parsimony, and *adequate* because there is no point in retaining an inadequate model that does not describe a significant fraction of the variation in the data. It is very important to understand that *there is not one model*; this is one of the common implicit errors involved in traditional regression and ANOVA, where the same models are used, often uncritically, over and over again. In most circumstances, there will be a large number of different, more or less plausible models that might be fitted to any given set of data. Part of the job of data analysis is to determine which, if any, of the possible models are adequate, and then, out of the set of adequate models, which is the minimal adequate model. In some cases there may be no single best model and a set of different models may all describe the data equally well (or equally poorly if the variability is great).

9.1   First things first

The commonest mistake is to try to do the statistical modelling straight away. The best thing is to spend a substantial amount of time, right at the outset, getting to know your data and what they show. This will help guide your thinking as to exactly what kind of statistical modelling is most appropriate.

- In a spreadsheet, make sure the dataframe is correct in structure and content:

  - Do *all* of the values of each variable appear in the same column?

  - Are all the zeros really 0, or should they be NA?

  - Does every row contain the same number of entries?

  - Are there any variable names with blank spaces in them?

- Read the dataframe into R using `read.table` (or `read.csv` if factor levels (like place names) contain blank spaces) (p. 139).

- Look at the `head` and the `tail` of the dataframe and check for mistakes (p. 161).

- Plot every one of the variables on its own to check for gross errors (`plot(x)`, `plot(y)` etc.; see p. 190).

- Look at the relationships between variables (use `tapply`, `plot`, `tree` and `gam`) (p. 768).

- Think about model choice (p. 1)

  ○ Which explanatory variables should be included?

  ○ What transformation of the response is most appropriate?

  ○ Which interactions should be included?

  ○ Which non-linear terms should be included?

  ○ Is there pseudoreplication, and if so, how should it be dealt with?

  ○ Should the explanatory variables be transformed?

- Try to use the simplest kind of analysis that is appropriate to your data and the question you are trying to answer (e.g. do a one-way ANOVA rather than a mixed-effects model) (p. 344).

- Fit a maximal model and simplify it by stepwise deletion (p. 391).

- Check the minimal adequate model for constancy of variance and normality of errors using `plot(model)` (p. 405).

- Emphasize the effect sizes and standard errors (`summary.lm`), and play down the analysis of deviance table (`summary.aov`) (p. 382).

- Document carefully what you have done, and explain all the steps you took. That way, you should be able to understand what you did and why you did it, when you return to the analysis in 6 months' time.

9.2 Maximum likelihood

What, exactly, do we mean when we say that the parameter values should afford the 'best fit of the model to the data'? The convention we adopt is that our techniques should lead to **unbiased, variance-minimizing estimators**. We define 'best' in terms of **maximum likelihood**. This notion may be unfamiliar, so it is worth investing some time to get a feel for it. This is how it works:

- given the data,

- and given our choice of model,

- what values of the parameters of that model

- make the observed data most likely?

We judge the model on the basis how likely the data would be if the model were correct.

9.3 The principle of parsimony (Occam's razor)

One of the most important themes running through this book concerns model simplification. The principle of parsimony is attributed to the early fourteenth-century English nominalist philosopher, William of Occam, who insisted that, given a set of equally good explanations for a given phenomenon, *the correct explanation is the simplest explanation*. It is called Occam's razor because he 'shaved' his explanations down to the bare minimum: his point was that in explaining something, assumptions must not be needlessly multiplied.

In particular, for the purposes of explanation, things not *known* to exist should not, unless it is absolutely necessary, be postulated as existing. For statistical modelling, the principle of parsimony means that:

- models should have as few parameters as possible;

- linear models should be preferred to non-linear models;

- experiments relying on few assumptions should be preferred to those relying on many;

- models should be pared down until they are *minimal adequate*;

- simple explanations should be preferred to complex explanations.

The process of model simplification is an integral part of hypothesis testing in R. In general, a variable is retained in the model only *if it causes a significant increase in deviance when it is removed from the current model*. Seek simplicity, then distrust it.

In our zeal for model simplification, however, we must be careful not to throw the baby out with the bathwater. Einstein made a characteristically subtle modification to Occam's razor. He said: 'A model should be as simple as possible. But no simpler.' Remember, too, what Oscar Wilde said: 'Truth is rarely pure, and never simple.'

9.4 Types of statistical model

Fitting models to data is the central function of R. The process is essentially one of exploration; there are no fixed rules and no absolutes. The object is to determine a minimal adequate model (see Table 9.1) from the large set of potential models that might be used to describe the given set of data. In this book we discuss five

Table 9.1. Statistical modelling involves the selection of a minimal adequate model from a potentially large set of more complex models, using stepwise model simplification.

Model	Interpretation
Saturated model	One parameter for every data point Fit: perfect Degrees of freedom: none Explanatory power of the model: none
Maximal model	Contains all (p) factors, interactions and covariates that might be of any interest. Many of the model's terms are likely to be insignificant Degrees of freedom: $n - p - 1$ Explanatory power of the model: it depends
Minimal adequate model	A simplified model with $1 \leq p' \leq p$ parameters Fit: less than the maximal model, but not significantly so Degrees of freedom: $n - p' - 1$ Explanatory power of the model: $r^2 = SSR/SSY$
Null model	Just one parameter, the overall mean $\bar{y}$ Fit: none; $SSE = SSY$ Degrees of freedom: $n - 1$ Explanatory power of the model: none

types of model:

- the null model;
- the minimal adequate model;
- the current model;
- the maximal model; and
- the saturated model.

The stepwise progression from the saturated model (or the maximal model, whichever is appropriate) through a series of simplifications to the minimal adequate model is made on the basis of **deletion tests**. These are F tests or chi-squared tests that assess the significance of the increase in deviance that results when a given term is removed from the current model.

Models are representations of reality that should be both accurate and convenient. However, it is impossible to maximize a model's realism, generality and holism simultaneously, and the principle of parsimony is a vital tool in helping to choose one model over another. Thus, we would only include an explanatory variable in a model if it significantly improved the fit of the model. The fact that we went to the trouble of measuring something does not mean we have to have it in our model. Parsimony says that, other things being equal, we prefer:

- a model with $n - 1$ parameters to a model with n parameters;
- a model with $k - 1$ explanatory variables to a model with k explanatory variables;
- a linear model to a model which is curved;
- a model without a hump to a model with a hump;
- a model without interactions to a model containing interactions between factors.

Other considerations include a preference for models containing explanatory variables that are easy to measure over variables that are difficult or expensive to measure. Also, we prefer models that are based on a sound mechanistic understanding of the process over purely empirical functions. Some variables are so important that we retain them in the model even though their parameters are not significantly different from zero (e.g. density dependence in population models).

Parsimony requires that the model should be as simple as possible. This means that the model should not contain any redundant parameters or factor levels. We achieve this by fitting a maximal model and then simplifying it by following one or more of these steps:

- remove non-significant interaction terms;
- remove non-significant quadratic or other non-linear terms;
- remove non-significant explanatory variables;
- group together factor levels that do not differ from one another;
- in ANCOVA, set non-significant slopes of continuous explanatory variables to zero.

All the above are subject, of course, to the caveats that the simplifications make good scientific sense and do not lead to significant reductions in explanatory power.

Just as there is no perfect model, so there may be no optimal scale of measurement for a model. Suppose, for example, we had a process that had Poisson errors with multiplicative effects amongst the explanatory variables. Then, we must choose between three different scales, each of which optimizes one of three different properties:

- the scale of $\sqrt{y}$ would give constancy of variance;
- the scale of $y^{2/3}$ would give approximately normal errors;
- the scale of $\ln(y)$ would give additivity.

Thus, any measurement scale is always going to be a compromise, and we should choose the scale that gives the best overall performance of the model.

9.5   Steps involved in model simplification

There are no hard-and-fast rules, but the procedure laid out in Table 9.2 works well in practice. With large numbers of explanatory variables, and many interactions and non-linear terms, the process of model simplification can take a very long time. But this is time well spent because it reduces the risk of overlooking an important aspect of the data. It is important to realize that there is no guaranteed way of finding all the important structures in a complex dataframe.

9.5.1   Caveats

Model simplification is an important process but it should not be taken to extremes. For example, care should be taken with the interpretation of deviances and standard errors produced with fixed parameters that have been estimated from the data. Again, the search for 'nice numbers' should not be pursued uncritically. Sometimes there are good scientific reasons for using a particular number (e.g. a power of 0.66 in an allometric

Table 9.2.   Model simplification process.

Step	Procedure	Explanation
1	Fit the maximal model	Fit all the factors, interactions and covariates of interest. Note the residual deviance. If you are using Poisson or binomial errors, check for overdispersion and rescale if necessary.
2	Begin model simplification	Inspect the parameter estimates using the R function `summary`. Remove the least significant terms first, using `update -`, starting with the highest-order interactions.
3	If the deletion causes an insignificant increase in deviance	Leave that term out of the model. Inspect the parameter values again. Remove the least significant term remaining.
4	If the deletion causes a significant increase in deviance	Put the term back in the model using `update +`. These are the statistically significant terms as assessed by deletion from the maximal model.
5	Keep removing terms from the model	Repeat steps 3 or 4 until the model contains nothing but significant terms. This is the minimal adequate model. If none of the parameters is significant, then the minimal adequate model is the null model.

relationship between respiration and body mass). It is much more straightforward, for example, to say that yield increases by 2 kg per hectare for every extra unit of fertilizer, than to say that it increases by 1.947 kg. Similarly, it may be preferable to say that the odds of infection increase 10-fold under a given treatment, than to say that the logits increase by 2.321; without model simplification this is equivalent to saying that there is a 10.186-fold increase in the odds. It would be absurd, however, to fix on an estimate of 6 rather than 6.1 just because 6 is a whole number.

9.5.2   Order of deletion

The data in this book fall into two distinct categories. In the case of planned experiments, all of the treatment combinations are equally represented and, barring accidents, there are no missing values. Such experiments are said to be **orthogonal**. In the case of observational studies, however, we have no control over the number of individuals for which we have data, or over the combinations of circumstances that are observed. Many of the explanatory variables are likely to be correlated with one another, as well as with the response variable. Missing treatment combinations are commonplace, and the data are said to be **non-orthogonal**. This makes an important difference to our statistical modelling because, in orthogonal designs, the variation that is attributed to a given factor is constant, and does not depend upon the order in which factors are removed from the model. In contrast, with non-orthogonal data, we find that the variation attributable to a given factor *does* depend upon the order in which factors are removed from the model. We must be careful, therefore, to judge the significance of factors in non-orthogonal studies, when they are *removed from the maximal model* (i.e. from the model including all the other factors and interactions with which they might be confounded). Remember that, *for non-orthogonal data, order matters*.

Also, if your explanatory variables are correlated with each other, then the significance you attach to a given explanatory variable will depend upon whether you delete it from a maximal model or add it to the null model. If you always test by model simplification then you will not fall into this trap.

The fact that you have laboured long and hard to include a particular experimental treatment does not justify the retention of that factor in the model if the analysis shows it to have no explanatory power. ANOVA tables are often published containing a mixture of significant and non-significant effects. This is not a problem in orthogonal designs, because sums of squares can be unequivocally attributed to each factor and interaction term. But as soon as there are missing values or unequal weights, then it is impossible to tell how the parameter estimates and standard errors of the significant terms would have been altered if the non-significant terms had been deleted. The best practice is as follows:

- Say whether your data are orthogonal or not.

- Explain any correlations amongst your explanatory variables.

- Present a minimal adequate model.

- Give a list of the non-significant terms that were omitted, and the deviance changes that resulted from their deletion.

If you do this, then readers can judge for themselves the relative magnitude of the non-significant factors, and the importance of correlations between the explanatory variables.

The temptation to retain terms in the model that are 'close to significance' should be resisted. The best way to proceed is this. If a result would have been *important* if it had been statistically significant, then it is worth repeating the experiment with higher replication and/or more efficient blocking, in order to demonstrate the importance of the factor in a convincing and statistically acceptable way.

9.6   Model formulae in R

The structure of the model is specified in the model formula like this:

$$\text{response variable} \sim \text{explanatory variable(s)}$$

where the tilde symbol $\sim$ reads 'is modelled as a function of' (see Table 9.3 for examples). So a simple linear regression of y on x would be written as

```
y~x
```

and a one-way ANOVA where sex is a two-level factor would be written as

```
y~sex
```

Table 9.3.   Examples of R model formulae. In a model formula, the function I (upper case 'I') stands for 'as is' and is used for generating sequences, `I(1:10)`, or calculating quadratic terms, `I(x^2)`.

Model	Model formula	Comments
Null	`y~1`	1 is the intercept in regression models, but here it is the overall mean y
Regression	`y~x`	x is a continuous explanatory variable
Regression through origin	`y~x-1`	Do not fit an intercept
One-way ANOVA	`y~sex`	sex is a two-level categorical variable
One-way ANOVA	`y~sex-1`	as above, but do not fit an intercept (gives two means rather than a mean and a difference)
Two-way ANOVA	`y~sex + genotype`	genotype is a four-level categorical variable
Factorial ANOVA	`y~N * P * K`	N, P and K are two-level factors to be fitted along with all their interactions
Three-way ANOVA	`y~N*P*K - N:P:K`	As above, but do not fit the three-way interaction
Analysis of covariance	`y~x + sex`	A common slope for y against x but with two intercepts, one for each sex
Analysis of covariance	`y~x * sex`	Two slopes and two intercepts
Nested ANOVA	`y~a/b/c`	Factor c nested within factor b within factor a
Split-plot ANOVA	`y~a*b*c+Error(a/b/c)`	A factorial experiment but with three plot sizes and three different error variances, one for each plot size
Multiple regression	`y~x + z`	Two continuous explanatory variables, flat surface fit
Multiple regression	`y~x * z`	Fit an interaction term as well (x + z + x:z)
Multiple regression	`y~x + I(x^2) + z + I(z^2)`	Fit a quadratic term for both x and z
Multiple regression	`y <- poly(x,2) + z`	Fit a quadratic polynomial for x and linear z
Multiple regression	`y~(x + z + w)^2`	Fit three variables plus all their interactions up to two-way
Non-parametric model	`y~s(x) + s(z)`	y is a function of smoothed x and z in a generalized additive model
Transformed response and explanatory variables	`log(y)~I(1/x) + sqrt(z)`	All three variables are transformed in the model

The right-hand side of the model formula shows:

- the number of explanatory variables and their identities – their attributes (e.g. continuous or categorical) are usually defined prior to the model fit;
- the interactions between the explanatory variables (if any);
- non-linear terms in the explanatory variables.

On the right of the tilde, one also has the option to specify offsets or error terms in some special cases. As with the response variable, the explanatory variables can appear as transformations, or as powers or polynomials.

It is very important to note that symbols are used differently in model formulae than in arithmetic expressions. In particular:

+	indicates inclusion of an explanatory variable in the model (not addition);
−	indicates deletion of an explanatory variable from the model (not subtraction);
*	indicates inclusion of explanatory variables and interactions (not multiplication);
/	indicates nesting of explanatory variables in the model (not division);
|	indicates conditioning (not 'or'), so that y~x | z is read as 'y as a function of x given z'.

There are several other symbols that have special meaning in model formulae. A colon denotes an interaction, so that A:B means the two-way interaction between A and B, and N:P:K:Mg means the four-way interaction between N, P, K and Mg.

Some terms can be written in an expanded form. Thus:

A*B*C	is the same as A+B+C+A:B+A:C+B:C+A:B:C;
A/B/C	is the same as A+B%in%A+C%in%B%in%A;
(A+B+C)^3	is the same as A*B*C;
(A+B+C)^2	is the same as A*B*C - A:B:C.

9.6.1 Interactions between explanatory variables

Interactions between two two-level categorical variables of the form A*B mean that two main effect means and one interaction mean are evaluated. On the other hand, if factor A has three levels and factor B has four levels, then seven parameters are estimated for the main effects (three means for A and four means for B). The number of interaction terms is $(a - 1)(b - 1)$, where a and b are the numbers of levels of the factors A and B, respectively. So in this case, R would estimate $(3 - 1)(4 - 1) = 6$ parameters for the interaction.

Interactions between two continuous variables are fitted differently. If x and z are two continuous explanatory variables, then x*z means fit x+z+x:z and the interaction term x:z behaves as if a new variable had been computed that was the pointwise product of the two vectors x and z. The same effect could be obtained by calculating the product explicitly,

```
product.xz <- x * z
```

then using the model formula y~x + z + product.xz. Note that the representation of the interaction by the *product* of the two continuous variables is an assumption, not a fact. The real interaction might be of an altogether different functional form (e.g. x * z^2).

Interactions between a categorical variable and a continuous variable are interpreted as an analysis of covariance; a separate slope and intercept are fitted for each level of the categorical variable. So y~A*x would fit three regression equations if the factor A had three levels; this would estimate six parameters from the data – three slopes and three intercepts.

The slash operator is used to denote nesting. Thus, with categorical variables A and B,

```
y ~ A/B
```

means fit 'A plus B within A'. This could be written in two other equivalent ways:

```
y ~ A + A:B
y ~ A + B %in% A
```

both of which alternatives emphasize that there is no point in attempting to estimate a main effect for B (it is probably just a factor label like 'tree number 1' that is of no scientific interest; see p. 681).

Some functions for specifying non-linear terms and higher-order interactions are useful. To fit a polynomial regression in x and z, we could write

```
y ~ poly(x,3) + poly(z,2)
```

to fit a cubic polynomial in x and a quadratic polynomial in z. To fit interactions, but only up to a certain level, the ^ operator is useful. The formula

```
y ~ (A + B + C)^2
```

fits all the main effects and two-way interactions (i.e. it excludes the three-way interaction that A*B*C would have included).

The I function (upper-case letter 'i') stands for 'as is'. It overrides the interpretation of a model symbol as a formula operator when the intention is to use it as an arithmetic operator. Suppose you wanted to fit $1/x$ as an explanatory variable in a regression. You might try

```
y ~ 1/x
```

but this actually does something very peculiar. It fits x nested within the intercept (whatever that might represent). When it appears in a model formula, the slash operator is assumed to imply nesting. To obtain the effect we want, we use I ('as is') to write

```
y ~ I(1/x)
```

We also need to use I when we want * to represent multiplication and ^ to mean 'to the power' rather than an interaction model expansion: thus to fit x and x^2 in a quadratic regression we would write

```
y~x+I(x^2)
```

9.6.2   Creating formula objects

You can speed up the creation of complicated model formulae using paste to create series of variable names and collapse to join the variable names together by symbols. Here, for instance, is a multiple regression formula with 25 continuous explanatory variables created using the as.formula function:

```
xnames  <-  paste("x", 1:25, sep="")

(model.formula <- as.formula(paste("y~", paste(xnames, collapse= "+"))))

y ~ x1 + x2 + x3 + x4 + x5 + x6 + x7 + x8 + x9 + x10 + x11 +
    x12 + x13 + x14 + x15 + x16 + x17 + x18 + x19 + x20 + x21 +
    x22 + x23 + x24 + x25
```

9.7   Multiple error terms

When there is nesting (e.g. split plots in a designed experiment; see p. 685) or temporal pseudoreplication (see p. 695) you can include an `Error` function as part of the model formula. Suppose you had a three-factor factorial experiment with categorical variables *A*, *B* and *C*. The twist is that each treatment is applied to plots of different sizes: *A* is applied to replicated whole fields, *B* is applied at random to half fields and *C* is applied to smaller split–split plots within each half-field. This is shown in a model formula like this:

```
y ~ A*B*C + Error(A/B/C)
```

Note that the terms within the model formula are separated by asterisks to show that it is a full factorial with all interaction terms included, whereas the terms are separated by slashes in the `Error` statement. There are as many terms in the `Error` statement as there are different sizes of plots – three in this case, although the smallest plot size (*C* in this example) can be omitted from the list – and the terms are listed left to right from the largest to the smallest plots; see p. 686 for details and examples.

For the more modern mixed-effects model using `lmer` (in package `lme4`), the preferred method is to create unique factor level names, rather than use slashes to indicate nesting. The colon operator is useful for this, when both of the arguments are factors. So if our nested factors *A*, *B* and *C* are numbers (say 1:2, 1:4 and 1:3, respectively)

```
A <- rep(1:2,each=12)
B <- rep(1:4,each=3,length=24)
C <- rep(1:3,length=24)
```

then we compute new factors *a*, *b*, and *c*

```
a <- factor(A)
b <- factor(A):factor(B)
c <- factor(A):factor(B):factor(C)
```

and then fit each as a separate random effect (see p. 692)

```
lmer(y ~ x + (1|a)+(1|b)+(1|c) )
```

9.8   The intercept as parameter 1

The simple command

```
y ~ 1
```

causes the null model to be fitted. This works out the grand mean (the overall average) of all the data, and the total deviance (or the total sum of squares, *SSY*, in models with normal errors and the identity link). In some cases, this may be the minimal adequate model; it is possible that none of the explanatory variables we have measured contribute anything significant to our understanding of the variation in the response variable. This is normally what you do not want to happen at the end of your three-year research project.

To remove the intercept (parameter 1) from a regression model (i.e. to force the regression line through the origin) you fit '–1' like this:

```
y ~ x - 1
```

You should not do this unless you know exactly what you are doing, and exactly why you are doing it. Removing the intercept from an ANOVA model where all the variables are categorical has a different effect:

```
y ~ sex - 1
```

This gives the mean for males and the mean for females in the summary table, rather than the mean for females and the difference in mean for males.

9.9 The `update` function in model simplification

In the `update` function used during model simplification, the dot '.' is used to specify 'what is there already' on either side of the tilde. So if your original model said

```
model <- lm(y~A*B)
```

then the `update` function to remove the interaction term `A:B` could be written like this:

```
model2 <- update(model,~.- A:B)
```

Note that there is no need to repeat the name of the response variable, and the punctuation 'tilde dot' means take model as it is, and remove from it the interaction term `A:B`.

9.10 Model formulae for regression

The important point to grasp is that model formulae look very like equations but there are important differences. Our simplest useful equation looks like this:

$$y = a + bx.$$

It is a two-parameter model with one parameter for the intercept, a, and another for the slope, b, of the graph of the continuous response variable y against a continuous explanatory variable x. The model formula for the same relationship looks like this:

```
y ~ x
```

The equals sign is replaced by a tilde, and all of the parameters are left out. It we had a multiple regression with two continuous explanatory variables x and z, the equation would be

```
y = a + bx + cz,
```

but the model formula is

```
y~x + z
```

It is all wonderfully simple. But just a minute. How does R know what parameters we want to estimate from the data? We have only told it the names of the explanatory variables. We have said nothing about how to fit them, or what sort of equation we want to fit to the data. The key to this is to understand what kind of explanatory variable is being fitted to the data. If the explanatory variable x specified on the right of the tilde is a continuous variable, then R *assumes* that you want to do a regression, and hence that you want to estimate two parameters in a linear regression whose equation is $y = a + bx$.

A common misconception is that linear models involve a straight-line relationship between the response variable and the explanatory variables. This is *not* the case, as you can see from these two linear models:

```
windows(7,4)
par(mfrow=c(1,2))
x <- seq(0,10,0.1)
plot(x,1+x-x^2/15,type="l",col="red")
plot(x,3+0.1*exp(x),type="l",col="red")
```

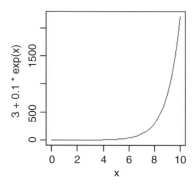

The definition of a linear model is an equation that contains mathematical variables, parameters and random variables and that is linear in the parameters and in the random variables. What this means is that if a, b and c are parameters then obviously

$$y = a + bx$$

is a linear model, but so is

$$y = a + bx - cx^2$$

because x^2 can be replaced by z which gives a linear relationship

$$y = a + bx + cz,$$

and so is

$$y = a + be^x$$

because we can create a new variable $z = \exp(x)$, so that

$$y = a + bz.$$

Some models are non-linear but can be readily linearized by transformation. For example,

$$y = \exp(a + bx)$$

is non-linear, but on taking logs of both sides, it becomes

$$\ln(y) = a + bx$$

If the equation you want to fit is more complicated than this, then you need to specify the form of the equation, and use non-linear methods (`nls` or `nlme`) to fit the model to the data (see p. 715).

9.11   Box–Cox transformations

Sometimes it is not clear from theory what the optimal transformation of the response variable should be. In these circumstances, the Box–Cox transformation offers a simple empirical solution. The idea is to find the power transformation, λ (lambda), that maximizes the likelihood when a specified set of explanatory variables is fitted to

$$\frac{y^\lambda - 1}{\lambda}$$

as the response. The value of lambda can be positive or negative, but it cannot be zero (you would get a zero-divide error when the formula was applied to the response variable, y). For the case $\lambda = 0$ the Box–Cox transformation is defined as $\log(y)$. Suppose that $\lambda = -1$. The formula now becomes

$$\frac{y^{-1} - 1}{-1} = \frac{1/y - 1}{-1} = 1 - \frac{1}{y},$$

and this quantity is regressed against the explanatory variables and the log-likelihood computed.

In this example, we want to find the optimal transformation of the response variable, which is timber volume:

```
data <- read.delim("c:\\temp\\timber.txt")
attach(data)
names(data)
```

```
[1]  "volume"  "girth"  "height"
```

We start by loading the MASS library of Venables and Ripley:

```
library(MASS)
```

The boxcox function is very easy to use: just specify the model formula, and the default options take care of everything else.

```
windows(7,7)
boxcox(volume~log(girth)+log(height))
```

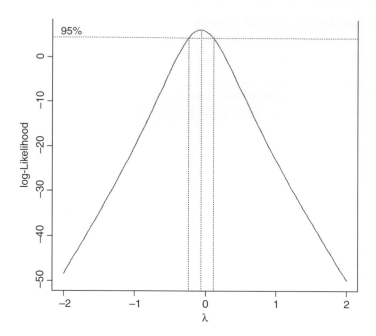

It is clear that the optimal value of lambda is close to zero (i.e. the log transformation). We can zoom in to get a more accurate estimate by specifying our own, non-default, range of lambda values. It looks as if it would be sensible to plot from −0.5 to +0.5:

```
boxcox(volume~log(girth)+log(height),lambda=seq(-0.5,0.5,0.01))
```

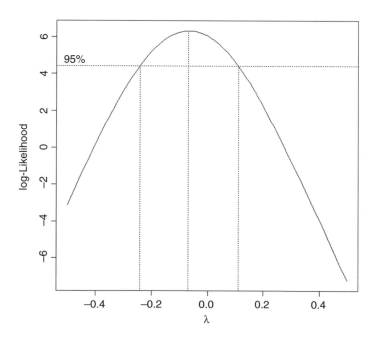

The likelihood is maximized at $\lambda \approx -0.08$, but the log-likelihood for $\lambda = 0$ is very close to the maximum. This also gives a much more straightforward interpretation, so we would go with that, and model `log(volume)` as a function of `log(girth)` and `log(height)` (see p. 262).

What if we had not log transformed the explanatory variables? What would have been the optimal transformation of volume in that case? To find out, we rerun the `boxcox` function, simply changing the model formula like this:

```
boxcox(volume~girth+height)
```

We can zoom in from 0.1 to 0.6 like this:

```
boxcox(volume~girth+height,lambda=seq(0.1,0.6,0.01))
```

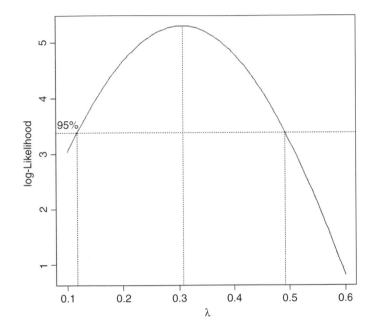

This suggests that the cube root transformation would be best ($\lambda = 1/3$). Again, this accords with dimensional arguments, since the response and explanatory variables would all have dimensions of length in this case.

9.12 Model criticism

There is a temptation to become personally attached to a particular model. Statisticians call this 'falling in love with your model'. It is as well to remember the following truths about models:

- All models are wrong.
- Some models are better than others.
- The correct model can never be known with certainty.
- The simpler the model, the better it is.

There are several ways that we can improve things if it turns out that our present model is inadequate:

- Transform the response variable.
- Transform one or more of the explanatory variables.
- Try fitting different explanatory variables if you have any.
- Use a different error structure.
- Use non-parametric smoothers instead of parametric functions.
- Use different weights for different y values.

All of these are investigated in the coming chapters. In essence, you need a set of tools to establish whether, and how, your model is inadequate. For example, the model might:

- predict some of the y values poorly;
- show non-constant variance;
- show non-normal errors;
- be strongly influenced by a small number of influential data points;
- show some sort of systematic pattern in the residuals;
- exhibit overdispersion.

9.13   Model checking

After fitting a model to data we need to investigate how well the model describes the data. In particular, we should look to see if there are any systematic trends in the goodness of fit. For example, does the goodness of fit increase with the observation number, or is it a function of one or more of the explanatory variables? We can work with the raw residuals:

$$residuals = y - \text{fitted values}.$$

For instance, we should routinely plot the residuals against:

- the fitted values (to look for heteroscedasticity);
- the explanatory variables (to look for evidence of curvature);
- the sequence of data collection (to look for temporal correlation);
- standard normal deviates (to look for non-normality of errors).

9.13.1   Heteroscedasticity

A good model must also account for the variance–mean relationship adequately and produce additive effects on the appropriate scale (as defined by the link function). A plot of standardized residuals against fitted values should look like the sky at night (points scattered at random over the whole plotting region), with no trend in the size or degree of scatter of the residuals. A common problem is that the variance increases with the mean, so that we obtain an expanding, fan-shaped pattern of residuals (right-hand panel):

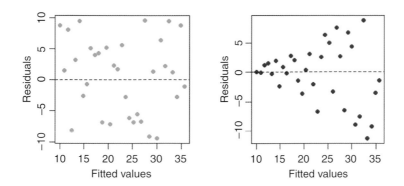

The plot on the left is what we want to see: no trend in the residuals with the fitted values. The plot on the right is a problem. There is a clear pattern of increasing residuals as the fitted values get larger. This is a picture of what heteroscedasticity looks like.

9.13.2 Non-normality of errors

Errors may be non-normal for several reasons. They may be skew, with long tails to the left or right. Or they may be kurtotic, with a flatter or more pointy top to their distribution. In any case, the theory is based on the assumption of normal errors, and if the errors are *not* normally distributed, then we shall not know how this affects our interpretation of the data or the inferences we make from it.

It takes considerable experience to interpret normal error plots. Here we generate a series of data sets where we introduce different but known kinds of non-normal errors. Then we plot them using a simple home-made function called mcheck (first developed by John Nelder in the original GLIM language; the name stands for 'model checking'). The idea is to see what patterns are generated in normal plots by the different kinds of non-normality. In real applications we would use the generic plot(model) rather than mcheck (see below). First, we write the function mcheck. The idea is to produce two plots, side by side: a plot of the residuals against the fitted values on the left, and a plot of the ordered residuals against the quantiles of the normal distribution on the right.

```
mcheck <- function (obj, ...){
       rs <- obj$resid
       fv <- obj$fitted
       windows(7,4)
       par(mfrow=c(1,2))
       plot(fv,rs,xlab="Fitted values",ylab="Residuals",pch=16,col="red")
       abline(h=0, lty=2)
       qqnorm(rs,xlab="Normal scores",ylab="Ordered residuals",main="",pch=16)
       qqline(rs,lty=2,col="green")
       par(mfrow=c(1,1))
       invisible(NULL)   }
```

Note the use of $ (component selection) to extract the residuals and fitted values from the model object which is passed to the function as obj (the expression x$name is the name *component* of x). The functions qqnorm and qqline are built-in functions to produce normal probability plots. It is good programming practice to set the graphics parameters back to their default settings before leaving the function.

The aim is to create a catalogue of some of the commonest problems that arise in model checking. We need a vector of x values for the following regression models:

```
x <- 0:30
```

Now we manufacture the response variables according to the equation

$$y = 10 + x + \varepsilon$$

where the errors, ε, have zero mean but are taken from different probability distributions in each case.

Normal errors

```
e <- rnorm(31,mean=0,sd=5)
yn <- 10+x+e

mn <- lm(yn~x)

mcheck(mn)
```

There is no suggestion of non-constant variance (left plot) and the normal plot (right) is reasonably straight. The judgement as to what constitutes an important departure from normality takes experience, and this is the reason for looking at some distinctly non-normal, but known, error structures next.

Uniform errors

```
eu <- 20*(runif(31)-0.5)
yu <- 10+x+eu

mu <- lm(yu~x)

mcheck(mu)
```

Uniform errors show up as a distinctly S-shaped pattern in the quantile–quantile plot on the right. The fit in the centre is fine, but the largest and smallest residuals are too small (they are constrained in this example to be ± 10).

Negative binomial errors

```
enb <- rnbinom(31,2,.3)
ynb <- 10+x+enb

mnb <- lm(ynb~x)

mcheck(mnb)
```

The large negative residuals are all above the line, but the most obvious feature of the plot is the single, very large positive residual (in the top right-hand corner). In general, negative binomial errors will produce a J-shape on the quantile–quantile plot. The biggest positive residuals are much too large to have come from a normal distribution. These values may turn out to be highly influential (see below).

Gamma errors and increasing variance

Here the shape parameter is set to 1 and the rate parameter to $1/x$, and the variance increases with the square of the mean:

```
eg <- rgamma(31,1,1/x)
yy <- 10+x+eg
```

```
mg <- lm(yg~x)
```

```
mcheck(mg)
```

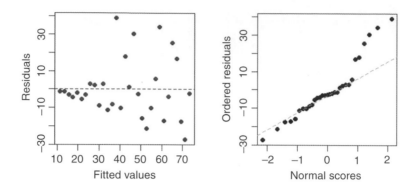

The left-hand plot shows the residuals increasing steeply with the fitted values, and illustrates an asymmetry between the size of the positive and negative residuals. The right-hand plot shows the highly non-normal distribution of errors.

9.14   Influence

One of the commonest reasons for a lack of fit is through the existence of outliers in the data. It is important to understand, however, that a point may *appear* to be an outlier because of misspecification of the model, and not because there is anything wrong with the data. It is important to understand that analysis of residuals is a very poor way of looking for influence. Precisely because a point is highly influential, it forces the regression line close to it, and hence the influential point may have a very small residual.

Take this circle of data that shows absolutely no relationship between *y* and *x*:

```
x <- c(2,3,3,3,4)
y <- c(2,3,2,1,2)
```

We want to draw two graphs side by side, and we want them to have the same axis scales:

```
windows(7,4)
par(mfrow=c(1,2))
```

```
plot(x,y,xlim=c(0,8),ylim=c(0,8))
```

Obviously, there is no relationship between *y* and *x* in the original data. But let us add an outlier at the point (7, 6) using concatenation c and see what happens:

```
x1 <- c(x,7)
y1 <- c(y,6)
plot(x1,y1,xlim=c(0,8),ylim=c(0,8))
abline(lm(y1~x1),col="blue")
```

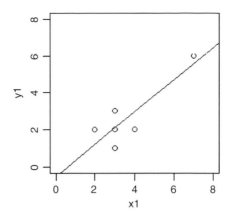

Now, there is a significant regression of y on x. The outlier is said to be highly *influential*. This makes our write-up much more complicated. We need to own up and show that the entire edifice depends upon the single point at $(7, 6)$. This requires an explanation of two models rather than one. We cannot pretend that the point $(7, 6)$ does not exist (that would be a scientific scandal), but we must describe just how influential it is.

Testing for the presence of influential points is an important part of statistical modelling. You cannot rely on analysis of the residuals, because by their very influence, these points force the regression line close to them:

```
reg <- lm(y1~x1)
summary(reg)

Call:
lm(formula = y1 ~ x1)

Residuals:
       1        2        3        4        5        6
 0.78261  0.91304 -0.08696 -1.08696 -0.95652  0.43478

Coefficients:
            Estimate Std. Error t value Pr(>|t|)
(Intercept)  -0.5217     0.9876  -0.528   0.6253
x1            0.8696     0.2469   3.522   0.0244 *
```

As you can see, the influential point (no. 6) has the second smallest residual (0.434 78). Instead, we look at the most extreme values of the explanatory variable, both to the left (extreme low values) and the right (extreme high values, as with point no. 6), as judged by $(x - \bar{x})^2$:

```
influence.measures(reg)

Influence measures of
         lm(formula = y1 ~ x1) :

    dfb.1  dfb.x1   dffit cov.r   cook.d   hat inf
1   0.687 -0.5287  0.7326 1.529  0.26791 0.348
2   0.382 -0.2036  0.5290 1.155  0.13485 0.196
3  -0.031  0.0165 -0.0429 2.199  0.00122 0.196
```

```
4 -0.496   0.2645  -0.6871 0.815 0.19111 0.196
5 -0.105  -0.1052  -0.5156 1.066 0.12472 0.174
6 -3.023   4.1703   4.6251 4.679 7.62791 0.891    *
```

You can see that point no. 6 is highlighted by an asterisk, drawing attention to its high influence. To extract the subscripts of the influential points, use the is.inf attribute like this:

```
influence.measures(reg)$is.inf
```

```
   dfb.1_  dfb.x1 dffit  cov.r  cook.d    hat
1  FALSE   FALSE  FALSE  FALSE   FALSE  FALSE
2  FALSE   FALSE  FALSE  FALSE   FALSE  FALSE
3  FALSE   FALSE  FALSE  FALSE   FALSE  FALSE
4  FALSE   FALSE  FALSE  FALSE   FALSE  FALSE
5  FALSE   FALSE  FALSE  FALSE   FALSE  FALSE
6   TRUE    TRUE   TRUE   TRUE    TRUE  FALSE
```

As you see, all of the influence measures (with the exception of hat), pick out point no. 6. For more detail, use lm.influence(reg) :

```
lm.influence(reg)
```

```
$hat
        1         2         3         4         5         6
0.3478261 0.1956522 0.1956522 0.1956522 0.1739130 0.8913043
```

```
$coefficients
   (Intercept)           x1
1   0.67826087 -0.130434783
2   0.37015276 -0.049353702
3  -0.03525264  0.004700353
4  -0.44065805  0.058754407
5  -0.10068650 -0.025171625
6  -2.52173913  0.869565217
```

```
$sigma
        1         2         3         4         5         6
0.9660918 0.9491580 1.1150082 0.8699177 0.9365858 0.8164966
```

```
$wt.res
         1          2          3          4          5          6
0.78260870 0.91304348 -0.08695652 -1.08695652 -0.95652174 0.43478261
```

The first component, $hat, is a vector containing the diagonal of the hat matrix. This is the orthogonal projector matrix onto the model space. Large values of elements of this vector mean that changing y_i will have a big impact on the fitted values (i.e. the hat diagonals are measures of the leverage of y_i).

Most interesting in the present context are the coefficients affecting the two parameters of the model (intercept and slope). The rows contain the *change* in the estimated coefficients which results when the ith case is dropped from the regression. Data in row 6 have much the biggest effect on both slope and intercept.

The third component, $sigma, is a vector whose ith element contains the estimate of the residual standard error obtained when the ith case is dropped from the regression; thus 0.816 496 6 is the residual standard error when point no. 6 is dropped, `lm(y1[-6]~x1[-6])`, and the error variance in this case is $0.816\ 496\ 6^2 = 0.666\ 666$, as you can see below:

```
summary.aov(lm(y1[-6]~x1[-6]))
```

```
           Df Sum Sq Mean Sq F value Pr(>F)
x1[-6]      1      0  0.0000       0      1
Residuals   3      2  0.6667
```

Finally, `$wt.res` is a vector of weighted residuals (or deviance residuals in a generalized linear model) or raw residuals if weights are not set (as in this example).

9.15   Summary of statistical models in R

Models are fitted using one of the following model-fitting functions:

lm   fits a linear model with normal errors and constant variance; generally this is used for regression analysis using continuous explanatory variables.

aov   fits analysis of variance with normal errors, constant variance and the identity link; generally used for categorical explanatory variables or ANCOVA with a mix of categorical and continuous explanatory variables.

glm   fits generalized linear models to data using categorical or continuous explanatory variables, by specifying one of a family of **error structures** (e.g. Poisson for count data or binomial for proportion data) and a particular **link function**.

gam   fits generalized additive models to data with one of a family of error structures (e.g. Poisson for count data or binomial for proportion data) in which the continuous explanatory variables can (optionally) be fitted as arbitrary smoothed functions using non-parametric smoothers rather than specific parametric functions.

lme   and `lmer` fit linear mixed-effects models with specified mixtures of fixed effects and random effects and allow for the specification of correlation structure among the explanatory variables and autocorrelation of the response variable (e.g. time series effects with repeated measures). `lmer` allows for non-normal errors and non-constant variance with the same error families as a GLM.

nls   fits a non-linear regression model via least squares, estimating the parameters of a specified non-linear function.

nlme   fits a specified non-linear function in a mixed-effects model where the parameters of the non-linear function are assumed to be random effects; it allows for the specification of correlation structure among the explanatory variables and autocorrelation of the response variable (e.g. time series effects with repeated measures).

loess   fits a local regression model with one or more continuous explanatory variables using non-parametric techniques to produce a smoothed model surface.

tree   and `rpart` fit a regression tree model using binary recursive partitioning whereby the data are successively split along coordinate axes of the explanatory variables so that at any node the split is chosen that maximally distinguishes the response variable in the left and right branches. With a categorical response variable, the tree is called a classification tree, and the model used for classification assumes that the response variable follows a multinomial distribution.

For most of these models, a range of **generic functions** can be used to obtain information about the model. The most important and most frequently used are as follows:

summary   produces parameter estimates and standard errors from lm, and ANOVA tables from aov; this will often determine your choice between lm and aov. For either lm or aov you can choose summary.aov or summary.lm to get the alternative form of output (an ANOVA table or a table of parameter estimates and standard errors; see p. 517).

plot   produces diagnostic plots for model checking, including residuals against fitted values, normality checks, influence tests, etc.

anova   is a wonderfully useful function for comparing different models and producing ANOVA tables.

update   is used to modify the last model fit; it saves both typing effort and computing time.

Other useful generic functions include the following:

coef   gives the coefficients (estimated parameters) from the model.

fitted   gives the fitted values, predicted by the model for the values of the explanatory variables included.

resid   gives the residuals (the differences between measured and predicted values of *y*).

predict   uses information from the fitted model to produce smooth functions for plotting a line through the scatterplot of your data. Make sure you provide a list or a dataframe containing all of the necessary information on each of the explanatory variables in your model to enable the prediction to be made.

9.16   Optional arguments in model-fitting functions

Unless you argue to the contrary, all of the rows in the dataframe will be used in the model fitting, there will be no offsets, and all values of the response variable will be given equal weight. Variables named in the model formula will come from the defined dataframe (data=mydata), the with function (p. 113) or from the attached dataframe (if there is one). Here we illustrate the following options:

- subset
- weights
- data
- offset
- na.action

We shall work with an example involving analysis of covariance (see p. 538 for details) where we have a mix of both continuous and categorical explanatory variables:

```
data <- read.table("c:\\temp\\ipomopsis.txt",header=T)
attach(data)
names(data)

[1]  "Root"  "Fruit"  "Grazing"
```

The response is seed production (Fruit) with a continuous explanatory variable (Root, Root diameter) and a two-level factor (Grazing, with levelss Grazed and Ungrazed).

9.16.1 Subsets

Perhaps the most commonly used modelling option is to fit the model to a subset of the data (e.g. fit the model to data from just the grazed plants). You could do this using subscripts on the response variable and all the explanatory variables:

```
model <- lm(Fruit[Grazing=="Grazed"]~Root[Grazing=="Grazed"])
```

but it is much more straightforward to use the subset argument, especially when there are lots of explanatory variables:

```
model <- lm(Fruit~Root, subset=(Grazing=="Grazed"))
```

The answer, of course, is the same in both cases, but the summary.lm and summary.aov tables are neater with subset. Note the round brackets used with the subset option (not the square brackets used with subscripts in the first example)

9.16.2 Weights

The default is for all the values of the response to have equal weights (all equal to 1)

```
weights = rep(1, n.observations)
```

Where data points are to be weighted unequally, the classical approach is to weight each value by the inverse of the variance of the distribution from which that point is drawn. This downplays the influence of highly variable data.

Instead of using initial root size as a covariate (as above) you could use Root as a weight in fitting a model with Grazing as the sole categorical explanatory variable:

```
model <- lm(Fruit~Grazing, weights=Root)
summary(model)

Call:
lm(formula = Fruit~Grazing, weights = Root)
Coefficients:
                   Estimate    Std. Error    t value    Pr(>|t|)
(Intercept)         70.725        4.849        14.59     <2e-16    ***
GrazingUngrazed    -16.953        7.469        -2.27      0.029      *

Residual standard error: 62.51 on 38 degrees of freedom
Multiple R-Squared: 0.1194,   Adjusted R-squared: 0.0962
F-statistic: 5.151 on 1 and 38 DF,  p-value: 0.02899
```

When weights (w) are specified the model is fitted using weighted least squares, in which the quantity to be minimized is $\sum w \times d^2$ (rather than $\sum d^2$), where d is the difference between the response variable and the fitted values predicted by the model. Needless to say, the use of weights alters the parameter estimates and their standard errors:

```
model <- lm(Fruit~Grazing)
summary(model)
```

```
Coefficients:
                  Estimate    Std. Error    t value    Pr(>|t|)
   (Intercept)      67.941          5.236     12.976    1.54e-15    ***
GrazingUngrazed    -17.060          7.404     -2.304      0.0268    *

Residual standard error: 23.41 on 38 degrees of freedom
Multiple R-Squared: 0.1226,  Adjusted R-squared: 0.09949
F-statistic: 5.309 on 1 and 38 DF,  p-value: 0.02678
```

Fitting root size as a statistical weight is scientifically wrong in this case: why should values from larger plants be given greater influence? Also, this analysis gives entirely the wrong interpretation of the data (ungrazed plants come out as being *less* fecund than the grazed plants). Analysis of covariance reverses this interpretation, showing that for a given root size, the grazed plants produced 36.013 *fewer* fruits than the ungrazed plants; the problem was that the big plants were almost all in the grazed treatment (see p. 538).

9.16.3 Missing values

What to do about missing values in the dataframe is an important issue (p. 172). Ideally, of course, there are no missing values, so you do not need to worry about what action to take (na.action). If there are missing values, you have two choices:

- leave out any row of the dataframe in which one or more variables are missing, then na.action = na.omit; or

- fail the fitting process, so na.action = na.fail.

If in doubt, you should specify na.action = na.fail because you will not get nasty surprises if unsuspected NAs in the dataframe cause strange (but unwarned) behaviour in the model.
Let us introduce a missing value into the initial root weights:

```
Root[37] <- NA

model <- lm(Fruit~Grazing*Root)
```

The model is fitted without comment, and the only thing you might notice is that the residual degrees of freedom is reduced from 36 to 35. If you want to be warned about missing values, then use the na.action option:

```
model <- lm(Fruit~Grazing*Root,na.action=na.fail)

Error in na.fail.default(list(Fruit = c(59.77, 60.98, 14.73, 19.28, 34.25,   :
    missing values in object
```

If you are carrying out regression with time series data that include missing values then you should use na.action = NULL so that residuals and fitted values are time series as well (if the missing values were omitted, then the resulting vector would not be a time series of the correct length).

9.16.4 Offsets

You would not use offsets with a linear model (you could simply subtract the offset from the value of the response variable, and work with the transformed values). But with generalized linear models you may want to specify part of the variation in the response using an offset (see p. 566 for details and examples).

9.16.5 Dataframes containing the same variable names

If you have several different dataframes containing the same variable names (say, x and y) then the simplest way to ensure that the correct variables are used in the modelling is to name the dataframe in the function call:

```
model <- lm(y~x,data=correct.frame)
```

The alternative is much more cumbersome to type:

```
model <- lm(correct.frame$y~correct.frame$x)
```

9.17 Akaike's information criterion

Akaike's information criterion (AIC) is known in the statistics trade as a **penalized log-likelihood**. If you have a model for which a log-likelihood value can be obtained (see Section 7.3.3), then

$$AIC = -2 \times \log\text{-likelihood} + 2(p + 1),$$

where p is the number of parameters in the model, and 1 is added for the estimated variance (you could call this another parameter if you wanted to). To demystify AIC let us calculate it by hand. These data show the relationship between growth and dietary tannin for caterpillars in a feeding experiment:

```
data <- read.table("c:\\temp\\regression.txt",header=T)
```

```
attach(data)
names(data)
```

```
[1] "growth" "tannin"
```

The regression model for these data is worked out, one term at a time, by hand in Chapter 10.

```
model <- lm(growth~tannin)
```

To calculate the log-likelihood we need three quantities (p. 282): the sample size, n; the error variance $s2 = \sigma^2$; and the sum of the squares of the residuals, $\text{sse} = \sum(y - \mu)^2$:

```
n <- length(growth)
sse <- sum((growth-fitted(model))^2)
s2 <- sse/(n-2)
s <- sqrt(s2)
```

Now we can compute the log-likelihood:

```
-(n/2)*log(2*pi)-n*log(s)-sse/(2*s2)
```

```
[1] -16.51087
```

There is an R function `logLik` to calculate the log likelihood from any appropriate model object directly:

```
logLik(model)
```

```
'log Lik.' -16.37995 (df=3)
```

The three degrees of freedom (`df`) refer to the slope, the intercept and the variance. The difference between the two estimates is just rounding error. Now we can compute AIC:

```
-2 * -16.37995 + 6
```

```
[1] 38.7599
```

Again, not surprisingly, there is an R function called `AIC` to compute the information criterion directly from the model object:

```
AIC(model)
```

```
[1] 38.7599
```

9.17.1 AIC as a measure of the fit of a model

The more parameters there are in the model, the better the fit. You could obtain a perfect fit if you had a separate parameter for every data point, but this model would have absolutely no explanatory power. There is always going to be a trade-off between the goodness of fit and the number of parameters required by parsimony. AIC is useful because it explicitly penalizes any superfluous parameters in the model, by adding $2(p + 1)$ to the deviance.

When comparing two models, the smaller the AIC, the better the fit. This is the basis of automated model simplification using `step`.

You can use the function `AIC` to compare two models, in exactly the same way as you can use `anova` (as explained on p. 415). Here we develop an analysis of covariance that is introduced on p. 538.

```
model.1 <- lm(Fruit~Grazing*Root)
model.2 <- lm(Fruit~Grazing+Root)
```

```
AIC(model.1, model.2)
```

```
        df      AIC
model.1  5 263.6269
model.2  4 261.7835
```

Because `model.2` has the *lower* AIC, we prefer it to `model.1`. The log-likelihood was penalized by $2 \times (4 + 1) = 10$ in `model.1` because that model contained 4 parameters (2 slopes and 2 intercepts) and by $2 \times (3 + 1)=8$ in `model.2` because that model had 3 parameters (two intercepts and a common slope). You can see where the two values of AIC come from by calculation:

```
-2*logLik(model.1)+2*(4+1)
```

```
[1] 263.6269
```

```
-2*logLik(model.2)+2*(3+1)
```

```
[1] 261.7835
```

If you want to compare many models, you can combine the models into a list,

```
models <- list (model1, model2, model3, model4, model5, model6)
```

then extract the AIC of each of them using `lapply` like this:

```
aic <- unlist(lapply(models, AIC))
```

where `aic` will be a vector of numbers in which you can search for the minimum.

9.18 Leverage

Points increase in influence to the extent that they lie on their own, a long way from the mean value of x (to either the left or right). To account for this, measures of leverage for a given data point y are proportional to $(x - \bar{x})^2$. Here are the x data from our earlier example:

```
x <- c(2,3,3,3,4,7)
```

The commonest measure of leverage is

$$h_i = \frac{1}{n} + \frac{(x_i - \bar{x})^2}{\Sigma(x_i - \bar{x})^2},$$

where the denominator is *SSX*. A good rule of thumb is that a point is highly influential if its

$$h_i > \frac{2p}{n},$$

where p is the number of parameters in the model.

We could easily calculate the leverage value of each point in our vector. It is more efficient, perhaps, to write a general function that could carry out the calculation of the h values for any vector of x values,

```
leverage <- function(x){1/length(x)+(x-mean(x))^2/sum((x-mean(x))^2)}
```

and then use this function with our vector of x values to produce a leverage plot:

```
plot(leverage(x),type="h",ylim=c(0,1),col="blue")
abline(h=4/6,lty=2,col="green")
```

As you can see, only the sixth=point shows more leverage than is reasonable (the horizontal green dashed line shows $2p/n = 4/6$ in this example). For built-in functions for checking influence, see p. 463.

9.19   Misspecified model

The model may have the wrong terms in it, or the terms may be included in the model in the wrong way. We deal with the selection of terms for inclusion in the minimal adequate model in Chapter 10. Here we simply note that **transformation of the explanatory variables** often produces improvements in model performance. The most frequently used transformations are logs, powers and reciprocals.

When both the error distribution and functional form of the relationship are unknown, there is no single specific rationale for choosing any given transformation in preference to another. The aim is pragmatic, namely to find a transformation that gives:

- constant error variance;

- approximately normal errors;

- additivity;

- a linear relationship between the response variables and the explanatory variables;

- straightforward scientific interpretation.

The choice is bound to be a compromise and, as such, is best resolved by quantitative comparison of the deviance produced under different model forms. Again, in testing for non-linearity in the relationship between y and x we might add a term in x^2 to the model; a significant parameter in the x^2 term indicates curvilinearity in the relationship between y and x.

A further element of misspecification can occur because of **structural non-linearity**. Suppose, for example, that we were fitting a model of the form

$$y = a + \frac{b}{x},$$

but the underlying process was really of the form

$$y = a + \frac{b}{c + x};$$

then the fit is going to be poor. Of course if we *knew* that the model structure was of this form, then we could fit it as a non-linear model (p. 715) or as a non-linear mixed-effects model (p. 722), but in practice this is seldom the case.

9.20   Model checking in R

The data we examine in this section are on the decay of a biodegradable plastic in soil: the response, y, is the mass of plastic remaining and the explanatory variable, x, is duration of burial:

```
Decay <- read.table("c:\\temp\\Decay.txt",header=T)
attach(Decay)
```

```
names(Decay)
```

```
[1] "time"     "amount"
```

For the purposes of illustration we shall fit a linear regression to these data and then use model-checking plots to investigate the adequacy of that model:

```
model <- lm(amount~time)
```

The basic model checking could not be simpler:

```
par(mfrow=c(2,2))
plot(model)
```

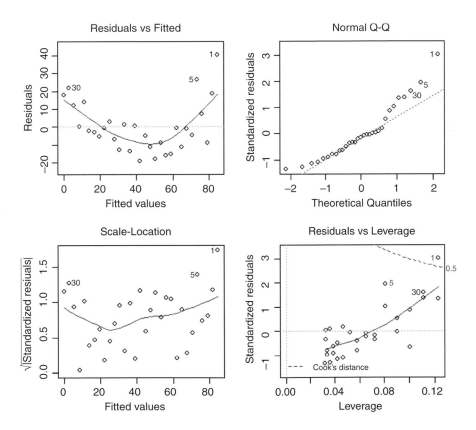

This one command produces a series of graphs, spread over four pages (here compressed to a single page by specifying par(mfrow=c(2,2)). The upper two graphs are the most important. First, you get a plot of the residuals against the fitted values (top left) which shows very pronounced curvature; most of the residuals for intermediate fitted values are negative, and the positive residuals are concentrated at the smallest and largest fitted values. Remember, this plot should look like the sky at night, with no pattern of any sort. This suggests systematic inadequacy in the structure of the model. Perhaps the relationship between y and x is non-linear rather than linear as we assumed here? Second (top right), you get a quunatile–quantile plot (p. 463) which indicates pronounced non-normality in the residuals (the line should be straight, not banana-shaped as here). The third graph is like a positive-valued version of the first graph; it is good for detecting non-constancy of variance (heteroscedasticity), which shows up as a triangular scatter (like a wedge of cheese) with an increasing red line through it. The fourth graph shows a pronounced pattern in the standardized residuals as a function of the leverage. The graph also shows Cook's distance, highlighting the identity of particularly influential data points.

Cook's distance is an attempt to combine leverage and residuals in a single measure. The absolute values of the deletion residuals $|r_i^*|$ are weighted as follows:

$$C_i = |r_i^*| \left(\frac{n-p}{p} \cdot \frac{h_i}{1 - h_i} \right)^{1/2}.$$

Data points 1, 5 and 30 are singled out as being influential, with point 1 especially so. When we were happier with other aspects of the model, we would repeat the modelling, leaving out each of these points in turn. Alternatively, we could jackknife the data, which involves leaving every data point out, one at a time, in turn. In any event, this is clearly *not* a good model for these data. The analysis is completed on p. 469, when we fit an exponential rather than a linear model to the data.

9.21   Extracting information from model objects

We often want to extract material from fitted models (e.g. slopes, residuals or *p* values) and there are three different ways of doing this:

- by name, e.g. `coef(model)`;
- with list subscripts, e.g. `summary(model)[[3]]`;
- using $ to name the component, e.g. `model$resid`.

The model object we use to demonstrate these techniques is the simple linear regression that was analysed in full by hand on p. 450.

```
data <- read.table("c:\\temp\\regression.txt",header=T)
attach(data)
names(data)
```

```
[1]  "growth"  "tannin"
```

```
model <- lm(growth~tannin)
summary(model)
```

```
Call:
lm(formula = growth ~ tannin)
```

```
Residuals:
    Min       1Q  Median       3Q      Max
-2.4556  -0.8889  -0.2389   0.9778   2.8944
```

```
Coefficients:
            Estimate Std. Error t value Pr(>|t|)
(Intercept)  11.7556     1.0408  11.295 9.54e-06 ***
tannin       -1.2167     0.2186  -5.565 0.000846 ***
```

```
Residual standard error: 1.693 on 7 degrees of freedom
```

```
Multiple R-squared: 0.8157,       Adjusted R-squared: 0.7893
F-statistic: 30.97 on 1 and 7 DF,  p-value: 0.0008461
```

9.21.1 Extracting information by name

You can extract the coefficients of the model, the fitted values, the residuals, the effect sizes and the variance–covariance matrix by name, as follows:

```
coef(model)
```

```
(Intercept)        tannin
  11.755556     -1.216667
```

gives the parameter estimates ('coefficients') for the intercept (a) and slope (b);

```
fitted(model)
```

```
[1]      1         2         3         4         5         6
11.755556 10.538889  9.322222  8.105556  6.888889  5.672222

[7]      7         8         9
 4.455556  3.238889  2.022222
```

gives the fitted values ($\hat{y} = a + bx$) of the model (its predictions) for each value of the explanatory variable(s);

```
resid(model)
```

```
[1]       1          2          3          4          5          6
 0.2444444 -0.5388889 -1.3222222  2.8944444 -0.8888889  1.3277778

[7]      7          8          9
-2.4555556 -0.2388889  0.9777778
```

gives the residuals (y minus fitted values) for the nine data points.

For a linear model fitted by `lm` or `aov` the effects are the uncorrelated single-degree-of-freedom values obtained by projecting the data onto the successive orthogonal subspaces generated by the QR decomposition during the fitting process. The first r (=2 in this case; the rank of the model) are associated with coefficients and the remainder span the space of residuals but are not associated with particular residuals. The name `effects` produces a numeric vector of the same length as residuals of class `coef`. The first two rows are labelled by the corresponding coefficients (intercept and slope), and the remaining seven rows are unlabelled.

```
vcov(model)
```

```
            (Intercept)        tannin
(Intercept)    1.083263  -0.19116402
tannin        -0.191164   0.04779101
```

This extracts the variance–covariance matrix of the model's parameters.

9.21.2 Extracting information by list subscripts

The two model summary objects `summary.aov(model)` and `summary.lm(model)` are lists with many components. Here each of them is investigated in turn.

Here is `summary.aov`:

```
summary.aov(model)
```

```
            Df Sum Sq Mean Sq F value    Pr(>F)
tannin       1  88.82   88.82   30.97 0.000846 ***
Residuals    7  20.07    2.87
```

The columns of the ANOVA table can be extracted one at a time:

```
summary.aov(model)[[1]][1]
```

```
            Df
tannin       1
Residuals    7
```

```
summary.aov(model)[[1]][2]
```

```
            Sum Sq
tannin      88.817
Residuals   20.072
```

```
summary.aov(model)[[1]][3]
```

```
            Mean Sq
tannin       88.817
Residuals     2.867
```

```
summary.aov(model)[[1]][4]
```

```
            F value
tannin       30.974
Residuals
```

```
summary.aov(model)[[1]][5]
```

```
               Pr(>F)
tannin      0.0008461 ***
Residuals
---
Signif. codes:   0 '***' 0.001 '**' 0.01 '*' 0.05 '.' 0.1 ' ' 1
```

It can be quite involved to extract the numerical values that you might want to use in subsequent work. For instance, to get the *F* ratio (30.974) out of the fourth element of the list, we need to `unlist` the object, then use `as.numeric`, and then add a further subscript:

```
as.numeric(unlist(summary.aov(model)[[1]][4]))[1]
```

```
[1] 30.97398
```

Here is `summary.lm`:

```
summary(model)
```

```
Call:
lm(formula = growth ~ tannin)
```

```
Residuals:
    Min      1Q  Median      3Q      Max
-2.4556 -0.8889 -0.2389  0.9778  2.8944

Coefficients:
            Estimate Std. Error t value Pr(>|t|)
(Intercept)  11.7556     1.0408  11.295 9.54e-06 ***
tannin       -1.2167     0.2186  -5.565 0.000846 ***

Residual standard error: 1.693 on 7 degrees of freedom
Multiple R-squared: 0.8157,    Adjusted R-squared: 0.7893
F-statistic: 30.97 on 1 and 7 DF,  p-value: 0.0008461
```

The first element of the list is the model formula (or Call) showing the response variable (growth) and the explanatory variable(s) (tannin):

```
summary(model)[[1]]
```

```
lm(formula = growth ~ tannin)
```

The second describes the attributes of the object called summary(model):

```
summary(model)[[2]]
```

```
growth ~ tannin
attr(,"variables")
list(growth, tannin)
attr(,"factors")
       tannin
growth      0
tannin      1
attr(,"term.labels")
[1] "tannin"
attr(,"order")
[1] 1
attr(,"intercept")
[1] 1
attr(,"response")
[1] 1
attr(,".Environment")
<environment: R_GlobalEnv>
attr(,"predvars")
list(growth, tannin)
attr(,"dataClasses")
   growth     tannin
"numeric"  "numeric"
```

The third gives the residuals for the nine data points:

```
summary(model)[[3]]
```

as shown above. The fourth gives the parameter table, including standard errors of the parameters, t values and p values. This is the really important information:

```
summary(model)[[4]]
```

```
              Estimate Std. Error   t value      Pr(>|t|)
(Intercept) 11.755556  1.0407991 11.294740 9.537315e-06
tannin      -1.216667  0.2186115 -5.565427 8.460738e-04
```

You will often want to extract information from parts of this table, using extra subscripts

```
summary(model)[[4]]  [1]
```

```
[1] 11.75556
```

```
summary(model)[[4]]  [2]
```

```
[1] -1.216667
```

```
summary(model)[[4]]  [3]
```

```
[1] 1.040799
```

```
summary(model)[[4]]  [4]
```

```
[1] 0.2186115
```

to extract the individual value of the intercept, slope, standard error of the intercept and standard error of the slope respectively. The p value for the slope is in `[[4]]  [8]`, for instance:

```
summary(model)[[4]]  [8]
```

```
[1] 0.0008460738
```

The fifth is concerned with whether the corresponding components of the fit (the model frame, the model matrix, the response or the QR decomposition) should be returned. The default is FALSE:

```
summary(model)[[5]]
```

```
(Intercept) tannin
   FALSE      FALSE
```

The sixth is the residual standard error: the square root of the error variance from the summary.aov table ($s^2 = 2.867$; see above):

```
summary(model)[[6]]
```

```
[1]  1.693358
```

The seventh shows the number of rows in the summary.lm table (showing two parameters to have been estimated from the data with this model, and the residual degrees of freedom (d.f. = 7):

```
summary(model)[[7]]
```

```
[1]  2  7  2
```

The eighth is $r^2 = SSR/SST$, the fraction of the total variation in the response variable that is explained by the model (see p. 456 for details):

```
summary(model)[[8]]
```

```
[1]  0.8156633
```

The ninth is the adjusted R^2, explained on p. 461 but seldom used in practice:

```
summary(model)[[9]]
```

```
[1]  0.7893294
```

The tenth gives F ratio information: the three values given here are the F ratio (30.973 98), the number of degrees of freedom in the model (i.e. in the numerator, numdf) and the residual degrees of freedom (i.e. in the denominator, dendf):

```
summary(model)[[10]]
```

```
   value      numdf      dendf
30.97398   1.00000   7.00000
```

The eleventh component is the correlation matrix of the parameter estimates:

```
summary(model)[[11]]
```

```
              (Intercept)        tannin

(Intercept)   0.37777778  -0.06666667
tannin       -0.06666667   0.01666667
```

9.21.3   Extracting components of the model using $

Another way to extract model components is to use the $ symbol. To get the intercept (a) and the slope (b) of the regression, type

```
model$coef
```

```
(Intercept)       tannin
11.755556     -1.216667
```

Finally, the residual degrees of freedom (9 points – 2 estimated parameters = 7 d.f.) are

```
model$df
```

```
[1]  7
```

9.21.4   Using lists with models

You might want to extract the coefficients from a series of related statistical models, and you want to avoid the use of a loop. One solution is to create a list and then employ lapply to do the work. Here are the data with y as a function of x:

```
x <- 0:100
y <- 17+0.2*x+3*rnorm(101)
```

Now create three linear models of increasing complexity:

```
model0 <- lm(y~1)
model1 <- lm(y~x)
model2 <- lm(y~x+I(x^2))
```

Make a list containing the three model objects:

```
models <- list(model0,model1,model2)
```

To obtain the coefficients from the three models, it is simple to use `lapply` on the list to apply the function `coef` to each element of the list:

```
lapply(models,coef)

[[1]]
(Intercept)
   26.90530

[[2]]
(Intercept)            x
 15.8267899    0.2215701

[[3]]
  (Intercept)           x          I(x^2)
1.593695e+01 2.148935e-01 6.676673e-05
```

To get a *vector* (rather than a list) as output, and to select only the three intercepts, we use subscripts `[c(1,2,4)]` with `unlist` and `as.vector` like this:

```
as.vector(unlist(lapply(models,coef)))[c(1,2,4)]

[1]  26.90530 15.82679 15.93695
```

This protocol can be useful in model selection. Here we extract the AIC of each model:

```
lapply(models,AIC)

[[1]]
[1]  672.7502

[[2]]
[1]  510.787

[[3]]
[1]  512.5231
```

Other things being equal, we would chose the model with the lowest AIC (the linear regression (`model1`) has AIC $= 510.787$).

9.22   The `summary` tables for continuous and categorical explanatory variables

It is important to understand the difference between `summary.lm` and `summary.aov` for the same model. Here is a one-way analysis of variance of the plant competition experiment (p. 511):

```
comp <- read.table("c:\\temp\\competition.txt",header=T)
attach(comp)
names(comp)

[1]  "biomass"  "clipping"
```

The categorical explanatory variable is `clipping` and it has five levels as follows:

```
levels(clipping)
[1]  "control"  "n25"  "n50"  "r10"  "r5"
```

The analysis of variance model is fitted like this:

```
model <- lm(biomass~clipping)
```

and we can obtain two different summaries of it.

Here is `summary.aov`:

```
summary.aov(model)
```

```
          Df  Sum Sq  Mean Sq  F value  Pr(>F)
clipping   4   85356    21339    4.302  0.00875 **
Residuals 25  124020     4961
```

showing one row for the treatment and one row for the residuals (the row for the total sum of squares is not printed in R), each row with degrees of freedom, sum of squares, variance (labelled 'Mean Square') and the F ratio, testing the null hypothesis of no significant differences between the treatment means. The only interesting things in `summary.aov` are the error variance ($s^2 = 4961$) which we use in calculating measures of unreliability, and the F ratio (4.302) showing that there are significant differences amongst the means to be explained.

Here is `summary.lm`:

```
summary.lm(model)
```

```
Call:
lm(formula = biomass ~ clipping)
```

```
Residuals:
     Min       1Q   Median       3Q      Max
 -103.333  -49.667    3.417   43.375  177.667
```

```
Coefficients:
             Estimate Std. Error t value Pr(>|t|)
(Intercept)    465.17      28.75  16.177  9.4e-15 ***
clippingn25     88.17      40.66   2.168  0.03987 *
clippingn50    104.17      40.66   2.562  0.01683 *
clippingr10    145.50      40.66   3.578  0.00145 **
clippingr5     145.33      40.66   3.574  0.00147 **
```

```
Residual standard error: 70.43 on 25 degrees of freedom
Multiple R-squared: 0.4077,      Adjusted R-squared: 0.3129
F-statistic: 4.302 on 4 and 25 DF,  p-value: 0.008752
```

The residuals are summarized first by their 'five numbers'. The coefficients table has as many rows as there are parameters in the model (five in this case, one for each factor level mean). The top row, labelled (`Intercept`), is the *only* mean value in the table: it is the mean for the factor level that comes first in the alphabet (`control` in this example). The other four rows are *differences between means* (each mean compared to the control mean in this example). The second column contains the unreliability estimates. The first row contains *the standard error of a mean* (28.75). The other four rows contain *the standard error of the difference between two means* (40.66). The significance stars are highly misleading in this example, suggesting wrongly that there are four significant contrasts for this model. The problem arises because the default 'Treatment contrasts' in R are not orthogonal. The four lower rows are being compared with the first row. As we shall see later, there is only one significant orthogonal contrast in this experiment (the control versus the other four treatments).

So where do the effect sizes come from? What is 465.17 and what is 88.17? To understand the answers to these questions, we need to know how the equation for the explanatory variables is structured in a linear model when the explanatory variable, as here, is categorical. To recap, the linear regression model is written as

```
lm(y~x)
```

which R interprets as the two-parameter linear equation. R knows this because x is a continuous variable, so the equation it invokes is

$$y = a + bx,$$

in which the values of the parameters a and b are to be estimated from the data. But what about our analysis of variance? We have one explanatory variable, $x = \text{clipping}$, but it is a categorical variable with five levels, control, n25, n50, r10 and r5. The aov model is exactly analogous to the regression model

```
lm(y~x)
```

but what is the associated equation? Let us look at the equation first, and try to understand it:

$$y = a + bx_1 + cx_2 + dx_3 + ex_4 + fx_5.$$

This looks just like a multiple regression, with five explanatory variables, $x_1, \ldots, x_5$. The key point to understand is that $x_1, \ldots, x_5$ are dummy variables representing the levels of the factor called x. The intercept, a, is the overall (or grand) mean for the whole experiment. The parameters $b, \ldots, f$ are differences between the grand mean and the mean for a given factor level. You will need to concentrate to understand this.

With a categorical explanatory variable, all the variables are coded as $x = 0$ except for the factor level that is associated with the y value in question, when x is coded as $x = 1$. You will find this hard to understand without a good deal of practice. Let us look at the first row of data in our dataframe:

```
comp[1,]
```

```
    biomass  clipping
1      551       n25
```

So the first biomass value (551) in the dataframe comes from clipping treatment n25 which, out of all the factor levels (above), comes second in the alphabet. This means that for this row of the dataframe $x_1 = 0$, $x_2 = 1$, $x_3 = 0$, $x_4 = 0$, $x_5 = 0$. The equation for the first row therefore looks like this:

$$y = a + b \times 0 + c \times 1 + d \times 0 + e \times 0 + f \times 0,$$

so the model for the fitted value at n25 is

$$\hat{y} = a + c;$$

and similarly for the other factor levels. The fitted value $\hat{y}$ is the sum of two parameters, a and c. The equation apparently does not contain an explanatory variable (there is no x in the equation as there would be in a regression equation, above). Note, too, how many parameters the full model contains: they are represented by the letters *a* to *f* and there are six of them. But we can only estimate five parameters in this experiment (one mean for each of the five factor levels). Our model contains one redundant parameter, and we need to deal with this. There are several sensible ways of doing this, and people differ in their opinions about what is the best way. The writers of R agree that *treatment contrasts* represent the best solution. This method does away with parameter *a*, the overall mean (in the jargon, this overall mean is intentionally aliased). The mean of the factor level that comes first in the alphabet (control, in our example) is promoted to pole position, and the other effects are shown as differences (contrasts) between this mean and the other four factor level means.

An example might help make this clearer. Here are our five means:

```
means <- tapply(biomass,clipping,mean)
means
```

```
 control        n25        n50         r10          r5
465.1667  553.3333  569.3333  610.6667  610.5000
```

In treatment contrasts, the control mean (465.1667) becomes the first parameter of the model (known as the intercept). The second parameter is the difference between the second mean (n25 = 553.3333) and the intercept:

```
means[2]-means[1]
```

```
     n25
88.16667
```

The third parameter is the difference between the third mean (n50 = 569.3333) and the intercept:

```
means[3]-means[1]
```

```
     n50
104.1667
```

The fourth parameter is the difference between the fourth mean (r10 = 610.6667) and the intercept:

```
means[4]-means[1]
```

```
  r10
145.5
```

The fifth parameter is the difference between the fifth mean (r5 = 610.5) and the intercept:

```
means[5]-means[1]
```

```
      r5
145.3333
```

So much for the effect sizes. What about their standard errors? The first row is a mean, so we need the standard error of one factor-level mean. This mean is based on six numbers in this example, so the standard error of the mean is $\sqrt{s^2/n}$ where the error variance, $s^2 = 4961$, is obtained from summary.aov(model) above:

```
sqrt(4961/6)
```

```
[1]  28.75471
```

All the other rows have the same standard error, but it is bigger than this. That is because the effects on the second and subsequent rows are *not* means, but *differences between means*. That means that the appropriate standard error is not the standard error of a mean, but rather *the standard error of the difference between two means*. When two samples are independent, the variance of their difference is the sum of their two variances. Thus, the formula for the standard error of a difference between two means is

$$se_{\text{diff}} = \sqrt{\frac{s_1^2}{n_1} + \frac{s_2^2}{n_2}}.$$

When the two variances and the two sample sizes are the same (as here, because our design is balanced and we are using the pooled error variance (4961) from the `summary.aov` table) the formula simplifies to $\sqrt{2 \times s^2 / n}$:

```
sqrt(2*4961/6)
```

```
[1]   40.6653
```

With some practice, that should demystify the origin of the numbers in the `summary.lm` table. But it does take lots of practice, and people do find this very difficult at first, so do not feel bad about it.

9.23   Contrasts

Contrasts are the essence of hypothesis testing and model simplification in analysis of variance and analysis of covariance. They are used to compare means or groups of means with other means or groups of means, in what are known as **single-degree-of-freedom comparisons**. There are two sorts of contrasts we might be interested in:

- contrasts we had planned to examine at the experimental design stage (these are referred to as ***a priori*** contrasts);

- contrasts that look interesting after we have seen the results (these are referred to as ***a posteriori*** contrasts).

Some people are very snooty about *a posteriori* contrasts, on the grounds that they were unplanned. You are not supposed to decide what comparisons to make *after* you have seen the analysis, but scientists do this all the time. The key point is that you should only do contrasts *after* the ANOVA has established that there really are significant differences to be investigated. It is not good practice to carry out tests to compare the largest mean with the smallest mean, if the ANOVA has failed to reject the null hypothesis (tempting though this may be).

There are two important points to understand about contrasts:

- there is a huge number of *possible* contrasts, and

- there are only $k - 1$ *orthogonal* contrasts,

where k is the number of factor levels. Two contrasts are said to be **orthogonal** to one another if the comparisons are statistically independent. Technically, two contrasts are orthogonal if the products of their contrast coefficients sum to zero (we shall see what this means in a moment).

Let us take a simple example. Suppose we have one factor with five levels and the factor levels are called a, b, c, d, e. Let us start writing down the possible contrasts. Obviously we could compare each mean singly with every other:

$$a \text{ vs. } b, a \text{ vs. } c, a \text{ vs. } d, a \text{ vs. } e, b \text{ vs. } c, b \text{ vs. } d, b \text{ vs. } e, c \text{ vs. } d, c \text{ vs. } e, d \text{ vs. } e.$$

But we could also compare pairs of means:

$$\{a, b\} \text{ vs. } \{c, d\}, \{a, b\} \text{ vs. } \{c, e\}, \{a, b\} \text{ vs. } \{d, e\}, \{a, c\} \text{ vs. } \{b, d\}, \{a, c\} \text{ vs. } \{b, e\}, \ldots$$

or triplets of means:

$$\{a, b, c\} \text{ vs. } d, \{a, b, c\} \text{ vs. } e, \{a, b, d\} \text{ vs. } c, \{a, b, d\} \text{ vs. } e, \{a, c, d\} \text{ vs. } b, \ldots$$

or groups of four means:

$$\{a, b, c, d\} \text{ vs. } e, \{a, b, c, e\} \text{ vs. } d, \{a, b, d, e\} \text{ vs. } c, \{a, c, d, e\} \text{ vs. } b, \{b, c, d, e\} \text{ vs. } a.$$

You doubtless get the idea. There are absolutely masses of possible contrasts. In practice, however, we should only compare things once, either directly or implicitly. So the two contrasts a vs. b and a vs. c implicitly contrast b vs. c. This means that if we have carried out the two contrasts a vs. b and a vs. c then the third contrast b vs. c is *not* an orthogonal contrast because you have already carried it out, implicitly. Which particular contrasts are orthogonal depends very much on your choice of the first contrast to make. Suppose there were good reasons for comparing $\{a, b, c, e\}$ vs. d. For example, d might be the placebo and the other four might be different kinds of drug treatment, so we make this our first contrast. Because $k - 1 = 4$ we only have three possible contrasts that are orthogonal to this. There may be *a priori* reasons to group $\{a, b\}$ and $\{c, e\}$, so we make this our second orthogonal contrast. This means that we have no degrees of freedom in choosing the last two orthogonal contrasts: they have to be a vs. b and c vs. e. Just remember that *with orthogonal contrasts you only compare things once*.

9.23.1 Contrast coefficients

Contrast coefficients are a numerical way of embodying the hypothesis we want to test. The rules for constructing contrast coefficients are straightforward:

- Treatments to be lumped together get the same sign (plus or minus).
- Groups of means to be to be contrasted get opposite sign.
- Factor levels to be excluded get a contrast coefficient of 0.
- The contrast coefficients, c, must add up to 0.

Suppose that with our five-level factor $\{a, b, c, d, e\}$ we want to begin by comparing the our levels $\{a, b, c, e\}$ with the single level d. All levels enter the contrast, so none of the coefficients is 0. The four terms $\{a, b, c, e\}$ are grouped together so they all get the same sign (minus, for example, although it makes no difference which sign is chosen). They are to be compared to d, so it gets the opposite sign (plus, in this case). The choice of what numeric values to give the contrast coefficients is entirely up to you. Most people use whole numbers rather than fractions, but it really does not matter. All that matters is that the coefficients sum to 0. The positive and negative coefficients have to add up to the same value. In our example, comparing four means with one mean, a natural choice of coefficients would be -1 for each of $\{a, b, c, e\}$ and $+4$ for d. Alternatively, with could select $+0.25$ for each of $\{a, b, c, e\}$ and -1 for d.

Factor level:	a	b	c	d	e
contrast 1 coefficients:	-1	-1	-1	4	-1

Suppose the second contrast is to compare $\{a, b\}$ with $\{c, e\}$. Because this contrast excludes d, we set its contrast coefficient to 0. $\{a, b\}$ get the same sign (say, plus) and $\{c, e\}$ get the opposite sign. Because the number of levels on each side of the contrast is equal (2 in both cases) we can use the name numeric value for all the coefficients. The value 1 is the most obvious choice (but you could use 13.7 if you wanted to be perverse):

Factor level:	a	b	c	d	e
Contrast 2 coefficients:	1	1	-1	0	-1

There are only two possibilities for the remaining orthogonal contrasts, a vs. b and c vs. e:

Factor level:	a	b	c	d	e
Contrast 3 coefficients:	1	-1	0	0	0
Contrast 4 coefficients:	0	0	1	0	-1

The variation in y attributable to a particular contrast is called the **contrast sum of squares**, SSC. The sums of squares of the $k-1$ orthogonal contrasts add up to the total treatment sum of squares, SSA ($\sum_{i=1}^{k-1} SSC_i = SSA$). The contrast sum of squares is computed like this:

$$SSC_i = \frac{\left(\sum (c_i T_i / n_i) \right)^2}{\sum (c_i^2 / n_i)},$$

where the c_i are the contrast coefficients (above), n_i are the sample sizes within each factor level and T_i are the totals of the y values within each factor level (often called the treatment totals). The significance of a contrast is judged by an F test, dividing the contrast sum of squares by the error variance. The F test has 1 degree of freedom in the numerator (because a contrast is a comparison of two means, and $2-1=1$) and $k(n-1)$ degrees of freedom in the denominator (the error variance degrees of freedom).

9.23.2   An example of contrasts in R

The following example comes from the competition experiment we analysed on p. 511, in which the biomass of control plants is compared to the biomass of plants grown in conditions where competition was reduced in one of four different ways. There are two treatments in which the roots of neighbouring plants were cut (to 5 cm or 10 cm depth) and two treatments in which the shoots of neighbouring plants were clipped (25% or 50% of the neighbours were cut back to ground level).

```
comp <- read.table("c:\\temp\\competition.txt",header=T)
attach(comp)
names(comp)}
```

```
[1] "biomass" "clipping"
```

We start with the one-way analysis of variance:

```
model1 <- aov(biomass~clipping)
summary(model1)
```

```
            Df Sum Sq Mean Sq F value  Pr(>F)
clipping     4  85356   21339   4.302 0.00875 **
Residuals   25 124020    4961
```

Clipping treatment has a highly significant effect on biomass. But have we fully understood the result of this experiment? Probably not. For example, which factor levels had the biggest effect on biomass, and were all of the competition treatments significantly different from the controls? To answer these questions, we need to use summary.lm:

```
summary.lm(model1)
```

```
Call:
aov(formula = biomass ~ clipping)
```

```
Residuals:
     Min       1Q   Median       3Q      Max
-103.333  -49.667    3.417   43.375  177.667
```

```
Coefficients:
             Estimate Std. Error t value Pr(>|t|)
(Intercept)    465.17      28.75  16.177  9.4e-15 ***
clippingn25     88.17      40.66   2.168  0.03987 *
clippingn50    104.17      40.66   2.562  0.01683 *
clippingr10    145.50      40.66   3.578  0.00145 **
clippingr5     145.33      40.66   3.574  0.00147 **
```

```
Residual standard error: 70.43 on 25 degrees of freedom
Multiple R-squared: 0.4077,     Adjusted R-squared: 0.3129
F-statistic: 4.302 on 4 and 25 DF,  p-value: 0.008752
```

This looks as if we need to keep all five parameters, because all five rows of the summary table have one or more significance stars. If fact, this is not the case. This example highlights the major shortcoming of treatment contrasts: they do not show how many significant factor levels we need to retain in the minimal adequate model because all of the rows are being compared with the intercept (with the controls in this case, simply because the factor level name for 'controls' comes first in the alphabet):

```
levels(clipping)
```

```
[1]  "control"  "n25"  "n50"  "r10"  "r5"
```

9.23.3 *A priori* contrasts

In this experiment, there are several planned comparisons we should like to make. The obvious place to start is by comparing the control plants, exposed to the full rigours of competition, with all of the other treatments. That is to say, we want to contrast the first level of clipping with the other four levels. The contrast coefficients, therefore, would be 4, −1, −1, −1, −1. The next planned comparison might contrast the shoot-pruned treatments (n25 and n50) with the root-pruned treatments (r10 and r5). Suitable contrast coefficients for this would be 0, 1, 1, −1, −1 (because we are ignoring the control in this contrast). A third contrast might compare the two depths of root pruning; 0, 0, 0, 1, −1. The last orthogonal contrast would therefore have to compare the two intensities of shoot pruning: 0, 1, −1, 0, 0. Because the factor called `clipping` has five levels there are only 5 − 1 = 4 orthogonal contrasts.

R is outstandingly good at dealing with contrasts, and we can associate these five user-specified *a priori* contrasts with the categorical variable called `clipping` like this:

```
contrasts(clipping) <-
cbind(c(4,-1,-1,-1,-1),c(0,1,1,-1,-1),c(0,0,0,1,-1),c(0,1,-1,0,0))
```

We can check that this has done what we wanted by typing:

```
clipping

attr(,"contrasts")
         [,1] [,2] [,3] [,4]
control     4    0    0    0
n25        -1    1    0    1
n50        -1    1    0   -1
r10        -1   -1    1    0
r5         -1   -1   -1    0

Levels: control n25 n50 r10 r5
```

which produces the matrix of contrast coefficients that we specified. One contrast is contained in each column. Note that all the columns add to zero (i.e. each set of contrast coefficients is correctly specified). Note also that the products of any two of the columns sum to zero (this shows that all the contrasts are orthogonal, as intended): for example, comparing contrasts 1 and 2 gives products 0 + (−1) + (−1) + 1 + 1 = 0.

Now we can refit the model and inspect the results of our specified contrasts, rather than the default treatment contrasts:

```
model2 <- aov(biomass~clipping)
summary.lm(model2)

Call:
aov(formula = biomass ~ clipping)

Residuals:
     Min       1Q   Median       3Q      Max
-103.333  -49.667    3.417   43.375  177.667

Coefficients:
              Estimate Std. Error t value Pr(>|t|)
(Intercept)  561.80000   12.85926  43.688  < 2e-16 ***
clipping1    -24.15833    6.42963  -3.757 0.000921 ***
clipping2    -24.62500   14.37708  -1.713 0.099128 .
clipping3      0.08333   20.33227   0.004 0.996762
clipping4     -8.00000   20.33227  -0.393 0.697313

Residual standard error: 70.43 on 25 degrees of freedom
Multiple R-squared: 0.4077,    Adjusted R-squared: 0.3129
F-statistic: 4.302 on 4 and 25 DF,  p-value: 0.008752
```

Instead of requiring five parameters (as suggested by our initial treatment contrasts), this analysis shows that we need only two parameters: the overall mean (561.8) and the contrast between the controls and the four competition treatments ($p = 0.000\ 921$). All the other contrasts are non-significant.

When we specify the contrasts, the intercept is the overall (grand) mean:

```
mean(biomass)
```

```
[1]   561.8
```

The second row, labelled `clipping1`, estimates, like all contrasts, the difference between two means. But which two means, exactly? The means for the different factor levels are:

```
tapply(biomass,clipping,mean)
```

```
  control         n25        n50        r10          r5
465.1667  553.3333  569.3333  610.6667  610.5000
```

Thus this first contrast compares the controls (with mean 465.1667) with the mean of the other four treatments. The simplest way to get this other mean is to create a new factor, c_1 that has value 1 for the controls and 2 for the rest:

```
c1 <- factor(1+(clipping!="control"))
tapply(biomass,c1,mean)
```

```
        1           2
465.1667  585.9583
```

The estimate reflecting the first contrast is the difference between the overall mean (561.8) and the mean of the four non-control treatments (585.9583):

```
mean(biomass)  -  tapply(biomass,c1,mean)[2]
```

```
        2
-24.15833
```

and you see the estimate in row 2 is –24.15833. What about the second contrast? This compares the root- and shoot-pruned treatments, and c_2 is a factor that lumps together the two root and two shoot treatments:

```
c2 <- factor(2*(clipping=="n25")+2*(clipping=="n50")+
                    (clipping=="r10")+(clipping=="r5"))
```

We can compute the mean biomass for the two treatments using `tapply`, then subtract the means from one another, then halve the differences:

```
(tapply(biomass,c2,mean)[3]- tapply(biomass,c2,mean)[2])/2
```

```
        2
-24.625
```

So the second contrast (–24.625) is *half the difference* between the root- and shoot-pruned treatments. What about the third contrast? This is between the two root-pruned treatments. We know their values already from `tapply`, above:

```
       r10               r5
610.6667    610.5000
```

The two means differ by 0.166666 so the third contrast is half the difference between the two means:

```
(610.666666-610.5)/2
```

```
[1]   0.083333
```

The final contrast compares the two shoot-pruning treatments, and the contrast is half the difference between these two means:

```
(553.3333-569.3333)/2
```

```
[1]   -8
```

To recap: the first contrast compares the overall mean with the mean of the four non-control treatments, the second contrast is half the difference between the root and shoot-pruned treatment means, the third contrast is half the difference between the two root-pruned treatments, and the fourth contrast is half the difference between the two shoot-pruned treatments.

It is important to note that the first four standard errors in the summary.lm table are all different from one another. As we have just seen, the estimate in the first row of the table is a mean, while all the other rows contain estimates that are *differences between means*. The overall mean on the top row is based on 30 numbers so the standard error of the mean is $se = \sqrt{s^2/30}$, where s^2 comes from the ANOVA table:

```
sqrt(4961/30)
```

```
[1]   12.85950
```

The small difference in the fourth decimal place is due to rounding errors in calling the variance 4961.0. The next row compares two means so we need the standard error of the difference between two means. The complexity comes from the fact that the two means are each based on *different numbers of numbers*. The overall mean is based on all five factor levels (30 numbers) while the non-control mean with which it is compared is based on four means (24 numbers). Each factor level has $n = 6$ replicates, so the denominator in the standard error formula is $5 \times 4 \times 6 = 120$. Thus, the standard error of the difference between the these two means is $se = \sqrt{s^2/(5 \times 4 \times 6)}$:

```
sqrt(4961/(5*4*6))
```

```
[1]   6.429749
```

For the second contrast, each of the means is based on 12 numbers so the standard error is $se = \sqrt{2 \times (s^2/12)}$ so the standard error of half the difference is:

```
sqrt(2*(4961/12))/2
```

```
[1]   14.37735
```

The last two contrasts are both between means based on six numbers, so the standard error of the difference is $se = \sqrt{2 \times (s^2/6)}$ and the standard error of half the difference is:

```
sqrt(2*(4961/6))/2
```

```
[1]   20.33265
```

The complexity of these calculations is another reason for preferring treatment contrasts rather than user-specified contrasts as the default. The advantage of orthogonal contrasts, however, is that the summary.lm table gives us a much better idea of the number of parameters required in the minimal adequate model (two in this case). Treatment contrasts had significance stars on all five rows (see below) because all the non-control treatments were compared to the controls (the intercept).

9.24 Model simplification by stepwise deletion

An alternative to specifying the contrasts ourselves (as above) is to aggregate non-significant factor levels in a stepwise *a posteriori* procedure. To demonstrate this, we revert to treatment contrasts. First, we switch off our user-defined contrasts:

```
contrasts(clipping) <- NULL
options(contrasts=c("contr.treatment","contr.poly"))
```

Now we fit the model with all five factor levels as a starting point:

```
model3 <- aov(biomass~clipping)
summary.lm(model3)
```

```
Coefficients:
             Estimate Std. Error t value Pr(>|t|)
(Intercept)    465.17      28.75  16.177  9.4e-15 ***
clippingn25     88.17      40.66   2.168  0.03987 *
clippingn50    104.17      40.66   2.562  0.01683 *
clippingr10    145.50      40.66   3.578  0.00145 **
clippingr5     145.33      40.66   3.574  0.00147 **
```

Looking down the list of parameter estimates, we see that the most similar are the effects of root pruning to 10 and 5 cm (145.5 vs. 145.33). We shall begin by simplifying these to a single root-pruning treatment called root. The trick is to use the gets arrow <- to change the names of the appropriate factor levels. Start by copying the original factor name:

```
clip2 <- clipping
```

Now inspect the level numbers of the various factor level names:

```
levels(clip2)
```

```
[1]  "control"  "n25"   "n50"   "r10"   "r5"
```

The plan is to lump together r10 and r5 under the same name, root. These are the fourth and fifth levels of clip2, so we write:

```
levels(clip2)[4:5] <-  "root"
```

Now if we type

```
levels(clip2)
```

```
[1] "control" "n25"       "n50"       "root"
```

we see that r10 and r5 have indeed been replaced by root.

The next step is to fit a new model with clip2 in place of clipping, and to test whether the new simpler model is significantly worse as a description of the data using anova:

```
model4 <- aov(biomass~clip2)
anova(model3,model4)
```

```
Analysis of Variance Table

Model 1: biomass ~ clipping
Model 2: biomass ~ clip2
  Res.Df      RSS Df Sum of Sq  F Pr(>F)
1     25  124020
2     26  124020 -1 -0.083333  0 0.9968
```

As we expected, this model simplification was completely justified.

The next step is to investigate the effects using summary.lm:

```
summary.lm(model4)
```

```
Coefficients:
            Estimate Std. Error t value Pr(>|t|)
(Intercept)   465.17      28.20  16.498 2.72e-15 ***
clip2n25       88.17      39.87   2.211 0.036029 *
clip2n50      104.17      39.87   2.612 0.014744 *
clip2root     145.42      34.53   4.211 0.000269 ***
```

It looks as if the two shoot-clipping treatments (n25 and n50) are not significantly different from one another (they differ by just $104.17 - 88.17 = 16.0$ with a standard error of 39.87). We can lump these together into a single shoot-pruning treatment as follows:

```
clip3 <- clip2
levels(clip3)[2:3] <- "shoot"
levels(clip3)
```

```
[1] "control" "shoot"    "root"
```

Then we fit a new model with clip3 in place of clip2:

```
model5 <- aov(biomass~clip3)
anova(model4,model5)
```

```
Analysis of Variance Table

Model 1: biomass ~ clip2
Model 2: biomass ~ clip3
  Res.Df      RSS Df Sum of Sq      F Pr(>F)
1     26  124020
2     27  124788 -1      -768 0.161 0.6915
```

Again, this simplification was fully justified. Do the root and shoot competition treatments differ?

```
clip4 <- clip3
levels(clip4)[2:3] <- "pruned"
levels(clip4)
```

```
[1]  "control"  "pruned"
```

Now fit a new model with clip4 in place of clip3:

```
model6 <- aov(biomass~clip4)
anova(model5,model6)
```

```
Analysis of Variance Table

Model 1: biomass ~ clip3
Model 2: biomass ~ clip4
  Res.Df     RSS Df Sum of Sq       F  Pr(>F)
1      27 124788
2      28 139342 -1    -14553 3.1489 0.08726 .
```

This simplification was close to significant, but we are ruthless ($p > 0.05$), so we accept the simplification. Now we have the minimal adequate model:

```
summary.lm(model6)

Call:
aov(formula = biomass ~ clip4)

Residuals:
     Min      1Q  Median      3Q      Max
-135.958  -49.667  -4.458  50.635  145.042

Coefficients:
               Estimate Std. Error t value Pr(>|t|)
(Intercept)       465.2       28.8  16.152 1.01e-15 ***
clip4pruned       120.8       32.2   3.751 0.000815 ***

Residual standard error: 70.54 on 28 degrees of freedom
Multiple R-squared: 0.3345,      Adjusted R-squared: 0.3107
F-statistic: 14.07 on 1 and 28 DF,  p-value: 0.0008149
```

It has just two parameters: the mean for the controls (465.2) and the difference between the control mean and the four treatment means ($465.2 + 120.8 = 586.0$):

```
tapply(biomass,clip4,mean)

  control    pruned
465.1667 585.9583
```

We know that these two means are significantly different because of the p value of 0.000 815, but just to show how it is done, we can make a final model7 that has no explanatory variable at all (it fits only the overall mean). This is achieved by writing $y \sim 1$ in the model formula:

```
model7 <- aov(biomass~1)
anova(model6,model7)

Analysis of Variance Table

Model 1: biomass ~ clip4
Model 2: biomass ~ 1
  Res.Df     RSS Df Sum of Sq       F     Pr(>F)
1      28 139342
2      29 209377 -1    -70035 14.073 0.0008149 ***
```

Note that the p value is exactly the same as in model6. The p values in R are calculated such that they avoid the need for this final step in model simplification: they are 'deletion p values'.

9.25   Comparison of the three kinds of contrasts

In order to show the differences between treatment, Helmert and sum contrasts, we shall reanalyse the competition experiment using each in turn. Contrasts are explained on p. 430. For present purposes, you need only know that R provides three types of contrasts that summarize the differences between parameter estimates in different ways. Treatment contrasts (Section 9.25.1) are more intuitive than Hermert (Section 9.25.2) or sum (Section 9.25.3) contrasts.

9.25.1   Treatment contrasts

This is the default in R. These are the contrasts you get, unless you explicitly choose otherwise.

```
options(contrasts=c("contr.treatment","contr.poly"))
```

Here are the contrast coefficients as set under treatment contrasts:

```
contrasts(clipping)
```

```
        n25 n50 r10 r5
control   0   0   0  0
n25       1   0   0  0
n50       0   1   0  0
r10       0   0   1  0
r5        0   0   0  1
```

Notice that the contrasts are *not* orthogonal (the products of the coefficients do not sum to zero).

```
output.treatment <- lm(biomass~clipping)
summary(output.treatment)
```

```
Coefficients:
               Estimate Std. Error t value Pr(>|t|)
(Intercept)     465.17     28.75   16.177  9.4e-15 ***
clippingn25      88.17     40.66    2.168  0.03987 *
clippingn50     104.17     40.66    2.562  0.01683 *
clippingr10     145.50     40.66    3.578  0.00145 **
clippingr5      145.33     40.66    3.574  0.00147 **
```

With treatment contrasts, the factor levels are arranged in alphabetical sequence, and the level that comes first in the alphabet is made into the intercept. In our example this is control, so we can read off the control mean as 465.17, and the standard error of the mean as 28.75. The remaining four rows are differences between means, and the standard errors are standard errors of differences. Thus, clipping neighbours back to 25 cm increases biomass by 88.17 over the controls and this difference is significant at $p = 0.039\,87$. And so on. The downside of treatment contrasts is that all the rows appear to be significant despite the fact that rows 2–5 are actually not significantly different from one another, as we saw earlier.

9.25.2   Helmert contrasts

This is the default in S-PLUS, so beware if you are switching back and forth between the two languages.

```
options(contrasts=c("contr.helmert","contr.poly"))
contrasts(clipping)
```

```
           [,1] [,2] [,3] [,4]
control     -1   -1   -1   -1
n25          1   -1   -1   -1
n50          0    2   -1   -1
r10          0    0    3   -1
r5           0    0    0    4
```

Notice that the contrasts are orthogonal (the products sum to zero) and their coefficients sum to zero, unlike treatment contrasts, above:

```
output.helmert <- lm(biomass~clipping)
summary(output.helmert)
```

```
Coefficients:
             Estimate Std. Error t value Pr(>|t|)
(Intercept)   561.800     12.859  43.688   <2e-16 ***
clipping1      44.083     20.332   2.168   0.0399 *
clipping2      20.028     11.739   1.706   0.1004
clipping3      20.347      8.301   2.451   0.0216 *
clipping4      12.175      6.430   1.894   0.0699 .
```

With Helmert contrasts, the intercept is the overall mean (561.8). The first contrast (labelled `clipping1`) compares the first mean in alphabetical sequence with the average of the first and second factor levels in alphabetical sequence (`control` plus n25; see above): its parameter value is the mean of the first two factor levels, minus the mean of the first factor level:

```
(465.16667+553.33333)/2-465.166667
```

```
[1]   44.08332
```

The second contrast (`clipping2`) compares the third factor level (n50) and the two levels already compared (`control` and n25): its value is the difference between the average of the first three factor levels and the average of the first two factor levels:

```
(465.16667+553.33333+569.333333)/3-(465.166667+553.3333)/2
```

```
[1]   20.02779
```

The third contrast (`clipping3`) compares the fourth factor level (r10) and the three levels already compared (`control`, n25 and n50): its value is the difference between the average of the first four factor levels and the average of the first three factor levels

```
(465.16667+553.33333+569.333333+610.66667)/4
                          -(553.3333+465.166667+569.3333)/3
```

```
[1]   20.34725
```

The fourth contrast (`clipping3`) compares the fifth factor level (r5) and the four levels already compared (`control`, n25, n50 and r10): its value is the difference between the average of the first five factor levels (the grand mean), and the average of the first four factor levels:

```
mean(biomass)-(465.16667+553.33333+569.333333+610.66667)/4
```

```
[1]   12.175
```

So much for the parameter estimates. Now look at the standard errors. We have seen rather few of these values in any of the analyses we have done to date. The standard error in row 1 is the standard error of the overall mean, with s^2 taken from the overall ANOVA table: $\sqrt{s^2/kn}$.

```
sqrt(4961/30)
```

```
[1]   12.85950
```

The standard error in row 2 is a comparison of a group of two means with a single mean ($2 \times 1 = 2$). Thus 2 is multiplied by the sample size n in the denominator: $\sqrt{s^2/2n}$.

```
sqrt(4961/(2*6))
```

```
[1]   20.33265
```

The standard error in row 3 is a comparison of a group of three means with a group of two means (so $3 \times 2 = 6$ in the denominator): $\sqrt{s^2/6n}$.

```
sqrt(4961/(3*2*6))
```

```
[1]   11.73906
```

The standard error in row 4 is a comparison of a group of four means with a group of three means (so $4 \times 3 = 12$ in the denominator): $\sqrt{s^2/12n}$.

```
sqrt(4961/(4*3*6))
```

```
[1]   8.30077
```

The standard error in row 5 is a comparison of a group of five means with a group of four means (so $5 \times 4 = 20$ in the denominator): $\sqrt{s^2/20n}$.

```
sqrt(4961/(5*4*6))
```

```
[1]   6.429749
```

It is true that the parameter estimates and their standard errors are much more difficult to understand in Helmert than in treatment contrasts. But the advantage of Helmert contrasts is that they give you proper orthogonal contrasts, and hence give a much clearer picture of which factor levels need to be retained in the minimal adequate model. They do not eliminate the need for careful model simplification, however. As we saw earlier, this example requires only two parameters in the minimal adequate model, but Helmert contrasts suggest the need for three (albeit only marginally significant) parameters.

9.25.3   Sum contrasts

Sum contrasts are the third option:

```
options(contrasts=c("contr.sum","contr.poly"))
output.sum <- lm(biomass~clipping)
summary(output.sum)
```

```
Coefficients:
            Estimate Std. Error t value Pr(>|t|)
(Intercept)  561.800     12.859  43.688  < 2e-16 ***
clipping1    -96.633     25.719  -3.757 0.000921 ***
clipping2     -8.467     25.719  -0.329 0.744743
```

```
clipping3       7.533      25.719     0.293 0.772005
clipping4      48.867      25.719     1.900 0.069019 .
```

As with Helmert contrasts, the first row contains the overall mean and the standard error of the overall mean. The remaining four rows are different: they are the differences between the grand mean and the first four factor means (`control`, `n25`, `n50` and `r10`):

```
tapply(biomass,clipping,mean) - 561.8
```

```
    control          n25          n50          r10           r5
-96.633333    -8.466667     7.533333    48.866667    48.700000
```

The standard errors are all the same (25.719) for all four contrasts. The contrasts compare the grand mean (based on 30 numbers) with a single treatment mean:

```
sqrt(4961/30+4961/10)
```

```
[1]     25.71899
```

9.26 Aliasing

Aliasing occurs when there is no information available on which to base an estimate of a parameter value. Parameters can be aliased for one of two reasons:

- there are no data in the dataframe from which to estimate the parameter (e.g. missing values, partial designs or correlation among the explanatory variables), or

- the model is structured in such a way that the parameter value cannot be estimated (e.g. over-specified models with more parameters than necessary).

Intrinsic aliasing occurs when it is due to the *structure of the model*. **Extrinsic aliasing** occurs when it is due to the *nature of the data*.

Suppose that in a factorial experiment all of the animals receiving level 2 of diet (factor A) and level 3 of temperature (factor B) have died accidentally as a result of attack by a fungal pathogen. This particular combination of diet and temperature contributes no data to the response variable, so the interaction term $A(2) : B(3)$ cannot be estimated. It is **extrinsically aliased**, and its parameter estimate is set to zero.

If one continuous variable is perfectly correlated with another variable that has already been fitted to the data (perhaps because it is a constant multiple of the first variable), then the second term is aliased and adds nothing to the model. Suppose that $x_2 = 0.5x_1$; then fitting a model with $x_1 + x_2$ will lead to x_2 being **intrinsically aliased** and given a zero parameter estimate.

If all the values of a particular explanatory variable are set to zero for a given level of a particular factor, then that level is **intentionally aliased**. This sort of aliasing is a useful programming trick in ANCOVA when we wish a covariate to be fitted to some levels of a factor but not to others.

9.27 Orthogonal polynomial contrasts: `contr.poly`

Here are the data from a randomized experiment with four levels of dietary supplement:

```
data <- read.table("c:\\temp\\poly.txt",header=T)
attach(data)
```

```
names(data)
```

```
[1]  "treatment"  "response"
```

We begin by noting that the factor levels are in alphabetical order (not in ranked sequence – none, low, medium, high – as we might prefer):

```
tapply(response,treatment,mean)
```

```
  high     low medium     none
  4.50    5.25   7.00     2.50
```

The summary.lm table from the one-way analysis of variance looks like this:

```
Coefficients:
            Estimate Std. Error t value Pr(>|t|)
(Intercept)   4.8125     0.1875  25.667 7.45e-12 ***
treatment1   -0.3125     0.3248  -0.962    0.355
treatment2    0.4375     0.3248   1.347    0.203
treatment3    2.1875     0.3248   6.736 2.09e-05 ***
```

and the summary.aov table looks like this:

```
summary.aov(model)
```

```
           Df Sum Sq Mean Sq F value    Pr(>F)
treatment   3  41.69  13.896    24.7 2.02e-05 ***
Residuals  12   6.75   0.563
```

We can see that treatment is a factor but it is not ordered:

```
is.factor(treatment)
```

```
[1]  TRUE
```

```
is.ordered(treatment)
```

```
[1]  FALSE
```

To convert it into an ordered factor, we use the ordered function like this:

```
treatment <- ordered(treatment,levels=c("none","low","medium","high"))
levels(treatment)
```

```
[1] "none"    "low"     "medium" "high"
```

Now the factor levels appear in their ordered sequence, rather than in alphabetical order.

Fitting the ordered factor makes no difference to the summary.aov table:

```
model2 <- lm(response~treatment)
summary.aov(model2)
```

```
           Df Sum Sq Mean Sq F value    Pr(>F)
treatment   3  41.69  13.896    24.7 2.02e-05 ***
Residuals  12   6.75   0.562
```

but the summary.lm table is fundamentally different when the factors are ordered. Now the contrasts are not contr.treatment but contr.poly (which stands for 'orthogonal polynomial contrasts'):

```
summary.lm(model2)

Call:
lm(formula = response ~ treatment)

Residuals:
   Min     1Q Median     3Q    Max
 -1.25  -0.50   0.00   0.50   1.00

Coefficients:
              Estimate Std. Error t value Pr(>|t|)
(Intercept)     4.8125     0.1875  25.667 7.45e-12 ***
treatment.L     1.7330     0.3750   4.621 0.000589 ***
treatment.Q    -2.6250     0.3750  -7.000 1.43e-05 ***
treatment.C    -0.7267     0.3750  -1.938 0.076520 .

Residual standard error: 0.75 on 12 degrees of freedom
Multiple R-squared: 0.8606,      Adjusted R-squared: 0.8258
F-statistic:  24.7 on 3 and 12 DF,  p-value: 2.015e-05
```

The levels of the factor called treatment are no longer labelled low, medium, none as with treatment contrasts (above). Instead they are labelled L, Q and C, which stand for 'linear', 'quadratic' and 'cubic' polynomial terms, respectively. But what are the coefficients, and why are they so difficult to interpret? The first thing you notice is that the intercept 4.8125 is no longer one of the treatment means:

```
tapply(response,treatment,mean)

 none    low medium   high
 2.50   5.25   7.00   4.50
```

You could fit a polynomial regression model to the mean values of the response with the four ordered levels of treatment represented by a continuous (dummy) explanatory variable (say, x <- c(1,2,3,4)), then fit terms for x, x^2 and x^3 independently (using the 'as is' function I in the model formula). This is what it would look like:

```
yv <- as.vector(tapply(response,treatment,mean))
x <- 1:4
model <- lm(yv~x+I(x^2)+I(x^3))
summary(model)

Call:
lm(formula = yv ~ x + I(x^2) + I(x^3))

Residuals:
ALL 4 residuals are 0: no residual degrees of freedom!

Coefficients:
              Estimate Std. Error t value Pr(>|t|)
(Intercept)     2.0000         NA      NA       NA
x              -1.7083         NA      NA       NA
I(x^2)          2.7500         NA      NA       NA
I(x^3)         -0.5417         NA      NA       NA
```

There are four data points and four estimated parameters, so there are no residual degrees of freedom. Thus the equation for y as a function of treatment (x) could be written

$$y = 2 - 1.7083x + 2.75x^2 - 0.5417x^3.$$

Notice that the intercept is *not* one of the factor-level means. To find the mean for factor level 1 (none), the equation is evaluated for $x = 1$ (namely $2 - 1.7083 + 2.75 - 0.5417 = 2.5$; exactly the correct answer, as we can see above). So why does R not do it this way? There are two main reasons: orthogonality and computational accuracy. If the linear, quadratic and cubic contrasts are *orthogonal* and fitted *stepwise*, then we can see whether adding an extra term produces significantly improved explanatory power in the model. In this case, for instance, there is no justification for retaining the cubic term ($p = 0.076\,52$). Computational accuracy can become a major problem when fitting many polynomial terms, because these terms are necessarily so highly correlated:

```
x <- 1:4
x2 <- x^2
x3 <- x^3
cor(cbind(x,x2,x3))
```

```
            x          x2         x3
x   1.0000000 0.9843740 0.9513699
x2  0.9843740 1.0000000 0.9905329
x3  0.9513699 0.9905329 1.0000000
```

Orthogonal polynomial contrasts fix both these problems simultaneously. Here is one way to obtain orthogonal polynomial contrasts for a factor with four levels. The contrasts (in the rows) will go up to polynomials of degree $= k - 1 = 4 - 1 = 3$.

term	x_1	x_2	x_3	x_4
linear	-3	-1	1	3
quadratic	1	-1	-1	1
cubic	-1	3	-3	1

Note that the linear x terms are equally spaced, and have a mean of zero (i.e. each point on the x axis is separated by 2). Also, note that all the rows sum to zero. The key point is that the pointwise products of the terms in any two rows also sum to zero: thus for the linear and quadratic terms we have products of $(-3, 1, -1, 3)$, for the linear and cubic terms $(3, -3, -3, 3)$ and for the quadratic and cubic terms $(-1, -3, 3, 1)$. In R, the orthogonal polynomial contrasts have different numerical values, but the same properties:

```
t(contrasts(treatment))
```

```
          [,1]        [,2]        [,3]       [,4]
.L -0.6708204 -0.2236068  0.2236068 0.6708204
.Q  0.5000000 -0.5000000 -0.5000000 0.5000000
.C -0.2236068  0.6708204 -0.6708204 0.2236068
```

If you wanted to be especially perverse, you could reconstruct the four estimated mean values from these polynomial contrasts and the treatment effects shown in summary.lm (above). The means for none, low, medium and high are respectively

```
4.8125 - 0.6708204*1.733 - 0.5*2.6250 + 0.2236068*0.7267
```

```
[1]   2.499963
```

```
4.8125 - 0.2236068*1.733+0.5*2.6250 - 0.6708204*0.7267
```

```
[1]   5.250004
```

```
4.8125 + 0.2236068*1.733 + 0.5*2.6250 + 0.6708204*0.7267
```

```
[1]   6.999996
```

```
4.8125 + 0.6708204*1.733 - 0.5*2.6250 - 0.2236068*0.7267
```

```
[1]   4.500037
```

in agreement (to 3 decimal places) with the four mean values (above). Thus, the parameters can be interpreted as the coefficients in a polynomial model of degree 3 ($= k - 1$ because there are $k = 4$ levels of the factor called treatment), but only so long as the factor levels are equally spaced (and we do not know whether that is true from the information in the current dataframe, because we know only the ranking) and the class sizes are equal (that is true in the present case where $n = 4$).

Because we have four data points (the treatment means) and four parameters, the fit of the model to the means is perfect (there are no residual degrees of freedom and no unexplained variation). We can see what the polynomial function looks like by drawing the smooth curve on top of a barplot for the means:

```
y <- as.vector(tapply(response,treatment,mean))
model <- lm(y~poly(x,3))
summary(model)
```

```
Coefficients:
             Estimate Std. Error t value Pr(>|t|)
(Intercept)    4.8125         NA      NA       NA
poly(x, 3)1    1.7330         NA      NA       NA
poly(x, 3)2   -2.6250         NA      NA       NA
poly(x, 3)3   -0.7267         NA      NA       NA
```

Now we can generate a smooth series of x values between 1 and 4 from which to predict the smooth polynomial function:

```
xv <- seq(1,4,0.1)
```

```
yv <- predict(model,list(x=xv))
```

The only slight difficulty is that the x axis values on the barplot do not scale exactly one-to-one with our x values, so we need to adjust the x-location of our smooth line from xv to xs $= -0.5 + 1.2$xv. The parameters -0.5 and 1.2 come from noting that the centres of the four bars are at 0.7, 1.9, 3.1 and 4.3:

```
(bar.x <- barplot(y))
```

```
         [,1]
[1,]   0.7
[2,]   1.9
[3,]   3.1
[4,]   4.3
```

```
barplot(y,names=levels(treatment))
xs   <-   -0.5 + 1.2 * xv
lines(xs,yv,col="red")
```

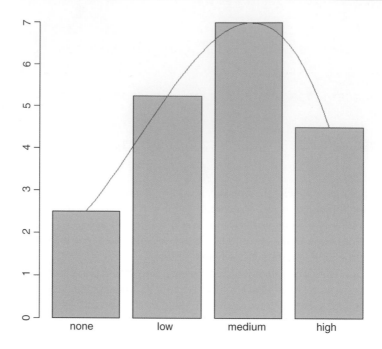

9.28   Summary of statistical modelling

The steps in the statistical analysis of data are always the same, and should always be done in the following order:

(1) data inspection (plots and tabular summaries, identifying errors and outliers);

(2) model specification (picking an appropriate model from many possibilities);

(3) ensure that there is no pseudoreplication, or specify appropriate random effects;

(4) fit a maximal model with an appropriate error structure;

(5) model simplification (by deletion from a complex initial model);

(6) model criticism (using diagnostic plots, influence tests, etc.);

(7) repeat steps 2 to 6 as often as necessary.

10

Regression

Regression analysis is the statistical method you use when both the response variable and the explanatory variable are continuous variables (i.e. real numbers with decimal places – things like heights, weights, volumes, or temperatures). Perhaps the easiest way of knowing when regression is the appropriate analysis is to see that a scatterplot is the appropriate graphic (in contrast to analysis of variance, say, where it would have been a box-and-whisker plot or a bar chart). We cover seven important kinds of regression analysis in this book:

- linear regression (the simplest, and much the most frequently used);
- polynomial regression (often used to test for non-linearity in a relationship);
- piecewise regression (two or more adjacent straight lines);
- robust regression (models that are less sensitive to outliers);
- multiple regression (where there are numerous explanatory variables);
- non-linear regression (to fit a specified non-linear model to data);
- non-parametric regression (used when there is no obvious functional form).

The first five cases are covered here, non-linear regression in Chapter 20 and non-parametric regression in Chapter 18 (where we deal with generalized additive models and non-parametric smoothing).

The essence of regression analysis is using sample data to estimate parameter values and their standard errors. First, however, we need to select a model which describes the relationship between the response variable and the explanatory variable(s). The simplest of all is the linear model

$$y = a + bx.$$

There are two variables and two parameters. The response variable is y, and x is a single continuous explanatory variable. The parameters are a and b: the intercept is a (the value of y when $x = 0$); and the slope is b (the change in y divided by the change in x which brought it about).

The R Book, Second Edition. Michael J. Crawley.
© 2013 John Wiley & Sons, Ltd. Published 2013 by John Wiley & Sons, Ltd.

10.1   Linear regression

Let us start with an example which shows the growth of caterpillars fed on experimental diets differing in their tannin content:

```
reg.data <- read.table("c:\\temp\\regression.txt",header=T)
attach(reg.data)
names(reg.data)

[1]  "growth"  "tannin"

plot(tannin,growth,pch=21,col="blue",bg="red")
```

The higher the percentage of tannin in the diet, the more slowly the caterpillars grew. You can get a crude estimate of the parameter values by eye. Tannin content increased by 8 units, in response to which growth declined from about 12 units to about 2 units, a change of −10 units of growth. The slope, b, is the change in y divided by the change in x, so

$$b \approx \frac{-10}{8} = -1.25.$$

The intercept, a, is the value of y when $x = 0$, and we see by inspection of the scatterplot that growth was close to 12 units when tannin was zero. Thus, our rough parameter estimates allow us to write the regression equation as

$$y \approx 12.0 - 1.25x.$$

Of course, different people would get different parameter estimates by eye. What we want is an objective method of computing parameter estimates from the data that are in some sense the 'best' estimates of the parameters for these data and this particular model. The convention in modern statistics is to use the **maximum**

likelihood estimates of the parameters as providing the 'best' estimates. That is to say that, given the data, and having selected a linear model, we want to find the values of the slope and intercept that make the data most likely. Keep re-reading this sentence until you understand what it is saying.

For the simple kinds of regression models with which we begin, we make several important assumptions:

- The variance in y is constant (i.e. the variance does not change as y gets bigger).

- The explanatory variable, x, is measured without error.

- The difference between a measured value of y and the value predicted by the model for the same value of x is called a residual.

- Residuals are measured on the scale of y (i.e. parallel to the y axis).

- The residuals are normally distributed.

```
model <- lm(growth~tannin)
abline(model,col="red")
yhat <- predict(model,tannin=tannin)
join <- function(i)
  lines(c(tannin[i],tannin[i]),c(growth[i],yhat[i]),col="green")
sapply(1:9,join)
```

Under these assumptions, the maximum likelihood is given by the **method of least squares**. The phrase 'least squares' refers to the residuals, as shown in the figure. The residuals are the vertical differences between the data (solid circles) and the fitted model (the straight line). Each of the residuals is a distance, d, between a data point, y, and the value predicted by the fitted model, $\hat{y}$, evaluated at the appropriate value of the explanatory variable, x:

$$d = y - \hat{y}.$$

Now we replace the predicted value $\hat{y}$ by its formula $\hat{y} = a + bx$, noting the change in sign:

$$d = y - a - bx.$$

Finally, our measure of lack of fit is the sum of the squares of these distances:

$$\sum d^2 = \sum (y - a - bx)^2.$$

The sum of the residuals will always be zero, because the positive and negative residuals cancel out, so $\sum d$ is no good as a measure of lack of fit (although $\sum |d|$ is useful in computationally intensive statistics; see p. 65). The best fit line is defined as passing through the point defined by the mean value of x ($\bar{x}$) and the mean value of y ($\bar{y}$). The large open circle marks the point($\bar{x}, \bar{y}$). You can think of maximum likelihood as working as follows. Imagine that the straight line is pivoted, so that it can rotate around the point($\bar{x}, \bar{y}$). When the line is too steep, some of the residuals are going to be very large. Likewise, if the line is too shallow, some of the residuals will again be very large. Now ask yourself what happens to the sum of the squares of the residuals as the slope is rotated from too shallow, through just right, to too steep. The sum of squares will be big at first, then decline to a minimum value, then increase again. A graph of the sum of squares against the value of the slope used in estimating it would look like this:

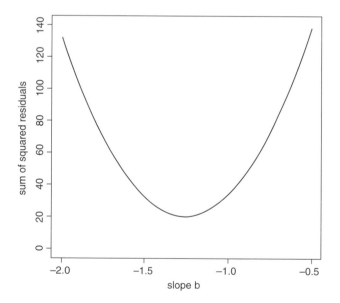

```
bs <- seq(-2,-0.5,0.01)
SSE <- function(i) sum((growth - 12 - bs[i]*tannin)^2)
plot(bs,sapply(1:length(bs),SSE),type="l",ylim=c(0,140),
          xlab="slope b",ylab="sum of squared residuals",col="blue")
```

The maximum likelihood estimate of the slope is the value of b associated with the minimum value of the sum of the squares of the residuals (i.e. close to −1.25). Ideally we want an analytic solution that gives the maximum likelihood of the slope directly (this is done using calculus in Box 10.1). It turns out, however, that the least-squares estimate of b can be calculated very simply from the covariance of x and y (which we met on p. 304).

10.1.1   The famous five in R

We want to find the minimum value of $\sum d^2$. To work this out we need the 'famous five': these are $\sum y^2$ and $\sum y$, $\sum x^2$ and $\sum x$, and the sum of products, $\sum xy$ (introduced on p. 331). The sum of products is worked out pointwise. You can calculate the numbers from the data the long way:

```
sum(tannin);sum(tannin^2);sum(growth);sum(growth^2);sum(tannin*growth)
```

```
[1]  36
[1]  204
[1]  62
[1]  536
[1]  175
```

Alternatively, as we saw on p. 332, you can create a matrix and use matrix multiplication:

```
XY <- cbind(1,growth,tannin)
t(XY) %*% XY
```

```
          growth tannin
       9      62     36
growth 62     536    175
tannin 36     175    204
```

10.1.2   Corrected sums of squares and sums of products

The next thing is to use the famous five to work out three essential 'corrected sums'. We are already familiar with corrected sums of squares, because these are used in calculating variance: s^2 is calculated as the corrected sum of squares divided by the degrees of freedom (p. 333). We shall need the corrected sums of squares of both the explanatory variable, SSX, and the response variable, SSY:

$$SSX = \sum x^2 - \frac{\left(\sum x\right)^2}{n},$$

$$SSY = \sum y^2 - \frac{\left(\sum y\right)^2}{n}.$$

The third term is the corrected sum of products, SSXY. The covariance of x and y is the expectation of the vector product $E\left[(x - \bar{x})(y - \bar{y})\right]$, and this depends on the value of the corrected sum of products (p. 334), which is given by

$$SSXY = \sum xy - \frac{\left(\sum x\right)\left(\sum y\right)}{n}.$$

If you look carefully you will see that the corrected sum of products has exactly the same kind of structure as SSY and SSX. For SSY, the first term is the sum of y times y and the second term contains the sum of y times the sum of y (and similarly for SSX). For SSXY, the first term contains the sum of x times y and the second term contains the sum of x times the sum of y.

Note that for accuracy within a computer program it is best *not* to use these shortcut formulae, because they involve differences (minus) between potentially very large numbers (sums of squares) and hence are potentially subject to rounding errors. Instead, when programming, use the following equivalent formulae:

$$SSY = \sum (y - \bar{y})^2,$$
$$SSX = \sum (x - \bar{x})^2,$$
$$SSXY = \sum (y - \bar{y})(x - \bar{x}).$$

The three key quantities *SSY, SSX* and *SSXY* can be computed the long way, substituting the values of the famous five:

$$SSY = 536 - \frac{62^2}{9} = 108.8889,$$

$$SSX = 204 - \frac{36^2}{9} = 60,$$

$$SSXY = 175 - \frac{36 \times 62}{9} = -73.$$

Alternatively, the matrix can be used (see p. 334).

The next question is how we use *SSX, SSY* and *SSXY* to find the maximum likelihood estimates of the parameters and their associated standard errors. It turns out that this step is much simpler than what has gone before. The maximum likelihood estimate of the slope, *b*, is just

$$b = \frac{SSXY}{SSX}$$

(the detailed derivation of this is in Box 10.1). So, for our example,

$$b = \frac{-73}{60} = -1.216667.$$

Compare this with our by-eye estimate of –1.25. Now that we know the value of the slope, we can use any point that we know to lie on the fitted straight line to work out the maximum likelihood estimate of the intercept, *a*. One part of the definition of the best-fit straight line is that it passes through the point $(\bar{x}, \bar{y})$ determined by the mean values of *x* and *y*. Since we know that $y = a + bx$, it must be the case that $\bar{y} = a + b\bar{x}$, and so

$$a = \bar{y} - b\bar{x} = \frac{\sum y}{n} - b \frac{\sum x}{n}$$

and, using R as a calculator, we get the value of the intercept as

```
mean(growth)+1.216667*mean(tannin)
```

```
[1]   11.75556
```

noting the change of sign. This is reasonably close to our original estimate by eye ($a \approx 12$).

The function for carrying out linear regression in R is `lm` (which stands for 'linear model'). The response variable comes first (`growth` in our example), then the tilde $\sim$, then the name of the continuous explanatory variable (`tannin`). R prints the values of the intercept and slope like this:

```
lm(growth~tannin)
```

```
Coefficients:
(Intercept)        tannin
    11.756        -1.217
```

We can now write the maximum likelihood equation like this:

$$\texttt{growth} = 11.755\ 56 - 1.216\ 667 \times \texttt{tannin}.$$

Box 10.1   The least-squares estimate of the regression slope, *b*

The **best fit** slope is found by rotating the line until the *error sum of squares, SSE,* is minimized, so we want to find the minimum of $\sum (y - a - bx)^2$. We start by finding the derivative of *SSE* with respect to *b*:

$$\frac{\mathrm{d}SSE}{\mathrm{d}b} = -2 \sum x(y - a - bx).$$

Now, multiplying through the bracketed term by *x* gives

$$\frac{\mathrm{d}SSE}{\mathrm{d}b} = -2 \sum xy - ax - bx^2.$$

Apply summation to each term separately, set the derivative to zero, and divide both sides by -2 to remove the unnecessary constant:

$$\sum xy - \sum ax - \sum bx^2 = 0.$$

We cannot solve the equation as it stands because there are two unknowns, *a* and *b*. However, we know that the value of *a* is $\bar{y} - b\bar{x}$. Also, note that $\sum ax$ can be written as $a\sum x$, so replacing *a* and taking both *a* and *b* outside their summations gives:

$$\sum xy - \left[\frac{\sum y}{n} - b\frac{\sum x}{n}\right]\sum x - b\sum x^2 = 0.$$

Now multiply out the bracketed term by $\sum x$ to get:

$$\sum xy - \frac{\sum x \sum y}{n} + b\frac{(\sum x)^2}{n} - b\sum x^2 = 0.$$

Next, take the two terms containing *b* to the right-hand side, and note their change of sign:

$$\sum xy - \frac{\sum x \sum y}{n} = b\sum x^2 - b\frac{(\sum x)^2}{n}.$$

Finally, divide both sides by $\sum x^2 - \left(\sum x\right)^2 / n$ to obtain the required estimate b:

$$b = \frac{\sum xy - \sum x \sum y / n}{\sum x^2 - \left(\sum x\right)^2 / n}.$$

Thus, the value of b that minimizes the sum of squares of the departures is given simply by:

$$b = \frac{SSXY}{SSX}.$$

This is the **maximum likelihood estimate of the slope** of the linear regression.

10.1.3   Degree of scatter

There is another very important issue that needs to be considered, because two data sets with exactly the same slope and intercept could look quite different:

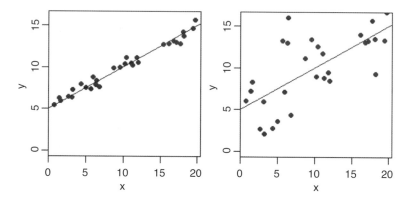

We need a way to quantify the degree of fit, so that the graph on the left has a high value and the graph on the right has a low value. It turns out that we already have the appropriate quantity: it is the sum of squares of the residuals (p. 338). This is referred to as the *error sum of squares*, *SSE*. Here, **error** does not mean 'mistake', but refers to residual variation or *unexplained variation*:

$$SSE = \sum (y - a - bx)^2.$$

Graphically, you can think of *SSE* as the sum of the squares of the lengths of the vertical residuals (the green lines) in the plot on p. 452. By tradition, however, when talking about the degree of scatter we actually quantify the *lack* of scatter, so the graph on the left, with a perfect fit (zero scatter) gets a value of 1, and the graph on the right, which shows no relationship at all between y and x (100% scatter), gets a value of 0. This quantity used to measure the lack of scatter is officially called the 'coefficient of determination', but everybody refers to it as 'r squared'. This is an important definition that you should try to memorize: r^2 is the fraction of the total variation in y that is explained by variation in x. We have already defined the total variation in the response variable as *SSY* (p. 454). The unexplained variation in the response variable is defined above as *SSE* (the error sum of squares) so the explained variation is simply *SSY* − *SSE*. Thus,

$$r^2 = \frac{SSY \quad SSE}{SSE}.$$

A value of $r^2 = 1$ means that all of the variation in the response variable is explained by variation in the explanatory variable (the left-hand graph below) while a value of $r^2 = 0$ means none of the variation in the response variable is explained by variation in the explanatory variable (the right-hand graph).

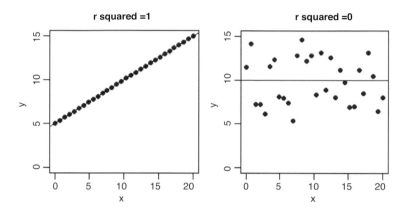

```
y <- 5+0.5*x
plot(x,y,pch=16,xlim=c(0,20),ylim=c(0,15),col="red",main="r squared = 1")
abline(5,0.5,col="blue")
y <- 5+runif(30)*10
plot(x,y,pch=16,xlim=c(0,20),ylim=c(0,15),col="red",main="r squared = 0")
abline(h=10,col="blue")
```

You can get the value of *SSY* the long way as on p. 454 (*SSY* = 108.8889), or using R to fit the null model in which growth is described by a single parameter, the intercept *a*. In R, the intercept is called parameter 1, so the null model is expressed as lm(growth~1). There is a function called deviance that can be applied to a linear model which returns the sum of the squares of the residuals (in this null case, it returns $\sum (y - \bar{y})^2$, which is *SSY* as we require):

```
deviance(lm(growth~1)
```

```
[1]   108.8889
```

The value of *SSE* is worked out longhand from $\sum (y - a - bx)^2$ but this is a pain, and the value can be extracted very simply from the regression model using deviance like this:

```
deviance(lm(growth~tannin))
```

```
[1]   20.07222
```

Now we can calculate the value of r^2:

$$r^2 = \frac{SSY - SSE}{SSY} = \frac{108.8889 - 20.072\,22}{108.8889} = 0.815\,663\,3.$$

You will not be surprised that the value of r^2 can be extracted from the model:

```
summary(lm(growth~tannin))[[8]]
```

```
[1] 0.8156633
```

The **correlation coefficient**, r, introduced on p. 373, is given by

$$r = \frac{SSXY}{\sqrt{SSX \times SSY}}.$$

Of course r is the square root of r^2, but we use the formula above so that we retain the sign of the correlation: $SSXY$ is positive for positive correlations between y and x and negative for negative correlations between y and x. For our example, the correlation coefficient is

$$r = \frac{-73}{\sqrt{60 \times 108.8889}} = -0.903\ 140\ 7.$$

10.1.4   Analysis of variance in regression: $SSY = SSR + SSE$

The idea is simple: we take the total variation in y, SSY, and partition it into components that tell us about the explanatory power of our model. The variation that is explained by the model is called the regression sum of squares (denoted by SSR), and the unexplained variation is called the error sum of squares (denoted by SSE). Then $SSY = SSR + SSE$. Now, in principle, we could compute SSE because we know that it is the sum of the squares of the deviations of the data points from the fitted model, $\sum d^2 = \sum (y - a - bx)^2$. Since we know the values of a and b, we are in a position to work this out. The formula is fiddly, however, because of all those subtractions, squarings and addings-up. Fortunately, there is a very simple shortcut that involves computing SSR, the explained variation, rather than SSE. This is because

$$SSR = b \times SSXY = \frac{SSXY^2}{SSX},$$

so we can immediately work out $SSR = -1.21667 \times -73 = 88.816\ 67$. And since $SSY = SSR + SSE$, we can get SSE by subtraction:

$$SSE = SSY - SSR = 108.8889 - 88.81667 = 20.07222.$$

Using R to do the calculations, we get:

```
(sse <- deviance(lm(growth~tannin)))
```

```
[1]  20.07222
```

```
(ssy <- deviance(lm(growth~1)))
```

```
[1]  108.8889
```

```
(ssr <- ssy-sse)
```

```
[1]  88.81667
```

We now have all of the sums of squares, and all that remains is to think about the degrees of freedom. We had to estimate one parameter, the overall mean, $\bar{y}$, before we could calculate $SSY = \sum (y - \bar{y})^2$, so the total degrees of freedom are $n - 1$. The error sum of squares was calculated only after two parameters had been estimated from the data (the intercept and the slope) since $SSE = \sum (y - bx)^2$, so the error degrees of freedom are $n - 2$. Finally, the regression model added just one parameter, the slope b, compared with the null model, so there is *one* regression degree of freedom. Thus, the ANOVA table looks like this:

Source	Sum of squares	Degrees of freedom	Mean squares	F ratio
Regression	88.817	1	88.817	30.974
Error	20.072	7	$s^2 = 2.867\,46$	
Total	108.889	8		

Notice that the component degrees of freedom add up to the total degrees of freedom (this is always true, in any ANOVA table, and is a good check on your understanding of the design of the experiment). The third column, headed 'Mean squares', contains the variances obtained by dividing the sums of squares by the degrees of freedom in the same row. In the row labelled 'Error' we obtain the very important quantity called the error variance, denoted by s^2, by dividing the error sum of squares by the error degrees of freedom. Obtaining the value of the error variance is the main reason for drawing up the ANOVA table. Traditionally, one does not fill in the bottom box (it would be the overall variance in y, $SSY/(n - 1)$, although this is the basis of the adjusted r^2 value; see p. 461). Finally, the ANOVA table is completed by working out the F ratio, which is a ratio between two variances. In most simple ANOVA tables, you divide the treatment variance in the numerator (the regression variance in this case) by the error variance s^2 in the denominator. The null hypothesis under test in a linear regression is that the slope of the regression line is zero (i.e. that there is no dependence of y on x). The two-tailed alternative hypothesis is that the slope is significantly different from zero (either positive or negative). In many applications it is not particularly interesting to reject the null hypothesis, because we are interested in the estimates of the slope and its standard error (we often know from the outset that the null hypothesis is false). To test whether the F ratio is sufficiently large to reject the null hypothesis, we compare the calculated value of F in the final column of the ANOVA table with the critical value of F, expected by chance alone (this is found from quantiles of the F distribution qf, with 1 d.f. in the numerator and $n - 2$ d.f. in the denominator, as described below). The table can be produced directly from the fitted model in R by using the anova function:

```
anova(lm(growth~tannin))
```

```
Analysis of Variance Table
```

```
Response: growth
          Df Sum Sq Mean Sq F value    Pr(>F)
tannin     1 88.817  88.817  30.974 0.0008461 ***
Residuals  7 20.072   2.867
```

The same output can be obtained using summary.aov(lm(growth~tannin)). The extra column given by R is the p value associated with the computed value of F.

There are two ways to assess our F ratio of 30.974. One way is to compare it with the critical value of F, with 1 d.f. in the numerator and 7 d.f. in the denominator. We have to decide on the level of uncertainty that we are willing to put up with; the traditional value for work like this is 5%, so our certainty is 0.95. Now we can use quantiles of the F distribution, qf, to find the critical value of F:

```
qf(0.95,1,7)
```

```
[1]  5.591448
```

Because our calculated value of F is much larger than this critical value, we can be confident in rejecting the null hypothesis. The other way, which is perhaps better than working rigidly at the 5% uncertainty level, is to ask what is the probability of getting a value for F as big as 30.974 or larger if the null hypothesis is true. For this we use 1-pf rather than qf:

```
1-pf(30.974,1,7)
```

```
[1]    0.0008460725
```

It is very unlikely indeed ($p < 0.001$). This value is in the last column of the R output. Note that the p value is *not* the probability that the null hypothesis is true. On the contrary, it is the probability, given that the null hypothesis *is* true, of obtaining a value of F this large or larger by chance alone.

10.1.5   Unreliability estimates for the parameters

Finding the least-squares values of slope and intercept is only half of the story, however. In addition to the parameter estimates, $a = 11.756$ and $b = -1.2167$, we need to measure the unreliability associated with each of the estimated parameters. In other words, we need to calculate the standard error of the intercept and the standard error of the slope. We have already met the standard error of the mean, and we used it in calculating confidence intervals (p. 122) and in doing Student's t test (p. 358). Standard errors of regression parameters are similar in so far as they are enclosed inside a big square root term (so that the units of the standard error are the same as the units of the parameter), and they have the error variance, s^2, from the ANOVA table (above) in the numerator. There are extra components, however, which are specific to the unreliability of a slope or an intercept (see Boxes 10.2 and 10.3 for details).

Box 10.2   Standard error of the slope

The uncertainty of the estimated slope increases with increasing variance and declines with increasing number of points on the graph. In addition, however, the uncertainty is greater when the range of x values (as measured by SSX) is small:

$$se_b = \sqrt{\frac{s^2}{SSX}}.$$

Box 10.3   Standard error of the intercept

The uncertainty of the estimated intercept increases with increasing variance and declines with increasing number of points on the graph. As with the slope, uncertainty is greater when the range of x values (as measured by SSX) is small. Uncertainty in the estimate of the intercept also increases with the square of the distance between the origin and the mean value of x (as measured by $\sum x^2$):

$$se_a = \sqrt{\frac{s^2 \sum x^2}{n \times SSX}}$$

Longhand calculation shows that the standard error of the slope is

$$se_b = \sqrt{\frac{s^2}{SSX}} = \sqrt{\frac{2.867}{60}} - 0.2186,$$

and the standard error of the intercept is

$$se_a = \sqrt{\frac{s^2 \Sigma x^2}{n \times SSX}} = \sqrt{\frac{2.867 \times 204}{9 \times 60}} = 1.0408.$$

However, in practice you would always use the `summary.lm` function applied to the fitted linear model like this:

```
summary(lm(growth~tannin))
```

```
Coefficients:
            Estimate Std. Error t value Pr(>|t|)
(Intercept)  11.7556     1.0408  11.295 9.54e-06 ***
tannin       -1.2167     0.2186  -5.565 0.000846 ***
```

I have stripped out the details about the residuals and the explanation of the significance stars in order to highlight the parameter estimates and their standard errors (as calculated above). The residual standard error is the square root of the error variance from the ANOVA table ($1.693 = \sqrt{2.867}$). Multiple R-squared is the fraction of the total variance explained by the model ($SSR/SSY = 0.8157$). The adjusted R-squared is close to, but different from, the value of r^2 we have just calculated. Instead of being based on the explained sum of squares, SSR, and the total sum of squares, SSY, it is based on the overall variance (a quantity we do not typically calculate), $s_T^2 = SSY/(n-1) = 13.611$, and the error variance s^2 (from the ANOVA table, $s^2 = 2.867$) and is worked out like this:

$$\text{adjusted} R\text{-squared} = \frac{s_T^2 - s^2}{s_T^2}.$$

So in this example, adjusted R-squared = $(13.611 - 2.867)/13.611 = 0.7893$. We discussed the F statistic and p value in the previous section.

The `summary.lm` table shows everything you need to know about the parameters and their standard errors, but there is a built-in function, `confint`, which produces 95% confidence intervals for the estimated parameters from the model directly like this:

```
confint(model)
```

```
               2.5 %     97.5 %
(Intercept)  9.294457  14.2166544
tannin      -1.733601  -0.6997325
```

These values are obtained by subtracting from, and adding to, each parameter estimate an interval which is the standard error times Student's t with 7 degrees of freedom (the appropriate value of t is given by `qt(.975,7)` = 2.364 624). The fact that neither interval includes 0 indicates that both parameter values are significantly different from zero, as established by the earlier F tests.

Of the two sorts of summary table, `summary.lm` is by far the more informative, because it shows the effect sizes (in this case the slope of the graph) and their unreliability estimates (the standard error of the slope). Generally, you should resist the temptation to put ANOVA tables in your written work. The important information such as the p value and the error variance can be put in the text, or in figure legends, much more efficiently. ANOVA tables put far too much emphasis on hypothesis testing, and show nothing directly about effect sizes.

Box 10.4   Standard error for a predicted value

The standard error of a predicted value $\hat{y}$ is given by:

$$se_{\hat{y}} = \sqrt{s^2 \left[\frac{1}{n} + \frac{(x - \bar{x})^2}{SSX} \right]}.$$

It increases with the *square* of the difference between mean x and the value of x at which the prediction is made. As with the standard error of the slope, the wider the range of x values, SSX, the lower the uncertainty. The bigger the sample size, n, the lower the uncertainty. Note that the formula for the standard error of the intercept is just the special case of this for $x = 0$ (you should check the algebra of this result as an exercise).

For predictions made on the basis of the regression equation we need to know the standard error for a predicted single sample of y,

$$se_y \sqrt{s^2 \left[1 + \frac{1}{n} + \frac{(x - \bar{x})^2}{SSX} \right]},$$

while the standard error for a predicted mean for k items at a given level of x_i is

$$se_{\bar{y}_i} = \sqrt{s^2 \left[\frac{1}{k} + \frac{1}{n} + \frac{(x - \bar{x})^2}{SSX} \right]}.$$

10.1.6   Prediction using the fitted model

It is good practice to save the results of fitting the model in a named object. Naming models is very much a matter of personal taste: some people like the name of the model to describe its structure, other people like the name of the model to be simple and to rely on the formula (which is part of the structure of the model) to describe what the model does. I like the second approach, so I might write

```
model <- lm(growth~tannin)
```

The object called `model` can now be used for all sorts of things. For instance, we can use the `predict` function to work out values for the response at values of the explanatory variable that we did not measure. Thus, we can ask for the predicted growth if tannin concentration was 5.5%. The value or values of the explanatory variable to be used for prediction are specified in a list like this:

```
predict(model,list(tannin=5.5))
```

```
[1]   5.063889
```

indicating a predicted growth rate of 5.06 if a tannin concentration of 5.5% had been applied. To predict growth at more than one level of tannin, the list of values for the explanatory variable is specified as a vector. Here are the predicted growth rates at 3.3, 4.4, 5.5 and 6.6% tannin:

```
predict(model,list(tannin=c(3.3,4.4,5.5,6.6)))
```

```
       1        2        3        4
7.740556 6.402222 5.063889 3.725556
```

For drawing smooth curves through a scatterplot we use `predict` with a vector of 100 or so closely-spaced *x* values, as illustrated on p. 207.

10.1.7 Model checking

The final thing you will want to do is to expose the model to critical appraisal. The assumptions we really want to be sure about are constancy of variance and normality of errors. The simplest way to do this is with model-checking plots. Six plots (selectable by `which`) are currently available: a plot of residuals against fitted values; a scale–location plot of $\sqrt{|\text{residuals}|}$ against fitted values; a normal qunatile–quantile plot; a plot of Cook's distances versus row labels; a plot of residuals against leverages; and a plot of Cook's distances against leverage/(1 – leverage). By default four plots are provided (the first three plus the fifth):

```
windows(7,7)
par(mfrow=c(2,2))
plot(model)
```

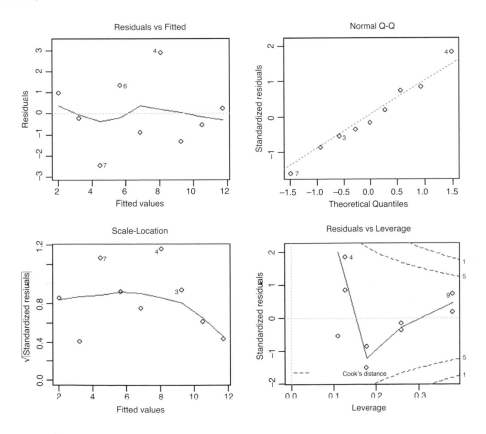

The first graph (top left) shows residuals on the y axis against fitted values on the x axis. It takes experience to interpret these plots, but what you *do not* want to see is lots of structure or pattern in the plot. Ideally, as here, the points should look like the sky at night. It is a major problem if the scatter increases as the fitted values get bigger; this would look like a wedge of cheese on its side (see p. 405). But in our present case, everything is OK on the constancy of variance front.

The next plot (top right) shows the normal `qqnorm` plot (p. 406) which should be a straight line if the errors are normally distributed. Again, the present example looks fine. If the pattern were S-shaped or banana-shaped, we would need to fit a different model to the data.

The third plot (bottom left) is a repeat of the first, but on a different scale; it shows the square root of the standardized residuals (where all the values are positive) against the fitted values. If there was a problem, such as the variance increasing with the mean, then the points would be distributed inside a triangular shape, with the scatter of the residuals increasing as the fitted values increase. The red line would then show a pronounced upward trend. But there is no such pattern here, which is good.

The fourth and final plot (bottom right) shows standardized residuals as a function of leverage, along with Cook's distance (p. 419) for each of the observed values of the response variable. The point of this plot is to highlight those y values that have the biggest effect on the parameter estimates (high influence; p. 409). You can see that point 9 has the highest leverage, but point 7 is quite influential (it is closest to the Cook's distance contour). You might like to investigate how much this influential point (6, 2) affected the parameter estimates and their standard errors. To do this, we repeat the statistical modelling but leave out the point in question, using `subset` like this (recall that `!=` means 'not equal to'):

```
model2 <- update(model,subset=(tannin != 6))
summary(model2)
```

```
Coefficients:
            Estimate Std. Error t value Pr(>|t|)
(Intercept)  11.6892     0.8963  13.042 1.25e-05 ***
tannin       -1.1171     0.1956  -5.712  0.00125 **
```

First of all, notice that we have lost one degree of freedom, because there are now eight values of y rather than nine. The estimate of the slope has changed from -1.2167 to -1.1171 (a difference of about 9%) and the standard error of the slope has changed from 0.2186 to 0.1956 (a difference of about 12%). What you do in response to this information depends on the circumstances. Here, we would simply note that point (6, 2) was influential and stick with our first model, using all the data. In other circumstances, a data point might be so influential that the structure of the model is changed completely by leaving it out. In that case, we might gather more data or, if the study was already finished, we might publish both results (with and without the influential point) so that the reader could make up their own mind about the interpretation. The important point is that we always do model checking; the `summary.lm(model)` table is not the end of the process of regression analysis.

You might also want to check for lack of serial correlation in the residuals (e.g. time series effects) using the `durbin.watson` function from the `car` package (see p. 484), but there are too few data to use it with this example.

10.2   Polynomial approximations to elementary functions

Elementary functions such sin(x), log(x) and exp(x) can be expressed as Maclaurin series:

$$\sin(x) = x - \frac{x^3}{3!} + \frac{x^5}{5!} - \frac{x^7}{7!} + \cdots,$$

$$\cos(x) = 1 - \frac{x^2}{2!} + \frac{x^4}{4!} - \frac{x^6}{6!} + \cdots,$$

$$\exp(x) = \frac{x^0}{0!} + \frac{x^1}{1!} + \frac{x^2}{2!} + \frac{x^3}{3!} + \cdots,$$

$$\log(x+1) = x - \frac{x^2}{2} + \frac{x^3}{3} - \frac{x^4}{4} + \frac{x^5}{5} - \cdots.$$

In fact, we can approximate any smooth continuous single-valued function by a polynomial of sufficiently high degree. To see this in action, consider the graph of sin(x) against x in the range $0 < x < \pi$ (where x is an angle measured in radians):

```
x <- seq(0,pi,0.01)
y <- sin(x)
plot(x,y,type="l",ylab="sin(x)")
```

Up to about $x = 0.3$ the very crude approximation sin(x) = x works reasonably well. The first approximation, including a single extra term for $-x^3/3!$, extends the reasonable fit up to about $x = 0.8$:

```
a1 <- x-x^3/factorial(3)
lines(x,a1,col="green")
```

Adding the term in $x^5/5!$ captures the first peak in sin(x) quite well. And so on.

```
a2 <- x-x^3/factorial(3)+x^5/factorial(5)
lines(x,a2,col="red")
```

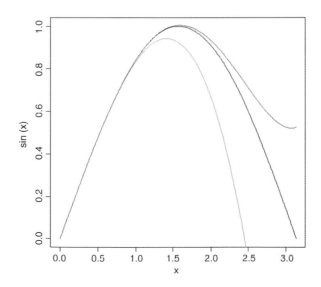

10.3   Polynomial regression

The relationship between y and x often turns out not to be a straight line. However, Occam's razor requires that we fit a straight-line model unless a non-linear relationship is significantly better at describing the data. So this begs the question: how do we assess the significance of departures from linearity? One of the simplest ways is to use polynomial regression.

   The idea of polynomial regression is straightforward. As before, we have just one continuous explanatory variable, x, but we can fit higher powers of x, such as x^2 and x^3, to the model in addition to x to explain curvature in the relationship between y and x. It is useful to experiment with the kinds of curves that can be generated with very simple models. Even if we restrict ourselves to the inclusion of a quadratic term, x^2, there are many curves we can describe, depending upon the signs of the linear and quadratic terms:

```
par(mfrow=c(2,2))
x <- seq(0,10,0.1)
y1 <- 4 + 2 * x - 0.1 * x^2
y2 <- 4 + 2 * x - 0.2 * x^2
y3 <- 12 - 4 * x + 0.35 * x^2
y4 <- 4 + 0.5 * x + 0.1 * x^2
plot(x,y1,type="l",ylim=c(0,15),ylab="y",col="red")
plot(x,y2,type="l",ylim=c(0,15),ylab="y",col="red")
plot(x,y3,type="l",ylim=c(0,15),ylab="y",col="red")
plot(x,y4,type="l",ylim=c(0,15),ylab="y",col="red")
```

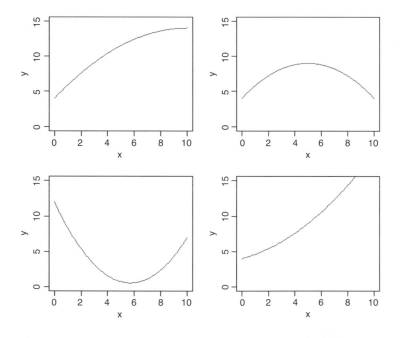

   In the top left-hand panel, there is a curve with positive but declining slope, with no hint of a hump ($y = 4 + 2x - 0.1x^2$). The top right-hand graph shows a curve with a clear maximum ($y = 4 + 2x - 0.2x^2$), and at bottom left we have a curve with a clear minimum ($y = 12 - 4x + 0.35x^2$). The bottom right-hand curve shows a positive association between y and x with the slope increasing as x increases ($y = 4 + 0.5x + 0.1x^2$). So you can see that a simple quadratic model with three parameters (an intercept, a slope for x, and a slope for x^2) is capable of describing a wide range of functional relationships between y and x. It is very important to understand that the quadratic model *describes* the relationship between y and x; it does not pretend to *explain* the mechanistic (or causal) relationship between y and x.

We can see how polynomial regression works by analysing an example where diminishing returns in output (yv) are suspected as inputs (xv) are increased:

```
poly <- read.table("c:\\temp\\diminish.txt",header=T)
attach(poly)
names(poly)
```

```
[1]   "xv"   "yv"
```

We begin by fitting a straight-line model to the data:

```
windows(7,4)
par(mfrow=c(1,2))
model1 <- lm(yv~xv)
plot(xv,yv,pch=21,col="brown",bg="yellow")
abline(model1,col="navy")
```

This is not a bad fit to the data ($r^2 = 0.8725$), but there is a distinct hint of curvature (diminishing returns in this case). Next, we fit a second explanatory variable which is the square of the x value (the so-called 'quadratic term'). Note the use of I (for 'as is') in the model formula; see p. 210.

```
model2 <- lm(yv~xv+I(xv^2))
```

Now we use model2 to predict the fitted values for a smooth range of x values between 0 and 90:

```
plot(xv,yv,pch=21,col="brown",bg="yellow")
x <- 0:90
y <- predict(model2,list(xv=x))
lines(x,y,col="navy")
```

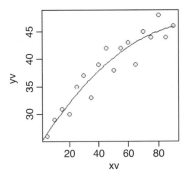

This looks like a slightly better fit than the straight line ($r^2 = 0.9046$), but we shall choose between the two models on the basis of an F test using anova:

```
anova(model1,model2)
```

```
Analysis of Variance Table
```

```
Model 1: yv ~ xv
Model 2: yv ~ xv + I(xv^2)
```

```
   Res.Df      RSS Df Sum of Sq        F Pr(>F)
1      16 91.057
2      15 68.143  1    22.915 5.0441 0.0402 *
```

The more complicated curved model is a significant improvement over the linear model ($p = 0.04$) so we accept that there is evidence of curvature in these data.

10.4   Fitting a mechanistic model to data

Rather than fitting some arbitrary model for curvature (as above, with a quadratic term for inputs), we sometimes have a mechanistic model relating the value of the response variable to the explanatory variable (e.g. a mathematical model of a physical process). In the following example we are interested in the decay of organic material in soil, and our mechanistic model is based on the assumption that the fraction of dry matter lost per year is a constant. This leads to a two-parameter model of exponential decay in which the amount of material remaining (y) is a function of time (t):

$$y = y_0 e^{-bt}.$$

Here y_0 is the initial dry mass (at time $t = 0$) and b is the decay rate (the parameter we want to estimate by linear regression). Taking logs of both sides, we get

$$\log(y) = \log(y_0) - bt.$$

Now you can see that we can estimate the parameter of interest, b, as the slope of a linear regression of $\log(y)$ on t (i.e. we log-transform the y axis but not the x axis) and the value of y_0 as the antilog of the intercept.

We begin by plotting our data:

```
data <- read.table("c:\\temp\\Decay.txt",header=T)
names(data)

[1]  "time"  "amount"

attach(data)
plot(time,amount,pch=21,col="blue",bg="brown")
abline(lm(amount~time),col="green")
```

The curvature in the relationship is clearly evident from the poor fit of the straight-line (green) model through the scatterplot (there are groups of positive residuals for low and high values of time, and a large group of negative residuals at intermediate times). Now we fit the linear model of log(amount) as a function of time:

```
model <- lm(log(amount)~time)
summary(model)

Call:
lm(formula = log(amount) ~ time)

Residuals:
   Min      1Q  Median      3Q     Max
-0.5935 -0.2043  0.0067  0.2198  0.6297
```

```
Coefficients:
             Estimate Std. Error t value Pr(>|t|)
(Intercept)  4.547386   0.100295   45.34  < 2e-16 ***
time        -0.068528   0.005743  -11.93 1.04e-12 ***

Residual standard error: 0.286 on 29 degrees of freedom
Multiple R-squared: 0.8308,      Adjusted R-squared: 0.825
F-statistic: 142.4 on 1 and 29 DF,  p-value: 1.038e-12
```

Thus, the slope is $-0.068\,528$ and y_0 is the antilog of the intercept: $y_0 = \exp(4.547\,386) = 94.385\,36$. The equation can now be parameterized (with standard errors in brackets) as

$$y = e^{4.5474(\pm 0.1003) - 0.0685(\pm 0.00574)t},$$

or written in its original form, without the uncertainty estimates, as

$$y = 94.385^{-0.0685t},$$

and we can draw the fitted line through the data, remembering to take the antilogs of the predicted values (the model predicts `log(amount)` and we want `amount`), like this:

```
ts <- seq(0,30,0.02)
left <- exp(predict(model,list(time=ts)))
plot(time,amount,pch=21,col="blue",bg="brown")
lines(ts,left,col="blue")
```

 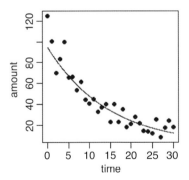

10.5   Linear regression after transformation

Many mathematical functions that are non-linear in their parameters can be linearized by transformation (see p. 258). The most frequent transformations (in order of frequency of use), are logarithms, antilogs and reciprocals. Here is an example of linear regression associated with a power law (p. 261):

$$y = ax^b.$$

This is a two-parameter function, where the parameter a describes the slope of the function for low values of x and b is the shape parameter. For $b = 0$ we have a horizontal relationship $y = a$, for $b = 1$ we have a straight line through the origin $y = ax$ with slope a, for $b > 1$ the slope is positive but increases with increasing x, for

$0 < b < 1$ the slope is positive but decreases with increasing x, while for $b < 0$ (negative powers) the curve is a negative hyperbola that is asymptotic to infinity as x approaches 0 and asymptotic to zero as x approaches infinity.

Let us load a new dataframe and plot the data:

```
power <- read.table("c:\\temp\\power.txt",header=T)
attach(power)
names(power)
```

```
[1] "area"      "response"
```

```
plot(area,response,pch=21,col="green",bg="orange")
abline(lm(response~area),col="blue")
plot(log(area),log(response),pch=21,col="green",bg="orange")
abline(lm(log(response)~log(area)),col="blue")
```

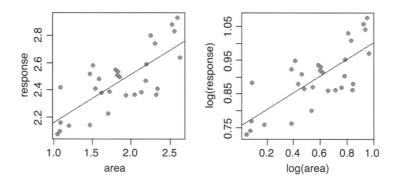

The two plots look very similar (this is not always the case), but we need to compare the two models:

```
model1 <- lm(response~area)
```

```
model2 <- lm(log(response)~log(area))
```

```
summary(model2)
```

```
Coefficients:
            Estimate Std. Error t value Pr(>|t|)
(Intercept)  0.75378    0.02613  28.843  < 2e-16 ***
log(area)    0.24818    0.04083   6.079 1.48e-06 ***
```

We need to do a t test to see whether the estimated shape parameter, $b = 0.248\,18$, is significantly less than $b = 1$ (a straight line):

$$t = \frac{|0.24818 - 1.0|}{0.04083} = 18.41342.$$

This is highly significant ($p < 0.0001$), so we conclude that there is a non-linear relationship between response and area.

Let us get a visual comparison of the two models:

```
windows(7,7)
plot(area,response,pch=21,col="green",bg="orange")
```

```
abline(lm(response~area),col="blue")
xv <- seq(1,2.7,0.01)
yv <- exp(0.75378)*xv^0.24818
lines(xv,yv,col="red")
```

This is a nice example of the distinction between statistical significance and scientific importance. The power law transformation shows that the curvature is highly significant ($b < 1$ with $p<0.0001$) but over the range of the data, and given the high variance in y, the effect of the curvature is very small; the straight line and the power function are very close to one another. However, the choice of model makes an enormous difference if the function is to be used for prediction. Here are the two functions over an extended range of values for x:

```
plot(area,response,xlim=c(0,5),ylim=c(0,4),pch=21,col="green",bg="orange")
abline(lm(response~area),col="blue")
xv <- seq(0,5,0.01)
yv <- exp(0.75378)*xv^0.24818
lines(xv,yv,col="red")
```

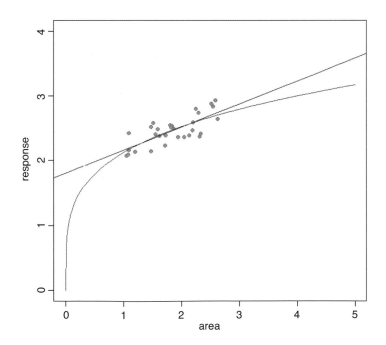

The moral is clear: you need to extremely careful when using regression models for prediction. If you know that response *must* be zero when area is zero (the graph has to pass through the origin) then obviously the power function is likely to be better for extrapolation to the left of the data. But if we have no information on non-linearity other than that contained within the data, then parsimony suggests that errors will be smaller using the simpler, linear model for prediction. Both models are equally good at describing the data (the linear model has $r^2 = 0.574$ and the power law model has $r^2 = 0.569$), but extrapolation beyond the range of the data is always fraught with difficulties. Targeted collection of new data for response at values of area close to 0 and close to 5 might resolve the issue.

10.6 Prediction following regression

The popular notion is that predicting the future is impossible, and that attempts at prediction are nothing more that crystal-gazing. However, all branches of applied science rely upon prediction. These predictions may be based on extensive experimentation (as in engineering or agriculture) or they may be based on detailed, long-term observations (as in astronomy or meteorology). In all cases, however, the main issue to be confronted in prediction is how to deal with uncertainty: uncertainty about the suitability of the fitted model, uncertainty about the representativeness of the data used to parameterize the model, and uncertainty about future conditions (in particular, uncertainty about the future values of the explanatory variables).

There are two kinds of prediction, and these are subject to very different levels of uncertainty. **Interpolation**, which is prediction *within* the measured range of the data, can often be very accurate and is not greatly affected by model choice. **Extrapolation**, which is prediction *beyond* the measured range of the data, is far more problematical, and model choice is a major issue. Choice of the wrong model can lead to wildly different predictions (see p. 471).

Here are two kinds of plots involved in prediction following regression: the first illustrates uncertainty in the parameter estimates; the second indicates uncertainty about predicted values of the response. We continue with the tannin example:

```
reg.data <- read.table("c:\\temp\\regression.txt",header=T)
attach(reg.data)
names(reg.data)
```

```
[1]  "growth"  "tannin"
```

```
plot(tannin,growth,pch=21,col="blue",bg="red")
```

```
model <- lm(growth~tannin)
abline(model,col="blue")
```

The first plot is intended to show the uncertainty associated with the estimate of the slope. It is easy to extract the slope from the vector of coefficients:

```
coef(model)[2]
```

```
    tannin
-1.216667
```

The standard error of the slope is a little trickier to find. After some experimentation, you will discover that it is in the fourth element of the list that is `summary(model)`:

```
summary(model)[[4]][4]
```

```
[1]   0.2186115
```

Here is a function that will add dotted lines showing two extra regression lines to our existing plot – the estimated slope plus and minus one standard error of the slope:

```
se.lines <- function(model){
        b1 <- coef(model)[2]+ summary(model)[[4]][4]
        b2 <- coef(model)[2]- summary(model)[[4]][4]
        xm <- sapply(model[[12]][2],mean)
        ym <- sapply(model[[12]][1],mean)
        a1 <- ym-b1*xm
        a2 <- ym-b2*xm
                abline(a1,b1,lty=2,col="blue")
                abline(a2,b2,lty=2,col="blue")
}
se.lines(model)
```

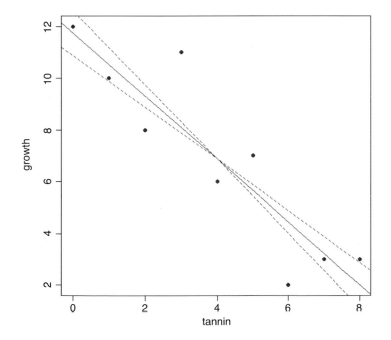

More often, however, we are interested in the uncertainty about predicted values (rather than uncertainty of parameter estimates, as above). We might want to draw the 95% confidence intervals associated with predictions of *y* at different values of *x*. As we saw on p. 460, uncertainty increases with the *square* of the difference between the mean value of *x* and the value of *x* at which the value of *y* is to be predicted. Before we can draw these lines we need to calculate a vector of *x* values; you need 100 or so values to make an attractively smooth curve. Then we need the value of Student's *t* (p. 122). Finally, we multiply Student's *t* by the standard error of the predicted value of *y* (p. 462) to get the confidence interval. This is added to the fitted values of *y* to get the upper limit and subtracted from the fitted values of *y* to get the lower limit. Here is the function:

```
ci.lines <- function(model){
      xm <- sapply(model[[12]][2],mean)
      n <- sapply(model[[12]][2],length)
      ssx <-   sum(model[[12]][2]^2)-sum(model[[12]][2])^2/n
      s.t <-   qt(0.975,(n-2))
xv <- seq(min(model[[12]][2]),max(model[[12]][2]),length=100)
yv <- coef(model)[1]+coef(model)[2]*xv
se <- sqrt(summary(model)[[6]]^2*(1/n+(xv-xm)^2/ssx))
ci <- s.t*se
      uyv <- yv+ci
      lyv <- yv-ci
      lines(xv,uyv,lty=2,col="blue")
      lines(xv,lyv,lty=2,col="blue")
}
```

We replot the linear regression, then overlay the confidence intervals (Box 10.4):

```
plot(tannin,growth,pch=21,col="blue",bg="red")
abline(model, col="blue")

ci.lines(model)
```

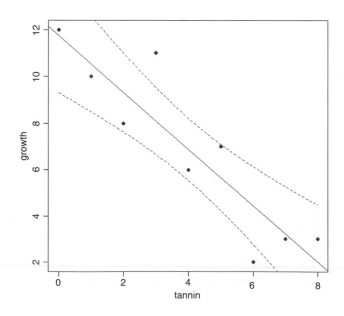

This draws attention to the points at `tannin = 3` and `tannin = 6` that fall outside the 95% confidence limits of our fitted values.

You can speed up this procedure by using the built-in ability to generate confidence intervals coupled with `matlines`. The familiar 95% confidence intervals are `int="c"`, while prediction intervals (fitted values plus or minus 2 standard deviations) are `int="p"` .

```
plot(tannin,growth,pch=16,ylim=c(0,15))
model <-lm(growth~tannin)
```

As usual, start by generating a series of *x* values for generating the curves, then create the scatterplot. The *y* values are predicted from the model, specifying `int="c"`, then `matlines` is used to draw the regression line (solid) and the two confidence intervals (dotted), producing exactly the same graph as our last plot (above) without writing a special function:

```
xv <- seq(0,8,0.1)
yv <- predict(model,list(tannin=xv),int="c")
matlines(xv,yv,lty=c(1,2,2),col="black")
```

A similar plot can be obtained using the `effects` library (see p. 968).

10.7 Testing for lack of fit in a regression

The unreliability estimates of the parameters explained in Boxes 10.2 and 10.3 draw attention to the important issues in optimizing the efficiency of regression designs. We want to make the error variance as small as possible (as always), but in addition, we want to make *SSX* as large as possible, by placing as many points as possible at the extreme ends of the *x* axis. Efficient regression designs allow for:

- replication of least some of the levels of *x*;
- a preponderance of replicates at the extremes (to maximize *SSX*);
- sufficient levels of *x* to allow testing for non-linearity;
- sufficient different values of *x* to allow accurate location of thresholds.

Here is an example where replication allows estimation of pure sampling error, and this in turn allows a test of the significance of the data's departure from linearity. As the concentration of an inhibitor is increased, the reaction rate declines:

```
data <- read.delim("c:\\temp\\lackoffit.txt")
attach(data)
names(data)
```

```
[1]   "conc"   "rate"
```

```
plot(conc,jitter(rate),pch=16,col="red",ylim=c(0,8),ylab="rate")
abline(lm(rate~conc),col="blue")
```

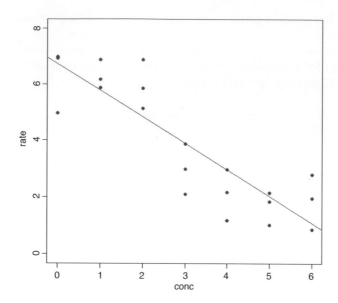

The linear regression does not look too bad, and the slope is highly significantly different from zero:

```
model.reg <- lm(rate~conc)
summary(model.reg)
```

```
Coefficients:
            Estimate Std. Error t value Pr(>|t|)
(Intercept)   6.7262     0.4559  14.755 7.35e-12 ***
conc         -0.9405     0.1264  -7.439 4.85e-07 ***
Residual standard error: 1.159 on 19 degrees of freedom
Multiple R-squared: 0.7444,      Adjusted R-squared: 0.7309
F-statistic: 55.33 on 1 and 19 DF,  p-value: 4.853e-07
```

Because there is replication at each level of x we can do something extra, compared with a typical regression analysis. We can estimate what is called the **pure error variance**. This is the sum of the squares of the differences between the y values and the *mean* values of y for the relevant level of x. This should sound somewhat familiar. In fact, it is the definition of *SSE* from a one-way analysis of variance (see p. 501). By creating a factor to represent the seven levels of x, we can estimate this *SSE* simply by fitting a one-way ANOVA:

```
fac.conc <- factor(conc)
model.aov <- aov(rate~fac.conc)
summary(model.aov)
```

```
          Df Sum Sq Mean Sq F value   Pr(>F)
fac.conc   6  87.81  14.635   17.07 1.05e-05 ***
Residuals 14  12.00   0.857
```

This shows that the pure error sum of squares is 12.0 on 14 degrees of freedom (three replicates, and hence 2 d.f., at each of seven levels of x). See if you can figure out why this sum of squares is less than the observed in the model.reg regression (25.512). If the means from the seven different concentrations all fell exactly on the same straight line then the two sums of squares would be identical. It is the fact that the means do *not* fall on the regression line that causes the difference. The difference between these two sums of squares

$(25.512 - 12.9 = 13.512)$ is a measure of lack of fit of the `rate` data to the straight-line model. We can compare the two models to see if they differ in their explanatory powers:

```
anova(model.reg,model.aov)
```

```
Analysis of Variance Table
Model 1: rate ~ conc
Model 2: rate ~ fac.conc

  Res.Df     RSS Df Sum of Sq        F  Pr(>F)
1     19  25.512
2     14  12.000  5    13.512 3.1528 0.04106 *
```

A single ANOVA table showing the lack-of-fit sum of squares on a separate line is obtained by fitting both the regression line (1 d.f.) and the lack of fit (5 d.f.) in the same model:

```
anova(lm(rate~conc+fac.conc))
```

```
Analysis of Variance Table

Response: rate
          Df Sum Sq Mean Sq F value     Pr(>F)
conc       1 74.298  74.298 86.6806 2.247e-07 ***
fac.conc   5 13.512   2.702  3.1528   0.04106 *
Residuals 14 12.000   0.857
```

To get a visual impression of this lack of fit we can draw vertical lines from the mean values to the fitted values of the linear regression for each level of x:

```
my <- as.vector(tapply(rate,fac.conc,mean))
for (i in 0:6)
 lines(c(i,i),c(my[i+1],predict(model.reg,list(conc=0:6))[i+1]),col="green")
points(0:6,my,pch=16,col="green")
```

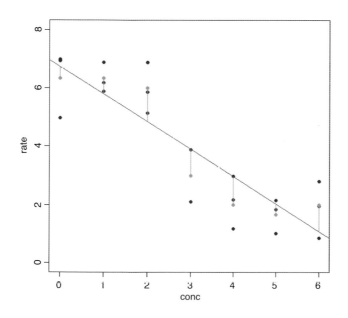

This significant lack of fit indicates that the straight-line model is *not* an adequate description of these data ($p < 0.05$). A negative S-shaped function is likely to fit the data better (see p. 301).

There is an R package called `lmtest` on CRAN, which is full of tests for linear models.

10.8   Bootstrap with regression

An alternative to estimating confidence intervals on the regression parameters from the pooled error variance in the ANOVA table (p. 459) is to use bootstrapping. There are two ways of doing this:

- sample cases with replacement, so that some points are left off the graph while others appear more than once in the dataframe;

- calculate the residuals from the fitted regression model, and randomize which fitted *y* values get which residuals.

In both cases, the randomization is carried out many times, the model fitted and the parameters estimated. The confidence interval is obtained from the quantiles of the distribution of parameter values (see p. 41).

The following dataframe contains a response variable (profit from the cultivation of a crop of carrots for a supermarket) and a single explanatory variable (the cost of inputs, including fertilizers, pesticides, energy and labour):

```
regdat <- read.table("c:\\temp\\regdat.txt",header=T)
attach(regdat)
names(regdat)

[1]  "explanatory"   "response"

plot(explanatory,response,pch=21,col="green",bg="red")
model <- lm(response~explanatory)
abline(model,col="blue")
```

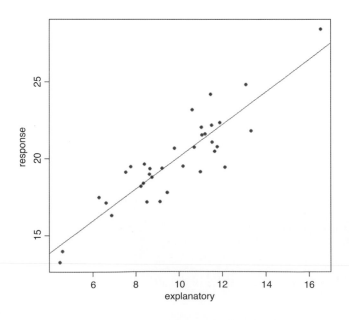

The response is a reasonably linear function of the explanatory variable, but the variance in the response is quite large. For instance, when the explanatory variable is about 12, the response variable ranges between less than 20 and more than 24.

```
model
```

```
Coefficients:
(Intercept)    explanatory
      9.630          1.051
```

Theory suggests that the slope should be 1.0, and our estimated slope is very close to this (1.051). We want to establish a 95% confidence interval on the estimate. Here is a home-made bootstrap which resamples the data points 10 000 times and gives a bootstrapped estimate of the slope:

```
b.boot <- numeric(10000)
```

```
for (i in 1:10000){
        indices <- sample(1:35,replace=T)
        xv <- explanatory[indices]
        yv <- response[indices]
        model <- lm(yv~xv)
        b.boot[i] <- coef(model)[2]
}
hist(b.boot,main="",col="green")
```

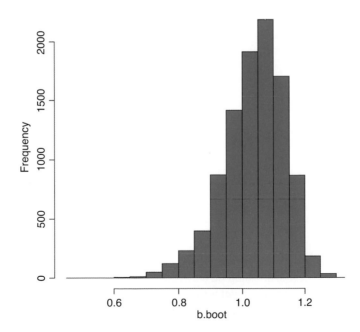

Here is the 95% interval for the bootstrapped estimate of the slope:

```
quantile(b.boot,c(0.025,0.975))
```

```
     2.5%       97.5%
0.8137637 1.1964226
```

Evidently, the bootstrapped data provide no support for the hypothesis that the slope is significantly greater than 1.0.

We now repeat the exercise, using the `boot` function from the `boot` package:

```
library(boot)
```

The first step is to write what is known as the 'statistic' function. This shows `boot` how to calculate the statistic we want from the resampled data (the slope in this case). The resampling of the data is achieved by a subscript provided by `boot` (here called `index`). The point is that every time the model is fitted within the bootstrap it uses a different data set (`yv` and `xv`): we need to describe how these data are constructed and how they are to be used in the model fitting:

```
reg.boot <- function(regdat, index){
     xv <- explanatory[index]
     yv <- response[index]
     model <- lm(yv~xv)
     coef(model)
}
```

Now we can run the `boot` function, then extract the intervals with the `boot.ci` function:

```
reg.model <- boot(regdat,reg.boot,R=10000)
boot.ci(reg.model,index=2)

BOOTSTRAP CONFIDENCE INTERVAL CALCULATIONS
Based on 10000 bootstrap replicates

CALL :
boot.ci(boot.out = reg.model, index = 2)

Intervals :
Level        Normal                    Basic
95%    ( 0.870,  1.254 )     ( 0.903,  1.287 )

Level       Percentile                 BCa
95%    ( 0.815,  1.198 )     ( 0.821,  1.202 )

Calculations and Intervals on Original Scale
Warning message:
In boot.ci(reg.model, index = 2) :
  bootstrap variances needed for studentized intervals
```

All the intervals are reasonably similar: statisticians typically prefer the bias-corrected, accelerated (BCa) intervals. These indicate that if we were to repeat the data-collection exercise we can be 95% confident that the regression slope for those new data would be between 0.821 and 1.202.

The other way of bootstrapping with a model is to randomize the allocation of the residuals to fitted y values estimated from the original regression model. We start by calculating the residuals and the fitted values:

```
model <- lm(response~explanatory)
fit <- fitted(model)
res <- resid(model)
```

What we intend to do is to randomize which of the `res` values is added to the `fit` values to get a reconstructed response variable, y, which we regress as a function of the original explanatory variable. Here is the statistic function to do this:

```
residual.boot <- function(res, index){
y <- fit+res[index]
model <- lm(y~explanatory)
coef(model) }
```

Note that the data passed to the statistic function are `res` in this case (rather than the original dataframe `regdat` as in the first example, above). Now use the `boot` function and the `boot.ci` function to obtain the 95% confidence intervals on the slope (this is `index=2`; the intercept is `index=1`):

```
res.model <- boot(res,residual.boot,R=10000)
boot.ci(res.model,index=2)

BOOTSTRAP CONFIDENCE INTERVAL CALCULATIONS
Based on 10000 bootstrap replicates

CALL :
boot.ci(boot.out = res.model, index = 2)

Intervals :
Level        Normal                  Basic
95%    ( 0.878,  1.224 )    ( 0.884,  1.225 )

Level       Percentile               BCa
95%    ( 0.876,  1.218 )    ( 0.872,  1.215 )

Calculations and Intervals on Original Scale
Warning message:
In boot.ci(res.model, index = 2) :
  bootstrap variances needed for studentized intervals
```

The BCa from randomizing the residuals is from 0.872 to 1.215, while from selecting random x and y points with replacement it was from 0.821 to 1.202 (above). The two rather different approaches to bootstrapping produce reassuringly similar estimates of the same parameter.

10.9   Jackknife with regression

A second alternative to estimating confidence intervals on regression parameters is to **jackknife** the data. Each point in the data set is left out, one at a time, and the parameter of interest is re-estimated. The `regdat` dataframe (above) has `length(response)` data points:

```
names(regdat)
```

```
[1]  "explanatory"   "response"
```

```
length(response)
```

```
[1]   35
```

We create a vector to contain the 35 different estimates of the slope:

```
jack.reg <- numeric(35)
```

Now carry out the regression 35 times, leaving out a different *x*, *y* pair each time:

```
for (i in 1:35) {
model <- lm(response[-i]~explanatory[-i])
jack.reg[i] <- coef(model)[2] }
```

Here is a histogram of the different estimates of the slope of the regression:

```
hist(jack.reg,main="",col="pink")
```

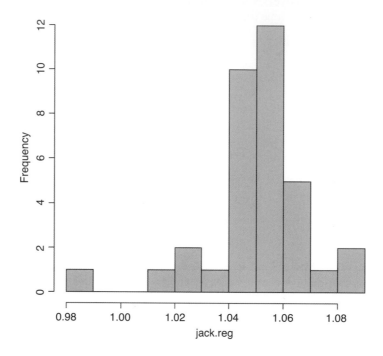

As you can see, the distribution is strongly skew to the left. The quantiles of `jack.reg` are not particularly informative because the sample is so small (just 35). However, the jackknife does draw attention to one particularly influential point (the extreme left-hand bar) which, when omitted from the dataframe, causes the estimated slope to fall below 1.0. We say the point is **influential** because it is the only one of the 35 points whose omission causes the estimated slope to fall below 1.0. But which data point is this? We extract Cook's distance `$infmat[,5]` from the influence matrix from the model (`influence.measures(model)$infmat`) and ask which data point has the maximum value of this influence measure:

```
model <- lm(response~explanatory)
which(influence.measures(model)$infmat[,5]
       == max(influence.measures(model)$infmat[,5]))
```

22

Now we can draw regression lines for the full data set (blue line) and for the model with the influential point number 22 omitted (red line) to see just how influential (or not) this point really is for the location of the line:

```
plot(explanatory,response,pch=21,col="green",bg="red")
abline(model,col="blue")
abline(lm(response[-22]~explanatory[-22]),col="red")
```

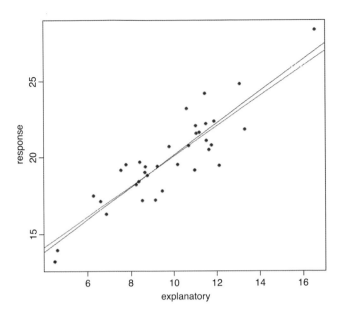

Neither model describes at all well the location of the response for the two lowest values of the explanatory variable (and the fit is worse with the most influential point removed).

10.10   Jackknife after bootstrap

The `jack.after.boot` function calculates the jackknife influence values from a bootstrap output object, and plots the corresponding jackknife-after-bootstrap plot. We illustrate its use with the `boot` object calculated earlier called `reg.model`. We are interested in the slope, which is `index=2`:

```
jack.after.boot(reg.model,index=2)
```

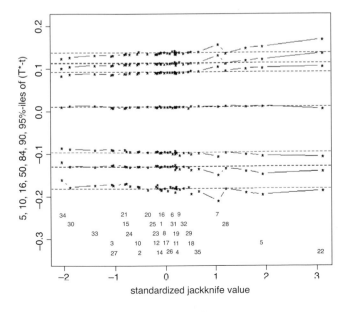

The centred jackknife quantiles for each observation are estimated from those bootstrap samples in which *the particular observation did not appear*. These are then plotted against the influence values. From the top downwards, the horizontal dotted lines show the 95th, 90th, 84th, 50th, 16th, 10th and 5th percentiles. The numbers at the bottom identify the 35 points by their index values within `regdat`. Again, the influence of point no. 22 shows up clearly (this time on the right-hand side), indicating that it has a strong positive influence on the slope, and the two left-hand outliers are identified as points nos 34 and 30.

10.11   Serial correlation in the residuals

The Durbin–Watson function is used for testing whether there is autocorrelation in the residuals from a linear model or a generalized linear model, and is implemented as part of the `car` package (see Fox, 2002):

```
library("car")
durbinWatsonTest(model)
```

```
lag Autocorrelation D-W Statistic p-value
  1      -0.07946739      2.049899   0.874
Alternative hypothesis: rho != 0
```

There is no evidence of serial correlation in these residuals ($p = 0.874$).

The `car` package also contains functions for drawing ellipses, including data ellipses, and confidence ellipses for linear and generalized linear models. Here is the `dataEllipse` function for the present example: by default, the ellipses are drawn at 50% and 90%:

```
dataEllipse(explanatory,response)
```

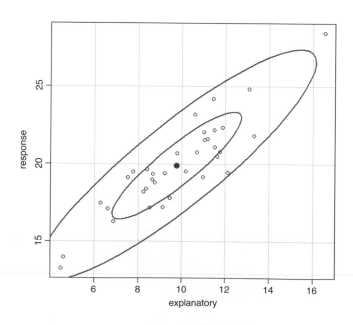

10.12 Piecewise regression

This kind of regression fits different functions over different ranges of the explanatory variable. For example, it might fit different linear regressions to the left and right hand halves of a scatterplot. Two important questions arise in piecewise regression:

- how many segments to divide the line into;
- where to position the break points on the x axis.

Suppose we want to do the simplest piecewise regression, using just two linear segments. Where do we break up the x values? A simple, pragmatic view is to divide the x values at the point where the piecewise regression best fits the response variable. Let us take an example using a linear model where the response is the log of a count (the number of species recorded) and the explanatory variable is the log of the size of the area searched for the species:

```
data <- read.table("c:\\temp\\sasilwood.txt",header=T)
attach(data)
names(data)
```

```
[1]   "Species"   "Area"
```

A quick scatterplot suggests that the relationship between `log(Species)` and `log(Area)` is not linear:

```
plot(log(Species)~log(Area),pch=21,col="red",bg="yellow")
```

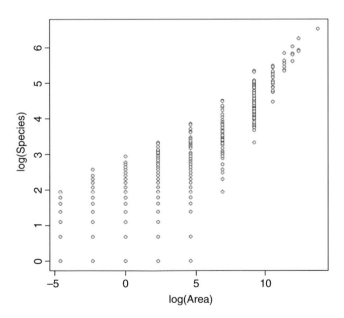

The slope appears to be shallower at small scales than at large. The overall regression highlights this at the model-checking stage:

```
model1 <- lm(log(Species)~log(Area))
par(mfrow=c(2,2))
plot(model1)
```

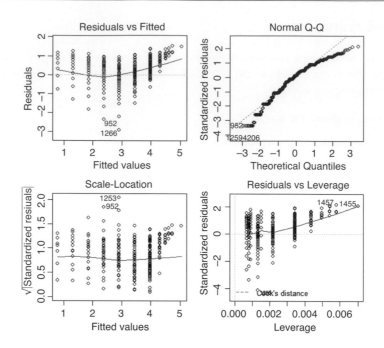

The residuals are very strongly U-shaped (this plot should look like the sky at night) and the errors are profoundly non-normal (the top right-hand line should be straight).

If we are to use piecewise regression, then we need to work out how many straight-line segments to use and where to put the breaks. Visual inspection of the scatterplot suggests that two segments would be an improvement over a single straight line and that the break point should be about `log(Area) = 5`. The choice of break point is made more objective by choosing a range of values for the break point and selecting the break that produces the minimum deviance. We should have a minimum of two x values for each of the pieces of the regression, so the areas associated with the first and last breaks can be obtained by examination of the table of x values:

```
table(Area)
```

```
Area
0.01    0.1      1      10     100    1000   10000  40000  90000 160000 250000  1e+06
 346    345    259     239      88      67    110     18      7      4      3      1
```

The leftmost break could be between areas 0.1 and 1, and the rightmost between 160 000 and 250 000 (i.e. between indices 2 and 3 and 10 and 11)

Piecewise regression is extremely simple in R: we just include a logical statement as part of the model formula, with as many logical statements as we want straight-line segments in the fit. In the present example with two linear segments, the two logical statements are `Area<Break` to define the left-hand regression and `Area>=Break` to define the right-hand regression. We want to fit the model for all values of `Break` between 1 and 250 000, so we create a vector of breaks like this:

```
Break <- sort(unique(Area))[3:11]
```

Now we use a loop to fit the two-segment piecewise model nine times and to store the value of the residual standard error in a vector called d. This quantity is the sixth element of the list that is the model summary object, `d[i] <- summary(model)[[6]]`:

```
d <- numeric(9)
for (i in 1:9) {
model <-
lm(log(Species)~(Area<Break[i])*log(Area)+(Area>=Break[i])*log(Area))
d[i] <- summary(model)[[6]] }
```

A plot shows where the minimum value of d occurs:

```
windows(7,4)
par(mfrow=c(1,2))
plot(log(Break),d,typ="l",col="red")
```

Where exactly does the minimum of d occur? We use the which function for this:

```
Break[which(d==min(d))]
```

```
[1] 100
```

The best piecewise regression will fit one line up to Area = 100 and a different line for Area > 100. The model formula looks like this:

```
model2 <- lm(log(Species)~log(Area)*(Area<100)+log(Area)*(Area>=100))
```

The piecewise regression is a massive improvement over the linear model:

```
anova(model1,model2)
```

```
Analysis of Variance Table
```

```
Model 1: log(Species) ~ log(Area)
Model 2: log(Species) ~ log(Area) * (Area < 100) + log(Area) * (Area >=
    100)
  Res.Df     RSS Df Sum of Sq       F    Pr(>F)
1   1485 731.98
2   1483 631.36  2    100.62 118.17 < 2.2e-16 ***
```

The summary of the piecewise regression takes some getting used to. We have fitted two linear regressions, so there are four parameters. Like an analysis of covariance, the table of coefficients contains one slope and one intercept, along with one difference between slopes and one difference between intercepts. The table has six rows because of the intentional aliasing, which we contrived by providing zeros for the explanatory variables where the two logical expressions evaluate to FALSE:

```
summary(model2)
```

```
Coefficients: (2 not defined because of singularities)
```

	Estimate	Std. Error	t value	Pr(>|t|)	
(Intercept)	0.61682	0.13059	4.723	2.54e-06	***
log(Area)	0.41019	0.01655	24.787	< 2e-16	***
Area < 100TRUE	1.07854	0.13246	8.143	8.12e-16	***
Area >= 100TRUE	NA	NA	NA	NA	
log(Area):Area < 100TRUE	-0.25611	0.01816	-14.100	< 2e-16	***
log(Area):Area >= 100TRUE	NA	NA	NA	NA	

```
Residual standard error: 0.6525 on 1483 degrees of freedom
```

```
Multiple R-squared: 0.724,        Adjusted R-squared: 0.7235
F-statistic:  1297 on 3 and 1483 DF,  p-value: < 2.2e-16
```

The intercept is for the factor level that comes first in the alphabet: this is the right-hand part of the graph `log(Area):Area < 100FALSE` and the slope (`log(Area)`) is for this factor level too. The difference between the two intercepts is labelled `Area < 100TRUE`. The next row is labelled `Area >= 100TRUE` and contains NAs because there were no x values (they were all zeros because logical FALSE was coerced to numeric zero by the multiplication). The difference between two slopes is labelled `log(Area):Area < 100TRUE` while the last row labelled `log(Area):Area >= 100TRUE` contains NAs because there were no x values. We cannot use `abline`, because we want two separate lines through different parts of the scatterplot (not two lines across the whole plotting area; try it and see). To make plotting the two lines easier it is a good idea to calculate the two slopes and two intercepts in advance. The parameters are in the fourth element of the list that makes up `summary(model2)`. It is worth looking at this separately:

```
summary(model2)[[4]]
```

	Estimate	Std. Error	t value	Pr(>|t|)
(Intercept)	0.6168168	0.13058554	4.723469	2.537983e-06
log(Area)	0.4101943	0.01654883	24.786903	9.081618e-114
Area < 100TRUE	1.0785395	0.13245572	8.142642	8.117406e-16
log(Area):Area < 100TRUE	-0.2561147	0.01816373	-14.100333	1.834740e-42

Note that the two rows with NAs have been excluded. Using subscripts, we extract the parameter estimates and calculate two intercepts (`a1` and `a2`) and two slopes (`b1` and `b2`) for the left and right hand pieces of the regression:

```
a1 <- summary(model2)[[4]][1]+summary(model2)[[4]][3]
a2 <- summary(model2)[[4]][1]
b1 <- summary(model2)[[4]][2]+summary(model2)[[4]][4]
b2 <- summary(model2)[[4]][2]
```

Finally, we need to decide on the x values between which to draw the two lines. Inspection of the scatterplot indicates that –5 would be a good minimum value and 15 would be a good maximum. The break point ($4.6 = \log(100)$) is the obvious point at which to stop the first line and start the second line.

```
plot(log(Area),log(Species),col="blue")
lines(c(-5,4.6),c(a1+b1*-5,a1+b1*4.6),col="red")
lines(c(4.6,15),c(a2+b2*4.6,a2+b2*15),col="red")
```

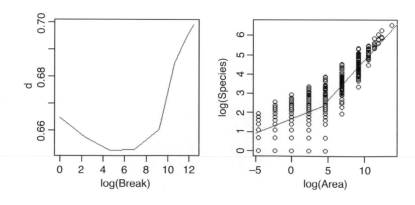

Of course, for spatial scales even smaller than studied here, the slope of the plot must go asymptotically to zero, because once the plot is so small that it can contain only one individual, making the plot even smaller is bound to contain that same individual (thus, species richness will be one for all subsequent smaller spatial scales).

10.13   Multiple regression

A multiple regression is a statistical model with two or more continuous explanatory variables. We contrast multiple regression with analysis of variance, where all the explanatory variables are categorical (Chapter 11) and analysis of covariance, where the explanatory variables are a mixture of continuous and categorical (Chapter 12). Multiple regressions models provide some of the most profound challenges faced by the analyst because of some crucial issues:

- over-fitting (we often have more explanatory variables than data points);

- parameter proliferation (we might want to fit parameters for curvature and interaction);

- correlation between explanatory variables (called collinearity);

- choice between contrasting models of roughly equal explanatory power.

The *principle of parsimony* (Occam's razor), discussed in Section 9.2, is again relevant here. It requires that the model should be as simple as possible. This means that the model should not contain any redundant parameters. Ideally, we achieve this by fitting a maximal model and then simplifying it by following one or more of these steps:

- Remove non-significant interaction terms.

- Remove non-significant quadratic or other non-linear terms.

- Remove non-significant explanatory variables.

- Amalgamate explanatory variables that have similar parameter values.

Of course, such simplifications must make good scientific sense, and must not lead to significant reductions in explanatory power. It is likely that many of the explanatory variables are correlated with each other, and so *the order in which variables are deleted from the model* will influence the explanatory power attributed to them. The thing to remember about multiple regression is that, in principle, there is no end to it. The number of combinations of interaction terms and curvature terms is endless. There are some simple rules (like parsimony) and some automated functions (like `step`) to help. But, in principle, you could spend a very great deal of time in modelling a single dataframe. There are no hard-and-fast rules about the best way to proceed, but we shall typically carry out simplification of a complex model by stepwise deletion: non-significant terms are left out, and significant terms are added back (see Chapter 9).

At the data inspection stage, there are many more kinds of plots we could do:

- Plot the response against each of the explanatory variables separately.

- Plot the explanatory variables against one another (e.g. `pairs`; see Section 10.13.1).

- Plot the response against pairs of explanatory variables in three-dimensional plots.

- Plot the response against explanatory variables for different combinations of other explanatory variables (e.g. conditioning plots, `coplot`; see p. 236).

- Fit non-parametric smoothing functions (e.g. using generalized additive models, to look for evidence of curvature).

- Fit tree models to investigate whether interaction effects are simple or complex.

10.13.1 The multiple regression model

There are several important issues involved in carrying out a multiple regression:

- which explanatory variables to include;

- curvature in the response to the explanatory variables;

- interactions between explanatory variables;

- correlation between explanatory variables;

- the risk of overparameterization.

The assumptions about the response variable are the same as with simple linear regression: the errors are normally distributed, the errors are confined to the response variable, and the variance is constant. The explanatory variables are assumed to be measured without error. The model for a multiple regression with two explanatory variables (x_1 and x_2) looks like this:

$$y_i = \beta_0 + \beta_1 x_{1i} + \beta_2 x_{2i} + \varepsilon_i.$$

The ith data point, y_i, is determined by the levels of the two continuous explanatory variables x_{1i} and x_{2i}, by the model's three parameters (the intercept β_0 and the two slopes β_1 and β_2), and by the residual ε_i of point i from the fitted surface. For each of the i rows of the dataframe, there are $k + 1$ parameters, β_j, so that

$$y_i = \sum_{j=0}^{k} \beta_j x_{ji} + \varepsilon_i,$$

where $x_{0i} = 1$.

Let us begin with an example from air pollution studies. How is ozone concentration related to wind speed, air temperature and the intensity of solar radiation?

```
ozone.pollution <- read.table("c:\\temp\\ozone.data.txt",header=T)
attach(ozone.pollution)
names(ozone.pollution)
```

```
[1]  "rad"  "temp"  "wind"  "ozone"
```

In multiple regression, it is always a good idea to use `pairs` to look at all the correlations:

```
pairs(ozone.pollution,panel=panel.smooth)
```

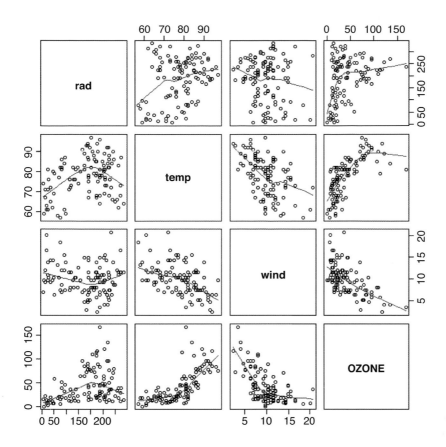

The response variable, ozone concentration, is shown on the y axis of the bottom row of panels: there is a strong negative relationship with wind speed, a positive correlation with temperature and a rather unclear, humped relationship with radiation.

A good way to tackle a multiple regression problem is using non-parametric smoothers in a generalized additive model like this:

```
library(mgcv)
par(mfrow=c(2,2))
model <- gam(ozone~s(rad)+s(temp)+s(wind))
plot(model)
```

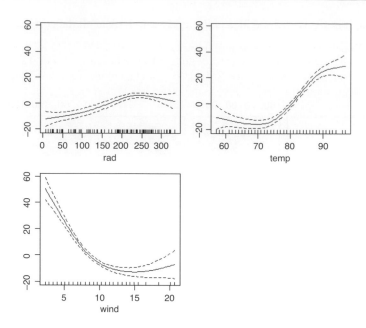

The confidence intervals are sufficiently narrow to suggest that the curvature in the relationships between ozone and temperature and ozone and wind are real, but the curvature of the relationship with solar radiation is marginal. The plots lead us to anticipate that quadratic terms for temperature and wind should be included in our initial model. What about interactions? This is where tree models can help:

```
library(tree)
model <- tree(ozone~.,data=ozone.pollution)
par(mfrow=c(1,1))
plot(model)
text(model)
```

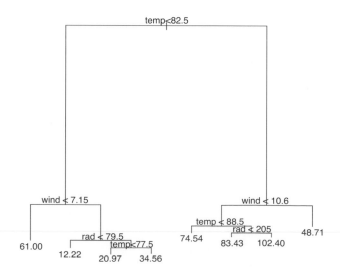

This shows that temperature is by far the most important factor affecting ozone concentration (the longer the branches in the tree, the greater the deviance explained). Wind speed is important at both high and low temperatures, with still air being associated with higher mean ozone levels (the figures at the ends of the branches). The interaction structure is relatively simple (compare with the other air pollution example on p. 768), but there is a hint of an interaction between wind and radiation and between wind and temperature. We could include these in an initial complex model, degrees of freedom permitting.

```
w2  <- wind^2
t2  <- temp^2
r2  <- rad^2
tw  <- temp*wind
wr  <- wind*rad
tr  <- temp*rad
wtr <- wind*temp*rad
```

Armed with this background information we can begin the linear modelling. We start with the most complicated model: this includes curvature terms for each variable, all three two-way interactions and a three-way interaction:

```
model1 <- lm(ozone~rad+temp+wind+t2+w2+r2+wr+tr+tw+wtr)
summary(model1)
```

```
Coefficients:
                Estimate Std. Error t value Pr(>|t|)
(Intercept)    5.683e+02  2.073e+02    2.741  0.00725 **
rad           -3.117e-01  5.585e-01   -0.558  0.57799
temp          -1.076e+01  4.303e+00   -2.501  0.01401 *
wind          -3.237e+01  1.173e+01   -2.760  0.00687 **
t2             5.833e-02  2.396e-02    2.435  0.01668 *
w2             6.106e-01  1.469e-01    4.157 6.81e-05 ***
r2            -3.619e-04  2.573e-04   -1.407  0.16265
wr             2.054e-02  4.892e-02    0.420  0.67552
tr             8.403e-03  7.512e-03    1.119  0.26602
tw             2.377e-01  1.367e-01    1.739  0.08519 .
wtr           -4.324e-04  6.595e-04   -0.656  0.51358
```

```
Residual standard error: 17.82 on 100 degrees of freedom
Multiple R-squared: 0.7394,      Adjusted R-squared: 0.7133
F-statistic: 28.37 on 10 and 100 DF,  p-value: < 2.2e-16
```

There looks to be rather little scope for model simplification, so we shall do it all by hand (rather than using step, for instance, not least because this is prone to remove main effects that are still present in interactions, or the linear parts of quadratic terms that would best be retained).

We start by removing the highest-order interaction. An excellent feature of R is that the p values are 'p values on deletion' so we do not have to use anova to compare the models produced by stepwise deletions:

```
model2 <- update(model1,~.-wtr)
summary(model2)
```

```
Coefficients:
                Estimate Std. Error t value Pr(>|t|)
(Intercept)    5.245e+02  1.957e+02    2.680  0.0086 **
```

rad	2.628e-02	2.142e-01	0.123	0.9026	
temp	-1.021e+01	4.209e+00	-2.427	0.0170	*
wind	-2.802e+01	9.645e+00	-2.906	0.0045	**
t2	5.953e-02	2.382e-02	2.499	0.0141	*
w2	6.173e-01	1.461e-01	4.225	5.25e-05	***
r2	-3.388e-04	2.541e-04	-1.333	0.1855	
wr	-1.127e-02	6.277e-03	-1.795	0.0756	.
tr	3.750e-03	2.459e-03	1.525	0.1303	
tw	1.734e-01	9.497e-02	1.825	0.0709	.

The least significant term is the quadratic term for radiation, so we remove that:

```
model3 <- update(model2,~.-r2)
summary(model3)
```

Coefficients:

	Estimate	Std. Error	t value	Pr(>|t|)	
(Intercept)	486.346603	194.333075	2.503	0.01392	*
rad	-0.043163	0.208535	-0.207	0.83644	
temp	-9.446780	4.185240	-2.257	0.02613	*
wind	-26.471461	9.610816	-2.754	0.00697	**
t2	0.056966	0.023835	2.390	0.01868	*
w2	0.599709	0.146069	4.106	8.14e-05	***
wr	-0.011359	0.006300	-1.803	0.07435	.
tr	0.003160	0.002428	1.302	0.19600	
tw	0.157637	0.094595	1.666	0.09869	.

The temperature by radiation interaction is not significant, so it goes next:

```
model4 <- update(model3,~.-tr)
summary(model4)
```

Coefficients:

	Estimate	Std. Error	t value	Pr(>|t|)	
(Intercept)	514.401470	193.783580	2.655	0.00920	**
rad	0.212945	0.069283	3.074	0.00271	**
temp	-10.654041	4.094889	-2.602	0.01064	*
wind	-27.391965	9.616998	-2.848	0.00531	**
t2	0.067805	0.022408	3.026	0.00313	**
w2	0.619396	0.145773	4.249	4.72e-05	***
wr	-0.013561	0.006089	-2.227	0.02813	*
tw	0.169674	0.094458	1.796	0.07538	.

The temperature by wind interaction is the next to go (it is marginally significant but we are ruthless):

```
model5 <- update(model4,~.-tw)
summary(model5)
```

Coefficients:

	Estimate	Std. Error	t value	Pr(>|t|)	
(Intercept)	223.573855	107.618223	2.077	0.040221	*
rad	0.173431	0.066398	2.612	0.010333	*

```
temp            -5.197139     2.775039   -1.873 0.063902   .
wind           -10.816032     2.736757   -3.952 0.000141   ***
t2               0.043640     0.018112    2.410 0.017731   *
w2               0.430059     0.101767    4.226 5.12e-05   ***
wr              -0.009819     0.005783   -1.698 0.092507   .
```

There is no place for the wind by rain interaction:

```
model6 <- update(model5,~.-wr)
summary(model6)
```

```
Coefficients:
                Estimate Std. Error t value Pr(>|t|)
(Intercept)  291.16758   100.87723    2.886  0.00473 **
rad            0.06586     0.02005    3.285  0.00139 **
temp          -6.33955     2.71627   -2.334  0.02150 *
wind         -13.39674     2.29623   -5.834 6.05e-08 ***
t2             0.05102     0.01774    2.876  0.00488 **
w2             0.46464     0.10060    4.619 1.10e-05 ***
```

The next job is to subject model6 to criticism:

```
par(mfrow=c(2,2))
plot(model6)
```

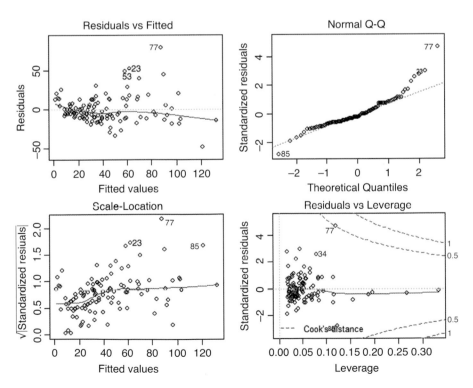

This is quite seriously badly behaved. The residuals increase with the fitted values (non-constant variance) and the errors are not normal. Let us try transforming the response variable. Having done this we need to start the modelling from scratch with all of the original explanatory variables included. Having transformed the response variable, we should expect that the curvature has been altered:

```
model7 <- lm(log(ozone) ~ rad+temp+wind+t2+w2+r2+wr+tr+tw+wtr)
summary(model7)
```

```
Coefficients:
              Estimate Std. Error t value Pr(>|t|)
(Intercept)  2.803e+00  5.676e+00   0.494   0.6225
rad          2.771e-02  1.529e-02   1.812   0.0729 .
temp        -3.018e-02  1.178e-01  -0.256   0.7983
wind        -9.812e-02  3.211e-01  -0.306   0.7605
t2           6.034e-04  6.559e-04   0.920   0.3598
w2           8.732e-03  4.021e-03   2.172   0.0322 *
r2          -1.489e-05  7.043e-06  -2.114   0.0370 *
wr          -2.001e-03  1.339e-03  -1.494   0.1382
tr          -2.507e-04  2.056e-04  -1.219   0.2256
tw          -1.985e-03  3.742e-03  -0.530   0.5971
wtr          2.535e-05  1.805e-05   1.404   0.1634
```

```
model8 <- update(model7,~.-wtr)
summary(model8)
```

```
model9 <- update(model8,~.-tr)
summary(model9)
```

```
model10 <- update(model9,~.-tw)
summary(model10)
```

```
model11 <- update(model10,~.-t2)
summary(model11)
```

```
model12 <- update(model11,~.-wr)
summary(model12)
```

```
Coefficients:
              Estimate Std. Error t value Pr(>|t|)
(Intercept)  7.724e-01  6.350e-01   1.216 0.226543
rad          7.466e-03  2.323e-03   3.215 0.001736 **
temp         4.193e-02  6.237e-03   6.723 9.52e-10 ***
wind        -2.211e-01  5.874e-02  -3.765 0.000275 ***
w2           7.390e-03  2.585e-03   2.859 0.005126 **
r2          -1.470e-05  6.734e-06  -2.183 0.031246 *
```

```
Residual standard error: 0.4851 on 105 degrees of freedom
Multiple R-squared: 0.7004,     Adjusted R-squared: 0.6861
F-statistic:  49.1 on 5 and 105 DF,  p-value: < 2.2e-16
```

```
plot(model12)
```

This is the minimum adequate model. It has five consequential parameters (the intercept of a multiple regression model is usually meaningless; it is the value of the response when every one of the explanatory variables is zero). As predicted by our initial plots, none of the interactions survived the model simplification.

The curvature on the scale of `log(ozone)` is different, of course (we plotted `ozone` against the explanatory variables, not `log(ozone)`). Log transformation of the response improved both the non-constancy of variance and the non-normality of errors. The model explains just over of 70% of the variation in log(ozone concentration).

10.13.2 Common problems arising in multiple regression

The following are some of the problems and difficulties that crop up when we do multiple regression:

- differences in the measurement scales of the explanatory variables, leading to large variation in the sums of squares and hence to an ill-conditioned matrix;

- multicollinearity, in which there is a near-linear relation between two of the explanatory variables, leading to unstable parameter estimates;

- parameter proliferation where quadratic and interaction terms soak up more degrees of freedom than our data can afford;

- rounding errors during the fitting procedure;

- non-independence of groups of measurements;

- temporal or spatial correlation amongst the explanatory variables;

- pseudoreplication.

Wetherill *et al.* (1986) give a detailed discussion of these problems. We shall encounter other examples of multiple regressions in the context of generalized linear models (Chapter 13), generalized additive models (Chapter 18), survival models (Chapter 27) and mixed-effects models (Chapter 19).

11

Analysis of Variance

Instead of fitting continuous, measured variables to data (as in regression), many experiments involve exposing experimental material to a range of discrete *levels* of one or more categorical variables known as *factors*. Thus, a factor might be drug treatment for a particular cancer, with five levels corresponding to a placebo plus four new pharmaceuticals. Alternatively, a factor might be mineral fertilizer, where the four levels represent four different mixtures of nitrogen, phosphorus and potassium. Factors are often used in experimental designs to represent statistical *blocks*; these are internally homogeneous units in which each of the experimental treatments is repeated. Blocks may be different fields in an agricultural trial, different genotypes in a plant physiology experiment, or different growth chambers in a study of insect photoperiodism.

It is important to understand that regression and analysis of variance (ANOVA) are identical approaches except for the nature of the explanatory variables. For example, it is a small step from having three levels of a shade factor (say light, medium and heavy shade cloths) then carrying out a one-way ANOVA, to measuring the light intensity in the three treatments and carrying out a regression with light intensity as the explanatory variable. As we shall see later on, some experiments combine regression and ANOVA by fitting a series of regression lines, one in each of several levels of a given factor (this is called analysis of covariance; see Chapter 12).

The emphasis in ANOVA was traditionally on hypothesis testing. Nowadays, the aim of an analysis of variance in R is to estimate means and standard errors of differences between means. Comparing two means by a t test involved calculating the difference between the two means, dividing by the standard error of the difference, and then comparing the resulting statistic with the value of Student's t from tables (or better still, using qt to calculate the critical value; see p. 287). The means are said to be significantly different when the calculated value of t is larger than the critical value. For large samples ($n > 30$) a useful rule of thumb is that a t value greater than 2 is significant. In ANOVA, we are concerned with cases where we want to compare three or more means. For the two-sample case, the t test and the ANOVA are identical, and the t test is to be preferred because it is simpler.

11.1 One-way ANOVA

There is a real paradox about analysis of variance, which often stands in the way of a clear understanding of exactly what is going on. The idea of ANOVA is to compare several means, but it does this by comparing variances. How can that work?

The R Book, Second Edition. Michael J. Crawley.
© 2013 John Wiley & Sons, Ltd. Published 2013 by John Wiley & Sons, Ltd.

A visual example should make this clear. To keep things simple, suppose we have just two levels of a single factor. We plot the data in the order in which they were measured: first for the first level of the factor and then for the second level. Draw the overall mean as a horizontal line through the data, and indicate the departures of each data point from the overall mean with a set of vertical lines:

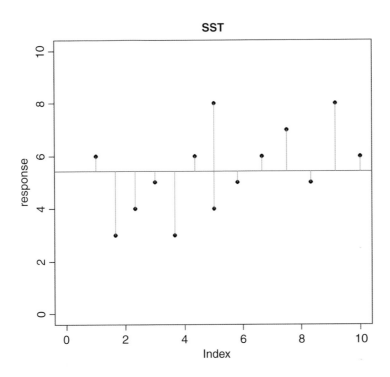

The green lines illustrate the total variation in the response. We shall call this quantity *SST* (the 'total sum of squares'). It is *the sum of the squares of the differences between the data, y, and the overall mean.* In symbols,

$$SST = \sum (y - \bar{\bar{y}})^2,$$

where $\bar{\bar{y}}$ ('y double bar') is the overall mean. Next we can fit each of the separate means, $\bar{y}_A$ (through the red points) and $\bar{y}_B$ (through the blue points), and consider the sum of squares of the differences between each *y* value and its own treatment mean (either the red line or the blue line). We call this *SSE* (the 'error sum of squares'), and calculate it like this:

$$SSE = \sum (y_A - \bar{y}_A)^2 + \sum (y_B - \bar{y}_B)^2$$

On the graph, the differences from which SSE is calculated look like this:

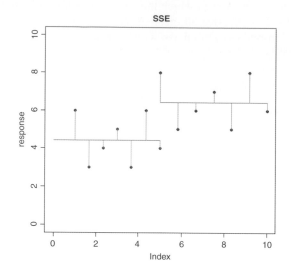

SSE is the sum of the squares of the green lines (the 'residuals', as they are known).

Now ask yourself this question. If the treatment means are different from the overall mean, what will be the relationship between *SST* and *SSE*? After a moment's thought you should have been able to convince yourself that if the means are the same, then *SSE* is the same as *SST*, because the two horizontal lines in the last plot would be in the same position as the single line in the earlier plot.

Now what if the means were significantly different from one another? What would be the relationship between *SSE* and *SST* in this case? Which would be the larger? Again, it should not take long for you to see that if the means *are* different, then *SSE* will be less than *SST*. Indeed, in the limit, *SSE* could be zero if the replicates from each treatment fell exactly on their respective means, like this:

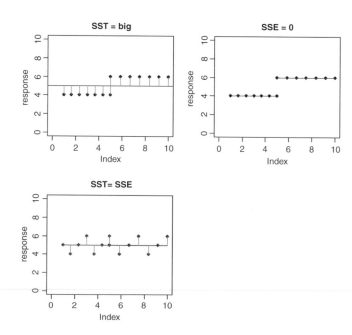

In the top row, there is a highly significant difference between the two means: *SST* is big but *SSE* is zero (all the replicates are identical). In the bottom row, the means are identical. *SST* is still big, but now $SSE = SST$. Once you have understood these three plots, you will see why *you can investigate differences between means by looking at variances*. This is how analysis of variance works.

We can calculate the difference between *SST* and *SSE*, and use this as a measure of the difference between the treatment means; this is traditionally called the *treatment sum of squares*, and is denoted by *SSA*:

$$SSA = SST - SSE.$$

When differences between means are significant, then *SSA* will be large relative to *SSE*. When differences between means are not significant, then *SSA* will be small relative to *SSE*. In the limit, *SSE* could be zero (top right in the last figure), so all of the variation in y is explained by differences between the means ($SSA = SST$). At the other extreme, when there is no difference between the means (bottom left), $SSA = 0$ and so $SSE = SST$.

The technique we are interested in, however, is analysis of variance, not analysis of sums of squares. We convert the sums of squares into variances by dividing by their degrees of freedom. In our example, there are two levels of the factor and so there is $2 - 1 = 1$ degree of freedom for *SSA*. In general, we might have k levels of any factor and hence $k - 1$ d.f. for treatments. If each factor level were replicated n times, then there would be $n - 1$ d.f. for error within each level (we lose one degree of freedom for each individual treatment mean estimated from the data). Since there are k levels, there would be $k(n - 1)$ d.f. for error in the whole experiment. The total number of numbers in the whole experiment is kn, so total d.f. is $kn - 1$ (the single degree is lost for our estimating the overall mean, $\bar{y}$). As a check in more complicated designs, it is useful to make sure that the individual component degrees of freedom add up to the correct total:

$$kn - 1 = k - 1 + k(n - 1) = k - 1 + kn - k.$$

The divisions for turning the sums of squares into variances are conveniently carried out in an ANOVA table:

Source	SS	d.f.	MS	F	Critical F
Treatment	SSA	$k - 1$	$MSA = \dfrac{SSA}{k - 1}$	$F = \dfrac{MSA}{s^2}$	qf(0.95, k-1,k(n-1))
Error	SSE	$k(n - 1)$	$s^2 = \dfrac{SSE}{k(n - 1)}$		
Total	SST	$kn - 1$			

Each element in the sums of squares column is divided by the number in the adjacent degrees of freedom column to give the variances in the mean square column (headed MS). The significance of the difference between the means is then assessed using an F test (a variance ratio test). The treatment variance *MSA* is divided by the error variance, s^2, and the value of this test statistic is compared with the critical value of F using qf (the quantiles of the F distribution, with $p = 0.95$, $k - 1$ degrees of freedom in the numerator, and $k(n - 1)$ degrees of freedom in the denominator). If you need to look up the critical value of F in tables, remember that you look up the numerator degrees of freedom (on top of the division) across the *top* of the table, and the denominator degrees of freedom down the rows. The null hypothesis, traditionally denoted as H_0, is stated as

$$H_0 : \text{nothing's happening.}$$

This does not imply that the sample means are exactly the same (the means will always differ from one another, simply because everything varies). In fact, the null hypothesis assumes that *the means are not significantly different from one another*. What this implies is that the differences between the sample means could have arisen by chance alone, through random sampling effects, despite the fact that the different factor levels have identical means.

If the test statistic is larger than the critical value we *reject* the null hypothesis and accept the alternative:

$$H_1 : \text{at least one of the means is significantly different from the others.}$$

If the test statistic is less than the critical value, then it could have arisen due to chance alone, and so we accept the null hypothesis.

Another way of visualizing the process of ANOVA is to think of the relative amounts of sampling variation between replicates receiving the same treatment (i.e. between individual samples in the same level), and between different treatments (i.e. between-level variation). When the variation between replicates within a treatment is large compared to the variation between treatments, we are likely to conclude that the difference between the treatment means is not significant. Only if the variation between replicates within treatments is relatively small compared to the differences between treatments will we be justified in concluding that the treatment means are significantly different.

11.1.1 Calculations in one-way ANOVA

The definitions of the various sums of squares can now be formalized, and ways found of calculating their values from samples. The total sum of squares, *SST*, is defined as:

$$SST = \sum y^2 - \frac{\left(\sum y\right)^2}{kn},$$

just as in regression (see Chapter 10). Note that we divide by the total number of numbers we added together to get $\sum y$ (the grand total of all the *y*s) which is *kn*. It turns out that the formula that we used to define *SSE* is rather difficult to calculate (see above), so we calculate the treatment sums of squares, *SSA*, and obtain *SSE* by difference. The treatment sum of squares, *SSA*, is calculated as:

$$SSA = \frac{\sum C^2}{n} - \frac{\left(\sum y\right)^2}{kn},$$

where the new term is *C*, the *treatment total*. This is the sum of all the *n* replicates within a given level. Each of the *k* different treatment totals is squared, added up, and then divided by *n* (the number of numbers added together to get the treatment total). The formula is slightly different if there is unequal replication in different treatments, as we shall see below. The meaning of *C* will become clear when we work through the example later on. Notice the symmetry of the equation. The second term on the right-hand side is also divided by the number of numbers that were added together (*kn*) to get the total ($\sum y$) which is squared in the numerator. Finally,

$$SSE = SST - SSA,$$

to give all the elements required for completion of the ANOVA table.

11.1.2 Assumptions of ANOVA

You should be aware of the assumptions underlying the analysis of variance. They are all important, but some are more important than others:

- random sampling;
- equal variances;
- independence of errors;
- normal distribution of errors;
- additivity of treatment effects.

11.1.3 A worked example of one-way ANOVA

To draw this background material together, we shall work through an example by hand. In so doing, it will become clear what R is doing during its analysis of the data. We have an experiment in which crop yields per unit area were measured from 10 randomly selected fields on each of three soil types. All fields were sown with the same variety of seed and provided with the same fertilizer and pest control inputs. The question is whether soil type significantly affects crop yield, and if so, to what extent.

```
results <- read.table("c:\\temp\\yields.txt",header=T)
attach(results)
names(results)

[1]    "sand"       "clay"        "loam"
```

Here are the data:

```
results

     sand clay loam
1       6   17   13
2      10   15   16
3       8    3    9
4       6   11   12
5      14   14   15
6      17   12   16
7       9   12   17
8      11    8   13
9       7   10   18
10     11   13   14
```

The function `sapply` is used to calculate the mean yields for the three soils (contrast this with `tapply`, below, where the response and explanatory variables are in adjacent columns in a dataframe):

```
sapply(list(sand,clay,loam),mean)

[1]    9.9    11.5    14.3
```

Mean yield was highest on loam (14.3) and lowest on sand (9.9).

It will be useful to have all of the yield data in a single vector called *y*. To create a dataframe from a spreadsheet like `results` where the values of the response are in multiple columns, we use the function called `stack` like this:

```
(frame <- stack(results))
     values  ind
1         6 sand
2        10 sand
3         8 sand
4         6 sand
...

...
27       17 loam
28       13 loam
29       18 loam
30       14 loam
```

You can see that the `stack` function has invented names for the response variable (`values`) and the explanatory variable (`ind`). We will always want to change these:

```
names(frame) <- c("yield","soil")
attach(frame)
head(frame)
  yield soil
1     6 sand
2    10 sand
3     8 sand
4     6 sand
5    14 sand
6    17 sand
```

That's more like it.

Before carrying out analysis of variance, we should check for constancy of variance (see p. 354) across the three soil types:

```
tapply(yield,soil,var)
      clay       loam       sand
15.388889   7.122222  12.544444
```

The variances differ by more than a factor of 2. But is this significant? We test for heteroscedasticity using the Fligner–Killeen test of homogeneity of variances:

```
fligner.test(y~soil)

        Fligner-Killeen test of homogeneity of variances
data: y by soil
Fligner-Killeen:med chi-squared = 0.3651, df = 2, p-value = 0.8332
```

We could have used `bartlett.test(y~soil)`, which gives $p = 0.5283$ (but this is more a test of non-normality than of equality of variances). Either way, there is no evidence of any significant difference in variance across the three samples, so it is legitimate to continue with our one-way analysis of variance.

Because the explanatory variable is categorical (three levels of soil type), initial data inspection involves a box-and-whisker plot of y against soil like this:

```
plot(yield~soil,col="green")
```

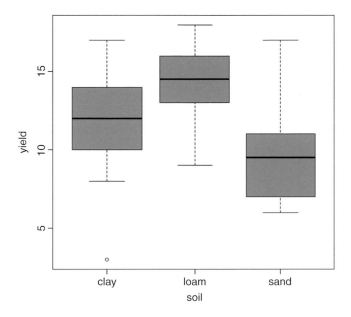

Median yield is lowest on sand and highest on loam, but there is considerable variation from replicate to replicate within each soil type (there is even a low outlier on clay). It looks as if yield on loam will turn out to be significantly higher than on sand (their boxes do not overlap) but it is not clear whether yield on clay is significantly greater than on sand or significantly lower than on loam. The analysis of variance will answer these questions.

The analysis of variance involves calculating the total variation in the response variable (yield in this case) and partitioning it ('analysing it') into informative components. In the simplest case, we partition the total variation into just two components, explained variation and unexplained variation:

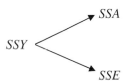

Explained variation is called the treatment sum of squares (*SSA*) and unexplained variation is called the error sum of squares (*SSE*, also known as the residual sum of squares), as defined earlier. Let us work through the numbers in R. From the formula for *SSY*, we can obtain the total sum of squares by finding the differences between the data and the overall mean:

```
sum((yield-mean(yield))^2)
```

```
[1] 414.7
```

The unexplained variation, *SSE*, is calculated from the differences between the yields and the mean yields *for that soil type*:

```
sand-mean(sand)
```

```
[1]   -3.9   0.1   -1.9   -3.9   4.1   7.1   -0.9   1.1   -2.9   1.1
```

```
clay-mean(clay)
```

```
[1]    5.5   3.5   -8.5   -0.5   2.5   0.5   0.5   -3.5   -1.5   1.5
```

```
loam-mean(loam)
```

```
[1]   -1.3   1.7   -5.3   -2.3   0.7   1.7   2.7   -1.3   3.7   -0.3
```

We need the sums of the squares of these differences:

```
sum((sand-mean(sand))^2)
```

```
[1]   112.9
```

```
sum((clay-mean(clay))^2)
```

```
[1]   138.5
```

```
sum((loam-mean(loam))^2)
```

```
[1]   64.1
```

To get the sum of these totals across all soil types, we can use `sapply` like this:

```
sum(sapply(list(sand,clay,loam),function (x)  sum((x-mean(x))^2) ))
```

```
[1]   315.5
```

So *SSE*, the unexplained (or residual, or error) sum of squares, is 315.5.

The extent to which *SSE* is less than *SSY* is a reflection of the magnitude of the differences between the means. The greater the difference between the mean yields on the different soil types, the greater will be the difference between *SSE* and *SSY*.

The treatment sum of squares, *SSA*, is the amount of the variation in yield that is explained by differences between the treatment means. In our example,

$$SSA = SSY - SSE = 414.7 - 315.5 = 99.2.$$

Now we can draw up the ANOVA table. There are six columns indicating, from left to right, the source of variation, the sum of squares attributable to that source, the degrees of freedom for that source, the variance for that source (traditionally called the mean square rather than the variance), the *F* ratio (testing the null hypothesis that this source of variation is not significantly different from zero) and the *p* value associated with that *F* value (if $p < 0.05$ then we reject the null hypothesis). We can fill in the sums of squares just calculated, then think about the degrees of freedom:

Source	Sum of squares	Degrees of freedom	Mean square	*F* ratio	*p* value
Soil type	99.2	2	49.6	4.24	0.025
Error	315.5	27	$s^2 = 11.685$		
Total	414.7	29			

There are 30 data points in all, so the total degrees of freedom are $30 - 1 = 29$. We lose 1 d.f. because in calculating SSY we had to estimate one parameter from the data in advance, namely the overall mean, $\bar{y}$, before we could calculate $SST = \sum (y - \bar{y})^2$. Each soil type has $n = 10$ replications, so each soil type has $10 - 1 = 9$ d.f. for error, because we estimated one parameter from the data *for each soil type*, namely the treatment means $\bar{y}_i$ in calculating SSE. Overall, therefore, the error has $3 \times 9 = 27$ d.f. There were three soil types, so there are $3 - 1 = 2$ d.f. for soil type.

The mean squares are obtained simply by dividing each sum of squares by its respective degrees of freedom (in the same row). The error variance, s^2, is the residual mean square (the mean square for the unexplained variation); this is sometimes called the 'pooled error variance' because it is calculated across all the treatments. The alternative would be to have three separate variances, one for each treatment:

```
tapply(yield,soil,var)
```

```
    clay       loam       sand
15.388889   7.122222  12.544444
```

```
mean(tapply(yield,soil,var))
```

```
[1] 11.68519
```

You will see that the pooled error variance $s^2 = 11.685$ is simply the mean of the three separate variances, because (in this case) there is equal replication in each soil type ($n = 10$).

By tradition, we do not calculate the total mean square, so the bottom cell of the fourth column of the ANOVA table is empty. The F ratio is the treatment variance divided by the error variance, testing the null hypothesis that the treatment means are not significantly different. If we reject this null hypothesis, we accept the alternative hypothesis that *at least one of the means is significantly different from the others*. The question naturally arises at this point as to whether 4.24 is a big number or not. If it is a big number then we reject the null hypothesis. If it is not a big number, then we accept the null hypothesis. As ever, we decide whether the test statistic $F = 4.24$ is big or small by comparing it with the *critical value* of F, given that there are 2 d.f. in the numerator and 27 d.f. in the denominator. Critical values in R are found from the function qf which gives us quantiles of the F distribution:

```
qf(.95,2,27)
```

```
[1]   3.354131
```

Our calculated test statistic of 4.24 is larger than the critical value of 3.35, so we reject the null hypothesis. At least one of the soils has a mean yield that is significantly different from the others. The modern approach is not to work slavishly at the 5% level but rather to calculate the p value associated with our test statistic of 4.24. Instead of using the function for quantiles of the F distribution, we use the function pf for cumulative probabilities of the F distribution like this:

```
1-pf(4.24,2,27)
```

```
[1]   0.02503987
```

The p value is 0.025, which means that a value of $F = 4.24$ or bigger would arise by chance alone when the null hypothesis was true about 25 times in 1000. This is a sufficiently small probability (i.e. it is less than 5%) for us to conclude that there is a significant difference between the mean yields (i.e. we reject the null hypothesis).

That was a lot of work. R can do the whole thing in a single line:

```
summary(aov(yield~soil))
```

```
          Df Sum Sq Mean Sq F value Pr(>F)
soil       2   99.2   49.60   4.245  0.025 *
Residuals 27  315.5   11.69
```

Here you see all the values that we calculated longhand. The error row is labelled `Residuals`. In the second and subsequent columns you see the degrees of freedom for treatment and error (2 and 27), the treatment and error sums of squares (99.2 and 315.5), the treatment mean square of 49.6, the error variance $s^2 = 11.685$, the F ratio and the p value (labelled `Pr(>F)`). The single asterisk next to the p value indicates that the difference between the soil means is significant at 5% (but not at 1%, which would have merited two asterisks). Notice that R does not print the bottom row of the ANOVA table showing the total sum of squares and total degrees of freedom.

The next thing we would do is to check the assumptions of the `aov` model. This is done using `plot` like this (see p. 419):

```
par(mfrow=c(2,2))
plot(aov(yield~soil))
```

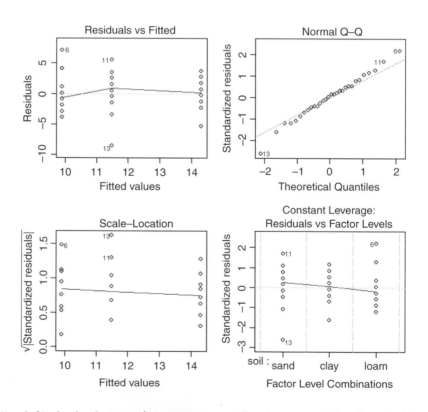

The first plot (top left) checks the most important assumption (constancy of variance); there should be no pattern in the residuals against the fitted values (the three treatment means) – and, indeed, there is none. The second plot (top right) tests the assumption of normality of errors: there should be a straight-line relationship between our standardized residuals and theoretical quantiles derived from a normal distribution. Points 6,

11 and 13 lie a little off the straight line, but this is nothing to worry about (see p. 405). The residuals are well behaved (bottom left) and there are no highly influential values that might be distorting the parameter estimates (bottom right).

11.1.4   Effect sizes

The best way to view the effect sizes graphically is to use `plot.design` (which takes a formula rather than a model object), but our current model with just one factor is perhaps too simple to get full value from this (`plot.design(yield~soil)`). To see the effect sizes in tabular form use `model.tables` (which takes a model object as its argument) like this:

```
model <- aov(yield~soil)
model.tables(model,se=T)
```

```
Tables of effects
 soil
soil
clay loam sand
-0.4  2.4 -2.0

Standard errors of effects
          soil
          1.081
replic.     10
```

The effects are shown as departures from the overall mean: soil 1 (sand) has a mean yield that is 2.0 below the overall mean, and soil 3 (loam) has a mean that is 2.4 above the overall mean. The standard error of effects is 1.081 on a replication of $n = 10$ (this is the standard error of a mean). You should note that this is *not* the appropriate standard error for comparing two means (see below). If you specify `"means"` you get:

```
model.tables(model,"means",se=T)
```

```
Tables of means
Grand mean

11.9

 soil
soil
clay loam sand
11.5 14.3   9.9

Standard errors for differences of means
          soil
          1.529
replic.     10
```

Now the three means are printed (rather than the effects) and the standard error of the difference of means is given (this *is* what you need for doing a *t* test to compare any two means).

Another way of looking at effect sizes is to use the `summary.lm` option for viewing the model, rather than `summary.aov` (as we used above):

```
summary.lm(model)
```

```
Call:
aov(formula = yield ~ soil)

Residuals:
   Min     1Q Median     3Q    Max
  -8.5   -1.8    0.3    1.7    7.1

Coefficients:
              Estimate Std. Error t value Pr(>|t|)
(Intercept)    11.500      1.081  10.638  3.7e-11 ***
soilloam        2.800      1.529   1.832   0.0781 .
soilsand       -1.600      1.529  -1.047   0.3046

Residual standard error: 3.418 on 27 degrees of freedom
Multiple R-squared: 0.2392,     Adjusted R-squared: 0.1829
F-statistic: 4.245 on 2 and 27 DF,  p-value: 0.02495
```

In regression analysis (p. 461) the `summary.lm` output was easy to understand because it gave us the intercept and the slope (the two parameters estimated by the model) and their standard errors. But this table has three rows. Why is that? What is an intercept in the context of analysis of variance? And why are the standard errors different for the intercept and for `soilsand`?

It will take a while before you feel at ease with `summary.lm` tables for analysis of variance. The details are explained on p. 424, but the central point is that all `summary.lm` tables have as many rows as there are parameters estimated from the data. There are three rows in this case because our `aov` model estimates three parameters: a mean yield for each of the three soil types. In the context of `aov`, an intercept is a mean value; in this case it is the mean yield for `clay` because this factor-level name comes first in the alphabet. So if `Intercept` is the mean yield for clay, what are the other two rows labelled `soilloam` and `soilsand`? This is the hardest thing to understand. All other rows in the `summary.lm` table for `aov` are *differences between means*. Thus row 2, labelled `soilloam`, is the difference between the mean yields on loam and clay, and row 3, labelled `soilsand`, is the difference between the mean yields of sand and clay.

The first row (`Intercept`) is a mean, so the standard error column in row 1 contains the standard error of a mean. Rows 2 and 3 are differences between means, so their standard error columns contain the standard error of the difference between two means (and this is a bigger number; see p. 358). The standard error of a mean is

$$se_{\text{mean}} = \sqrt{\frac{s^2}{n}} = \sqrt{\frac{11.685}{10}} = 1.081,$$

whereas the standard error of the difference between two means is

$$se_{\text{diff}} = \sqrt{2\frac{s^2}{n}} = \sqrt{2 \times \frac{11.685}{10}} = 1.529.$$

The `summary.lm` table shows that neither loam nor sand produces a significantly higher yield than clay (none of the p-values is less than 0.05, despite the fact that the ANOVA table showed $p = 0.025$). But what about the contrast in the yields from loam and sand? To assess this we need to do some arithmetic of our own. The two parameters differ by 2.8 + 1.6 = 4.4 (take care with the signs). The standard error of the difference is 1.529, so the *t* value is 2.88. This is much greater than 2 (our rule of thumb for *t*) so the mean yields of loam and sand are significantly different. To find the precise value of Student's *t* with 10 replicates in each treatment, the critical value of *t* is given by the function qt with 18 d.f. (we have lost two degrees of freedom for the two treatment means we have estimated from the data):

```
qt(0.975,18)
```

```
[1]     2.100922
```

Alternatively we can work out the *p* value associated with our calculated $t = 2.88$:

```
2*(1 - pt(2.88, df = 18))
```

```
[1]  0.009966426
```

We multiply by 2 because this is a two-tailed test (see p. 293); we did not know in advance that loam would outyield sand under the particular circumstances of this experiment.

The residual standard error in the summary.lm output is the square root of the error variance from the ANOVA table: $\sqrt{11.685} = 3.418$. *R*-squared is the fraction of the total variation in yield that is explained by the model (adjusted *R*-squared are explained on p. 461). The *F* statistic and the *p* value come from the last two columns of the ANOVA table.

So there it is. That is how analysis of variance works. When the means are significantly different, then the sum of squares computed from the individual treatment means will be significantly smaller than the sum of squares computed from the overall mean. We judge the significance of the difference between the two sums of squares using analysis of variance.

11.1.5 Plots for interpreting one-way ANOVA

There are two traditional ways of plotting the results of ANOVA:

- box-and-whisker plots;
- barplots with error bars.

Here is an example to compare the two approaches. We have an experiment on plant competition with one factor and five levels. The factor is called clipping and the five levels consist of control (i.e. unclipped), two intensities of shoot pruning and two intensities of root pruning:

```
comp <- read.table("c:\\temp\\competition.txt",header=T)
attach(comp)
names(comp)
```

```
[1]    "biomass"    "clipping"
```

```
plot(clipping,biomass,xlab="Competition treatment",
                          ylab="Biomass",col="yellow")
```

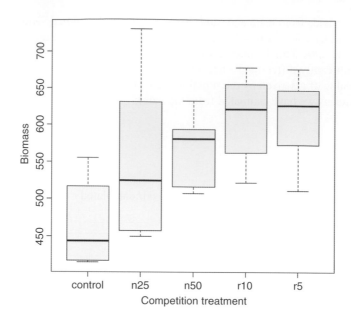

The box-and-whisker plot is good at showing the nature of the variation within each treatment, and also whether there is skew within each treatment (e.g. for the control plots, there is a wider range of values between the median and third quartile than between the median and first quartile). No outliers are shown above the whiskers, so the tops and bottoms of the bars are the maxima and minima within each treatment. The medians for the competition treatments are all higher than the third quartile of the controls, suggesting that they may be significantly different from the controls, but there is little to suggest that any of the competition treatments are significantly different from one another (see below for the analysis).

Barplots with error bars are preferred by many journal editors, and some people think that they make hypothesis testing easier. We shall see. Unlike S-PLUS, R does not have a built-in function called error.bar, so we shall have to write our own. Here is a very simple version without any bells or whistles. We shall call it error.bars to distinguish it from the much more general S-PLUS function:

```
error.bars <- function(yv,z,nn)
{xv <- barplot(yv,ylim=c(0,(max(yv)+max(z))),
             col="green",names=nn,ylab=deparse(substitute(yv)))
for (i in 1:length(xv)) {
arrows(xv[i],yv[i]+z[i],xv[i],yv[i]-z[i],angle=90,code=3,length=0.15)
}}
```

To use this function we need to decide what kind of values (z) to use for the lengths of the bars. Let us use the standard error of a mean based on the pooled error variance from the ANOVA, then return to a discussion of the pros and cons of different kinds of error bars later. Here is the one-way analysis of variance:

```
model <- aov(biomass~clipping)
summary(model)
```

	Df	Sum Sq	Mean Sq	F value	Pr(>F)	
clipping	4	85356	21339	4.302	0.00875	**
Residuals	25	124020	4961			

From the ANOVA table we can see that the pooled error variance $s^2 = 4961$. Now we need to know how many numbers were used in the calculation of each of the five means:

```
table(clipping)
```

```
clipping
control     n25       n50       r10       r5
     6       6         6         6         6
```

There was equal replication (which makes life easier), and each mean was based on six replicates, so the standard error of a mean is $\sqrt{s^2/n} = \sqrt{4961/6} = 28.75$. We shall draw an error bar up 28.75 from each mean and down by the same distance, so we need five values for z, one for each bar, each of 28.75:

```
se <- rep(28.75,5)
```

We need to provide labels for the five different bars – the factor levels should be good for this:

```
labels <- levels(clipping)
```

Now we work out the five mean values which will be the heights of the bars, and save them as a vector called ybar:

```
ybar <- tapply(biomass,clipping,mean)
```

Finally, we can create the barplot with error bars (the function is defined above):

```
error.bars(ybar,se,labels)
```

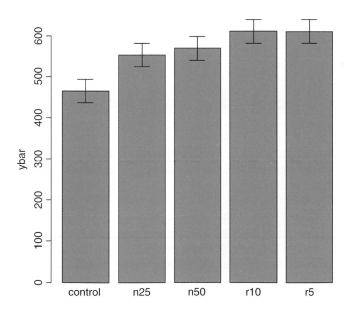

We do not get the same feel for the distribution of the values *within* each treatment as was obtained by the box-and-whisker plot, but we can certainly see clearly which means are not significantly different. If, as here, we use ± 1 standard error of the mean as the length of the error bars, then *when the bars overlap this implies that the two means are not significantly different*. Remember the rule of thumb for t: significance requires 2 or more standard errors, and if the bars overlap it means that the difference between the means is less than 2 standard errors. There is another issue, too. For comparing means, we should use the standard error of the difference between two means (not the standard error of one mean) in our tests (see p. 358); these bars would be about 1.4 times as long as the bars we have drawn here. So while we can be sure that the two root-pruning treatments are not significantly different from one another, and that the two shoot-pruning treatments are not significantly different from one another (because their bars overlap), we cannot conclude from this plot (although we do know it from the ANOVA table above; $p = 0.008\ 75$) that the controls have significantly lower biomass than the rest (because the error bars are not the correct length for testing differences between means).

An alternative graphical method is to use 95% confidence intervals for the lengths of the bars, rather than standard errors of means. This is easy to do: we multiply our standard errors by Student's t, qt(.975,5) = 2.570 582, to get the lengths of the confidence intervals:

```
error.bars(ybar,2.570582*se,labels)
```

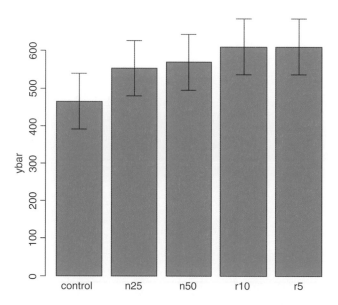

Now, all of the error bars overlap, implying visually that there are no significant differences between the means. But we know that this is not true from our analysis of variance, in which we rejected the null hypothesis that all the means were the same at $p = 0.008\ 75$. If it were the case that the bars did *not* overlap when we are using confidence intervals (as here), then that would imply that the means differed by more than 4 standard errors, and this is a much greater difference than is required to conclude that the means are significantly different. So this is not perfect either. With standard errors we could be sure that the means were *not* significantly different when the bars *did* overlap. And with confidence intervals we can be sure that the means *are* significantly different when the bars *do not* overlap. But the alternative cases are not clear-cut for either type of bar. Can we somehow get the best of both worlds, so that the means *are* significantly different when the bars *do not* overlap, and the means are *not* significantly different when the bars *do* overlap?

The answer is yes, we can, if we use least significant difference (LSD) bars. Let us revisit the formula for Student's t test:

$$t = \frac{\text{a difference}}{\text{standard error of the diffference}}.$$

We say that the difference is significant when $t > 2$ (by the rule of thumb, or $t > $ qt(0.975,df) if we want to be more precise). We can rearrange this formula to find the smallest difference that we would regard as being significant. We can call this the least significant difference:

$$LSD = \text{qt}(0.975,df) \times \text{standard error of a difference} \approx 2 \times se_{\text{diff}}.$$

In our present example this is

```
qt(0.975,10)*sqrt(2*4961/6)
```

```
[1]    90.60794
```

because a difference is based on $12 - 2 = 10$ degrees of freedom. What we are saying is the two means would be significantly different if they differed by 90.61 or more. How can we show this graphically? We want overlapping bars to indicate a difference less than 90.61, and non-overlapping bars to represent a difference greater than 90.61. With a bit of thought you will realize that we need to draw bars that are *LSD*/2 in length, up and down from each mean. Let us try it with our current example:

```
lsd <- qt(0.975,10)*sqrt(2*4961/6)
lsdbars <- rep(lsd,5)/2
error.bars(ybar,lsdbars,labels)
```

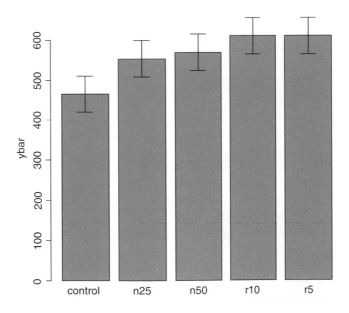

Now we can interpret the significant differences visually. The control biomass is significantly lower than any of the four treatments, but none of the four treatments is significantly different from any other. The statistical analysis of this contrast is explained in detail in Section 9.23 (p. 430). Sadly, most journal editors insist on error bars of 1 standard error of the mean. It is true that there are complicating issues to do with LSD bars (not least the vexed question of multiple comparisons; see p. 531), but at least they do what was intended by the error plot (i.e. overlapping bars means non-significance and non-overlapping bars means significance); neither standard errors nor confidence intervals can say that. A better option might be to use box-and-whisker plots with the notch=T option to indicate significance (see p. 213).

11.2 Factorial experiments

A factorial experiment has two or more factors, each with two or more levels, plus replication for each combination of factors levels. This means that we can investigate statistical interactions, in which *the response to one factor depends on the level of another factor*. Our example comes from a farm-scale trial of animal diets. There are two factors: diet and supplement. Diet is a factor with three levels: barley, oats and wheat. Supplement is a factor with four levels: agrimore, control, supergain and supersupp. The response variable is weight gain after 6 weeks.

```
weights <- read.table("c:\\temp\\growth.txt",header=T)
attach(weights)
```

Data inspection is carried out using `barplot` (note the use of `beside=T` to get the bars in adjacent clusters rather than vertical stacks):

```
barplot(tapply(gain,list(diet,supplement),mean),
        beside=T,ylim=c(0,30),col=c("orange","yellow","cornsilk"))
```

Note that the second factor in the list (`supplement`) appears as groups of bars from left to right in alphabetical order by factor level, from agrimore to supersupp. The first factor (`diet`) appears as three levels within each group of bars: orange = `barley`, yellow = `oats`, cornsilk = `wheat`, again in alphabetical order by factor level. We should really add a key to explain the levels of `diet`. Use `locator(1)` to find the coordinates for the *top left* corner of the box around the legend. You need to increase the default scale on the *y* axis to make enough room for the legend box.

```
labs <- c("Barley","Oats","Wheat")
legend(locator(1),labs,fill= c("orange","yellow","cornsilk"))
```

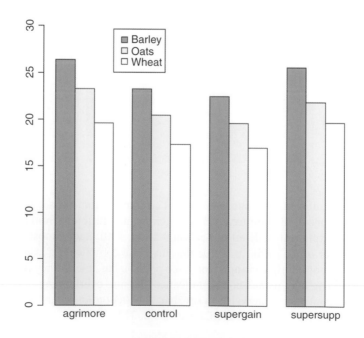

We inspect the mean values using `tapply` as usual:

```
tapply(gain,list(diet,supplement),mean)
```

```
        agrimore  control supergain supersupp
barley 26.34848 23.29665  22.46612  25.57530
oats   23.29838 20.49366  19.66300  21.86023
wheat  19.63907 17.40552  17.01243  19.66834
```

Now we use `aov` or `lm` to fit a factorial analysis of variance (the choice affects only whether we get an ANOVA table or a list of parameters estimates as the default output from `summary`). We estimate parameters for the main effects of each level of `diet` and each level of `supplement`, plus terms for the interaction between `diet` and `supplement`. Interaction degrees of freedom are the product of the degrees of freedom of the component terms (i.e. $(3 - 1) \times (4 - 1) = 6$). The model is `gain~diet+supplement+diet:supplement`, but this can be simplified using the asterisk notation like this:

```
model <- aov(gain~diet*supplement)
summary(model)
```

```
                Df Sum Sq Mean Sq F value   Pr(>F)
diet             2 287.17  143.59   83.52 3.00e-14 ***
supplement       3  91.88   30.63   17.82 2.95e-07 ***
diet:supplement  6   3.41    0.57    0.33    0.917
Residuals       36  61.89    1.72
```

The ANOVA table shows that there is no hint of any interaction between the two explanatory variables ($p = 0.917$); evidently the effects of diet and supplement are additive. The disadvantage of the ANOVA table is that it does not show us the effect sizes, and does not allow us to work out how many levels of each of the two factors are significantly different.

As a preliminary to model simplification, `summary.lm` is often more useful than `summary.aov`:

```
summary.lm(model)
```

```
Coefficients:
                            Estimate Std. Error t value Pr(>|t|)
(Intercept)                  26.3485     0.6556  40.191  < 2e-16 ***
dietoats                     -3.0501     0.9271  -3.290 0.002248 **
dietwheat                    -6.7094     0.9271  -7.237 1.61e-08 ***
supplementcontrol            -3.0518     0.9271  -3.292 0.002237 **
supplementsupergain          -3.8824     0.9271  -4.187 0.000174 ***
supplementsupersupp          -0.7732     0.9271  -0.834 0.409816
dietoats:supplementcontrol    0.2471     1.3112   0.188 0.851571
dietwheat:supplementcontrol   0.8183     1.3112   0.624 0.536512
dietoats:supplementsupergain  0.2470     1.3112   0.188 0.851652
dietwheat:supplementsupergain 1.2557     1.3112   0.958 0.344601
dietoats:supplementsupersupp -0.6650     1.3112  -0.507 0.615135
dietwheat:supplementsupersupp 0.8024     1.3112   0.612 0.544381

Residual standard error: 1.311 on 36 degrees of freedom
Multiple R-squared: 0.8607,      Adjusted R-squared: 0.8182
F-statistic: 20.22 on 11 and 36 DF,  p-value: 3.295e-12
```

This is a rather complex model, because there are 12 estimated parameters (the number of rows in the table): six main effects and six interactions. Remember that the parameter labelled `Intercept` is the mean with both factor levels set to their first in the alphabet (`diet=barley` and `supplement=agrimore`). All other rows are differences between means. The output re-emphasizes that none of the interaction terms is even close to significant, but it suggests that the minimal adequate model will require five parameters: an intercept, a difference due to `oats`, a difference due to `wheat`, a difference due to `control` and difference due to `supergain` (these are the five rows with significance stars). This draws attention to the main shortcoming of using treatment contrasts as the default. If you look carefully at the table, you will see that the effect sizes of two of the supplements, `control` and `supergain`, are not significantly different from one another. You need lots of practice at doing t tests in your head, to be able to do this quickly. Ignoring the signs (because the signs are negative for both of them), we have 3.05 vs. 3.88, a difference of 0.83. But look at the associated standard errors (both 0.927); the difference is less than 1 standard error of a difference between two means. For significance, we would need roughly 2 standard errors (remember the rule of thumb, in which $t \geq 2$ is significant; see p. 292). The rows get starred in the significance column because treatments contrasts compare all the main effects in the rows with the intercept (where each factor is set to its first level in the alphabet, namely `agrimore` and `barley` in this case). When, as here, several factor levels are different from the intercept, but not different from one another, they all get significance stars. This means that you cannot count up the number of rows with stars in order to determine the number of significantly different factor levels.

We first simplify the model by leaving out the interaction terms:

```
model <- aov(gain~diet+supplement)
summary.lm(model)
```

```
Coefficients:
                    Estimate Std. Error t value Pr(>|t|)
(Intercept)         26.1230    0.4408   59.258  < 2e-16 ***
dietoats            -3.0928    0.4408   -7.016 1.38e-08 ***
dietwheat           -5.9903    0.4408  -13.589  < 2e-16 ***
supplementcontrol   -2.6967    0.5090   -5.298 4.03e-06 ***
supplementsupergain -3.3815    0.5090   -6.643 4.72e-08 ***
supplementsupersupp -0.7274    0.5090   -1.429     0.16
```

It is clear that we need to retain all three levels of diet (`oats` differs from `wheat` by $5.99 - 3.09 = 2.90$ with a standard error of 0.44). But it is *not* clear that we need four levels of supplement: `supersupp` is not obviously different from `agrimore` (0.727 with standard error 0.509). Nor is `supergain` obviously different from the unsupplemented `control` animals ($3.38 - 2.70 = 0.68$). We shall try a new two-level factor to replace the four-level supplement, and see if this significantly reduces the model's explanatory power: `agrimore` and `supersupp` are recoded as 'best' and `control` and `supergain` as 'worst':

```
supp2 <- factor(supplement)
levels(supp2)

[1] "agrimore"  "control"  "supergain"  "supersupp"

levels(supp2)[c(1,4)] <- "best"
levels(supp2)[c(2,3)] <- "worst"
levels(supp2)

[1]   "best"       "worst"
```

Now we can compare the two models:

```
model2 <- aov(gain~diet+supp2)
anova(model,model2)

Analysis of Variance Table

Model 1: gain ~ diet + supplement
Model 2: gain ~ diet + supp2
  Res.Df    RSS Df Sum of Sq      F Pr(>F)
1     42 65.296
2     44 71.284 -2   -5.9876 1.9257 0.1584
```

The simpler model2 has saved two degrees of freedom and is not significantly worse than the more complex model ($p = 0.1584$). This is the minimal adequate model: all of the parameters are significantly different from zero and from one another:

```
summary.lm(model2)

Coefficients:
              Estimate Std. Error t value Pr(>|t|)
(Intercept)    25.7593     0.3674  70.106  < 2e-16 ***
dietoats       -3.0928     0.4500  -6.873 1.76e-08 ***
dietwheat      -5.9903     0.4500 -13.311  < 2e-16 ***
supp2worst     -2.6754     0.3674  -7.281 4.43e-09 ***
```

Model simplification has reduced our initial 12-parameter model to a four-parameter model.

11.3 Pseudoreplication: Nested designs and split plots

The model-fitting functions aov, lme and lmer have the facility to deal with complicated error structures, and it is important that you can recognize such error structures, and hence avoid the pitfalls of pseudoreplication. There are two general cases:

- nested sampling, as when repeated measurements are taken from the same individual, or observational studies are conducted at several different spatial scales (mostly random effects);

- split-plot analysis, as when designed experiments have different treatments applied to plots of different sizes (mostly fixed effects).

11.3.1 Split-plot experiments

In a split-plot experiment, different treatments are applied to plots of different sizes. Each different plot size is associated with its own error variance, so instead of having one error variance (as in all the ANOVA tables up to this point), we have as many error terms as there are different plot sizes. The analysis is presented as a series of component ANOVA tables, one for each plot size, in a hierarchy from the largest plot size with the lowest replication at the top, down to the smallest plot size with the greatest replication at the bottom.

The following example refers to a designed field experiment on crop yield with three treatments: irrigation (with two levels, irrigated or not), sowing density (with three levels, low, medium and high), and fertilizer application (with three levels, low, medium and high).

```
yields <- read.table("c:\\temp\\splityield.txt",header=T)
attach(yields)
names(yields)
```

```
[1] "yield"      "block"      "irrigation" "density"     "fertilizer"
```

The largest plots were the four whole fields (block), each of which was split in half, and irrigation was allocated at random to one half of the field. Each irrigation plot was split into three, and one of three different seed-sowing densities (low, medium or high) was allocated at random (independently for each level of irrigation and each block). Finally, each density plot was divided into three, and one of three fertilizer nutrient treatments (N, P, or N and P together) was allocated at random.

The issue with split-plot experiments is pseudoreplication. Think about the irrigation experiment. There were four blocks, each split in half, with one half irrigated and the other as a control. The dataframe for an analysis of this experiment should therefore contain just 8 rows (not 72 rows as in the present case). There would be seven degrees of freedom in total, three for blocks, one for irrigation and just $7 - 3 - 1 = 3$ d.f. for error. If you did not spot this, the model could be run with 51 d.f. representing massive pseudoreplication (the correct p value for the irrigation treatment is 0.0247, but for the pseudoreplicated mistaken analysis $p = 6.16 \times 10^{-10}$).

The model formula is specified as a factorial, using the asterisk notation. The error structure is defined in the Error term, with the plot sizes listed from left to right, from largest to smallest, with each variable separated by the slash operator /. Note that the smallest plot size, fertilizer, does not need to appear in the Error term:

```
model <-
    aov(yield~irrigation*density*fertilizer+Error(block/irrigation/density))
summary(model)
```

```
Error: block
          Df Sum Sq Mean Sq F value Pr(>F)
Residuals  3  194.4   64.81
```

```
Error: block:irrigation
           Df Sum Sq Mean Sq F value Pr(>F)
irrigation  1   8278    8278   17.59 0.0247 *
Residuals   3   1412     471
```

```
Error: block:irrigation:density
                  Df Sum Sq Mean Sq F value Pr(>F)
density            2   1758   879.2   3.784 0.0532 .
irrigation:density 2   2747  1373.5   5.912 0.0163 *
Residuals         12   2788   232.3
```

```
Error: Within
                             Df Sum Sq Mean Sq F value   Pr(>F)
fertilizer                    2 1977.4   988.7  11.449 0.000142 ***
irrigation:fertilizer         2  953.4   476.7   5.520 0.008108 **
density:fertilizer            4  304.9    76.2   0.883 0.484053
irrigation:density:fertilizer 4  234.7    58.7   0.680 0.610667
Residuals                    36 3108.8    86.4
```

Here you see the four ANOVA tables, one for each plot size: blocks are the biggest plots, half blocks get the irrigation treatment, one third of each half block gets a sowing density treatment, and one third of a sowing density treatment gets each fertilizer treatment. Note that the non-significant main effect for density ($p = 0.053$) does *not* mean that density is unimportant, because density appears in a significant interaction with irrigation (the density terms cancel out, when averaged over the two irrigation treatments; see below). The best way to understand the two significant interaction terms is to plot them using `interaction.plot` like this:

```
interaction.plot(fertilizer,irrigation,yield)
```

Irrigation increases yield proportionately more on the N-fertilized plots than on the P-fertilized plots. The irrigation–density interaction is more complicated:

```
interaction.plot(density,irrigation,yield)
```

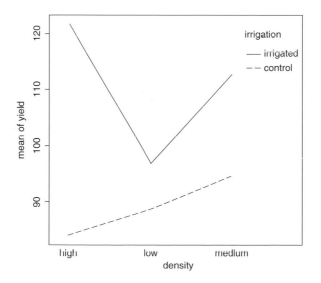

On the irrigated plots, yield is lowest on the low-density plots, but on control plots yield is lowest on the high-density plots. Alternatively, you could use the `effects` package which takes a model object (a linear model or a generalized linear model) and provides attractive trellis plots of specified interaction effects (p. 968).

When there are one or more missing values (`NA`), then factors have effects in more than one stratum and the same main effect turns up in more than one ANOVA table. In such a case, use `lme` or `lmer` rather than `aov`. The output of `aov` is not to be trusted under these circumstances.

11.3.2   Mixed-effects models

Mixed-effects models are so called because the explanatory variables are a mixture of fixed effects and random effects:

- fixed effects influence only the *mean* of y;

- random effects influence only the *variance* of y.

A random effect should be thought of as coming from a population of effects: the existence of this population is an extra assumption. We speak of **prediction** of random effects, rather than estimation: we **estimate** fixed effects from data, but we intend to make predictions about the population from which our random effects were sampled. Fixed effects are unknown constants to be estimated from the data. Random effects govern the variance–covariance structure of the response variable. The fixed effects are often experimental treatments that were applied under our direction, and the random effects are either categorical or continuous variables that are distinguished by the fact that we are typically not interested in the parameter values, but only in the variance they explain.

One of more of the explanatory variables might represent **grouping** in time or in space. Random effects that come from the same group will be correlated, and this contravenes one of the fundamental assumptions of standard statistical models: **independence of errors**. Mixed-effects models take care of this non-independence of errors by modelling the covariance structure introduced by the grouping of the data.

A major benefit of random-effects models is that they economize on the number of degrees of freedom used up by the factor levels. Instead of estimating a mean for every single factor level, the random-effects model estimates the distribution of the means (usually as the standard deviation of the differences of the factor-level means around an overall mean). Mixed-effects models are particularly useful in cases where there is temporal pseudoreplication (repeated measurements) and/or spatial pseudoreplication (e.g. nested designs or split-plot experiments). These models can allow for:

- spatial autocorrelation between neighbours;

- temporal autocorrelation across repeated measures on the same individuals;

- differences in the mean response between blocks in a field experiment;

- differences between subjects in a medical trial involving repeated measures.

The point is that we really do not want to waste precious degrees of freedom in estimating parameters for each of the separate levels of the categorical random variables. On the other hand, we do want to make use of the all measurements we have taken, but because of the pseudoreplication we want to take account of both the

- correlation structure, used to model within-group correlation associated with temporal and spatial dependencies, using **correlation**, and

- variance function, used to model non-constant variance in the within-group errors using **weights**.

11.3.3 Fixed effect or random effect?

It is difficult without lots of experience to know when to use a categorical explanatory variable as a fixed effect or as a random effect. Some guidelines are given below.

- Am I interested in the effect sizes? Yes means fixed effects.

- Is it reasonable to suppose that the factor levels come from a population of levels? Yes means random effects.

- Are there enough levels of the factor in the dataframe on which to base an estimate of the variance of the population of effects? No means fixed effects.

- Are the factor levels informative? Yes means fixed effects.

- Are the factor levels just numeric labels? Yes means random effects.

- Am I mostly interested in making inferences about the distribution of effects, based on the random sample of effects represented in the dataframe? Yes means random effects.

- Is there hierarchical structure? Yes means we need to ask whether the data are experimental or observations.

- Is it a hierarchical experiment, where the factor levels are experimental manipulations? Yes means fixed effects in a split-plot design (see p. 519)

- Is it a hierarchical observational study? Yes means random effects, perhaps in a variance components analysis (see p. 524).

- When the model contains both fixed and random effects, use mixed-effects models.

- If the model structure is linear, use linear mixed effects, `lme` or `lmer`.

- Otherwise, specify the model equation and use non-linear mixed effects, `nlme`.

11.3.4 Removing the pseudoreplication

If you are principally interested in the fixed effects, then the best response to pseudoreplication in a data set is simply to eliminate it. Spatial pseudoreplication can be averaged away. You will always get the correct effect size and p value from the reduced, non-pseudoreplicated dataframe. Note also that you should not use `anova` to compare different models for the fixed effects when using `lme` or `lmer` with REML (see p. 688). Temporal pseudoreplication can be dealt with by carrying out carrying out separate ANOVAs, one at each time (or just one at the end of the experiment). This approach, however, has two weaknesses:

- It cannot address questions about treatment effects that relate to the longitudinal development of the mean response profiles (e.g. differences in growth rates between successive times).

- Inferences made with each of the separate analyses are not independent, and it is not always clear how they should be combined.

The key feature of longitudinal data is that the same individuals are measured repeatedly through time. This would represent temporal pseudoreplication if the data were used uncritically in regression or ANOVA. The set of observations on one individual subject will tend to be positively correlated, and this correlation needs to be taken into account in carrying out the analysis. The alternative is a cross-sectional study, with all the data gathered at a single point in time, in which each individual contributes a single data point. The advantage of longitudinal studies is that they are capable of separating *age effects* from *cohort effects*; these are inextricably confounded in cross-sectional studies. This is particularly important when differences between years mean that cohorts originating at different times experience different conditions, so that individuals of the same age in different cohorts would be expected to differ.

There are two extreme cases in longitudinal studies:

- a few measurements on a large number of individuals;

- a large number of measurements on a few individuals.

In the first case it is difficult to fit an accurate model for change within individuals, but treatment effects are likely to be tested effectively. In the second case, it is possible to get an accurate model of the way that individuals change though time, but there is less power for testing the significance of treatment effects, especially if variation from individual to individual is large. In the first case, less attention will be paid to estimating the correlation structure, while in the second case the covariance model will be the principal focus of attention. The aims are:

- to estimate the average time course of a process;

- to characterize the degree of heterogeneity from individual to individual in the rate of the process;

- to identify the factors associated with both of these, including possible cohort effects.

The response is not the individual measurement, but the *sequence of measurements* on an individual subject. This enables us to distinguish between age effects and year effects; see Diggle et al. (1994) for details.

11.3.5   Derived variable analysis

The idea here is to get rid of the pseudoreplication by reducing the repeated measures into a set of summary statistics (slopes, intercepts or means), then *analyse these summary statistics* using standard parametric techniques such as ANOVA or regression. The technique is weak when the values of the explanatory variables change through time. Derived variable analysis makes most sense when it is based on the parameters of scientifically interpretable non-linear models from each time sequence. However, the best model from a theoretical perspective may not be the best model from the statistical point of view.

There are three qualitatively different sources of random variation:

- **random effects**, where experimental units differ (e.g. genotype, history, size, physiological condition) so that there are intrinsically high responders and other low responders;

- **serial correlation**, where there may be time-varying stochastic variation within a unit (e.g. market forces, physiology, ecological succession, immunity) so that correlation depends on the time separation of pairs of measurements on the same individual, with correlation weakening with the passage of time;

- **measurement error**, where the assay technique may introduce an element of correlation (e.g. shared bioassay of closely spaced samples; different assay of later specimens).

11.4   Variance components analysis

For random effects we are often more interested in the question of how much of the variation in the response variable can be attributed to a given factor, than we are in estimating means or assessing the significance of differences between means. This procedure is called **variance components analysis**.

The following classic example of spatial pseudoreplication comes from Snedecor and Cochran (1980):

```
rats <- read.table("c:\\temp\\rats.txt",header=T)
attach(rats)
names(rats)
```

```
[1] "Glycogen"   "Treatment" "Rat"          "Liver"
```

Three experimental treatments were administered to rats, and the glycogen content of the rats' livers was analysed as the response variable. There were two rats per treatment, so the total sample was $n = 3 \times 2 = 6$. The tricky bit was that after each rat was killed, its liver was cut up into three pieces: a left-hand bit, a central bit and a right-hand bit. So now there are six rats each producing three bits of liver, for a total of $6 \times 3 = 18$ numbers. Finally, two separate preparations were made from each macerated bit of liver, to assess the measurement error associated with the analytical machinery. At this point there are $2 \times 18 = 36$ numbers in the dataframe as a whole. The factor levels are numbers, so we need to declare the explanatory variables to be categorical before we begin:

```
Treatment <- factor(Treatment)
Rat <- factor(Rat)
Liver <- factor(Liver)
```

Here is the analysis done the *wrong* way:

```
model <- aov(Glycogen~Treatment)
summary(model)
```

```
            Df Sum Sq Mean Sq F value   Pr(>F)
Treatment    2   1558   778.8    14.5 3.03e-05 ***
Residuals   33   1773    53.7
```

A massively significant effect or treatment, right? Wrong. This result is due entirely to pseudoreplication. With just six rats in the whole experiment, there should be just three degrees of freedom for error, not 33.

The simplest way to do the analysis properly is to average away the pseudoreplication. Here are the mean glycogen values for the six rats:

```
(means <- tapply(Glycogen,list(Treatment,Rat),mean))
```

```
          1         2
1 132.5000  148.5000
2 149.6667  152.3333
3 134.3333  136.0000
```

We need a new variable to represent the treatments associated with each of these rats. The 'generate levels' function gl is useful here:

```
treat <- gl(3,1,length=6)
```

Now we can fit the non-pseudoreplicated model with the correct error degrees of freedom (3 d.f., not 33):

```
model <- aov(as.vector(means)~treat)
summary(model)
```

```
            Df Sum Sq Mean Sq F value Pr(>F)
treat        2  259.6  129.80   2.929  0.197
Residuals    3  132.9   44.31
```

As you can see, the treatment effect falls well short of significance ($p = 0.197$).

There are two different ways of doing the analysis properly in R: ANOVA with multiple error terms (aov) or linear mixed-effects models (lmer). The problem is that the bits of the same liver are pseudoreplicates because they are spatially correlated (they come from the same rat); they are not independent, as required if they are to be true replicates. Likewise, the two preparations from each liver bit are very highly correlated (the livers were macerated before the preparations were taken, so they are essentially the same sample (certainly not independent replicates of the experimental treatments).

Here is the correct analysis using aov with multiple error terms. In the Error term we start with the largest scale (treatment), then rats within treatments, then liver bits within rats within treatments. Finally, there were replicated measurements (two preparations) made for each bit of liver.

```
model2 <- aov(Glycogen~Treatment+Error(Treatment/Rat/Liver))
summary(model2)

Error: Treatment
          Df Sum Sq Mean Sq
Treatment  2   1558   778.8

Error: Treatment:Rat
          Df Sum Sq Mean Sq F value Pr(>F)
Residuals  3  797.7   265.9

Error: Treatment:Rat:Liver
          Df Sum Sq Mean Sq F value Pr(>F)
Residuals 12    594    49.5

Error: Within
          Df Sum Sq Mean Sq F value Pr(>F)
Residuals 18    381   21.17
```

You can do the correct, non-pseudoreplicated analysis of variance from this output (Box 11.1).

Box 11.1   Sums of squares in hierarchical designs

The trick to understanding these sums of squares is to appreciate that with nested categorical explanatory variables (random effects) the correction factor, which is subtracted from the sum of squared subtotals, is *not* the conventional $(\sum y)^2 / kn$. Instead, the correction factor is the uncorrected sum of squared subtotals from the level in the hierarchy immediately above the level in question. This is very hard to see without lots of practice. The total sum of squares, *SSY*, and the treatment sum of squares, *SSA*, are computed in the usual way (see p. 499):

$$SSY = \sum y^2 - \frac{\left(\sum y\right)^2}{n},$$

$$SSA = \frac{\sum_{i=1}^{k} C_i^2}{n} - \frac{\left(\sum y\right)^2}{kn}.$$

The analysis is easiest to understand in the context of an example. For the rats data, the treatment totals were based on 12 numbers (two rats, three liver bits per rat and two preparations per liver bit). In this case, in the formula for SSA, above, $n = 12$ and $kn = 36$. We need to calculate sums of squares for rats within treatments, SS_{Rats}, liver bits within rats within treatments, $SS_{Liverbits}$, and preparations within liver bits within rats within treatments, $SS_{Preparations}$:

$$SS_{Rats} = \frac{\sum R^2}{6} - \frac{\sum C^2}{12},$$

$$SS_{Liverbits} = \frac{\sum L^2}{2} - \frac{\sum R^2}{6},$$

$$SS_{Preparations} = \frac{\sum y^2}{1} - \frac{\sum L^2}{2}.$$

The correction factor at any level is the *uncorrected sum of squares from the level above*. The last sum of squares could have been computed by difference:

$$SS_{Preparations} = SSY - SSA - SS_{Rats} - SS_{Liverbits}.$$

The F test for equality of the treatment means is the treatment variance divided by the 'rats within treatment variance' from the row immediately beneath: $F = 778.78/265.89 = 2.928\,956$, with 2 d.f. in the numerator and 3 d.f. in the denominator (as we obtained in the correct ANOVA, above).

To turn this into a variance components analysis we need to do a little work. The mean squares are converted into variance components like this:

- residuals = preparations within liver bits: unchanged = 21.17,

- liver bits within rats within treatments: $(49.5 - 21.17)/2 = 14.165$,

- rats within treatments: $(265.89 - 49.5)/6 = 36.065$.

You divide the difference in variance in going from one spatial scale to the next, by the number of numbers in the level below (i.e. two preparations per liver bit, and six preparations per rat, in this case). Variance components analysis typically expresses these variances as percentages of the total:

```
varcomps <- c(21.17,14.165,36.065)
100*varcomps/sum(varcomps)

[1]  29.64986 19.83894 50.51120
```

illustrating that more than 50% of the random variation is accounted for by differences between the rats. Repeating the experiment using more than six rats would make much more sense than repeating it by cutting up the livers into more pieces. Analysis of the rats data using `lmer` is explained on p. 703.

11.5   Effect sizes in ANOVA: `aov` or `lm`?

The difference between `lm` and `aov` is mainly in the form of the output: the `summary` table with `aov` is in the traditional form for analysis of variance, with one row for each categorical variable and each interaction term. On the other hand, the `summary` table for `lm` produces one row per estimated parameter (i.e. one row for each factor level and one row for each interaction level). If you have multiple error terms (spatial pseudoreplication) then you must use aov because lm does not support the Error term.

Here is a three-way analysis of variance fitted first using aov then using lm:

```
daphnia <- read.table("c:\\temp\\Daphnia.txt",header=T)
attach(daphnia)
names(daphnia)
```

```
[1] "Growth.rate" "Water"          "Detergent"     "Daphnia"
```

```
model1 <- aov(Growth.rate~Water*Detergent*Daphnia)
summary(model1)
```

	Df	Sum Sq	Mean Sq	F value	Pr(>F)	
Water	1	1.99	1.985	2.850	0.097838	.
Detergent	3	2.21	0.737	1.059	0.375478	
Daphnia	2	39.18	19.589	28.128	8.23e-09	***
Water:Detergent	3	0.17	0.058	0.084	0.968608	
Water:Daphnia	2	13.73	6.866	9.859	0.000259	***
Detergent:Daphnia	6	20.60	3.433	4.930	0.000532	***
Water:Detergent:Daphnia	6	5.85	0.975	1.399	0.234324	
Residuals	48	33.43	0.696			

All three factors are likely to stay in the model because each is involved in at least one significant interaction. We must not be misled by the apparently non-significant main effect for detergent. The three-way interaction is clearly non-significant and can be deleted ($p = 0.234$). Here is the output from the same analysed using the linear model function:

```
model2 <- lm(Growth.rate~Water*Detergent*Daphnia)
summary(model2)
```

```
Coefficients:
```

	Estimate	Std. Error	t value	Pr(>|t|)	
(Intercept)	2.81126	0.48181	5.835	4.48e-07	***
WaterWear	-0.15808	0.68138	-0.232	0.81753	
DetergentBrandB	-0.03536	0.68138	-0.052	0.95883	
DetergentBrandC	0.47626	0.68138	0.699	0.48794	
DetergentBrandD	-0.21407	0.68138	-0.314	0.75475	
DaphniaClone2	0.49637	0.68138	0.728	0.46986	
DaphniaClone3	2.05526	0.68138	3.016	0.00408	**
WaterWear:DetergentBrandB	0.46455	0.96361	0.482	0.63193	
WaterWear:DetergentBrandC	-0.27431	0.96361	-0.285	0.77712	
WaterWear:DetergentBrandD	0.21729	0.96361	0.225	0.82255	
WaterWear:DaphniaClone2	1.38081	0.96361	1.433	0.15835	
WaterWear:DaphniaClone3	0.43156	0.96361	0.448	0.65627	
DetergentBrandB:DaphniaClone2	0.91892	0.96361	0.954	0.34506	
DetergentBrandC:DaphniaClone2	-0.16337	0.96361	-0.170	0.86609	
DetergentBrandD:DaphniaClone2	1.01209	0.96361	1.050	0.29884	
DetergentBrandB:DaphniaClone3	-0.06490	0.96361	-0.067	0.94658	
DetergentBrandC:DaphniaClone3	-0.80789	0.96361	-0.838	0.40597	
DetergentBrandD:DaphniaClone3	-1.28669	0.96361	-1.335	0.18809	
WaterWear:DetergentBrandB:DaphniaClone2	-1.26380	1.36275	-0.927	0.35837	
WaterWear:DetergentBrandC:DaphniaClone2	1.35612	1.36275	0.995	0.32466	

```
WaterWear:DetergentBrandD:DaphniaClone2  0.77616    1.36275   0.570  0.57164
WaterWear:DetergentBrandB:DaphniaClone3 -0.87443    1.36275  -0.642  0.52414
WaterWear:DetergentBrandC:DaphniaClone3 -1.03019    1.36275  -0.756  0.45337
WaterWear:DetergentBrandD:DaphniaClone3 -1.55400    1.36275  -1.140  0.25980
Residual standard error: 0.8345 on 48 degrees of freedom
Multiple R-squared: 0.7147,       Adjusted R-squared: 0.578
F-statistic: 5.227 on 23 and 48 DF,  p-value: 7.019e-07
```

Note that the two significant interactions from the aov table do *not* show up in the summary.lm table (Water–Daphnia and Detergent–Daphnia). This is because summary.lm shows treatment contrasts, comparing everything to the Intercept, rather than orthogonal contrasts (see p. 430). This draws attention to the importance of model simplification rather than per-row *t* tests in assessing statistical significance (i.e. removing the non-significant three-way interaction term in this case). In the aov table, the *p* values are 'on deletion' *p* values, which is a big advantage.

The main difference is that there are eight rows in the summary.aov table (three main effects, three two-way interactions, one three-way interaction and an error term) but there are 24 rows in the summary.lm table (four levels of detergent by three levels of daphnia clone by two levels of water). You can easily view the output of model1 in linear model layout, or model2 as an ANOVA table using the opposite summary options:

```
summary.lm(model1)
summary.aov(model2)
```

In complicated designed experiments, it is easiest to summarize the effect sizes with plot.design and model.tables functions. For main effects, use

```
plot.design(Growth.rate~Water*Detergent*Daphnia)
```

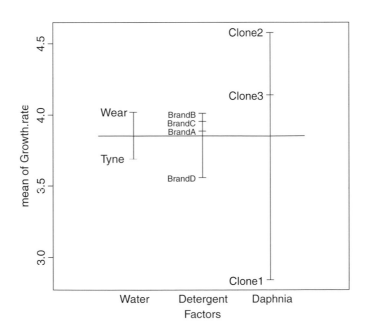

This simple graphical device provides a very clear summary of the three sets of main effects. It is no good, however, at illustrating the interactions. The `model.tables` function takes the name of the fitted model object as its first argument, and you can specify whether you want the standard errors (as you typically would):

```
model.tables(model1, "means", se = TRUE)
```

```
Tables of means
Grand mean
3.851905

 Water
Water
 Tyne  Wear
3.686 4.018

 Detergent
Detergent
BrandA BrandB  BrandC  BrandD
 3.885   4.010   3.955   3.558

 Daphnia
Daphnia
Clone1 Clone2 Clone3
 2.840   4.577   4.139

 Water:Detergent
        Detergent
Water  BrandA BrandB BrandC BrandD
  Tyne 3.662   3.911   3.814   3.356
  Wear 4.108   4.109   4.095   3.760

 Water:Daphnia
        Daphnia
Water  Clone1 Clone2 Clone3
  Tyne 2.868   3.806   4.383
  Wear 2.812   5.348   3.894

 Detergent:Daphnia
          Daphnia
Detergent Clone1 Clone2 Clone3
   BrandA 2.732   3.919   5.003
   BrandB 2.929   4.403   4.698
   BrandC 3.071   4.773   4.019
   BrandD 2.627   5.214   2.834

 Water:Detergent:Daphnia
, , Daphnia = Clone1

        Detergent
Water  BrandA BrandB BrandC BrandD
  Tyne 2.811   2.776   3.288   2.597
  Wear 2.653   3.082   2.855   2.656
```

```
, , Daphnia = Clone2

        Detergent
Water  BrandA BrandB BrandC BrandD
  Tyne 3.308  4.191  3.621  4.106
  Wear 4.530  4.615  5.925  6.322

, , Daphnia = Clone3

        Detergent
Water  BrandA BrandB BrandC BrandD
  Tyne 4.867  4.766  4.535  3.366
  Wear 5.140  4.630  3.504  2.303
```

```
Standard errors for differences of means
         Water Detergent Daphnia Water:Detergent Water:Daphnia Detergent:Daphnia Water:Detergent:Daphnia
        0.1967    0.2782  0.2409          0.3934        0.3407            0.4818                  0.6814
replic.     36        18      24               9            12                 6                       3
```

Note how the standard errors of the differences between two means increase as the replication declines. All the standard errors use the same pooled error variance $s^2 = 0.696$ (see above). For instance, the three-way interactions have $se = \sqrt{2 \times 0.696/3} = 0.681$.and the daphnia main effects have $se = \sqrt{2 \times 0.696/24} = 0.2409$.

Attractive plots of effect sizes can be obtained using the `effects` library (p. 968).

11.6 Multiple comparisons

One of the cardinal sins is to take a set of samples, search for the sample with the largest mean and the sample with the smallest mean, and then do a t test to compare them. You should not carry out contrasts until the analysis of variance, calculated over the whole set of samples, has indicated that there are significant differences present (i.e. until after the null hypothesis has been rejected). Also, bear in mind that there are just $k - 1$ orthogonal contrasts when you have a categorical explanatory variable with k levels, so do not carry out more than $k - 1$ comparisons of means (see p. 430 for discussion of these ideas).

When comparing the multiple means across the levels of a factor, a simple comparison using multiple t tests will inflate the probability of declaring a significant difference when there is none. This is because the intervals are calculated with a given coverage probability for each *interval* but the interpretation of the coverage is usually with respect to *the entire family of intervals* (i.e. for the factor as a whole).

If you follow the protocol of model simplification recommended in this book, then issues of multiple comparisons will not arise very often. An occasional significant t test amongst a bunch of non-significant interaction terms is not likely to survive a deletion test (see p. 437). Again, if you have factors with large numbers of levels you might consider using mixed-effects models rather than ANOVA (i.e. treating the factors as random effects rather than fixed effects; see p. 681).

John Tukey introduced intervals based on the range of the sample means rather than the individual differences; nowadays, these are called Tukey's honest significant differences. The intervals returned by the `TukeyHSD` function are based on Studentized range statistics. Technically the intervals constructed in this way would only apply to balanced designs where the same number of observations is made at each level of the factor. This function incorporates an adjustment for sample size that produces sensible intervals for mildly unbalanced designs.

The following example concerns the yield of fungi gathered from 16 different habitats:

```
data <- read.table("c:\\temp\\Fungi.txt",header=T)
attach(data)
names(data)
```

```
[1]  "Habitat"         "Fungus.yield"
```

First we establish whether there is any variation in fungus yield to explain:

```
model <- aov(Fungus.yield~Habitat)
summary(model)
```

```
            Df  Sum Sq  Mean Sq  F value  Pr(>F)
Habitat     15    7527    501.8    72.14  <2e-16 ***
Residuals  144    1002      7.0
```

Yes, there is ($p < 0.000\,001$). But this is not of much real interest, because it just shows that some habitats produce more fungi than others. We are likely to be interested in *which* habitats produce significantly more fungi than others. Multiple comparisons are an issue because there are 16 habitats and so there are (16 × 15)/2 = 120 possible pairwise comparisons. There are two options:

- apply the function TukeyHSD to the model to get Tukey's honest significant differences;

- use the function pairwise.t.test to get adjusted p values for all comparisons.

Here is Tukey's test in action: it produces a table of p values by default:

```
TukeyHSD(model)
```

```
   Tukey multiple comparisons of means
      95% family-wise confidence level

Fit: aov(formula = Fungus.yield ~ Habitat)

$Habitat
                        diff          lwr           upr         p adj
Ash-Alder          3.53292777   -0.5808096    7.6466651   0.1844088
Aspen-Alder       12.78574402    8.6720067   16.8994814   0.0000000
Beech-Alder       12.32365349    8.2099161   16.4373908   0.0000000
Birch-Alder       14.11348150    9.9997441   18.2272189   0.0000000
. . .

. . .
Willow-Rowan      -3.51860059   -7.6323379    0.5951368   0.1896363
Sycamore-Spruce    4.96019563    0.8464583    9.0739330   0.0044944
Willow-Spruce      4.92754623    0.8138089    9.0412836   0.0049788
Willow-Sycamore   -0.03264941   -4.1463868    4.0810879   1.0000000
```

You can plot the confidence intervals if you prefer (or do both, of course):

```
plot(TukeyHSD(model))
```

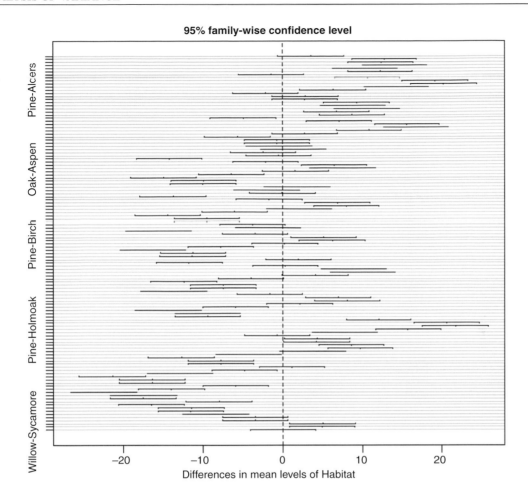

95% family-wise confidence level

Habitats on opposite sides of the dotted line and not overlapping it are significantly different from one another.

Alternatively, you can use the `pairwise.t.test` function in which you specify the response variable, and then the categorical explanatory variable containing the factor levels you want to be compared, separated by a comma (not a tilde):

```
pairwise.t.test(Fungus.yield,Habitat)
```

```
        Pairwise comparisons using t tests with pooled SD
data:   Fungus.yield and Habitat
```

	Alder	Ash	Aspen	Beech	Birch	Cherry	Chestnut	Holmoak	Hornbeam	Lime	Oak	Pine
Ash	0.10011	-	-	-	-	-	-	-	-	-	-	-
Aspen	< 2e-16	6.3e-11	-	-	-	-	-	-	-	-	-	-
Beech	< 2e-16	5.4e-10	1.00000	-	-	-	-	-	-	-	-	-
Birch	< 2e-16	1.2e-13	1.00000	1.00000	-	-	-	-	-	-	-	-
Cherry	4.7e-13	2.9e-06	0.87474	1.00000	0.04943	-	-	-	-	-	-	-
Chestnut	< 2e-16	7.8e-10	1.00000	1.00000	1.00000	1.00000	-	-	-	-	-	-
Holmoak	1.00000	0.00181	< 2e-16	< 2e-16	< 2e-16	3.9e-16	< 2e-16	-	-	-	-	-
Hornbeam	1.1e-13	8.6e-07	1.00000	1.00000	1.00000	0.10057	1.00000	< 2e-16	-	-	-	-
Lime	< 2e-16	< 2e-16	1.1e-05	1.9e-06	0.00131	3.3e-10	1.4e-06	< 2e-16	1.3e-09	-	-	-
Oak	< 2e-16	< 2e-16	1.4e-07	2.0e-08	2.7e-05	1.9e-12	1.5e-08	< 2e-16	8.4e-12	1.00000	-	-

```
Pine      < 2e-16 3.9e-14 1.00000 1.00000 1.00000 0.02757 1.00000   < 2e-16 0.05975   0.00253 6.1e-05 -
Rowan     1.8e-05 0.51826 8.5e-06 4.7e-05 3.9e-08 0.03053 6.2e-05   5.3e-08 0.01380   < 2e-16 < 2e-16 1.5e-08
P value adjustment method: holm
```

As you see, the default method of adjustment of the p values is holm, but other adjustment methods include hochberg, hommel, bonferroni, BH, BY, fdr and none. Without adjustment of the p values, the rowan–willow comparison looks highly significant ($p = 0.003\,35$), as you can see if you try

```
pairwise.t.test(Fungus.yield,Habitat,p.adjust.method="none")
```

I like TukeyHSD because it is conservative without being ridiculously so (in contrast to Bonferroni). For instance, Tukey gives the birch–cherry comparison as non-significant ($p = 0.101\,102\,7$), while Holm makes this difference significant ($p = 0.049\,43$). Tukey has the comparison between willow and Holm oak as significant ($p = 0.038\,091\,0$), whereas Bonferroni throws this baby out with the bathwater ($p = 0.056\,72$). You need to decide how circumspect you want to be in the context of your particular question.

There is a useful package for multiple comparisons called multcomp:

```
install.packages("multcomp")
```

You can see at once how contentious the issue of multiple comparisons is, just by looking at the length of the list of different multiple comparisons methods supported in this package:

- the many-to-one comparisons of Dunnett
- the all-pairwise comparisons of Tukey
- Sequen
- AVE
- changepoint
- Williams
- Marcus
- McDermott
- Tetrade
- Bonferroni correction
- Holm
- Hochberg
- Hommel
- Benjamini–Hochberg
- Benjamini–Yekutieli

The old-fashioned Bonferroni correction is highly conservative, because the p values are multiplied by the number of comparisons. Instead of using the usual Bonferroni and Holm procedures, the adjustment methods include less conservative corrections that take the exact correlations between the test statistics into account by use of the multivariate t distribution. The resulting procedures are therefore substantially more powerful (the Bonferroni and Holm adjusted p values are reported for reference). There seems to be no reason to use the unmodified Bonferroni correction because it is dominated by Holm's method, which is valid under arbitrary assumptions.

The tests are designed to suit multiple comparisons within the general linear model, so they allow for covariates, nested effects, correlated means and missing values. The first four methods are designed to give strong control of the familywise error rate. The methods of Benjamini, Hochberg, and Yekutieli control the false discovery rate, which is the expected proportion of false discoveries amongst the rejected hypotheses. The false discovery rate is a less stringent condition than the familywise error rate, so these methods are more powerful than the others.

11.7 Multivariate analysis of variance

Two or more response variables are sometimes measured in the same experiment. Of course you can analyse each response variable separately, and that is the typical way to proceed. But there are occasions where you want to treat the group of response variables as one multivariate response. The function for this is manova, the multivariate analysis of variance. Note that manova does not support multi-stratum analysis of variance, so the formula must not include an Error term.

```
data <- read.table("c:\\temp\\manova.txt",header=T)
attach(data)
names(data)
```

```
[1]   "tear"      "gloss"     "opacity"    "rate"      "additive"
```

First, create a multivariate response variable, Y, by binding together the three separate response variables (tear, gloss and opacity), like this:

```
Y <- cbind(tear, gloss, opacity)
```

Then fit the multivariate analysis of variance using the manova function:

```
model <- manova(Y~rate*additive)
```

There are two ways to inspect the output. First, as a multivariate analysis of variance:

```
summary(model)
```

	Df	Pillai	approx F	num Df	den Df	Pr(>F)	
rate	1	0.61814	7.5543	3	14	0.003034	**
additive	1	0.47697	4.2556	3	14	0.024745	*
rate:additive	1	0.22289	1.3385	3	14	0.301782	
Residuals	16						

This shows significant main effects for both rate and additive, but no interaction. Note that the F tests are based on 3 and 14 degrees of freedom (not 1 and 16). The default method in summary.manova is the Pillai–Bartlett statistic. Other options include Wilks, Hotelling–Lawley and Roy. Second, you will want to look at each of the three response variables separately:

```
summary.aov(model)
```

```
Response tear :
```

	Df	Sum Sq	Mean Sq	F value	Pr(>F)	
rate	1	1.7405	1.74050	15.7868	0.001092	**
additive	1	0.7605	0.76050	6.8980	0.018330	*

```
rate:additive  1 0.0005 0.00050   0.0045 0.947143
Residuals     16 1.7640 0.11025
```

```
Response gloss :
              Df Sum Sq Mean Sq F value  Pr(>F)
rate           1 1.3005 1.30050  7.9178 0.01248 *
additive       1 0.6125 0.61250  3.7291 0.07139 .
rate:additive  1 0.5445 0.54450  3.3151 0.08740 .
Residuals     16 2.6280 0.16425
```

```
Response opacity :
              Df Sum Sq Mean Sq F value Pr(>F)
rate           1  0.421  0.4205  0.1036 0.7517
additive       1  4.901  4.9005  1.2077 0.2881
rate:additive  1  3.961  3.9605  0.9760 0.3379
Residuals     16 64.924  4.0578
```

Notice that one of the three response variables, opacity, is not significantly associated with either of the explanatory variables.

12

Analysis of Covariance

Analysis of covariance (ANCOVA) combines elements from regression and analysis of variance. The response variable is continuous, and there is at least one continuous explanatory variable and at least one categorical explanatory variable. The procedure works like this:

- Fit two or more linear regressions of y against x (one for each level of the factor).
- Estimate different slopes and intercepts for each level.
- Use model simplification (deletion tests) to eliminate unnecessary parameters.

For example, we could use ANCOVA in a medical experiment where the response variable was 'days to recovery' and the explanatory variables were 'smoker or not' (categorical) and 'blood cell count' (continuous). In economics, local unemployment rate might be modelled as a function of country (categorical) and local population size (continuous). Suppose we are modelling weight (the response variable) as a function of sex and age. Sex is a factor with two levels (male and female) and age is a continuous variable. The maximal model therefore has four parameters: two slopes (a slope for males and a slope for females) and two intercepts (one for males and one for females) like this:

$$weight_{\text{male}} = a_{\text{male}} + b_{\text{male}} \times age,$$
$$weight_{\text{female}} = a_{\text{female}} + b_{\text{female}} \times age.$$

The R Book, Second Edition. Michael J. Crawley.
© 2013 John Wiley & Sons, Ltd. Published 2013 by John Wiley & Sons, Ltd.

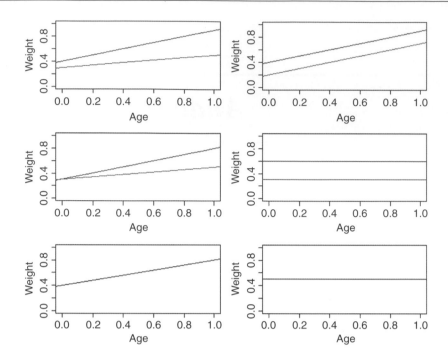

This maximal model is shown in the top left-hand panel. Model simplification is an essential part of analysis of covariance, because the principle of parsimony requires that we keep as few parameters in the model as possible.

There are six possible models in this case, and the process of model simplification begins by asking whether we need all four parameters (top left). Perhaps we could make do with two intercepts and a common slope (top right), or a common intercept and two different slopes (centre left). There again, age may have no significant effect on the response, so we only need two parameters to describe the main effects of sex on weight; this would show up as two separated, horizontal lines in the plot (one mean weight for each sex; centre right). Alternatively, there may be no effect of sex at all, in which case we only need two parameters (one slope and one intercept) to describe the effect of age on weight (bottom left). In the limit, neither the continuous nor the categorical explanatory variables might have any significant effect on the response, in which case model simplification will lead to the one-parameter null model $\hat{y} = \bar{y}$ (a single horizontal line; bottom right).

12.1   Analysis of covariance in R

We could use either lm or aov; the choice affects only the format of the summary table. We shall use both and compare their output. Our worked example concerns an experiment on the impact of grazing on the seed production of a biennial plant. Forty plants were allocated to two treatments, grazed and ungrazed, and the grazed plants were exposed to rabbits during the first two weeks of stem elongation. They were then protected from subsequent grazing by the erection of a fence and allowed to regrow. Because initial plant size was thought likely to influence fruit production, the diameter of the top of the rootstock was measured before each plant was potted up. At the end of the growing season, the fruit production (dry weight in milligrams) was recorded on each of the 40 plants, and this forms the response variable in the following analysis.

```
regrowth <- read.table("c:\\temp\\ipomopsis.txt",header=T)
attach(regrowth)
names(regrowth)
```

```
[1]  "Root"  "Fruit"  "Grazing"
```

   The object of the exercise is to estimate the parameters of the minimal adequate model for these data. We begin by inspecting the data with a plot of fruit production against root size for each of the two treatments separately:

```
plot(Root,Fruit,pch=16,col=c("blue","red")[as.numeric(Grazing)])
```

where red dots represent the ungrazed plants and blue dots represent the grazed plants. Note the use of as.numeric to select the plotting colours. How are the grazing treatments reflected in the factor levels?

```
levels(Grazing)
```

```
[1]  "Grazed"  "Ungrazed"
```

Now we can use logical subscripts (p. 39) to draw linear regression lines for the two grazing treatments separately, using abline (we could have used subset instead):

```
abline(lm(Fruit[Grazing=="Grazed"]~Root[Grazing=="Grazed"]),col="blue")
abline(lm(Fruit[Grazing=="Ungrazed"]~Root[Grazing=="Ungrazed"]),col="red")
```

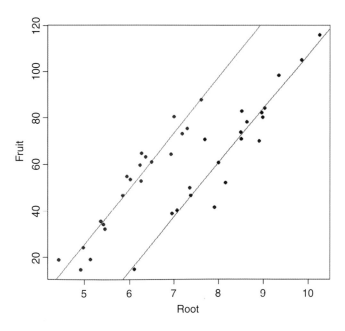

The odd thing about these data is that grazing seems to *increase* fruit production, which is a highly counter-intuitive result:

```
tapply(Fruit,Grazing,  mean)
```

```
Grazed      Ungrazed
67.9405      50.8805
```

This difference is statistically significant ($p = 0.027$) if you do a *t* test (although this is the *wrong* thing to do in this case, as explained below):

```
t.test(Fruit~Grazing)

        Welch Two Sample t-test
data: Fruit by Grazing
t = 2.304, df = 37.306, p-value = 0.02689
alternative hypothesis: true difference in means is not equal to 0
95 percent confidence interval:
2.061464   32.058536
sample estimates:
mean in group Grazed mean in group Ungrazed
            67.9405                 50.8805
```

Several important points are immediately apparent from this initial analysis:

- Different sized plants were allocated to the two treatments.
- The grazed plants were bigger at the outset.
- The regression line for the ungrazed plants is above the line for the grazed plants.
- The regression lines are roughly parallel.
- The intercepts (not shown to the left) are likely to be significantly different.

Each of these points will be addressed in detail.

To understand the output from analysis of covariance it is useful to work through the calculations by hand. We start by working out the sums, sums of squares and sums of products for the whole data set combined (40 pairs of numbers), and then for each treatment separately (20 pairs of numbers). We shall fill in a table of totals, because it helps to be really well organized for these calculations. Check to see where (and why) the sums and the sums of squares of the root diameters (the *x* values) and the fruit yields (the *y* values) have gone in the table. First, we shall work out the overall totals based on all 40 data points.

```
sum(Root);sum(Root^2)
```

```
[1]   287.246
[1]   2148.172
```

```
sum(Fruit);sum(Fruit^2)
```

```
[1]   2376.42
[1]   164928.1
```

```
sum(Root*Fruit)
```

```
[1]   18263.16
```

These are the famous five, which we shall make use of shortly, and complete the overall data summary. Now we select the root diameters for the grazed and ungrazed plants and then the fruit yields for the grazed and ungrazed plants:

```
sum(Root[Grazing=="Grazed"]);sum(Root[Grazing=="Grazed"]^2)
```

```
[1]   166.188
```

[1]   1400.834

sum(Root[Grazing=="Ungrazed"]);sum(Root[Grazing=="Ungrazed"]^2)

[1]   121.058
[1]   747.3387

sum(Fruit[Grazing=="Grazed"]);sum(Fruit[Grazing=="Grazed"]^2)

[1]   1358.81
[1]   104156.0

sum(Fruit[Grazing=="Ungrazed"]);sum(Fruit[Grazing=="Ungrazed"]^2)

[1]   1017.61
[1]   60772.11

Finally, we want the sums of products: first for the grazed plants and then for the ungrazed plants:

sum(Root[Grazing=="Grazed"]*Fruit[Grazing=="Grazed"])

[1]   11753.64

sum(Root[Grazing=="Ungrazed"]*Fruit[Grazing=="Ungrazed"])

[1]   6509.522

Here is our table:

	Sums	Squares and products
x ungrazed	121.058	747.3387
y ungrazed	1017.61	60772.11
xy ungrazed		6509.522
x grazed	166.188	1400.834
y grazed	1358.81	104156.0
xy grazed		11753.64
x overall	287.246	2148.172
y overall	2376.42	164928.1
xy overall		18263.16

Now we have all of the information necessary to carry out the calculations of the corrected sums of squares and products, SSY, SSX and $SSXY$, for the whole data set ($n = 40$) and for the two separate treatments (with 20 replicates in each). To get the right answer you will need to be extremely methodical, but there is nothing mysterious or difficult about the process. First, calculate the regression statistics for the whole experiment, ignoring the grazing treatment, using the famous five which we have just calculated:

$$SSY = 164\,928.1 - \frac{2376.42^2}{40} = 23\,743.84,$$

$$SSX = 2148.172 - \frac{287.246^2}{40} = 85.4158,$$

$$SSXY = 18\,263.16 - \frac{287.246 \times 2376.42}{40} = 1197.731,$$

$$SSR = \frac{1197.731^2}{85.4158} = 16\,795,$$

$$SSE = 23\,743.84 - 16\,795 = 6948.835.$$

The effect of differences between the two grazing treatments, SSA, is

$$SSA = \frac{1358.81^2 + 1017.61^2}{20} - \frac{2376.42^2}{40} = 2910.436.$$

Next calculate the regression statistics for each of the grazing treatments separately. First, for the grazed plants:

$$SSY_g = 104\,156 - \frac{1358.81^2}{20} = 11\,837.79,$$

$$SSX_g = 1400.834 - \frac{166.188^2}{20} = 19.9111,$$

$$SSXY_g = 11\,753.64 - \frac{1358.81 \times 166.188}{20} = 462.7415,$$

$$SSR_g = \frac{462.7415^2}{19.9111} = 10\,754.29,$$

$$SSE_g = 11\,837.79 - 10\,754.29 = 1083.509,$$

so the slope of the graph of Fruit against Root for the grazed plants is given by

$$b_g = \frac{SSXY_g}{SSX_g} = \frac{462.7415}{19.9111} = 23.240.$$

Now for the ungrazed plants:

$$SSY_u = 60\,772.11 - \frac{1017.61^2}{20} = 8995.606,$$

$$SSX_u = 747.3387 - \frac{121.058^2}{20} = 14.586\,77,$$

$$SSXY_u = 6509.522 - \frac{121.058 \times 1017.61}{20} = 350.0302,$$

$$SSR_u = \frac{350.0302^2}{14.586\,77} = 8399.466,$$

$$SSE_u = 8995.606 - 8399.466 = 596.1403,$$

so the slope of the graph of Fruit against Root for the ungrazed plants is given by

$$b_u = \frac{SSXY_u}{SSX_u} = \frac{350.0302}{14.586\,77} = 23.996.$$

Now add up the regression statistics across the factor levels (grazed and ungrazed):

$$SSY_{g+u} = 11\ 837.79 + 8995.606 = 20\ 833.4,$$

$$SSX_{g+u} = 19.9111 + 14.586\ 77 = 34.497\ 88,$$

$$SSXY_{g+u} = 462.7415 + 350.0302 = 812.7717,$$

$$SSR_{g+u} = 10\ 754.29 + 8399.436 = 19\ 153.75,$$

$$SSE_{g+u} = 1083.509 + 596.1403 = 1684.461$$

The SSR for a model with a single common slope is given by

$$SSR_c = \frac{(SSXY_{g+u})^2}{SSX_{g+u}} = \frac{812.7717^2}{34.497\ 88} = 19\ 148.94,$$

and the value of the single common slope is

$$b = \frac{SSXY_{g+u}}{SSX_{g+u}} = \frac{812.7717}{34.497\ 88} = 23.560.$$

The difference between the two estimates of SSR ($SSR_{\text{diff}} = SSR_{g+u} - SSR_c = 19\ 153.75 - 19\ 148.94 = 4.81$) is a measure of the significance of the difference between the two slopes estimated separately for each factor level. Finally, SSE is calculated by difference:

$$SSE = SSY - SSA - SSR_c - SSR_{\text{diff}}$$
$$= 23\ 743.84 - 2910.44 - 19\ 148.94 - 4.81 = 1679.65.$$

Now we can complete the ANOVA table for the full model:

Source	SS	d.f.	MS	F
Grazing	2910.44	1		
Root	19148.94	1		
Different slopes	4.81	1	4.81	n.s.
Error	1679.65	36	46.66	
Total	23743.84	39		

Degrees of freedom for error are $40 - 4 = 36$ because we have estimated four parameters from the data: two slopes and two intercepts. So the error variance is 46.66 ($= SSE/36$). The difference between the slopes is clearly not significant ($F = 4.81/46.66 = 0.10$) so we can fit a simpler model with a common slope of 23.56. The sum of squares for differences between the slopes (4.81) now becomes part of the error sum of squares:

Source	SS	d.f.	MS	F
Grazing	2910.44	1	2910.44	63.9291
Root	19148.94	1	19148.94	420.6156
Error	1684.46	37	45.526	
Total	23743.84	39		

This is the minimal adequate model. Both of the terms are highly significant and there are no redundant factor levels.

The next step is to calculate the intercepts for the two parallel regression lines. This is done exactly as before, by rearranging the equation of the straight line to obtain $a = y - bx$. For each line we can use the mean values of x and y, with the common slope in each case. Thus:

$$a_1 = \bar{Y}_1 - b\bar{X}_1 = 50.88 - 23.56 \times 6.0529 = -91.7261,$$

$$a_2 = \bar{Y}_2 - b\bar{X}_2 = 67.94 - 23.56 \times 8.309 = -127.8294.$$

This demonstrates that the grazed plants produce, on average, $127.83 - 91.73 = 36.1$ mg of fruit *less* than the ungrazed plants.

Finally, we need to calculate the standard errors for the common regression slope and for the difference in mean fecundity between the treatments, based on the error variance in the minimal adequate model, given in the table above:

$$s^2 = \frac{1684.46}{37} = 45.526.$$

The standard errors are obtained as follows. The standard error of the common slope is found in the usual way:

$$se_b = \sqrt{\frac{s^2}{SSX_{g+u}}} = \sqrt{\frac{45.526}{19.9111 + 14.456\ 67}} = 1.149.$$

The standard error of the intercept of the regression for the grazed treatment is also found in the usual way:

$$se_a = \sqrt{s^2\left[\frac{1}{n} + \frac{(0 - \bar{x})^2}{SSX_{g+u}}\right]} = \sqrt{45.526\left[\frac{1}{20} + \frac{8.3094^2}{34.498}\right]} = 9.664.$$

It is clear that the intercept of -127.829 is very significantly less than zero ($t = 127.829/9.664 = 13.2$), suggesting that there is a threshold rootstock size before reproduction can begin. The standard error of the difference between the elevations of the two lines (the grazing effect) is given by

$$se_{\hat{y}_u - \hat{y}_g} = \sqrt{s^2\left[\frac{2}{n} + \frac{(\bar{x}_1 - \bar{x}_2)^2}{SSX_{g+u}}\right]}$$

which, substituting the values for the error variance and the mean rootstock sizes of the plants in the two treatments, becomes:

$$se_{\hat{y}_u - \hat{y}_g} = \sqrt{45.526\left[\frac{2}{20} + \frac{(6.0529 - 8.3094)^2}{34.498}\right]} = 3.357.$$

This suggests that any lines differing in elevation by more than about $2 \times 3.357 = 6.71$ mg dry weight would be regarded as significantly different. Thus, the present difference of 36.09 clearly represents a highly significant reduction in fecundity caused by grazing ($t = 10.83$).

The hand calculations were convoluted, but ANCOVA is exceptionally straightforward in R, using `lm`. The response variable is fecundity (`Fruit`), and there is one experimental factor (`Grazing`) with two levels (`Ungrazed` and `Grazed`) and one covariate (initial rootstock diameter, `Root`). There are 40 values for each of these variables. As we saw earlier, the largest plants were allocated to the grazed treatments, but for a *given* rootstock diameter (say, 7 mm) the scatterplot shows that the grazed plants produced *fewer* fruits than the ungrazed plants (not more, as a simple comparison of the means suggested). This is an excellent example of the value of analysis of covariance. Here, the correct analysis using ANCOVA completely reverses our interpretation of the data.

The analysis proceeds in the following way. We fit the most complicated model first, then simplify it by removing non-significant terms until we are left with a minimal adequate model, in which all the parameters are significantly different from zero. For ANCOVA, the most complicated model has different slopes and intercepts for each level of the factor. Here we have a two-level factor (`Grazed` and `Ungrazed`) and we are fitting a linear model with two parameters ($y = a + bx$) so the most complicated mode has four parameters (two slopes and two intercepts). To fit different slopes and intercepts we use the asterisk * notation:

```
ancova <- lm(Fruit~Grazing*Root)
```

You should realize that *order matters*: we would get a different output if the model had been written `Fruit ~ Root * Grazing` (more of this on p. 555).

```
summary(ancova)
```

```
Coefficients:
                      Estimate    Std. Error    t value    Pr (>|t|)
(Intercept)           -125.173    12.811        -9.771     1.15e-11    ***
GrazingUngrazed        30.806     16.842         1.829     0.0757      .
Root                   23.240      1.531        15.182     < 2e-16     ***
GrazingUngrazed:Root    0.756      2.354         0.321     0.7500
```

This shows that initial root size has a massive effect on fruit production ($t = 15.182$), but there is no indication of any difference in the slope of this relationship between the two grazing treatments (this is the `Grazing` by `Root` interaction with $t = 0.321$, $p \gg 0.05$). The ANOVA table for the maximal model looks like this:

```
anova(ancova)
```

```
Analysis of Variance Table
Response:        Fruit
               Df     Sum Sq    Mean Sq    F value      Pr(>F)
Grazing         1     2910.4     2910.4    62.3795    2.262e-09    ***
Root            1    19148.9    19148.9   410.4201    < 2.2e-16    ***
Grazing:Root    1        4.8        4.8     0.1031       0.75
Residuals      36     1679.6       46.7
```

The next step is to delete the non-significant interaction term from the model. We can do this manually or automatically: here we shall do both for the purposes of demonstration. The function for manual model simplification is `update`. We update the current model (here called `ancova`) by deleting terms from it. The syntax is important: the punctuation reads 'comma tilde dot minus'. We define a new name for the simplified model:

```
ancova2 <- update(ancova, ~ . - Grazing:Root)
```

Now we compare the simplified model with just three parameters (one slope and two intercepts) with the maximal model using anova like this:

```
anova(ancova,ancova2)
Analysis of Variance Table
Model 1: Fruit ~ Grazing * Root
Model 2: Fruit ~ Grazing + Root
      Res.Df      RSS         Df    Sum of Sq       F           Pr (>F)
1       36      1679.65
2       37      1684.46       -1      -4.81       0.1031        0.75
```

This says that model simplification was justified because it caused a negligible reduction in the explanatory power of the model ($p = 0.75$; to retain the interaction term in the model we would need $p < 0.05$).

The next step in model simplification involves testing whether or not grazing had a significant effect on fruit production once we control for initial root size. The procedure is similar: we define a new model and use update to remove Grazing from ancova2 like this:

```
ancova3 <- update(ancova2, ~ . - Grazing)
```

Now we compare the two models using anova:

```
anova(ancova2,ancova3)
Analysis of Variance Table
Model 1: Fruit ~ Grazing + Root
Model 2: Fruit ~ Root
      Res.Df     RSS        Df    Sum of Sq      F          Pr(>F)
1       37     1684.5
2       38     6948.8       -1     -5264.4      115.63     6.107e-13     ***
```

This model simplification is a step too far. Removing the Grazing term causes a massive reduction in the explanatory power of the model, with an F value of 115.63 and a vanishingly small p value. The effect of grazing in reducing fruit production is highly significant and needs to be retained in the model. Thus ancova2 is our minimal adequate model, and we should look at its summary table to compare with our earlier calculations carried out by hand:

```
summary(ancova2)

Coefficients:
                Estimate    Std. Error    t value    Pr (>| t |)
(Intercept)     -127.829       9.664       -13.23     1.35e-15     ***
GrazingUngrazed   36.103       3.357        10.75     6.11e-13     ***
Root              23.560       1.149        20.51     < 2e-16      ***

Residual standard error: 6.747 on 37 degrees of freedom
Multiple R-Squared: 0.9291, Adjusted R-squared: 0.9252
F-statistic: 242.3 on 2 and 37 DF, p-value: < 2.2e-16
```

You know when you have got the minimal adequate model, because every row of the coefficients table has one or more significance stars (three in this case, because the effects are all so strong). In contrast to our initial interpretation based on mean fruit production, grazing is associated with a 36.103 mg *reduction* in fruit production.

```
anova(ancova2)
```

```
Analysis of Variance Table
Response: Fruit
            Df    Sum Sq     Mean Sq     F value      Pr(>F)
Grazing      1    2910.4     2910.4      63.929       1.397e-09     ***
Root         1    19148.9    19148.9     420.616      < 2.2e-16     ***
Residuals   37    1684.5     45.5
```

These are the values we obtained the long way on p. 540.

Now we repeat the model simplification using the automatic model-simplification function called `step`. It could not be easier to use. The full model is called `ancova`:

```
step(ancova)
```

This function causes all the terms to be tested to see whether they are needed in the minimal adequate model. The criterion used is Akaike's information criterion (AIC, p. 415). In the jargon, this is a 'penalized log-likelihood'. What this means in simple terms is that it weighs up the inevitable trade-off between degrees of freedom and fit of the model. You can have a perfect fit if you have a parameter for every data point, but this model has zero explanatory power. Thus *deviance goes down as degrees of freedom in the model go up*. The AIC adds 2 times the number of parameters in the model to the deviance (to penalize it). Deviance, you will recall, is twice the log-likelihood of the current model. Anyway, AIC is a measure of lack of fit; big AIC is bad, small AIC is good. The full model (four parameters: two slopes and two intercepts) is fitted first, and AIC calculated as 157.5:

```
step(ancova)
```

```
Start:  AIC=157.5
Fruit ~ Grazing * Root

                 Df Sum of Sq     RSS     AIC
- Grazing:Root    1    4.8122  1684.5  155.61
<none>                          1679.7  157.50

Step:  AIC=155.61
Fruit ~ Grazing + Root

                Df Sum of Sq      RSS     AIC
<none>                         1684.5  155.61
- Grazing        1    5264.4   6948.8  210.30
- Root           1   19148.9  20833.4  254.22

Call:
lm(formula = Fruit ~ Grazing + Root)

Coefficients:
   (Intercept)    GrazingUngrazed              Root
       -127.83              36.10             23.56
```

Then `step` tries removing the most complicated term (the `Grazing` by `Root` interaction). This reduces AIC to 155.61 (an improvement, so the simplification is justified). No further simplification is possible (as we saw when we used `update` to remove the `Grazing` term from the model) because AIC goes up to

210.3 when `Grazing` is removed and up to 254.2 if `Root` is removed. Thus, `step` has found the minimal adequate model (it does not always do so, as we shall see later; it is good, but not perfect).

12.2 ANCOVA and experimental design

There is an extremely important general message in this example for experimental design. No matter how carefully we randomize at the outset, our experimental groups are likely to be heterogeneous. Sometimes, as in this case, we may have made initial measurements that we can use as covariates later on, but this will not always be the case. There are bound to be important factors that we did not measure. If we had not measured initial root size in this example, we would have come to entirely the wrong conclusion about the impact of grazing on plant performance.

A far better design for this experiment would have been to measure the rootstock diameters of all the plants at the beginning of the experiment (as was done here), but then to place the plants in matched pairs with rootstocks of similar size. Then, one of the plants would be picked at random and allocated to one of the two grazing treatments (e.g. by tossing a coin); the other plant of the pair then receives the unallocated gazing treatment. Under this scheme, the size ranges of the two treatments would overlap, and the analysis of covariance would be unnecessary.

12.3 ANCOVA with two factors and one continuous covariate

The following experiment, with weight as the response variable, involved genotype and sex as two categorical explanatory variables and age as a continuous covariate. There are six levels of genotype and two levels of sex.

```
Gain <-  read.table("c:\\temp\\Gain.txt",header=T)
attach(Gain)
names(Gain)
```

```
[1]  "Weight"  "Sex"  "Age"  "Genotype"  "Score"
```

We begin by fitting the maximal model with its 24 parameters: there are different slopes and intercepts for every combination of sex and genotype ($2 \times 6 \times 2 = 24$).

```
m1 <-  lm(Weight~Sex*Age*Genotype)
summary(m1)
```

```
Coefficients:
```

	Estimate	Std. Error	t value	Pr(>|t|)	
(Intercept)	7.80053	0.24941	31.276	< 2e-16	***
Sexmale	-0.51966	0.35272	-1.473	0.14936	
Age	0.34950	0.07520	4.648	4.39e-05	***
GenotypeCloneB	1.19870	0.35272	3.398	0.00167	**
GenotypeCloneC	-0.41751	0.35272	-1.184	0.24429	
GenotypeCloneD	0.95600	0.35272	2.710	0.01023	*
GenotypeCloneE	-0.81604	0.35272	-2.314	0.02651	*
GenotypeCloneF	1.66851	0.35272	4.730	3.41e-05	***
Sexmale:Age	-0.11283	0.10635	-1.061	0.29579	
Sexmale:GenotypeCloneB	-0.31716	0.49882	-0.636	0.52891	

Sexmale:GenotypeCloneC	-1.06234	0.49882	-2.130	0.04010	*
Sexmale:GenotypeCloneD	-0.73547	0.49882	-1.474	0.14906	
Sexmale:GenotypeCloneE	-0.28533	0.49882	-0.572	0.57087	
Sexmale:GenotypeCloneF	-0.19839	0.49882	-0.398	0.69319	
Age:GenotypeCloneB	-0.10146	0.10635	-0.954	0.34643	
Age:GenotypeCloneC	-0.20825	0.10635	-1.958	0.05799	.
Age:GenotypeCloneD	-0.01757	0.10635	-0.165	0.86970	
Age:GenotypeCloneE	-0.03825	0.10635	-0.360	0.72123	
Age:GenotypeCloneF	-0.05512	0.10635	-0.518	0.60743	
Sexmale:Age:GenotypeCloneB	0.15469	0.15040	1.029	0.31055	
Sexmale:Age:GenotypeCloneC	0.35322	0.15040	2.349	0.02446	*
Sexmale:Age:GenotypeCloneD	0.19227	0.15040	1.278	0.20929	
Sexmale:Age:GenotypeCloneE	0.13203	0.15040	0.878	0.38585	
Sexmale:Age:GenotypeCloneF	0.08709	0.15040	0.579	0.56616	

```
Residual standard error: 0.2378 on 36 degrees of freedom
Multiple R-squared: 0.9742,     Adjusted R-squared: 0.9577
F-statistic: 59.06 on 23 and 36 DF,  p-value: < 2.2e-16
```

You should work through this output slowly and make sure that you can see what each term means. Remember that the intercept is for all the factor levels that come first in the alphabet (females of CloneA) and the slope (Age) likewise. There is only one intercept and one slope in this entire output. Everything else is either a difference between intercepts, or a difference between slopes. Names involving Age are differences between slopes; names *not* involving Age are differences between intercepts.

There are one or two significant parameters, but it is not at all clear that the three-way or two-way interactions need to be retained in the model. As a first pass, let us use step to see how far it gets with model simplification:

```
m2 <- step(m1)
summary(m2)
```

```
Coefficients:
               Estimate Std. Error t value Pr(>|t|)
(Intercept)     7.93701    0.10066  78.851  < 2e-16 ***
Sexmale        -0.83161    0.05937 -14.008  < 2e-16 ***
Age             0.29958    0.02099  14.273  < 2e-16 ***
GenotypeCloneB  0.96778    0.10282   9.412 8.07e-13 ***
GenotypeCloneC -1.04361    0.10282 -10.149 6.21e-14 ***
GenotypeCloneD  0.82396    0.10282   8.013 1.21e-10 ***
GenotypeCloneE -0.87540    0.10282  -8.514 1.98e-11 ***
GenotypeCloneF  1.53460    0.10282  14.925  < 2e-16 ***
```

We definitely do not need the three-way interaction, despite the effect of Sexmale:Age: Genotype-CloneC which gave a significant *t* test on its own. How about the three two-way interactions? The step function leaves out Sex by Genotype and then assesses the other two. No need for Age by Genotype. Try removing Sex by Age. Nothing. What about the main effects? They are all highly significant. This is R's idea of the minimal adequate model: three main effects but no interactions. That is to say, the slope of the graph of weight gain against age does not vary with sex or genotype, but the intercepts *do* vary.

This is where Helmert contrasts would actually come in handy (see p. 553). Everything is three-star significantly different from Genotype[1] Sex[1], but it is not obvious that the intercepts for genotypes

B and D need different values ($+0.96$ and $+0.82$ above genotype A with $se_{diff} = 0.1028$), nor is it obvious that C and E have different intercepts (-1.043 and -0.875). Perhaps we could reduce the number of factor levels of Genotype from the present six to four without any loss of explanatory power ?

We create a new categorical variable called newGenotype with separate levels for clones A and F, and for B and D combined and C and E combined:

```
newGenotype <- Genotype
levels(newGenotype)
```

```
[1]   "CloneA"   "CloneB"   "CloneC"   "CloneD"   "CloneE"   "CloneF"
```

```
levels(newGenotype)[c(3,5)] <- "ClonesCandE"
levels(newGenotype)[c(2,4)] <- "ClonesBandD"
levels(newGenotype)
```

```
[1]   "CloneA"   "ClonesBandD"   "ClonesCandE"   "CloneF"
```

Then we redo the modelling with newGenotype (four levels) instead of Genotype (six levels),

```
m3 <- lm(Weight~Sex+Age+newGenotype)
```

and check that the simplification was justified:

```
anova(m2,m3)
```

```
Analysis of Variance Table
Model 1: Weight ~ Sex + Age + Genotype
Model 2: Weight ~ Sex + Age + newGenotype
      Res.Df      RSS         Df    Sum of Sq          F     Pr(>F)
1         52    2.74890
2         54    2.99379     -2     -0.24489     2.3163     0.1087
```

Yes, the simplification was justified (p value $= 0.1087$) so we accept the simpler model m3:

```
summary(m3)
```

```
Coefficients:
                          Estimate    Std. Error    t value    Pr(> | t |)
(Intercept)                7.93701       0.10308     76.996      < 2e-16      ***
Sexmale                   -0.83161       0.06080    -13.679      < 2e-16      ***
Age                        0.29958       0.02149     13.938      < 2e-16      ***
newGenotypeClonesBandD     0.89587       0.09119      9.824     1.28e-13      ***
newGenotypeClonesCandE    -0.95950       0.09119    -10.522     1.10e-14      ***
newGenotypeCloneF          1.53460       0.10530     14.574      < 2e-16      ***
```

```
Residual standard error: 0.2355 on 54 degrees of freedom
Multiple R-Squared: 0.962, Adjusted R-squared: 0.9585
F-statistic: 273.7 on 5 and 54 DF, p-value: < 2.2e-16
```

After an analysis of covariance, it is useful to draw the fitted lines through a scatterplot, with each factor level represented by different plotting symbols and line types (see p. 198):

```
plot(Age,Weight,type="n")
```

```
colours <- c("green","red","black","blue")
lines <- c(1,2)
```

```
symbols <- c(16,17)

points(Age,Weight,pch=symbols[as.numeric(Sex)],
                    col=colours[as.numeric(newGenotype)])

xv <- c(1,5)
for (i in 1:2) {
for (j in 1:4) {
        a <- coef(m3)[1]+(i>1)* coef(m3)[2]+(j>1)*coef(m3)[j+2]
        b <- coef(m3)[3]
        yv <- a+b*xv
lines(xv,yv,lty=lines[i],col=colours[j]) } }
```

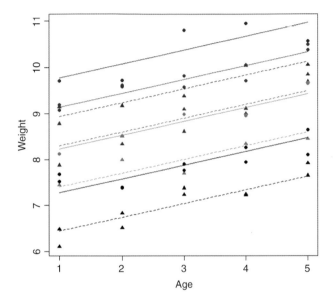

Note the use of colour to represent the four genotypes and plotting symbols and line types to represent the two sexes. You can see that the males (circles and solid lines) are heavier than the females (triangles and dashed lines) in all of the genotypes. Other functions to be considered in plotting the results of ANCOVA are split and augPred in lattice graphics.

12.4 Contrasts and the parameters of ANCOVA models

In analysis of covariance, we estimate a slope and an intercept for each level of one or more factors. Suppose we are modelling weight (the response variable) as a function of sex and age, as illustrated on p. 538. The difficulty arises because there are several different ways of expressing the values of the four parameters in the summary.lm table:

- two slopes, and two intercepts (as in the equations on p. 537);

- one slope and one difference between slopes, and one intercept and one difference between intercepts;

- the overall mean slope and the overall mean intercept, and one difference between slopes and one difference between intercepts.

In the second case (two estimates and two differences) a decision needs to be made about which factor level to associate with the estimate, and which level with the difference (e.g. whether males should be expressed as the intercept and females as the difference between intercepts, or vice versa) .

When the factor levels are unordered (the typical case), then R takes the factor level that comes first in the alphabet as the estimate and the others are expressed as differences. In our example, the parameter estimates would be female, and male parameters would be expressed as differences from the female values, because 'f' comes before 'm' in the alphabet. This should become clear from an example:

```
Ancovacontrasts <-  read.table("c:\\temp\\Ancovacontrasts.txt",header=T)
attach(Ancovacontrasts)
names(Ancovacontrasts)
```

```
[1]  "weight"  "sex"  "age"
```

First we work out the two regressions separately so that we know the values of the two slopes and the two intercepts (note the use of subscripts or subsets to select the sexes):

```
lm(weight[sex=="male"]~age[sex=="male"])
```

```
Coefficients:
     (Intercept)   age[sex == "male"]
           3.115                 1.561
```

```
lm(weight~age,subset=(sex=="female"))
```

```
Coefficients:
(Intercept)             age
     1.9663          0.9962
```

So the intercept for males is 3.115 and the intercept for females is 1.9663. The difference between the first (female) and second (male) intercepts is therefore

$$3.115 - 1.9663 = +1.1487$$

and the difference between the two slopes is

$$1.561 - 0.9962 = +0.5648$$

Now we can do an overall regression, ignoring gender:

```
lm(weight~age)
```

```
Call:
lm(formula = weight ~ age)
```

```
Coefficients:
(Intercept)             age
      2.541           1.279
```

This tells us that the average intercept is 2.541 and the average slope is 1.279.

Next we can carry out an analysis of covariance and compare the output produced by each of the three different **contrast options** allowed by R: treatment (the default in R and Glim), Helmert (the default in S-PLUS), and sum. First, the analysis using **treatment contrasts** as used by R and Glim:

```
options(contrasts=c("contr.treatment", "contr.poly"))
model1 <- lm(weight~age*sex)
summary(model1)
```

```
Coefficients:
              Estimate Std. Error t value Pr(>|t|)
(Intercept)     1.9663     0.6268   3.137  0.00636 **
age             0.9962     0.1010   9.862 3.33e-08 ***
sexmale         1.1489     0.8864   1.296  0.21331
age:sexmale     0.5646     0.1429   3.952  0.00114 **
```

The intercept (1.9663) is the intercept for females (because 'f' comes before 'm' in the alphabet). The age parameter (0.9962) is the slope of the graph of weight against age for females. The sex parameter (1.1489) is the difference between the (female) intercept and the male intercept $(1.9663 + 1.1489 = 3.1152)$. The age–sex interaction term is the difference between slopes of the female and male graphs $(0.9962 + 0.5646 = 1.5608)$. So with treatment contrasts, the parameters (in order 1 to 4) are an intercept, a slope, a difference between two intercepts, and a difference between two slopes. In the standard error column we see, from row 1 downwards, the standard error of an intercept for a regression with females only (0.6268 with $n = 10$, $\Sigma x^2 = 385$ and $SSX = 82.5$), the standard error of a slope for females only (0.1010, with $SSX = 82.5$), the standard error of the difference between two intercepts each based on $n = 10$ data points $(\sqrt{2 \times 0.6268^2} = 0.8864)$ and the standard error of the difference between two slopes each based on $n = 10$ data points $(\sqrt{2 \times 0.1010^2} = 0.1429)$. The formulas for these standard errors are on p. 554. Many people are more comfortable with this method of presentation than they are with Helmert or sum contrasts.

We now turn to the analysis using **Helmert contrasts**:

```
options(contrasts=c("contr.helmert", "contr.poly"))
model2 <- lm(weight~age*sex)
summary(model2)
```

```
Coefficients:
              Estimate Std. Error t value Pr(>|t|)
(Intercept)    2.54073    0.44319   5.733 3.08e-05 ***
age            1.27851    0.07143  17.899 5.26e-12 ***
sex1           0.57445    0.44319   1.296  0.21331
age:sex1       0.28230    0.07143   3.952  0.00114 **
```

Let us see if we can work out what the four parameter values represent. The first parameter, 2.540 73 (labelled `Intercept`), is the intercept of the overall regression, ignoring sex (see above). The parameter labelled `age` (1.278 51) is a *slope* because age is our continuous explanatory variable. Again, you will see that it is the slope for the regression of weight against age, ignoring sex. The third parameter, labelled `sex1` (0.574 45), must have something to do with intercepts because sex is our categorical variable. If we want to reconstruct the second intercept (for males) we need to add 0.574 45 to the overall intercept: $2.540\ 73 + 0.574\ 45 = 3.115\ 18$. To get the intercept for females we need to subtract it: $2.540\ 73 - 0.574\ 45 = 1.966\ 28$. The fourth parameter (0.282 30), labelled `age:sex1`, is the difference between the overall mean slope (1.278 51) and the male slope: $1.278\ 51 + 0.282\ 30 = 1.560\ 81$. To get the slope of weight against age for females we need to subtract the interaction term from the age term: $1.278\ 51 - 0.282\ 30 = 0.996\ 21$.

In the standard errors column, from the top row downwards, you see the standard error of an intercept based on a regression with all 20 points (the overall regression, ignoring sex, 0.443 19) and the standard error of a slope based on a regression with all 20 points (0.071 43). The standard errors of differences (both intercept and slope) involve *half* the difference between the male and female values, because with Helmert contrasts the difference is between the male value and the overall value, rather than between the male and female values. Thus the third row has the standard error of a difference between the overall intercept and the intercept for males based on a regression with 10 points (0.443 19 = 0.8864/2), and the bottom row has the standard error of a difference between the overall slope and the slope for males, based on a regression with 10 points (0.1429/2 = 0.071 43). Thus the values in the bottom two rows of the Helmert table are simply half the values in the same rows of the treatment table.

The advantage of Helmert contrasts is in hypothesis testing in more complicated models than this, because it is easy to see which terms we need to retain in a simplified model by inspecting their significance levels in the summary.lm table. The disadvantage is that it is much harder to reconstruct the slopes and the intercepts from the estimated parameters values (see also p. 440).

Finally, we look at the third option, which is **sum contrasts**:

```
options(contrasts=c("contr.sum", "contr.poly"))
model3 <- lm(weight~age*sex)
summary(model3)
```

```
Coefficients:
             Estimate Std. Error t value Pr(>|t|)
(Intercept)   2.54073    0.44319   5.733 3.08e-05 ***
age           1.27851    0.07143  17.899 5.26e-12 ***
sex1         -0.57445    0.44319  -1.296  0.21331
age:sex1     -0.28230    0.07143  -3.952  0.00114 **
```

The first two estimates are the same as those produced by Helmert contrasts: the overall intercept and slope of the graph relating weight to age, ignoring sex. The sex parameter (–0.574 45) is *sign reversed* compared with the Helmert option: it shows how to calculate the female (the *first*) intercept from the overall intercept 2.540 73– 0.574 45 = 1.966 28. The interaction term also has reversed sign: to get the slope for females, add the interaction term to the slope for age: 1.278 51– 0.282 30 = 0.996 21.

The four standard errors for the sum contrasts are exactly the same as those for Helmert contrasts (explained above).

12.5 Order matters in summary.aov

People are often disconcerted by the ANOVA table produced by summary.aov in analysis of covariance. Compare the tables produced for these two models:

```
summary.aov(lm(weight~sex*age))
```

```
          Df Sum Sq Mean Sq F value   Pr(>F)
sex        1  90.49   90.49 107.50 1.66e-08 ***
age        1 269.71  269.71 320.39 5.26e-12 ***
sex:age    1  13.15   13.15  15.62  0.00114 **
Residuals 16  13.47    0.84
```

```
summary.aov(lm(weight~age*sex))
```

	Df	Sum Sq	Mean Sq	F value	Pr(>F)	
age	1	269.71	269.71	320.39	5.26e-12	***
sex	1	90.49	90.49	107.50	1.66e-08	***
age:sex	1	13.15	13.15	15.62	0.00114	**
Residuals	16	13.47	0.84			

Exactly the same sums of squares and *p* values. No problem. But look at these two models from the plant compensation example analysed in detail earlier (p. 545):

```
summary.aov(lm(Fruit~Grazing*Root))
```

	Df	Sum Sq	Mean Sq	F value	Pr(>F)	
Grazing	1	2910	2910	62.380	2.26e-09	***
Root	1	19149	19149	410.420	< 2e-16	***
Grazing:Root	1	5	5	0.103	0.75	
Residuals	36	1680	47			

```
summary.aov(lm(Fruit~Root*Grazing))
```

	Df	Sum Sq	Mean Sq	F value	Pr(>F)	
Root	1	16795	16795	359.968	< 2e-16	***
Grazing	1	5264	5264	112.832	1.21e-12	***
Root:Grazing	1	5	5	0.103	0.75	
Residuals	36	1680	47			

In this case the order of variables within the model formula has a huge effect: it changes the sum of squares associated with the two main effects (root size is continuous and grazing is categorical, grazed or ungrazed) and alters their *p* values. The interaction term, the residual sum of squares and the error variance are unchanged. So what is the difference between the two cases?

In the first example, where order was irrelevant, the *x* values for the continuous variable (age) were identical for both sexes (there is one male and one female value at each of the 10 experimentally controlled ages). In the second example, the *x* values (root size) were different in the two treatments, and mean root size was greater for the grazed plants than for the ungrazed ones:

```
tapply(Root,Grazing, mean)
```

Grazed	Ungrazed
8.3094	6.0529

Whenever the *x* values are different in different factor levels, and/or there is different replication in different factor levels, then *SSX* and *SSXY* will vary from level to level and this will affect the way the sum of squares is distributed across the main effects. It is of no consequence in terms of your interpretation of the model, however, because the effect sizes and standard errors in the summary.lm table are completely unaffected:

```
summary(lm(Fruit~Root*Grazing))
```

Coefficients:

	Estimate	Std. Error	t value	Pr(>|t|)	
(Intercept)	-109.770	8.421	-13.035	3.36e-15	***
Root	23.618	1.177	20.066	< 2e-16	***
Grazing1	-15.403	8.421	-1.829	0.0757	.
Root:Grazing1	-0.378	1.177	-0.321	0.7500	

```
summary(lm(Fruit~Grazing*Root))
```

Coefficients:

	Estimate	Std. Error	t value	Pr(>|t|)	
(Intercept)	-109.770	8.421	-13.035	3.36e-15	***
Grazing1	-15.403	8.421	-1.829	0.0757	.
Root	23.618	1.177	20.066	< 2e-16	***
Grazing1:Root	-0.378	1.177	-0.321	0.7500	

The p values that are shown in summary.lm are the deletion p values that we are accustomed to. The p values in the full ANOVA table, however, are not. Once the interaction term is deleted, the problem disappears, and the p values, once again, are the familiar deletion p values.

13

Generalized Linear Models

We can use generalized linear models (GLMs) – pronounced 'glims' – when the variance is not constant, and/or when the errors are not normally distributed. Certain kinds of response variables invariably suffer from these two important contraventions of the standard assumptions, and GLMs are excellent at dealing with them. Specifically, we might consider using GLMs when the response variable is:

- count data expressed as proportions (e.g. logistic regressions);
- count data that are not proportions (e.g. log-linear models of counts);
- binary response variables (e.g. dead or alive);
- data on time to death where the variance increases faster than linearly with the mean (e.g. time data with gamma errors).

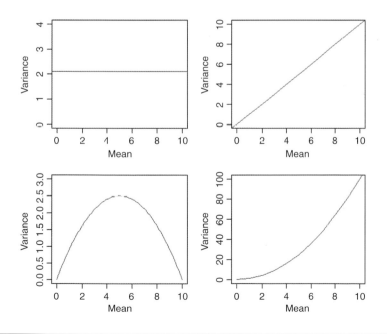

The R Book, Second Edition. Michael J. Crawley.
© 2013 John Wiley & Sons, Ltd. Published 2013 by John Wiley & Sons, Ltd.

The central assumption that we have made up to this point is that variance was constant (top left-hand graph). In count data, however, where the response variable is an integer and there are often lots of zeros in the dataframe, the variance may increase linearly with the mean (top tight). With proportion data, where we have a count of the number of failures of an event as well as the number of successes, the variance will be an inverted U-shaped function of the mean (bottom left). Where the response variable follows a gamma distribution (as in time-to-death data) the variance increases faster than linearly with the mean (bottom right). Many of the basic statistical methods such as regression and Student's t test assume that variance is constant, but in many applications this assumption is untenable. Hence the great utility of GLMs.

A GLM has three important properties:

- the error structure;

- the linear predictor;

- the link function.

These are all likely to be unfamiliar concepts. The ideas behind them are straightforward, however, and it is worth learning what each of the concepts involves.

13.1   Error structure

Up to this point, we have dealt with the statistical analysis of data with normal errors. In practice, however, many kinds of data have non-normal errors: for example:

- errors that are strongly skewed;

- errors that are kurtotic;

- errors that are strictly bounded (as in proportions);

- errors that cannot lead to negative fitted values (as in counts).

In the past, the only tools available to deal with these problems were transformation of the response variable or the adoption of non-parametric methods. A GLM allows the specification of a variety of different error distributions:

- Poisson errors, useful with count data;

- binomial errors, useful with data on proportions;

- gamma errors, useful with data showing a constant coefficient of variation;

- exponential errors, useful with data on time to death (survival analysis).

The **error structure** is defined by means of the `family` directive, used as part of the model formula. Examples are

```
glm(y ~ z, family = poisson)
```

which means that the response variable y has Poisson errors, and

```
glm(y ~ z, family = binomial)
```

which means that the response is binary, and the model has binomial errors. As with previous models, the explanatory variable z can be continuous (leading to a regression analysis) or categorical (leading to an ANOVA-like procedure called analysis of deviance, as described below).

13.2 Linear predictor

The structure of the model relates each observed y value to a predicted value. The predicted value is obtained by transformation of the value emerging from the **linear predictor**. The linear predictor, η (eta), is a linear sum of the effects of one or more explanatory variables, x_j,

$$\eta_i = \sum_{j=1}^{p} x_{ij}\beta_j,$$

where the xs are the values of the p different explanatory variables, and the βs are the (usually) unknown parameters to be estimated from the data. The right-hand side of the equation is called the **linear structure**.

There are as many terms in the linear predictor as there are parameters, p, to be estimated from the data. Thus, with a simple regression, the linear predictor is the sum of two terms whose parameters are the intercept and the slope. With a one-way ANOVA with four treatments, the linear predictor is the sum of four terms leading to the estimation of the mean for each level of the factor. If there are covariates in the model, they add one term each to the linear predictor (the slope of each relationship). Interaction terms in a factorial ANOVA add one or more parameters to the linear predictor, depending upon the degrees of freedom of each factor (e.g. there would be three extra parameters for the interaction between a two-level factor and a four-level factor, because $(2-1) \times (4-1) = 3$).

To determine the fit of a given model, a GLM evaluates the linear predictor for each value of the response variable, then compares the predicted value with a *transformed* value of y. The transformation to be employed is specified in the link function, as explained below. The fitted value is computed by applying the reciprocal of the link function, in order to get back to the original scale of measurement of the response variable.

13.3 Link function

One of the difficult things to grasp about GLMs is the relationship between the values of the response variable (as measured in the data and predicted by the model in fitted values) and the linear predictor. The thing to remember is that the **link function** relates the mean value of y to its linear predictor. In symbols, this means that

$$\eta = g(\mu),$$

which is simple, but needs thinking about. The linear predictor, η, emerges from the linear model as a sum of the terms for each of the p parameters. This is not a value of y (except in the special case of the **identity link** that we have been using (implicitly) up to now). The value of η is obtained by transforming the value of y by the link function, and the predicted value of y is obtained by applying the inverse link function to η.

The most frequently used link functions are shown below. An important criterion in the choice of link function is to ensure that the fitted values stay within reasonable bounds. We would want to ensure, for example, that counts were all greater than or equal to 0 (negative count data would be nonsense). Similarly, if the response variable was the proportion of individuals that died, then the fitted values would have to lie between 0 and 1 (fitted values greater than 1 or less than 0 would be meaningless). In the first case, a log link is appropriate because the fitted values are antilogs of the linear predictor, and all antilogs are greater than or equal to 0. In the second case, the logit link is appropriate because the fitted values are calculated as the antilogs of the log odds, $\log(p/q)$.

By using different link functions, the performance of a variety of models can be compared directly. The total deviance is the same in each case, and we can investigate the consequences of altering our assumptions about precisely how a given change in the linear predictor brings about a response in the fitted value of y. The most appropriate link function is the one which produces the minimum residual deviance.

13.3.1 Canonical link functions

The canonical link functions are the default options employed when a particular error structure is specified in the `family` directive in the model formula. Omission of a `link` directive means that the following settings are used:

Error	Canonical link
normal	identity
poisson	log
binomial	logit
Gamma	reciprocal

You should try to memorize these canonical links and to understand why each is appropriate to its associated error distribution. Note that only gamma errors have a capital initial letter in R.

Choosing between using a link function (e.g. log link) and transforming the response variable (i.e. having $\log(y)$ as the response variable rather than y) takes a certain amount of experience. The decision is usually based on *whether the variance is constant* on the original scale of measurement. If the variance was constant, you would use a link function. If the variance increased with the mean, you would be more likely to log-transform the response.

13.4 Proportion data and binomial errors

Proportion data have three important properties that affect the way the data should be analysed:

- the data are strictly bounded;
- the variance is non-constant;
- errors are non-normal.

You cannot have a proportion greater than 1 or less than 0. This has obvious implications for the kinds of functions fitted and for the distributions of residuals around these fitted functions. For example, it makes no sense to have a linear model with a negative slope for proportion data because there would come a point, with high levels of the x variable, where negative proportions would be predicted. Likewise, it makes no sense to have a linear model with a positive slope for proportion data because there would come a point, with high levels of the x variable, where proportions greater than 1 would be predicted.

With proportion data, if the probability of success is 0, then there will be no successes in repeated trials, all the data will be zeros and hence the variance will be zero. Likewise, if the probability of success is 1, then there will be as many successes as there are trials, and again the variance will be 0. For proportion data, therefore, the variance increases with the mean up to a maximum (when the probability of success is 0.5) then declines again towards zero as the mean approaches 1. The variance–mean relationship is humped, rather than constant as assumed in the classical tests.

The final assumption is that the errors (the differences between the data and the fitted values estimated by the model) are normally distributed. This cannot be so in proportional data because the data are bounded above and below: no matter how big a negative residual might be at high predicted values, $\hat{y}$, a positive residual cannot be bigger than $1 - \hat{y}$. Similarly, no matter how big a positive residual might be for low predicted values $\hat{y}$, a negative residual cannot be greater than $\hat{y}$ (because you cannot have negative proportions). This means that confidence intervals must be asymmetric whenever $\hat{y}$ takes large values (close to 1) or small values (close to 0).

All these issues (boundedness, non-constant variance, non-normal errors) are dealt with by using a generalized linear model with a binomial error structure. It could not be simpler to deal with this. Instead of using a linear model and writing

```
lm(y~x)
```

we use a generalized linear model and specify that the error family is binomial like this:

```
glm(y~x,family=binomial)
```

That's all there is to it. In fact, it is even easier than that, because we do not even need to write `family=`:

```
glm(y~x,binomial)
```

13.5   Count data and Poisson errors

Count data have a number of properties that need to be considered during modelling:

- Count data are bounded below (you cannot have counts less than zero).

- Variance is not constant (variance increases with the mean).

- Errors are not normally distributed.

- The fact that the data are whole numbers (integers) affects the error distribution.

It is very simple to deal with all these issues by using a GLM. All we need to write is

```
glm(y~x,poisson)
```

and the model is fitted with a log link (to ensure that the fitted values are bounded below) and Poisson errors (to account for the non-normality). If, having fitted the minimal adequate model, we discover that the residual deviance is greater than the residual degrees of freedom, then we have contravened an important assumption of the model. This is called overdispersion, and we can correct for it by specifying `quasipoisson` errors like this:

```
glm(y~x,quasipoisson)
```

It is important to understand that Poisson errors are an assumption, not a fact. Many of the count data you encounter in practice will have variance–mean ratios greater than 1, and in these cases you will need to correct for overdispersion.

13.6   Deviance: Measuring the goodness of fit of a GLM

The fitted values produced by the model are most unlikely to match the values of the data perfectly. The size of the discrepancy between the model and the data is a measure of the inadequacy of the model; a small discrepancy may be tolerable, but a large one will not be. The measure of discrepancy in a GLM to assess the goodness of fit of the model to the data is called the **deviance**. Deviance is defined as –2 times the difference in log-likelihood between the current model and a saturated model (i.e. a model that fits the data perfectly). Because the latter does not depend on the parameters of the model, minimizing the deviance is the same as maximizing the likelihood.

Deviance is estimated in different ways for different families within glm (Table 13.1). Numerical examples of the calculation of deviance for different glm families are given in Chapters 14 (Poisson errors), 16 (binomial errors), and 27 (gamma errors). Where there is grouping structure in the data, leading to spatial or temporal pseudoreplication, you will want to use generalized mixed models (lmer) with one of these error families (p. 710).

13.7   Quasi-likelihood

The precise relationship between the variance and the mean is well established for all the GLM error families (Table 13.1). In some cases, however, we may be uneasy about specifying the precise form of the error distribution. We may know, for example, that it is not normal (e.g. because the variance increases with the mean), but we do not know with any confidence that the underlying distribution is, say, negative binomial.

There is a very simple and robust alternative known as quasi-likelihood, introduced by Wedderburn (1974), which uses only the most elementary information about the response variable, namely the variance–mean relationship (see Taylor's power law, p. 262). It is extraordinary that this information alone is often sufficient to retain close to the full efficiency of maximum likelihood estimators.

Suppose that we know that the response is always positive, the data are invariably skew to the right, and the variance increases with the mean. This does not enable us to specify a particular distribution (e.g. it does not discriminate between Poisson or negative binomial errors), and hence we cannot use techniques like maximum likelihood or likelihood ratio tests. Quasi-likelihood frees us from the need to specify a particular

Table 13.1. Deviance formulae for different GLM families, where y is observed data, $\bar{y}$ the mean value of y, μ are the fitted values of y from the maximum likelihood model, and n is the binomial denominator in a binomial GLM.

Family (error structure)	Deviance	Variance function
gaussian	$\sum (y - \bar{y})^2$	1
poisson	$2 \sum y \ln(y/\mu) - (y - \mu)$	μ
binomial	$2 \sum y \ln(y/\mu) + (n - y) \ln(n - y)/(n - \mu)$	$\dfrac{\mu(n - \mu)}{n}$
Gamma	$2 \sum (y - \mu)/y - \ln(y/\mu)$	μ^2
inverse.gaussian	$\sum (y - \mu)^2/(\mu^2 y)$	μ^3

distribution, and requires us only to specify the mean–variance relationship up to a proportionality constant, which can be estimated from the data:

$$\text{var}(y_i) \propto v(\mu_i).$$

An example of the principle at work compares quasi-likelihood with maximum likelihood in the case of Poisson errors (full details are in McCulloch and Searle, 2001). This means that the maximum quasi-likelihood (MQL) equations for β are

$$\frac{\partial}{\partial \beta} \sum (y_i \log \mu_i - \mu_i) = 0,$$

exactly the same as the maximum likelihood equation for the Poisson (see p. 314). In this case, MQL and maximum likelihood give precisely the same estimates, and MQL would therefore be fully efficient. Other cases do not work out quite as elegantly as this, but MQL estimates are generally robust and efficient. Their great virtue is the simplicity of the central premise that $\text{var}(y_i) \propto v(\mu_i)$, and the lack of the need to assume a specific distributional form.

If we take the original GLM density function and find the derivative of the log-likelihood with respect to the mean,

$$\frac{\partial l_i}{\partial \mu_i} = \frac{\partial l_i}{\partial \theta_i} \frac{\partial \theta_i}{\partial \mu_i} = \frac{y_i b'(\theta_i)}{a_i(\phi)} \frac{1}{b''(\theta_i)} = \frac{y_i - \mu_i}{\text{var}(y_i)}$$

(where the primes denote differentiation), the quasi-likelihood Q is defined as

$$Q(y, \mu) = \int_y^\mu \frac{y - \mu}{\phi V(\mu)} d\mu.$$

Here, the denominator is the variance of y, $\text{var}(y) = \phi V(\mu)$, where ϕ is called the scale parameter (or the dispersion parameter) and $V(\mu)$ is the variance function. We need only specify the two moments (mean μ and variance $\phi V(\mu)$) and maximize Q to find the MQL estimates of the parameters.

The scale parameter is estimated from the generalized Pearson statistic rather than from the residual deviance (as when correcting for overdispersion with Poisson or binomial errors):

$$\hat{\phi} = \frac{\sum_i \{(y_i - \hat{\mu}_i)/V_i(\hat{\mu}_i)\}}{n - p} = \frac{\chi^2}{n - p}.$$

For normally distributed data, the residual sum of squares SSE is chi-squared distributed.

13.8   The quasi family of models

Instead of transforming the response variable in different ways, we can analyse the *untransformed* response variable but specify different link functions. This has the considerable advantage that the models are comparable by anova because they all have the same response variable. Recall that if we transform the response variable in different ways (e.g. log in one case, cube root in another case) then we cannot compare the resulting models using `anova`. Here are the timber data that we analysed on p. 401:

```
data<-read.table("c:\\temp\\timber.txt",header=T)
attach(data)
head(data)
```

```
  volume girth height
1 0.7458 66.23   21.0
2 0.7458 68.62   19.5
3 0.7386 70.22   18.9
4 1.1875 83.79   21.6
5 1.3613 85.38   24.3
6 1.4265 86.18   24.9
```

Models 1 and 2 fit the same explanatory variables,

```
volume~girth+height
```

but use a different link function in each case: `model1` uses the cube root link (this is specified as `power(0.333)`), `model2` uses the `log` link. These different link functions are fitted in `glm` using the `quasi` family (this stands for 'quasi-likelihood'):

```
model1 <- glm(volume~girth+height,family=quasi(link=power(0.3333)))
model2 <- glm(volume~girth+height,family=quasi(link=log))
```

The beauty of this approach is that we can now compare the models using `anova` because they both have the same response variable (i.e. untransformed volume):

```
anova(model1,model2)

Analysis of Deviance Table

Model 1: volume ~ girth + height
Model 2: volume ~ girth + height
  Resid. Df Resid. Dev Df Deviance
1        28    0.96554
2        28    1.42912  0 -0.46357
```

Both models have the same number of degrees of freedom, so there is no F test to do, but the `anova` function shows us the difference in deviance between the two models. The cube root link is better than the log link (residual deviance of 0.966 compared with 1.429). Note that the `anova` table gives us no indication of how the two models differ from one another in their link functions, so we need to be very careful in our book-keeping. We should assess the constancy-of-variance and normality-of-errors assumptions of the two models:

```
plot(model1)
```

```
plot(model2)
```

It is clear that `model1` much better behaved in terms of its assumptions as well as having the lower deviance.

13.9   Generalized additive models

Generalized additive models (GAMs) are like GLMs in that they can have different error structures and different link functions to deal with count data or proportion data. What makes them different is that the shape of the relationship between *y* and a continuous variable *x* is not specified by some explicit functional form. Instead, non-parametric smoothers are used to describe the relationship. This is especially useful for relationships that exhibit complicated shapes, such as hump-shaped curves (see p. 675). The model looks just like a GLM, except that the relationships we want to be smoothed are prefixed by s: thus, if we had a three-variable multiple regression (with three continuous explanatory variables *w*, *x* and *z*) on count data and we wanted to smooth all three explanatory variables, we would write:

```
model <- gam(y~s(w)+s(x)+s(z),poisson)
```

These are hierarchical models, so the inclusion of a high-order interaction (such as *A:B:C*) necessarily implies the inclusion of all the lower-order terms marginal to it (i.e. *A:B*, *A:C* and *B:C*, along with main effects for *A*, *B* and *C*).

Because the models are nested, the more complicated model will necessarily explain at least as much of the variation as the simpler model (and usually more). What we want to know is whether the extra parameters in the more complex model are justified in the sense that they add significantly to the models explanatory power. If they do not, then parsimony requires that we accept the simpler model.

13.10   Offsets

An **offset** is a component of the linear predictor that is known in advance (typically from theory, or from a mechanistic model of the process) and, because it is known, requires no parameter to be estimated from the data. For linear models with normal errors an offset is redundant, since you can simply subtract the offset from the values of the response variable, and work with the residuals instead of the y values. For GLMs, however, it is necessary to specify the offset; this is held constant while other explanatory variables are evaluated. Here is an example from the famous timber data.

The background theory is simple. We assume the logs are roughly cylindrical (i.e. that taper is negligible between the bottom and the top of the log). Then volume, v, in relation to girth, g, and height, h, is given by

$$v = \frac{g^2}{4\pi} h.$$

Taking logarithms gives

$$\log(v) = \log\left(\frac{1}{4\pi}\right) + 2\log(g) + \log(h).$$

We would expect, therefore, that if we did a multiple linear regression of $\log(v)$ on $\log(h)$ and $\log(g)$ we would get estimated slopes of 1.0 for $\log(h)$ and 2.0 for $\log(g)$. Let us see what happens:

```
data <- read.delim("c:\\temp\\timber.txt")
attach(data)
names(data)

[1] "volume" "girth" "height"
```

The girths are in centimetres but all the other data are in metres, so we convert the girths to metres at the outset:

```
girth <- girth/100
```

Now fit the model:

```
model1 <- glm(log(volume)~log(girth)+log(height))
summary(model1)

Coefficients:
            Estimate Std. Error t value Pr(>|t|)
(Intercept) -2.89938    0.63767   -4.547 9.56e-05 ***
log(girth)   1.98267    0.07503   26.426  < 2e-16 ***
log(height)  1.11714    0.20448    5.463 7.83e-06 ***
```

The estimates are reasonably close to expectation (1.117 14 rather than 1.0 for $\log(h)$ and 1.982 67 rather than 2.0 for $\log(g)$).

Now we shall use offset to specify the theoretical response of log(*v*) to log(*h*), i.e. a slope of 1.0 rather than the estimated 1.117 14:

```
model2 <- glm(log(volume)~log(girth)+offset(log(height)))
summary(model2)
```

```
Coefficients:
            Estimate Std. Error t value Pr(>|t|)
(Intercept) -2.53419    0.01457  -174.0   <2e-16 ***
log(girth)   2.00545    0.06287    31.9   <2e-16 ***
```

Naturally the residual deviance is greater, but only by a very small amount. The AIC has gone down from –62.697 to –64.336, so the model simplification was justified.

Let us try including the theoretical slope (2.0) for log(*g*) in the offset as well:

```
model3 <- glm(log(volume)~1+offset(log(height)+2*log(girth)))
summary(model3)
```

```
Coefficients:
            Estimate Std. Error t value Pr(>|t|)
(Intercept) -2.53403    0.01421  -178.3   <2e-16 ***

(Dispersion parameter for gaussian family taken to be 0.006259057)

    Null deviance: 0.18777  on 30  degrees of freedom
Residual deviance: 0.18777  on 30  degrees of freedom
AIC: -66.328
```

Again, the residual deviance is only marginally greater, and AIC is smaller, so the simplification is justified.

What about the intercept? If our theoretical model of cylindrical logs is correct then the intercept should be

```
log(1/(4*pi))
```

```
[1] -2.531024
```

This is almost exactly the same as the intercept estimated by GLM in model3, so we are justified in putting the entire model in the offset and informing GLM not to estimate an intercept from the data (y ~ -1):

```
model4 <- glm(log(volume) ~ offset(log(1/(4*pi))+log(height)+2*log(girth))-1)
summary(model4)
```

```
Call:
glm(formula = log(volume) ~ offset(log(1/(4 * pi)) + log(height) +
    2 * log(girth)) - 1)

Deviance Residuals:
      Min        1Q     Median        3Q       Max
 -0.171513  -0.050378  -0.006514  0.063356  0.133482

No Coefficients

(Dispersion parameter for gaussian family taken to be 0.006066198)

    Null deviance: 0.18805  on 31  degrees of freedom
Residual deviance: 0.18805  on 31  degrees of freedom
AIC: -68.282
```

This is a rather curious model with no estimated parameters, but it has a residual deviance of just 0.188 05 (compared with `model1`, where all three parameters were estimated from the data, which had a deviance of 0.185 55). Because we were saving one degree of freedom with each step in the procedure, AIC became smaller with each step, justifying all of the model simplifications. These logs are cylindrical, not tapered as alleged in some analyses of these data.

13.11 Residuals

After fitting a model to data, we should investigate how well the model describes the data. In particular, we should look to see if there are any systematic trends in the goodness of fit. For example, does the goodness of fit increase with the observation number, or is it a function of one or more of the explanatory variables? We can work with the raw residuals:

$$\text{residuals} = \text{response variable} - \text{fitted values}.$$

With normal errors, the identity link, equal weights and the default scale factor, the raw and standardized residuals are identical. The standardized residuals are required to correct for the fact that with non-normal errors (like count or proportion data) we violate the fundamental assumption that the variance is constant (p. 490) because the residuals tend to change in size as the mean value the response variable changes.

For Poisson errors, the standardized residuals are

$$\frac{y - \text{fitted values}}{\sqrt{\text{fitted values}}}.$$

For binomial errors they are

$$\frac{y - \text{fitted values}}{\sqrt{\text{fitted values} \times \left[1 - \dfrac{\text{fitted values}}{\text{binomial denominator}}\right]}}$$

where the binomial denominator is the size of the sample from which the y successes were drawn. For gamma errors they are

$$\frac{y - \text{fitted values}}{\text{fitted values}}.$$

In general, we can use several kinds of standardized residuals

$$\text{standardized residuals} = (y - \text{fitted values})\sqrt{\frac{\text{prior weight}}{\text{scale parameter} \times \text{variance funcion}}}.$$

where the prior weights are optionally specified by you to give individual data points more or less influence (see p. 463), the scale parameter measures the degree of overdispersion (see p. 592), and the variance function describes the relationship between the variance and the mean (e.g. equality for a Poisson process; see Table 13.1).

13.11.1 Misspecified error structure

A common problem with real data is that the variance increases with the mean. The assumption in previous chapters has been of normal errors with constant variance at all values of the response variable. For continuous measurement data with non-constant errors we can specify a generalized linear model with gamma errors. These are discussed in Chapter 27 along with worked examples, and we need only note at this stage that they assume a **constant coefficient of variation** (see Taylor's power law, p. 262).

With count data, we often assume Poisson errors, but the data may exhibit overdispersion (see below and p. 592), so that the variance is actually greater than the mean (rather than equal to it, as assumed by the Poisson distribution). An important distribution for describing aggregated data is the *negative binomial*. While R has no direct facility for specifying negative binomial errors, we can use quasi-likelihood to specify the variance function in a GLM with `family = quasi` (see p. 564).

13.11.2 Misspecified link function

Although each error structure has a canonical link function associated with it (see p. 560), it is quite possible that a different link function would give a better fit for a particular model specification. For example, in a GLM with normal errors we might try a log link or a reciprocal link using `quasi` to improve the fit (for examples, see p. 563). Similarly, with binomial errors we might try a complementary log-log link instead of the default logit link function (see p. 651).

An alternative to changing the link function is to transform the values of the response variable. The important point to remember here is that changing the scale of y will alter the error structure and this will make it difficult to compare one model with another. Thus, if you take logs of y and carry out regression with normal errors, then you will be assuming that the errors in y were lognormally distributed. This may well be a sound assumption, but a bias will have been introduced if the errors really were additive on the original scale of measurement. If, for example, theory suggests that there is an exponential relationship between y and x,

$$y = ae^{bx},$$

then it would be reasonable to suppose that the log of y would be linearly related to x:

$$\ln y = \ln a + bx.$$

Now suppose that the errors ε in y are multiplicative with a mean of 0 and constant variance, like this:

$$y = ae^{bx}(1 + \varepsilon)$$

Then they will also have a mean of 0 in the transformed model. But if the errors are additive,

$$y = ae^{bx} + \varepsilon,$$

then the error variance in the transformed model will depend upon the expected value of y. In a case like this, it is much better to analyse the untransformed response variable and to employ the log link function, because this retains the assumption of additive errors.

When both the error distribution and functional form of the relationship are unknown, there is no single specific rationale for choosing any given transformation in preference to another. The aim is pragmatic, namely to find a transformation that gives:

- constant error variance;

- approximately normal errors;

- additivity;

- a linear relationship between the response variables and the explanatory variables;

- straightforward scientific interpretation.

The choice is bound to be a compromise and, as such, is best resolved by quantitative comparison of the deviance produced under different model forms (see p. 270).

13.12   Overdispersion

Overdispersion is the polite statistician's version of Murphy's law: if something can go wrong, it will. Overdispersion can be a problem when working with Poisson or binomial errors, and tends to occur because you have not measured one or more of the factors that turn out to be important. It may also result from the underlying distribution being non-Poisson or non-binomial. This means that the probability you are attempting to model is not constant within each cell, but behaves like a random variable. This, in turn, means that the residual deviance is inflated. In the worst case, all the predictor variables you have measured may turn out to be unimportant so that you have no information at all on any of the genuinely important predictors. In this case, the minimal adequate model is just the overall mean, and all your 'explanatory' variables provide no extra information.

The techniques of dealing with overdispersion are discussed in detail when we consider Poisson errors (p. 592) and binomial errors (p. 664). Here it is sufficient to point out that there are two general techniques available to us:

- use *F* tests with an empirical scale parameter instead of chi-squared;

- use quasi-likelihood to specify a more appropriate variance function.

It is important, however, to stress that these techniques introduce another level of uncertainty into the analysis. Overdispersion happens for real, scientifically important reasons, and these reasons may throw doubt upon our ability to interpret the experiment in an unbiased way. It means that something we did not measure turned out to have an important impact on the results. If we did not measure this factor, then we have no confidence that our randomization process took care of it properly and we may have introduced an important bias into the results.

13.13   Bootstrapping a GLM

There are two contrasting ways of using bootstrapping with statistical models:

- Fit the model lots of times by selecting cases for inclusion at random with replacement, so that some data points are excluded and others appear more than once in any particular model fit.

- Fit the model once and calculate the residuals and the fitted values, then shuffle the residuals lots of times and add them to the fitted values in different permutations, fitting the model to the many different data sets.

In both cases, you will obtain a distribution of parameter values for the model from which you can derive confidence intervals. Here we use the timber data (a multiple regression with two continuous explanatory variables, introduced on p. 566) to illustrate the two approaches (see p. 349 for an introduction to the bootstrap).

```
library(boot)
```

The GLM model with its parameter estimates and standard errors is on p. 566. The hard part of using `boot` is writing the sampling function correctly. It has at least two arguments: the first *must* be the data on which the resampling is to be carried out (in this case, the whole dataframe called `trees`), and the second *must* be the index (the randomized subscripts showing which data values are to be used in a given realization; some cases will be repeated, others will be omitted). Inside the function we create a new dataframe based on the randomly selected indices, then fit the model to this new data set. Finally, the function should return the coefficients of the model. Here is the 'statistic' function in full:

```
model.boot <- function(data,indices){
sub.data <- data[indices,]
model <- glm(log(volume)~log(girth)+log(height),data=sub.data)
coef(model)  }
```

Now run the bootstrap for 2000 resamplings using the `boot` function:

```
glim.boot <- boot(trees,model.boot,R=2000)
glim.boot
```

```
ORDINARY NONPARAMETRIC BOOTSTRAP
Call:
boot(data = trees, statistic = model.boot, R = 2000)
Bootstrap Statistics:
          original              bias        std. error
t1*      -2.899379        -0.046089511       0.6452832
t2*       1.982665        -0.001071986       0.0603073
t3*       1.117138         0.014858487       0.2082793
```

There is very little bias in any of the three parameters, and the bootstrapped standard errors are close to their parametric estimates.

The other way of bootstrapping with a model, mentioned above, is to include all the original cases (rather than a subset of them with repeats, as we did above) but to randomize the residuals that are associated with each case. The raw residuals are `y - fitted(model)` and it is these values that are shuffled and allocated to cases at random. The model is then refitted and the coefficients extracted. The new `y` values, therefore, are

```
fitted(model)+ sample(y - fitted(model))
```

Here is a home-made version:

```
model <- glm(log(volume)~log(girth)+log(height))
yhat <- fitted(model)
residuals <- log(volume)  - yhat
```

```
coefs <- numeric(6000)
coefs <- matrix(coefs,nrow=2000)
```

We shuffle the residuals 2000 times to get different vectors of y values:

```
for (i in 1:2000){
y <- yhat+sample(residuals)
boot.model <- glm(y~log(girth)+log(height))
coefs[i,] <- coef(boot.model) }
```

Extracting the means and standard deviations of the coefficients using `apply` gives:

```
apply(coefs,2,mean)
```

```
[1] -2.898088 1.982693 1.116724
```

```
apply(coefs,2,sd)
```

```
[1] 0.60223281 0.07231379 0.19317107
```

These values are close to the estimates obtained by other means earlier. Next, we use the packaged `boot` function to carry out the same method. The preliminaries involve fitting the GLM and extracting the fitted values (`yhat`), which will be the same each time, and the residuals (`resids`), which will be independently shuffled each time:

```
model <- glm(log(volume)~log(girth)+log(height))
yhat <- fitted(model)
resids <- resid(model)
```

Now make a dataframe that will be fed into the bootstrap, containing the residuals to be shuffled, along with the two explanatory variables:

```
res.data <- data.frame(resids,girth,height)
```

Now for the only hard part: writing the 'statistic' function to do the work within `boot`. The first argument is always the dataframe and the second is always the index `i`, which controls the shuffling:

```
bf <- function(res.data,i) {
y <- yhat+res.data[i,1]
nd <- data.frame(y,girth,height)
model <- glm(y~log(girth)+log(height),data=nd)
coef(model) }
```

Inside the function we create a particular vector of y values by adding the shuffled residuals `res.data[i,1]` to the fitted values, then put this vector, y, along with the explanatory variables into a new dataframe nd that will be different each time GLM the is fitted. The function returns the three coefficients from the particular fitted model, `coef(model)`; the coefficients are the 'statistics' of the bootstrap, hence the name of the function.

Finally, because we want to shuffle the residuals rather than sample them with replacement, we specify `sim="permutation"` in the call to the `boot` function:

```
boot(res.data, bf, R=2000, sim="permutation")
```

```
DATA PERMUTATION
Call:
```

```
boot(data = res.data, statistic = bf, R = 2000, sim = "permutation")
Bootstrap Statistics :
            original             bias         std. error
t1*        -2.899379        0.014278399       0.62166875
t2*         1.982665        0.001601178       0.07064475
t3*         1.117138       -0.004586529       0.19938992
```

Again, the parameter values and their standard errors are very close to those obtained by our other boot-strapping methods. Here are the confidence intervals for the three parameters, specified by index=1 for the intercept, index=2 for the slope of the regression on $\log(g)$ and index=3 for the slope of the regression on $\log(h)$:

```
perms <-  boot(res.data, bf, R=2000, sim="permutation")
boot.ci(perms,index=1)

BOOTSTRAP CONFIDENCE INTERVAL CALCULATIONS
Based on 2000 bootstrap replicates
CALL :
boot.ci(boot.out = perms, index = 1)
Intervals :
Level               Normal                        Basic
95%            (-4.117, -1.692)            (-4.118, -1.680)
Level              Percentile                      BCa
95%            (-4.119, -1.681)            (-4.302, -1.784)
Calculations and Intervals on Original Scale
There were 32 warnings (use warnings() to see them)

boot.ci(perms,index=2)

BOOTSTRAP CONFIDENCE INTERVAL CALCULATIONS
Based on 2000 bootstrap replicates
CALL :
boot.ci(boot.out = perms, index = 2)
Intervals :
Level               Normal                        Basic
95%            ( 1.837, 2.125 )           ( 1.836, 2.124 )
Level              Percentile                      BCa
95%            ( 1.841, 2.129 )           ( 1.827, 2.115 )
Calculations and Intervals on Original Scale
There were 32 warnings (use warnings() to see them)
boot.ci(perms,index=3)
BOOTSTRAP CONFIDENCE INTERVAL CALCULATIONS
Based on 2000 bootstrap replicates
CALL :
boot.ci(boot.out = perms, index = 3)
Intervals :
Level               Normal                        Basic
95%            ( 0.730, 1.508 )           ( 0.726, 1.509 )
Level              Percentile                      BCa
95%            ( 0.725, 1.508 )           ( 0.758, 1.566 )
```

```
Calculations and Intervals on Original Scale
There were 32 warnings (use warnings() to see them)
```

You can see that all the intervals for the slope on log(g) include the value 2.0 and all the intervals for the slope on log(h) include 1.0, consistent with the theoretical expectation that the logs are cylindrical, and that the volume of usable timber can be estimated from the length of the log and the square of its girth.

13.14   Binomial GLM with ordered categorical variables

Ordered factors are introduced on p. 443. Here is a worked example using the built-in oesophageal cancer dataframe called esoph. The response is the number of cancer cases and the matching number of non-cancer patients (controls), with three categorical explanatory variables: age group (agegp, with six ordered levels each spanning 10 years), alcohol consumption (alcgp, with four ordered levels) and tobacco consumption (tobgp, with four ordered levels). There are too few cases to fit a full factorial of agegp*tobgpacco*alcgp, so we start with a maximal model that has a main effect for age and an interaction between tobacco and alcohol:

```
model1 <- glm(cbind(ncases,ncontrols)~agegp+alcgp*tobgp,binomial,data=esoph)
summary(model1)
```

```
Call:
glm(formula = cbind(ncases, ncontrols) ~ agegp + alcgp * tobgp,
    family = binomial, data = esoph)

Deviance Residuals:
    Min       1Q   Median       3Q      Max
-1.8895  -0.5317  -0.2304   0.2704   2.0724

Coefficients:
                 Estimate Std. Error z value Pr(>|z|)
(Intercept)      -1.75985    0.19822  -8.878  < 2e-16 ***
agegp.L           2.99646    0.65386   4.583 4.59e-06 ***
agegp.Q          -1.35008    0.59197  -2.281  0.02257 *
agegp.C           0.13436    0.45056   0.298  0.76554
agegp^4           0.07098    0.30974   0.229  0.81875
agegp^5          -0.21347    0.19627  -1.088  0.27676
alcgp.L           1.37077    0.21136   6.485 8.85e-11 ***
alcgp.Q          -0.14913    0.19645  -0.759  0.44778
alcgp.C           0.22823    0.18203   1.254  0.20990
tobgp.L           0.63846    0.19710   3.239  0.00120 **
tobgp.Q           0.02922    0.19617   0.149  0.88159
tobgp.C           0.15607    0.19796   0.788  0.43044
alcgp.L:tobgp.L  -0.70426    0.41128  -1.712  0.08683 .
alcgp.Q:tobgp.L   0.12948    0.38889   0.333  0.73917
alcgp.C:tobgp.L  -0.16118    0.36697  -0.439  0.66051
alcgp.L:tobgp.Q   0.12225    0.42044   0.291  0.77122
alcgp.Q:tobgp.Q  -0.44527    0.39224  -1.135  0.25629
alcgp.C:tobgp.Q   0.04843    0.36211   0.134  0.89361
alcgp.L:tobgp.C  -0.29187    0.42939  -0.680  0.49668
```

```
alcgp.Q:tobgp.C -0.05205     0.39538  -0.132  0.89527
alcgp.C:tobgp.C -0.13905     0.35754  -0.389  0.69734

(Dispersion parameter for binomial family taken to be 1)

    Null deviance: 227.241  on 87  degrees of freedom
Residual deviance:  47.484  on 67  degrees of freedom
AIC: 236.96

Number of Fisher Scoring iterations: 6
```

The good news is that there is no overdispersion (residual deviance 47.484 on 67 d.f.) so we can begin model simplification. You may not be familiar with the way that ordered factor levels are displayed in the summary.glm table: L means 'linear' testing whether there is evidence for a straight-line relationship with the response variable (look at the sign to see if it is increasing or decreasing); Q means 'quadratic' testing whether there is evidence for curvature in the response (look at the sign to see if the curvature is U-shaped or upside-down U-shaped); C means 'cubic' testing whether there is evidence for a point of inflection in the relationship; and numbers like ^4, ^5 (etc) test for higher-order polynomial effects, like local maxima and local minima in the relationship. We shall not interpret the output until we have finished with model simplification. There is no indication of an interaction between smoking and drinking, so we remove this:

```
model2<-glm(cbind(ncases,ncontrols)~agegp+alcgp+tobgp,binomial,data=esoph)
anova(model1,model2)

Analysis of Deviance Table

Model 1: cbind(ncases, ncontrols) ~ agegp + alcgp * tobgp
Model 2: cbind(ncases, ncontrols) ~ agegp + alcgp + tobgp
  Resid. Df Resid. Dev Df Deviance
1       67     47.484
2       76     53.973 -9  -6.4895
```

The critical value of chi-squared with 9 d.f. is 16.92:

```
qchisq(.95,9)

[1] 16.91898
```

Our relatively low value of 6.4895 therefore indicates that this model simplification was justified. What next?

```
summary(model2)

Call:
glm(formula = cbind(ncases, ncontrols) ~ agegp + alcgp + tobgp,
    family = binomial, data = esoph)

Deviance Residuals:
    Min       1Q    Median       3Q       Max
-1.6891  -0.5618  -0.2168   0.2314   2.0642

Coefficients:
            Estimate Std. Error z value Pr(>|z|)
(Intercept) -1.77997    0.19796  -8.992  < 2e-16 ***
agegp.L      3.00534    0.65215   4.608 4.06e-06 ***
```

```
agegp.Q      -1.33787       0.59111   -2.263   0.02362  *
agegp.C       0.15307       0.44854    0.341   0.73291
agegp^4       0.06410       0.30881    0.208   0.83556
agegp^5      -0.19363       0.19537   -0.991   0.32164
alcgp.L       1.49185       0.19935    7.484  7.23e-14 ***
alcgp.Q      -0.22663       0.17952   -1.262   0.20680
alcgp.C       0.25463       0.15906    1.601   0.10942
tobgp.L       0.59448       0.19422    3.061   0.00221  **
tobgp.Q       0.06537       0.18811    0.347   0.72823
tobgp.C       0.15679       0.18658    0.840   0.40071
```

```
(Dispersion parameter for binomial family taken to be 1)

    Null deviance: 227.241  on 87  degrees of freedom
Residual deviance:  53.973  on 76  degrees of freedom
AIC: 225.45

Number of Fisher Scoring iterations: 6
```

There is strong evidence for linear effects of alcohol and tobacco consumption, with a quadratic (decelerating positive) effect of age on cancer risk. The quadratic and higher-order polynomials are not significant for either tobacco or alcohol, so this is a good stage at which to consider factor-level reduction.

Let us look at the data as proportions:

```
p<-ncases/(ncases+ncontrols)
```

Now plot this against the three explanatory variables

```
par(mfrow=c(2,2))
plot(p~alcgp,col="red")
plot(p~tobgp,col="blue")
plot(p~agegp,col="green")
```

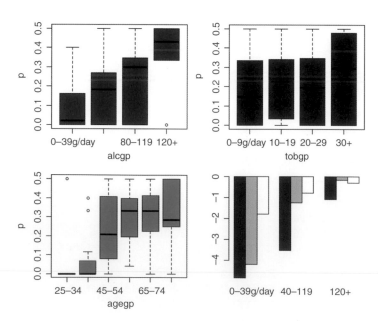

The plot suggests some sensible model simplifications (ignore the grey barplot for the moment). The alcohol response (red) looks to be linear, so we leave that as it is. The tobacco response (blue) could probably be simplified by combining the two intermediate smoking rates:

```
tob2<-tobgp
levels(tob2)[2:3]<-"10-30"
levels(tob2)
```

```
[1] "0-9g/day" "10-30"     "30+"
```

The age effect is most complicated, but a three-level factor might work just as well with a young group (under age 45), an intermediate group (age between 45 and 54) and an older group (55+):

```
age2<-agegp
levels(age2)[4:6]<-"55+"
levels(age2)[1:2]<-"under45"
levels(age2)
```

```
[1] "under45" "45-54"     "55+"
```

We can start the modelling again with these new ordered factors:

```
model3 <- glm(cbind(ncases,ncontrols)~age2*alcgp*tob2,binomial,data=esoph)
model4<-step(model3)
model5<-update(model4,~.-age2:alcgp)
anova(model4,model5,test="Chi")
```

```
Analysis of Deviance Table
```

```
Model 1: cbind(ncases, ncontrols) ~ age2 + alcgp + tob2 + age2:alcgp
Model 2: cbind(ncases, ncontrols) ~ age2 + alcgp + tob2
  Resid. Df Resid. Dev Df Deviance Pr(>Chi)
1        74     44.677
2        80     60.388 -6  -15.711  0.01539 *
```

At this stage, the ordering of the factors is more of a hindrance than a help, because it is so hard to interpret the significant age by alcohol interaction. Let us also get rid of the apparently problematical alcohol group (look at the size of its standard error in `summary(model4)`),

```
levels(alc3)[2:3]<-"40-119"
alc3<-factor(alc3,ordered=FALSE)
levels(alc3)
```

```
1] "0-39g/day" "40-119"     "120+"
```

and unorder the other two factors for good measure:

```
age3<-factor(age2,ordered=FALSE)
tob3<-factor(tob2,ordered=FALSE)
```

We start the modelling again with a full factorial of the three unordered factors, then use `step` to simplify it:

```
model6 <- glm(cbind(ncases,ncontrols)~age3*alc3*tob3,binomial,data=esoph)
model7<-step(model6)
summary(model7)
```

```
Coefficients:
                        Estimate Std. Error z value Pr(>|z|)
(Intercept)              -5.1929     1.0066  -5.159 2.49e-07 ***
age345-54                 0.6559     1.4217   0.461 0.644562
age355+                   3.0628     1.0248   2.989 0.002801 **
alc340-119                1.3133     1.1248   1.168 0.242985
alc3120+                  3.8060     1.1313   3.364 0.000767 ***
tob310-30                 0.3293     0.1798   1.832 0.066980 .
tob330+                   0.7837     0.2715   2.887 0.003893 **
age345-54:alc340-119      1.6374     1.5225   1.075 0.282189
age355+:alc340-119       -0.2974     1.1504  -0.259 0.795984
age345-54:alc3120+        0.2203     1.5625   0.141 0.887895
age355+:alc3120+         -2.2843     1.1780  -1.939 0.052494 .

(Dispersion parameter for binomial family taken to be 1)

    Null deviance: 227.241  on 87  degrees of freedom
Residual deviance:  48.559  on 77  degrees of freedom
AIC: 218.04

Number of Fisher Scoring iterations: 6
```

The interaction between age and alcohol is not obviously significant, so we test it by deletion:

```
model8<-update(model7,~.-age3:alc3)
anova(model7,model8,test="Chi")
```

```
Analysis of Deviance Table
```

```
Model 1: cbind(ncases, ncontrols) ~ age3 + alc3 + tob3 + age3:alc3
Model 2: cbind(ncases, ncontrols) ~ age3 + alc3 + tob3
  Resid. Df Resid. Dev Df Deviance Pr(>Chi)
1        77     48.559
2        81     62.468 -4  -13.909  0.00759 **
```

It turns out that the interaction is highly significant ($p < 0.008$). To visualize the interaction, we plot the predicted means of the logits by age and alcohol consumption:

```
barplot(tapply(predict(model7),list(age3,alc3),mean),beside=T)
```

This is the grey-scale barplot in the lower right-hand panel, above. The young and middle-aged subjects both have low rates of cancer at low alcohol consumption rates, but the middle-aged subjects have proportionally higher rates at intermediate alcohol consumption and, by the highest rates of alcohol consumption, the two older age classes have equally high rates of cancer. Again, the interaction has become clear only after model simplification.

14

Count Data

Up to this point, the response variables have all been continuous measurements such as weights, heights, lengths, temperatures and growth rates. A great deal of the data collected by scientists, medical statisticians and economists, however, is in the form of **counts** (whole numbers or integers). The number of individuals who died, the number of firms going bankrupt, the number of days of frost, the number of red blood cells on a microscope slide, and the number of craters in a sector of lunar landscape are all potentially interesting variables for study. With count data, the number 0 often appears as a value of the response variable (consider, for example, what a 0 would mean in the context of the examples just listed). In this chapter we deal with data on **frequencies**, where we count how many times something happened, but we have no way of knowing how often it did *not* happen (e.g. lightning strikes, bankruptcies, deaths, births). This is in contrast to count data on **proportions**, where we know the number doing a particular thing, but also the number not doing that thing (e.g. the proportion dying, sex ratios at birth, proportions of different groups responding to a questionnaire).

Straightforward linear regression methods (assuming constant variance, normal errors) are not appropriate for count data for four main reasons:

- The linear model might lead to the prediction of negative counts.

- The variance of the response variable is likely to increase with the mean.

- The errors will not be normally distributed.

- Zeros are difficult to handle in transformations.

In R, count data are handled very elegantly in a generalized linear model by specifying family=poisson which sets errors = Poisson and link = log (see p. 558). The log link ensures that all the fitted values are positive, while the Poisson errors take account of the fact that the data are integer and have variances that are equal to their means.

14.1 A regression with Poisson errors

The following example has a count (the number of reported cancer cases per year per clinic) as the response variable, and a single continuous explanatory variable (the distance from a nuclear plant to the clinic in kilometres). The question is whether or not proximity to the reactor affects the number of cancer cases.

```
clusters<-read.table("c:\\temp\\clusters.txt",header=T)
attach(clusters)
names(clusters)
```

```
[1] "Cancers"  "Distance"
```

There seems to be a downward trend in cancer cases with distance (see the plot below). But is the trend significant? We do a regression of cases against distance, using a GLM with Poisson errors:

```
model1<-glm(Cancers~Distance,poisson)
summary(model1)
```

```
Call:
glm(formula = Cancers ~ Distance, family = poisson)

Deviance Residuals:
    Min       1Q    Median        3Q       Max
-1.5504   -1.3491   -1.1553    0.3877    3.1304

Coefficients:
              Estimate Std. Error z value Pr(>|z|)
(Intercept)   0.186865   0.188728   0.990   0.3221
Distance     -0.006138   0.003667  -1.674   0.0941 .

(Dispersion parameter for poisson family taken to be 1)

    Null deviance: 149.48  on 93  degrees of freedom
Residual deviance: 146.64  on 92  degrees of freedom
AIC: 262.41
```

The trend does not look to be significant, but look at the residual deviance. Under Poisson errors, it is assumed that residual deviance is equal to the residual degrees of freedom (because the variance and the mean should be the same). The fact that residual deviance is larger than residual degrees of freedom indicates that we have overdispersion (extra, unexplained variation in the response). We compensate for the overdispersion by refitting the model using quasi-Poisson rather than Poisson errors:

```
model2<-glm(Cancers~Distance,quasipoisson)
summary(model2)
```

```
Call:
glm(formula = Cancers ~ Distance, family = quasipoisson)

Deviance Residuals:
    Min       1Q    Median        3Q       Max
-1.5504   -1.3491   -1.1553    0.3877    3.1304

Coefficients:
              Estimate Std. Error t value Pr(>|t|)
(Intercept)   0.186865   0.235364   0.794    0.429
Distance     -0.006138   0.004573  -1.342    0.183

(Dispersion parameter for quasipoisson family taken to be 1.555271)

    Null deviance: 149.48  on 93  degrees of freedom
Residual deviance: 146.64  on 92  degrees of freedom
```

Compensating for the overdispersion has increased the *p* value to 0.183, so there is no compelling evidence to support the existence of a trend in cancer incidence with distance from the nuclear plant. To draw the fitted model through the data, you need to understand that the GLM with Poisson errors uses the log link, so the parameter estimates and the predictions from the model (the 'linear predictor') are in logs, and need to be antilogged exp (yv) before the (non-significant) fitted line is drawn.

```
xv <- seq(0,100)
yv <- predict(model2,list(Distance=xv))

plot(Cancers~Distance,pch=21,col="red",bg="orange")
lines(xv,exp(yv),col="blue")
```

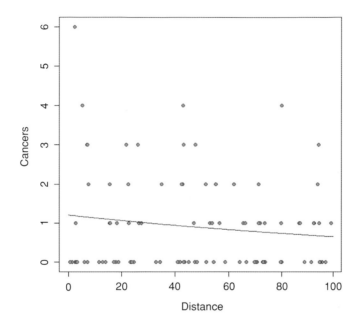

Note how odd the scatterplot looks when we have count data as the response. The values of the response variable are in rows (because they are all whole numbers), and there are data points all over the place (not clustered around the regression line, as they were in previous regressions).

14.2 Analysis of deviance with count data

In our next example the response variable is a count of infected blood cells per square millimetre on microscope slides prepared from randomly selected individuals. The explanatory variables are smoker (logical: yes or no), age (three levels: under 20, 21 to 59, 60 and over), sex (male or female) and body mass score (three levels: normal, overweight, obese).

```
count<-read.table("c:\\temp\\cellcounts.txt",header=T)
attach(count)
names(count)

[1] "cells"  "smoker" "age"    "sex"     "weight"
```

It is always a good idea with count data to get a feel for the overall frequency distribution of counts using `table`:

```
table(cells)

cells
   0    1    2    3    4    5    6    7
 314   75   50   32   18   13    7    2
```

Most subjects (314 of them) showed no damaged cells, and the maximum of 7 was observed in just two patients.

We begin data inspection by tabulating the main effect means:

```
tapply(cells,smoker,mean)

     FALSE        TRUE
 0.5478723  1.9111111

tapply(cells,weight,mean)

    normal      obese       over
 0.5833333  1.2814371  0.9357143

tapply(cells,sex,mean)

    female       male
 0.6584507  1.2202643

tapply(cells,age,mean)

       mid        old      young
 0.8676471  0.7835821  1.2710280
```

It looks as if smokers had a substantially higher mean count than non-smokers, overweight and obese subjects had higher counts than those of normal weight, males had a higher count than females, and young subjects had a higher mean count than middle-aged or older people. We need to test whether any of these differences are significant and to assess whether there are interactions between the explanatory variables.

```
model1<-glm(cells~smoker*sex*age*weight,poisson)
summary(model1)

        Null deviance: 1052.95 on 510 degrees of freedom
Residual deviance: 736.33 on 477 degrees of freedom
AIC: 1318
Number of Fisher Scoring iterations: 6
```

The residual deviance (736.33) is much greater than the residual degrees of freedom (477), indicating overdispersion, so before interpreting any of the effects, we should refit the model using quasi-Poisson errors:

```
model2<-glm(cells~smoker*sex*age*weight,quasipoisson)
summary(model2)

Coefficients: (2 not defined because of singularities)
                              Estimate Std. Error t value Pr(>|t|)
(Intercept)                    -0.8329     0.4307  -1.934   0.0537 .
smokerTRUE                     -0.1787     0.8057  -0.222   0.8246
```

sexmale	0.1823	0.5831	0.313	0.7547	
ageold	-0.1830	0.5233	-0.350	0.7267	
ageyoung	0.1398	0.6712	0.208	0.8351	
weightobese	1.2384	0.8965	1.381	0.1678	
weightover	-0.5534	1.4284	-0.387	0.6986	
smokerTRUE:sexmale	0.8293	0.9630	0.861	0.3896	
smokerTRUE:ageold	-1.7227	2.4243	-0.711	0.4777	
smokerTRUE:ageyoung	1.1232	1.0584	1.061	0.2892	
sexmale:ageold	-0.2650	0.9445	-0.281	0.7791	
sexmale:ageyoung	-0.2776	0.9879	-0.281	0.7788	
smokerTRUE:weightobese	3.5689	1.9053	1.873	0.0617	.
smokerTRUE:weightover	2.2581	1.8524	1.219	0.2234	
sexmale:weightobese	-1.1583	1.0493	-1.104	0.2702	
sexmale:weightover	0.7985	1.5256	0.523	0.6009	
ageold:weightobese	-0.9280	0.9687	-0.958	0.3386	
ageyoung:weightobese	-1.2384	1.7098	-0.724	0.4693	
ageold:weightover	1.0013	1.4776	0.678	0.4983	
ageyoung:weightover	0.5534	1.7980	0.308	0.7584	
smokerTRUE:sexmale:ageold	1.8342	2.1827	0.840	0.4011	
smokerTRUE:sexmale:ageyoung	-0.8249	1.3558	-0.608	0.5432	
smokerTRUE:sexmale:weightobese	-2.2379	1.7788	-1.258	0.2090	
smokerTRUE:sexmale:weightover	-2.5033	2.1120	-1.185	0.2365	
smokerTRUE:ageold:weightobese	0.8298	3.3269	0.249	0.8031	
smokerTRUE:ageyoung:weightobese	-2.2108	1.0865	-2.035	0.0424	*
smokerTRUE:ageold:weightover	1.1275	1.6897	0.667	0.5049	
smokerTRUE:ageyoung:weightover	-1.6156	2.2168	-0.729	0.4665	
sexmale:ageold:weightobese	2.2210	1.3318	1.668	0.0960	.
sexmale:ageyoung:weightobese	2.5346	1.9488	1.301	0.1940	
sexmale:ageold:weightover	-1.0641	1.9650	-0.542	0.5884	
sexmale:ageyoung:weightover	-1.1087	2.1234	-0.522	0.6018	
smokerTRUE:sexmale:ageold:weightobese	-1.6169	3.0561	-0.529	0.5970	
smokerTRUE:sexmale:ageyoung:weightobese	NA	NA	NA	NA	
smokerTRUE:sexmale:ageold:weightover	NA	NA	NA	NA	
smokerTRUE:sexmale:ageyoung:weightover	2.4160	2.6846	0.900	0.3686	

(Dispersion parameter for quasipoisson family taken to be 1.854815)

```
    Null deviance: 1052.95  on 510  degrees of freedom
Residual deviance:  736.33  on 477  degrees of freedom
AIC: NA

Number of Fisher Scoring iterations: 6
```

The first thing to understand is what the NAs in the coefficients table mean. This is a sign of aliasing: there is no information in the dataframe from which to estimate this particular interaction term. There is an apparently significant three-way interaction between smoking, age and obesity ($p = 0.0424$). There were too few subjects to assess the four-way interaction (see the NAs in the table), so we begin model simplification by removing the highest-order interaction:

```
model3<-update(model2, ~. -smoker:sex:age:weight)
summary(model3)
```

Coefficients:

	Estimate	Std. Error	t value	Pr(>|t|)	
(Intercept)	-0.897195	0.436988	-2.053	0.04060	*
smokerTRUE	0.030263	0.735386	0.041	0.96719	
sexmale	0.297192	0.570009	0.521	0.60234	
ageold	-0.118726	0.528165	-0.225	0.82224	
ageyoung	0.289259	0.639618	0.452	0.65130	
weightobese	1.302660	0.898307	1.450	0.14768	
weightover	-0.005052	1.027198	-0.005	0.99608	
smokerTRUE:sexmale	0.527345	0.867294	0.608	0.54345	
smokerTRUE:ageold	-0.566584	1.700590	-0.333	0.73915	
smokerTRUE:ageyoung	0.757297	0.939746	0.806	0.42073	
sexmale:ageold	-0.379884	0.935365	-0.406	0.68483	
sexmale:ageyoung	-0.610703	0.920969	-0.663	0.50758	
smokerTRUE:weightobese	3.924591	1.475476	2.660	0.00808	**
smokerTRUE:weightover	1.192159	1.259888	0.946	0.34450	
sexmale:weightobese	-1.273202	1.040701	-1.223	0.22178	
sexmale:weightover	0.154097	1.098781	0.140	0.88853	
ageold:weightobese	-0.993355	0.970484	-1.024	0.30656	
ageyoung:weightobese	-1.346913	1.459454	-0.923	0.35653	
ageold:weightover	0.454217	1.090260	0.417	0.67715	
ageyoung:weightover	-0.483955	1.300866	-0.372	0.71004	
smokerTRUE:sexmale:ageold	0.771116	1.451512	0.531	0.59549	
smokerTRUE:sexmale:ageyoung	-0.210317	1.140384	-0.184	0.85376	
smokerTRUE:sexmale:weightobese	-2.500668	1.369941	-1.825	0.06857	.
smokerTRUE:sexmale:weightover	-1.110222	1.217531	-0.912	0.36230	
smokerTRUE:ageold:weightobese	-0.882951	1.187871	-0.743	0.45766	
smokerTRUE:ageyoung:weightobese	-2.453315	1.047067	-2.343	0.01954	*
smokerTRUE:ageold:weightover	0.823018	1.528233	0.539	0.59045	
smokerTRUE:ageyoung:weightover	0.040795	1.223664	0.033	0.97342	
sexmale:ageold:weightobese	2.338617	1.324805	1.765	0.07816	.
sexmale:ageyoung:weightobese	2.822032	1.623849	1.738	0.08288	.
sexmale:ageold:weightover	-0.442066	1.545451	-0.286	0.77497	
sexmale:ageyoung:weightover	0.357807	1.291194	0.277	0.78181	

(Dispersion parameter for quasipoisson family taken to be 1.847991)

```
    Null deviance: 1052.95  on 510   degrees of freedom
Residual deviance:  737.87  on 479   degrees of freedom
```

The remaining model simplification is left to you as an exercise. Your minimal adequate model might look something like this:

```
newWt<-weight
levels(newWt)[c(1,3)]<-"not"
summary(model15)
```

```
Call:
glm(formula = cells ~ smoker + newWt + smoker:newWt, family = quasipoisson)
```

```
Deviance Residuals:
    Min       1Q    Median        3Q       Max
-2.6511   -1.1742   -0.9709    0.5217    3.8157
Coefficients:
                       Estimate Std. Error t value Pr(>|t|)
(Intercept)             -0.7522     0.1279  -5.883 7.31e-09 ***
smokerTRUE               1.0523     0.1740   6.048 2.84e-09 ***
newWtobese               0.3803     0.1924   1.977   0.0486 *
smokerTRUE:newWtobese    0.5764     0.2573   2.240   0.0255 *
```

This model shows a highly significant interaction between smoking and weight in determining the number of damaged cells, but there are no convincing effects of age or sex. In a case like this, it is useful to produce a summary table to highlight the effects:

```
tapply(cells,list(smoker,weight),mean)
```

```
            normal       obese        over
FALSE    0.4184397   0.6893939   0.5436893
TRUE     0.9523810   3.5142857   2.0270270
```

The interaction arises because the response to smoking depends on body weight: smoking adds a mean of about 0.5 damaged cells for individuals with normal body weight, but adds 2.8 damaged cells for obese people.

It is straightforward to turn the summary table into a barplot:

```
barplot(tapply(cells,list(smoker,weight),mean),col=c("wheat2","wheat4"),
              beside=T,ylab="damaged cells",xlab="body mass")
legend(1.2,3.4,c("non-smoker","smoker"),fill=c("wheat2","wheat4"))
```

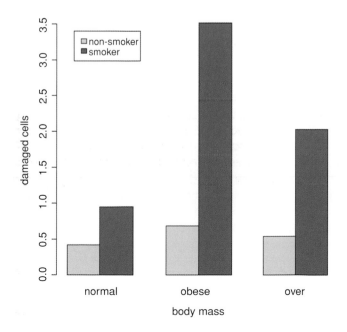

14.3 Analysis of covariance with count data

In this next example the response is a count of the number of plant species on plots that have different biomass (a continuous explanatory variable) and different soil pH (a categorical variable with three levels: high, mid and low).

```
species<-read.table("c:\\temp\\species.txt",header=T)
attach(species)
names(species)
[1] "pH" "Biomass" "Species"
plot(Biomass,Species,type="n")
spp<-split(Species,pH)
bio<-split(Biomass,pH)
points(bio[[1]],spp[[1]],pch=16,col="red")
points(bio[[2]],spp[[2]],pch=16,col="green")
points(bio[[3]],spp[[3]],pch=16,col="blue")
legend(locator(1),legend=c("high","low","medium"),
                pch=c(16,16,16),col=c("red","green","blue"),title="pH")
```

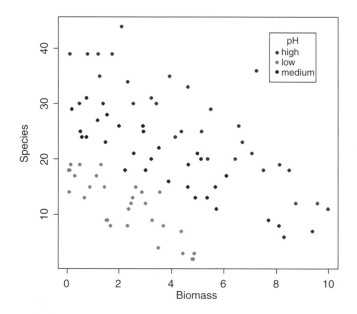

Note the use of `split` to create separate lists of plotting coordinates for the three levels of pH. It is clear that species declines with biomass, and that soil pH has a big effect on species, but does the slope of the relationship between species and biomass depend on pH? The lines look reasonably parallel from the scatterplot. This is a question about interaction effects, and in analysis of covariance, interaction effects are about differences between slopes:

```
model1<-glm(Species~ Biomass*pH,poisson)
summary(model1)
```

```
Coefficients:
                 Estimate Std. Error z value Pr(>|z|)
(Intercept)       3.76812    0.06153  61.240  < 2e-16 ***
Biomass          -0.10713    0.01249  -8.577  < 2e-16 ***
pHlow            -0.81557    0.10284  -7.931 2.18e-15 ***
pHmid            -0.33146    0.09217  -3.596 0.000323 ***
Biomass:pHlow    -0.15503    0.04003  -3.873 0.000108 ***
Biomass:pHmid    -0.03189    0.02308  -1.382 0.166954

(Dispersion parameter for poisson family taken to be 1)

    Null deviance: 452.346  on 89  degrees of freedom
Residual deviance:  83.201  on 84  degrees of freedom
AIC: 514.39
```

There is no evidence of overdispersion (residual deviance = 83.2 on 89 d.f.). We can test for the need for different slopes by comparing this maximal model (with six parameters) with a simpler model with different intercepts but the same slope (four parameters):

```
model2<-glm(Species~Biomass+pH,poisson)
anova(model1,model2,test="Chi")
```

```
Analysis of Deviance Table

Model 1: Species ~ Biomass * pH
Model 2: Species ~ Biomass + pH
  Resid. Df Resid. Dev Df Deviance  Pr(>Chi)
1        84     83.201
2        86     99.242 -2   -16.04 0.0003288 ***
```

The slopes are very significantly different ($p = 0.000\,33$), so we are justified in retaining the more complicated model1.

Finally, we draw the fitted lines through the scatterplot, using predict We need to specify values for all of the explanatory variables in the model, namely biomass (a continuous variable) and soil pH (a three-level categorical variable). First, the continuous variable for the x axis:

```
xv<-seq(0,10,0.1)
```

Next we need to provide a vector of factor levels for soil pH and this vector must be exactly the same length as the vector of x values (length(xv) = 101 in this case). It is simplest to use the factor levels for pH in the order in which they appear:

```
levels(pH)
```

```
[1] "high" "low" "mid"
```

We shall draw the line for the high pH first, remembering to antilog the predictions:

```
pHs<-factor(rep("high",101))
xv<-seq(0,10,0.1)

yv<-predict(model1,list(Biomass=xv,pH=pHs))
```

```
lines(xv,exp(yv),col="red")
pHs<-factor(rep("low",101))

yv<-predict(model1,list(Biomass=xv,pH=pHs))

lines(xv,exp(yv),col="green")
pHs<-factor(rep("mid",101))

yv<-predict(model1,list(Biomass=xv,pH=pHs))
lines(xv,exp(yv),col="blue")
```

You could make the R code more elegant by writing a function to plot any number of lines, depending on the number of levels of the factor (three levels of pH in this case).

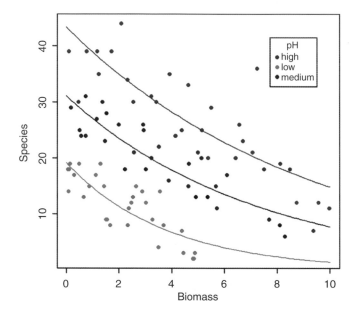

14.4   Frequency distributions

Here are data on the numbers of bankruptcies in 80 districts. The question is whether there is any evidence that some districts show greater than expected numbers of cases. What would we expect? Of course we should expect some variation, but how much, exactly? Well that depends on our model of the process. Perhaps the simplest model is that absolutely nothing is going on, and that every singly bankruptcy case is absolutely independent of every other. That leads to the prediction that the numbers of cases per district will follow a Poisson process, a distribution in which the variance is equal to the mean (see p. 314). Let us see what the data show.

```
case.book<-read.table("c:\\temp\\cases.txt",header=T)
attach(case.book)
names(case.book)
```

```
[1] "cases"
```

First we need to count the numbers of districts with no cases, one case, two cases, and so on. The R function that does this is called `table`:

```
frequencies<-table(cases)
frequencies
```

```
cases
 0  1  2  3  4  5  6  7  8  9 10
34 14 10  7  4  5  2  1  1  1  1
```

There were no cases at all in 34 districts, but one district had 10 cases. A good way to proceed is to compare our distribution (called `frequencies`) with the distribution that would be observed if the data really did come from a Poisson distribution as postulated by our model. We can use the R function `dpois` to compute the probability density of each of the 11 frequencies from 0 to 10 (we multiply the probability produced by `dpois` by the total sample of 80 to obtain the predicted frequencies). We need to calculate the mean number of cases per district – this is the Poisson distribution's only parameter:

```
mean(cases)
```

```
[1] 1.775
```

The plan is to draw two distributions side by side, so we set up the plotting region:

```
windows(7,4)
par(mfrow=c(1,2))
```

Now we plot the observed frequencies in the left-hand panel and the predicted, Poisson frequencies in the right-hand panel:

```
barplot(frequencies,ylab="Frequency",xlab="Cases",col="green4", main="Cases")
barplot(dpois(0:10,1.775)*80,names=as.character(0:10),
                 ylab="Frequency",xlab="Cases",col="green3",main="Poisson")
```

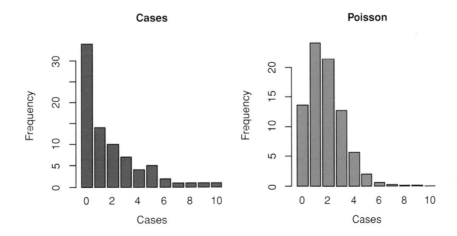

The distributions are very different: the mode of the observed data is 0, but the mode of the Poisson distribution with the same mean is 1; the observed data contained examples of 8, 9 and 10 cases, but these would be highly unlikely under a Poisson process. We would say that the observed data are highly **aggregated** – they have a variance–mean ratio much greater than 1 (the Poisson distribution, of course, has a variance–mean ratio of 1):

```
var(cases)/mean(cases)
```

```
[1] 2.99483
```

So, if the data are not Poisson distributed, how are they distributed? A good candidate distribution where the variance–mean ratio is this big (around 3.0) is the negative binomial distribution (see p. 315). This is a two-parameter distribution: the first parameter is the mean number of cases (1.775), and the second is called the clumping parameter, k (measuring the degree of aggregation in the data: small values of k ($k < 1$) show high aggregation, while large values of k ($k > 5$) show randomness). We can get an approximate estimate of the magnitude of k from

$$\hat{k} = \frac{\bar{x}^2}{s^2 - \bar{x}}.$$

We can work this out:

```
mean(cases)^2/(var(cases)-mean(cases))
```

```
[1] 0.8898003
```

so we shall work with $k = 0.89$. How do we compute the expected frequencies? The density function for the negative binomial distribution is dnbinom and it has three arguments: the frequency for which we want the probability (in our case 0 to 10), the number of successes (in our case 1), and the mean number of cases (1.775); we multiply by the total number of cases (80) to obtain the expected frequencies

```
exp<-dnbinom(0:10,1,mu=1.775)*80
```

We will draw a single figure in which the observed and expected frequencies are drawn side by side. The trick is to produce a new vector (called both) which is twice as long as the observed and expected frequency vectors ($2 \times 11 = 22$). Then, we put the observed frequencies in the odd-numbered elements (using modulo 2 to calculate the values of the subscripts), and the expected frequencies in the even-numbered elements:

```
both<-numeric(22)
both[1:22 %% 2 != 0]<-frequencies
both[1:22 %% 2 == 0]<-exp
```

On the x axis, we intend to label only every other bar:

```
labels<-character(22)
labels[1:22 %% 2 == 0]<-as.character(0:10)
```

Now we can produce the barplot, using dark red for the observed frequencies and dark blue for the negative binomial frequencies ('expected'):

```
windows(7,7)
barplot(both,col=rep(c("red4","blue4"),11),names=labels,ylab="Frequency",
                xlab="Cases")
```

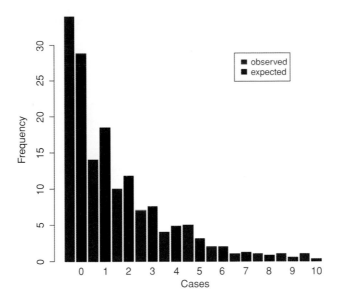

Now we need to add a legend to show what the two colours of the bars mean. You can locate the legend by trial and error, or by left-clicking the mouse when the cursor is in the correct position, using the `locator(1)` function to fix the lop left corner of the legend box (see p. 194):

```
legend(locator(1),c("observed","expected") ,fill=c("red4","blue4"))
```

The fit to the negative binomial distribution is much better than it was with the Poisson distribution, especially in the right-hand tail. But the observed data have too many 0s and too few 1s to be represented perfectly by a negative binomial distribution. If you want to quantify the lack of fit between the observed and expected frequency distributions, you can calculate Pearson's chi-squared $\sum (O - E)^2 / E$ based on the number of comparisons that have expected frequency greater than 4:

```
exp
```

```
[1]  28.8288288  18.4400617  11.7949944   7.5445460   4.8257907   3.0867670

[7]   1.9744185   1.2629164   0.8078114   0.5167082   0.3305070
```

If we accumulate the rightmost six frequencies, then all the values of `exp` will be bigger than 4. The degrees of freedom are then given by the number of legitimate comparisons (6) minus the number of parameters estimated from the data (2 in our case) minus 1 (for contingency, because the total frequency must add up to 80), i.e. 3 d.f. We reduce the lengths of the observed and expected vectors, creating an upper interval called 5+ for '5 or more':

```
cs<-factor(0:10)
levels(cs)[6:11]<-"5+"
levels(cs)
```

```
[1]  "0"  "1"  "2"  "3"  "4"  "5+"
```

Now make the two shorter vectors `of` and `ef` (for 'observed' and 'expected frequencies'):

```
ef<-as.vector(tapply(exp,cs,sum))
```

```
of<-as.vector(tapply(frequencies,cs,sum))
```

Finally, we can compute the chi-squared value measuring the difference between the observed and expected frequency distributions, and use `1-pchisq` to work out the *p* value:

```
sum((of-ef)^2/ef)
```

```
[1]  3.594145
```

```
1-pchisq(3.594145,3)
```

```
[1]  0.3087555
```

We conclude that a negative binomial description of these data is reasonable (the observed and expected distributions are not significantly different, $p = 0.31$).

14.5   Overdispersion in log-linear models

The data analysed in this section refer to children from Walgett, New South Wales, Australia, who were classified by sex (with two levels: male (M) and female (F)), culture (also with two levels: Aboriginal (A) and not (N)), age group (with four levels: F0 (primary), F1, F2 and F3) and learner status (with two levels: average (AL) and slow (SL)). The response variable is a count of the number of days absent from school in a particular school year (Days).

```
library(MASS)
data(quine)
attach(quine)
names(quine)
```

```
[1]  "Eth"  "Sex"  "Age"  "Lrn"  "Days"
```

We begin with a log-linear model for the counts, and fit a maximal model containing all the factors and all their interactions:

```
model1<-glm(Days~Eth*Sex*Age*Lrn,poisson)
summary(model1)
```

```
(Dispersion parameter for poisson family taken to be 1)

       Null deviance: 2073.5 on 145 degrees of freedom
Residual deviance: 1173.9 on 118 degrees of freedom
AIC: 1818.4
```

Next, we check the residual deviance to see if there is overdispersion. Recall that the residual deviance should be equal to the residual degrees of freedom if the Poisson errors assumption is appropriate. Here it is 1173.9 on 118 d.f., indicating overdispersion by a factor of roughly 10. This is much too big to ignore, so before embarking on model simplification we try a different approach, using quasi-Poisson errors to account for the overdispersion:

```
model2<-glm(Days~Eth*Sex*Age*Lrn,quasipoisson)
summary(model2)
```

```
Call:
glm(formula = Days ~ Eth * Sex * Age * Lrn, family = quasipoisson)
```

```
Deviance Residuals:
     Min        1Q     Median        3Q       Max
 -7.3872   -2.5129   -0.4205    1.7424    6.6783

Coefficients: (4 not defined because of singularities)
                       Estimate Std. Error t value Pr(>|t|)
(Intercept)              3.0564     0.3346   9.135 2.22e-15 ***
EthN                    -0.1386     0.4904  -0.283   0.7780
SexM                    -0.4914     0.5082  -0.967   0.3356
AgeF1                   -0.6227     0.5281  -1.179   0.2407
AgeF2                   -2.3632     2.2066  -1.071   0.2864
AgeF3                   -0.3784     0.4296  -0.881   0.3802
LrnSL                   -1.9577     1.8120  -1.080   0.2822
EthN:SexM               -0.7524     0.8272  -0.910   0.3649
EthN:AgeF1               0.1029     0.7427   0.139   0.8901
EthN:AgeF2              -0.5546     3.8094  -0.146   0.8845
EthN:AgeF3               0.0633     0.6194   0.102   0.9188
SexM:AgeF1               0.4092     0.9372   0.437   0.6632
SexM:AgeF2               3.1098     2.2506   1.382   0.1696
SexM:AgeF3               1.1145     0.6173   1.806   0.0735 .
EthN:LrnSL               2.2588     1.9474   1.160   0.2484
SexM:LrnSL               1.5900     1.9448   0.818   0.4152
AgeF1:LrnSL              2.6421     1.8688   1.414   0.1601
AgeF2:LrnSL              4.8585     2.8413   1.710   0.0899 .
AgeF3:LrnSL                  NA         NA      NA       NA
EthN:SexM:AgeF1         -0.3105     1.6756  -0.185   0.8533
EthN:SexM:AgeF2         0.3469      3.8928   0.089   0.9291
EthN:SexM:AgeF3         0.8329      0.9629   0.865   0.3888
EthN:SexM:LrnSL        -0.1639      2.1666  -0.076   0.9398
EthN:AgeF1:LrnSL       -3.5493      2.0712  -1.714   0.0892 .
EthN:AgeF2:LrnSL       -3.3315      4.2739  -0.779   0.4373
EthN:AgeF3:LrnSL            NA          NA      NA       NA
SexM:AgeF1:LrnSL       -2.4285      2.1901  -1.109   0.2697
SexM:AgeF2:LrnSL       -4.1914      2.9472   1.422   0.1576
SexM:AgeF3:LrnSL            NA          NA      NA       NA
EthN:SexM:AgeF1:LrnSL   2.1711      2.7527   0.789   0.4319
EthN:SexM:AgeF2:LrnSL   2.1029      4.4203   0.476   0.6351
EthN:SexM:AgeF3:LrnSL       NA          NA      NA       NA

(Dispersion parameter for quasipoisson family taken to be 9.514226)

    Null deviance: 2073.5  on 145  degrees of freedom
Residual deviance: 1173.9  on 118  degrees of freedom
```

Notice that certain interactions are aliased (shown by rows with NA). These have not been estimated because of missing factor-level combinations, as indicated by the zeros in the following table:

```
ftable(table(Eth,Sex,Age,Lrn))
```

```
Eth Sex Age      AL SL
A   F   F0       4  1
        F1       5 10
        F2       1  8
        F3       9  0
    M   F0       5  3
        F1       2  3
        F2       7  4
        F3       7  0
N   F   F0       4  1
        F1       6 11
        F2       1  9
        F3      10  0
    M   F0       6  3
        F1       2  7
        F2       7  3
        F3       7  0
```

Most of this occurs because slow learners never get into Form 3.

Unfortunately, Akaike's information citerion is not defined for this model, so we cannot automate the simplification using `step` or `stepAIC`. We need to do the model simplification longhand, therefore, remembering to do F tests (not chi-squared) because of the overdispersion. Here is the last step of the simplification before obtaining the minimal adequate model. Do we need the age by learning interaction?

```
model4<-update(model3,~. - Age:Lrn)
anova(model3,model4,test="F")
```

```
Analysis of Deviance Table
    Resid. Df     Res.Dev   Df    Deviance   F     Pr(> -F)
1   127   1280.52
2   129   1301.08               -2    -20.56  1.0306  0.3598
```

No, we do not. So here is the minimal adequate model with quasi-Poisson errors:

```
summary(model4)
Coefficients:
```

	Estimate	Std. Error	t value	Pr(> | t |)	
(Intercept)	2.83161	0.30489	9.287	4.98e-16	***
EthN	0.09821	0.38631	0.254	0.79973	
SexM	-0.56268	0.38877	-1.447	0.15023	
AgeF1	-0.20878	0.35933	-0.581	0.56223	
AgeF2	0.16223	0.37481	0.433	0.66586	
AgeF3	-0.25584	0.37855	-0.676	0.50036	
LrnSL	0.50311	0.30798	1.634	0.10479	
EthN:SexM	-0.24554	0.37347	-0.657	0.51206	
EthN:AgeF1	-0.68742	0.46823	-1.468	0.14450	
EthN:AgeF2	-1.07361	0.42449	-2.529	0.01264	*
EthN:AgeF3	0.01879	0.42914	0.044	0.96513	
EthN:LrnSL	-0.65154	0.45857	-1.421	0.15778	
SexM:AgeF1	-0.26358	0.50673	-0.520	0.60385	

SexM:AgeF2	0.94531	0.43530	2.172	0.03171	*
SexM:AgeF3	1.35285	0.42933	3.151	0.00202	**
SexM:LrnSL	-0.29570	0.41144	-0.719	0.47363	
EthN:SexM:LrnSL	1.60463	0.57112	2.810	0.00573	**

```
(Dispersion parameter for quasipoisson family taken to be 9.833426)
      Null deviance: 2073.5 on 145 degrees of freedom
Residual deviance: 1301.1 on 129 degrees of freedom
```

There is a very significant three-way interaction between ethnic origin, sex and learning difficulty; non-Aboriginal slow-learning boys were more likely to be absent than non-Aboriginal boys without learning difficulties.

```
ftable(tapply(Days,list(Eth,Sex,Lrn),mean))
```

```
          AL        SL

A F   14.47368  27.36842
  M   22.28571  20.20000
N F   13.14286   7.00000
  M   13.36364  17.00000
```

Note, however, that among the pupils without learning difficulties it is the Aboriginal boys who miss the most days, and it is Aboriginal girls with learning difficulties who have the highest rate of absenteeism overall.

14.6   Negative binomial errors

Instead of using quasi-Poisson errors (as above) we could use a negative binomial model. This is in the MASS library and involves the function glm.nb. The modelling proceeds in exactly the same way as with a typical GLM:

```
model.nb1<-glm.nb(Days~Eth*Sex*Age*Lrn)
summary(model.nb1,cor=F)
```

```
Call:
glm.nb(formula = Days ~ Eth * Sex * Age * Lrn,
          init.theta = 1.92836014510701, link - log)

(Dispersion parameter for Negative Binomial(1.9284) family taken to be 1)

    Null deviance:     272.29 on 145 degrees of freedom
Residual deviance:     167.45 on 118 degrees of freedom
AIC: 1097.3
              Theta: 1.928
          Std. Err.: 0.269
    2 x log-likelihood: -1039.324
```

The output is slightly different than for a conventional GLM: you see the estimated negative binomial parameter (here called Theta, but known to us as k, and equal to 1.928) and its approximate standard error (0.269) and 2 times the log-likelihood (contrast this with the residual deviance from our quasi-Poisson model, which was 1301.1; see above). Note that the residual deviance in the negative binomial model (167.45) is not 2 times the log-likelihood.

An advantage of the negative binomial model over the quasi-Poisson is that we can automate the model simplification with `stepAIC`:

```
model.nb2<-stepAIC(model.nb1)
summary(model.nb2,cor=F)
```

```
Coefficients: (3 not defined because of singularities)
                 Estimate Std. Error z value Pr(>|z|)
(Intercept)        3.1693     0.3411   9.292  < 2e-16 ***
EthN              -0.3560     0.4210  -0.845 0.397848
SexM              -0.6920     0.4138  -1.672 0.094459 .
AgeF1             -0.6405     0.4638  -1.381 0.167329
AgeF2             -2.4576     0.8675  -2.833 0.004612 **
AgeF3             -0.5880     0.3973  -1.480 0.138885
LrnSL             -1.0264     0.7378  -1.391 0.164179
EthN:SexM         -0.3562     0.3854  -0.924 0.355364
EthN:AgeF1         0.1500     0.5644   0.266 0.790400
EthN:AgeF2        -0.3833     0.5640  -0.680 0.496746
EthN:AgeF3         0.4719     0.4542   1.039 0.298824
SexM:AgeF1         0.2985     0.6047   0.494 0.621597
SexM:AgeF2         3.2904     0.8941   3.680 0.000233 ***
SexM:AgeF3         1.5412     0.4548   3.389 0.000702 ***
EthN:LrnSL         0.9651     0.7753   1.245 0.213255
SexM:LrnSL         0.5457     0.8013   0.681 0.495873
AgeF1:LrnSL        1.6231     0.8222   1.974 0.048373 *
AgeF2:LrnSL        3.8321     1.1054   3.467 0.000527 ***
AgeF3:LrnSL            NA         NA      NA       NA
EthN:SexM:LrnSL    1.3578     0.5914   2.296 0.021684 *
EthN:AgeF1:LrnSL  -2.1013     0.8728  -2.408 0.016058 *
EthN:AgeF2:LrnSL  -1.8260     0.8774  -2.081 0.037426 *
EthN:AgeF3:LrnSL      NA         NA      NA       NA
SexM:AgeF1:LrnSL  -1.1086     0.9409  -1.178 0.238671
SexM:AgeF2:LrnSL  -2.8800     1.1550  -2.493 0.012651 *
SexM:AgeF3:LrnSL      NA         NA      NA       NA

(Dispersion parameter for Negative Binomial(1.8653) family taken to be 1)

    Null deviance: 265.27  on 145  degrees of freedom
Residual deviance: 167.44  on 123  degrees of freedom
AIC: 1091.4

Number of Fisher Scoring iterations: 1

          Theta:   1.865
      Std. Err.:   0.258

2 x log-likelihood:  -1043.409
```

We take up the model simplification where AIC leaves off:

```
model.nb3<-update(model.nb2,~. - Sex:Age:Lrn)
anova(model.nb3,model.nb2)
```

```
Likelihood ratio tests of Negative Binomial Models

      theta Resid. df    2 x log-lik.    Test     df LR stat.    Pr(Chi)
1 1.789507         125      -1049.111
2 1.865343         123      -1043.409 1 vs 2      2 5.701942 0.05778817
```

Because we are so ruthless, the marginally significant sex by age by learning interaction does not survive a deletion test ($p = 0.058$), nor do ethnic origin by age by learning ($p = 0.115$) nor age by learning ($p = 0.150$):

```
model.nb4<-update(model.nb3,~. - Eth:Age:Lrn)
anova(model.nb3,model.nb4)

Likelihood ratio tests of Negative Binomial Models

   Resid. df    2 x log-lik.    Test     df LR stat.    Pr(Chi)
1        127      -1053.431
2        125      -1049.111 1 vs 2      2 4.320086 0.1153202

model.nb5<-update(model.nb4,~. - Age:Lrn)
anova(model.nb4,model.nb5)

Likelihood ratio tests of Negative Binomial Models

    2 x log-lik.    Test     df LR stat.    Pr(Chi)
1      -1057.219
2      -1053.431 1 vs 2      2 3.787823 0.150482

summary(model.nb5,cor=F)

Coefficients:
                 Estimate Std. Error z value Pr(>|z|)
(Intercept)       2.91755    0.32626   8.942  < 2e-16 ***
EthN              0.05666    0.39515   0.143  0.88598
SexM             -0.55047    0.39014  -1.411  0.15825
AgeF1            -0.32379    0.38373  -0.844  0.39878
AgeF2            -0.06383    0.42046  -0.152  0.87933
AgeF3            -0.34854    0.39128  -0.891  0.37305
LrnSL             0.57697    0.33382   1.728  0.08392 .
EthN:SexM        -0.41608    0.37491  -1.110  0.26708
EthN:AgeF1       -0.56613    0.43162  -1.312  0.18965
EthN:AgeF2       -0.89577    0.42950  -2.086  0.03702 *
EthN:AgeF3        0.08467    0.44010   0.192  0.84744
SexM:AgeF1       -0.08459    0.45324  -0.187  0.85195
SexM:AgeF2        1.13752    0.45192   2.517  0.01183 *
SexM:AgeF3        1.43124    0.44365   3.226  0.00126 **
EthN:LrnSL       -0.78724    0.43058  -1.828  0.06750 .
SexM:LrnSL       -0.47437    0.45908  -1.033  0.30147
EthN:SexM:LrnSL   1.75289    0.58341   3.005  0.00266 **

(Dispersion parameter for Negative Binomial(1.6786) family taken to be 1)

    Null deviance: 243.98  on 145  degrees of freedom
Residual deviance: 168.03  on 129  degrees of freedom
AIC: 1093.2
```

```
Number of Fisher Scoring iterations: 1

              Theta:   1.679
          Std. Err.:   0.227

2 x log-likelihood:   -1057.219
```

The minimal adequate model, therefore, contains exactly the same terms as we obtained with quasi-Poisson, but the significance levels are higher (e.g. the three-way interaction has $p = 0.002\,66$ compared with $p = 0.005\,73$). We need to plot the model to check assumptions:

```
plot(model.nb5)
```

The variance is well behaved and the residuals are close to normally distributed. The combination of low p values and the ability to use stepAIC makes glm.nb a very useful modelling function for count data such as these.

15

Count Data in Tables

The analysis of count data with categorical explanatory variables comes under the heading of contingency tables. The general method of analysis for contingency tables involves log-linear modelling, but the simplest contingency tables are often analysed by Pearson's chi-squared, Fisher's exact test or tests of binomial proportions (see p. 365).

15.1   A two-class table of counts

You count 47 animals and find that 29 of them are males and 18 are females. Are these data sufficiently male-biased to reject the null hypothesis of an even sex ratio? With an even sex ratio the expected number of males and females is $47/2 = 23.5$. The simplest test is Pearson's chi-squared in which we calculate

$$\chi^2 = \sum \frac{(\text{observed} - \text{expected})^2}{\text{expected}}.$$

Substituting our observed and expected values, we get

$$\chi^2 = \frac{(29 - 23.5)^2 + (18 - 23.5)^2}{23.5} = 2.574\,468.$$

This is less than the critical value for chi-squared with 1 degree of freedom (3.841), so we conclude that the sex ratio is not significantly different from 50:50. There is a built-in function for this:

```
observed <- c(29,18)
chisq.test(observed)
```

```
        Chi-squared test for given probabilities
data: observed
X-squared = 2.5745, df = 1, p-value = 0.1086
```

The R Book, Second Edition. Michael J. Crawley.
© 2013 John Wiley & Sons, Ltd. Published 2013 by John Wiley & Sons, Ltd.

which indicates that a sex ratio of this size or more extreme than this would arise by chance alone about 10% of the time ($p = 0.1086$). Alternatively, you could carry out a binomial test:

```
binom.test(observed)
```

```
        Exact binomial test
data: observed
number of successes = 29, number of trials = 47, p-value = 0.1439
alternative hypothesis: true probability of success is not equal to 0.5
95 percent confidence interval:
0.4637994   0.7549318
sample estimates:
probability of success
        0.6170213
```

You can see that the 95% confidence interval for the proportion of males (0.46, 0.75) contains 0.5, so there is no evidence against a 50:50 sex ratio in these data. The p value is slightly different than it was in the chi-squared test, but the interpretation is exactly the same.

15.2   Sample size for count data

How many samples do you need before you have any chance of detecting a significant departure from equality? Suppose you are studying sex ratios in families. How many female children would you need to discover in a family with no males before you could conclude that a father's sex-determining chromosomes were behaving oddly? What about five females and no males? This is not significant because it can occur by chance when $p = 0.5$ with probability $2 \times 0.5^5 = 0.0625$ (note that this is a two-tailed test). The smallest sample that gives significance is a family of six children, all of one sex: $2 \times 0.5^6 = 0.031\,25$. How big would the sample need to be to reject the null hypothesis if one of the children was of the opposite sex? One out of seven is no good, as is one out of eight. You need a sample of at least nine children before you can reject the hypothesis that $p = 0.5$ when one of the children is of the opposite sex. Here is that calculation using the binom.test function:

```
binom.test(1,9)
```

```
        Exact binomial test
data: 1 and 9
number of successes = 1, number of trials = 9, p-value = 0.03906
alternative hypothesis: true probability of success is not equal to 0.5
95 percent confidence interval:
0.002809137   0.482496515
sample estimates:
probability of success
        0.1111111
```

15.3   A four-class table of counts

Mendel's famous peas produced 315 yellow round phenotypes, 101 yellow wrinkled, 108 green round and 32 green wrinkled offspring (a total of 556):

```
observed <- c(315,101,108,32)
```

The question is whether these data depart significantly from the 9:3:3:1 expectation that would arise if there were two independent 3:1 segregations (with round seeds dominating wrinkled, and yellow seeds dominating green).

Because the null hypothesis is not a 25:25:25:25 distribution across the four categories, we need to calculate the expected frequencies explicitly:

```
(expected <- 556*c(9,3,3,1)/16)

312.75 104.25 104.25 34.75
```

The expected frequencies are very close to the observed frequencies in Mendel's experiment, but we need to quantify the difference between them and ask how likely such a difference is to arise by chance alone:

```
chisq.test(observed,p=c(9,3,3,1),rescale.p=TRUE)

        Chi-squared test for given probabilities
data: observed
X-squared = 0.47, df = 3, p-value = 0.9254
```

Note the use of different probabilities for the four phenotypes: $p=c(9,3,3,1)$. Because these values do not sum to 1.0, we require the extra argument `rescale.p=TRUE`. A difference as big as or bigger than the one observed will arise by chance alone in more than 92% of cases and is clearly not statistically significant. The Pearson's chi-squared value is

```
sum((observed-expected)^2/expected)

[1]   0.470024
```

and the p-value comes from the right-hand tail of the cumulative probability function of the chi-squared distribution `1-pchisq` with 3 degrees of freedom (4 comparisons minus 1 for contingency; the total count must be 556)

```
1-pchisq(0.470024,3)

[1]   0.9254259
```

exactly as we obtained using the built-in `chisq.test` function, above.

15.4   Two-by-two contingency tables

Count data are often classified by more than one categorical explanatory variable. When there are two explanatory variables and both have just two levels, we have the famous 2×2 contingency table (see p. 365). We can return to the example of Mendel's peas. We need to convert the vector of observed counts into a matrix with two rows:

```
observed <- matrix(observed,nrow=2)
observed

        [,1]  [,2]
[1,]   315   108
[2,]   101    32
```

Fisher's exact test (p. 371) can take such a matrix as its sole argument:

```
fisher.test(observed)
```

```
        Fisher's Exact Test for Count Data
data: observed
p-value = 0.819
alternative hypothesis: true odds ratio is not equal to 1
95 percent confidence interval:
0.5667874  1.4806148
sample estimates:
odds ratio
0.9242126
```

Alternatively we can use Pearson's chi-squared test with Yates' continuity correction:

```
chisq.test(observed)
```

```
          Pearson's Chi-squared test with Yates' continuity correction
data: observed
X-squared = 0.0513, df = 1, p-value = 0.8208
```

Again, the *p*-values are different with different tests, but the interpretation is the same: these pea plants behave in accordance with Mendel's predictions of two independent traits, coat colour and seed shape, each segregating 3:1.

15.5   Using log-linear models for simple contingency tables

It is worth repeating these simple examples with a log-linear model so that when we analyse more complex cases you have a feel for what the GLM is doing. Recall that the deviance for a log-linear model of count data (p. 562) is

$$\text{deviance} = 2 \sum O \ln\left(\frac{O}{E}\right),$$

where O is a vector of observed counts and E is a vector of expected counts. Our first example had 29 males and 18 females and we wanted to know if the sex ratio was significantly male-biased:

```
observed <- c(29,18)
summary(glm(observed~1,poisson))
```

```
Coefficients:
            Estimate Std. Error z value Pr(>|z|)
(Intercept)   3.1570     0.1459   21.64   <2e-16 ***
(Dispersion parameter for poisson family taken to be 1)

    Null deviance: 2.5985  on 1  degrees of freedom
Residual deviance: 2.5985  on 1  degrees of freedom
AIC: 14.547
```

Only the bottom part of the summary table is informative in this case. The residual deviance is compared to the critical value of chi-squared in tables with 1 d.f.:

```
1-pchisq(2.5985,1)

[1]  0.1069649
```

We accept the null hypothesis that the sex ratio is 50:50 ($p = 0.106\,96$).

In the case of Mendel's peas we had a four-level categorical variable (i.e. four phenotypes) and the null hypothesis was a 9:3:3:1 distribution of traits:

```
observed <- c(315,101,108,32)
```

We need vectors of length 4 for the two seed traits, shape and colour:

```
shape <- factor(c("round","round","wrinkled","wrinkled"))
colour <- factor(c("yellow","green","yellow","green"))
```

Now we fit a saturated model (model1) and a model without the interaction term (model2) and compare the two models using anova with a chi-squared test:

```
model1 <- glm(observed~shape*colour,poisson)
model2 <- glm(observed~shape+colour,poisson)
anova(model1,model2,test="Chi")

Analysis of Deviance Table
Model 1: observed ~ shape * colour
Model 2: observed ~ shape + colour
     Resid. Df       Resid. Dev       Df      Deviance    P(> | Chi |)
1            0         1.021e-14
2            1           0.11715      -1      -0.11715         0.73215
```

There is no interaction between seed colour and seed shape ($p = 0.732\,15$) so we conclude that the two traits are independent and the phenotypes are distributed 9:3:3:1 as predicted. The p value is slightly different because the ratios of the two dominant traits are not exactly 3:1 in the data: round to wrinkled is $\exp(1.089\,04) = 2.971\,42$ and yellow to green is $\exp(1.157\,02) = 3.180\,441$:

```
summary(model2)

Coefficients:
                Estimate    Std. Error    z value    Pr(> | z |)
(Intercept)      4.60027       0.09013      51.04        <2e-16    ***
shapewrinkled   -1.08904       0.09771     -11.15        <2e-16    ***
colouryellow     1.15702       0.09941      11.64        <2e-16    ***
```

To summarize, the log-linear model involves fitting a saturated model with zero residual deviance (a parameter is estimated for every row of the dataframe) and then simplifying the model by removing the highest-order interaction term. The increase in deviance gives the chi-squared value for testing the hypothesis of independence. The minimal model must contain all the nuisance variables necessary to constrain the marginal totals (i.e. main effects for shape and colour in this example), as explained on p. 604.

15.6   The danger of contingency tables

We have already dealt with simple contingency tables and their analysis using Fisher's exact test or Pearson's chi-squared (see p. 365). But there is an important further issue to be dealt with. In observational studies we quantify only a limited number of explanatory variables. It is inevitable that we shall fail to measure a number of factors that have an important influence on the behaviour of the system in question. That's life, and given that we make every effort to note the important factors, there is little we can do about it. The problem comes when we ignore factors that have an important influence on the response variable. This difficulty can be particularly acute if we aggregate data over important explanatory variables. An example should make this clear.

Suppose we are carrying out a study of induced defences in trees. A preliminary trial has suggested that early feeding on a leaf by aphids may cause chemical changes in the leaf which reduce the probability of that leaf being attacked later in the season by hole-making insects. To this end we mark a large cohort of leaves, then score whether they were infested by aphids early in the season and whether they were holed by insects later in the year. The work was carried out on two different trees and the results were as follows:

Tree	Aphids	Holed	Intact	Total leaves	Proportion holed
Tree 1	Absent	35	1750	1785	0.0196
	Present	23	1146	1169	0.0197
Tree 2	Absent	146	1642	1788	0.0817
	Present	30	333	363	0.0826

There are four variables: the response variable, count, with eight values (in columns 3 and 4, with row totals in column 5), a two-level factor for late season feeding by caterpillars (holed or intact), a two-level factor for early season aphid feeding (aphids present or absent) and a two-level factor for tree (the observations come from two separate trees, imaginatively named tree 1 and tree 2).

```
induced <- read.table("C:\\temp\\induced.txt",header=T)
attach(induced)
names(induced)
```

```
[1]"Tree"   "Aphid"   "Caterpillar"   "Count"
```

We begin by fitting what is known as a **saturated model**. This is a curious thing, which has as many parameters as there are values of the response variable. The fit of the model is perfect, so there are no residual degrees of freedom and no residual deviance. The reason why we fit a saturated model is that it is always the best place to start modelling complex contingency tables. If we fit the saturated model, then there is no risk that we inadvertently leave out important interactions between the so-called 'nuisance variables'. These are the parameters that need to be in the model to ensure that the marginal totals (the row and column totals, for instance) are properly constrained.

```
model <- glm(Count~Tree*Aphid*Caterpillar,family=poisson)
```

The asterisk notation ensures that the saturated model is fitted, because all of the main effects and two-way interactions are fitted, along with the three-way `Tree` by `Aphid` by `Caterpillar` interaction. The model fit involves the estimation of $2 \times 2 \times 2 = 8$ parameters, and exactly matches the eight values of the response variable, `Count`. Looking at the saturated model in any detail serves no purpose, because the reams of information it contains are all superfluous.

The first real step in the modelling is to use `update` to remove the three-way interaction from the saturated model, and then to use `anova` to test whether the three-way interaction is significant or not:

```
model2 <- update(model, ~ . - Tree:Aphid:Caterpillar)
```

The punctuation here is very important (it is 'comma, tilde, dot, minus'), and note the use of colons rather than asterisks to denote interaction terms rather than main effects plus interaction terms. Now we can see whether the three-way interaction was significant by specifying `test="Chi"` like this:

```
anova(model,model2,test="Chi")

Analysis of Deviance Table

Model 1: Count ~ Tree * Aphid * Caterpillar
Model 2: Count ~ Tree + Aphid + Caterpillar + Tree:Aphid +
    Tree:Caterpillar + Aphid:Caterpillar
  Resid. Df Resid. Dev Df    Deviance Pr(>Chi)
1         0 0.00000000
2         1 0.00079137 -1 -0.00079137   0.9776
```

This shows very clearly that the interaction between caterpillar attack and leaf holing does not differ from tree to tree ($p = 0.977\,56$). Note that if this interaction had been significant, then we would have stopped the modelling at this stage. But it was not, so we leave it out and continue.

What about the main question? Is there an interaction between aphid attack and leaf holing? To test this we delete the `Caterpillar` by `Aphid` interaction from the model, and assess the results using `anova`:

```
model3 <- update(model2, ~ . - Aphid:Caterpillar)
anova(model3,model2,test="Chi")

Analysis of Deviance Table

Model 1: Count ~ Tree + Aphid + Caterpillar + Tree:Aphid + Tree:Caterpillar
Model 2: Count ~ Tree + Aphid + Caterpillar + Tree:Aphid + Tree:Caterpillar +
    Aphid:Caterpillar
  Resid. Df Resid. Dev Df Deviance Pr(>Chi)
1         2  0.0040853
2         1  0.0007914  1 0.003294   0.9542
```

There is absolutely no hint of an interaction ($p = 0.954$). The interpretation is clear: this work provides no evidence at all for induced defences caused by early season caterpillar feeding.

But look what happens when we do the modelling the wrong way. Suppose we went straight for the interaction of interest, `Aphid` by `Caterpillar`. We might proceed like this:

```
wrong <- glm(Count~Aphid*Caterpillar,family=poisson)
wrong1 <- update (wrong,~. - Aphid:Caterpillar)
anova(wrong,wrong1,test="Chi")

Analysis of Deviance Table

Model 1: Count ~ Aphid * Caterpillar
Model 2: Count ~ Aphid + Caterpillar
  Resid. Df Resid. Dev Df Deviance Pr(>Chi)
1         4     550.19
2         5     556.85 -1  -6.6594 0.009864 **
```

The `Aphid` by `Caterpillar` interaction is highly significant ($p = 0.01$), providing strong evidence for induced defences. This is wrong! By failing to include `Tree` in the model we have omitted an important explanatory variable. As it turns out, and as we should really have determined by more thorough preliminary analysis, the trees differ enormously in their average levels of leaf holing:

```
as.vector(tapply(Count,list(Caterpillar,Tree),sum))[1]/tapply(Count,Tree,sum)   [1]

        Tree1
0.01963439
```

```
as.vector(tapply(Count,list(Caterpillar,Tree),sum))[3]/tapply(Count,Tree,sum)   [2]

        Tree2
0.08182241
```

`Tree2` has more than four times the proportion of its leaves holed by caterpillars. If we had been paying more attention when we did the modelling the wrong way, we should have noticed that the model containing only `Aphid` and `Caterpillar` had massive overdispersion, and this should have alerted us that all was not well.

The moral is simple and clear. Always fit a saturated model first, containing all the variables of interest and all the interactions involving the nuisance variables (`Tree` in this case). Only delete from the model those interactions that involve the variables of interest (`Aphid` and `Caterpillar` in this case). Main effects are meaningless in contingency tables (they do nothing more than constrain the marginal totals), as are the model summaries. Always test for overdispersion. It will never be a problem if you follow the advice of simplifying down from a saturated model, because you only ever leave out non-significant terms, and you never delete terms involving any of the nuisance variables.

15.7   Quasi-Poisson and negative binomial models compared

The data on red blood cell counts are read from a file:

```
data <- read.table("c:\\temp\\bloodcells.txt",header=T)
attach(data)
names(data)

[1] "count"
```

Now we need to create a vector for gender containing 5000 repeats of 'female' and then 5000 repeats of 'male':

```
gender <- factor(rep(c("female","male"),c(5000,5000)))
```

The idea is to test the significance of the difference in mean cell counts for the two genders, which is slightly higher in males than in females:

```
tapply(count,gender,mean)

female      male
1.1986    1.2408
```

We begin with the simplest log-linear model – a GLM with Poisson errors:

```
model <- glm (count~gender,poisson)
summary(model)
```

You should check for overdispersion before drawing any conclusions about the significance of the gender effect. It turns out that there is substantial overdispersion (scale parameter $= 23\,154/9998 = 2.315\,863$), so we repeat the modelling using quasi-Poisson errors instead:

```
model <- glm(count~gender,quasipoisson)
summary(model)

Coefficients:
                Estimate      Std. Error    t value   Pr(>| t |)
(Intercept)     0.18115       0.02167       8.360     <2e-16      ***
gendermale      0.03460       0.03038       1.139     0.255

(Dispersion parameter for quasipoisson family taken to be 2.813817)

    Null deviance:    23158    on    9999    degrees    of    freedom
Residual deviance:    23154    on    9998    degrees    of    freedom
AIC: NA
Number of Fisher Scoring iterations: 6
```

As you see, the gender effect falls well short of significance ($p = 0.255$).

Alternatively, you could use a GLM with negative binomial errors. The function is in the MASS library:

```
library(MASS)
model <- glm.nb(count~gender)
summary(model)

Call:
glm.nb(formula = count ~ gender, init.theta = 0.6676246007, link = log)

Deviance Residuals:
    Min        1Q    Median        3Q       Max
-1.1842   -1.1716   -1.1716    0.3503    3.1522

Coefficients:
            Estimate Std. Error z value Pr(>|z|)
(Intercept)  0.18115    0.02160   8.388   <2e-16 ***
gendermale   0.03460    0.03045   1.136    0.256

(Dispersion parameter for Negative Binomial(0.6676) family taken to be 1)

    Null deviance: 9610.8  on 9999  degrees of freedom
Residual deviance: 9609.5  on 9998  degrees of freedom
AIC: 30362

Number of Fisher Scoring iterations: 1

            Theta:   0.6676
        Std. Err.:   0.0185

2 x log-likelihood:   -30355.6010
```

You would come to the same conclusion, although the p value is slightly different ($p = 0.256$).

15.8 A contingency table of intermediate complexity

We start with a three-dimensional table of count data from college records. It is a contingency table with two levels of year (freshman and sophomore), two levels of discipline (arts and science), and two levels of gender (male and female):

```
numbers <- c(24,30,29,41,14,31,36,35)
```

The statistical question is whether the relationship between gender and discipline varies between freshmen and sophomores (i.e. we want to know the significance of the three-way interaction between year, discipline and gender).

The first task is to define the dimensions of `numbers` using the `dim` function:

```
dim(numbers) <- c(2,2,2)
numbers
```

```
, , 1

      [,1]    [,2]
[1,]    24      29
[2,]    30      41
, , 2

      [,1]    [,2]
[1,]    14      36
[2,]    31      35
```

The top table refers to the males `[,,1]` and the bottom table to the females `[,,2]`. Within each table, the rows are the year groups and the columns are the disciplines. It would make the table much easier to understand if we provided these dimensions with names using the `dimnames` function:

```
dimnames(numbers)[[3]] <- list("male", "female")
dimnames(numbers)[[2]] <- list("arts", "science")
dimnames(numbers)[[1]] <- list("freshman", "sophomore")
```

To see this as a flat table, use the `ftable` function like this:

```
ftable(numbers)
```

```
                    male female
freshman     arts     24     14
             science  29     36
sophomore    arts     30     31
             science  41     35
```

The thing to understand is that the `dimnames` are the factor levels (e.g. male or female), not the names of the factors (e.g. gender).

We convert this table into a dataframe using the `as.data.frame.table` function. This saves us from having to create separate vectors to describe the levels of gender, year and discipline associated with each count:

```
as.data.frame.table(numbers)
```

```
       Var1      Var2   Var3 Freq
1  freshman      arts   male   24
```

```
2 sophomore     arts    male    30
3  freshman science    male    29
4 sophomore science    male    41
5  freshman    arts female    14
6 sophomore    arts female    31
7  freshman science female    36
8 sophomore science female    35
```

You can see that R has generated reasonably sensible variable names for the four columns, but we want to use our own names:

```
frame <- as.data.frame.table(numbers)
names(frame) <- c("year","discipline","gender","count")
frame
```

```
          year discipline gender count
1   freshman         arts   male    24
2 sophomore         arts   male    30
3   freshman      science   male    29
4 sophomore      science   male    41
5   freshman         arts female    14
6 sophomore         arts female    31
7   freshman      science female    36
8 sophomore      science female    35
```

Now we can do the statistical modelling. The response variable is count, and we begin by fitting a saturated model with eight estimated parameters (i.e. the model generates the observed counts exactly, so the deviance is zero and there are no degrees of freedom):

```
attach(frame)
model1 <- glm(count~year*discipline*gender,poisson)
```

We test for the significance of the year by discipline by gender interaction by deleting the year by discipline by gender interaction from model1 to make model2 using update:

```
model2 <- update(model1,~. - year:discipline:gender)
```

then comparing model1 and model2 using anova with a chi-squared test:

```
anova(model1,model2,test="Chi")
```

```
Analysis of Deviance Table
Model 1: count ~ year * discipline * gender
Model 2: count ~ year + discipline + gender + year:discipline +
year:gender +discipline:gender
```

	Resid. Df	Resid. Dev	Df	Deviance	P(> | Chi |)
1	0	-5.329e-15			
2	1	3.08230	-1	-3.08230	0.07915

The interaction is not significant ($p = 0.079$), indicating similar gender by discipline relationships in the two year groups. We finish the analysis at this point because we have answered the question that we were asked to address.

15.9   Schoener's lizards: A complex contingency table

In this section we are interested in whether lizards show any niche separation across various ecological factors and, in particular, whether there are any interactions – for example, whether they show different habitat separation at different times of day:

```
lizards <- read.table("c:\\temp\\lizards.txt",header=T)
attach(lizards)
names(lizards)
```

```
[1]"n" "sun" "height" "perch" "time" "species"
```

The response variable is n, the count for each contingency. The explanatory variables are all categorical: sun is a two-level factor (Sun and Shade within the bush), height is a two-level factor (High and Low within the bush), perch is a two-level factor (Broad and Narrow twigs), time is a three-level factor (Afternoon, Mid.day and Morning), and there are two lizard species both belonging to the genus *Anolis* (*A. grahamii* and *A. opalinus*). As usual, we begin by fitting a saturated model, fitting all the interactions and main effects:

```
model1 <- glm(n~sun*height*perch*time*species,poisson)
```

Model simplification begins with removal of the highest-order interaction effect: the sun by height by perch by time by species interaction:

```
model2 <- update(model1, ~.- sun:height:perch:time:species)
anova(model1,model2,test="Chi")
```

```
Analysis of Deviance Table

  Resid. Df Resid. Dev Df    Deviance Pr(>Chi)
1         0 3.3472e-10
2         2 2.1807e-10 -2 1.1665e-10
```

When the change in deviance is so close to zero, R does not print a *p* value. Also, you need to ignore the warning messages in the early stages of model simplification with complex contingency tables. It is a considerable relief that this interaction is not significant (imagine trying to explain what it meant in the Discussion section of your paper). The key point to understand in this kind of analysis is that the only interesting terms are interactions involving species. All of the other interactions and main effects are nuisance variables that have to be retained in the model to constrain the marginal totals (see p. 368 for an explanation of what this means).

There are four four-way interactions of interest – species by sun by height by perch, species by sun by height by time, species by sun by perch by time, and species by height by perch by time – and we should test their significance by deleting them from model2 which contains all of the four-way interactions. Here goes:

```
model3 <- update(model2, ~.-sun:height:perch:species)
anova(model2,model3,test="Chi")
```

```
Analysis of Deviance Table

  Resid. Df Resid. Dev Df Deviance Pr(>Chi)
1         2     0.0000
2         3     2.7088 -1  -2.7088   0.0998 .
```

Close, but not significant ($p = 0.0998$). As ever, we are ruthless, so we shall leave it out.

```
model4 <- update(model2, ~.-sun:height:time:species)
anova(model2,model4,test="Chi")
```

```
Analysis of Deviance Table

  Resid. Df Resid. Dev Df Deviance Pr(>Chi)
1         2    0.00000
2         4    0.44164 -2 -0.44164   0.8019
```

Nothing at all ($p = 0.802$).

```
model5 <- update(model2, ~.-sun:perch:time:species)
anova(model2,model5,test="Chi")
```

```
Analysis of Deviance Table

  Resid. Df Resid. Dev Df Deviance Pr(>Chi)
1         2    0.00000
2         4    0.81008 -2 -0.81008    0.667
```

Again, nothing there ($p = 0.667$). Finally:

```
model6 <- update(model2, ~.-height:perch:time:species)
anova(model2,model6,test="Chi")
```

```
Analysis of Deviance Table

  Resid. Df Resid. Dev Df Deviance Pr(>Chi)
1         2     0.0000
2         4     3.2217 -2  -3.2217   0.1997
```

This means that none of the four-way interactions involving species need be retained.

Now we have to assess all six of the three-way interactions involving species: species by height by perch, species by height by time, species by perch by time, species by sun by height, species by sun by time, and species by sun by perch. Can we speed this up by using automatic deletion? Yes and no. Yes, we can use step, to remove terms assessed by AIC to be non-significant (p. 547). No, unless we are very careful. We must not allow step to remove any interactions that do not involve species (because these are the essential nuisance variables). We do this with the lower argument:

```
model7 <- step(model1,lower=~sun*height*perch*time)
```

```
Start:  AIC=259.25
n ~ sun * height * perch * time * species

                               Df   Deviance     AIC
- sun:height:perch:time:species  2 2.1807e-10  255.25
<none>                             3.3472e-10  259.25

Step:  AIC=255.25

                          Df Deviance     AIC
- sun:height:time:species  2   0.4416  251.69
- sun:perch:time:species   2   0.8101  252.06
- height:perch:time:species 2  3.2217  254.47
```

```
<none>                                   0.0000 255.25
- sun:height:perch:species      1     2.7088 255.96
- sun:height:perch:time         2     4.7901 256.04

Step:   AIC=251.69

                                Df Deviance      AIC
- sun:perch:time:species         2    1.0713 248.32
<none>                                0.4416 251.69
- height:perch:time:species      2    4.6476 251.90
- sun:height:perch:time          2    4.9482 252.20
- sun:height:perch:species       1    3.1113 252.36

Step:   AIC=248.32

                                Df Deviance      AIC
- sun:time:species               2    3.3403 246.59
<none>                                1.0713 248.32
- sun:height:perch:time          2    5.1261 248.38
- sun:height:perch:species       1    3.3016 248.55
- height:perch:time:species      2    5.7906 249.04

Step:   AIC=246.59

                                Df Deviance      AIC
<none>                                3.3403 246.59
- sun:height:perch:time          2    7.5288 246.78
- sun:height:perch:species       1    5.8273 247.08
- height:perch:time:species      2    8.5418 247.79
```

You can see that `step` has been very forgiving, and has left two of the four-way interactions involving species in the model. What we can do next is to take out all of the four-way interactions and start `step` off again with this simpler starting point. We want to start at the lower model plus all the three-way interactions involving species with sun, height, perch and time:

```
model8 <-
glm(n ~sun*height*perch*time+(species+sun+height+perch+time)^3,poisson)

model9 <- step(model8,lower= ~sun*height*perch*time)


Start:  AIC=249.82

                            Df Deviance     AIC
- perch:time:species         2    8.5912 245.84
- height:time:species        2    9.9709 247.22
- sun:perch:species          1    8.5874 247.84
- height:perch:species       1    8.8752 248.12
- sun:time:species           2   10.9032 248.15
<none>                            8.5730 249.82
- sun:height:perch:time      2   13.2313 250.48
- sun:height:species         1   11.8672 251.12
```

```
Step:   AIC=245.84

                            Df  Deviance      AIC
- height:time:species        2    9.9867   243.24
- sun:perch:species          1    8.6053   243.85
- height:perch:species       1    8.8891   244.14
- sun:time:species           2   10.9090   244.16
<none>                               8.5912   245.84
- sun:height:perch:time      2   13.2426   246.49
- sun:height:species         1   11.8675   247.12

Step:   AIC=243.24

                            Df  Deviance      AIC
- sun:time:species           2   11.7667   241.01
- sun:perch:species          1    9.9930   241.24
- height:perch:species       1   10.2429   241.49
<none>                               9.9867   243.24
- sun:height:species         1   12.4873   243.74
- sun:height:perch:time      2   14.8884   244.14

Step:   AIC=241.02

                            Df  Deviance      AIC
- sun:perch:species          1   11.783   239.03
- height:perch:species       1   11.979   239.23
<none>                              11.767   241.01
- sun:height:species         1   13.848   241.10
- sun:height:perch:time      2   16.911   242.16
- time:species               2   23.245   248.49

Step:   AIC=239.03

                            Df  Deviance      AIC
- height:perch:species       1   11.984   237.23
<none>                              11.783   239.03
- sun:height:species         1   13.923   239.17
- sun:height:perch:time      2   16.949   240.20
- time:species               2   23.385   246.63

Step:   AIC=237.23

                            Df  Deviance      AIC
<none>                              11.984   237.23
- sun:height:species         1   14.205   237.45
- sun:height:perch:time      2   17.188   238.44
- time:species               2   23.713   244.96
- perch:species              1   24.921   248.17
```

Again, we need to be harsh and to test whether these terms really to deserve to stay in model9. The most complex term is the interaction sun by height by perch by time, but we do not want to remove this because it is a nuisance variable (the interaction does not involve species). We should start by removing sun by height by species:

```
model10 <- update(model9, ~.-sun:height:species)
anova(model9,model10,test="Chi")
```

```
Analysis of Deviance Table

  Resid. Df Resid. Dev Df Deviance Pr(>Chi)
1        17    11.984
2        18    14.205 -1  -2.2203   0.1362
```

This interaction is not significant, so we leave it out. Our model10 contains no three- or four-way interactions involving species.

Let us try deleting the two-way interactions in turn from model10:

```
model11 <- update(model10, ~.-sun:species)
model12 <- update(model10, ~.-height:species)
model13 <- update(model10, ~.-perch:species)
model14 <- update(model10, ~.-time:species)
anova(model10,model11,test="Chi")
```

```
Analysis of Deviance Table
      Resid. Df    Resid. Dev    Df    Deviance    P(> | Chi |)
1           18       14.2046
2           19       21.8917    -1     -7.6871          0.0056
```

We need to retain a main effect for sun ($p = 0.0056$).

```
anova(model10,model12,test="Chi")
```

```
Analysis of Deviance Table
      Resid. Df    Resid. Dev    Df    Deviance    P(> | Chi |)
1           18       14.205
2           19       36.271     -1     -22.066        2.634e-06
```

We need to retain a main effect for height ($p < 0.0001$).

```
anova(model10,model13,test="Chi")
```

```
Analysis of Deviance Table
      Resid. Df    Resid. Dev    Df    Deviance    P(> | Chi |)
1           18       14.2046
2           19       27.3346    -1     -13.1300         0.0003
```

We need to retain a main effect for perch ($p = 0.0003$).

```
anova(model10,model14,test="Chi")
```

```
Analysis of Deviance Table
        Resid. Df    Resid. Dev    Df    Deviance  P(> | Chi |)
1             18       14.205
2             20       25.802     -2     -11.597         0.003
```

We need to retain a main effect for time of day ($p = 0.003$).

To see where we are, we should produce a summary table of the counts:

```
ftable(tapply(n,list(species,sun,height,perch,time),sum))
```

				Afternoon	Mid.day	Morning
grahamii	Shade	High	Broad	4	1	2
			Narrow	3	1	3
		Low	Broad	0	0	0
			Narrow	1	0	0
	Sun	High	Broad	10	20	11
			Narrow	8	32	15
		Low	Broad	3	4	5
			Narrow	4	5	1
opalinus	Shade	High	Broad	4	8	20
			Narrow	5	4	8
		Low	Broad	12	8	13
			Narrow	1	0	6
	Sun	High	Broad	18	69	34
			Narrow	8	60	17
		Low	Broad	13	55	31
			Narrow	4	21	12

The modelling has indicated that species differ in their responses to all four explanatory variables, but that there are no interactions between the factors. The only remaining question for model simplification is whether we need to keep all three levels for time of day, or whether two levels would do just as well (we lump together Mid.day and Morning):

```
tod <- factor(1+(time=="Afternoon"))
model15 <- update(model10, ~.-species:time+species:tod)
anova(model10,model15,test="Chi")
```

```
Analysis of Deviance Table
      Resid. Df   Resid. Dev    Df    Deviance     P(> | Chi |)
1           18      14.2046
2           19      15.0232    -1     -0.8186            0.3656
```

That simplification was justified, so we keep time in the model but as a two-level factor.

That was hard, I think you will agree. You need to be extremely well organized to do this sort of analysis without making any mistakes. A high degree of serenity is required throughout. What makes it difficult is keeping track of the interactions that are in the model and those that have been excluded, and making absolutely sure that no nuisance variables have been omitted unintentionally. It turns out that life can be made much more straightforward if the analysis can be reformulated as an exercise in proportions rather than counts, because if it can, then all of the problems with nuisance variables disappear. On p. 643 the example is reanalysed with the response variable as a proportion in a GLM with binomial errors. This is possible because we have just two species, so we can reformulate the response as the proportion of all lizards that are *A. opalinus*. This is a big advantage because it does away with the need to retain any of the nuisance variables.

15.10   Plot methods for contingency tables

The departures from expectations of the observed frequencies in a contingency table can be regarded as $(O - E)/\sqrt{E}$. The R function called `assocplot` produces a Cohen–Friendly association plot indicating deviations from independence of rows and columns in a two-dimensional contingency table.

Here are data on hair colour and eye colour:

```
data(HairEyeColor)
(x <- margin.table(HairEyeColor, c(1, 2)) )
```

```
       Eye
Hair     Brown Blue Hazel Green
  Black     68   20    15     5
  Brown    119   84    54    29
  Red       26   17    14    14
  Blond      7   94    10    16
```

```
assocplot(x, main = "Relation between hair and eye color")
```

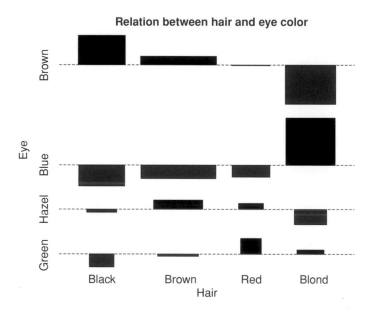

The plot shows the excess (black bars) of people with black hair who have brown eyes, the excess of people with blond hair who have blue eyes, and the excess of redheads who have green eyes. The red bars show categories where fewer people were observed than expected under the null hypothesis of independence of hair colour and eye colour.

Here are the same data plotted as a mosaic plot:

```
mosaicplot(HairEyeColor, shade = TRUE)
```

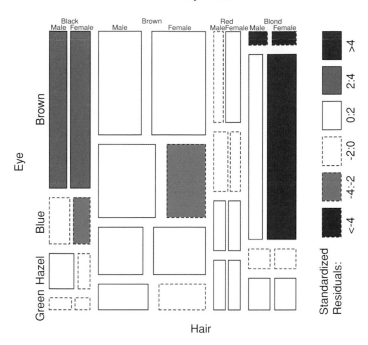

HairEyeColor

The plot indicates that there are significantly more blue-eyed blond females than expected in the case of independence, and too few brown-eyed blond females. Extended mosaic displays show the standardized residuals of a log-linear model of the counts by the colour and outline of the mosaic's tiles. Negative residuals are drawn in shades of red and with broken outlines, while positive residuals are drawn in shades of blue with solid outlines.

Where there are multiple 2×2 tables (dataframes with three or more categorical explanatory variables), then `fourfoldplot` might be useful. It allows a visual inspection of the association between two dichotomous variables in one or several populations (known as strata).

A classic example of contingency table data comes built in with R: `UCBAdmissions` describes the admissions policy of different departments at the University of California at Berkeley in relation to gender:

```
data(UCBAdmissions)
head(UCBAdmissions)
```

```
[1]  512 313   89   19  353 207
```

```
str(UCBAdmissions)
```

```
table [1:2, 1:2, 1:6]  512 313 89 19 353 207 17 8 120 205 . . .
- attr(*, "dimnames")=List of 3
 ..$ Admit : chr [1:2] "Admitted" "Rejected"
 ..$ Gender: chr [1:2] "Male" "Female"
 ..$ Dept  : chr [1:6] "A" "B" "C" "D" . . .
```

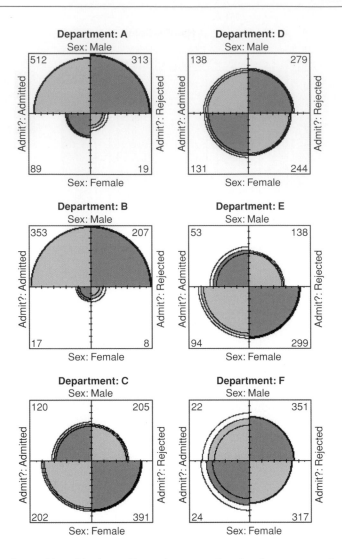

You see that the object is a table with three dimensions: two levels of status, two levels of gender and six departments. Here are the college admissions data plotted for each department separately:

```
x <- aperm(UCBAdmissions, c(2, 1, 3))
names(dimnames(x)) <- c("Sex", "Admit?", "Department")
ftable(x)
```

```
                  Department   A    B    C    D    E    F
Sex     Admit?
Male    Admitted             512  353  120  138   53   22
        Rejected             313  207  205  279  138  351
Female  Admitted              89   17  202  131   94   24
        Rejected              19    8  391  244  299  317

fourfoldplot(x, margin = 2)
```

You will need to compare the graphs with the frequency table (above) to see what is going on. The central questions are whether the rejection rate for females is different from the rejection rate for males, and whether any such difference varies from department to department. The log-linear model suggests that the difference docs vary with department ($p - 0.0011$; see below). That Department B attracted a smaller number of female applicants is very obvious. What is less clear (but in many ways more interesting) is that they rejected proportionally fewer of the female applicants (32%) than the male applicants (37%). You may find these plots helpful, but I must admit that I do not.

Here we use gl to generate factor levels for department, sex and admission, then fit a saturated contingency table model for the counts, x. We then use anova with test="Chi" to assess the significance of the three-way interaction:

```
dept <- gl(6,4)
sex <- gl(2,1,24)
admit <- gl(2,2,24)
model1 <- glm(as.vector(x) ~dept*sex*admit,poisson)
model2 <- update(model1, ~. -dept:sex:admit)
anova(model1,model2,test="Chi")
```

```
Analysis of Deviance Table
```

```
Model 1: as.vector(x) ~ dept * sex * admit
Model 2: as.vector(x) ~ dept + sex + admit + dept:sex + dept:admit +
sex:admit
  Resid. Df Resid. Dev Df Deviance Pr(>Chi)
1         0      0.000
2         5     20.204 -5  -20.204 0.001144 **
```

The interaction is highly significant, indicating that the admission rates of the two sexes differ substantially from department to department. There are five degrees of freedom for the interaction: (departments − 1) × (genders − 1). The remaining parameters are nuisance variables of no interest in this analysis.

Another way to do the same test is to turn the three-dimensional contingency table into a dataframe:

```
admissions <- as.data.frame(UCBAdmissions)
admissions
```

	Admit	Gender	Dept	Freq
1	Admitted	Male	A	512
2	Rejected	Male	A	313
3	Admitted	Female	A	89
4	Rejected	Female	A	19
5	Admitted	Male	B	353
6	Rejected	Male	B	207
7	Admitted	Female	B	17
8	Rejected	Female	B	8
9	Admitted	Male	C	120
10	Rejected	Male	C	205
11	Admitted	Female	C	202
12	Rejected	Female	C	391
13	Admitted	Male	D	138
14	Rejected	Male	D	279
15	Admitted	Female	D	131

```
16 Rejected Female    D   244
17 Admitted   Male    E    53
18 Rejected   Male    E   138
19 Admitted Female    E    94
20 Rejected Female    E   299
21 Admitted   Male    F    22
22 Rejected   Male    F   351
23 Admitted Female    F    24
24 Rejected Female    F   317
```

There is a useful function called xtabs ('cross tabulations') which creates a contingency table from cross-classifying factors using a formula interface (it has the added advantage of allowing you to specify the name of the dataframe with which to work). Here are the total numbers of applicants by gender and department:

```
xtabs(Freq~Gender+Dept,admissions)
```

```
        Dept
Gender    A    B    C    D    E    F
  Male   825  560  325  417  191  373
  Female 108   25  593  375  393  341
```

Notice that the proportion of female applicants was very low in some departments (e.g. A and B) but high in others (e.g. Department E had a female-biased application rate).

The formula interface allows the dot '.' convention, which means 'fit all the explanatory variables'; i.e. everything else in the dataframe except the response. The summary option with xtabs calculates the total number of applications, and tests for independence:

```
summary(xtabs(Freq ~ ., admissions))
```

```
Call: xtabs(formula = Freq ~ ., data = admissions)
Number of cases in table: 4526
Number of factors: 3
Test for independence of all factors:
        Chisq = 2000.3, df = 16, p-value = 0
```

Clearly there is a highly significant difference in the proportion of female applicants admitted across the different departments.

To turn the counts into proportions we can extract parts of the summary tables produced by xtabs, using subscripts. We start by calculating the total numbers of female applicants to each department:

```
xtabs(Freq~Admit+Dept+Gender,admissions)
```

```
, , Gender = Male

          Dept
Admit       A    B    C    D    E    F
  Admitted 512  353  120  138   53   22
  Rejected 313  207  205  279  138  351

, , Gender = Female

          Dept
Admit       A    B    C    D    E    F
  Admitted  89   17  202  131   94   24
  Rejected  19    8  391  244  299  317
```

Because this table is three-dimensional, we need to specify three subscripts separated by two commas. We want the column totals (`colSums`) from the lower half of the table (i.e. only for the females), for which the appropriate index is `[,,2]`:

```
females <- colSums(xtabs(Freq~Admit+Dept+Gender,admissions)[,,2])
females
```

```
  A   B   C   D   E   F
108  25 593 375 393 341
```

Now we want to extract the numbers of females admitted to each department, which is the top row `[1,]` of the lower half-table `[,,2]`:

```
admitted.females <- xtabs(Freq~Admit+Dept+Gender,admissions)[,,2][1,]
```

The proportion of female applicants admitted by departments is then simply:

```
(female.success <- admitted.females/females)
```

```
         A          B          C          D          E          F
0.82407407 0.68000000 0.34064081 0.34933333 0.23918575 0.07038123
```

It is no wonder that the interaction term was so significant ($p = 0.001\ 144$): the success rate varies from a low of 7% in Department F to a high of 82% in Department A.

15.11   Graphics for count data: Spine plots and spinograms

Here is an illustration of what a spine plot does. Suppose you have three treatments (placebo, drug A and drug B) and the response variable is a five-level categorical variable (much worse, worse, no change, better, much better). The data, one row per patient, consist of their current condition and the treatment they were given:

```
data <- read.table("c:\\temp\\spino.txt",header=T)
attach(data)
head(data)
```

```
   condition treatment
1  no.change    drug.A
2     better    drug.B
3     better    drug.B
4  no.change   placebo
5  no.change    drug.B
6  no.change    drug.B
```

The plot will be easier to interpret if we specify the order of the factor levels (they would be in alphabetic order by default):

```
condition<-factor(condition,c("much.worse","worse","no.change","better","much.better"))
treatment<-factor(treatment,c("placebo","drug.A","drug.B"))
```

Now we can use `spineplot` (either in (x, y) format or in $(y \sim x)$ format) like this:

```
spineplot(condition~treatment)
```

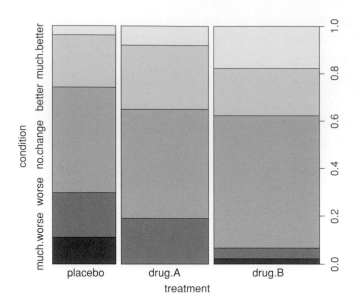

There are two things to notice about the spine plot. The grey-scale partitions within the bars are the proportions of a given treatment in each of the five conditions, i.e. the conditional relative frequencies of y in every x group (these are labelled on the left axis and quantified on the right axis). Where a category is empty (as in the 'much worse' level with drug A) the labels on the left can be confusing. The widths of the bars reflect the total sample sizes (there were more patients getting drug B than the placebo). It looks like drug B is better than the placebo, but the efficacy of drug A is less clear. Here are the counts:

```
table(condition,treatment)
```

```
             treatment
condition     placebo drug.A drug.B
  much.worse        3      0      1
  worse             5      7      2
  no.change        12     17     25
  better            6     10      9
  much.better       1      3      8
```

With so few patients showing changes in condition, we are going to struggle to find significant effects in this data set. To do the stats, we need to create a dataframe of these counts with matching columns, one to show the level of treatment and one to show the level of condition. The tool for this is as.data.frame.table:

```
as.data.frame.table(table(condition,treatment))
```

```
     condition treatment Freq
1    much.worse   placebo    3
2         worse   placebo    5
3     no.change   placebo   12
4        better   placebo    6
5   much.better   placebo    1
6    much.worse    drug.A    0
```

```
7          worse      drug.A     7
8      no.change      drug.A    17
9         better      drug.A    10
10  much.better       drug.A     3
11   much.worse       drug.B     1
12        worse       drug.B     2
13    no.change       drug.B    25
14       better       drug.B     9
15  much.better       drug.B     8
```

You might have thought about using `aggregate` to do this, where the function we want to apply would be `length` (for instance, to count how many patients receiving the placebo got much worse; we can see from the table above that the answer is `length=3` in this case)

```
aggregate(data,data,length)
```

```
      condition treatment condition treatment
1        better     drug.A        10        10
2   much.better     drug.A         3         3
3     no.change     drug.A        17        17
4         worse     drug.A         7         7
5        better     drug.B         9         9
6   much.better     drug.B         8         8
7    much.worse     drug.B         1         1
8     no.change     drug.B        25        25
9         worse     drug.B         2         2
10       better    placebo         6         6
11  much.better    placebo         1         1
12   much.worse    placebo         3         3
13    no.change    placebo        12        12
14        worse    placebo         5         5
```

As you can see, the problem is that `aggregate` leaves out rows from the dataframe when there were zero cases (i.e. no patients receiving drug A got much worse), so there are only 14 rows in the dataframe, not the 15 we want for doing the statistics.

As usual, we start by fitting a saturated model, then remove the highest-order interaction:

```
new <- as.data.frame.table(table(condition,treatment))
```

```
model1 <- glm(Freq~condition*treatment,poisson,data=new)
model2 <- glm(Freq~condition+treatment,poisson,data=new)
```

Then we compare the two models using `anova` with a chi-squared test:

```
anova(model1,model2,test="Chi")
```

```
Analysis of Deviance Table

Model 1: Freq ~ condition * treatment
Model 2: Freq ~ condition + treatment
  Resid. Df Resid. Dev Df Deviance P(>|Chi|)
1         0      0.000
2         8     15.133 -8  -15.133   0.05661 .
```

As we suspected, given the low replication, there is no significant interaction between treatment and condition ($p > 0.05$; close, but not significant).

In a spinogram, the response is categorical but the explanatory variable is continuous. The following data show parasitism (a binary response, parasitized or not) as a function of host population density:

```
wasps<-read.table("c:\\temp\\para.txt",header=T)
attach(wasps)
head(wasps)
```

```
  density fate
1       1  not
2       2  not
3       2  not
4       4  not
5       4  not
6       4  not
```

```
table(density,fate)
```

```
        fate
density not parasitized
      1   1           0
      2   2           0
      4   3           1
      8   6           2
     16  11           5
     32  12          20
     64  12          52
```

Apparently, the proportion of hosts parasitized increases as host density is increased. To visualize this, we use `spineplot` like this:

```
spineplot(fate~density)
```

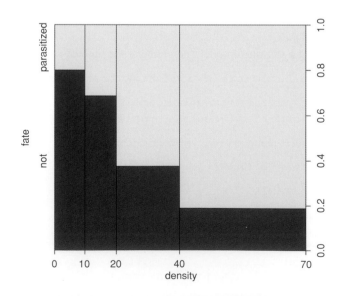

The trend of increasing parasitism with density is very clear. In these plots, the width of the sector indicates how many of the data fell in this range of population densities; there were equal numbers of hosts in the first two bins, but twice as many in the highest density category than in the category below, with a peak of just over 80% parasitized.

Alternatively, if you want a smooth curve you can use the conditional density plot `cdplot` like this:

```
cdplot(fate~density)
```

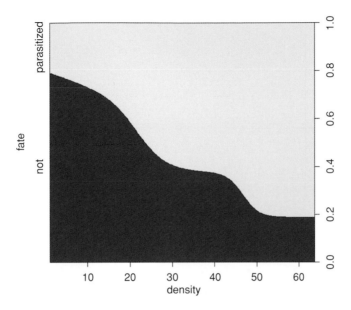

The trend is quantified using logistic regression like this:

```
model1<-glm(fate~density,binomial)
model2<-glm(fate~log(density),binomial)
AIC(model1,model2)
```

```
        df      AIC
model1  2 144.3978
model2  2 143.0790
```

We choose the log-transformed explanatory variable because this gives a lower AIC:

```
summary(model2)
```

```
Coefficients:
              Estimate Std. Error z value Pr(>|z|)
(Intercept)    -4.0230     1.0585  -3.801 0.000144 ***
log(density)    1.3062     0.2942   4.440    9e-06 ***

(Dispersion parameter for binomial family taken to be 1)

    Null deviance: 167.39  on 126  degrees of freedom
Residual deviance: 139.08  on 125  degrees of freedom
AIC: 143.08
```

The data and the fitted regression line are plotted like this:

```
plot(as.numeric(fate)-1~log(density),pch=16,ylab="parasitized")
xv<-seq(0,4.5,0.01)
yv<-1/(1+1/exp(-4.023+1.3062*xv))
lines(xv,yv)
```

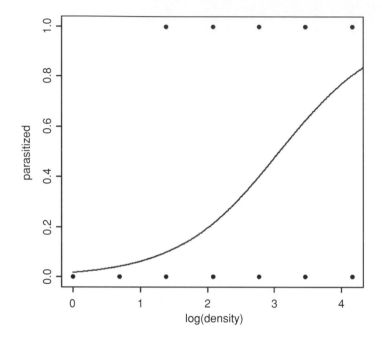

The spinogram is much better at illustrating the pattern in the data than are the non-jittered 0s and 1s of the binary response plot. The logistic plot is improved by overlaying the empirical frequencies, rather than by showing the raw data as 0s and 1s. We might choose four bins in an example like this, averaging the four lowest density classes, and using the counts data from the three highest classes (16, 32 and 64):

```
den <- c(3.75,16,32,64)
pd  <- c(3/15,5/16,20/32,52/64)
```

Now redraw the axes leaving out the 0s and 1s (`type="n"`) then add the logistic trend line and overlay the empirical frequencies as larger solid circles (`cex=2`):

```
plot(as.numeric(fate)-1~log(density),pch=16,ylab="parasitized",type="n")
lines(xv,yv)
points(log(den),pd,pch=16,cex=2)
```

To add error bars (eb) to show plus and minus one standard error of the estimated proportion, $SE_p = \sqrt{p(1-p)/n}$, we put the code to draw lines up and down from each point in a loop:

```
eb<-sqrt(pd*(1-pd)/den)
for (i in 1:4) lines(c(log(den)[i],log(den)[i]),c(pd[i]+eb[i],pd[i]-eb[i]))
```

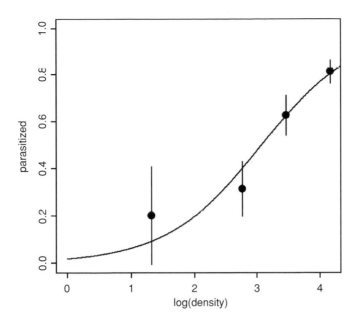

This has the virtue of illustrating the excellent fit of the model at high densities, but the rather poor fit at lower densities.

16

Proportion Data

An important class of problems involves count data on proportions such as:

- studies on death rates,

- infection rates of diseases,

- answers to questionnaires,

- proportion responding to clinical treatment,

- proportion admitting to particular voting intentions,

- sex ratios, or

- data on proportional response to an experimental treatment.

What all these have in common is that we know how many of the experimental objects are in one category (dead, insolvent, male or infected) and we also know how many are in another (alive, solvent, female or uninfected). This contrasts with Poisson count data, where we knew how many times an event occurred, but *not* how many times it did not occur (p. 579).

We model processes involving proportional response variables in R by specifying a generalized linear model with `family=binomial`. The only complication is that whereas with Poisson errors we could simply specify `family=poisson`, with binomial errors we must give the number of failures as well as the numbers of successes in a two-vector response variable. To do this we bind together two vectors using `cbind` into a single object, *y*, comprising the numbers of successes and the number of failures. The **binomial denominator**, *n*, is the total sample, and

```
number.of.failures <- binomial.denominator - number.of.successes
y <- cbind(number.of.successes, number.of.failures)
```

The old fashioned way of modelling this sort of data was to use the percentage mortality as the response variable. There are four problems with this:

- The errors are not normally distributed.

- The variance is not constant.

The R Book, Second Edition. Michael J. Crawley.
© 2013 John Wiley & Sons, Ltd. Published 2013 by John Wiley & Sons, Ltd.

- The response is bounded (by 1 above and by 0 below).

- By calculating the percentage, we lose information on the size of the sample, n, from which the proportion was estimated.

R carries out weighted regression, using the individual sample sizes as weights, and the logit link function to ensure linearity. There are some kinds of proportion data, such as **percentage cover**, which are best analysed using conventional linear models (assuming normal errors and constant variance) following **arcsine trans-formation**. The response variable, y, measured in radians, is $\sin^{-1}\sqrt{0.01 \times p}$, where p is percentage cover. If, however, the response variable takes the form of a **percentage change** in some continuous measurement (such as the percentage change in weight on receiving a particular diet), then rather than arcsine-transforming the data, it is usually better treated by either

- analysis of covariance (see p. 537), using final weight as the response variable and initial weight as a covariate, or

- by specifying the response variable as a relative growth rate, measured as log(final weight/initial weight),

both of which can be analysed as linear models with normal errors without further transformation.

16.1 Analyses of data on one and two proportions

For comparisons of one binomial proportion with a constant, use `binom.test` (see p. 600). For comparison of two samples of proportion data, use `prop.test` (see p. 365). The methods of this chapter are required only for more complex models of proportion data, including regression and contingency tables, where GLMs are used.

16.2 Count data on proportions

The traditional transformations of proportion data were arcsine and probit. The arcsine transformation took care of the error distribution, while the probit transformation was used to linearize the relationship between percentage mortality and log dose in a bioassay. There is nothing wrong with these transformations, and they are available within R, but a simpler approach is often preferable, and is likely to produce a model that is easier to interpret.

The major difficulty with modelling proportion data is that the responses are *strictly bounded*. There is no way that the percentage dying can be greater than 100% or less than 0%. But if we use simple techniques such as regression or analysis of covariance, then the fitted model could quite easily predict negative values or values greater than 100%, especially if the variance was high and many of the data were close to 0 or close to 100%.

The **logistic** curve is commonly used to describe data on proportions, because, unlike the straight-line model, it asymptotes at 0 and 1 so that negative proportions and responses of more than 100% cannot be predicted. Throughout this discussion we shall use p to describe the proportion of individuals observed to respond in a given way. Because much of their jargon was derived from the theory of gambling, statisticians call these **successes**, although to a demographer measuring death rates this may seem somewhat macabre. The proportion of individuals that respond in other ways (the statistician's **failures**) is therefore $1 - p$, and we shall call this proportion q. The third variable is the size of the sample, n, from which p was estimated (this is the binomial denominator, and the statistician's **number of attempts**).

An important point about the binomial distribution is that the variance is not constant. In fact, the variance of a binomial distribution with mean np is

$$s^2 = npq,$$

so that the variance changes with the mean like this:

The variance is low when p is very high or very low, and the variance is greatest when $p = q = 0.5$. As p gets smaller, so the binomial distribution gets closer and closer to the Poisson distribution. You can see why this is so by considering the formula for the variance of the binomial (above). Remember that for the Poisson, the variance is equal to the mean: $s^2 = np$. Now, as p gets smaller, so q gets closer and closer to 1, so the variance of the binomial converges to the mean:

$$s^2 = npq \approx np \qquad (q \approx 1).$$

16.3   Odds

The logistic model for p as a function of x is given by

$$p = \frac{e^{a+bx}}{1 + e^{a+bx}},$$

and there are no prizes for realizing that the model is not linear. But if $x = -\infty$ then $p = 0$, and if $x = +\infty$ then $p = 1$, so the model is strictly bounded. If $x = 0$, then $p = \exp(a)/[1 + \exp(a)]$. The trick of linearizing the logistic model actually involves a very simple transformation. You may have come across the way in which bookmakers specify probabilities by quoting the **odds** against a particular horse winning a race (they might give odds of 2 to 1 on a reasonably good horse or 25 to 1 on an outsider). This is a rather different way of presenting information on probabilities than scientists are used to dealing with. Thus, where the scientist might state a proportion as 0.333 (one chance of winning in three), the bookmaker would give odds of 2 to 1 (based on the counts of outcomes: one success against two failures). In symbols, this is the difference between the scientist stating the probability p, and the bookmaker stating the odds p/q. Now if we take the odds p/q and substitute this into the formula for the logistic, we get

$$\frac{p}{q} = \frac{e^{a+bx}}{1 + e^{a+bx}} \left[1 - \frac{e^{a+bx}}{1 + e^{a+bx}} \right]^{-1}$$

which looks awful. But a little algebra shows that

$$\frac{p}{q} = \frac{e^{a+bx}}{1 + e^{a+bx}} \left[\frac{1}{1 + e^{a+bx}} \right]^{-1} = e^{a+bx}.$$

Taking natural logs and recalling that $\ln(e^x) = x$ will simplify matters even further, so that

$$\ln\left(\frac{p}{q}\right) = a + bx.$$

This gives a **linear predictor**, $a + bx$, not for p but for the **logit** transformation of p, namely $\ln(p/q)$. In the jargon of R, the logit is the **link function** relating the linear predictor to the value of p.

Here are p as a function of x (left panel) and logit(p) as a function of x (right panel) for the logistic model with $a = 1$ and $b = 0.1$:

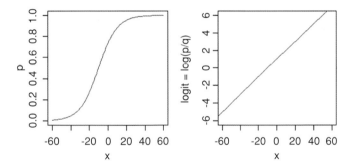

You might ask at this stage: 'why not simply do a linear regression of $\ln(p/q)$ against the explanatory x-variable?' GLM with binomial errors has three great advantages here:

- It allows for the non-constant binomial variance.

- It deals with the fact that logits for ps near 0 or 1 are infinite.

- It allows for differences between the sample sizes by weighted regression.

16.4 Overdispersion and hypothesis testing

That the errors are binomially distributed is an assumption, not a fact. When we have overdispersion, this assumption is wrong and we need to deal with it.

All the different statistical procedures that we have met in earlier chapters can also be used with data on proportions. Factorial analysis of variance, multiple regression, and a variety of models in which different regression lines are fitted in each of several levels of one or more factors, can be carried out. The only difference is that we assess the significance of terms on the basis of chi-squared – the increase in scaled deviance that results from removal of the term from the current model.

The important point to bear in mind is that hypothesis testing with binomial errors is less clear-cut than with normal errors. While the chi-squared approximation for changes in scaled deviance is reasonable for large samples (i.e. larger than about 30), it is poorer with small samples. Most worrying is the fact that the degree to which the approximation is satisfactory is itself unknown. This means that considerable care must be exercised in the interpretation of tests of hypotheses on parameters, especially when the parameters are marginally significant or when they explain a very small fraction of the total deviance. With binomial or Poisson errors we cannot hope to provide exact p values for our tests of hypotheses.

When we have obtained the minimal adequate model, *the residual scaled deviance should be roughly equal to the residual degrees of freedom*. Overdispersion occurs when the residual deviance is larger than the residual degrees of freedom. There are two possibilities: either the model is misspecified, or the probability of success, p, is not constant within a given treatment level. The effect of randomly varying p is to increase the binomial variance from npq to

$$s^2 = npq + n(n-1)\sigma^2,$$

leading to a large residual deviance. This occurs even for models that would fit well if the random variation were correctly specified.

One simple solution is to assume that the variance is not npq but $npq\phi$, where ϕ is an unknown *scale parameter* ($\phi > 1$). We obtain an estimate of the scale parameter by dividing the Pearson chi-squared by the degrees of freedom, and use this estimate of ϕ to compare the resulting scaled deviances. To accomplish this, we use `family=quasibinomial` rather than `family=binomial` when there is overdispersion.

The most important points to emphasize in modelling with binomial errors are as follows:

- Create a two-column object for the response, using `cbind` to join together the two vectors containing the counts of success and failure.

- Check for overdispersion (residual deviance greater than the residual degrees of freedom), and correct for it by using `family=quasibinomial` rather than `family=binomial` if necessary.

- Remember that you do not obtain exact p values with binomial errors; the chi-squared approximations are sound for large samples, but small samples may present a problem.

- The fitted values are two sets of counts, like the response variable.

- The linear predictor is in logits (the log of the odds $= \ln(p/q)$).

- You can back-transform from logits (z) to proportions (p) by $p = 1/[1 + 1/\exp(z)]$.

16.5   Applications

You can do as many kinds of modelling in a GLM as in a linear model. Here we show examples of:

- regression with binomial errors (continuous explanatory variables);

- analysis of deviance with binomial errors (categorical explanatory variables);

- analysis of covariance with binomial errors (both kinds of explanatory variables).

16.5.1 Logistic regression with binomial errors

This example concerns sex ratios in insects (the proportion of all individuals that are males). In the species in question, it has been observed that the sex ratio is highly variable, and an experiment was set up to see whether population density was involved in determining the fraction of males.

```
numbers <- read.table("c:\\temp\\sexratio.txt",header=T)
attach(numbers)
head(numbers)
```

```
  density females males
1       1       1     0
2       4       3     1
3      10       7     3
4      22      18     4
5      55      22    33
6     121      41    80
```

It certainly looks as if there are proportionally more males at high density, but we should plot the data as proportions to see this more clearly:

```
windows(7,4)
par(mfrow=c(1,2))

p <- males/(males+females)

plot(density,p,ylab="Proportion male",pch=16,col="blue")
plot(log(density),p,ylab="Proportion male",pch=16,col="blue")
```

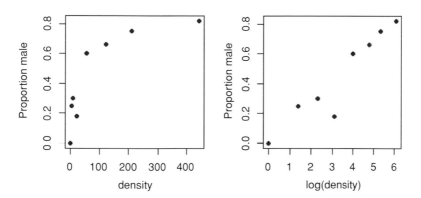

Evidently, a logarithmic transformation of the explanatory variable is likely to improve the model fit. We shall see in a moment.

The question is whether increasing population density leads to a significant increase in the proportion of males in the population – or, more briefly, whether the sex ratio is density-dependent. It certainly looks from the plot as if it is.

The response variable is a matched pair of counts that we wish to analyse as proportion data using a GLM with binomial errors. First, we use cbind to bind together the vectors of male and female counts into a single object that will be the response in our analysis:

```
y <- cbind(males,females)
```

This means that y will be interpreted in the model as the proportion of all individuals that were male. The model is specified like this:

```
model <- glm(y~density,binomial)
```

This says that the object called model gets a generalized linear model in which y (the sex ratio) is modelled as a function of a single continuous explanatory variable (called density), using an error distribution from the binomial family. The output looks like this:

```
summary(model)

Call:
glm(formula = y ~ density, family = binomial)

Deviance Residuals:
    Min        1Q    Median         3Q        Max
-3.4619   -1.2760   -0.9911     0.5742     1.8795

Coefficients:
                Estimate Std. Error z value Pr(>|z|)
(Intercept) 0.0807368  0.1550376    0.521    0.603
density     0.0035101  0.0005116    6.862 6.81e-12 ***

(Dispersion parameter for binomial family taken to be 1)

    Null deviance: 71.159  on 7  degrees of freedom
Residual deviance: 22.091  on 6  degrees of freedom
AIC: 54.618

Number of Fisher Scoring iterations: 4
```

The model table looks just as it would for a straightforward regression. The first parameter in the Coefficients table is the intercept and the second is the slope of the graph of sex ratio against population density. The slope is highly significantly steeper than zero (proportionately more males at higher population density; $p = 6.81 \times 10^{-12}$), but there is substantial overdispersion (residual deviance = 22.091 is much greater than residual d.f. = 6) . We can see if log transformation of the explanatory variable improves this:

```
model2 <- glm(y~log(density),binomial)
summary(model2)

Coefficients:
                Estimate Std. Error z value Pr(>|z|)
(Intercept)   -2.65927    0.48758   -5.454 4.92e-08 ***
log(density)   0.69410    0.09056    7.665 1.80e-14 ***

(Dispersion parameter for binomial family taken to be 1)

    Null deviance: 71.1593  on 7  degrees of freedom
Residual deviance:  5.6739  on 6  degrees of freedom
AIC: 38.201
```

This is a big improvement, so we shall adopt it. In the model with log(density) there is no evidence of overdispersion (residual deviance = 5.67 on 6 d.f.), whereas the lack of fit introduced by the curvature in our first model caused substantial overdispersion (residual deviance = 22.09 on 6 d.f.).

Model checking involves the use of `plot (model2)`. As you will see, there is no pattern in the residuals against the fitted values, and the normal plot is reasonably linear. Point no. 4 is highly influential (it has a large Cook's distance), but the model is still significant with this point omitted.

We conclude that the proportion of animals that are males increases significantly with increasing density, and that the logistic model is linearized by logarithmic transformation of the explanatory variable (population density). We finish by drawing the fitted line though the scatterplot:

```
windows(7,7)
xv <- seq(0,6,0.01)

yv <- predict(model2,list(density=exp(xv)),type="response")

plot(log(density),p,ylab="Proportion male",pch=16,col="blue")
lines(xv,yv,col="red")
```

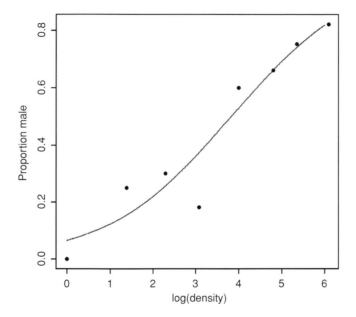

Note the use of `type="response"` to back-transform from the logit scale to the S-shaped proportion scale.

16.5.2   Estimating LD50 and LD90 from bioassay data

The data consist of numbers dead and initial batch size for five doses of pesticide application, and we wish to know what dose kills 50% of the individuals (or 90% or 95%, as required). The tricky statistical issue is that one is using a value of y (50% dead) to predict a value of x (the relevant dose) and to work out a standard error on the x axis.

```
data <- read.table("c:\\temp\\bioassay.txt",header=T)
attach(data)
names(data)

[1]   "dose"   "dead"   "batch"
```

The logistic regression is carried out in the usual way:

```
y <- cbind(dead,batch-dead)

model <- glm(y~log(dose),binomial)
```

Then the function dose.p from the MASS library is run with the model object, specifying the proportions killed for which we want the predicted log(doses) ($p = 0.5$ is the default for LD50):

```
library(MASS)
dose.p(model,p=c(0.5,0.9,0.95))

              Dose           SE
p = 0.50: 2.306981  0.07772065
p = 0.90: 3.425506  0.12362080
p = 0.95: 3.805885  0.15150043
```

Despite the label 'Dose', the output shows the logs of the doses associated with kills of LD50, LD90 and LD95, along with their standard errors.

16.5.3   Proportion data with categorical explanatory variables

This next example concerns the germination of seeds of two genotypes of the parasitic plant *Orobanche* and two extracts from host plants (bean and cucumber) that were used to stimulate germination. It is a two-way factorial analysis of deviance.

```
germination <- read.table("c:\\temp\\germination.txt",header=T)
attach(germination)
names(germination)

[1]  "count"   "sample"   "Orobanche"   "extract"
```

The count is the number of seeds that germinated out of a batch of size = sample. So the number that did not germinate is sample - count, and we construct the response vector like this:

```
y <- cbind(count, sample-count)
```

Each of the categorical explanatory variables has two levels:

```
levels(Orobanche)

[1]   "a73"   "a75"

levels(extract)

[1]   "bean"   "cucumber"
```

We want to test the hypothesis that there is no interaction between *Orobanche* genotype (a73 or a75) and plant extract (bean or cucumber) on the germination rate of the seeds. This requires a factorial analysis using the asterisk * operator like this:

```
model <- glm(y ~ Orobanche * extract, binomial)
summary(model)

Call:
glm(formula = y ~ Orobanche * extract, family = binomial)
```

```
Deviance Residuals:
      Min        1Q     Median        3Q        Max
 -2.01617  -1.24398    0.05995    0.84695    2.12123

Coefficients:
                          Estimate Std. Error  z value Pr(>|z|)
(Intercept)                -0.4122     0.1842   -2.238   0.0252 *
Orobanchea75               -0.1459     0.2232   -0.654   0.5132
extractcucumber             0.5401     0.2498    2.162   0.0306 *
Orobanchea75:extractcucumber 0.7781    0.3064    2.539   0.0111 *

(Dispersion parameter for binomial family taken to be 1)

    Null deviance: 98.719  on 20  degrees of freedom
Residual deviance: 33.278  on 17  degrees of freedom
AIC: 117.87
```

At first glance, it looks as if there is a highly significant interaction ($p = 0.0111$). But we need to check that the model is sound. The first thing is to check for is overdispersion. The residual deviance is 33.278 on 17 d.f., so the model is quite badly overdispersed:

```
33.279 / 17

[1]   1.957588
```

The overdispersion factor is almost 2. The simplest way to take this into account is to use what is called an 'empirical scale parameter' to reflect the fact that the errors are not binomial as we assumed, but were larger than this (i.e. overdispersed) by a factor of 1.9576. We refit the model using quasi-binomial errors to account for the overdispersion:

```
model <- glm(y ~ Orobanche * extract, quasibinomial)
```

Then we use update to remove the interaction term in the normal way:

```
model2 <- update(model, ~ . - Orobanche:extract)
```

The only difference is that we use an F test instead of a chi-squared test to compare the original and simplified models because now we have estimated two parameters from the model (the mean plus the empirical scale parameter):

```
anova(model,model2,test="F")

Analysis of Deviance Table

Model 1: y ~ Orobanche * extract
Model 2: y ~ Orobanche + extract
  Resid. Df Resid. Dev Df Deviance      F   Pr(>F)
1        17     33.278
2        18     39.686 -1  -6.4081 3.4418 0.08099 .
```

Now you see that the interaction is not significant ($p = 0.081$). There is no compelling evidence that different genotypes of *Orobanche* respond differently to the two plant extracts.

The next step is to see if any further model simplification is possible:

```
anova(model2,test="F")
```

```
Analysis of Deviance Table
```

```
Model: quasibinomial, link: logit
```

	Df	Deviance	Resid. Df	Resid. Dev	F	Pr(>F)
NULL			20	98.719		
Orobanche	1	2.544	19	96.175	1.1954	0.2887
extract	1	56.489	18	39.686	26.5412	6.692e-05 ***

There is a highly significant difference between the two plant extracts on germination rate, but it is not obvious that we need to keep *Orobanche* genotype in the model. We try removing it:

```
model3 <- update(model2, ~ . - Orobanche)
anova(model2,model3,test="F")
```

```
Analysis of Deviance Table
```

```
Model 1: y ~ Orobanche + extract
Model 2: y ~ extract
```

	Resid. Df	Resid. Dev	Df	Deviance	F	Pr(>F)
1	18	39.686				
2	19	42.751	-1	-3.065	1.4401	0.2457

There is no justification for retaining *Orobanche* in the model. So the minimal adequate model contains just two parameters:

```
coef(model3)
```

```
  (Intercept)  extractcucumber
   -0.5121761        1.0574031
```

What, exactly, do these two numbers mean? Remember that the coefficients are from the linear predictor. They are on the transformed scale, so because we are using quasi-binomial errors, they are in logits ($\ln(p/(1-p))$). To turn them into the germination rates for the two plant extracts requires a little calculation.

To go from a logit x to a proportion p, you need to calculate

$$p = \frac{1}{1 + 1/e^x}.$$

So our first x value is -0.5122 and we calculate

```
1/(1+1/(exp(-0.5122)))
```

```
[1]  0.3746779
```

This says that the mean germination rate of the seeds with the first plant extract (bean) was 37%. What about the parameter for extract (1.0574)? Remember that with categorical explanatory variables the parameter values are *differences between means*. So to get the second germination rate we add 1.057 to the intercept before back-transforming:

```
1/(1+1/(exp(-0.5122+1.0574)))
```

```
[1]  0.6330212
```

This says that the germination rate was nearly twice as great (63%) with the second plant extract (cucumber). Obviously we want to generalize this process, and also to speed up the calculations of the estimated mean proportions. We can use `predict` to help here, because `type="response"` makes predictions on the back-transformed scale automatically:

```
tapply(predict(model3,type="response"),extract,mean)
```

```
      bean   cucumber
0.3746835  0.6330275
```

It is interesting to compare these figures with the averages of the raw proportions. First we need to calculate the proportion germinating, p, in each sample:

```
p <- count/sample
```

Then we can find the average germination rates for each extract:

```
tapply(p,extract,mean)
```

```
      bean   cucumber
0.3487189  0.6031824
```

You see that this gives different answers. Not too different in this case, but different none the less. The correct way to average proportion data is to add up the total counts for the different levels of abstract, and only then to turn them into proportions:

```
tapply(count,extract,sum)
```

```
   bean cucumber
    148      276
```

This means that 148 seeds germinated with bean extract and 276 with cucumber. But how many seeds were involved in each case?

```
tapply(sample,extract,sum)
```

```
   bean cucumber
    395      436
```

This means that 395 seeds were treated with bean extract and 436 seeds were treated with cucumber. So the answers we want are 148/395 and 276/436 (i.e. the correct mean proportions). We can automate the calculation like this:

```
as.vector(tapply(count,extract,sum))/as.vector(tapply(sample,extract,sum))
```

```
[1]  0.3746835  0.6330275
```

These are the correct mean proportions that were produced by the GLM. The moral here is that you calculate the average of proportions by using total counts and total samples and not by averaging the raw proportions.

16.6   Averaging proportions

Here is an example of what *not* to do. We have four proportions:

$$0.2, 0.17, 0.2, 0.53.$$

So surely we just add them up and divide by 4. This gives 1.1/4 = 0.275. Wrong! And not by just a little bit. We need to look at the counts on which the proportions were based. These turn out to be:

$$1/5, 1/6, 2/10, 53/100.$$

The correct way to average proportions is to add up the total count of successes $(1 + 1 + 2 + 53 = 57)$ and divide this by the total number of samples $(5 + 6 + 10 + 100 = 121)$. The correct mean proportion is $57/121 = 0.4711$. This is nearly double our incorrect answer (above).

16.7 Summary of modelling with proportion count data

- Make a two-column response vector containing the successes and failures.

- Use `glm` with `family=binomial` (you can omit `family=`).

- Fit the maximal model.

- Test for overdispersion.

- If you find overdispersion then use `quasibinomial` rather than `binomial` errors.

- Begin model simplification by removing interaction terms.

- Remove non-significant main effects

- Use `plot` to obtain your model-checking diagnostics.

- Back-transform using `predict` with the option `type="response"` to obtain means.

16.8 Analysis of covariance with binomial data

We now turn to an example concerning flowering in five varieties of perennial plant. Replicated individuals in a fully randomized design were sprayed with one of six doses of a controlled mixture of growth promoters. After 6 weeks, plants were scored as flowering or not flowering. The count of flowering individuals forms the response variable. This is an ANCOVA because we have both continuous (dose) and categorical (variety) explanatory variables. We use logistic regression because the response variable is a count (flowered) that can be expressed as a proportion (flowered/number).

```
props <- read.table("c:\\temp\\flowering.txt",header=T)
attach(props)
names(props)

[1]  "flowered"  "number"  "dose"  "variety"

y <- cbind(flowered,number-flowered)
pf <- flowered/number
pfc <- split(pf,variety)
dc <- split(dose,variety)

plot(dose,pf,type="n",ylab="Proportion flowered")
points(jitter(dc[[1]]),jitter(pfc[[1]]),pch=21,col="blue",bg="red")
points(jitter(dc[[2]]),jitter(pfc[[2]]),pch=21,col="blue",bg="green")
```

```
points(jitter(dc[[3]]),jitter(pfc[[3]]),pch=21,col="blue",bg="yellow")
points(jitter(dc[[4]]),jitter(pfc[[4]]),pch=21,col="blue",bg="green3")
points(jitter(dc[[5]]),jitter(pfc[[5]]),pch=21,col="blue",bg="brown")
```

Note the use of split to separate the different varieties, so that we can plot them with different symbols, and of jitter to stop repeated values hiding one another.

There is clearly a substantial difference between the plant varieties in their response to the flowering stimulant. The modelling proceeds in the normal way. We begin by fitting the maximal model with different slopes and intercepts for each variety (estimating ten parameters in all):

```
model1 <- glm(y~dose*variety,binomial)
summary(model1)
```

```
Coefficients:
                Estimate Std. Error z value Pr(>|z|)
(Intercept)     -4.59165    1.03215  -4.449 8.64e-06 ***
dose             0.41262    0.10033   4.113 3.91e-05 ***
varietyB         3.06197    1.09317   2.801 0.005094 **
varietyC         1.23248    1.18812   1.037 0.299576
varietyD         3.17506    1.07516   2.953 0.003146 **
varietyE        -0.71466    1.54849  -0.462 0.644426
dose:varietyB   -0.34282    0.10239  -3.348 0.000813 ***
dose:varietyC   -0.23039    0.10698  -2.154 0.031274 *
dose:varietyD   -0.30481    0.10257  -2.972 0.002961 **
dose:varietyE   -0.00649    0.13292  -0.049 0.961057
```

(Dispersion parameter for binomial family taken to be 1)

```
    Null deviance: 303.350  on 29  degrees of freedom
Residual deviance:  51.083  on 20  degrees of freedom
AIC: 123.55
```

The model exhibits substantial overdispersion, but this is probably due to poor model selection rather than extra, unmeasured variability. Let us investigate this by plotting the fitted curves through the scatterplot.

```
xv <- seq(0,35,0.1)
vn <- rep("A",length(xv))
yv <- predict(model1,list(variety=factor(vn),dose=xv),type="response")
lines(xv,yv,col="red")

vn <- rep("B",length(xv))
yv <- predict(model1,list(variety=factor(vn),dose=xv),type="response")
lines(xv,yv,col="green")

vn <- rep("C",length(xv))
yv <- predict(model1,list(variety=factor(vn),dose=xv),type="response")
lines(xv,yv,col="yellow")

vn <- rep("D",length(xv))
yv <- predict(model1,list(variety=factor(vn),dose=xv),type="response")
lines(xv,yv,col="green3")
```

```
vn <- rep("E",length(xv))
yv <- predict(model1,list(variety=factor(vn),dose=xv),type="response")
lines(xv,yv,col="brown")

legend(locator(1),legend=c("A","B","C","D","E"),title="variety",
            lty=rep(1,5),col=c("red","green","yellow","green3","brown"))
```

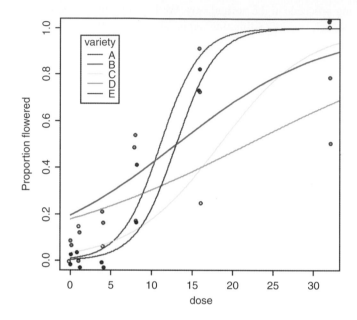

As you can see, the model is reasonable for two of the genotypes (A and E, represented by red and brown lines, respectively), moderate for one genotype (C, yellow) but very poor for two of them, B (lower, light green line) and D (upper, dark green line). For both of the latter, the model greatly overestimates the proportion flowering at zero dose, and for genotype B there seems to be some inhibition of flowering at the highest dose because the graph falls from 90% flowering at dose 16 to just 50% at dose 32. Variety D appears to be asymptoting at less than 100% flowering.

```
tapply(pf,list(dose,variety),mean)
```

```
            A          B          C          D          E
0   0.0000000 0.08333333 0.00000000 0.06666667 0.0000000
1   0.0000000 0.00000000 0.14285714 0.11111111 0.0000000
4   0.0000000 0.20000000 0.06666667 0.15789474 0.0000000
8   0.4000000 0.50000000 0.17647059 0.53571429 0.1578947
16  0.8181818 0.90000000 0.25000000 0.73076923 0.7500000
32  1.0000000 0.50000000 1.00000000 0.77777778 1.0000000
```

These failures of the model should focus attention for future work.

The moral is that the fact that we have proportion data does not mean that the data will necessarily be well described by the logistic model. For instance, in order to describe the response of genotype B, the model would need to have a hump, rather than to asymptote at $p = 1$ for large doses. It is essential to look closely at the data, both with plots and with tables, before accepting the model output. Model choice is very big deal indeed. The logistic was a poor choice for two of the five varieties in this case.

16.9   Converting complex contingency tables to proportions

In this section we show how to remove the need for all of the nuisance variables that are involved in complex contingency table modelling (the row and column totals that have to be included in all of the models). We can do this when the response can be restated as a binary proportion (a choice of one out of two contingencies). For instance, in the case of Schoener's lizards which proved so tricky to analyse as a complex contingency table (see p. 610), we can work with *the proportion of all lizards* that are *Anolis grahamii* as the response variable, instead of analysing the *counts* of the numbers of *A. grahamii* and *A. opalinus* separately. This has the huge advantage of requiring none of the nuisance variables to be included in the model. The technique would not work if we had three lizard species, however. Then we would have to stick with the complex contingency table modelling.

```
lizards <- read.table("c:\\temp\\lizards.txt",header=T)
attach(lizards)
names(lizards)

[1] "n"        "sun"       "height"  "perch"    "time"     "species"

head(lizards)

   n   sun height   perch    time   species
1 20 Shade   High   Broad Morning opalinus
2 13 Shade    Low   Broad Morning opalinus
3  8 Shade   High  Narrow Morning opalinus
4  6 Shade    Low  Narrow Morning opalinus
5 34   Sun   High   Broad Morning opalinus
6 31   Sun    Low   Broad Morning opalinus
```

First, we need to make absolutely sure that all the explanatory variables are in exactly the same order for both species of lizards. The reason for this is that we are going to cbind the counts for one of the lizard species onto the half dataframe containing the other species counts and all of the explanatory variables. Any mistakes here would be disastrous because the count would be lined up with the wrong combination of explanatory variables, and the analysis would be wrong and utterly meaningless.

```
sorted <- lizards[order(species,sun,height,perch,time),]
head(sorted)

   n   sun height   perch      time  species
41 4 Shade   High   Broad Afternoon grahamii
33 1 Shade   High   Broad   Mid.day grahamii
25 2 Shade   High   Broad   Morning grahamii
43 3 Shade   High  Narrow Afternoon grahamii
35 1 Shade   High  Narrow   Mid.day grahamii
27 3 Shade   High  Narrow   Morning grahamii
```

Next we need to extract the top half of this dataframe (i.e. rows 1–24):

```
short <- sorted[1:24,]
```

Note that this operation has lost all of the data for *A. opalinus*. Also, the name for the left-hand variable, *n*, is no longer appropriate. It is the count for *A. grahamii*, so we should rename it Ag, say (with the intention of adding another column called Ao in due course to contain the counts of *A. opalinus*):

```
names(short)[1] <- "Ag"

names(short)

[1] "Ag"    "sun"    "height"    "perch"    "time"    "species"
```

The right-hand variable, species, is redundant now (all the entries are *A. grahamii*), so we should drop it:

```
short <- short[,-6]
head(short)

   Ag   sun height   perch       time
41  4 Shade   High   Broad  Afternoon
33  1 Shade   High   Broad    Mid.day
25  2 Shade   High   Broad    Morning
43  3 Shade   High  Narrow  Afternoon
35  1 Shade   High  Narrow    Mid.day
27  3 Shade   High  Narrow    Morning
```

The counts for each row of *A. opalinus* are in the variable called n in the bottom half of the dataframe called sorted. We extract them like this:

```
sorted$n[25:48]


 [1]  4  8 20  5  4  8 12  8 13  1  0  6 18 69 34  8 60 17 13 55 31  4 21 12
```

The idea is to create a new dataframe with these counts for *A. opalinus* lined up alongside the matching counts for *A. grahamii*:

```
new.lizards <- data.frame(sorted$n[25:48], short)
```

The first variable needs an informative name, like Ao:

```
names(new.lizards)[1] <- "Ao"
head(new.lizards)

   Ao Ag   sun height   perch       time
41  4  4 Shade   High   Broad  Afternoon
33  8  1 Shade   High   Broad    Mid.day
25 20  2 Shade   High   Broad    Morning
43  5  3 Shade   High  Narrow  Afternoon
35  4  1 Shade   High  Narrow    Mid.day
27  8  3 Shade   High  Narrow    Morning
```

That completes the editing of the dataframe. Notice, however, that we have got three dataframes, all of different configurations, but each containing the same variable names (sun, height, perch and time) – look at objects() and search(). We need to do some housekeeping:

```
detach(lizards)
rm(short,sorted)
attach(new.lizards)
```

16.9.1 Analysing Schoener's lizards as proportion data

With the foregoing preliminaries, here are the variable names:

```
names(new.lizards)
```

```
[1]   "Ao"   "Ag"   "sun"   "height"   "perch"   "time"
```

The response variable is a two-column object containing the counts of the two species:

```
y <- cbind(Ao,Ag)
```

We begin by fitting the saturated model containing all possible interactions:

```
model1 <- glm(y~sun*height*perch*time,binomial)
```

Since there are no nuisance variables, we can use `step` directly to begin the model simplification (compare this with p. 611 with a log-linear model of the same data):

```
model2 <- step(model1)
```

```
Start:   AIC=102.82
y ~ sun * height * perch * time

                         Df    Deviance    AIC
- sun:height:perch:time  1 2.1797e-10 100.82
<none>                     3.5825e-10 102.82
```

Out goes the four-way interaction (with a sigh of relief):

```
Step:   AIC=100.82

                    Df Deviance      AIC
- sun:height:time    2   0.4416   97.266
- sun:perch:time     2   0.8101   97.634
- height:perch:time  2   3.2217 100.046
<none>                   0.0000 100.824
- sun:height:perch   1   2.7088 101.533
```

Next, we wave goodbye to three of the three-way interactions:

```
Step:   AIC=97.27

                    Df Deviance     AIC
- sun:perch:time     2   1.0713 93.896
<none>                   0.4416 97.266
- height:perch:time  2   4.6476 97.472
- sun:height:perch   1   3.1113 97.936
```

```
Step:   AIC=93.9

                    Df Deviance     AIC
- sun:time           2   3.3403 92.165
<none>                   1.0713 93.896
- sun:height:perch   1   3.3016 94.126
- height:perch:time  2   5.7906 94.615
```

Then we discard the two-way interaction of sun by time:

```
Step:   AIC=92.16

                     Df Deviance    AIC
<none>                   3.3403 92.165
- sun:height:perch    1  5.8273 92.651
- height:perch:time   2  8.5418 93.366
```

We have seen that AIC is very generous in leaving terms in the model that we would ruthlessly eliminate. To begin with, we need to test whether we would have kept the two three-way interactions and the five two-way interactions:

```
model3 <- update(model2,~. - height:perch:time)
model4 <- update(model2,~. - sun:height:perch)
anova(model2,model3,test="Chi")

Analysis of Deviance Table
```

Model 1: y ~ sun + height + perch + time + sun:height + sun:perch + height:perch +
    height:time + perch:time + sun:height:perch + height:perch:time
Model 2: y ~ sun + height + perch + time + sun:height + sun:perch + height:perch +
    height:time + perch:time + sun:height:perch

```
  Resid. Df Resid. Dev Df Deviance Pr(>Chi)
1         7     3.3403
2         9     8.5418 -2  -5.2014  0.07422 .
```

That was close, but not significant. So out it goes.

```
anova(model2,model4,test="Chi")
Analysis of Deviance Table
```

Model 1: y ~ sun + height + perch + time + sun:height + sun:perch + height:perch +
    height:time + perch:time + sun:height:perch + height:perch:time
Model 2: y ~ sun + height + perch + time + sun:height + sun:perch + height:perch +
    height:time + perch:time + height:perch:time

```
  Resid. Df Resid. Dev Df Deviance Pr(>Chi)
1         7     3.3403
2         8     5.8273 -1   -2.487   0.1148
```

No. We would not keep either of those three-way interactions. What about the two-way interactions? We need to start with a simpler base model than model2:

```
model5 <- glm(y~(sun+height+perch+time)^2-sun:time,binomial)
```

We shall remove each of the two-way interactions separately, comparing each to model5 which contains all of the two-way interactions:

```
model6 <- update(model5,~. - sun:height)
anova(model5,model6,test="Chi")
```

```
Analysis of Deviance Table

  Resid. Df Resid. Dev Df Deviance Pr(>Chi)
1        10     10.903
2        11     13.254 -1  -2.3511    0.1252

model7 <- update(model5,~. - sun:perch)
anova(model5,model7,test="Chi")

Analysis of Deviance Table

  Resid. Df Resid. Dev Df  Deviance Pr(>Chi)
1        10     10.903
2        11     10.927 -1 -0.023597   0.8779

model8 <- update(model5,~. - height:perch)
anova(model5,model8,test="Chi")

Analysis of Deviance Table

  Resid. Df Resid. Dev Df Deviance Pr(>Chi)
1        10     10.903
2        11     11.143 -1 -0.24006    0.6242

model9 <- update(model5,~. - time:perch)
anova(model5,model9,test="Chi")

Analysis of Deviance Table

  Resid. Df Resid. Dev Df   Deviance Pr(>Chi)
1        10     10.903
2        12     10.909 -2 -0.0058263    0.9971

model10 <- update(model5,~. - time:height)
anova(model5,model10,test="Chi")

Analysis of Deviance Table

  Resid. Df Resid. Dev Df Deviance Pr(>Chi)
1        10     10.903
2        12     11.760 -2 -0.85679    0.6516
```

So we do not need any of the two-way interactions. What about the main effects?

```
model11 <- glm(y~sun+height+perch+time,binomial)
summary(model11)

Call:
glm(formula = y ~ sun + height + perch + time, family = binomial)

Deviance Residuals:
     Min        1Q     Median        3Q       Max
-1.66015  -0.37800    0.04488   0.62644   1.48717

Coefficients:
            Estimate Std. Error z value Pr(>|z|)
```

```
(Intercept)     1.2079      0.3536    3.416 0.000634 ***
sunSun         -0.8473      0.3224   -2.628 0.008585 **
heightLow       1.1300      0.2571    4.395 1.11e-05 ***
perchNarrow    -0.7626      0.2113   -3.610 0.000306 ***
timeMid.day     0.9639      0.2816    3.423 0.000619 ***
timeMorning     0.7368      0.2990    2.464 0.013730 *

(Dispersion parameter for binomial family taken to be 1)

    Null deviance: 70.102  on 22  degrees of freedom
Residual deviance: 14.205  on 17  degrees of freedom
AIC: 83.029
```

All the main effects are significant and so must be retained.

Just one last point. We might not need all three levels for time, since the summary suggests that Mid.day and Morning are not significantly different (parameter difference of $0.9639 - 0.7368 = 0.2271$, with a standard error of the difference of 0.29). We lump them together in a new factor called t2:

```
t2 <- time
levels(t2)[c(2,3)] <- "other"
levels(t2)
```

```
[1] "Afternoon" "other"
```

```
model12 <- glm(y~sun+height+perch+t2,binomial)
anova(model11,model12,test="Chi")
```

```
Analysis of Deviance Table

Model 1: y ~ sun + height + perch + time
Model 2: y ~ sun + height + perch + t2
  Resid. Df Resid. Dev Df Deviance Pr(>Chi)
1        17     14.205
2        18     15.023 -1 -0.81863   0.3656
```

A model with just two times of day is not significantly worse than a model with three.

```
summary(model12)
```

```
Call:
glm(formula = y ~ sun + height + perch + t2, family = binomial)

Deviance Residuals:
    Min        1Q    Median       3Q       Max
-1.59707  -0.37407   0.06965   0.64616   1.53004

Coefficients:
            Estimate Std. Error z value Pr(>|z|)
(Intercept)   1.1595     0.3484   3.328 0.000874 ***
sunSun       -0.7872     0.3159  -2.491 0.012722 *
heightLow     1.1188     0.2566   4.360  1.3e-05 ***
perchNarrow  -0.7485     0.2104  -3.557 0.000375 ***
t2other       0.8717     0.2611   3.338 0.000844 ***
```

```
(Dispersion parameter for binomial family taken to be 1)

    Null deviance: 70.102  on 22  degrees of freedom
Residual deviance: 15.023  on 18  degrees of freedom
AIC: 81.847
```

All the parameters are significant, so this is the minimal adequate model. There is no evidence of overdispersion. There are just five parameters, and the model contains no nuisance variables (compare this with the massive contingency table model `summary.lm(model1)`). The ecological interpretation is straightforward: the two lizard species differ significantly in their niches on all the niche axes that were measured. However, there were no significant interactions (nothing subtle was happening such as swapping perch sizes at different times of day).

17

Binary Response Variables

Many statistical problems involve binary response variables. For example, we often classify individuals as

- dead or alive,

- occupied or empty,

- healthy or diseased,

- wilted or turgid,

- male or female,

- literate or illiterate,

- mature or immature,

- solvent or insolvent, or

- employed or unemployed.

It is interesting to understand the factors that are associated with an individual being in one class or the other. Binary analysis will be a useful option when at least one of your explanatory variables is continuous (rather than categorical). In a study of company insolvency, for instance, the data would consist of a list of measurements made on the insolvent companies (their age, size, turnover, location, management experience, workforce training, and so on) and a similar list for the solvent companies. The question then becomes which, if any, of the explanatory variables increase the probability of an individual company being insolvent.

The response variable contains only 0s and 1s; for example, 0 to represent dead individuals and 1 to represent live ones. Thus, there is only a single column of numbers for the response, in contrast to proportion data where two vectors (successes and failures) were bound together to form the response (see Chapter 16). The way that R treats binary data is to assume that the 0s and 1s come from *a binomial trial with sample size 1*. If the probability that an individual is dead is p, then the probability of obtaining y (where y is either dead or alive, 0 or 1) is given by an abbreviated form of the binomial distribution with $n = 1$, known as the Bernoulli distribution:

$$P(y) = p^y (1 - p)^{(1-y)}.$$

The R Book, Second Edition. Michael J. Crawley.
© 2013 John Wiley & Sons, Ltd. Published 2013 by John Wiley & Sons, Ltd.

The random variable y has a mean of p and a variance of $p(1 - p)$, and the objective is to determine how the explanatory variables influence the value of p. The trick to using binary response variables effectively is to know when it is worth using them, and when it is better to lump the successes and failures together and analyse the *total counts* of dead individuals, occupied patches, insolvent firms or whatever. The question you need to ask yourself is: do I have unique values of one or more explanatory variables for each and every individual case?

If the answer is 'yes', then analysis with a binary response variable is likely to be fruitful. If the answer is 'no', then there is nothing to be gained, and you should reduce your data by aggregating the counts to the resolution at which each count *does* have a unique set of explanatory variables. For example, suppose that all your explanatory variables were categorical – sex (male or female), employment (employed or unemployed) and region (urban or rural). In this case there is nothing to be gained from analysis using a binary response variable because none of the individuals in the study have *unique* values of any of the explanatory variables. It might be worthwhile if you had each individual's body weight, for example; then you could ask whether, when you control for sex and region, heavy people are more likely to be unemployed than light people. In the absence of *unique* values for any explanatory variables, there are two useful options:

- Analyse the data as a contingency table using Poisson errors, with the count of the total number of individuals in each of the eight contingencies ($2 \times 2 \times 2$) as the response variable (see Chapter 15) in a dataframe with just eight rows.

- Decide which of your explanatory variables is the key (perhaps you are interested in gender differences), then express the data as proportions (the number of males and the number of females) and recode the binary response as a count of a two-level factor. The analysis is now of proportion data (the proportion of all individuals that are female, for instance) using binomial errors (see Chapter 16).

If you *do* have unique measurements of one or more explanatory variables for each individual, these are likely to be continuous variables such as body weight, income, medical history, distance to the nuclear reprocessing plant, geographic isolation, and so on. This being the case, successful analyses of binary response data tend to be multiple regression analyses or complex analyses of covariance, and you should consult Chapters 10 and 12 for details on model simplification and model criticism.

In order to carry out modelling on a binary response variable we take the following steps:

- Create a single vector containing 0s and 1s as the response variable.

- Use `glm` with `family=binomial`.

- Consider changing the link function from the default logit to complementary log-log.

- Fit the model in the usual way.

- Test significance by deletion of terms from the maximal model, and compare the change in deviance with chi-squared.

Note that there is no such thing as overdispersion with a binary response variable, and hence no need to change to using `quasibinomial` when the residual deviance is large. The choice of link function is generally made by trying both links and selecting the link that gives the lowest deviance. The logit link that we used earlier is symmetric in p and q, but the complementary log-log link is asymmetric. You may also improve the fit by transforming one or more of the explanatory variables. Bear in mind that you can fit non-parametric smoothers to binary response variables using generalized additive models (as described in Chapter 18) instead of carrying out parametric logistic regression.

17.1 Incidence functions

In this example, the response variable is called `incidence`; a value of 1 means that an island was occupied by a particular species of bird, and 0 means that the bird did not breed there. The explanatory variables are the area of the island (km^2) and the isolation of the island (distance from the mainland, km).

```
island <- read.table("c:\\temp\\isolation.txt",header=T)
attach(island)
names(island)
```

```
[1] "incidence" "area" "isolation"
```

There are two continuous explanatory variables, so the appropriate analysis is multiple regression. The response is binary, so we shall do logistic regression with binomial errors.

We begin by fitting a complex model involving an interaction between `isolation` and `area`:

```
model1 <- glm(incidence~area*isolation,binomial)
```

Then we fit a simpler model with only main effects for `isolation` and `area`:

```
model2 <- glm(incidence~area+isolation,binomial)
```

We now compare the two models using ANOVA:

```
anova(model1,model2,test="Chi")
```

```
Analysis of Deviance Table

Model 1: incidence ~ area * isolation
Model 2: incidence ~ area + isolation
  Resid. Df Resid. Dev Df Deviance Pr(>Chi)
1        46     28.252
2        47     28.402 -1 -0.15043   0.6981
```

The simpler model is not significantly worse, so we accept this for the time being, and inspect the parameter estimates and standard errors:

```
summary(model2)
```

```
Call:
glm(formula = incidence ~ area + isolation, family = binomial)

Deviance Residuals:
    Min        1Q    Median        3Q       Max
-1.8189   -0.3089    0.0490    0.3635    2.1192

Coefficients:
            Estimate Std. Error z value Pr(>|z|)
(Intercept)   6.6417     2.9218   2.273  0.02302 *
area          0.5807     0.2478   2.344  0.01909 *
isolation    -1.3719     0.4769  -2.877  0.00401 **

(Dispersion parameter for binomial family taken to be 1)

    Null deviance: 68.029  on 49  degrees of freedom
Residual deviance: 28.402  on 47  degrees of freedom
AIC: 34.402
```

The estimates and their standard errors are in logits. We see that `area` has a significant positive effect (larger islands are more likely to be occupied), but `isolation` has a very strong negative effect (isolated islands are much less likely to be occupied). This is the minimal adequate model. We should plot the fitted model through the scatterplot of the data. It is much easier to do this for each variable separately, like this:

```
modela <- glm(incidence~area,binomial)
modeli <- glm(incidence~isolation,binomial)

windows(7,4)
par(mfrow=c(1,2))

xv <- seq(0,9,0.01)
yv <- predict(modela,list(area=xv),type="response")
plot(area,incidence)
lines(xv,yv, col="red")

xv2 <- seq(0,10,0.01)
yv2 <- predict(modeli,list(isolation=xv2),type="response")
plot(isolation,incidence)
lines(xv2,yv2, col="red")
```

 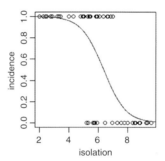

17.2   Graphical tests of the fit of the logistic to data

The logistic plots above are all well and good, but it is very difficult to know how good the fit of the model is when the data are shown only as 0s or 1s. Some people have argued for putting histograms instead of rugs on the top and bottom axes, but there are issues here about the arbitrary location of the bins (see p. 279). **Rugs** are a one-dimensional addition to the bottom (or top) of the plot showing the locations of the data points along the x axis. The idea is to indicate the extent to which the values are clustered at certain values of the explanatory variable, rather than evenly spaced out along it. If there are many values at the same value of x, it will be useful to use the `jitter` function to spread them out (by randomly selected small distances from x).

A different tack is to cut the data into a number of sectors and plot empirical probabilities (ideally with their standard errors) as a guide to the fit of the logistic curve, but this, too, can be criticized on the arbitrariness of the boundaries to do the cutting, coupled with the fact that there are often too few data points to give acceptable precision to the empirical probabilities and standard errors in any given group.

For what it is worth, here is an example of this approach. The response is occupation of territories and the explanatory variable is resource availability in each territory:

```
occupy <- read.table("c:\\temp\\occupation.txt",header=T)
attach(occupy)
names(occupy)
```

```
[1]  "resources"  "occupied"
```

```
plot(resources,occupied,type="n")
rug(jitter(resources[occupied==0]))
rug(jitter(resources[occupied==1]),side=3)
```

Now fit the logistic regression and draw the line:

```
model  <-  glm(occupied~resources,binomial)
xv  <-  0:1000
yv  <-  predict(model,list(resources=xv),type="response")
lines(xv,yv,col="red")
```

The idea is to cut up the ranked values on the *x* axis (resources) into five categories and then work out the mean and the standard error of the proportions in each group:

```
cutr  <-  cut(resources,5)
tapply(occupied,cutr,sum)
```

```
(13.2,209]    (209,405]    (405,600]    (600,796]    (796,992]
         0           10           25           26           31
```

If you have not met the `cut` function before, you will be impressed. It has taken the continuous variable called `resources`, and cut it up into five bins in creating a factor called `cutr`. The margins of the bins are defined within curved and square brackets which are read as follows: (13.2, 209] means 'from, but not including, 13.2 to, and including, 209'. So the figure next to the round bracket is excluded from this bin and is included in the adjacent bin (to the left in this case). This option is called `right=TRUE` and is the default for `cut`. We use the `table` function to count the number of cases in each bin:

```
table(cutr)
```

```
cutr
(13.2,209]    (209,405]    (405,600]    (600,796]    (796,992]
        31           29           30           29           31
```

So the empirical probabilities are given by:

```
probs  <-  tapply(occupied,cutr,sum)/table(cutr)
probs
```

```
(13.2,209]    (209,405]    (405,600]    (600,796]    (796,992]
 0.0000000    0.3448276    0.8333333    0.8965517    1.0000000
```

```
probs  <-  as.vector(probs)
resmeans  <-  tapply(resources,cutr,mean)
resmeans  <-  as.vector(resmeans)
```

We can plot these as big points on the graph – the closer they fall to the line, the better the fit of the logistic model to the data:

```
points(resmeans,probs,pch=16,cex=2,col="blue")
```

We need to add a measure of unreliability to the points. The standard error of a binomial proportion will do: $se = \sqrt{p(1-p)/n}$.

```
se  <-  sqrt(probs*(1-probs)/table(cutr))
```

Finally, draw lines up and down from each point indicating one standard error:

```
up <- probs+as.vector(se)
down <- probs-as.vector(se)

for (i in 1:5) {
  lines(c(resmeans[i],resmeans[i]),c(up[i],down[i]), col="blue")  }
```

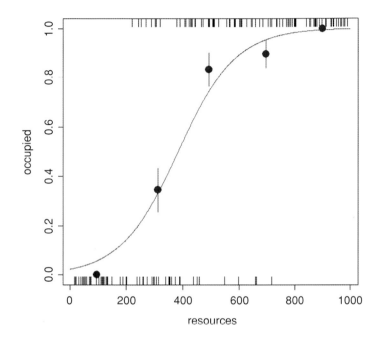

Evidently, the logistic is a good fit to the data above resources of 800 (not surprising, though, given that there were no unoccupied patches in this region), but it is rather a poor fit for resources between 400 and 800, as well as below 200, despite the fact that there were no occupied patches in the latter region (empirical $p = 0$).

17.3 ANCOVA with a binary response variable

In our next example the binary response variable is parasite infection (infected or not) and the explanatory variables are weight and age (continuous) and sex (categorical). We begin with data inspection:

```
infection <- read.table("c:\\temp\\infection.txt",header=T)
attach(infection)
names(infection)

[1] "infected" "age"      "weight"   "sex"

windows(7,4)
par(mfrow=c(1,2))
plot(infected,weight,xlab="Infection",ylab="Weight",col="green")
plot(infected,age,xlab="Infection",ylab="Age",col="green4")
```

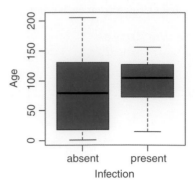

Infected individuals are substantially lighter than uninfected individuals, and occur in a much narrower range of ages. To see the relationship between infection and gender (both categorical variables) we can use `table`:

```
table(infected,sex)
```

```
          sex
infected  female male
  absent      17   47
  present     11    6
```

This indicates that the infection is much more prevalent in females (11/28) than in males (6/53).

We now proceed, as usual, to fit a maximal model with different slopes for each level of the categorical variable:

```
model <- glm(infected~age*weight*sex,family=binomial)
summary(model)
```

```
Coefficients:
                      Estimate Std. Error z value Pr(>|z|)
(Intercept)          -0.109124   1.375388  -0.079    0.937
age                   0.024128   0.020874   1.156    0.248
weight               -0.074156   0.147678  -0.502    0.616
sexmale              -5.969109   4.278066  -1.395    0.163
age:weight           -0.001977   0.002006  -0.985    0.325
age:sexmale           0.038086   0.041325   0.922    0.357
weight:sexmale        0.213830   0.343265   0.623    0.533
age:weight:sexmale   -0.001651   0.003419  -0.483    0.629
```

```
(Dispersion parameter for binomial family taken to be 1)
```

```
    Null deviance: 83.234  on 80  degrees of freedom
Residual deviance: 55.706  on 73  degrees of freedom
AIC: 71.706
```

It certainly does not look as if any of the high-order interactions are significant. Instead of using `update` and `anova` for model simplification, we can use `step` to compute the AIC for each term in turn:

```
model2 <- step(model)
```

```
Start:  AIC=71.71
```

First, it tests whether the three-way interaction is required:

```
infected ~ age * weight * sex

                   Df Deviance   AIC
- age:weight:sex   1    55.943 69.943
<none>                  55.706 71.706
```

This causes a reduction in AIC of just $71.7 - 69.9 = 1.8$ and hence is not significant. Next, it looks at the three two-way interactions and decides which to delete first:

```
Step:  AIC=69.94

                Df Deviance    AIC
- weight:sex    1   56.122  68.122
- age:sex       1   57.828  69.828
<none>              55.943  69.943
- age:weight    1   58.674  70.674
```

Only the removal of the weight–sex interaction causes a reduction in AIC, so this interaction is deleted and the other two interactions are retained. Let us see if we would have been this lenient:

```
Step:  AIC=68.12

              Df Deviance    AIC
<none>            56.122  68.122
- age:sex     1   58.142  68.142
- age:weight  1   58.899  68.899
```

```
summary(model2)
```

```
Coefficients:
              Estimate Std. Error z value Pr(>|z|)
(Intercept) -0.391566   1.265230  -0.309   0.7570
age          0.025764   0.014921   1.727   0.0842 .
weight      -0.036494   0.128993  -0.283   0.7772
sexmale     -3.743771   1.791962  -2.089   0.0367 *
age:weight  -0.002221   0.001365  -1.627   0.1038
age:sexmale  0.020464   0.015232   1.343   0.1791

(Dispersion parameter for binomial family taken to be 1)

    Null deviance: 83.234  on 80  degrees of freedom
Residual deviance: 56.122  on 75  degrees of freedom
AIC: 68.122
```

Neither of the two interactions retained by step would figure in our model ($p > 0.10$). We shall use update to simplify model2:

```
model3 <- update(model2,~.-age:weight)
anova(model2,model3,test="Chi")
```

```
Analysis of Deviance Table

Model 1: infected ~ age + weight + sex + age:weight + age:sex
Model 2: infected ~ age + weight + sex + age:sex
```

```
   Resid. Df Resid. Dev Df Deviance Pr(>Chi)
1        75        56.122
2        76        58.899 -1   -2.777  0.09562 .
```

So there is no really persuasive evidence of an age–weight term ($p = 0.096$).

```
model4 <- update(model2,~.-age:sex)
anova(model2,model4,test="Chi")
```

Analysis of Deviance Table

```
Model 1: infected ~ age + weight + sex + age:weight + age:sex
Model 2: infected ~ age + weight + sex + age:weight
  Resid. Df Resid. Dev Df Deviance Pr(>Chi)
1        75        56.122
2        76        58.142 -1   -2.0203   0.1552
```

Note that we are testing all the two-way interactions by deletion from the model that contains all two-way interactions (model2): $p = 0.1552$, so nothing there, then.

What about the three main effects?

```
model5 <- glm(infected~age+weight+sex,family=binomial)
summary(model5)
```

Coefficients:

```
              Estimate Std. Error z value Pr(>|z|)
(Intercept)   0.609369   0.803288   0.759 0.448096
age           0.012653   0.006772   1.868 0.061701 .
weight       -0.227912   0.068599  -3.322 0.000893 ***
sexmale      -1.543444   0.685681  -2.251 0.024388 *
```

(Dispersion parameter for binomial family taken to be 1)

```
    Null deviance: 83.234  on 80  degrees of freedom
Residual deviance: 59.859  on 77  degrees of freedom
AIC: 67.859
```

Weight is highly significant, as we expected from the initial boxplot, sex is quite significant, and age is marginally significant. It is worth establishing whether there is any evidence of non-linearity in the response of infection to weight or age. We might begin by fitting quadratic terms for the two continuous explanatory variables:

```
model6 <- glm(infected~age+weight+sex+I(weight^2)+I(age^2),family=binomial)
summary(model6)
```

Coefficients:

```
              Estimate Std. Error z value Pr(>|z|)
(Intercept) -3.4475839  1.7978359  -1.918   0.0552 .
age          0.0829364  0.0360205   2.302   0.0213 *
weight       0.4466284  0.3372352   1.324   0.1854
sexmale     -1.2203683  0.7683288  -1.588   0.1122
I(weight^2) -0.0415128  0.0209677  -1.980   0.0477 *
I(age^2)    -0.0004009  0.0002004  -2.000   0.0455 *
```

```
(Dispersion parameter for binomial family taken to be 1)

    Null deviance: 83.234  on 80  degrees of freedom
Residual deviance: 48.620  on 75  degrees of freedom
AIC: 60.62
```

Evidently, both relationships are significantly curvilinear. It is worth looking at these non-linearities in more detail, to see if we can do better with other kinds of models (e.g. non-parametric smoothers, piecewise linear models or step functions). A generalized additive model is often a good way to start when we have continuous covariates:

```
library(mgcv)
model7 <- gam(infected~sex+s(age)+s(weight),family=binomial)
windows(7,4)
par(mfrow=c(1,2))
plot.gam(model7)
```

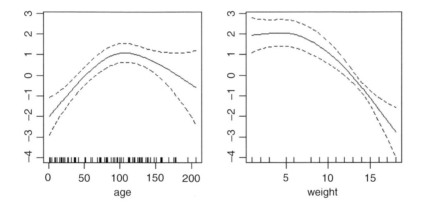

These non-parametric smoothers are excellent at showing the humped relationship between infection and age, and at highlighting the possibility of a threshold at weight ≈ 8 in the relationship between weight and infection. We can now return to a GLM to incorporate these ideas. We shall fit age and age^2 as before, but try a piecewise linear fit for weight, estimating the threshold weight at a range of values (say 8–14) and selecting the threshold that gives the lowest residual deviance; this turns out to be a threshold of 12 (rather higher than suggested by the gam plot above). The piecewise regression is specified by the term:

```
I((weight - 12) * (weight > 12))
```

The I ('as is') is necessary to stop the * being evaluated as an interaction term in the model formula. What this expression says is 'regress infection on the value of weight - 12, but only do this when weight > 12 is true' (see p. 25). Otherwise, assume that infection is independent of weight.

```
model8 <- glm(infected~sex+age+I(age^2)
                 +I((weight-12)*(weight>12)),family=binomial)
summary(model8)

Coefficients:
                         Estimate Std. Error z value Pr(>|z|)
(Intercept)            -2.7511382  1.3678824  -2.011   0.0443 *
sexmale                -1.2864683  0.7349201  -1.750   0.0800 .
```

```
age                                 0.0798629  0.0348184    2.294    0.0218 *
I(age^2)                           -0.0003892  0.0001955   -1.991    0.0465 *
I((weight - 12) * (weight > 12)) -1.3547520  0.5350853   -2.532    0.0113 *
(Dispersion parameter for binomial family taken to be 1)
    Null deviance: 83.234  on 80   degrees of freedom
Residual deviance: 48.687  on 76   degrees of freedom
AIC: 58.687
```

The effect of sex on infection is not quite significant ($p = 0.071$ for a chi-squared test on deletion), so we leave it out. The quadratic term for age does not look highly significant here, but a deletion test gives $p = 0.011$, so we retain it. The minimal adequate model is therefore

```
model9 <- glm(infected~age+I(age^2)+I((weight-12)*(weight>12)),family=binomial)
summary(model9)
```

```
Coefficients:
                               Estimate  Std. Error  z value  Pr(>|z|)
(Intercept)                    -3.1207552  1.2665593   -2.464    0.0137 *
age                             0.0765784  0.0323376    2.368    0.0179 *
I(age^2)                       -0.0003843  0.0001846   -2.081    0.0374 *
I((weight - 12) * (weight > 12)) -1.3511706  0.5134681   -2.631    0.0085 **
(Dispersion parameter for binomial family taken to be 1)
    Null deviance: 83.234  on 80   degrees of freedom
Residual deviance: 51.953  on 77   degrees of freedom
AIC: 59.953
```

We conclude that there is a humped relationship between infection and age, and a threshold effect of weight on infection. The effect of sex is marginal, but might repay further investigation ($p = 0.071$).

17.4   Binary response with pseudoreplication

In the `bacteria` dataframe, which is part of the `MASS` library, we have repeated assessment of bacterial infection (yes or no, coded as y or n) in a series of patients allocated at random to one of three treatments: placebo, drug and drug plus supplement (drug+). The trial lasted for 11 weeks and different patients were assessed on different numbers of occasions. The question is whether the two treatments significantly reduced bacterial infection.

```
library(MASS)
attach(bacteria)
names(bacteria)
```

```
[1]  "y"   "ap"   "hilo"   "week"   "ID"   "trt"
```

```
table(y)
```

```
y
   n    y
  43  177
```

The data are binary, so we need to use `family=binomial`. There is temporal pseudoreplication (repeated measures on the same patients) so we cannot use `glm`. The ideal solution is the generalized mixed models function `lmer`. Like `glm`, the `lmer` function can take text (e.g. a two-level factor like `y`) as the response variable. We start by looking at the data:

```
table(y,trt)
```

```
    trt
y   placebo drug drug+
  n      12   18    13
  y      84   44    49
```

Preliminary data inspection suggests that the drug might be effective because only 12 out of 96 patient visits were bacteria-free in the placebos, compared with 31 out of 124 for the treated individuals. We shall see. The modelling goes like this: the `lmer` function is in the `lme4` package. The random effects appear in the same formula as the fixed effects, but defined by the round brackets and the 'given' operator | to separate the continuous random effect (`week`) from the categorical random effect (patient `ID`):

```
library(lme4)

model1 <- lmer(y~trt+(week|ID),family=binomial)
summary(model1)

Generalized linear mixed model fit by the Laplace approximation
Formula: y ~ trt + (week|ID)
   AIC   BIC logLik deviance
 209.2 229.6  -98.6    197.2
Random effects:
 Groups Name        Variance Std.Dev. Corr
 ID     (Intercept) 0.147815 0.38447
        week        0.062371 0.24974  1.000
Number of obs: 220, groups: ID, 50

Fixed effects:
            Estimate Std. Error z value Pr(>|z|)
(Intercept)   2.6195     0.4894   5.352  8.7e-08 ***
trtdrug      -1.2185     0.6588  -1.850   0.0644 .
trtdrug+     -0.5290     0.6991  -0.757   0.4492
```

Variation in intercepts across the patients (0.148) explained roughly twice as much variation in infection as did random variation in slopes (0.062). The fixed effects are not significant.

We can simplify the model by removing the dependence of infection on week, retaining only the intercept as a random effect `+ (1|ID)`:

```
model2 <- lmer(y~trt+(1|ID),family=binomial)
anova(model1,model2)

Data:
Models:
model2: y ~ trt + (1|ID)
model1: y ~ trt + (week|ID)
       Df    AIC    BIC  logLik  Chisq Chi Df Pr(>Chisq)
model2  4 214.32 227.90 -103.162
model1  6 209.21 229.57  -98.603 9.1184      2    0.01047 *
```

The simpler `model2` is significantly worse ($p = 0.010\ 47$) so we accept `model1` (it has a lower AIC than `model2`), retaining the random effect for `week`.

There is a question about the factor levels: perhaps the drug effect would be more significant if we combine the `drug` and `drug+` treatments?

```
drugs <- factor(1+(trt!="placebo"))
table(y,drugs)

    drugs
y    1  2
  n 12 31
  y 84 93

model3 <- lmer(y~drugs+(week|ID),family=binomial)
summary(model3)

Generalized linear mixed model fit by the Laplace approximation
Formula: y ~ drugs + (week|ID)
   AIC   BIC logLik deviance
 208.2 225.2 -99.12    198.2
Random effects:
 Groups Name        Variance Std.Dev. Corr
 ID     (Intercept) 0.196499 0.44328
        week        0.059128 0.24316  1.000
Number of obs: 220, groups: ID, 50

Fixed effects:
            Estimate Std. Error z value Pr(>|z|)
(Intercept)   2.6419     0.4958   5.329 9.89e-08 ***
drugs2       -0.8990     0.5977  -1.504    0.133
```

The interpretation is straightforward: there is no evidence in this experiment that either treatment significantly reduces bacterial infection. Note that this is not the same as saying that the drug does not work. It is simply that this trial is too small to demonstrate the significance of its efficacy.

It is also important to appreciate the importance of the pseudoreplication. If we had ignored the fact that there were multiple measures per patient we should have concluded wrongly that the drug effect was significant. Here are the raw data on the counts:

```
table(y,trt)

    trt
y    placebo drug drug+
  n       12   18    13
  y       84   44    49
```

Here is the *wrong* way of testing for the significance of the treatment effect:

```
prop.test(c(12,18,13),c(96,62,62))

        3-sample test for equality of proportions without continuity
correction

data:  c(12, 18, 13) out of c(96, 62, 62)
X-squared = 6.6585, df = 2, p-value = 0.03582
```

```
alternative hypothesis: two.sided
sample estimates:
    prop 1      prop 2      prop 3
0.1250000 0.2903226 0.2096774
```

It appears that the drug has increased the rate of non-infection from 0.125 in the placebos to 0.29 in the treated patients, and that this effect is significant ($p = 0.035\,82$). As we have seen, however, when we remove the pseudoreplication by using the appropriate mixed model with `lmer` the response is non-significant.

Another way to get rid of the pseudoreplication is to restrict the analysis to the patients that were there at the end of the experiment. We just use `subset=(week==11)` and this removes all the pseudoreplication because no subjects were measured twice within any week – we can check this with the `any` function:

```
any(table(ID,week) >1)
```

```
[1]   FALSE
```

The model is a straightforward GLM with a binary response variable and a single explanatory variable (the three-level factor called `trt`):

```
model <- glm(y~trt,binomial,subset=(week==11))
summary(model)
```

```
Coefficients:
              Estimate Std. Error z value Pr(>|z|)
(Intercept)    1.3863     0.5590   2.480   0.0131 *
trtdrug       -0.6931     0.8292  -0.836   0.4032
trtdrug+      -0.6931     0.8292  -0.836   0.4032
```

```
(Dispersion parameter for binomial family taken to be 1)
```

```
    Null deviance: 51.564  on 43  degrees of freedom
Residual deviance: 50.569  on 41  degrees of freedom
AIC: 56.569
```

Neither drug treatment has anything approaching a significant effect in lowering bacterial infection rates compared with the placebos ($p = 0.4032$). The supplement was expected to *increase* bacterial control over the drug treatment, so perhaps the interpretation will be modified lumping together the two drug treatments:

```
drugs <- factor(1+(trt=="placebo"))
```

Here are placebos plus patients getting one drug treatment or the other:

```
table(drugs[week==11])
```

```
 1   2
24  20
```

Thus there were 24 patients receiving one drug or the other, and 20 placebos (at 11 weeks).

```
model <- glm(y~drugs,binomial,subset=(week==11))
summary(model)
```

```
Coefficients:
              Estimate Std. Error z value Pr(>|z|)
(Intercept)    0.6931     0.4330   1.601    0.109
drugs2         0.6931     0.7071   0.980    0.327
```

```
(Dispersion parameter for binomial family taken to be 1)

    Null deviance: 51.564   on 43   degrees of freedom
Residual deviance: 50.569   on 42   degrees of freedom
AIC: 54.569
```

Clearly, there is no convincing effect for the drug treatments on bacterial infection when we use this subset of the data ($p = 0.327$).

An alternative way of analysing all the data (including the pseudoreplication) is to ask what proportion of tests on each patient scored positive for the bacteria. The response variable now becomes a proportion, and the pseudoreplication disappears because we only have one number for each patient (i.e. a count of the number of occasions on which each patient scored positive for the bacteria, with the binomial denominator as the total number of tests on that patient).

There are some preliminary data-shortening tasks. We need to create a vector of length 50 containing the drug treatments of each patient (tss) and a table (ys, with elements of length 50) scoring how many times each patient was infected and uninfected by bacteria. Finally, we use cbind to create a two-column response variable, yv:

```
dss <- data.frame(table(trt,ID))
head(dss)

      trt  ID Freq
1 placebo X01    4
2    drug X01    0
3   drug+ X01    0
4 placebo X02    0
5    drug X02    0
6   drug+ X02    4
```

We need to find out the treatments of the patients that scored Freq > 0:

```
tss <- dss[dss[,3]>0,]$trt

ys <- table(y,ID)

yv <- cbind(ys[2,],ys[1,])
```

Now we can fit a very simple model for the binomial response (glm with binomial errors):

```
model <- glm(yv~tss,binomial)
summary(model)

Coefficients:
            Estimate Std. Error z value Pr(>|z|)
(Intercept)   1.9459     0.3086   6.306 2.87e-10 ***
tssdrug      -1.0521     0.4165  -2.526   0.0115 *
tssdrug+     -0.6190     0.4388  -1.411   0.1583

(Dispersion parameter for binomial family taken to be 1)

    Null deviance: 86.100   on 49   degrees of freedom
Residual deviance: 79.444   on 47   degrees of freedom
AIC: 130.9
```

Drug looks to be significant here, but note that the residual deviance is much bigger than the residual degrees of freedom, so we should correct for overdispersion by using quasi-binomial instead of binomial errors (recall that the response is now binomial rather than binary; see p. 651):

```
model <- glm(yv~tss,quasibinomial)
summary(model)
```

```
Coefficients:
            Estimate Std. Error t value Pr(>|t|)
(Intercept)   1.9459     0.3837   5.071 6.62e-06 ***
tssdrug      -1.0521     0.5180  -2.031   0.0479 *
tssdrug+     -0.6190     0.5457  -1.134   0.2624
```

```
(Dispersion parameter for quasibinomial family taken to be 1.546375)

    Null deviance: 86.100  on 49  degrees of freedom
Residual deviance: 79.444  on 47  degrees of freedom
AIC: NA
```

There is a marginally significant effect of drug, but no significant difference between the two drug treatments, so we aggregate them into a single drug treatment:

```
tss2 <- factor(1+(tss=="placebo"))
model <- glm(yv~tss2,quasibinomial)
summary(model)
```

```
Coefficients:
            Estimate Std. Error t value Pr(>|t|)
(Intercept)   1.0986     0.2582   4.255 9.64e-05 ***
tss22         0.8473     0.4629   1.830   0.0734 .
```

```
(Dispersion parameter for quasibinomial family taken to be 1.55006)

    Null deviance: 86.100  on 49  degrees of freedom
Residual deviance: 80.523  on 48  degrees of freedom
AIC: NA
```

Again, the treatment effect is not significant, in agreement with the generalized mixed-effects model (p. 662). This example re-emphasizes the importance of correcting for pseudoreplication and overdispersion. Had we not made allowance for these we would have concluded (wrongly) that the drug brought about significant a reduction in infection.

18

Generalized Additive Models

Up to this point, continuous explanatory variables have been added to models as linear functions, linearized parametric transformations, or through various link functions. In all cases, an explicit or implicit assumption was made about the parametric form of the function to be fitted to the data (whether quadratic, logarithmic, exponential, logistic, reciprocal or whatever). In many cases, however, you have one or more continuous explanatory variables, but you have no *a priori* reason to choose one particular parametric form over another for describing the shape of the relationship between the response variable and the explanatory variable(s). Generalized additive models (GAMs) are useful in such cases because they allow you to capture the shape of a relationship between y and x without prejudging the issue by choosing a particular parametric form.

Generalized additive models (implemented in R by the gam function) extend the range of application of generalized linear models (glm) by allowing non-parametric smoothers in addition to parametric forms, and these can be associated with a range of link functions. All of the error families allowed with glm are available with gam (binomial, poisson, Gamma, etc.). Indeed, gam has many of the attributes of both glm and lm, and the output can be modified using update. You can use all of the familiar methods such as print, plot, summary, anova, predict and fitted after a GAM has been fitted to data. The gam function used in this book is in the mgcv package contributed by Simon Wood:

```
library(mgcv)
```

There are many ways of specifying the model in a GAM: all of the continuous explanatory variables x, w and z can enter the model as non-parametrically smoothed functions like this:

```
y~s(x) + s(w) + s(z)
```

Alternatively, the model can contain a mix of parametrically estimated parameters (x and z) and smoothed variables $s(w)$:

```
y~x + s(w) + z
```

Formulae can involve nested (two-dimensional) terms in which the smoothing s() terms have more than one argument, implying an isotropic smooth:

```
y~s(x) + s(z) + s(x,z)
```

Alternatively the smoothers can have overlapping terms such as

```
y~s(x,z) + s(z,w)
```

The R Book, Second Edition. Michael J. Crawley.
© 2013 John Wiley & Sons, Ltd. Published 2013 by John Wiley & Sons, Ltd.

The user has a high degree of control over the way that interactions terms can be fitted, and te() smooths are provided as an effective means for modelling smooth interactions of any number of variables via scale-invariant tensor product smooths. Here is an example of a model formula with a fully nested tensor product te(x,z,k=6):

```
y ~ s(x,bs="cr",k=6) + s(z,bs="cr",k=6) + te(x,z,k=6)
```

The optional arguments to the smoothers are bs="cr",k=6, where bs indicates the basis to use for the smooth ("cr" is a cubic regression spline; the default is thin plate bs="tp"), and k is the dimension of the basis used to represent the smooth term (it defaults to k = 10*3(d-1) where d is the number of covariates for this term).

18.1 Non-parametric smoothers

You can see non-parametric smoothers in action for fitting a curve through a scatterplot in Chapter 10 (p. 491). Here we are concerned with using non-parametric smoothers in statistical modelling where the object is to assess the relative merits of a range of different models in explaining variation in the response variable. One of the simplest model-fitting functions is loess (which replaces its predecessor called lowess).

The following example shows population change, $Delta = \log(N(t+1)/N(t))$, as a function of population density ($N(t)$) in an investigation of density dependence in a sheep population. This is what the data look like:

```
soay <- read.table("c:\\temp\\soaysheep.txt",header=T)
attach(soay)
names(soay)

[1] "Year" "Population" "Delta"

plot(Population,Delta,pch=21,col="green",bg="red")
```

Broadly speaking, population change is positive at low densities (Delta > 0) and negative at high densities (Delta < 0) but there is a great deal of scatter, and it is not at all obvious what shape of smooth function would best describe the data. Here is the default loess:

```
model <- loess(Delta~Population)
summary(model)

Call:
loess(formula = Delta~Population)
Number of Observations: 44
Equivalent Number of Parameters: 4.66
Residual Standard Error: 0.2616
Trace of smoother matrix: 5.11
Control settings:
    normalize:      TRUE
    span :     0.75
    degree : 2
    family : gaussian
    surface :interpolate cell = 0.2
```

Now draw the smoothed line using `predict` to extract the predicted values from `model`:

```
xv <- seq(600,2000,1)
yv <- predict(model,data.frame(Population=xv))
lines(xv,yv col="red")
```

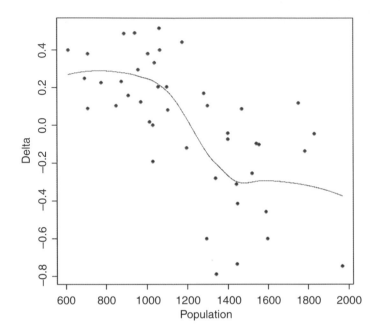

The smooth curve looks rather like a step function. We can compare this smooth function with a step function, using a tree model (p. 768) as an objective way of determining the threshold for splitting the data into low- and high-density parts:

```
library(tree)
thresh <- tree(Delta~Population)
print(thresh)
```

The threshold for the first split of the tree model is at `Population = 1289.5`, so we define this as the threshold density:

```
th <- 1289.5
```

Then we can use this threshold to create a two-level factor for fitting two constant rates of population change using aov:

```
model2 <- aov(Delta~(Population>th))
summary(model2)
```

```
                Df  Sum Sq  Mean Sq  F value    Pr(>F)
Population > th  1   2.810   2.810    47.63   2.01e-08  ***
Residuals       42   2.477   0.059
```

showing a residual error variance of 0.059. This compares with the residual of $0.2616^2 = 0.068$ from the `loess` (above). To draw the step function we need the average low-density population increase and the average high-density population decline:

```
tapply(Delta[-45],(Population[-45]>th),mean)
```

```
     FALSE          TRUE
0.2265084  -0.2836616
```

Note the use of negative subscripts to drop the NA from the last value of `Delta`. Then use these figures to draw the step function:

```
lines(c(600,th),c(0.2265,0.2265),lty=2, col="blue")
lines(c(th,2000),c(-0.2837,-0.2837),lty=2, col="blue")
lines(c(th,th),c(-0.2837,0.2265),lty=2, col="blue")
```

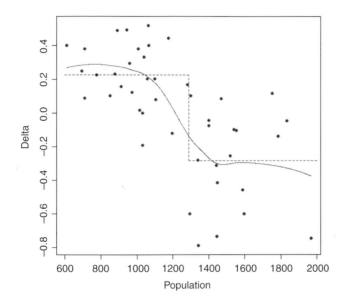

It is a moot point which of these two models is the most realistic scientifically, but the step function involved three estimated parameters (two averages and a threshold), while the `loess` is based on 4.66 degrees of freedom, so parsimony favours the step function (it also has a slightly lower residual sum of squares).

18.2 Generalized additive models

This dataframe contains measurements of radiation, temperature, wind speed and ozone concentration. We want to model ozone concentration as a function of the three continuous explanatory variables using non-parametric smoothers rather than specified nonlinear functions (the parametric multiple regression analysis of these data is on p. 490):

```
ozone.data <- read.table("c:\\temp\\ozone.data.txt",header=T)
attach(ozone.data)
names(ozone.data)
```

```
[1] "rad" "temp" "wind" "ozone"
```

For data inspection we use `pairs` with a non-parametric smoother, `lowess`:

```
pairs(ozone.data, panel=function(x,y) { points(x,y); lines(lowess(x,y))} )
```

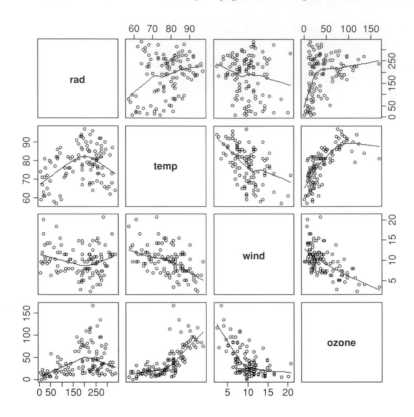

Now fit all three explanatory variables using the non-parametric smoother `s()`:

```
model <- gam(ozone~s(rad)+s(temp)+s(wind))
summary(model)

Family: gaussian
Link function: identity

Formula:
ozone ~ s(rad) + s(temp) + s(wind)

Parametric coefficients:
            Estimate Std. Error t value Pr(>|t|)
(Intercept)    42.10       1.66   25.36   <2e-16 ***
Approximate significance of smooth terms:
          edf Ref.df      F  p-value
s(rad)  2.763  3.451  3.964  0.00736 **
s(temp) 3.841  4.762 11.612 1.28e-08 ***
s(wind) 2.918  3.666 13.770 1.53e-08 ***

R-sq.(adj) =  0.724   Deviance explained = 74.8%
GCV score =      338   Scale est. = 305.96    n = 111
```

Note that the intercept is estimated as a parametric coefficient (42.10; upper table) and the three explanatory variables are fitted as smooth terms. All three are significant, but radiation is the least significant at $p = 0.007\,36$. We can compare a GAM with and without a term for radiation using ANOVA in the normal way:

```
model2 <- gam(ozone~s(temp)+s(wind))
anova(model,model2,test="F")

Analysis of Deviance Table

Model 1: ozone ~ s(rad) + s(temp) + s(wind)
Model 2: ozone ~ s(temp) + s(wind)
  Resid. Df Resid. Dev      Df Deviance      F    Pr(>F)
1    100.48       30742
2    102.85       34885 -2.3672  -4142.2 5.7192 0.002696 **
```

Clearly, radiation should remain in the model, since deletion of radiation caused a highly significant increase in deviance ($p = 0.0027$), emphasizing the fact that deletion is a better test than inspection of parameters (the p values in the full model table were *not* deletion p values).

We should investigate the possibility that there is an interaction between wind and temperature:

```
model3 <- gam(ozone~s(temp)+s(wind)+s(rad)+s(wind,temp))
summary(model3)

Family: gaussian
Link function: identity

Formula:
ozone ~ s(temp) + s(wind) + s(rad) + s(wind, temp)

Parametric coefficients:
            Estimate Std. Error t value Pr(>|t|)
(Intercept)   42.099      1.361   30.92   <2e-16 ***

Approximate significance of smooth terms:
                edf Ref.df      F  p value
s(temp)       1.000  1.000  0.000   0.9892
s(wind)       5.613  6.482  2.492   0.0255 *
s(rad)        1.389  1.667  4.694   0.0164 *
s(wind,temp) 18.246 22.061  3.202 6.52e-05 ***

R-sq.(adj) =  0.814   Deviance explained = 85.9%
GCV score = 272.66  Scale est. = 205.72    n = 111
```

The interaction appears to be highly significant, but the main effect of temperature is cancelled out. We can inspect the fit of model3 like this:

```
par(mfrow-c(2,2))
plot(model3,residuals=T,pch=16)
```

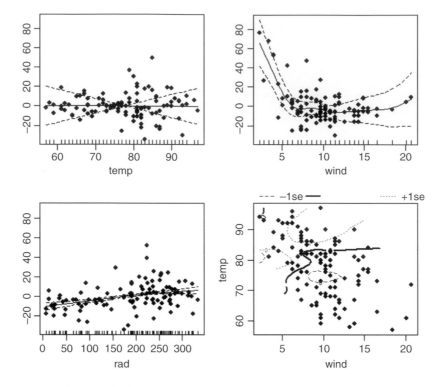

The etchings on the x axis are called rugs (see Section 17.2) and indicate the locations of measurements of x values on each axis. The default option is rug=T. The bottom right-hand plot shows the complexity of the interaction between temperature and wind speed.

18.2.1 Technical aspects

The degree of smoothness of model terms is estimated as part of fitting; isotropic or scale-invariant smooths of any number of variables are available as model terms. Confidence or credible intervals are readily available for any quantity predicted using a fitted model. In mgcv, gam solves the smoothing parameter estimation problem by using the generalized cross validation (GCV) criterion

$$GCV = \frac{nD}{(n - \text{d.f.})^2}$$

or an unbiased risk estimator (UBRE) criterion

$$UBRE = \frac{D}{n} + 2\phi \frac{\text{d.f.}}{n} - \phi,$$

where D is the deviance, n the number of data, ϕ the scale parameter and d.f. the effective degrees of freedom of the model. Notice that UBRE is effectively just AIC rescaled, but is only used when ϕ is known. It is also possible to replace D by the Pearson statistic (see ?gam.method), but this can lead to oversmoothing. Smoothing parameters are chosen to minimize the GCV or UBRE score for the model, and the main computational challenge solved by the mgcv package is to do this efficiently and reliably. Various alternative numerical methods are provided: see ?gam.method. Smooth terms are represented using penalized regression splines (or similar smoothers) with smoothing parameters selected by GCV or UBRE or by regression splines with fixed degrees of freedom (mixtures of the two are permitted). Multi-dimensional smooths are available using penalized thin plate regression splines (isotropic) or tensor product splines (when an isotropic smooth is inappropriate).

This gam function is *not* a clone of what S-PLUS provides – there are three major differences. First, by default, estimation of the degree of smoothness of model terms is part of model fitting. Second, a Bayesian approach to variance estimation is employed that makes for easier confidence interval calculation (with good coverage probabilities). Third, the facilities for incorporating smooths of more than one variable are different.

If absolutely any smooth functions were allowed in model fitting then maximum likelihood estimation of such models would invariably result in complex overfitting estimates of the smoothed functions $s(x)$ and $s(z)$. For this reason the models are usually fitted by penalized likelihood maximization, in which the model deviance (negative log-likelihood) is modified by the addition of a penalty for each smooth function, penalizing what the author of gam, Simon Wood, calls its 'wiggliness'. To control the trade-off between penalizing wiggliness and penalizing badness of fit, each penalty is multiplied by an associated smoothing parameter: how to estimate these parameters and how to practically represent the smooth functions are the main statistical questions introduced by moving from GLMs to GAMs.

The built-in alternatives for univariate smooths terms are: a conventional penalized cubic regression spline basis, parameterized in terms of the function values at the knots; a cyclic cubic spline with a similar parameterization; and thin plate regression splines. The cubic spline bases are computationally very efficient, but require knot locations to be chosen (automatically by default). The thin plate regression splines are optimal low-rank smooths which do not have knots, but are computationally more costly to set up. Smooths of several variables can be represented using thin plate regression splines, or tensor products of any available basis, including user-defined bases (tensor product penalties are obtained automatically form the marginal basis penalties).

Thin plate regression splines are constructed by starting with the basis for a full thin plate spline and then truncating this basis in an optimal manner, to obtain a low-rank smoother. Details are given in Wood (2003). One key advantage of the approach is that it avoids the knot placement problems of conventional regression spline modelling, but it also has the advantage that smooths of lower rank are nested within smooths of higher rank, so that it is legitimate to use conventional hypothesis testing methods to compare models based on pure regression splines. The thin plate regression spline basis can become expensive to calculate for large data sets. In this case the user can supply a reduced set of knots to use in basis construction (see knots in the argument list), or use tensor products of cheaper bases. In the case of the cubic regression spline basis, knots of the spline are placed evenly throughout the covariate values to which the term refers. For example, if fitting 101 data points with an 11-knot spline of x then there would be a knot at every 10th (ordered) x value. The parameterization used represents the spline in terms of its values at the knots. The values at neighbouring knots are connected by sections of cubic polynomial constrained to be continuous up to and including second derivatives at the knots. The resulting curve is a natural cubic spline through the values at the knots (given two extra conditions specifying that the second derivative of the curve should be zero at the two end knots). This parameterization gives the parameters a nice interpretability. Details of the underlying fitting methods are given in Wood (2000, 2004).

You must have more unique combinations of covariates than the model has total parameters (total parameters being the sum of basis dimensions plus the sum of non-spline terms less the number of spline terms.). Automatic smoothing parameter selection is not likely to work well when fitting models to very few response data. With large data sets (more than a few thousand data) the tp basis gets very slow to use: use the knots argument as discussed above and shown in the examples. Alternatively, for low-density smooths you can use the cr basis and for multi-dimensional smooths use te smooths.

For data with many zeros clustered together in the covariate space it is quite easy to set up GAMs which suffer from identifiability problems, particularly when using Poisson or binomial families. The problem is that with log or logit links, for example, mean value zero corresponds to an infinite range on the linear predictor scale.

Another situation that occurs quite often is the one in which we would like to find out if the model

$$E(y) = f(x, z)$$

is really necessary, or whether

$$E(y) = f_1(x) + f_2(z)$$

would not do just as well. One way to do this is to look at the results of fitting

```
y~s(x)+s(z)+s(x,z)
```

gam automatically generates side conditions to make this model identifiable. You can also estimate overlapping models such as

```
y~s(x,z)+s(z,v)
```

Sometimes models of the form

$$E(y) = b_0 + f(x)z$$

need to be estimated (where f is a smooth function, as usual). The appropriate formula is

```
y~z+s(x,by=z)
```

where the by argument ensures that the smooth function gets multiplied by covariate z, but GAM smooths are centred (average value zero), so the parametric term for z is needed as well (f is being represented by a constant plus a centred smooth). If we wanted

$$E(y) = f(x)z$$

then the appropriate formula would be

```
y~z+s(x,by=z)-1
```

The by mechanism also allows models to be estimated in which the form of a smooth depends on the level of a factor, but to do this the user must generate the dummy variables for each level of the factor. Suppose, for example, that fac is a factor with three levels 1, 2, 3, and at each level of this factor the response depends smoothly on a variable x in a manner that is level-dependent. Three dummy variables, fac.1, fac.2, fac.3, can be generated for the factor (e.g. fac.1 <- as.numeric(fac==1)). Then the model formula would be:

```
y~fac+s(x,by=fac.1)+s(x,by=fac.2)+s(x,by=fac.3)
```

In the above examples the smooths of more than one covariate have all employed single-penalty thin plate regression splines. These isotropic smooths are not always appropriate: if variables are not naturally well scaled relative to each other then it is often preferable to use tensor product smooths, with a wiggliness penalty for each covariate of the term. See ?te for examples.

The most logically consistent method to use for deciding which terms to include in the model is to compare GCV/UBRE scores for models with and without the term. More generally, the score for the model with a smooth term can be compared to the score for the model with the smooth term replaced by appropriate parametric terms. Candidates for removal can be identified by reference to the approximate *p* values provided by summary.gam. Candidates for replacement by parametric terms are smooth terms with estimated degrees of freedom close to their minimum possible.

18.3   An example with strongly humped data

The ethanol dataframe contains 88 sets of measurements for variables from an experiment in which ethanol was burned in a single cylinder automobile test engine. The response variable, NOx, is the concentration of nitric oxide (NO) and nitrogen dioxide (NO_2) in engine exhaust, normalized by the work done by the engine, and the two continuous explanatory variables are C (the compression ratio of the engine), and E (the equivalence ratio at which the engine was run, which is a measure of the richness of the air–ethanol mix).

```
install.packages("SemiPar")
library(SemiPar)
data(ethanol)
attach(ethanol)
head(ethanol)
```

```
     NOx  C      E
1 3.741 12 0.907
2 2.295 12 0.761
3 1.498 12 1.108
4 2.881 12 1.016
5 0.760 12 1.189
6 3.120  9 1.001
```

Because NOx is such a strongly humped function of the equivalence ratio, E, we start with a model, NOx~s(E)+C, that fits this as a smoothed term and estimates a parametric term for the compression ratio:

```
model <- gam(NOx~s(E)+C)
windows(7,4)
par(mfrow=c(1,2))
plot.gam(model,residuals=T,pch=16,all.terms=T)
```

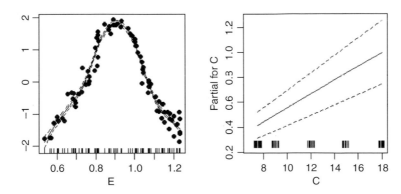

The coplot function is helpful in showing where the effect of C on NOx was most marked:

```
coplot(NOx~C|E,panel=panel.smooth)
```

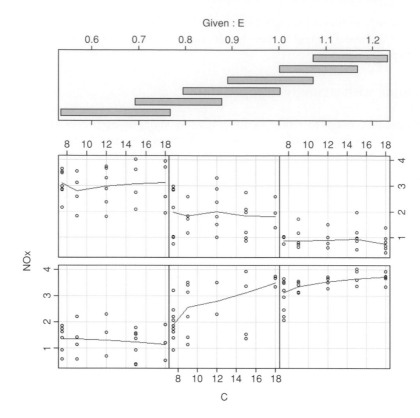

There is a pronounced positive effect of C on NOx only in panel 2 (ethanol $0.7 < E < 0.9$ from the shingles in the upper panel), but only slight effects elsewhere (most of the red lines are roughly horizontal). You can estimate the interaction between E and C from the product of the two variables:

```
CE <- E*C
model2 <- gam(NOx~s(E)+s(CE))
windows(7,4)
par(mfrow=c(1,2))

plot.gam(model2,residuals=T,pch=16,all.terms=T)
```

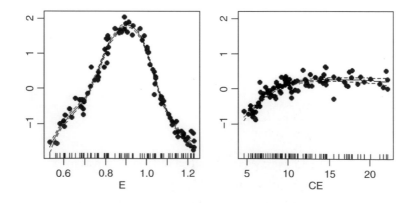

```
summary(model2)

Family: gaussian
Link function: identity

Formula:
NOx ~ s(E) + s(CE)

Parametric coefficients:
            Estimate Std. Error t value Pr(>|t|)
(Intercept)  1.95737    0.02126   92.07   <2e-16 ***

Approximate significance of smooth terms:
        edf Ref.df      F  p-value
s(E)  7.636  8.509 270.13  < 2e-16 ***
s(CE) 4.261  5.224  25.25 3.65e-15 ***

R-sq.(adj) =  0.969   Deviance explained = 97.3%
GCV score = 0.0466   Scale est. = 0.039771  n = 88
```

The summary of this GAM shows highly significant terms for both smoothed terms: the effect of ethanol, s(E), on 7.6 estimated degrees of freedom, and the interaction between E and C, s(CE), on 4.3 estimated degrees of freedom. The model explains a highly impressive 97.3% of the deviance in NOx concentration.

18.4   Generalized additive models with binary data

GAMs are particularly valuable with binary response variables (for background, see p. 650). To illustrate the use of gam for modelling binary response data, we return to the example analysed by logistic regression on p. 652. We want to understand how the isolation of an island and its area influence the probability that the island is occupied by our study species.

```
island <- read.table("c:\\temp\\isolation.txt",header=T)
attach(island)
names(island)

[1] "incidence" "area" "isolation"
```

In the logistic regression, isolation had a highly significant negative effect on the probability that an island will be occupied by our species ($p = 0.004$), and area (island size) had a significant positive effect on the likelihood of occupancy ($p = 0.019$). But we have no *a priori* reason to believe that the logit of the probability should be linearly related to either of the explanatory variables. We can try using a GAM to fit smoothed functions to the incidence data:

```
model3 <- gam(incidence~s(area)+s(isolation),binomial)
summary(model3)

Family: binomial
Link function: logit

Formula:
incidence ~ s(area) + s(isolation)
```

```
Parametric coefficients:
            Estimate Std. Error z value Pr(>|z|)
(Intercept)    1.6371     0.9898    1.654    0.0981 .

Approximate significance of smooth terms:
                edf Ref.df Chi.sq p-value
s(area)       2.429  3.066  3.623 0.31522
s(isolation) 1.000  1.000  7.480 0.00624 **

R-sq.(adj) =   0.63    Deviance explained = 63.1%
UBRE score = -0.32096  Scale est. = 1          n = 50
```

This indicates a highly significant effect of isolation on occupancy ($p = 0.006\,24$) but no effect of area ($p = 0.315\,22$). We plot the model to look at the residuals:

```
windows(7,4)
par(mfrow=c(1,2))
plot.gam(model3,residuals=T,pch=16)
```

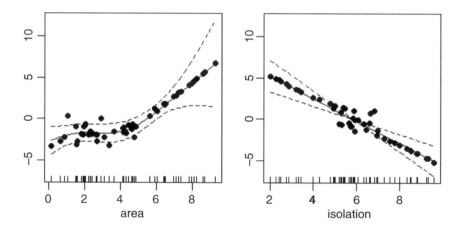

This suggests a strong effect of area, with very little scatter, but only above a threshold of about area = 5. We assess the significance of area by deletion and compare a model containing s(area) + s(isolation) with a model containing s(isolation) alone:

```
model4 <- gam(incidence~s(isolation),binomial)
anova(model3,model4,test="Chi")
```

```
Analysis of Deviance Table

Model 1: incidence ~ s(area) + s(isolation)
Model 2: incidence ~ s(isolation)
  Resid. Df Resid. Dev      Df Deviance Pr(>Chi)
1    45.571    25.094
2    45.799    29.127 -0.22824   -4.033 0.006461 **
```

This shows the effect of area to be highly significant ($p = 0.006\,461$), despite the non-significant p value in the summary table of model3. An alternative is to fit area as a parametric term and isolation as a smoothed term:

```
model5 <- gam(incidence~area+s(isolation),binomial)
summary(model5)

Family: binomial
Link function: logit

Formula:
incidence ~ area + s(isolation)

Parametric coefficients:
            Estimate Std. Error z value Pr(>|z|)
(Intercept)  -1.3928     0.9002  -1.547   0.1218
area          0.5807     0.2478   2.344   0.0191 *

Approximate significance of smooth terms:
              edf Ref.df Chi.sq p-value
s(isolation)    1      1  8.275 0.00402 **

R-sq.(adj) =   0.597   Deviance explained = 58.3%
UBRE score = -0.31196  Scale est. = 1          n = 50
```

Again, this shows a significant effect of area on occupancy. The lesson here is that a term can appear to be significant when entered into the model as a parametric term (area has $p = 0.019$ in model5) but not come close to significance when entered as a smoothed term (s(area) has $p = 0.275$ in model3). Also, the comparison of model3 and model4 draws attention to the benefits of using deletion with anova in assessing the significance of model terms.

18.5 Three-dimensional graphic output from gam

Here is an example by Simon Wood which shows the kind of three-dimensional graphics that can be obtained from gam using vis.gam when there are two continuous explanatory variables. Note that in this example the smoother works on both variables together, y~s(x,z):

```
windows(7,7)
test1 <- function(x,z,sx=0.3,sz=0.4)
    {(pi**sx*sz)*(1.2*exp(-(x-0.2)^2/sx^2-(z-0.3)^2/sz^2)+
    0.8*exp(-(x-0.7)^2/sx^2-(z-0.8)^2/sz^2))
    }
n <- 500
x <- runif(n);z <- runif(n);
y <- test1(x,z)+rnorm(n)*0.1
b4 <- gam(y~s(x,z))

vis.gam(b4)
```

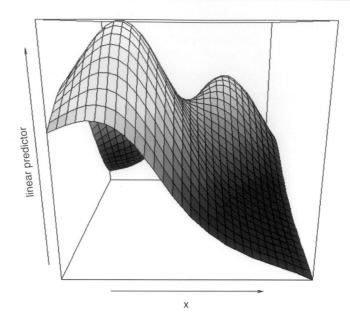

Note also that the vertical scale of the graph is the linear predictor, not the response.

19

Mixed-Effects Models

Up to this point, we have treated all categorical explanatory variables as if they were the same. This is certainly what R.A. Fisher had in mind when he invented the analysis of variance in the 1920s and 1930s. It was Eisenhart (1947) who realized that there were actually two fundamentally different sorts of categorical explanatory variables: he called these **fixed effects** and **random effects**. It will take a good deal of practice before you are confident in deciding whether a particular categorical explanatory variable should be treated as a fixed effect or a random effect, but in essence:

- fixed effects influence only the **mean** of y;

- random effects influence only the **variance** of y.

Fixed effects are unknown constants to be estimated from the data. Random effects govern the variance–covariance structure of the response variable (see p. 519). Nesting (or hierarchical structure) of random effects is a classic source of pseudoreplication, so it important that you are able to recognize it and hence not fall into its trap. Random effects that come from the same group will be correlated, and this contravenes one of the fundamental assumptions of standard statistical models: *independence of errors*. Random effects occur in two contrasting kinds of circumstances:

- observational studies with hierarchical structure;

- designed experiments with different spatial or temporal scales.

Fixed effects have informative factor levels, while random effects often have uninformative factor levels. The distinction is best seen by an example. In most mammal species the categorical variable sex has two levels: male and female. For any individual that you find, the knowledge that it is, say, female conveys a great deal of information about the individual, and this information draws on experience gleaned from many other individuals that were female. A female will have a whole set of attributes (associated with her being female) no matter what population that individual was drawn from. Take a different categorical variable like genotype. If we have two genotypes in a population we might label them A and B. If we take two more genotypes from a *different* population we might label them A and B as well. In a case like this, the label A does not convey any information at all about the genotype, other than that it is probably different from genotype B. In the case of sex, the factor level (male or female) is informative: sex is a fixed effect. In the case of genotype, the factor level (A or B) is uninformative: genotype is a random effect.

The R Book, Second Edition. Michael J. Crawley.
© 2013 John Wiley & Sons, Ltd. Published 2013 by John Wiley & Sons, Ltd.

Random effects have factor levels that are drawn from a large (potentially very large) population in which the individuals differ in many ways, but we do not know exactly how or why they differ. To get a feel for the difference between fixed effects and random effects here are some more examples:

Fixed effects	Random effects
Drug administered or not	Genotype
Insecticide sprayed or not	Brood
Nutrient added or not	Block within a field
One country versus another	Split plot within a plot
Male or female	History of development
Upland or lowland	Household
Wet versus dry	Individuals with repeated measures
Light versus shade	Family
One age versus another	Parent

The important point is that because the random effects come from a large population, there is not much point in concentrating on estimating means of our small subset of factor levels, and no point at all in comparing individual pairs of means for different factor levels. Much better to recognize them for what they are, random samples from a much larger population, and to concentrate on their variance. This is the *added* variation caused by differences between the levels of the random effects.

Variance components analysis is all about estimating the size of this variance, and working out its percentage contribution to the overall variation. There are five fundamental assumptions of linear mixed-effects models:

- Within-group errors are independent with mean zero and variance σ^2.

- Within-group errors are independent of the random effects.

- The random effects are normally distributed with mean zero and covariance matrix Ψ.

- The random effects are independent in different groups.

- The covariance matrix does not depend on the group.

The validity of these assumptions needs to be tested by employing a series of plotting methods involving the residuals, the fitted values and the predicted random effects. The tricks with mixed-effects models are:

- learning which variables are random effects;

- specifying the fixed and random effects in the model formula;

- getting the nesting structure of the random effects right;

- remembering to get library(lme4) or library(nlme) at the outset.

The issues fall into two broad categories: questions about experimental design and the management of experimental error (e.g. where does most of the variation occur, and where would increased replication be most profitable?); and questions about hierarchical structure, and the relative magnitude of variation at different levels within the hierarchy (e.g. studies on the genetics of individuals within families, families within parishes, and parishes with counties, to discover the relative importance of genetic and phenotypic variation).

Most ANOVA models are based on the assumption that there is a single error term. But in hierarchical studies and nested experiments, where the data are gathered at two or more different spatial scales, there is *a different error variance for each different spatial scale*. There are two reasonably clear-cut sets of circumstances where your first choice would be to use a linear mixed-effects model: you want to do variance components analysis because all your explanatory variables are categorical random effects and you do not have any fixed effects; or you do have fixed effects, but you also have pseudoreplication of one sort or another (e.g. temporal pseudoreplication resulting from repeated measurements on the same individuals; see p. 699). To test whether one should use a model with mixed effects or just a plain old linear model, Douglas Bates wrote in the R help archive: 'I would recommend the likelihood ratio test against a linear model fit by `lm`. The *p*-value returned from this test will be conservative because you are testing on the boundary of the parameter space.'

19.1 Replication and pseudoreplication

To qualify as replicates, measurements must have the following properties:

- They must be independent.
- They must not form part of a time series (data collected from the same place on successive occasions are not independent).
- They must not be grouped together in one place (aggregating the replicates means that they are not spatially independent).
- They must be of an appropriate spatial scale;
- Ideally, one replicate from each treatment ought to be grouped together into a block, and the whole experiment repeated in many different blocks.
- Repeated measures (e.g. from the same individual or the same spatial location) are not replicates (this is probably the commonest cause of pseudoreplication in statistical work).

Pseudoreplication occurs when you analyse the data as if you had more degrees of freedom than you really have. There are two kinds of pseudoreplication:

- temporal pseudoreplication, involving repeated measurements from the same individual;
- spatial pseudoreplication, involving several measurements taken from the same vicinity.

Pseudoreplication is a problem because one of the most important assumptions of standard statistical analysis is *independence of errors*. Repeated measures through time on the same individual will have non-independent errors because peculiarities of the individual will be reflected in all of the measurements made on it (the repeated measures will be temporally correlated with one another). Samples taken from the same vicinity will have non-independent errors because peculiarities of the location will be common to all the samples (e.g. yields will all be high in a good patch and all be low in a bad patch).

Pseudoreplication is generally quite easy to spot. The question to ask is this. How many degrees of freedom for error does the experiment really have? If a field experiment appears to have lots of degrees of freedom, it is probably pseudoreplicated. Take an example from pest control of insects on plants. There are 20 plots, 10 sprayed and 10 unsprayed. Within each plot there are 50 plants. Each plant is measured five times during the growing season. Now this experiment generates $20 \times 50 \times 5 = 5000$ numbers. There are two spraying treatments, so there must be 1 degree of freedom for spraying and 4998 degrees of freedom for error. Or must there? Count up the replicates in this experiment. Repeated measurements on the same plants (the five sampling occasions) are certainly not replicates. The 50 individual plants within each quadrat are not replicates either. The reason for this is that conditions within each quadrat are quite likely to be unique, and so all 50 plants will experience more or less the same unique set of conditions, irrespective of the spraying treatment they receive. In fact, there are 10 replicates in this experiment. There are 10 sprayed plots and 10 unsprayed plots, and each plot will yield only one independent datum for the response variable (the mean proportion of leaf area consumed by insects, for example). Thus, there are 9 degrees of freedom within each treatment, and $2 \times 9 = 18$ degrees of freedom for error in the experiment as a whole. It is not difficult to find examples of pseudoreplication on this scale in the literature (Hurlbert 1984). The problem is that it leads to the reporting of masses of spuriously significant results (with 4998 degrees of freedom for error, it is almost impossible *not* to have significant differences). The first skill to be acquired by the budding experimenter is the ability to plan an experiment that is properly replicated. There are various things that you can do when your data are pseudoreplicated:

- Average away the pseudoreplication and carry out your statistical analysis on the means.

- Carry out separate analyses for each time period.

- Use proper time series analysis or mixed-effects models.

19.2   The `lme` and `lmer` functions

Most of the examples in this chapter use the linear mixed model formula `lme`. This is to provide compatibility with the excellent book by Pinheiro and Bates (2000) on *Mixed-Effects Models in S and S-PLUS*. More recently, however, Douglas Bates has released the generalized mixed model function `lmer` as part of the `lme4` package, and you may prefer to use this in your own work, especially for nested count data or proportion data. To begin with, I provide a simple comparison of the basic syntax of the two functions.

19.2.1   `lme`

Specifying the fixed and random effects in the model formula is done with two formulae. Suppose that there are no fixed effects, so that all of the categorical variables are random effects. Then the fixed effect simply estimates the intercept (parameter 1):

```
fixed = y ~ 1
```

The fixed effect (a compulsory part of the `lme` structure) is just the overall mean value of the response variable y ~ 1. The `fixed` = part of the formula is optional if you put this object first. The random effects show the identities of the random variables and their relative locations in the hierarchy. The three random effects (a, b, and c) are specified like this:

```
random = ~ 1 | a/b/c
```

and in this case the phrase `random` = is *not* optional. An important detail to notice is that the name of the response variable (y) is not repeated in the random-effects formula: there is a blank space to the left of the tilde ~. In most mixed-effects models we assume that the random effects have a mean of zero and that we are interested in quantifying variation in the intercept caused by differences between the factor levels of the random effects. After the intercept comes the vertical bar | which is read as 'given the following spatial arrangement of the random variables'. In this example there are three random effects with 'c nested within b which in turn is nested within a'. The factors are separated by forward slash characters, and the variables are listed from left to right in declining order of spatial (or temporal) scale. This will only become clear with practice, but it is a simple idea. The formulae are put together like this:

```
lme(fixed = y ~ 1, random = ~ 1 | a/b/c)
```

19.2.2 `lmer`

There is just one formula in `lmer`, not separate formulae for the fixed and random effects. The fixed effects are specified first, to the right of the tilde, in the normal way. Next comes a plus sign, then one or more random terms enclosed in parentheses (in this example there is just one random term, but we might want separate random terms for the intercept and for the slopes, for instance). R can identify the random terms because they must contain a 'given' symbol |, to the right of which are listed the random effects in the usual way, from largest to smallest scale, left to right. So the `lmer` formula for this example is:

```
lmer(y ~ 1+( 1 | a/b/c ))
```

19.3 Best linear unbiased predictors

In `aov`, the effect size for treatment i is defined as $\bar{y}_i - \mu$, where μ is the overall mean. In mixed-effects models, however, correlation between the pseudoreplicates within a group causes what is called **shrinkage**. The best linear unbiased predictors (BLUPs, denoted by a_i) are smaller than the effect sizes ($\bar{y}_i - \mu$), and are given by

$$a_i = (\bar{y}_i - \mu)\left(\frac{\sigma_a^2}{\sigma_a^2 + \sigma^2/n}\right),$$

where σ^2 is the residual variance and σ_a^2 is the between-group variance which introduces the correlation between the pseudoreplicates within each group. Thus, the parameter estimate a_i is 'shrunk' compared to the fixed effect size ($\bar{y}_i - \mu$). When σ_a^2 is estimated to be large compared with the estimate of σ^2/n (i.e. when most of the variation is between classes and there is little variation within classes), the fixed effects and the BLUP are similar. On the other hand, when σ_a^2 is estimated to be small compared with the estimate of σ^2/n, then the fixed effects and the BLUP can be very different.

19.4 Designed experiments with different spatial scales: Split plots

The important distinction in models with categorical explanatory variables is between cases where the data come from a designed experiment, in which treatments were allocated to locations or subjects at random, and cases where the data come from an observational study in which the categorical variables are associated with an observation before the study. Here, we call the first case split-plot experiments and the second case hierarchical designs. The point is that their dataframes look identical, so it is easy to analyse one case wrongly as if it were the other. You need to be able to distinguish between fixed effects and random effects in both cases.

Here is the linear model for a split-plot experiment analysed in Chapter 11 by `aov` (see p. 519):

```
yields <- read.table("c:\\temp\\splityield.txt",header=T)
attach(yields)
names(yields)

[1] "yield" "block" "irrigation" "density" "fertilizer"

library(nlme)
```

The fixed-effects part of the model is specified in just the same way as in a straightforward factorial experiment: `yield ~ irrigation*density*fertilizer`. The random-effects part of the model says that we want the random variation to enter via effects on the intercept as `random=~1`. Finally, we define the spatial structure of the random effects after the 'given' symbol | as: `block/irrigation/density` reflecting the progressively smaller plot sizes. There is no need to specify the smallest spatial scale (fertilizer plots in this example).

```
model <- lme(yield ~ irrigation*density*fertilizer,random= ~
1|block/irrigation/density)
summary(model)

Linear mixed-effects model fit by REML
 Data: NULL
       AIC       BIC    logLik
  481.6212  525.3789 -218.8106

Random effects:
 Formula: ~1 | block
        (Intercept)
StdDev: 0.000660972

 Formula: ~1 | irrigation %in% block
        (Intercept)
StdDev:    1.982463

 Formula: ~1 | density %in% irrigation %in% block
        (Intercept) Residual
StdDev:    6.975553 9.292805
```

Fixed effects: yield ~ irrigation * density * fertilizer

	Value	Std.Error	DF	t-value	p-value
(Intercept)	80.50	5.893741	36	13.658558	0.0000
irrigationirrigated	31.75	8.335008	3	3.809234	0.0318
densitylow	5.50	8.216281	12	0.669403	0.5159
densitymedium	14.75	8.216281	12	1.795216	0.0978
fertilizerNP	5.50	6.571005	36	0.837010	0.4081
fertilizerP	4.50	6.571005	36	0.684827	0.4978
irrigationirrigated:densitylow	-39.00	11.619577	12	-3.356405	0.0057
irrigationirrigated:densitymedium	-22.25	11.619577	12	-1.914872	0.0796
irrigationirrigated:fertilizerNP	13.00	9.292805	36	1.398932	0.1704
irrigationirrigated:fertilizerP	5.50	9.292805	36	0.591856	0.5576
densitylow:fertilizerNP	3.25	9.292805	36	0.349733	0.7286
densitymedium:fertilizerNP	-6.75	9.292805	36	-0.726368	0.4723
densitylow:fertilizerP	-5.25	9.292805	36	-0.564953	0.5756
densitymedium:fertilizerP	-5.50	9.292805	36	-0.591856	0.5576
irrigationirrigated:densitylow:fertilizerNP	7.75	13.142011	36	0.589712	0.5591
irrigationirrigated:densitymedium:fertilizerNP	3.75	13.142011	36	0.285344	0.7770
irrigationirrigated:densitylow:fertilizerP	20.00	13.142011	36	1.521837	0.1368
irrigationirrigated:densitymedium:fertilizerP	4.00	13.142011	36	0.304367	0.7626

This output suggests that the only significant effects are the main effect of irrigation ($p = 0.0318$) and the irrigation by density interaction ($p = 0.0057$). The three-way interaction is not significant so we remove it, fitting all terms up to two-way interactions:

```
model <-  lme(yield~(irrigation+density+fertilizer)^2,
                        random=~1|block/irrigation/density)
summary(model)
```

```
Linear mixed-effects model fit by REML
 Data: NULL
      AIC       BIC     logLik
  503.1256 540.2136 -233.5628

Random effects:
 Formula: ~1 | block
         (Intercept)
StdDev: 0.0005634512

 Formula: ~1 | irrigation %in% block
         (Intercept)
StdDev:     1.982562

 Formula: ~1 | density %in% irrigation %in% block
         (Intercept) Residual
StdDev:     7.041303 9.142696
```

Fixed effects: yield ~ (irrigation + density + fertilizer)^2

	Value	Std.Error	DF	t-value	p-value
(Intercept)	82.47222	5.443438	40	15.150760	0.0000
irrigationirrigated	27.80556	7.069256	3	3.933307	0.0293
densitylow	0.87500	7.256234	12	0.120586	0.9060
densitymedium	13.45833	7.256234	12	1.854727	0.0884
fertilizerNP	3.58333	5.278538	40	0.678850	0.5011
fertilizerP	0.50000	5.278538	40	0.094723	0.9250
irrigationirrigated:densitylow	-29.75000	8.800165	12	-3.380618	0.0055
irrigationirrigated:densitymedium	-19.66667	8.800165	12	-2.234807	0.0452
irrigationirrigated:fertilizerNP	16.83333	5.278538	40	3.189014	0.0028
irrigationirrigated:fertilizerP	13.50000	5.278538	40	2.557526	0.0144
densitylow:fertilizerNP	7.12500	6.464862	40	1.102112	0.2770
densitymedium:fertilizerNP	-4.87500	6.464862	40	-0.754076	0.4552
densitylow:fertilizerP	4.75000	6.464862	40	0.734741	0.4668
densitymedium:fertilizerP	-3.50000	6.464862	40	-0.541388	0.5912

The fertilizer by density interaction is not significant, so we remove it:

```
model <-  lme(yield~irrigation*density+irrigation*fertilizer,
                        random=~1|block/irrigation/density)
summary(model)
```

```
Linear mixed-effects model fit by REML
Data: NULL
      AIC       BIC     logLik
  519.9035 549.6834 -245.9517

Random effects:
 Formula: ~1 | block
```

```
            (Intercept)
StdDev: 0.0005566885

  Formula: ~1 | irrigation %in% block
            (Intercept)
StdDev:    1.982614

  Formula: ~1 | density %in% irrigation %in% block
            (Intercept) Residual
StdDev:    7.057132 9.105995

Fixed effects: yield ~ irrigation * density + irrigation * fertilizer
                                    Value Std.Error DF   t-value p-value
(Intercept)                      82.08333  4.994999 44 16.433103  0.0000
irrigationirrigated              27.80556  7.063995  3  3.936236  0.0292
densitylow                        4.83333  6.222653 12  0.776732  0.4524
densitymedium                    10.66667  6.222653 12  1.714167  0.1122
fertilizerNP                      4.33333  3.717507 44  1.165656  0.2500
fertilizerP                       0.91667  3.717507 44  0.246581  0.8064
irrigationirrigated:densitylow   -29.75000  8.800161 12 -3.380620  0.0055
irrigationirrigated:densitymedium -19.66667  8.800161 12 -2.234808  0.0452
irrigationirrigated:fertilizerNP 16.83333  5.257349 44  3.201867  0.0025
irrigationirrigated:fertilizerP  13.50000  5.257349 44  2.567834  0.0137
```

Both the `irrigation` by `fertilizer` and `irrigation` by `density` interactions are now highly significant. The apparently non-significant main effect of `density` is spurious because `density` appears in a significant interaction with `irrigation`. The moral is that you must do the model simplification to get the appropriate *p* values.

Remember, too, that if you want to use `anova` to compare mixed models with different fixed-effects structures, then you must use maximum likelihood (`method = "ML"` in `lme` but `REML = FALSE` in `lmer`) rather than the default restricted maximum likelihood (REML). Here is the analysis again, but this time using `anova` to compare models with progressively simplified fixed effects:

```
model.lme <- lme(yield~irrigation*density*fertilizer,
              random=~ 1| block/irrigation/density,method="ML")
model.lme.2 <- update(model.lme,~. - irrigation:density:fertilizer)
anova(model.lme,model.lme.2)


            Model df     AIC      BIC     logLik    Test  L.Ratio p-value
model.lme       1 22 573.5108 623.5974 -264.7554
model.lme.2     2 18 569.0046 609.9845 -266.5023 1 vs 2 3.493788  0.4788


model.lme.3 <- update(model.lme.2,~. - density:fertilizer)
anova(model.lme.3,model.lme.2)


            Model df     AIC      BIC     logLik    Test  L.Ratio p-value
model.lme.3     1 14 565.1933 597.0667 -268.5967
model.lme.2     2 18 569.0046 609.9845 -266.5023 1 vs 2 4.188774  0.3811


model.lme.4 <- update(model.lme.3,~. - irrigation:fertilizer)
anova(model.lme.3,model.lme.4)
```

```
         Model df      AIC      BIC    logLik    Test  L.Ratio p-value
model.lme.3   1 14 565.1933 597.0667 -268.5967
model.lme.4   2 12 572.3373 599.6573 -274.1687 1 vs 2 11.14397  0.0038

model.lme.5 <- update(model.lme.2,~. - irrigation:density)
anova(model.lme.5,model.lme.2)

         Model df      AIC      BIC    logLik    Test  L.Ratio p-value
model.lme.5   1 16 576.7134 613.1400 -272.3567
model.lme.2   2 18 569.0046 609.9845 -266.5023 1 vs 2 11.70883  0.0029
```

The irrigation by fertilizer interaction is more significant ($p = 0.0038$ compared to $p = 0.0081$) under this mixed-effects model than it was in the linear model earlier, as is the irrigation by density interaction ($p = 0.0029$ compared to $p = 0.01633$). You need to do the model simplification in lme to uncover the significance of the main effect and interaction terms, but it is worth it, because the lme analysis can be more powerful. The minimal adequate model under the lme is:

```
summary(model.lme.3)
Linear mixed-effects model fit by maximum likelihood
 Data: NULL
       AIC      BIC    logLik
 565.1933 597.0667 -268.5967

Random effects:
 Formula: ~1 | block
        (Intercept)
StdDev: 0.0005260885

 Formula: ~1 | irrigation %in% block
        (Intercept)
StdDev:    1.716888

 Formula: ~1 | density %in% irrigation %in% block
         (Intercept) Residual
StdDev:    5.722413 8.718327

Fixed effects: yield ~ irrigation + density + fertilizer
                    + irrigation:density + irrigation:fertilizer

                                   Value Std.Error DF    t-value p-value
(Intercept)                     82.08333  4.756285 44  17.257867  0.0000
irrigationirrigated             27.80556  6.726403  3   4.133793  0.0257
densitylow                       4.83333  5.807347 12   0.832279  0.4215
densitymedium                   10.66667  5.807347 12   1.836754  0.0911
fertilizerNP                     4.33333  3.835552 44   1.129781  0.2647
fertilizerP                      0.91667  3.835552 44   0.238992  0.8122
irrigationirrigated:densitylow   -29.75000  8.212829 12  -3.622382  0.0035
irrigationirrigated:densitymedium -19.66667  8.212829 12  -2.394628  0.0338
irrigationirrigated:fertilizerNP  16.83333  5.424290 44   3.103325  0.0033
irrigationirrigated:fertilizerP   13.50000  5.424290 44   2.488805  0.0167
```

You should pay special attention to the degrees of freedom column. Note that the degrees of freedom are not pseudoreplicated: there are only 3 d.f. for testing the `irrigation` main effect; 12 d.f. for testing the `irrigation` by `density` interaction and 44 d.f. for `irrigation` by `fertilizer` (this is $36 + 4 + 4 = 44$ after model simplification). Also, remember that you must do your model simplification using maximum likelihood (`method = "ML"`) because you cannot use `anova` to compare models with different fixed-effect structures using REML.

Model-checking plots show that the residuals are well behaved:

```
plot(model.lme.3)
```

the response variable is a reasonably linear function of the fitted values:

```
plot(model.lme.3,yield~fitted(.))
```

and the errors are reasonably close to normally distributed in all four blocks:

```
qqnorm(model.lme.3,~ resid(.)| block)
```

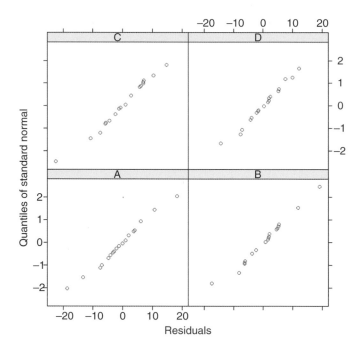

When, as here, the experiment is balanced and there are no missing values, then it is much simpler to interpret the `aov` using an `Error` term to describe the structure of the spatial pseudoreplication (p. 526), not least because it produces separate ANOVA tables for each of the spatial scales, which makes it very easy to check that these is no pseudoreplication. Without balance, however, you will need to use `lme` or `lmer` and to use model simplification to estimate the p values of the significant interaction terms.

If you do this example using `lmer`, you will want to switch off the matrix of correlations for the fixed effects. You do this with the `print(model,cor=F)` option (rather than `summary`):

```
library(lme4)

b <- block
bi <- block:irrigation
```

```
bid <- block:irrigation:density

model1 <-
lmer(yield~irrigation*density*fertilizer+(1|b)+(1|bi)+(1|bid),REML=FALSE)

print(model1,cor=F)


Linear mixed model fit by maximum likelihood
Formula: yield ~ irrigation * density * fertilizer + (1|b) + (1|bi) + (1|bid)
        AIC   BIC  logLik deviance REMLdev
      573.5 623.6  -264.8    529.5   437.6
Random effects:
Groups    Name        Variance    Std.Dev.
bid       (Intercept) 3.6493e+01  6.0410e+00
bi        (Intercept) 2.9479e+00  1.7169e+00
b         (Intercept) 8.9145e-13  9.4417e-07
Residual              6.4767e+01  8.0478e+00
Number of obs: 72, groups: bid, 24; bi, 8; b, 4
Fixed effects:
                                                    Estimate Std. Error t value
(Intercept)                                           80.500      5.104  15.772
irrigationirrigated                                   31.750      7.218   4.399
densitylow                                             5.500      7.115   0.773
densitymedium                                         14.750      7.115   2.073
fertilizerNP                                           5.500      5.691   0.966
fertilizerP                                            4.500      5.691   0.791
irrigationirrigated:densitylow                       -39.000     10.063  -3.876
irrigationirrigated:densitymedium                    -22.250     10.063  -2.211
irrigationirrigated:fertilizerNP                      13.000      8.048   1.615
irrigationirrigated:fertilizerP                        5.500      8.048   0.683
densitylow:fertilizerNP                                3.250      8.048   0.404
densitymedium:fertilizerNP                            -6.750      8.048  -0.839
densitylow:fertilizerP                                -5.250      8.048  -0.652
densitymedium:fertilizerP                             -5.500      8.048  -0.683
irrigationirrigated:densitylow:fertilizerNP            7.750     11.381   0.681
irrigationirrigated:densitymedium:fertilizerNP         3.750     11.381   0.329
irrigationirrigated:densitylow:fertilizerP            20.000     11.381   1.757
irrigationirrigated:densitymedium:fertilizerP          4.000     11.381   0.351
```

As before, it requires model simplification before the significant interactions become evident.

19.5 Hierarchical sampling and variance components analysis

Hierarchical data are often encountered in observational studies where information is collected at a range of different spatial scales. The principal aim is to discover the scale at which most of the variation is generated. This information would then allow a closer focus on mechanisms operating at this scale in subsequent more detailed studies. The following study involves a test with a mean score of 100 administered to children in four British towns. Each town was divided into districts by postcodes, and six districts were selected at random. Within districts, 10 streets were selected at random, and within streets, four households were selected at random. Naturally, different households had different numbers of children (childless households were excluded from the study) and there was no control over sex ratio of children within household.

```
library(lme4)
```

```
data <- read.table("c:\\temp\\childfull.txt",header=T)
attach(data)
head(data)
```

```
  childID child house street district  town response gender
1       1     1 door1        1        A Leeds 83.88773   male
2       1     1 door2        1        A Leeds 99.96294   male
3       1     3 door3        1        A Leeds 87.20253 female
4       2     3 door3        1        A Leeds 89.37665   male
5       3     3 door3        1        A Leeds 92.01751 female
6       1     5 door4        1        A Leeds 87.12672 female
```

You can see that the factor levels are not unique: for instance, there is a street 1 in each district of Leeds (and of every other town). We use the colon operator to create unique factor levels for each random effect: district within town (d), street within district within town (s) and household within street within district within town (h). Each household has one or more children (maximum = 8, mostly in Leeds) but the sex ratio varies from house to house.

```
d <- town:district
s <- town:district:factor(street)
h <- town:district:factor(street):house
```

The mixed effects model has one fixed effect (gender) and four nested random effects:

```
model <- lmer(response~gender+(1|town)+(1|d)+(1|s)+(1|h))
summary(model)
```

```
Linear mixed model fit by REML
Formula: response ~ gender + (1 | town) + (1 | d) + (1 | s) + (1 | h)
   AIC   BIC logLik deviance REMLdev
 19878 19920  -9932    19868   19864
Random effects:
Groups   Name         Variance Std.Dev.
h        (Intercept)    4.0817   2.0203
s        (Intercept)   15.6747   3.9591
d        (Intercept)  168.3451  12.9748
town     (Intercept)   36.9802   6.0811
Residual               36.2405   6.0200

Number of obs: 2972, groups: h, 960; s, 240; d, 24; town, 4

Fixed effects:
            Estimate Std. Error t value
(Intercept)  97.8965     4.0424  24.218
gendermale    0.5368     0.2363   2.272

Correlation of Fixed Effects:
           (Intr)
gendermale -0.030
```

The fixed effect is significant ($t = 2.272$) but small (0.537) compared to the means for towns, which varied by as much as 20 (e.g. Coventry vs. Derby). Of the random effects, most of the variation in the response is between districts within towns (64%) and least is between households within streets (4%). You get the percentage variance components like this:

```
vc <- c(36.2405,4.0817,15.6747,168.3451,36.9802)
vc <- 100*c(36.2405,4.0817,15.6747,168.3451,36.9802)/sum(vc)
vc
```

```
[1] 13.868129  1.561942  5.998227 64.420512 14.151190
```

The key feature of these data, however, is the substantial variation between children within the same family (i.e. even after you have controlled for family and environment (town and district within town)).

There are issues about the appropriate graphics in studies like this, simply because there are so many combinations of factor levels, and the nested factor levels typically make sense only in the context of their higher-level associations. Here is one way of showing the variation:

```
par(mfrow=c(2,2))
hist(response[town=="Coventry"],main="Coventry",breaks=seq(40,150,5),
    xlab="response")
hist(response[town=="Derby"],main="Derby",breaks=seq(40,150,5),
    xlab="response")
hist(response[town=="Leeds"],main="Leeds",breaks=seq(40,150,5),
    xlab="response")
hist(response[town=="Norwich"],main="Norwich",breaks=seq(40,150,5),
    xlab="response")
```

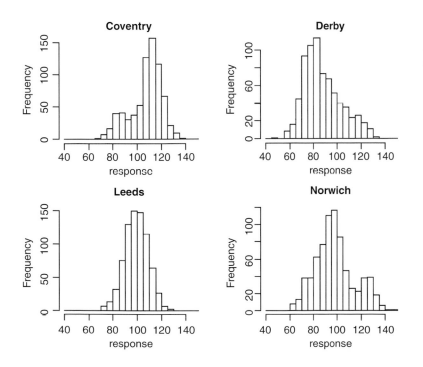

This is good at highlighting non-normality (e.g. the bimodal distribution of the response in Norwich, and the opposite skew in Coventry and Derby). Here is an alternative using box-and-whisker plots, with subset to chose the towns rather than subscripts:

```
plot(response~district,subset=(town=="Coventry"),main="Coventry")
plot(response~district,subset=(town=="Derby"),main="Derby")
plot(response~district,subset=(town=="Leeds"),main="Leeds")
plot(response~district,subset=(town=="Norwich"),main="Norwich")
```

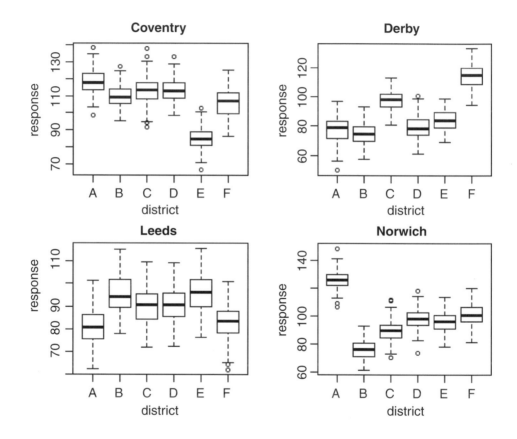

This is particularly good for drawing attention to districts with strikingly different mean scores (e.g. high ones like district A in Norwich and low ones like district E in Coventry). For looking at interactions, coplot is often useful:

```
coplot(response~district|town)
```

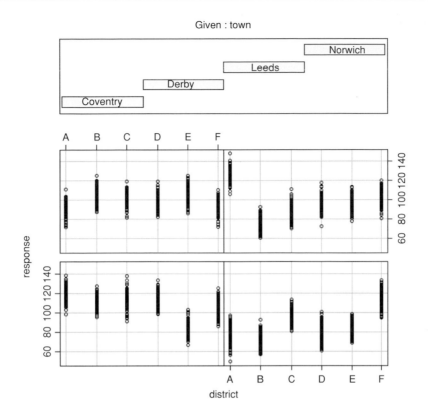

In practice, you will want to try many different kinds of plots across all of the spatial scales.

19.6 Mixed-effects models with temporal pseudoreplication

A common cause of temporal pseudoreplication in growth experiments with fixed effects is when each individual is measured several times as it grows during the course of an experiment. The next example is as simple as possible: we have a single fixed effect (a two-level categorical variable, with fertilizer added or not) and six replicate plants in each treatment, with each plant measured on five occasions (after 2, 4, 6, 8 or 10 weeks of growth). The response variable is root length, measured non-destructively through a glass panel, which is opened to the light only when the root length measurements are being taken. The fixed-effect formula looks like this:

```
fixed = root~fertilizer
```

The random-effects formula needs to indicate that the week of measurement (a continuous random effect) represents pseudoreplication within each individual plant:

```
random = ~week|plant
```

Because we have a continuous random effect (weeks) we write ~week in the random-effects formula rather than the ~1 that we used with categorical random effects (above). Here are the data:

```
results <- read.table("c:\\temp\\fertilizer.txt",header=T)
attach(results)
```

```
names(results)
```

```
[1]   "root"   "week"   "plant"   "fertilizer"
```

We begin with data inspection. For the kind of data involved in mixed-effects models there are some excellent built-in plotting functions (variously called panel plots, trellis plots, or lattice plots).

```
library(nlme)
```

```
library(lattice)
```

To use trellis plotting, we begin by turning our dataframe called `results` into a `groupedData` object (p. 957). To do this we specify the nesting structure of the random effects, and indicate the fixed effect by defining `fertilizer` as `outer` to this nesting (a fixed effect):

```
results <- groupedData(root~week|plant,outer = ~ fertilizer,results)
```

Because results is now a `groupedData` object, the plotting is fantastically simple:

```
plot(results)
```

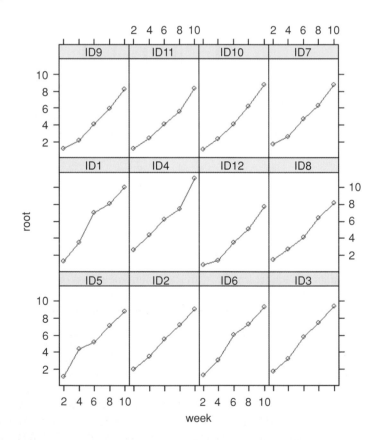

Here you get separate time series plots for each of the individual plants (created, in this case, by joining the dots, which is the default option), with plant identities ranked from bottom left (ID5) to top right (ID7) on the basis of mean root length. In terms of understanding the fixed effects, it is often more informative to group together the six replicates within each treatment, and to have one panel for each of the treatment levels (i.e.

one for the fertilized plants and one for the controls in this case). This is very straightforward, using `outer` to indicate the grouping:

```
plot(results,outer=T)
```

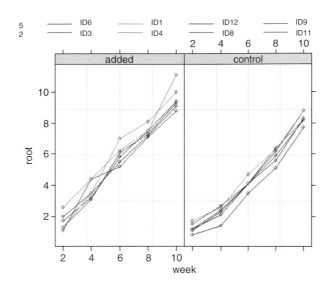

You can see that by week 10 there is virtually no overlap between the two treatment groups. The largest control plant has about the same root length as the smallest fertilized plant (about 9 cm).

Now for the statistical modelling. Ignoring the pseudoreplication, we should have 1 d.f. for `fertilizer` and $2 \times (6-1) = 10$ d.f. for error.

```
model <- lme(root~fertilizer,random=~week|plant)
summary(model)
```

```
Linear mixed-effects model fit by REML
 Data: NULL
       AIC       BIC    logLik
  171.0236 183.3863 -79.51181

Random effects:
 Formula: ~week | plant
 Structure: General positive-definite, Log-Cholesky parametrization
             StdDev    Corr
(Intercept) 2.8639832 (Intr)
week        0.9369412 -0.999
Residual    0.4966308

Fixed effects: root ~ fertilizer
                     Value Std.Error DF   t-value p-value
(Intercept)       2.799710 0.1438367 48 19.464499    0e+00
fertilizercontrol -1.039383 0.2034158 10 -5.109645    5e-04
 Correlation:
                  (Intr)
fertilizercontrol -0.707
```

```
Standardized Within-Group Residuals:
        Min          Q1         Med          Q3         Max
-1.9928118 -0.6586834 -0.1004301   0.6949714   2.0225381
```

```
Number of Observations: 60
Number of Groups: 12
```

The output looks dauntingly complex, but once you learn your way around it, the essential information is relatively easy to extract. The mean reduction in root size associated with the unfertilized controls is −1.039 383 and this has a standard error of 0.203 415 8 based on the correct 10 residual d.f. (six replicates per factor level). Can you see why the intercept has 48 d.f.? (Hint: ask yourself how many graphs have been fitted to the data.)

Here is a simple one-way ANOVA for the non-pseudoreplicated data taken from the end of the experiment in week 10:

```
model2 <- aov(root~fertilizer,subset=(week==10))
summary(model2)
```

```
            Df Sum Sq Mean Sq F value Pr(>F)
fertilizer   1  4.941   4.941   11.49 0.0069 **
Residuals   10  4.302   0.430
```

```
summary.lm(model2)
```

```
Call:
aov(formula = root ~ fertilizer, subset = (week == 10))
```

```
Residuals:
    Min      1Q  Median      3Q     Max
-0.8167 -0.3667 -0.1333  0.4042  1.4833
```

```
Coefficients:
                   Estimate Std. Error t value Pr(>|t|)
(Intercept)          9.6167     0.2678  35.915 6.65e-12 ***
fertilizercontrol   -1.2833     0.3787  -3.389   0.0069 **
```

```
Residual standard error: 0.6559 on 10 degrees of freedom
Multiple R-squared: 0.5346,      Adjusted R-squared: 0.488
F-statistic: 11.49 on 1 and 10 DF,  p-value: 0.006897
```

We can compare this with the output from the lme. The effect size in the lme is slightly smaller (−1.039 393 compared to −1.2833) but the standard error is appreciably lower (0.203 415 8 compared to 0.3787), so the significance of the result is higher in the lme than in the aov. You get increased statistical power as a result of going to the trouble of fitting the mixed-effects model. And, crucially, you do not need to make potentially arbitrary judgements about which time period to select for the non-pseudoreplicated analysis. You use all of the data in the model, and you specify its structure appropriately so that the hypotheses are tested with the correct degrees of freedom (10 in this case, not 48).

The reason why the effect sizes are different in the lm and lme models is that linear models use maximum likelihood estimates of the parameters based on arithmetic means. The linear mixed models, however, use the wonderfully named BLUPs (see p. 685).

19.7 Time series analysis in mixed-effects models

It is common to have repeated measures on subjects in observational studies, where we would expect that the observation on an individual at time $t + 1$ would be quite strongly correlated with the observation on the same individual at time t. This contravenes one of the central assumptions of linear models (p. 503), that the within-group errors are independent. However, we often observe significant serial correlation in data such as these.

The following example comes from Pinheiro and Bates (2000) and forms part of the nlme library. The data refer to the numbers of ovaries observed in repeated measures on 11 mares (their oestrus cycles have been scaled such that ovulation occurred at time 0 and at time 1). The issue is how best to model the correlation structure of the data. We know from previous work that the fixed effect can be modelled as a three-parameter sine–cosine function of time x:

$$y = a + b \sin(2\pi x) + d \cos(2\pi x) + \varepsilon_{ij},$$

and we want to assess different structures for modelling the within-class correlation.

The dataframe is of class groupedData which makes the plotting and error checking much simpler.

```
data(Ovary)
attach(Ovary)
names(Ovary)

[1]   "Mare"   "Time"   "follicles"

plot(Ovary)
```

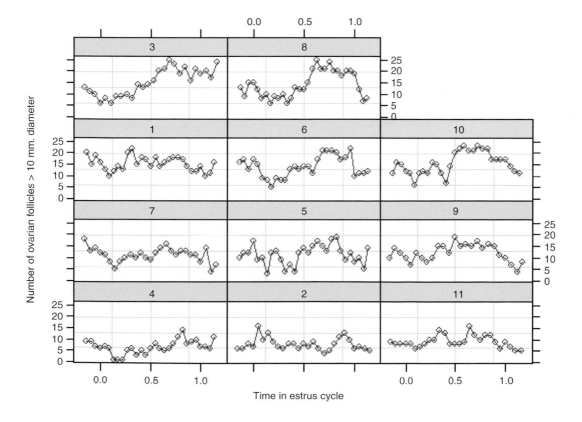

The panel plot has ranked the horses from bottom left to top right on the basis of their mean number of ovules (mare 4 with the lowest number, mare 8 with the highest). Some animals show stronger cyclic behaviour than others.

We begin by fitting a mixed-effects model making no allowance for the correlation structure, and investigate the degree of autocorrelation that is exhibited by the residuals (recall that the assumption of the model is that there is no correlation):

```
model <- lme(follicles~sin(2*pi*Time)+cos(2*pi*Time),
                                data=Ovary,random=~ 1| Mare)
summary(model)

Linear mixed-effects model fit by REML
 Data: Ovary
      AIC      BIC      logLik
  1669.36  1687.962  -829.6802

Random effects:
 Formula: ~1 | Mare
        (Intercept) Residual
StdDev:    3.041344 3.400466

Fixed effects: follicles ~ sin(2 * pi * Time) + cos(2 * pi * Time)
                      Value Std.Error  DF    t-value  p-value
(Intercept)        12.182244 0.9390009 295  12.973623  0.0000
sin(2 * pi * Time) -3.339612 0.2894013 295 -11.539727  0.0000
cos(2 * pi * Time) -0.862422 0.2715987 295  -3.175353  0.0017

Correlation:
                   (Intr) s(*p*T
sin(2 * pi * Time)  0.00
cos(2 * pi * Time) -0.06    0.00

Standardized Within-Group Residuals:
       Min         Q1        Med         Q3        Max
-2.4500138 -0.6721813 -0.1349236  0.5922957  3.5506618

Number of Observations: 308
Number of Groups: 11
```

The function ACF allows us to calculate the empirical autocorrelation structure of the residuals from this model:

```
plot(ACF(model),alpha=0.05)
```

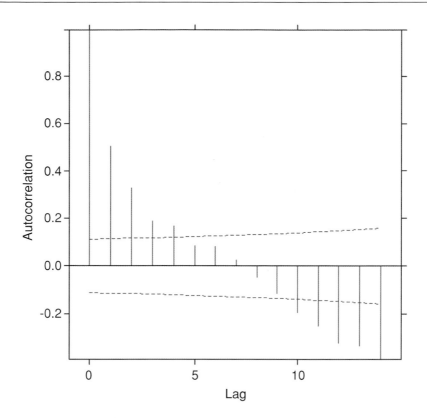

You can see that there is highly significant autocorrelation at lags 1 and 2 and marginally significant autocorrelation at lags 3 and 4. We model the autocorrelation structure using one of the standard corStruct classes (p. 863). For time series data like this, we typically choose between 'moving average', 'autoregressive' or 'autoregressive moving average' classes. Again, experience with horse biology suggests that a simple moving average model might be appropriate, so we start with this. The class is called corARMA and we need to specify the order of the model (the lag of the moving average part). The simplest assumption is that only the first two lags exhibit non-zero correlations (q=2):

```
model2 <- update(model,correlation=corARMA(q=2))
anova(model,model2)
```

	Model	df	AIC	BIC	logLik	Test	L.Ratio	p-value
model	1	5	1669.360	1687.962	-829.6802			
model2	2	7	1574.895	1600.937	-780.4476	1 vs 2	98.4652	<.0001

This is a great improvement over the original model, which assumed no correlation in the residuals. But what about a different time series assumption? Let us compare the moving average assumption with a simple first-order autoregressive model corAR1():

```
model3 <- update(model2,correlation=corAR1())
anova(model2,model3)
```

	Model	df	AIC	BIC	logLik	Test	L.Ratio	p-value
model2	1	7	1574.895	1600.937	-780.4476			
model3	2	6	1562.447	1584.769	-775.2233	1 vs 2	10.4484	0.0012

This is a very significant improvement, $p = 0.0012$, so we choose the `corAR1()` because it has the lowest AIC (it also uses fewer degrees of freedom, d.f. $= 6$). Error checking on `model3` might proceed like this:

```
plot(model3,resid(.,type="p")~fitted(.)|Mare)
```

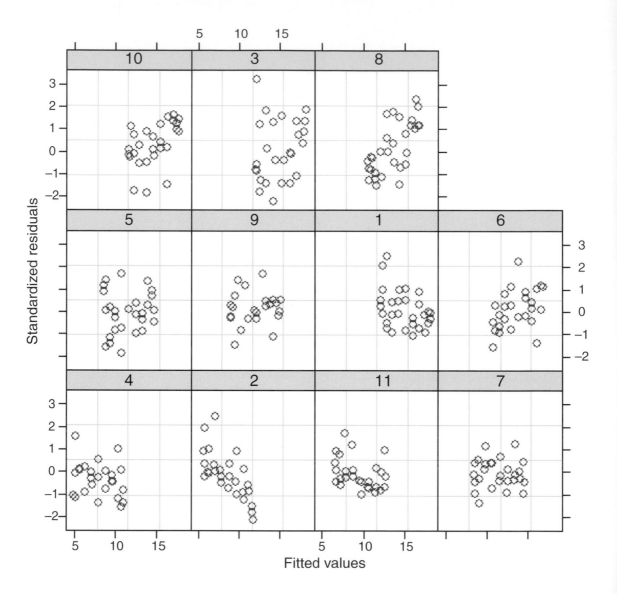

which shows that residuals are reasonably well behaved. And the normality assumption?

```
qqnorm(model3,~resid(.)|Mare)
```

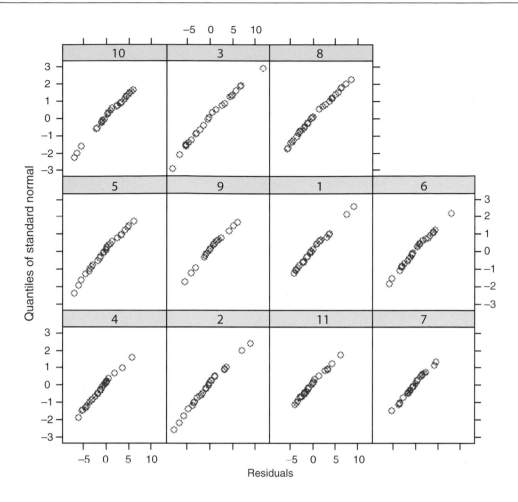

The errors are close to normally distributed for all of the mares. The model is well behaved, so we accept a first-order autocorrelation structure `corAR1()`.

19.8 Random effects in designed experiments

The rats example, studied by `aov` with an `Error` term on p. 526, can be repeated as a linear mixed-effects model, but only if we recode the factor levels. This example works much better with `lmer` than with `lme`.

```
dd <- read.table("c:\\temp\\rats.txt",h=T)
attach(dd)
names(dd)
```

```
[1]  "Glycogen"  "Treatment" "Rat"   "Liver"
```

```
Treatment <- factor(Treatment)
Liver <- factor(Liver)
Rat <- factor(Rat)
```

There is a single fixed effect (Treatment), and pseudoreplication enters the dataframe because each rat's liver is cut into three pieces and each liver bit is macerated and divided into separate aliquots to produce two readings. First, compute unique factor levels for each rat and each liver bit:

```
rat <- Treatment:Rat
liver <- Treatment:Rat:Liver
```

Then use these as random effects in the lmer model:

```
model <- lmer(Glycogen~Treatment+(1|rat)+(1|liver))
summary(model)

Linear mixed model fit by REML
Formula: Glycogen ~ Treatment + (1 | rat) + (1 | liver)
   AIC   BIC logLik deviance REMLdev
 231.6 241.1 -109.8    234.3   219.6

Random effects:
 Groups   Name        Variance Std.Dev.
 liver    (Intercept) 14.167   3.7639
 rat      (Intercept) 36.065   6.0054
 Residual             21.167   4.6007

Number of obs: 36, groups: liver, 18; rat, 6

Fixed effects:
            Estimate Std. Error t value
(Intercept)  140.500      4.707  29.850
Treatment2    10.500      6.656   1.577
Treatment3    -5.333      6.656  -0.801
```

You can see that the treatment effect is correctly interpreted as being non-significant (both *t* values are less than 2 in absolute value). The variance components (p. 527) can be extracted by expressing the variances as percentages:

```
vars <- c(14.167,36.065,21.167)

100*vars/sum(vars)

[1]   19.84201   50.51191   29.64607
```

Thus 50% of the variation is between rats within treatments, 19.8% is between liver bits within rats and 29.6% is between readings within liver bits within rats. If you are interested principally in the fixed effects, then much the best way to proceed is to average away the pseudoreplication and do a one-way ANOVA with 3 d.f. for error (see p. 525).

19.9 Regression in mixed-effects models

The next example involves a regression of plant size against local point measurements of soil nitrogen (N) at five places within each of 24 farms. It is expected that plant size and soil nitrogen will be positively correlated. There is only one measurement of plant size and soil nitrogen at any given point (i.e. there is no temporal pseudoreplication; cf. p. 695):

```
yields <- read.table("c:\\temp\\farms.txt",header=T)
attach(yields)
```

```
names(yields)
```

```
[1]  "N"   "size"   "farm"
```

Here are the data in aggregate, with different plotting colours and symbols for each farm:

```
plot(N,size,pch=rep(16:19,each=40),col=farm)
```

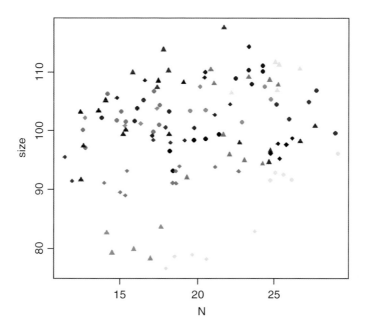

The most obvious pattern is that there is substantial variation in mean values of both soil nitrogen and plant size across the farms: the minimum-yielding fields (yellow) have a mean *y* value of less than 80, while the maximum (red) fields have a mean *y* value above 110. Note that because there are more farms (24) than colours (8), we need to use different plotting symbols as well as different colours to distinguish the five points from each farm.

The key distinction to understand is between fitting lots of linear regression models (one for each farm) and fitting one mixed-effects model, taking account of the differences between farms in terms of their contribution to the variance in response as measured by a standard deviation in intercept and a standard deviation in slope. We investigate these differences by contrasting the two fitting functions, lmList and lme. We begin by fitting 24 separate linear models, one for each farm:

```
linear.models <- lmList(size~N|farm,yields)
coef(linear.models)
```

```
   (Intercept)            N
1     67.46260   1.5153805
2    118.52443  -0.5550273
3     91.58055   0.5551292
4     87.92259   0.9212662
5     92.12023   0.5380276
6     97.01996   0.3045431
7     68.52117   0.9339957
8     91.54383   0.8220482
```

```
9        92.04667    0.8842662
10       85.08964    1.4676459
11      114.93449   -0.2689370
12       82.56263    1.0138488
13       78.60940    0.1324811
14       80.97221    0.6551149
15       84.85382    0.9809902
16       87.12280    0.3699154
17       52.31711    1.7555136
18       83.40400    0.8715070
19       88.91675    0.2043755
20       93.08216    0.8567066
21       90.24868    0.7830692
22       78.30970    1.1441291
23       59.88093    0.9536750
24       89.07963    0.1091016
```

You can see very substantial variations in the value of the intercept from 118.52 on farm 2 to 52.32 on farm 17. Slopes are also dramatically different, from negative –0.555 on farm 2 to steep and positive 1.7555 on farm 17. This is a classic problem in regression analysis when (as here) the intercept is a long way from the average value of x (see p. 460); large values of the intercept are almost bound to be correlated with low values of the slope.

Here are the slopes and intercepts from the model specified entirely in terms of random effects: a population of regression slopes predicted within each farm with nitrogen as the continuous explanatory variable, and a population of intercepts for each farm:

```
random.model <- lme(size~1,random=~N|farm)
coef(random.model)
```

```
     (Intercept)             N
1        85.98139    0.574205332
2       104.67366   -0.045401474
3        95.03442    0.331080929
4        98.62679    0.463579847
5        95.00270    0.407906220
6        99.82294    0.207203700
7        85.57345    0.285520355
8        96.09461    0.520896479
9        95.22186    0.672262931
10       93.14157    1.017995748
11      108.27200    0.015213748
12       87.36387    0.689406424
13       80.83933    0.003617022
14       89.84309    0.306402254
15       93.37050    0.636778731
16       92.10914    0.145772156
17       94.93395    0.084935464
18       85.90160    0.709943272
19       92.00628    0.052485987
```

20	95.26296	0.738029408
21	93.35069	0.591151964
22	87.66161	0.673119289
23	70.57827	0.432993929
24	90.29151	0.036747129

Differences between the intercepts explain 97.26% of the variance, differences in slope a mere 0.245%, with a residual variance of 2.49% (see the `summary` table). The thing you notice is that the random effects are less extreme (i.e. closer to the mean) than the fixed effects. This is an example of shrinkage (p. 685), and is clearest from a graphical comparison of the coefficients of the linear and mixed models:

```
mm <- coef(random.model)
ll <- coef(linear.models)
windows(7,4)
par(mfrow=c(1,2))
plot(ll[,1],mm[,1],pch=16,xlab="linear",ylab="random effects")
abline(0,1)
plot(ll[,2],mm[,2],pch=16,xlab="linear",ylab="random effects")
abline(0,1)
```

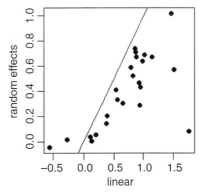

Most of the random-effects intercepts (left) are greater than their linear model equivalents (they are above the 45 degree line) while most of the random-effects slopes (right) are shallower than their linear model equivalents (i.e. below the line). For farm 17 the linear model had an intercept of 52.317 11 while the random-effects model had an intercept of 94.933 95. Likewise, the linear model for farm 17 had a slope of 1.755 513 6 while the random-effects model had a slope of 0.084 935 465.

We can fit a mixed model with both fixed and random effects. Here is a model in which size is modelled as a function of nitrogen and farm as fixed effects, with farm as a random effect. Because we intend to compare models with different fixed effect structures we need to specify `method="ML"` in place of the default REML.

```
farm <- factor(farm)
mixed.model1 <- lme(size~N*farm,random=~1|farm,method="ML")
mixed.model2 <- lme(size~N+farm,random=~1|farm,method="ML")
mixed.model3 <- lme(size~N,random=~1|farm,method="ML")
mixed.model4 <- lme(size~1,random=~1|farm,method="ML")
anova(mixed.model1,mixed.model2,mixed.model3,mixed.model4)
```

	Model	df	AIC	BIC	logLik	Test	L.Ratio	p-value
mixed.model1	1	50	542.9035	682.2781	-221.4518			
mixed.model2	2	27	524.2971	599.5594	-235.1486	1 vs 2	27.39359	0.2396
mixed.model3	3	4	614.3769	625.5269	-303.1885	2 vs 3	136.07981	<.0001
mixed.model4	4	3	658.0058	666.3683	-326.0029	3 vs 4	45.62892	<.0001

The first model contains a full factorial, with different slopes and intercepts for each of the 25 farms (using up 50 degrees of freedom). The second model has a common slope but different intercepts for the 25 farms (using 27 degrees of freedom); model2 does not have significantly lower explanatory power than model1 ($p = 0.2396$). The main effects of farm and of nitrogen application (model3 and model4) are both highly significant ($p < 0.0001$), so we select model2 because it has the lowest AIC.

Finally, we could do an old-fashioned analysis of covariance, fitting a different two-parameter model to each and every farm without any random effects:

```
model <- lm(size~N*factor(farm))
summary(model)

Call:
lm(formula = size ~ N * factor(farm))

Residuals:
    Min      1Q  Median      3Q     Max
-3.6077 -1.2947  0.0479  1.0732  4.1297

Coefficients:
                  Estimate Std. Error t value Pr(>|t|)
(Intercept)       67.46260   14.43749   4.673 1.35e-05 ***
N                  1.51538    0.73395   2.065   0.0426 *
factor(farm)2     51.06183   22.86930   2.233   0.0287 *
factor(farm)3     24.11794   16.54029   1.458   0.1492
factor(farm)4     20.45999   34.59610   0.591   0.5561
factor(farm)5     24.65762   17.29578   1.426   0.1583
factor(farm)6     29.55736   17.74007   1.666   0.1000
factor(farm)7      1.05856   20.53771   0.052   0.9590
factor(farm)8     24.08122   16.23722   1.483   0.1424
factor(farm)9     24.58407   15.45967   1.590   0.1162
factor(farm)10    17.62703   16.68467   1.056   0.2943
factor(farm)11    47.47189   18.24214   2.602   0.0112 *
factor(farm)12    15.10002   15.77085   0.957   0.3415
factor(farm)13    11.14680   17.82896   0.625   0.5338
factor(farm)14    13.50961   19.36739   0.698   0.4877
factor(farm)15    17.39122   20.74850   0.838   0.4047
factor(farm)16    19.66019   18.72739   1.050   0.2973
factor(farm)17   -15.14550   49.01250  -0.309   0.7582
factor(farm)18    15.94140   15.15371   1.052   0.2963
factor(farm)19    21.45414   17.99214   1.192   0.2370
factor(farm)20    25.61956   15.50019   1.653   0.1027
factor(farm)21    22.78608   15.65699   1.455   0.1499
factor(farm)22    10.84710   17.69820   0.613   0.5419
factor(farm)23    -7.58167   16.89435  -0.449   0.6549
```

```
factor(farm)24        21.61703   17.28697    1.250    0.2152
N:factor(farm)2       -2.07041    0.98369   -2.105    0.0388  *
N:factor(farm)3       -0.96025    0.89786   -1.069    0.2884
N:factor(farm)4       -0.59411    1.52204   -0.390    0.6974
N:factor(farm)5       -0.97735    0.84718   -1.154    0.2525
N:factor(farm)6       -1.13084    0.97207   -1.163    0.2485
N:factor(farm)7       -0.58138    0.92164   -0.631    0.5302
N:factor(farm)8       -0.69333    0.87773   -0.790    0.4322
N:factor(farm)9       -0.63111    0.81550   -0.774    0.4415
N:factor(farm)10      -0.04773    0.86512   -0.055    0.9562
N:factor(farm)11      -1.78432    0.87838   -2.031    0.0459  *
N:factor(farm)12      -0.50153    0.84820   -0.591    0.5562
N:factor(farm)13      -1.38290    0.98604   -1.402    0.1651
N:factor(farm)14      -0.86027    0.89294   -0.963    0.3386
N:factor(farm)15      -0.53439    0.94640   -0.565    0.5741
N:factor(farm)16      -1.14547    0.91070   -1.258    0.2125
N:factor(farm)17       0.24013    1.97779    0.121    0.9037
N:factor(farm)18      -0.64387    0.79080   -0.814    0.4182
N:factor(farm)19      -1.31100    0.90886   -1.442    0.1535
N:factor(farm)20      -0.65867    0.78956   -0.834    0.4069
N:factor(farm)21      -0.73231    0.81990   -0.893    0.3747
N:factor(farm)22      -0.37125    0.89597   -0.414    0.6798
N:factor(farm)23      -0.56171    0.85286   -0.659    0.5122
N:factor(farm)24      -1.40628    0.95103   -1.479    0.1436
```

```
Residual standard error: 1.978 on 72 degrees of freedom
Multiple R-squared: 0.9678,       Adjusted R-squared: 0.9468
F-statistic: 46.07 on 47 and 72 DF,  p-value: < 2.2e-16
```

There is a marginally significant overall effect of soil nitrogen on plant size (N has $p = 0.0426$) and (compared to farm 1) farms 2 and 11 have significantly higher intercepts and shallower slopes. The problem, of course, is that this model, with its 24 slopes and 24 intercepts, is vastly overparameterized. Let us fit a greatly simplified model with a common slope but different intercepts for the different farms:

```
model2 <- lm(size~N+factor(farm))
anova(model,model2)
```

```
Analysis of Variance Table
```

```
Model 1: size ~ N * factor(farm)
Model 2: size ~ N + factor(farm)
  Res.Df    RSS  Df Sum of Sq       F Pr(>F)
1     72 281.60
2     95 353.81 -23   -72.212 0.8028  0.717
```

This analysis provides no support for any significant differences between slopes. What about differences between farms in their intercepts?

```
model3 <- lm(size~N)
anova(model2,model3)
```

```
Analysis of Variance Table
```

```
Model 1: size ~ N + factor(farm)
Model 2: size ~ N
  Res.Df    RSS  Df Sum of Sq       F    Pr(>F)
1     95  353.8
2    118 8454.9 -23   -8101.1 94.574 < 2.2e-16 ***
```

This shows that there are highly significant differences in intercepts between farms. The interpretation of the analysis of covariance is exactly the same as the interpretation of the mixed model in this case where there is balanced structure and equal replication, but lme is vastly superior to the linear model when there is unequal replication.

19.10   Generalized linear mixed models

Pseudoreplicated data with non-normal errors lead to a choice of generalized linear mixed-effects models using lmer with a specified error family. These were previously handled by the glmmPQL function which is part of the MASS library (see Venables and Ripley, 2002). That function fitted a generalized linear mixed model with multivariate normal random effects, using penalized quasi-likelihood (hence the 'PQL'). The default method for a generalized linear model fit with lmer has been switched from PQL to the more reliable Laplace method. The lmer function can deal with the same error structures as a generalized linear model, namely Poisson (for count data), binomial (for binary data or proportion data) or gamma (for continuous data where the variance increase with the square of the mean). The model call is just like a mixed-effects model but with the addition of the name of the error family, like this:

```
lmer(y~fixed+(time | random), family=binomial)
```

For a worked example with binary data, involving patients who were tested for the presence of a bacterial infection on a number of occasions (the number varying somewhat from patient to patient), see pp. 660–665. The response variable is binary: yes for infected patients or no for patients not scoring as infected, so the family is binomial. There is a single categorical explanatory variable (a fixed effect) called treatment, which has three levels: drug, drug plus supplement, and placebo. The week numbers in which the repeated assessments on each patient were made is also recorded.

19.10.1   Hierarchically structured count data

This is an example of lmer with Poisson errors. Beetles were collected in pitfall traps laid out in a grid of two columns and five rows (10 pitfalls per quadrat) in each of two quadrats, randomly located within each of 3 randomly-located blocks within a field. On each of 4 farms there were 5 protocols of hedgerow management allocated at random to each of fields within the farm (1 = uncut control, 2 = grass cut, 3 = grass cut twice, 4 = hedge cut, 5 = grass and hedge cut). Here are the 1200 counts:

```
data<-read.table("c:\\temp\\nested2.txt",header=T)
attach(data)
head(data)

count farm field block quadrat
1   1    1    1     1        1
2   0    1    1     1        1
3   0    1    1     1        1
4   1    1    1     1        1
```

```
5    1    1    1    1       1
6    0    1    1    1       1
```

```
farm<-factor(farm)
field <- factor(field)
block <- factor(block)
quadrat <- factor(quadrat)
library(lme4)
```

This is what the data look like, classified by farms:

```
par(mfrow=c(2,2))
hist(count[farm==1],breaks=-0.5:16.5,
          main="Branton",col="wheat1",xlab="")
hist(count[farm==2],breaks=-0.5:16.5,
          main="Fawdon",col="wheat2",xlab="")
hist(count[farm==3],breaks=-0.5:16.5,
          main="Ingram",col="wheat3",xlab="")
hist(count[farm==4],breaks=-0.5:16.5,
          main="Powburn",col="wheat4",xlab="")
```

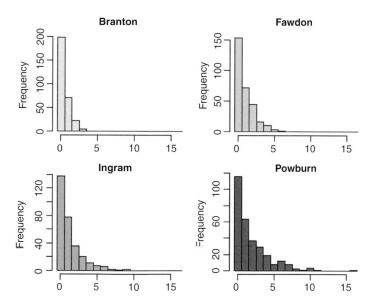

Most of the pitfall traps contained no beetles at all sites, and the maximum number caught was 16 on one occasion at Powburn. There are large differences between the farms in mean beetle counts, but despite this, the five hedgerow management treatments showed substantial differences:

```
tapply(count,list(farm,field),mean)
```

	1	2	3	4	5
1	0.4000000	0.4666667	0.4500000	0.5833333	0.5833333
2	0.6833333	0.7666667	0.7833333	1.2000000	1.1333333
3	0.8500000	0.9166667	1.6666667	1.2500000	1.3833333
4	1.5000000	2.2166667	1.8000000	1.4500000	2.2000000

We need to establish whether the treatment differences (as shown here between the column means) are significant. Because of the massive pseudoreplication, we need to analyse the counts very carefully. We shall use a generalized mixed effects model with Poisson errors, treating field as a fixed effect (the `field` codes refer to the five hedgerow management treatments) with the other factors as nested random effects (quadrats within blocks, within fields, within farms):

```
model <- lmer(count~field+(1|farm/field/block/quadrat),family=poisson)
summary(model)

Generalized linear mixed model fit by the Laplace approximation
Formula: count ~ field + (1 | farm/field/block/quadrat)
  AIC  BIC logLik deviance
 2216 2262  -1099     2198

Random effects:
 Groups                        Name        Variance Std.Dev.
 quadrat:(block:(field:farm))  (Intercept) 0.133785 0.36577
 block:(field:farm)            (Intercept) 0.017054 0.13059
 field:farm                    (Intercept) 0.000000 0.00000
 farm                          (Intercept) 0.216319 0.46510

Number of obs: 1200, groups: quadrat:(block:(field:farm)), 120;
block:(field:farm), 60; field:farm, 20; farm, 4

Fixed effects:
            Estimate Std. Error z value Pr(>|z|)
(Intercept)  -0.3322     0.2586  -1.285  0.19889
field2        0.2202     0.1561   1.411  0.15830
field3        0.2851     0.1552   1.836  0.06632 .
field4        0.3205     0.1546   2.074  0.03811 *
field5        0.4402     0.1533   2.872  0.00407 **
```

As you can see, management treatments 4 and 5 produced significantly higher beetle counts than the controls. Among the random effects you will see that `field:farm` (the fixed effect) registers as zero. Much the biggest cause of variation in beetle numbers was differences between the four farms (variance = 0.216) , then differences between the quadrats within a block (0.134), with only very modest differences from block to block within each field (0.017).

If, instead of fitting the management treatment as a fixed effect, we were to fit is as a random effect, then the model changes as follows:

```
model2 <- lmer(count~1+(1|farm/field/block/quadrat),family=poisson)
summary(model2)

Generalized linear mixed model fit by the Laplace approximation
Formula: count ~ 1 + (1 | farm/field/block/quadrat)
  AIC  BIC logLik deviance
 2216 2242  -1103     2206

Random effects:
 Groups                        Name        Variance Std.Dev.
 quadrat:(block:(field:farm))  (Intercept) 0.133591 0.365501
 block:(field:farm)            (Intercept) 0.035101 0.187354
```

```
field:farm                        (Intercept) 0.001888 0.043451
farm                              (Intercept) 0.213685 0.462260
```

Number of obs: 1200, groups: quadrat:(block:(field:farm)), 120;
block:(field:farm), 60; field:farm, 20; farm, 4

Fixed effects:
```
                Estimate Std. Error z value Pr(>|z|)
(Intercept) -0.07779      0.23703  -0.328     0.743
```

The variance component attributable to differences between blocks has almost doubled (this is the spatial scale immediately below fields, which was the scale at which the fixed effects treatments were applied), but the farm-scale variation (larger scale) and quadrat-scale variation (smaller scale) are almost unaffected. So the difference between fields which looks so small as a random effect (0.001 888) is clearly significant when the term is fitted as a fixed effect (as was appropriate in this example).

Here is the much simpler analysis with the pseudoreplication averaged away. Now, there are only 20 numbers in the dataframe (5 numbers (the five treatment mean counts) from each of 4 farms), and we use these to create a new response variable:

```
y <- as.vector(tapply(count,list(farm,field),mean))
y
```

```
 [1] 0.4000000 0.6833333 0.8500000 1.5000000 0.4666667 0.7666667 0.9166667
 [8] 2.2166667 0.4500000 0.7833333 1.6666667 1.8000000 0.5833333 1.2000000
[15] 1.2500000 1.4500000 0.5833333 1.1333333 1.3833333 2.2000000
```

We also need new shorter vectors of categorical explanatory variables for the farm (which now serves as a four-level block in this two-way, non-replicated analysis) and the five-level hedgerow management treatment. The 'generate factor levels' function, gl, is useful here. For farm blocks (fblock) we read the instruction to gl like this: 'generate levels up to 4, each with a repeat of 1, to a total length of 20'. For hedge it says 'generate levels up to 5, each with a repeat of 4'.

```
fblock <- gl(4,1,20)
hedge <- gl(5,4)
```

We cannot fit an exactly analogous model because we no longer have count data (the average beetle numbers are all real numbers) so we cannot use glm with Poisson errors. We can get close enough, however, by using a linear model with the logarithms of the mean counts as the response variable like this:

```
model3 <- lm(log(y)~fblock+hedge)
summary(model3)
```

```
Call:
lm(formula = log(y) ~ fblock + hedge)
Residuals:
     Min       1Q    Median       3Q       Max
-0.29679 -0.08192 -0.01837   0.05106   0.31607
```

```
Coefficients:
              Estimate Std. Error t value Pr(>|t|)
(Intercept)   -0.9555     0.1179  -8.102 3.30e-06 ***
fblock2        0.5943     0.1179   5.039  0.00029 ***
fblock3        0.8728     0.1179   7.400 8.27e-06 ***
```

```
fblock4          1.3008      0.1179   11.030 1.23e-07 ***
hedge2           0.1838      0.1319    1.394  0.18859
hedge3           0.2775      0.1319    2.105  0.05708 .
hedge4           0.3230      0.1319    2.450  0.03060 *
hedge5           0.4383      0.1319    3.324  0.00606 **

Residual standard error: 0.1865 on 12 degrees of freedom
Multiple R-squared: 0.9214,       Adjusted R-squared: 0.8756
F-statistic:  20.1 on 7 and 12 DF,  p-value: 9.875e-06
```

The p values are slightly different, but the interpretation is exactly the same. After model simplification, it turns out that there is one highly significant contrast, between the last two hedge management treatments (4 and 5) and the rest ($p = 0.0102$). It is hedge cutting, not grass cutting, that makes the difference in beetle counts in this case.

An alternative method of producing the reduced dataframe (i.e. eliminating the pseudoreplication) is to use aggregate like this:

```
d2<-aggregate(data,list(farm,field),mean)
model<-lm(log(count)~factor(farm)+factor(field),data=d2)
summary(model)
```

This produces the same output with less than half the R code.

20

Non-Linear Regression

Sometimes we have a mechanistic model for the relationship between y and x, and we want to estimate the parameters and standard errors of the parameters of a specific non-linear equation from data. Some frequently used non-linear models are shown in Table 20.1. What we mean in this case by 'non-linear' is not that the relationship is curved (it was curved in the case of polynomial regressions, but these were linear models), but that the relationship cannot be linearized by transformation of the response variable or the explanatory variable (or both). Here is an example: it shows jaw bone length as a function of age in deer. Theory indicates that the relationship is an asymptotic exponential with three parameters:

$$y = a - be^{-cx}.$$

In R, the main difference between linear models and non-linear models is that we have to tell R the exact nature of the equation as part of the model formula when we use non-linear modelling. In place of `lm` we write `nls` (this stands for 'non-linear least squares'). Then, instead of `y~x`, we write `y~a-b*exp(-c*x)` to spell out the precise non-linear model we want R to fit to the data.

The slightly tedious thing is that R requires us to specify initial guesses for the values of the parameters a, b and c (note, however, that some common non-linear models have 'self-starting' versions in R which bypass this step; see p. 728). Let us plot the data to work out sensible starting values. It always helps in cases like this to work out the equation's 'behaviour at the limits' – that is to say, to find the values of y when $x = 0$ and when $x = \infty$ (p. 258). For $x = 0$, we have exp(–0) which is 1, and $1 \times b = b$, so $y = a - b$. For $x = \infty$, we have exp(–∞) which is 0, and $0 \times b = 0$, so $y = a$. That is to say, the asymptotic value of y is a, and the intercept is $a - b$.

```
deer <- read.table("c:\\temp\\jaws.txt",header=T)
attach(deer)
names(deer)
```

```
[1] "age" "bone"
```

```
plot(age,bone,pch=21,col="purple",bg="green")
```

The R Book, Second Edition. Michael J. Crawley.
© 2013 John Wiley & Sons, Ltd. Published 2013 by John Wiley & Sons, Ltd.

Table 20.1. Useful non-linear functions.

Name	Equation
Asymptotic functions	
Michaelis–Menten	$y = \dfrac{ax}{1 + bx}$
2-parameter asymptotic exponential	$y = a(1 - e^{-bx})$
3-parameter asymptotic exponential	$y = a - be^{-cx}$
S-shaped functions	
2-parameter logistic	$y = \dfrac{e^{a+bx}}{1 + e^{a+bx}}$
3-parameter logistic	$y = \dfrac{a}{1 + be^{-cx}}$
4-parameter logistic	$y = a + \dfrac{b - a}{1 + e^{(c-x)/d}}$
Weibull	$y = a - be^{-(cx^d)}$
Gompertz	$y = ae^{-be^{-cx}}$
Humped curves	
Ricker curve	$y = axe^{-bx}$
First-order compartment	$y = k\exp(-\exp(a)x) - \exp(-\exp(b)x)$
Bell-shaped	$y = a\exp(-|bx|^2)$
Biexponential	$y = ae^{bx} - ce^{-dx}$

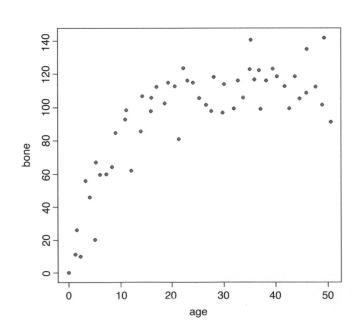

Inspection suggests that a reasonable estimate of the asymptote is $a \approx 120$ and intercept ≈ 10, so $b = 120 - 10 = 110$. Our guess at the value of c is slightly harder. Where the curve is rising most steeply, jaw length is about 40 where age is 5. Rearranging the equation gives

$$c = -\frac{\log((a-y)/b)}{x} = -\frac{\log(120-40)/110)}{5} = 0.063\,690\,75$$

Now that we have the three parameter estimates, we can provide them to R as the starting conditions as part of the `nls` call like this:

```
model <- nls(bone~a-b*exp(-c*age),start=list(a=120,b=110,c=0.064))
summary(model)
```

```
Formula: bone ~ a - b * exp(-c * age)

Parameters:
   Estimate Std. Error t value Pr(>|t|)
a 115.2528     2.9139   39.55  < 2e-16 ***
b 118.6875     7.8925   15.04  < 2e-16 ***
c   0.1235     0.0171    7.22 2.44e-09 ***

Residual standard error: 13.21 on 51 degrees of freedom

Number of iterations to convergence: 5
Achieved convergence tolerance: 2.381e-06
```

All the parameters appear to be significantly different from zero at $p < 0.001$. Beware, however. This does not necessarily mean that all the parameters need to be retained in the model. In this case, $a = 115.2528$ with standard error 2.9139 is clearly not significantly different from $b = 118.6875$ with standard error 7.8925 (they would need to differ by more than 2 standard errors to be significant). So we should try fitting the simpler two-parameter model

$$y = a(1 - e^{-cx}).$$

```
model2 <- nls(bone~a*(1-exp(-c*age)),start=list(a=120,c=0.064))
anova(model,model2)
```

```
Analysis of Variance Table

Model 1: bone ~ a - b * exp(-c * age)
Model 2: bone ~ a * (1 - exp(-c * age))

  Res.Df Res.Sum Sq Df  Sum Sq F value Pr(>F)
1     51     8897.3
2     52     8929.1 -1 -31.843  0.1825  0.671
```

Model simplification was clearly justified ($p = 0.671$), so we accept the two-parameter version, `model2`, as our minimal adequate model. We finish by plotting the curve through the scatterplot. The age variable needs to go from 0 to 50 in smooth steps:

```
av <- seq(0,50,0.1)
```

and we use `predict` with `model2` to generate the predicted bone lengths:

```
bv <- predict(model2,list(age=av))
lines(av,bv,col="red")
```

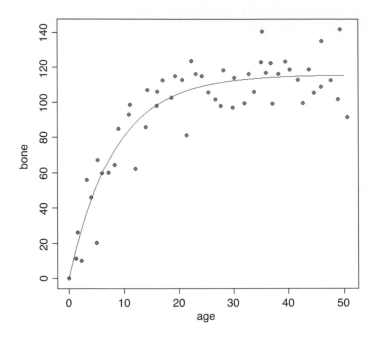

The parameters of this curve are obtained from `model2`:

```
summary(model2)

Formula: bone ~ a * (1 - exp(-c * age))

Parameters:
    Estimate Std. Error t value Pr(>|t|)
a 115.58056    2.84365  40.645  < 2e-16 ***
c   0.11882    0.01233   9.635 3.69e-13 ***

Residual standard error: 13.1 on 52 degrees of freedom

Number of iterations to convergence: 5
Achieved convergence tolerance: 1.356e-06
```

which we could write as $y = 115.58(1 - e^{-0.1188x})$ or as $y = 115.58(1 - \exp(-0.1188x))$ according to taste or journal style. If you want to present the standard errors as well as the parameter estimates, you could write: 'The model $y = a (1 - \exp(-bx))$ had $a = 115.58 \pm 2.84$ (1 standard error) and $b = 0.1188 \pm 0.0123$ (1 standard error, $n = 54$) and explained 84.87% of the total variation in bone length.' Note that because there are only two parameters in the minimal adequate model, we have called them a and b (rather than a and c as in the original formulation).

The variation explained needs to be calculated by comparing the residual variation (13.1, above) with the total variation. There are two steps to this. First, work out `sse`, the variation not explained by the regression.

We extract the residual standard error from the model summary (13.1), square it, then multiply by the degrees of freedom (52) to obtain the sum of squares:

```
sse <- as.vector((summary(model2)[[3]])^2*52)
sse

[1] 8929.143
```

Second, we work out the total variation in jaw bone length by fitting a null model, estimating only the overall mean, then extract the total sums of squares, sst, from the model summary:

```
null <- lm(bone~1)
sst <- as.vector(unlist(summary.aov(null)[[1]][2]))
sst

[1] 59007.99
```

Now, the percentage variation explained is simply:

```
100*(sst-sse)/sst

[1] 84.86791
```

20.1   Comparing Michaelis–Menten and asymptotic exponential

Model choice is always an important issue in curve fitting. We shall compare the fit of the asymptotic exponential (above) with a Michaelis–Menten with parameter values estimated from the same deer jaws data. As to starting values for the parameters, it is clear that a reasonable estimate for the asymptote would be 100 (this is a/b; see p. 264). The curve passes close to the point (5, 40) so we can guess a value of a of $40/5 = 8$ and hence $b = 8/100 = 0.08$. Now use nls to estimate the parameters:

```
(model3 <- nls(bone~a*age/(1+b*age),start=list(a=8,b=0.08)))

Nonlinear regression model
  model:  bone ~ a * age/(1 + b * age)
   data:  parent.frame()
       a       b
18.725   0.136
 residual sum-of-squares: 9854

Number of iterations to convergence: 7
Achieved convergence tolerance: 1.522e-06
```

Finally, we can add the line for Michaelis–Menten to the original plot, using predict with the model name, with a list to allocate av values to age:

```
yv <- predict(model3, list(age=av))
lines(av,yv,col="blue")
```

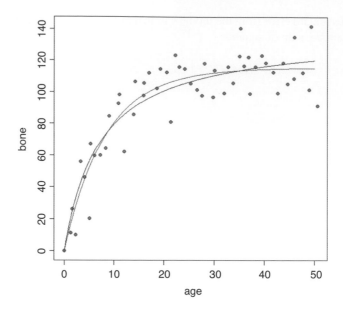

You can see that the asymptotic exponential (red line) tends to get to its asymptote first, and that the Michaelis–Menten (blue line) continues to increase. Model choice, therefore would be enormously important if you intended to use the model for prediction to ages much greater than 50 months.

20.2   Generalized additive models

Sometimes we can see that the relationship between y and x is non-linear but we do not have any theory or any mechanistic model to suggest a particular functional form (mathematical equation) to describe the relationship. In such circumstances, generalized additive models (GAMs) are particularly useful because they fit non-parametric smoothers to the data without requiring us to specify any particular mathematical model to describe the non-linearity (background and more examples are given in Chapter 18).

```
humped <- read.table("c:\\temp\\hump.txt",header=T)
attach(humped)
names(humped)
[1] "y" "x"
plot(x,y,pch=21,col="brown",bg="orange")
library(mgcv)
```

The model is specified very simply by showing which explanatory variables (in this case just x) are to be fitted as smoothed functions using the notation `y~s(x)`:

```
model <- gam(y~s(x))
```

Now we can use `predict` in the normal way to fit the curve estimated by `gam`:

```
xv <- seq(0.5,1.3,0.01)
yv <- predict(model,list(x=xv))
lines(xv,yv)
```

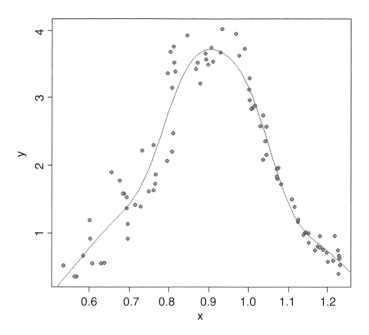

```
summary(model)

Family: gaussian
Link function: identity

Formula:
y ~ s(x)

Parametric coefficients:
            Estimate Std. Error t value Pr(>|t|)
(Intercept)  1.95737    0.03446    56.8   <2e-16 ***

Approximate significance of smooth terms:
       edf Ref.df      F p-value
s(x) 7.452  8.403 116.7  <2e-16 ***
R-sq.(adj) =  0.919    Deviance explained = 92.6%
GCV score = 0.1156  Scale est. = 0.1045    n = 88
```

Fitting the curve uses up 7.452 degrees of freedom (i.e. it is quite expensive) but the resulting fit is excellent and the model explains more than 92.6% of the deviance in y.

20.3 Grouped data for non-linear estimation

Here is a dataframe containing experimental results on reaction rates as a function of enzyme concentration for five different bacterial strains, with reaction rate measured just once for each strain at each of ten enzyme concentrations. The idea is to fit a family of five Michaelis–Menten functions with parameter values depending on the strain.

```
reaction <- read.table("c:\\temp\\reaction.txt",header=T)
attach(reaction)
names(reaction)
```

```
[1] "strain" "enzyme" "rate"
```

```
plot(enzyme,rate,pch=20+as.numeric(strain),bg=1+as.numeric(strain))
```

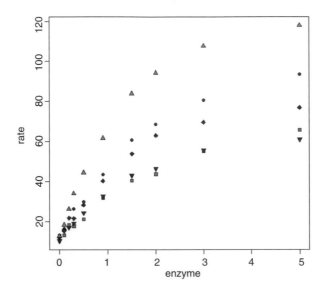

Clearly the different strains will require different parameter values, but there is a reasonable hope that the same functional form will describe the response of the reaction rate of each strain to enzyme concentration.

```
library(nlme)
```

The function we need is `nlsList`, which fits the same functional form to a group of subjects (as indicated by the 'given' operator $|$):

```
model <- nlsList(rate~c+a*enzyme/(1+b*enzyme)|strain,
                        data=reaction,start=c(a=20,b=0.25,c=10))
```

Note the use of the `groupedData` style formula `rate~enzyme|strain`.

```
summary(model)
```

```
Call:
  Model: rate ~ c + a * enzyme/(1 + b * enzyme)|strain
   Data: reaction
Coefficients:
  a
  Estimate  Std. Error    t value      Pr(>|t|)
A 51.79746    4.093791  12.652687  1.943004e-06
B 26.05893    3.063474   8.506335  2.800344e-05
C 51.86774    5.086678  10.196781  7.842356e-05
D 94.46245    5.813975  16.247481  2.973298e-06
E 37.50984    4.840749   7.748767  6.462817e-06
```

```
  b
  Estimate Std. Error    t value       Pr(>|t|)
A 0.4238572 0.04971637   8.525506 2.728564e-05
B 0.2802433 0.05761532   4.864041 9.173722e-04
C 0.5584897 0.07412454   7.534479 5.150212e-04
D 0.6560539 0.05207362  12.598586 1.634553e-05
E 0.5253479 0.09354863   5.615774 5.412405e-05
  c
  Estimate Std. Error    t value       Pr(>|t|)
A 11.46498   1.194155   9.600916 1.244488e-05
B 11.73312   1.120452  10.471780 7.049414e-06
C 10.53219   1.254928   8.392663 2.671651e-04
D 10.40964   1.294447   8.041767 2.909373e-04
E 10.30139   1.240664   8.303123 4.059887e-06

Residual standard error: 1.81625 on 35 degrees of freedom
```

There is substantial variation from strain to strain in the values of a and b, but we should test whether a model with a common intercept of, say, 11.0 might not fit equally well.

The plotting is made much easier if we convert the dataframe to a `groupedData` object:

```
reaction <- groupedData(rate~enzyme|strain,data=reaction)
plot(reaction)
```

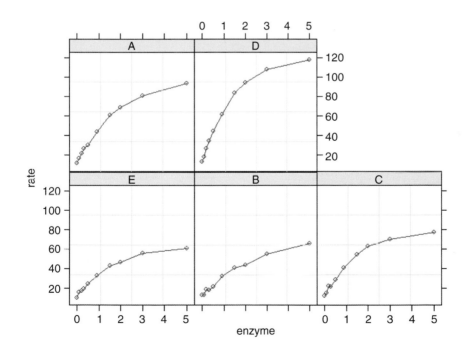

This default plot has just joined the dots, but we want to fit the separate non-linear regressions. To do this we fit a non-linear mixed-effects model with `nlme`, rather than use `nlsList`:

```
model2 <- nlme(rate~c+a*enzyme/(1+b*enzyme),fixed=a+b+c~1,
       random=a+b+c~1|strain,data=reaction,start=c(a=20,b=0.25,c=10))
```

Now we can employ the very powerful `augPred` function to fit the curves to each panel:

```
plot(augPred(model2))
```

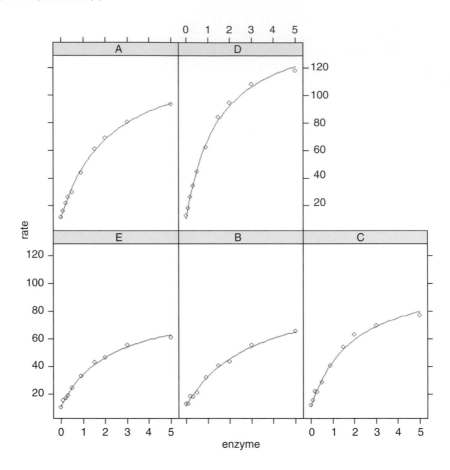

Here is the summary of the non-linear mixed model:

```
summary(model2)
```

```
Nonlinear mixed-effects model fit by maximum likelihood
  Model: rate ~ c + a * enzyme/(1 + b * enzyme)
 Data: reaction
       AIC        BIC      logLik
  253.4794 272.5997 -116.7397

Random effects:
 Formula: list(a ~ 1, b ~ 1, c ~ 1)
 Level: strain
 Structure: General positive-definite, Log-Cholesky parametrization
        StdDev      Corr
a       22.9144053  a        b
b        0.1132369  0.876
c        0.4231029 -0.536 -0.875
```

```
Residual  1.7105816

Fixed effects: a + b + c ~ 1
      Value Std.Error DF    t-value p-value
a 51.59873 10.741019 43   4.803895       0
b  0.47665  0.058786 43   8.108321       0
c 10.98537  0.556462 43 19.741461       0
 Correlation:
  a       b
b  0.843
c -0.313 -0.543

Standardized Within-Group Residuals:
        Min             Q1          Med           Q3          Max
-1.79191571 -0.65655493   0.05675519   0.74272994   2.02715761

Number of Observations: 50
Number of Groups: 5
```

The fixed effects in this model are the means of the parameter values. To see the separate parameter estimates for each strain, use `coef`:

```
coef(model2)
```

```
          a           b          c
E 34.09244 0.4533741 10.81642
B 28.01211 0.3238606 11.54848
C 49.64062 0.5193966 10.67127
A 53.20342 0.4426117 11.23660
D 93.04507 0.6440231 10.65410
```

Note that the rows of this table are no longer in alphabetical order but sequenced in the way they appeared in the panel plot (i.e. ranked by their maximum values). The parameter estimates are close to, but not equal to, the values estimated by `nlsList` (above) as a result of 'shrinkage' in the restricted maximum likelihood estimates (see p. 685). The efficiency of the random effects model in terms of degrees of freedom is illustrated by contrasting the numbers of parameters estimated by `nlsList` (15 = 5 values for each of 3 parameters) and by `nlme` (7 = 3 parameters plus 4 variances), giving residual degrees of freedom of 35 and 43, respectively.

If you want to draw the fitted lines yourself, then repeat the scatterplot we did earlier:

```
plot(enzyme,rate,pch=20+as.numeric(strain),bg=1+as.numeric(strain))
```

and use a loop to extract the three parameters from the coefficients table `coef(model)` and to copy the colours for the fitted lines:

```
for(i in 1:5){
yv <- coef(model)[i,3]+coef(model)[i,1]*xv/(1+coef(model)[i,2]*xv)
lines(xv,yv,col=(i+1))   }
```

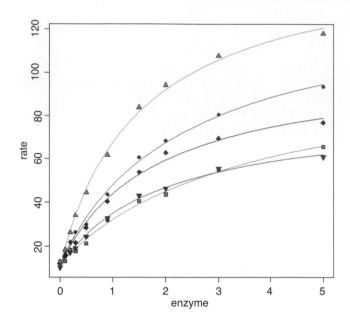

20.4 Non-linear time series models (temporal pseudoreplication)

The previous example was a designed experiment in which there was no pseudoreplication. However, we often want to fit non-linear models to growth curves where there is temporal pseudoreplication across a set of subjects, each providing repeated measures on the response variable. In such a case we shall want to model the temporal autocorrelation.

```
nl.ts <- read.table("c:\\temp\\nonlinear.txt",header=T)
attach(nl.ts)
names(nl.ts)
```

```
[1] "time" "dish" "isolate" "diam"
```

```
growth <- groupedData(diam~time|dish,data=nl.ts)
```

Here, we model the temporal autocorrelation as first-order autoregessive, `corAR1()`:

```
model <- nlme(diam~a+b*time/(1+c*time),
      fixed=a+b+c~1,
      random=a+b+c~1,
      data=growth,
      correlation=corAR1(),
      start=c(a=0.5,b=5,c=0.5))
summary(model)
```

```
Nonlinear mixed-effects model fit by maximum likelihood
  Model: diam ~ a + b * time/(1 + c * time)
 Data: growth
       AIC      BIC    logLik
  129.7693 158.3157 -53.88467
```

```
Random effects:
 Formula: list(a ~ 1, b ~ 1, c ~ 1)
 Level: dish
 Structure: General positive-definite, Log-Cholesky parametrization
         StdDev     Corr
a        0.1014481 a         b
b        1.2060389 -0.557
c        0.1095792 -0.958   0.772
Residual 0.3150068

Correlation Structure: AR(1)
 Formula: ~1 | dish
 Parameter estimate(s):
       Phi
-0.0334494
Fixed effects: a + b + c ~ 1
     Value Std.Error DF  t-value p-value
a 1.288262 0.1086391 88 11.85817       0
b 5.215252 0.4741958 88 10.99810       0
c 0.498222 0.0450644 88 11.05577       0
Correlation:
  a        b
b -0.506
c -0.542  0.823

Standardized Within-Group Residuals:
        Min           Q1          Med          Q3          Max
-1.74222876 -0.64713734 -0.03349903  0.70298704  2.24686642

Number of Observations: 99
Number of Groups: 9

coef(model)


          a        b         c
5 1.288831 3.348753 0.4393774
4 1.235631 5.075223 0.5373954
1 1.252722 5.009540 0.5212435
3 1.285846 4.843222 0.4885947
9 1.111130 7.171316 0.7061061
7 1.272569 5.361570 0.5158167
6 1.435789 4.055236 0.3397506
2 1.348524 5.440492 0.4553721
8 1.363314 6.631915 0.4803382
```

It could not be simpler to plot the family of non-linear models in a panel of scatterplots. We just use augPred like this:

```
plot(augPred(model))
```

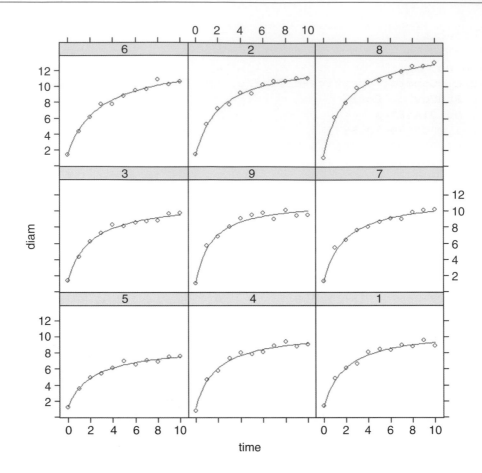

20.5   Self-starting functions

One of the most likely things to go wrong in non-linear least squares is that the model fails because your initial guesses for the starting parameter values were too far off. The simplest solution is to use one of R's 'self-starting' models, which work out the starting values for you automatically. These are the most frequently used self-starting functions:

SSasymp	asymptotic regression model;
SSasympOff	asymptotic regression model with an offset;
SSasympOrig	asymptotic regression model through the origin;
SSbiexp	biexponential model;
SSfol	first-order compartment model;
SSfpl	four-parameter logistic model;
SSgompertz	Gompertz growth model;
SSlogis	logistic model;
SSmicmen	Michaelis–Menten model;
SSweibull	Weibull growth curve model.

20.5.1 Self-starting Michaelis–Menten model

In our next example, reaction rate is a function of enzyme concentration; reaction rate increases quickly with concentration at first but asymptotes once the reaction rate is no longer enzyme-limited. R has a self-starting version called SSmicmen, parameterized as

$$y = \frac{ax}{b + x},$$

where the two parameters are a (the asymptotic value of y) and b (which is the x value at which half of the maximum response, $a/2$, is attained). In the field of enzyme kinetics b is called the Michaelis parameter (see p. 264; in R help the two parameters are called Vm and K, respectively).

Here is SSmicmen in action:

```
data <- read.table("c:\\temp\\mm.txt",header=T)
attach(data)
names(data)
```

```
[1] "conc" "rate"
```

```
plot(rate~conc,pch=16)
```

To fit the non-linear model, just put the name of the response variable (rate) on the left of the tilde ~ then put SSmicmen(conc,a,b)) on the right of the tilde, with the name of your explanatory variable first in the list of arguments (conc in this case), then your names for the two parameters (a and b, as defined above):

```
model <- nls(rate~SSmicmen(conc,a,b))
summary(model)
```

```
Formula: rate ~ SSmicmen(conc, a, b)
```

```
Parameters:
    Estimate Std. Error t value Pr(>|t|)
a 2.127e+02  6.947e+00  30.615 3.24e-11 ***
b 6.412e-02  8.281e-03   7.743 1.57e-05 ***
```

```
Residual standard error: 10.93 on 10 degrees of freedom
```

```
Number of iterations to convergence: 0
Achieved convergence tolerance: 1.93e-06
```

So the equation is

$$y = \frac{212.7x}{0.064\,12 + x},$$

and we can plot it like this:

```
xv <- seq(0,1.2,.01)
yv <- predict(model,list(conc=xv))
lines(xv,yv,col="blue")
```

20.5.2   Self-starting asymptotic exponential model

The three-parameter asymptotic exponential is usually written like this:

$$y = a - be^{-cx}.$$

In R's self-starting version, SSasymp, the parameters are as follows:

- a is the horizontal asymptote on the right-hand side (called Asym in R help);

- $b = a - R0$ where R0 is the intercept (the response when x is zero);

- c is the rate constant (the log of lrc in R help).

   Here is SSasymp applied to the jaws data (p. 209):

```
deer <- read.table("c:\\temp\\jaws.txt",header=T)
attach(deer)
names(deer)
```

```
[1] "age" "bone"
```

```
model <- nls(bone~SSasymp(age,a,b,c))
plot(age,bone,pch=16)
xv <- seq(0,50,0.2)
yv <- predict(model,list(age=xv))
lines(xv,yv)
```

```
summary(model)
```

```
Formula: bone ~ SSasymp(age, a, b, c)
```

```
Parameters:
  Estimate Std. Error t value Pr(>|t|)
a 115.2527    2.9139  39.553   <2e-16 ***
b  -3.4348    8.1961  -0.419    0.677
c  -2.0915    0.1385 -15.101   <2e-16 ***
```

```
Residual standard error: 13.21 on 51 degrees of freedom
```

```
Number of iterations to convergence: 0
Achieved convergence tolerance: 2.472e-07
```

The plot of this fit is on p. 720 along with the simplified model without the non-significant parameter b.
   Alternatively, one can use the two-parameter form that passes through the origin, SSasympOrig, which fits the function $y = a(1-\exp(-bx))$. The final form of the asymptotic exponential allows one to specify the function with an offset, d, on the x values, using SSasympOff, which fits the function $y = a - b \exp(-c(x - d))$.

20.5.3   Self-starting logistic

This is one of the most commonly used three-parameter growth models, producing a classic S-shaped curve:

```
sslogistic <- read.table("c:\\temp\\sslogistic.txt",header=T)
attach(sslogistic)
```

```
names(sslogistic)
```

```
[1] "density" "concentration"
```

```
plot(density~log(concentration),pch=16,col="green3")
```

We estimate the three parameters (*a*, *b*, *c*) using the self-starting function SSlogis:

```
model <- nls(density ~ SSlogis(log(concentration), a, b, c))
```

Now draw the fitted line using predict (note the antilog of xv in list):

```
xv <- seq(-3,3,0.1)
yv <- predict(model,list(concentration=exp(xv)))
lines(xv,yv,col="red")
```

The fit is excellent, and the parameter values and their standard errors are given by:

```
summary(model)
```

```
Formula: density ~ SSlogis(log(concentration), a, b, c)
```

```
Parameters:
  Estimate Std. Error t value Pr(>|t|)
a  2.34518    0.07815   30.01 2.17e-13 ***
b  1.48309    0.08135   18.23 1.22e-10 ***
c  1.04146    0.03227   32.27 8.51e-14 ***
```

```
Residual standard error: 0.01919 on 13 degrees of freedom
```

```
Number of iterations to convergence: 0
Achieved convergence tolerance: 3.284e-06
```

Here *a* is the asymptotic value, *b* is the mid-value of *x* when *y* is *a*/2, and *c* is the scale.

20.5.4   Self-starting four-parameter logistic

This model allows a lower asymptote (the fourth parameter) as well as an upper:

```
data <- read.table("c:\\temp\\chicks.txt",header=T)
attach(data)
names(data)
```

```
[1] "weight" "Time"
```

```
model <- nls(weight~SSfpl(Time, a, b, c, d))
xv <- seq(0,22,.2)
yv <- predict(model,list(Time=xv))
plot(weight·Time,pch=21,col="red",bg="green4")
lines(xv,yv,col="navy")
```

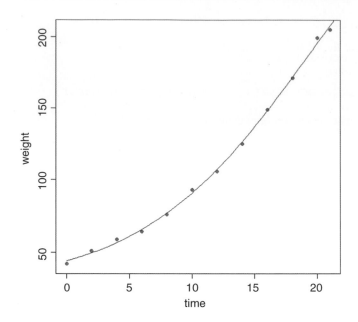

```
summary(model)
```

```
Formula: weight ~ SSfpl(Time, a, b, c, d)
```

```
Parameters:
  Estimate Std. Error t value Pr(>|t|)
a   27.453        6.601   4.159 0.003169 **
b  348.971       57.899   6.027 0.000314 ***
c   19.391        2.194   8.836 2.12e-05 ***
d    6.673        1.002   6.662 0.000159 ***
```

```
Residual standard error: 2.351 on 8 degrees of freedom
```

```
Number of iterations to convergence: 0
Achieved convergence tolerance: 3.324e-07
```

The four-parameter logistic is given by

$$y = A + \frac{B - A}{1 + e^{(D-x)/C}}.$$

This is the same formula as we used in Chapter 7, but note that C above is $1/c$ on p. 267. A is the horizontal asymptote on the left (for low values of x), B is the horizontal asymptote on the right (for large values of x), D is the value of x at the point of inflection of the curve (represented by xmid in our model for the chicks data), and C is a numeric scale parameter on the x axis (represented by scal). The parameterized model would be written like this:

$$y = 27.453 + \frac{348.971 - 27.453}{1 + \exp((19.391 - x)/6.673)}.$$

20.5.5 Self-starting Weibull growth function

R's parameterization of the Weibull growth function is

```
Asym-Drop*exp(-exp(lrc)*x^pwr)
```

where `Asym` is the horizontal asymptote on the right, `Drop` is the difference between the asymptote and the intercept (the value of y at $x = 0$), `lrc` is the natural logarithm of the rate constant, and `pwr` is the power to which x is raised.

```
weights <- read.table("c:\\temp\\weibull.growth.txt",header=T)
attach(weights)
names(weights)
```

```
[1] "weight" "time"
```

```
model <- nls(weight ~ SSweibull(time, Asym, Drop, lrc, pwr))
summary(model)
```

```
Formula: weight ~ SSweibull(time, Asym, Drop, lrc, pwr)

Parameters:
      Estimate Std. Error t value Pr(>|t|)
Asym 158.5012     1.1769  134.67 3.28e-13 ***
Drop 110.9971     2.6330   42.16 1.10e-09 ***
lrc   -5.9934     0.3733  -16.05 8.83e-07 ***
pwr    2.6461     0.1613   16.41 7.62e-07 ***

Residual standard error: 2.061 on 7 degrees of freedom

Number of iterations to convergence: 0
Achieved convergence tolerance: 5.702e-06
```

```
xt <- seq(2,22,0.1)
yw <- predict(model,list(time=xt))
plot(time,weight,pch=21,col="blue",bg="orange")
lines(xt,yw,col="blue2")
```

The fit is good, but the model cannot accommodate a drop in *y* values once the asymptote has been reached (you would need to fit some kind of humped function).

20.5.6 Self-starting first-order compartment function

In the following model, the response is drug concentration in the blood, which is to be plotted as a function of time after the dose was administered. There are three parameters (*a*, *b*, *c*) to be estimated:

```
foldat <- read.table("c:\\temp\\fol.txt",header=T)
attach(foldat)
names(foldat)

[1] "Wt"    "Dose" "Time" "conc"
```

The model looks like this: the response (*y*) is conc and the explanatory (*x*) is Time,

$$y = k \exp(- \exp(a)x) - \exp(- \exp(b)x),$$

where $k = \text{Dose} \times \exp(a + b - c)/(\exp(b) - \exp(a))$ and Dose is a vector of identical values provided in the dataframe (4.02 in this example):

```
model <- nls(conc~SSfol(Dose, Time, a, b, c))
summary(model)

Formula: conc ~ SSfol(Dose, Time, a, b, c)

Parameters:
  Estimate Std. Error t value Pr(>|t|)
a  -2.9196     0.1709 -17.085 1.40e-07 ***
b   0.5752     0.1728   3.328   0.0104 *
c  -3.9159     0.1273 -30.768 1.35e-09 ***

Residual standard error: 0.732 on 8 degrees of freedom

Number of iterations to convergence: 8
Achieved convergence tolerance: 4.907e-06

xv <- seq(0,25,0.1)
yv <- predict(model,list(Time=xv))
plot(conc~Time,pch=21,col="blue",bg="red")
lines(xv,yv,col="green4")
```

As you can see, this is a rather poor model for predicting the value of the peak concentration, but a reasonable description of the ascending and declining sections.

20.6   Bootstrapping a family of non-linear regressions

There are two broad applications of bootstrapping to the estimation of parameters in non-linear models:

- Select certain of the data points at random with replacement, so that, for any given model fit, some data points are duplicated and others are left out.

- Fit the model and estimate the residuals, then allocate the residuals at random, adding them to different fitted values in different simulations.

Our next example involves the viscosity data from the MASS library, where sinking time ($y = $ Time) is measured for three different weights ($w = $ Wt) in fluids of nine different viscosities ($x = $ Viscosity):

$$y = \frac{bx}{W - c}.$$

We need to estimate the two parameters b and c and their standard errors.

```
library(MASS)
data(stormer)
attach(stormer)
```

Here are the results of the straightforward non-linear regression:

```
model <- nls(Time~b*Viscosity/(Wt-c),start=list(b=29,c=2))
summary(model)
```

Formula: Time ~ b * Viscosity/(Wt - c)

Parameters:
  Estimate Std. Error t value Pr(>|t|)
b 29.4013   0.9155 32.114 < 2e-16 ***
c   2.2182   0.6655   3.333 0.00316 **

Residual standard error: 6.268 on 21 degrees of freedom

Number of iterations to convergence: 2
Achieved convergence tolerance: 8.959e-06

```
plot(Viscosity,Time,pch=16,col=1+as.numeric(factor(Wt)))
xv <- 0:300
yv <- predict(model,list(Wt=20,Viscosity=xv))
lines(xv,yv,col=2)
yv <- predict(model,list(Wt=50,Viscosity=xv))
lines(xv,yv,col=3)
yv <- predict(model,list(Wt=100,Viscosity=xv))
lines(xv,yv,col=4)
```

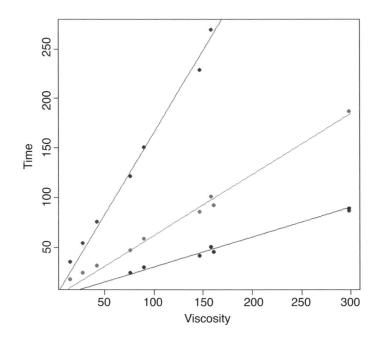

Here is a home-made bootstrap which leaves out cases at random. The idea is to sample the indices (subscripts) of the 23 cases at random with replacement:

```
sample(1:23,replace=T)
```

```
[1]  4  4 10 10 12  3 23 22 21 13  9 14  8  5 15 14 21 14 12  3 20 14 19
```

In this realization cases 1 and 2 were left out, cases 4 and 10 appeared twice, and so on. We call the subscripts `ss` as follows, and use the subscripts to select values for the response (*y*) and the two explanatory variables (`x1` and `x2`) like this:

```
ss <- sample(1:23,replace=T)
y <- Time[ss]
x1 <- Viscosity[ss]
x2 <- Wt[ss]
```

Now we put this in a loop and fit the model:

```
model <- nls(y~b*x1/(x2-c),start=list(b=29,c=2))
```

one thousand times, storing the coefficients in vectors called *bv* and *cv*:

```
bv <- numeric(1000)
cv <- numeric(1000)
for(i in 1:1000){
ss <- sample(1:23,replace=T)
y <- Time[ss]
x1 <- Viscosity[ss]
x2 <- Wt[ss]
model <- nls(y~b*x1/(x2-c),start=list(b=29,c=2))
bv[i] <- coef(model)[1]
cv[i] <- coef(model)[2] }
```

This took 2 seconds for 1000 iterations. The 95% confidence intervals for the two parameters are obtained using the `quantile` function:

```
quantile(bv,c(0.025,0.975))

    2.5%     97.5%
27.80559 30.70029
```

```
quantile(cv,c(0.025,0.975))

    2.5%      97.5%
0.7798505 3.7625656
```

Alternatively, you can randomize the locations of the residuals while keeping all the cases in the model for every simulation. We use the built-in functions in the `boot` package to illustrate this procedure.

```
library(boot)
```

The `boot` package (Canty and Ripley, 2012) allows both parametric and non-parametric resampling. For the non-parametric bootstrap, possible resampling methods are the ordinary bootstrap, the balanced bootstrap, antithetic resampling, and permutation. For non-parametric multi-sample problems stratified resampling is used: this is specified by including a vector of strata in the call to `boot`. Importance resampling weights may be specified. The function generates a specified number bootstrap replicates (R) of a `statistic` applied to data. The tricky part is in understanding how to write the `statistic` function. This is a function which when applied to data returns a vector containing the statistic(s) of interest.

First, we need to calculate the residuals and the fitted values from the `nls` model we fitted on p. 735:

```
rs <- resid(model)
fit <- fitted(model)
```

We make the fit along with the two explanatory variables `Viscosity` and `Wt` into a new dataframe called `storm` that will be used inside the `statistic` function:

```
storm <- data.frame(fit,Viscosity,Wt)
```

Next, we need to write the `statistic` function to describe the model fitting. We add the randomized residuals to the fitted values, then extract the coefficients from the non-linear model fitted to these new data:

```
statistic <- function(rs,i){
storm$y <- storm$fit+rs[i]
coef(nls(y~b*Viscosity/(Wt-c),data=storm,start=coef(model)))}
```

The two arguments to `statistic` are the vector of residuals, `rs`, and the randomized indices, `i`. Now we can run the `boot` function over 1000 iterations:

```
boot.model <- boot(rs,statistic,R=1000)
boot.model

ORDINARY NONPARAMETRIC BOOTSTRAP

Call:
boot(data = rs, statistic = statistic, R = 1000)
Bootstrap Statistics :
     original        bias   std. error
t1* 29.401294  0.6714883    0.8565776
t2*  2.218247 -0.2528173    0.6177332
```

The parametric estimates for b (t1) and c (t2) in `boot.model` are reasonably unbiased, and the bootstrap standard errors are slightly smaller than when we used `nls`. We get the bootstrapped confidence intervals with the `boot.ci` function: b is index=1 and c is index=2:

```
boot.ci(boot.model,index=1)

BOOTSTRAP CONFIDENCE INTERVAL CALCULATIONS
Based on 1000 bootstrap replicates

CALL :
boot.ci(boot.out = boot.model, index = 1)

Intervals :
Level      Normal                  Basic
95%    (27.05, 30.41 )     (27.10, 30.42 )

Level      Percentile              BCa
95%    (28.39, 31.70 )     (27.63, 30.39 )
Calculations and Intervals on Original Scale

boot.ci(boot.model,index=2)
```

```
BOOTSTRAP CONFIDENCE INTERVAL CALCULATIONS
Based on 1000 bootstrap replicates

CALL :
boot.ci(boot.out = boot.model, index = 2)

Intervals :
Level       Normal                  Basic
95%    ( 1.260,   3.682 )     ( 1.311,   3.735 )

Level       Percentile              BCa
95%    ( 0.701,   3.126 )     ( 1.250,   3.479 )
Calculations and Intervals on Original Scale
```

For comparison, here are the parametric confidence intervals from our home-made bootstrap: for b, from 27.80 to 30.70; and for c, from 0.7798 to 3.7625.

21

Meta-Analysis

There is a compelling case to be made that analysts should look at the whole body of evidence, rather than trying to understand individual studies in isolation. The systematic review of a body of evidence is known as **meta-analysis**. The idea is to draw together all of the appropriate studies that have addressed the same question, and calculate an overall effect and an overall measure of uncertainly for that effect. Of course, there are plenty of issues with this, among them the following:

- What do we mean by 'the same question'?

- What is an appropriate study?

- What makes a study inappropriate ?

- How different can a study be, and still be worth including in the meta-analysis?

- What are the publication biases in the various studies?

- Are the effects fixed or random?

The central idea is to remove as much as possible of the subjectivity that was such a feature of old-fashioned narrative reviews. In an ideal world, we should be able to extract from every published study the exact question addressed, the effect size, the variance of that effect, the replication, and enough detail on the methods used to be confident that the study was comparable with the others that we have already included. Then we simply calculate a weighted average effect size and an appropriate weighted average measure of the unreliability of that overall effect. And that would be it.

If the effect is consistent across experiments, this gives considerable confidence in its generality, and the size of the effect is a useful measure of its importance. On the other hand, if the effect is found to be inconsistent across experiments, then its generality is called into question, and further studies are required to understand why the effect was so variable.

Meta-analysis requires:

- clear rules for the selection of studies;

- clear rules for the omission of studies;

- simple protocols for assigning a statistical weight to each study;

The R Book, Second Edition. Michael J. Crawley.
© 2013 John Wiley & Sons, Ltd. Published 2013 by John Wiley & Sons, Ltd.

- agreed methods for calculating an appropriate measure of uncertainly for the effect;
- sufficient detail in the Methods section that another analyst would get exactly the same result, and come to exactly the same conclusion, if they were to repeat the analysis using the same group of studies.

A meta-analysis of existing studies would be a useful component of the Introduction of any scientific paper, because it would show what the question was, and why it was interesting. Likewise, the Discussion section would benefit from a meta-analysis of existing studies, by showing what the current study means and where it fits into existing knowledge.

It is very important that biases are investigated and reported. We can never know how many experiments were carried out but never published. We can make an educated guess, but we can never know for certain. It is highly likely that proportionately more studies go unpublished if they failed to find a significant effect. It is possible (but regrettable) that more manuscripts were rejected during the refereeing process if they reported non significant effects. All of the effect sizes from all of the relevant studies must be included in the analysis (not just the most pleasing ones). The variation in the effect sizes might be due to sampling effects alone, or there may be genuine differences in the effects reported by different studies.

The idea is to calculate an effect size and a variance for each study. The summary of the meta-analysis is then just a weighted average of these effect sizes. More precise studies are given more weight than less precise ones in calculating the summary effect.

21.1 Effect size

Meta-analysis can work with a variety of kinds of effects. The simplest effects would be arithmetic means. Or we might choose to work with the differences between control and treatment means. Alternatively, we could use the scaled difference (the raw difference between the treatment and control means, divided by the pooled standard deviation). In other circumstances, we might look at the ratio between the two means. This is called a response ratio, R, and it is often used in cases where the response variable is a continuous measurement (like yield in an agricultural trial). R is the mean value for the treatment divided by the mean value of the control. Often in such cases, the effect size we analyse is the logarithm of this ratio, because this is symmetric above and below zero, and is hence more likely to have normal errors.

For count data, proportion data and binary data (dead or alive, recovered or ill) we cannot use linear models, because we shall have non-constant variance and non-normal errors (see Chapter 13). In such cases, we might use the risk ratio, the odds ratio or the risk difference. For studies reporting correlation coefficients we might do a meta-analysis of the correlation coefficients. In any event, the sampling distribution of the effect size should be known, so that variances and confidence intervals can be computed.

21.2 Weights

More precise studies should be given higher weights. Typically this means that studies with higher replication get higher weights, but the details depend on our assumptions about the distribution of the true effects. Many meta-analyses use the inverse of the variances as weights (the less variable a study, the more weight is given to that effect).

21.3 Fixed versus random effects

If you can convince yourself that there is a single true effect size (for instance, if many studies are trying to estimate the same physical constant) then the fixed-effect model is appropriate. It is much more likely in practice, however, that the effect being estimated varies with the context (e.g. with location, genetics or environmental conditions). In such cases, it is appropriate to do a random-effects meta-analysis. In this case, the measured effects are assumed to represent a random sample from a distribution of effect sizes.

21.3.1 Fixed-effect meta-analysis of scaled differences

This example comes from Borenstein et al. (2009) and concerns six studies each with a treatment and a control group, with replication varying from 40 to 200 across studies:

```
data <- read.table("c:\\temp\\metadata.txt",header=T)
attach(data)
head(data)
```

```
  study meanT sdT  nT meanC sdC  nC
1     A    94  22  60    92  20  60
2     B    98  21  65    92  22  65
3     C    98  28  40    88  26  40
4     D    94  19 200    82  17 200
5     E    98  21  50    88  22  45
6     F    96  21  85    92  22  85
```

The effect size for each study in this case is going to be the scaled difference between the treatment and control means:

```
d <- meanT-meanC
d
```

```
[1] 0.0951303 0.2789943 0.3701166 0.6656402 0.4655727 0.1859962
```

Table 21.1. Calculations for a fixed-effects meta-analysis. The individual effects (Y) have variances (V) reflecting sampling error alone. Study weights (W) are the inverse of the study variances. The assumption is that there is a common overall fixed effect, of which M is our best estimate. Nomenclature follows Borenstein et al. (2009).

Term	Formula
Weight	$W = \dfrac{1}{V}$
Weighted mean summary effect	$M = \dfrac{\sum WY}{\sum W}$
Variance of the summary effect	$VM = \dfrac{2}{\sum W}$
Standard error of the summary effect	$SEM = \sqrt{VM}$
Lower limit on the summary effect	$M - 1.96\,SEM$
Upper limit on the summary effect	$M + 1.96\,SEM$
Test statistic	$z = \dfrac{M}{SEM}$

Work out the pooled within-study standard deviation, using the formulae in Table 21.1. Note the use of round brackets to print the answer, avoiding typing `swithin` again:

```
(swithin <- sqrt(((nT-1)*sdT^2+(nC-1)*sdC^2)/(nT+nC-2)))
```

```
[1] 21.02380 21.50581 27.01851 18.02776 21.47892 21.50581
```

Scale the difference between the means by the pooled within-study standard deviation – this will be the effect size in our meta-analysis:

```
(d <- d/swithin)
```

```
[1] 0.0951303 0.2789943 0.3701166 0.6656402 0.4655727 0.1859962
```

Calculate the variance of the scaled difference:

```
(Vd <- (nC+nT)/(nC*nT) + d^2/(2*(nC+nT)))
```

```
[1] 0.03337104 0.03106861 0.05085616 0.01055385 0.04336305 0.02363116
```

Compute Hedges' g (g) and its variance (Vg) of the bias-corrected mean difference:

```
(J <- 1-3/(4*(nC+nT-2)-1))
```

```
[1] 0.9936306 0.9941292 0.9903537 0.9981144 0.9919137 0.9955291
```

```
(g <- J*d)
```

```
[1] 0.09452437 0.27735640 0.36654635 0.66438510 0.46180798 0.18516464
```

```
(Vg <- J^2*Vd)
```

```
[1] 0.03294729 0.03070488 0.04987975 0.01051408 0.04266460 0.02342033
```

Confidence intervals (int) for the scaled differences (lower limit ll, and upper limit, ul) are computed next:

```
(int <- 1.96*sqrt(Vg))
```

```
[1] 0.3557672 0.3434470 0.4377420 0.2009749 0.4048461 0.2999525
```

```
(ll <- d-int)
```

```
[1] -0.26063689 -0.06445270 -0.06762538  0.46466536  0.06072666 -0.11395631
```

```
(ul <- d+int)
```

```
[1] 0.4508975 0.6224414 0.8078586 0.8666151 0.8704188 0.4859488
```

As a flourish, we compute the total samples for each experiment (ns) to use in making the square symbols in the forest plot:

```
(ns <- nT+nC)
```

```
[1] 120 130  80 400  95 170
```

Now for the summary effect. First, work out the weights (the reciprocals of the variances)

```
(W <- 1/Vg)
```

```
[1] 30.35151 32.56811 20.04822 95.11053 23.43864 42.69795
```

Now calculate the sum of the products of the weight and the effects:

```
(WY <- W*g)
```

```
[1]   2.868958   9.032975   7.348601 63.190019 10.824149   7.906151
```

Here are the calculations so far – effect size, variance within, weight, and weight by effect:

```
data.frame(d,Vd,W,WY)
```

```
          d          Vd          W          WY
1 0.0951303 0.03337104 30.35151   2.868958
2 0.2789943 0.03106861 32.56811   9.032975
3 0.3701166 0.05085616 20.04822   7.348601
4 0.6656402 0.01055385 95.11053 63.190019
5 0.4655727 0.04336305 23.43864 10.824149
6 0.1859962 0.02363116 42.69795   7.906151
```

The summary effect is the sum of the products WY divided by the sum of the weights W:

```
(M <- sum(WY)/sum(W))
```

```
[1] 0.4142697
```

The variance of the summary effect is just the reciprocal of the sum of the weights:

```
(VM <- 1/sum(W))
```

```
[1] 0.004094753
```

Finally, we want the standard error of the mean effect which is the square root of VM, and the value of the test statistic, z:

```
(SEM <- sqrt(VM))
```

```
[1] 0.06399026
```

```
(z <- M/SEM)
```

```
[1] 6.473949
```

```
(ci <- 1.96*SEM)
```

```
[1] 0.1254209
```

The forest plot needs to show the six studies (labelled A–F) and their unreliability estimates, with the summary effect and its unreliability estimate shown as a diamond at the bottom (in a different colour for emphasis):

```
plot(c(-1,1),c(0,8),type="n",xlab="",yaxt="n",ylab="")
points(d,8-(1:6),pch=15,cex=10*ns/sum(ns))
for (i in 1:7) lines(c(ll[i],ul[i]),c(8-i,8-i))
polygon(c(M-ci,M,M+ci,M),c(1,1.1,1,0.9),col="red")
abline(v=0,lty=2)
text(rep(-0.7,7),1:7,c("summary effect","F","E","D","C","B","A"))
```

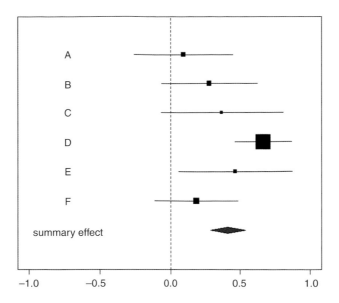

It is important to note that so-called 'vote counting' gets the wrong answer: only two significant results (D and E) out of six suggests non-significance overall. Not so. The summary effect is highly significant, and the lower bound of the red diamond comes nowhere near the vertical dashed line showing no effect. The sizes of the squares show the fraction of all replicates (across the six studies) that come from each study (the relative weights = ns/sum(ns)).

As an exercise, you might like to convert the above code into a simple general function that will work with any number of studies. It should take six vectors as its arguments (mean, standard deviation and sample size for treatment, then the same for control, for each of the studies in a meta-analysis).

If you are doing a meta-analysis in earnest, you are likely to use the meta package (Schwarzer, 2012). Here is the last example, repeated using metacont (this stands for 'meta-analysis for continuous variables'):

```
install.packages("meta")
library(meta)
metacont(nT,meanT,sdT,nC,meanC,sdC)
```

	MD	95%-CI	%W(fixed)	%W(random)
1	2	[-5.5231; 9.5231]	11.01	15.46
2	6	[-1.3937; 13.3937]	11.40	15.78
3	10	[-1.8412; 21.8412]	4.45	8.14
4	12	[8.4666; 15.5334]	49.93	29.36
5	10	[1.3283; 18.6717]	8.29	12.88
6	4	[-2.4656; 10.4656]	14.91	18.37

Number of studies combined: k=6

	MD	95%-CI	z	p.value
Fixed effect model	8.7666	[6.2698; 11.2634]	6.8817	< 0.0001
Random effects model	7.6172	[3.8233; 11.4110]	3.9351	< 0.0001

Quantifying heterogeneity:
tau^2 = 9.5103; H = 1.35 [1; 2.14]; I^2 = 44.9% [0%; 78.2%]

```
Test of heterogeneity:
   Q d.f.  p.value
9.07    5   0.1063

Details on meta-analytical method:
- Inverse variance method
- DerSimonian-Laird estimator for tau^2
```

There was no significant heterogeneity across the treatments ($p = 0.1063$). Here is its forest plot:

```
forest(metacont(nT,meanT,sdT,nC,meanC,sdC))
```

Study	Experimental Total	Mean	SD	Control Total	Mean	SD	Mean difference	MD	95%-CI	W(fixed)	W(random)
1	60	94	22	60	92	20		2.00	[−5.52; 9.52]	11.0%	15.5%
2	65	98	21	65	92	22		6.00	[−1.39; 13.39]	11.4%	15.8%
3	40	98	28	40	88	26		10.00	[−1.84; 21.84]	4.4%	8.1%
4	200	94	19	200	82	17		12.00	[−8.47; 15.53]	49.9%	29.4%
5	50	98	21	45	88	22		10.00	[1.33; 18.67]	8.3%	12.9%
6	85	96	21	85	92	22		4.00	[−2.47; 10.47]	14.9%	18.4%
Fixed effect model	**500**			**495**				8.77	[6.27; 11.26]	100%	--
Random effects model								7.62	[3.82; 11.41]	--	100%

Heterogeneity: I-squared=44.9%, tau-squared=9.51, p=0.1063

```
           −20  −10    0    10   20
```

This output shows both the fixed-effect and the random-effect analysis of these data: in this case, the interpretation is the same under both models, but this need not be so. Note that the mean effect size is smaller and the significance lower under the random-effects model.

21.3.2   Random effects with a scaled mean difference

We repeat the analysis using a random effects model, using the formulae in Table 21.2. We start by computing the components Q, d.f. and C, from which we shall calculate the between study variance, tau-squared.

```
WY2 <- W*g^2
W2 <- W^2
Q <- sum(WY2)-(sum(WY)^2/sum(W))
Q

[1] 12.00325

df <- 6-1
df

[1] 5

C <- sum(W)-(sum(W2)/sum(W))
C

[1] 187.6978

T2 <- (Q-df)/C
T2

[1] 0.03731131
```

Table 21.2. Calculations for a random effects meta-analysis. The individual effects (Y) have variances (V) reflecting sampling error alone as in Table 21.1. The total variance $V*$ now has two components: the within study variance (V) and the between study variance T^2 (tau squared). The method of moments (DerSimonian and Laird method) calculates T^2 from three quantities: Q (a corrected weighted sum of squares of the individual effects), C (a corrected sum of weights) and d.f. (degrees of freedom, one less than the number of studies k), where W and Y are as defined in Table 21.1. If all the studies had the same sample variance σ^2 and were based on the same sample size n, then the standard error of the summary effect would be $SEM^* = \sqrt{\frac{\sigma^2}{kn} + \frac{\tau^2}{k}}$. Nomenclature follows Borenstein et al. (2009).

Term	Formula
Q	$Q = \sum WY^2 - \dfrac{\left(\sum WY\right)^2}{\sum W}$
C	$C = \sum W - \dfrac{\sum W^2}{\sum W}$
Degrees of freedom	$\text{d.f.} = k - 1$
Between-study variance (tau squared)	$T^2 = \dfrac{Q - \text{d.f.}}{C}$
Total variance	$V^* = V + T^2$
Weight	$W^* = \dfrac{1}{V^*}$
Weighted mean summary effect	$M^* = \dfrac{\sum W^* Y}{\sum W^*}$
Variance of the summary effect	$VM^* = \dfrac{1}{\sum W^*}$
Standard error of the summary effect	$SEM^* = \sqrt{VM^*}$

So the DerSimonian and Laird estimate of between-study variance, tau-squared, is 0.0373. For the random-effects estimate of the summary effect, we need to allow for shrinkage (see p. 685). The weights are the reciprocals of the total variance (*within* Vg plus *between* T2):

```
Wstar   <- 1/(Vg+T2)
Wstar
```

```
[1]  14.23313 14.70238 11.46907 20.90940 12.50377 16.46588
```

```
Mstar <- sum(Wstar*g)/sum(Wstar)
Mstar
```

```
[1]  0.3582294
```

```
Vstar <- 1/sum(Wstar)
Vstar
```

```
[1]  0.01107621
```

```
SEMstar <- sqrt(Vstar)
SEMstar
```

```
[1]  0.1052436
```

```
LLstar <- Mstar-1.96*SEMstar
ULstar <- Mstar+1.96*SEMstar
LLstar
```

```
[1]  0.1519521
```

```
ULstar
```

```
[1]  0.5645068
```

```
Zstar <- Mstar/SEMstar
Zstar
```

```
[1]  3.403813
```

Because of shrinkage, the random effects estimate of the mean effect (`Mstar` = 0.358) is smaller than the fixed effects estimate (M = 0.414), as illustrated in the forest plot from `metacont` (above).

21.4   Random-effects meta-analysis of binary data

Next we demonstrate a random effects meta-analysis on count data. Instead of a difference between means, the effect in this analysis is the log-odds ratio, based on proportion data from six different studies (A–F) each with two treatments (control, C, and treated, T). The count data are the number of deaths (success) and number of survivors (failure; the data come from Borenstein et al., 2009):

```
data <- read.table("c:\\temp\\metadata2.txt",header=T)
attach(data)
data
```

```
  study successT failureT  nT successC failureC  nC
1     A       12       53  65       16       49  65
2     B        8       32  40       10       30  40
3     C       14       66  80       19       61  80
4     D       25      375 400       80      320 400
5     E        8       32  40       11       29  40
6     F       16       49  65       18       47  65
```

We should get a quick overview of the effects before starting the calculations. In the treatment group, the total number of successes was 83 and the total number of failures was 607, so the odds of success were

```
sum(successT)/sum(failureT)
```

```
[1]  0.1367381
```

In the control group, the total number of successes was 154 and the total number of failures was 536, so the odds of success were

```
sum(successC)/sum(failureC)
```

```
[1]  0.2873134
```

Now a rough and ready estimate of the overall effect size is given by the odds ratio, which is the treatment odds divided by control odds:

```
(sum(successT)/sum(failureT))/(sum(successC)/sum(failureC))
```

```
[1] 0.4759195
```

indicating that treatment has reduced the effect by roughly 50% (47.6%) taking the six studies together. The meta-analysis proceeds as follows, using the formulae in Table 21.2.

First, use the raw count data to compute a vector of odds ratios (or), one for each study (the number of successes divided by number of failures for the treated group, divided by the same quantity for the control group):

```
(or <- successT*failureC/(failureT*successC))
```

```
[1] 0.6933962 0.7500000 0.6810207 0.2666667 0.6590909 0.8526077
```

The response variable in this kind of meta-analysis is the logarithm of the odds ratio (lor)

```
(lor <- log(or))
```

```
[1] -0.3661537 -0.2876821 -0.3841625 -1.3217558 -0.4168938 -0.1594557
```

The variance of the log-odds ratio for each of the six studies is the sum of the reciprocals of the four counts of success and failure:

```
(vlor <- 1/successT + 1/failureT + 1/successC + 1/failureC)
```

```
[1] 0.18510942 0.28958333 0.15560511 0.05829167 0.28164185 0.15974031
```

As usual in meta-analysis, the weight given to each study is the reciprocal of its variance:

```
(W <- 1/vlor)
```

```
[1]  5.402210  3.453237  6.426524 17.155111  3.550609  6.260160
```

Note the very high weight given to study 4 (17.16) because of its large sample size.

In order to carry out the meta-analysis using the random-effects model, we need to calculate a set of products and sums of products (Borenstein et al., 2009): WY is the vector of products of the weights times the effect sizes, and WY2 is the vector of products of the weights times the squares of the effect sizes:

```
WY <- W*lor
WY2 <- W*lor^2
```

The fundamental difference with the random-effects model is that we compute a variance in the true effect (traditionally denoted by tau squared, τ^2) as well as the separate, per-study variances. For this we use the DerSimonian and Laird (1986) method $\tau^2 = (Q - \text{d.f.})/C$, where the quantities Q, d.f. and C are computed as follows:

```
Q <- sum(WY2)-(sum(WY)^2/sum(W))
df <- length(nC)-1
C <- sum(W)-(sum(W^2)/sum(W))
T2 <- (Q-df)/C
```

Here is a summary of the results so far:

```
(res <- data.frame(sum(WY),sum(WY2),Q,df,C,T2))

    sum.WY.  sum.WY2.        Q df        C        T2
1 -30.59362   32.7054 10.55115  5 32.10525 0.1729048
```

The subsequent random-effects analysis is based on the assumption that the variance for each study is the sum of the within-study variance (`vlor`) and the between-study variance (`T2`). The new weights to be given to each study (`Wstar`) are therefore calculated like this:

```
(Wstar <- 1/(T2+vlor))

[1] 2.793185 2.162218 3.044048 4.325325 2.199994 3.006207
```

Note that although D is still the most highly weighted of the studies, the relative size of its weight is substantially reduced in the random effects model (down from 17.16 to 4.33). To calculate the summary effect size we need the sum of the weights and the sum of the products of the weights times the effect sizes:

```
(Mstar <- sum(Wstar*lor)/sum(Wstar))

[1] -0.5662959
```

This is the summary of the log-odds ratio across the six studies. We back-transform to the odds ratio by taking the antilog of `Mstar`:

```
exp(Mstar)

[1] 0.5676241
```

Note that the summary effect is considerably larger than the rough preliminary calculation we made earlier (0.4759, above). The unreliability estimates associated with the summary effect are calculated like this (V = variance, `SE` = standard error, `ll` = lower limit, `ul` = upper limit):

```
VMstar <- 1/sum(Wstar)
SEMstar <- sqrt(VMstar)
llMstar <- Mstar-1.96*SEMstar
ulMstar <- Mstar+1.96*SEMstar
Z <- Mstar/SEMstar
p <- 2*(1-pnorm(-Z))
```

Here is a summary of the random-effects model output:

```
(res2 <- data.frame(Mstar,SEMstar,llMstar,ulMstar,Z,p))

      Mstar   SEMstar   llMstar     ulMstar         Z          p
1 -0.5662959 0.2388344 -1.034411 -0.09818041 -2.371081 0.01773612
```

The forest plot looks like this:

```
ll<-lor-1.96*vlor
ul<-lor+1.96*vlor

plot(c(-2,0.3),c(0,8),type="n",xlab="",yaxt="n",ylab="")
points(lor,8-(1:6),pch=15,cex=10*Wstar/sum(Wstar))
for (i in 1:7) lines(c(ll[i],ul[i]),c(8-i,8-i))
polygon(c(llMstar,Mstar,ulMstar,Mstar),c(1,1.2,1,0.8),col="red")
```

```
abline(v=0,lty=2)
text(rep(-1.8,7),1:7,c("summary effect","F","E","D","C","B","A"))
```

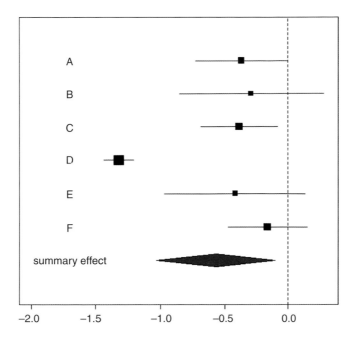

The summary effect is highly significant (note the log-odds scale), with zero effect marked as the vertical dashed line at 0. As before, note that vote counting would have led to the wrong conclusion (only two significant studies out of six).

22

Bayesian Statistics

Instead of asking 'what do my data show?', the Bayesian analyst asks 'how do my data alter our view of the world?'. It may not sound like much, but it is a fundamental change of outlook. The idea is that the results of the new study are assessed in the light of the existing data, to establish an updated assessment of parameter values and their uncertainties. There are now two models rather than one. There is a model for what we know already: this is called the **prior**. Then there is a model that we fit to our data: this is the **likelihood**. The two models combine to give us an estimate of the **posterior**. We use the posterior distribution to make statistical inferences. Under its Bayesian interpretation, probability measures our confidence that something is true.

When the new study is small, and existing knowledge is extensive, then we should not expect our work to make much difference to overall understanding. For instance, if we studied 150 subjects and found that age at death was not affected by their smoking habits, we would be ill advised to conclude that this result had great generality. This is because we know from studies of many thousands of subjects over many decades that smoking leads to a reduction of about 14 years in mean age at death. Small studies can be highly informative, of course, especially when they are original and address important questions. But the larger and better designed the new study, the more that study should be capable of altering our prior knowledge, and if the study is big enough and sufficiently well conducted, then it could turn prior expectation on its head. A Bayesian analysis works in three steps:

- using all the currently available information to create a full probability model;

- forming a posterior inference by conditioning the full probability model on the new data;

- criticizing the fit of the model to the data and evaluating the predictions it makes.

You should read *Bayesian Data Analysis* by Gelman et al. (2004) for background, examples and computational methods on each of these steps.

One of the strengths of the Bayesian approach is that it does not rely on two of the most peculiar aspects of the frequentist approach. Beginners often think that:

- if I reject the null hypothesis at the 5% level, then there is a 5% chance that the null hypothesis is true;

- if I establish a 95% confidence interval for the value of a parameter, then the value of the parameter lies within these bounds with probability 0.95.

Of course, both these assertions are wrong, but they are wrong in ways that are genuinely hard for beginners to understand. In the first case, the p value (say 0.05) means that a *test statistic* as large, or larger, than the one we observed, is expected to occur by chance alone with probability 0.05 when the null hypothesis is true. True, it is a measure of the plausibility of the null hypothesis, but it is *not* the probability that it is true, because we calculated the test statistic on the basis that the null hypothesis really *was* true. In the second case, we are asked to do an imaginary experiment in which we repeat our data collection exercise many times. The confidence interval indicates the distribution of these hypothetical repeated estimates of the parameter. It does *not* mean that the probability that our estimated parameter value lies outside these bounds is 0.05. But this is precisely what a Bayesian credible interval means.

Bayesians talk naturally about the probability of hypotheses:

- the unconditional probability of an hypothesis (its *prior probability*);

- the conditional probability, given some new evidence (its *posterior probability*).

The advantages of Bayesian analysis include the following:

- All sources of variation can be modelled (fixed effects, random effects, measurement error, etc.).

- All unknowns are treated as random variables.

- Confidence intervals are intuitive. The highest posterior density (HPD, also known as the credible interval) refers to the observed data on which the distribution of the parameter is conditioned – this is a much more intuitive concept than the conventional 95% confidence interval, which is based on some hypothetical replications of the sample.

- Explanatory variables *known to be of substantive importance* are included in the model even if they appear not to be statistically significant (e.g. density dependence in population models).

- Models can be as complex as necessary to describe the question in hand.

- Interpretation of complex models is much more straightforward.

- The uncertainly associated with missing values (so-called multiple imputation) is dealt with more satisfactorily.

- Multiple comparisons are dealt with intuitively, including comparisons of structurally different models.

- The very term 'hypothesis testing' dictates that it should be the hypothesis that is tested, given the data, not the other way around.

However, Bayesian analysis has a number of disadvantages:

- You need to know much more maths and statistics to do it competently.

- The solution to realistically complicated models can only be achieved through iterative simulation, drawing random samples from the posterior distribution.

- There are many ways of carrying out these simulations and it is not obvious in advance which method will be most effective.

- The choice of priors can be both controversial and consequential, especially with small sample sizes.

- Simulation is not exact, and it may not be obvious that you have the best possible model for the posterior.

To illustrate what I mean by needing to be better at statistics, consider these simple questions. What, precisely does 'knowing nothing' mean in terms of choosing between different priors? Should we assume that different parameters in the prior are independent of one another? What number of new observations do you reckon your prior information is worth ? How do I ensure that priors are conjugate? I think you will see what I mean.

Analytical solutions to simple Bayesian problems are relatively straightforward, so long as we stick to well-known distributions like the normal or the Poisson and use conjugate priors. But for realistically complicated models, it is essential to resort to numerical simulation to work out the parameter values involved in quantifying the posterior distribution.

It is not so much the priors that represent the principal motivation for using Bayesian models these days, but rather the potential to include all of the important sources of variation in a straightforward way. This has revolutionized statistical modelling in many fields. Estimating parameter values and their credible intervals by numerical simulation from the posterior distribution represents a quantum step in dealing with the most complicated models, including models with a high ratio of parameters to data points, and hierarchical models with complex spatial or temporal structure.

22.1   Background

We need to revise some simple probability theory. Let us define E as an event, and H as an hypothesis (a model) about the likelihood of E. The probability of an event E, given the hypothesis H, is written as $P(E|H)$, which is read 'the probability of E given H'. Some straightforward rules apply:

- $P(E|H) \geq 0$ for all E and all H;
- $P(H|H) = 1$ for all H;
- $P(E|FH) \times P(F|H) = P(EF|H)$.

The first just says that you cannot have negative probabilities. The second just says that the hypothesis needs to encompass all possible outcomes. The third is the subtle one. It says that given *independent* events E and F, then the probability of getting E and F given the hypothesis (the right-hand side), is the product of the probability of F given the hypothesis, times the probability of E given both F and H. Take the example of twins. They can derive from splitting of a single fertilized egg (monozygotic twins), or from two separate fertilizations of different eggs (dizygotic twins). Not all monozygotic twins look 'identical', and some dizygotic individuals can look very alike. This means that you cannot reliably tell monozygotic from dizygotic twins just by looking at them. But there is a reliable way of estimating the *proportions* of monozygotic and dizygotic twins in the *population*. The key fact is that while dizygotic twins can be 2 boys or 2 girls or 1 boy and 1 girl, monozygotic twins are *always* of the same sex. So if B is the event of a twin being a boy, and G is the event of a twin being a girl, then there are three outcomes for any batch of twins: BB, GG or BG. Let us define the second event as either monozygotic (M) or dizygotic (D). The probabilities for the dizygotic twins are easy:

$$P(GG|D) = \tfrac{1}{4}, \quad P(BB|D) = \tfrac{1}{4}, \quad P(BG|D) = \tfrac{1}{2}.$$

The probabilities for monozygotic twins include the fact that P(BG|M) is impossible:

$$P(BB|M) = \tfrac{1}{2}, \quad P(GG|M) = \tfrac{1}{2}, \quad P(BG|M) = 0.$$

We want to know $P(M)$. Using what we have learned about conditional probability, we can write down the probability of getting two girls if we know $P(M)$ and $P(D)$.

$$P(GG) = P(GG|M) \times P(M) + P(GG|D) \times P(D).$$

Half of all monozygotic twins will be GG but only one quarter of dizygotic twins will be GG:

$$P(GG) = \tfrac{1}{2}P(M) + \tfrac{1}{4}(1 - P(M)),$$

from which we can extract $P(M)$ by multiplying through the bracket by $\tfrac{1}{4}$, subtracting $\tfrac{1}{4}P(M)$ from $\tfrac{1}{2}P(M)$, then subtracting $\tfrac{1}{4}$ from both sides to leave

$$P(GG) - \tfrac{1}{4} = \tfrac{1}{4}P(M),$$

so the answer is $P(M) = 4P(GG) - 1$. In a town where 45% of twins are girl–girl, the proportion of twins that are monozygotic is $100 \times ((4 \times 0.45) - 1) = 80\%$.

22.2   A continuous response variable

Suppose that our data are X and our model contains the parameters θ. We might ask what is the probability of observing our data given that the model is true? This is called the **frequentist** approach to maximum likelihood (see p. 390):

$$p(X|\theta).$$

Alternatively, because we are usually a lot more certain about our data than we are about the truth of our model, we might ask what is the likelihood of our model, given the data:

$$l(\theta|X).$$

These two quantities are different ways of expressing the same idea, but they embody a fundamentally different approach. You need to think about this last paragraph until the penny drops.

The fundamental part of Bayesian statistics is that the posterior distribution $p(\theta|X)$ is proportional to the product of the prior and the likelihood:

$$p(\theta|X) \propto p(\theta) \times l(\theta|X).$$

This tells us how to modify our existing beliefs in the light of the newly available data. Note the proportionality: we can multiply the likelihood by any constant without affecting the posterior, but we need our posterior probability distribution to integrate to 1.

22.3   Normal prior and normal likelihood

The simplest Bayesian calculations are in cases where both the prior and the likelihood are normal. Assume that we have a single unknown parameter θ for which our prior is expressed in terms of a normal distribution for which we know both the mean (θ_0) and the variance (φ_0). We have collected new data X with sample mean μ and known variance φ_1.

The traditional way to measure precision is by the inverse of the variance (i.e. posterior precision = prior precision + data precision). We start by getting the overall estimate of the posterior variance: this is the reciprocal of the sum of the reciprocals of the prior variance and the sample variance:

$$v_{\text{posterior}} = \frac{1}{v_{\text{prior}}^{-1} + v_{\text{data}}^{-1}}.$$

The posterior mean is the weighted mean of the prior and the data, using the inverse variances as the weights:

$$m_{\text{posterior}} = v_{\text{posterior}} \left(\frac{m_{\text{prior}}}{v_{\text{prior}}} + \frac{m_{\text{data}}}{v_{\text{data}}} \right).$$

Now, of course, there are many assumptions here. The most important is that the two variances and the prior mean are known without error, and that the unknown parameter is normally distributed. These assumptions are all relaxed in more advanced Bayesian models.

Let us take a simple example, just to illustrate the concept of shrinkage. Suppose that Smith was the first to estimate the productivity of this particular kind of grassland, and she found a mean of 500 units per unit area with a standard deviation of 25. Our new data give a mean of 420 with a standard deviation of 10. If we use Smith's findings as our prior, we get a posterior variance, v, of roughly 86 like this:

```
(v <- 1/(1/25^2+1/10^2))
```

```
[1] 86.2069
```

The posterior mean is given by

```
(v*(500/25^2 + 420/10^2))
```

```
[1] 431.0345
```

So the **prior** was $N(500, 25^2)$, the **likelihood** from the new data was $N(420, 10^2)$ and the **posterior** was $N(431, 9^2)$. The new mean is closer to the likelihood than to the prior, and the standard deviation is substantially lower.

Jones, working a decade later than Smith, found a mean of 400 with a standard deviation of 20. How much difference would it have made if we had taken Jones rather than Smith for our estimate of the prior? Now the posterior variance would be

```
(v2 <- 1/(1/20^2+1/10^2))
```

```
[1] 80
```

and the new posterior mean would be

```
(v2*(400/20^2 + 420/10^2))
```

```
[1] 416
```

The posterior variance is roughly the same as before, but the posterior mean is slightly lower: $N(416, 9^2)$ compared to $N(431, 9^2)$. So our choice of priors does matter, and the smaller our new sample, the more that choice matters. With three studies, we might have done a meta-analysis (see Chapter 21).

22.4 Priors

Your prior beliefs can be specified as a point estimate (a single value) or as a distribution (many values, each with a different probability density). My prior for tossing this fair coin is a point estimate: the probability of heads is $\theta = \frac{1}{2}$. My prior for the yield of a new crop is 10% higher than the yield of the conventional crop ($\mu = 1.1$), but with a similar variance ($\sigma^2 = k$): this could be expressed as a normal prior $\theta \sim N(\mu, \sigma^2)$.

22.4.1 Conjugate priors

The likelihood function is $l(\theta|x)$ and the prior is $p(\theta)$. If the posterior density

$$p(\theta|x) \propto p(\theta) \times l(\theta|x).$$

is in the same *family of distributions* as the prior, then the class of prior distributions is said to form a conjugate family. The prior is called a **conjugate prior** for the likelihood. In most cases, this makes the maths much more straightforward, since the posterior has the same algebraic form as the prior. A classic example comes from the case where the response variable follows a binomial distribution with a single parameter, π, the probability of success in a single trial. The choice of a prior for π is best if $\pi \sim B(\alpha, \beta)$, where B is a beta distribution, because the family of beta distributions is conjugate to a binomial likelihood. Roughly speaking, you can choose α and β so that $\alpha + \beta$ represents the number of observations that you reckon your prior information is worth, while the mean of the beta distribution is equal to $\alpha/(\alpha + \beta)$ (Lee, 2012). The conjugate prior distribution for a response variable that has a Poisson likelihood (e.g. count data) is the gamma distribution. And so on. All the likelihood functions belonging to the exponential family have conjugate priors that are also in the exponential family (see Chapter 27).

Conjugate priors are useful if you are aiming for an analytical solution, but not essential if you intend to investigate your posterior distribution numerically. It is impossibly slow to calculate the probability associated with every combination of circumstances and every value of every parameter of the current model. We need a cleverer way of homing in on the right answer. It is in these cases that Markov chain Monte Carlo (MCMC) methods come into their own.

22.5 Bayesian statistics for realistically complicated models

Model choice is a very important part of Bayesian data analysis. Not only do we have to select a deterministic structure for the likelihood (which explanatory variables to include, and how to relate them to the response variable and to one another), but also specify a joint probability distribution for all the observable and unobservable quantities involved in the problem. This is known as the full probability model. Then, given the data, we need to compute the appropriate posterior distribution: this is the conditional probability distribution of all the unobserved quantities, given the observed data. Finally, we must subject the model to criticism: how good is the fit to the data, are the conclusions reasonable, and how sensitive is the interpretation to the assumptions that are embodied in the model choice? We cannot expect to be able to do these things analytically. The most popular methods for solving these issues numerically involve simulations known collectively as **Markov chain Monte Carlo** methods.

The Monte Carlo part of the name refers to random draws (the gambling part). The Markov chain part of the name refers to a series of events where what happens next depends only on current status (and not on any memory of historical status). The idea is step-by-step to better approximate the target posterior distribution (hill climbing is a useful analogy). Samples are drawn sequentially, with the distribution of the sample draws depending only on the last sample drawn, such that the approximate distributions are improved at every step. With time, the simulation converges on the target posterior distribution (Gelman et al., 2004).

The Metropolis algorithm is a kind of random walk that uses an acceptance/rejection rule to converge on a specified target distribution. It uses what is called a jumping distribution that either stays where it is or moves to a new place, with the probability of movement to a new place depending on the ratio of probability densities between the new place and the existing place. It always accepts steps that increase the density, but occasionally accepts downward steps. This ensures that the solution does not get stuck on small, local maxima that are some way distant from the global maximum. The Metropolis–Hastings algorithm differs in two ways: the jumping rules no longer need to be symmetric, and ratio of densities is modified to account for this asymmetry. The Gibbs sampling algorithm generates an instance from the distribution of each variable in turn, conditional on the current values of the other variables.

22.6　Practical considerations

MCMC modelling involves two important practical considerations: the burn-in period and thinning. By throwing away the early values from the simulation, the effects of the initial conditions will have died away, leaving a better estimate of the posterior distribution. The burn-in period (from which the results are discarded) is often set to half of the chain.

Because the hill-climbing process is based on a Markov chain, successive values of the parameters show strong serial correlations, so successive values typically give little extra information about the shape of the posterior distribution. You might choose to take one point in every hundred or so, to reduce the serial correlation. This is the thinning rate.

22.7　Writing BUGS models

BUGS stands for 'Bayesian inference Using Gibbs Sampling' (Lunn et al., 2009). You can read about the history of BUGS at the OpenBUGS website http://www.openbugs.info/w/.

The trick is to learn how to express your particular model in BUGS code. The code looks superficially like R, but it is fundamentally different. You do *not* type the code into R, but you write it in an editor, and save it as an ASCII file outside of R. The name of the file containing your BUGS model is provided as an argument to the MCMC function inside R. You can find lots of clear examples of the way that different kinds of models are expressed in BUGS code on the website for WinBUGS (Spiegelhalter et al., 2003) at http://www.mrc-bsu.cam.ac.uk/bugs/winbugs under the headings 'Volume 1', 'Volume 2' and 'Volume 3' at the bottom of 'Introduction to WinBUGS' on the home page. You should spend time browsing through these examples to find the one closest to the problem you are trying to solve, then edit the code to tailor it to your specific requirements. Three examples are described in detail below (a simple regression, a study with temporal pseudoreplication, and an experiment involving proportion data with overdispersion).

22.8　Packages in R for carrying out Bayesian analysis

There is a huge amount of information, and a great many computing resources for Bayesian analysis available on the CRAN website. This is summarized in the Bayesian Inference Task View written by Jong Hee Park (University of Chicago, USA), Andrew D. Martin (Washington University, St. Louis, MO, USA), and Kevin M. Quinn (University of California, Berkeley, USA); see http://cran.r-project.org/web/views/Bayesian.html. The Task View subdivides the packages under five headings:

- Bayesian packages for general model fitting;

- Bayesian packages for specific models or methods;

- post-estimation tools (like coda);

- packages for learning Bayesian statistics;

- packages that link R to other sampling engines (like `R2jags`, `R2WinBUGS`, `R2OpenBUGS`).

Applied researchers interested in Bayesian statistics are increasingly attracted to R because of the ease of which one can code algorithms to sample from posterior distributions as well as the significant number of packages contributed to CRAN. In particular, there are several choices for MCMC sampling. For many years the most popular of these was WinBUGS (Spiegelhalter et al., 2003), and this can still be run from R using the function R2WinBUGS. This is not used here because WinBUGS does not run on a Mac, and the software is no longer being developed. The final manifestation of WinBUGS, frozen at version 1.4.3, is still perfectly functional, but it has been replaced as an evolving framework by OpenBUGS. This means that the main choice is between OpenBUGS and JAGS. I have chosen to illustrate this chapter using JAGS because people who use it a lot speak very highly of it, and because, unlike WinBUGS, it runs on Macs as well as PC and Linux. It may well be that OpenBUGS will become the standard in future, and if it does, it will be simple to switch from JAGS.

22.9   Installing JAGS on your computer

JAGS stands for 'Just Another Gibbs Sampler' (Plummer, 2012). It is a program for analysis of Bayesian hierarchical models using MCMC simulation. It is very like BUGS in spirit and language, and was written with three aims in mind:

- to have a cross-platform engine for the BUGS language;

- to be extensible, allowing users to write their own functions, distributions and samplers;

- to be a platform for experimentation with ideas in Bayesian modelling.

JAGS is licensed under the GNU General Public License. First, you need to install JAGS on your machine. You do this by visiting http://mcmc-jags.sourceforge.net/. Click on 'files page' under Downloads, then after "Looking for the latest version?" click on Download JAGS. Then run the program and chose all the default options that are offered.

22.10   Running JAGS in R

The next thing you have to do is install the R2jags package that allows R to communicate with JAGS and vice versa. Inside R, while running R as administrator, install the package in the usual way:

```
install.packages("R2jags")
```

Now you are ready to start Bayesian modelling. The first thing to appreciate is that you do most of the hard work *outside* R. You have to use the BUGS language to write down your model of the likelihood and the priors, incorporating all of the important details about hierarchical structure, pseudoreplication, and so forth. Having sketched out the model, you write it in a text editor and save it as an ASCII file (*.txt file). Only now do you go into R to start the modelling. This is the sequence of events:

- Use read.table to enter your data into a dataframe in the familiar way.

- attach the dataframe and make a list of the variable names that need to be passed into the BUGS code.

- Work out the initial conditions (if any) that you want to specify.

- Load the jags package using library(jags).

- Run the JAGS model by specifying the name of the list of variables, the initial conditions (optional), the path and name of the file where the BUGS code is to be found, and the number of Markov chains that you want to run (a popular choice is three).

With any luck the JAGS model will run, and its progress is indicated by a slowly moving horizontal bar. Once the model has finished, you can inspect the parameter estimates and their uncertainty measures, and create various plots.

22.11   MCMC for a simple linear regression

In our example (analysed in detail as a linear model on p. 450) the response variable is growth and the continuous explanatory variable is the concentration of tannin in the diet. We start by reading the data into R:

```
data2 <- read.table("d:\\temp\\regression.txt",header=T)
attach(data2)
head(data2)
```

```
  growth tannin
1     12      0
2     10      1
3      8      2
4     11      3
5      6      4
6      7      5
```

Here is a reminder of the output of the simple linear regression for comparison with the JAGS output in due course:

```
summary(lm(growth~tannin))
```

```
             Estimate Std. Error t value Pr(>|t|)
(Intercept)   11.7556     1.0408  11.295 9.54e-06 ***
tannin        -1.2167     0.2186  -5.565 0.000846 ***
```

The intercept is 11.7556 (± 1.0408) and the slope is -1.2167 (± 0.2186).

Outside R, write the BUGS model and save it in a text file. This is the part that you will find difficult at first. The model contains the information on the structure of the model relating the response variable to the explanatory variables (growth = a + b * tannin), the nature of the priors (normal), and the assumed distributions (again, normal). You need to specify the number of rows in the dataframe ($N = 9$), the formula for the likelihood, and the probability density functions from which the priors and the likelihood are to be selected. This is what the contents of the ASCII file look like:

```
model {
for (i in 1:N) {
growth[i] ~ dnorm(mu[i], tau)
mu[i] <- a + b * tannin[i]
}
a ~ dnorm(0.0, 1.0E-4)
b ~ dnorm(0.0, 1.0E-4)
sigma <- 1.0/sqrt(tau)
tau ~ dgamma(1.0E-3, 1.0E-3)
}
```

The model contains a mixture of deterministic and stochastic elements. The deterministic components are indicated by 'gets' `<-` symbols: `mu[i] <- a + b * tannin[i]` and `sigma <- 1.0/sqrt(tau)`. The stochastic components are indicated by a tilde (~): the likelihood `growth[i] ~ dnorm(mu[i], tau)`, the prior for the intercept `a ~ dnorm(0.0, 1.0E-4)`, the uninformative prior for the slope `b ~ dnorm(0.0, 1.0E-4)`, while tau is sampled from a gamma distribution `tau ~ dgamma(1.0E-3, 1.0E-3)`. The code is saved to a file called `c:\\temp\\regression.bugs.txt` in this example. Now go back into R.

We need to open the library to connect our R session with the JAGS program:

```
library(R2jags)
```

You need to tell the `jags` function several important things:

- the names of the variables containing the data to which the model will be fitted;
- initial values for the parameters (you can omit this, and let JAGS figure them out);
- the names of the parameters to save for interpretation and inference;
- the name of the path and the ASCII file where the code for the model is to be found;
- the number of chains you want to simulate;
- the number of iterations per chain (by default, the burn-in is half this number).

Tell `jags` the names of the variables containing the data:

```
data.jags <- list("growth", "tannin", "N")
```

Finally, run the `jags` function to produce the model:

```
model <- jags(data=data.jags, parameters.to.save=c("a", "b", "tau"),
  n.iter=100000, model.file="C:/temp/regression.bugs.txt", n.chains=3)

Compiling model graph
   Resolving undeclared variables
   Allocating nodes
   Graph Size: 46
```

We inspect the model output like this:

```
model
```

```
Inference for Bugs model at "C:/temp/regression.bugs.txt", fit using jags,
 3 chains, each with 1e+05 iterations (first 50000 discarded), n.thin = 50
 n.sims = 3000 iterations saved
```

	mu.vect	sd.vect	2.5%	25%	50%	75%	97.5%	Rhat	n.eff
a	11.799	1.268	9.305	11.051	11.808	12.518	14.377	1.002	2000
b	-1.230	0.263	-1.781	-1.384	-1.226	-1.067	-0.712	1.002	2000
tau	0.351	0.195	0.080	0.211	0.313	0.451	0.826	1.001	3000
deviance	36.508	3.152	33.054	34.217	35.682	37.923	44.825	1.002	2100

```
For each parameter, n.eff is a crude measure of effective sample size,
and Rhat is the potential scale reduction factor (at convergence, Rhat=1).

DIC info (using the rule, pD = var(deviance)/2)
```

```
pD = 5.0 and DIC = 41.5
DIC is an estimate of expected predictive error (lower deviance is better).
```

As you can see, the parameter estimates are very close to those obtained by the linear model (intercept = 11.799 ± 1.268 rather than 11.7556 ± 1.0408; slope $= -1.230 \pm 0.263$ rather than -1.2167 ± 0.2186). The unreliability estimates are slightly greater than in the linear model, and the deviance is substantially greater (36.51 rather than 20.07), but the interpretation is unaffected. A plot of the $\texttt{jags}$ model object produces strip diagrams with credible interval bars for the parameters, the deviance and tau, with the results from the different chains in different colours:

```
plot(model)
```

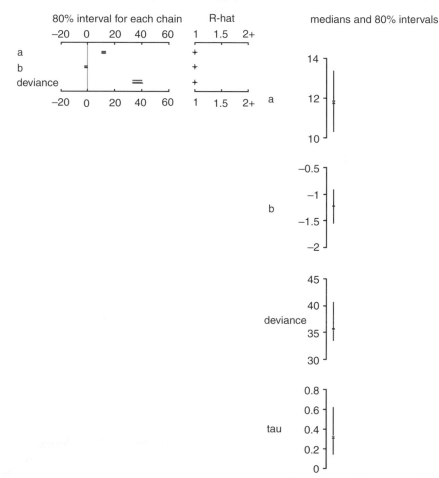

Bugs model at "C:/temp/regression.bugs.txt", fit using jags, 3 chains, each with 1e+05 iterations (first 50000 discarded)

You can obtain some attractive graphical output using functions from the $\texttt{coda}$ package:

```
model.mcmc <- as.mcmc(model)
densityplot(model.mcmc)
```

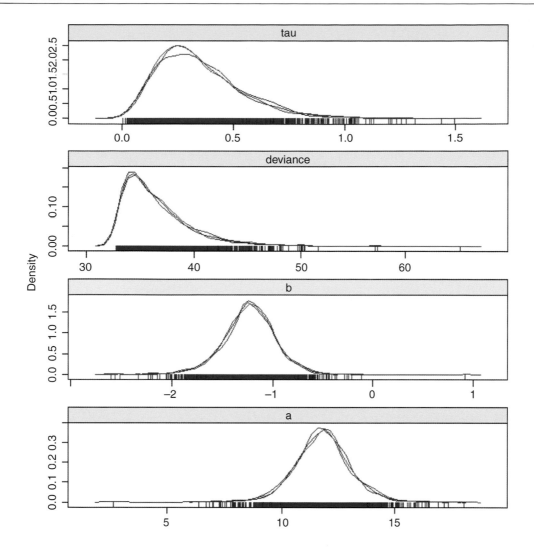

This shows each of the three chains in a different colour, and gives a clear visual impression of the unreliability of the estimates of the two model parameters (intercept *a* and slope *b*) and tau (the reciprocal of the error variance). Note the strong positive skew in the distribution of deviance.

22.12 MCMC for a model with temporal pseudoreplication

The root growth of N individuals was measured over T time periods. The temporal pseudoreplication is built into the BUGS model in the form of nested loops (time j within individual i). Mean size (mu) is assumed to be a linear function of time (x) with intercept alpha and slope beta. The data are plotted on p. 695, where they are analysed with a linear mixed effects model:

```
data <- read.table("d:\\temp\\fertilizer.txt",header=T)
attach(data)
head(data)
```

```
   root week plant fertilizer
1   1.3    2   ID1      added
2   3.5    4   ID1      added
3   7.0    6   ID1      added
4   8.1    8   ID1      added
5  10.0   10   ID1      added
6   2.0    2   ID2      added
```

Write the BUGS model and save it to an ASCII file called `c:\\temp\\bayes.lme.txt`:

```
model
{
for( i in 1 : N ) {
for( j in 1 : T ) {
Y[i , j] ~ dnorm(mu[i , j],tau.c)
mu[i , j] <- alpha[i] + beta[i] * (x[j])
}
alpha[i] ~ dnorm(alpha.c,alpha.tau)
beta[i] ~ dnorm(beta.c,beta.tau)
}
tau.c ~ dgamma(0.001,0.001)
sigma <- 1 / sqrt(tau.c)
alpha.c ~ dnorm(0.0,1.0E-6)
alpha.tau ~ dgamma(0.001,0.001)
beta.c ~ dnorm(0.0,1.0E-6)
beta.tau ~ dgamma(0.001,0.001)
alpha0 <- 0
}
```

The issues in this example concern the shape of the data. The response (root length) needs to be a matrix `Y` with the individuals as the rows (not a single vector as `root` is at present):

```
Y <- root
dim(Y) <- c(5,12)
Y <- t(Y)
```

The explanatory variable (`week`) needs to be a vector `x` of length $T = 5$ (not length 60 as at present):

```
x <- week[1:5]
```

We need to provide `jags` with the names of the variables containing the data:

```
data.jags <- list("Y", "x", "N","T")
```

Finally, we can run the `jags` model like this:

```
model <- jags(data=data.jags,
      parameters.to.save=c("alpha", "beta", "tau.c","alpha.c",
      "alpha.tau","beta.c","beta.tau"), n.iter=100000,
      model.file="C:/temp/bayes.lme.txt", n.chains=3)
```

This may take a few seconds to execute, but then we can investigate the posterior estimates of the parameters and their associated uncertainty measures:

```
model

Inference for Bugs model at "C:/temp/bayes.lme.txt", fit using jags,
 3 chains, each with 1e+05 iterations (first 50000 discarded), n.thin = 50
 n.sims = 3000 iterations saved
            mu.vect sd.vect    2.5%     25%     50%     75%    97.5%  Rhat n.eff
alpha[1]     -0.302   0.365  -0.942  -0.564  -0.329  -0.053    0.421 1.003   780
alpha[2]     -0.308   0.332  -0.907  -0.540  -0.321  -0.095    0.405 1.001  3000
alpha[3]     -0.396   0.326  -0.982  -0.626  -0.416  -0.186    0.271 1.001  2200
alpha[4]      0.038   0.452  -0.772  -0.309   0.052   0.377    0.873 1.003   780
alpha[5]     -0.392   0.316  -0.965  -0.611  -0.408  -0.188    0.283 1.001  3000
alpha[6]     -0.480   0.311  -1.075  -0.693  -0.495  -0.275    0.166 1.002  1100
alpha[7]     -0.651   0.311  -1.270  -0.850  -0.647  -0.455   -0.039 1.001  3000
alpha[8]     -0.723   0.317  -1.371  -0.927  -0.708  -0.513   -0.120 1.003   780
alpha[9]     -0.926   0.348  -1.626  -1.159  -0.911  -0.674   -0.318 1.001  3000
alpha[10]    -0.926   0.337  -1.593  -1.155  -0.911  -0.683   -0.307 1.001  3000
alpha[11]    -0.879   0.346  -1.578  -1.106  -0.868  -0.636   -0.242 1.001  3000
alpha[12]    -1.233   0.454  -2.103  -1.554  -1.238  -0.881   -0.417 1.001  3000
alpha.c      -0.598   0.212  -1.019  -0.734  -0.595  -0.461   -0.188 1.002  1100
alpha.tau    43.945 189.030   1.050   2.501   4.427  10.190  415.840 1.001  3000
beta[1]       1.038   0.058   0.928   0.997   1.039   1.080    1.151 1.001  2400
beta[2]       0.948   0.049   0.853   0.915   0.948   0.981    1.046 1.002  1800
beta[3]       0.978   0.049   0.884   0.944   0.978   1.011    1.075 1.002  1400
beta[4]       1.029   0.067   0.909   0.980   1.025   1.080    1.160 1.003   970
beta[5]       0.943   0.048   0.843   0.912   0.944   0.976    1.036 1.001  3000
beta[6]       0.980   0.048   0.884   0.947   0.980   1.013    1.073 1.003   850
beta[7]       0.913   0.048   0.818   0.882   0.913   0.945    1.008 1.001  3000
beta[8]       0.888   0.049   0.793   0.856   0.888   0.922    0.982 1.002  1800
beta[9]       0.884   0.054   0.781   0.847   0.885   0.921    0.986 1.001  2800
beta[10]      0.918   0.050   0.822   0.884   0.919   0.953    1.017 1.001  2900
beta[11]      0.878   0.053   0.776   0.841   0.878   0.915    0.982 1.001  3000
beta[12]      0.848   0.066   0.723   0.800   0.849   0.895    0.970 1.001  3000
beta.c        0.937   0.034   0.869   0.915   0.937   0.959    1.004 1.001  2300
beta.tau    270.178 345.016  40.808  98.051 165.446 307.572 1128.382 1.002  1200
tau.c         4.416   0.979   2.757   3.737   4.332   5.016    6.589 1.001  3000
deviance     82.675   7.429  69.787  77.350  82.132  87.391   97.923 1.000  3000

For each parameter, n.eff is a crude measure of effective sample size,
and Rhat is the potential scale reduction factor (at convergence, Rhat=1).

DIC info (using the rule, pD = var(deviance)/2)
pD = 27.6 and DIC = 110.3
DIC is an estimate of expected predictive error (lower deviance is better).
```

As you can see, there is convincing evidence of non-linearity here, because several of the individuals (7–12) have intercepts significantly less than zero. There is also significant variation in the slopes (from 1.038 down to 0.848, with standard deviation less than 0.067).

22.13 MCMC for a model with binomial errors

We analysed this model for the percentage germination of seeds from a factorial experiment involving two genotypes of *Orobanche* and two extracts, as a GLM with quasi-binomial errors on p. 636. Here are the data: the response, `count`, is the number germinating out of an initial `sample` of seeds (i.e. 10 germinated out of 39 seeds in the first case):

```
data <- read.table("c:\\temp\\germination.txt",header=T)
attach(data)
head(data)

  count sample Orobanche  extract
1    10     39      a75     bean
2    23     62      a75     bean
3    23     81      a75     bean
4    26     51      a75     bean
5    17     39      a75     bean
6     5      6      a75 cucumber
```

Write the BUGS model and save it in as ASCII file called `c:\\temp\\bayes.glm.txt`:

```
model
{
for( i in 1 : N ) {
r[i] ~ dbin(p[i],n[i])
b[i] ~ dnorm(0.0,tau)
logit(p[i]) <- alpha0 + alpha1 * x1[i] + alpha2 * x2[i] +
alpha12 * x1[i] * x2[i] + b[i]
}
alpha0 ~ dnorm(0.0,1.0E-6)
alpha1 ~ dnorm(0.0,1.0E-6)
alpha2 ~ dnorm(0.0,1.0E-6)
alpha12 ~ dnorm(0.0,1.0E-6)
tau ~ dgamma(0.001,0.001)
sigma <- 1 / sqrt(tau)
}
```

The deterministic part of the model shows the prediction of `logit(p)` as a function of the factorial combination of extract and genotype with two main effects (`x1` and `x2`) and one interaction term (the product `x1` by `x2`). The stochastic terms involve the number of germinating seedlings drawn from a binomial distribution (`dbin`) with parameters p and n, `r[i] ~ dbin(p[i],n[i])`, the residuals on the logit scale with normal errors `b[i] ~ dnorm(0.0,tau)`, with independent, non-informative priors for the intercept, the two main effects and the interaction term (`alpha0`, `alpha1`, `alpha2` and `alpha12`), with `tau` selected from a gamma distribution.

The data we need to provide to the model are:

```
N <- 21
n <- sample
r <- count
x1 <- Orobanche
```

```
x2 <- extract
data.jags <- list("r", "n", "x1", "x2", "N")
```

Now run the model:

```
model <- jags(data=data.jags,
parameters.to.save=c("alpha0", "alpha1","alpha2","alpha12","tau"),
n.iter=100000, model.file="C:/temp/bayes.glm.txt", n.chains=3)
```

To inspect the output, write:

```
model
```

```
Inference for Bugs model at "C:/temp/bayes.glm.txt", fit using jags,
 3 chains, each with 1e+05 iterations (first 50000 discarded), n.thin = 50
 n.sims = 3000 iterations saved
          mu.vect sd.vect   2.5%    25%     50%     75%    97.5%  Rhat n.eff
alpha0     -0.077   1.174 -2.447 -0.820  -0.079   0.666   2.201 1.001  2700
alpha1     -0.913   0.700 -2.285 -1.345  -0.915  -0.475   0.465 1.001  3000
alpha12     0.831   0.443 -0.018  0.549   0.815   1.104   1.763 1.001  3000
alpha2     -0.309   0.739 -1.823 -0.780  -0.293   0.172   1.114 1.001  2200
tau        76.086 439.882  2.808  7.077  13.364  31.802 630.682 1.001  2300
deviance  102.119   7.249 89.962 96.770 101.592 107.349 116.046 1.001  3000
```

```
For each parameter, n.eff is a crude measure of effective sample size,
and Rhat is the potential scale reduction factor (at convergence, Rhat=1).

DIC info (using the rule, pD = var(deviance)/2)
pD = 26.3 and DIC = 128.4
DIC is an estimate of expected predictive error (lower deviance is better).
```

As we saw with the GLM with overdispersion, the interaction term (alpha12) falls short of significance and should be removed. Rewriting the simpler model is left as an exercise.

23

Tree Models

Tree models are computationally intensive methods that are used in situations where there are many explanatory variables and we would like guidance about which of them to include in the model. Often there are so many explanatory variables that we simply could not test them all, even if we wanted to invest the huge amount of time that would be necessary to complete such a complicated multiple regression exercise. Tree models are particularly good at tasks that might in the past have been regarded as the realm of multivariate statistics (e.g. classification problems). The great virtues of tree models are as follows:

- They are very simple.

- They are excellent for initial data inspection.

- They give a very clear picture of the structure of the data.

- They provide a highly intuitive insight into the kinds of interactions between variables.

It is best to begin by looking at a tree model in action, before thinking about how it works. Here is an air pollution example that we might want to analyze as a multiple regression. We begin by using `tree`, then illustrate the more modern function `rpart` (which stands for 'recursive partitioning')

```
install.packages("tree")
library(tree)
Pollute <- read.table("c:\\temp\\Pollute.txt",header=T)
attach(Pollute)
names(Pollute)

[1] "Pollution" "Temp" "Industry" "Population"  "Wind"
[6] "Rain"       "Wet.days"

model <- tree(Pollute)
plot(model)
text(model)
```

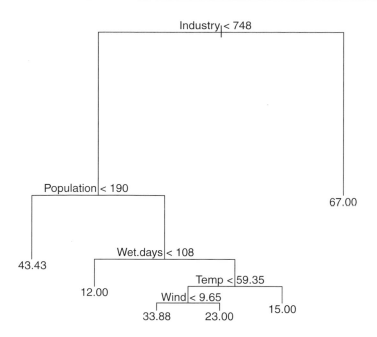

You follow a path from the top of the tree (called, in defiance of gravity, the **root**) and proceed to one of the terminal nodes (called a **leaf**) by following a succession of rules (called **splits**). The numbers at the tips of the leaves are the mean values in that subset of the data (mean SO$_2$ concentration in this case). The details are explained below.

23.1   Background

The model is fitted using **binary recursive partitioning**, whereby the data are successively split along coordinate axes of the explanatory variables so that, at any node, the split which maximally distinguishes the response variable in the left and the right branches is selected. Splitting continues until nodes are pure or the data are too sparse (fewer than six cases, by default; see Breiman et al., 1984).

   Each explanatory variable is assessed in turn, and the variable explaining the greatest amount of the deviance in y is selected. Deviance is calculated on the basis of a threshold in the explanatory variable; this threshold produces two mean values for the response (one mean above the threshold, the other below the threshold).

```
low <- (Industry<748)
tapply(Pollution,low,mean)

    FALSE      TRUE
67.00000 24.91667

plot(Industry,Pollution,pch=16)
abline(v=748,lty=2)
lines(c(0,748),c(24.92,24.92))
lines(c(748,max(Industry)),c(67,67))
```

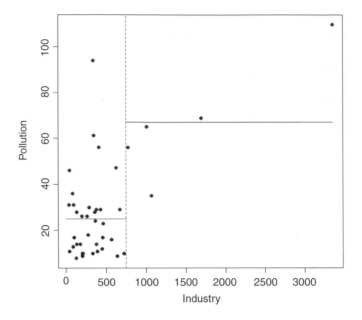

The procedure works like this. For a given explanatory variable (say, Industry above):

- Select a threshold value of the explanatory variable (the vertical dotted line at Industry = 748).

- Calculate the mean value of the response variable above and below this threshold (the two horizontal solid lines).

- Use the two means to calculate the deviance (as with *SSE*, see p. 500).

- Go through all possible values of the threshold (values on the *x* axis).

- Look to see which value of the threshold gives the lowest deviance.

- Split the data into high and low subsets on the basis of the threshold for this variable.

- Repeat the whole procedure on each subset of the data on either side of the threshold.

- Keep going until no further reduction in deviance is obtained, or there are too few data points to merit further subdivision (e.g. the right-hand side of the Industry split, above, is too sparse to allow further subdivision).

The deviance is defined as

$$D = \sum_j (y_j - \mu_{[j]})^2,$$

where $\mu_{[j]}$ is the mean of all the values of the response variable assigned to node j and this sum of squares is add up over all the nodes. The *value* of any split is defined as the reduction in this residual sum of squares. The probability model used in R is that the values of the response variable are normally distributed within each leaf of the tree with mean μ_i and variance σ^2. Note that because this assumption applies to the terminal nodes, the interior nodes represent a mixture of different normal distributions, so the deviance is only appropriate at the terminal nodes (i.e. for the leaves).

If the twigs of the tree are categorical (i.e. levels of a factor like names of particular species) then we have a **classification tree**. On the other hand, if the terminal nodes of the tree are predicted values of a continuous variable, then we have a **regression tree**.

The key questions are these:

- Which variables to use for the division.

- How best to achieve the splits for each selected variable.

It is important to understand that tree models have a tendency to over-interpret the data: for instance, the occasional 'ups' in a generally negative correlation probably do not mean anything substantial.

23.2 Regression trees

In this case the response variable is a continuous measurement, but the explanatory variables can be any mix of continuous and categorical variables. You can think of regression trees as analogous to multiple regression models. The difference is that a regression tree works by forward selection of variables, whereas we have been used to carrying out regression analysis by deletion (backward selection).

For our air pollution example, the regression tree is fitted by stating that the continuous response variable Pollution is to be estimated as a function of *all* of the explanatory variables in the dataframe called Pollute by use of the 'tilde dot' notation like this:

```
model <- tree(Pollution ~ . , Pollute)
```

For a regression tree, the print method produces the following kind of output:

```
print(model)

node), split, n, deviance, yval
      * denotes terminal node

1) root 41 22040 30.05
  2) Industry < 748 36 11260 24.92
    4) Population < 190 7  1096 43.43 *
    5) Population > 190 29  1187 20.45
     10) Wet.days < 108 11    96 12.00 *
     11) Wet.days > 108 18  2826 25.61
       22) Temp < 59.35 13  1895 29.69
         44) Wind < 9.65 8  1213 33.88 *
         45) Wind > 9.65 5   318 23.00 *
       23) Temp > 59.35 5   152 15.00 *
  3) Industry > 748 5  3002 67.00 *
```

The terminal nodes (the leaves) are denoted by * (there are six of them). The node number is on the left, labelled by the variable on which the split at that node was made. Next comes the 'split criterion' which shows the threshold value of the variable that was used to create the split. The number of cases going *into* the split (or into the terminal node) comes next. The penultimate figure is the deviance at that node. Notice how the deviance goes down as non-terminal nodes are split. In the root, based on all $n = 41$ data points, the deviance is *SSY* (see p. 499) and the y value is the overall mean for Pollution. The last figure on the right is the mean value of the response variable within that node or at that that leaf. The highest mean pollution (67.00) was in node 3 and the lowest (12.00) was in node 10.

Note how the nodes are nested: within node 2, for example, node 4 is terminal but node 5 is not; within node 5 node 10 is terminal but node 11 is not; within node 11, node 23 is terminal but node 22 is not, and so on.

Tree models lend themselves to circumspect and critical analysis of complex dataframes. In the present example, the aim is to understand the causes of variation in air pollution levels from case to case. The interpretation of the regression tree would proceed something like this:

- The five most extreme cases of Industry stand out (mean = 67.00) and need to be considered separately.

- For the rest, Population is the most important variable but, interestingly, it is low populations that are associated with the highest levels of pollution (mean = 43.43). Ask yourself which might be cause, and which might be effect.

- For high levels of population (greater than 190), the number of wet days is a key determinant of pollution; the places with the fewest wet days (less than 108 per year) have the lowest pollution levels of anywhere in the dataframe (mean = 12.00).

- For those places with more than 108 wet days, it is temperature that is most important in explaining variation in pollution levels; the warmest places have the lowest air pollution levels (mean = 15.00).

- For the cooler places with lots of wet days, it is wind speed that matters: the windier places are less polluted than the still places.

This kind of complex and contingent explanation is much easier to see, and to understand, in tree models than in the output of a multiple regression.

23.3   Using `rpart` to fit tree models

The newer function rpart differs from tree in the way it handles surrogate variables, but for the most part, it follows Breiman et al. (1984) quite closely. The name of the function stands for 'recursive partitioning'. We can compare the outputs of rpart (left) and tree (right) for the pollution data:

```
Pollute<-read.table("c:\\temp\\Pollute.txt",header=T)
attach(Pollute)
names(Pollute)
par(mfrow=c(1,2))

library(rpart)
model<-rpart(Pollution~.,data=Pollute)
plot(model)
text(model)

library(tree)
model<-tree(Pollute)
plot(model)
text(model)
```

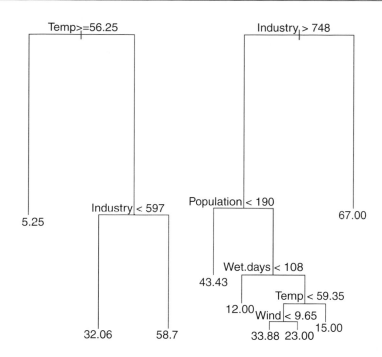

The new function `rpart` is much better at anticipating the results of model simplification, because it carries out analysis of variance with the two-level factors associated with each split. Thus, for temperature and industry

```
t2<-factor(Temp>=56.25)
i2<-factor(Industry<597)
model<-lm(Pollution~t2*i2)
summary(model)


Call:
lm(formula = Pollution ~ t2 * i2)

Residuals:
    Min      1Q  Median      3Q     Max
-29.714  -8.071  -3.071   6.286  61.944

Coefficients:
              Estimate Std. Error t value Pr(>|t|)
(Intercept)     58.714      6.984   8.407 4.15e-10 ***
t2TRUE         -49.214     14.816  -3.322  0.00202 **
i2TRUE         -26.659      8.231  -3.239  0.00254 **
t2TRUE:i2TRUE   33.230     16.213   2.050  0.04754 *

Residual standard error: 18.48 on 37 degrees of freedom
Multiple R-squared: 0.4267,     Adjusted R-squared: 0.3802
F-statistic:  9.18 on 3 and 37 DF,  p-value: 0.0001132
```

it produces a significant interaction (shown by the split on right branch the tree diagram) and this model does not allow the inclusion of any other significant terms. If population is added, it is marginally significant, but the original interaction between temperature and industry disappears.

```
Call:
lm(formula = Pollution ~ t2 * i2 + Population)

Residuals:
    Min      1Q   Median      3Q      Max
-25.169   -8.664   -3.351    8.142   64.778

Coefficients:
                Estimate Std. Error t value Pr(>|t|)
(Intercept)     39.263869   11.402820    3.443   0.00147 **
t2TRUE         -44.112844   14.378926   -3.068   0.00408 **
i2TRUE         -12.514816   10.351335   -1.209   0.23454
Population       0.013817    0.006564    2.105   0.04234 *
t2TRUE:i2TRUE   27.406065   15.754854    1.740   0.09049 .

Residual standard error: 17.68 on 36 degrees of freedom
Multiple R-squared: 0.4895,      Adjusted R-squared: 0.4328
F-statistic: 8.631 on 4 and 36 DF,  p-value: 5.432e-05

Call:
lm(formula = Pollution ~ t2 + Population)

Residuals:
    Min      1Q   Median      3Q      Max
-24.880  -12.076   -2.849    9.006   63.098

Coefficients:
                Estimate Std. Error t value Pr(>|t|)
(Intercept)     27.763869    4.888500    5.679 1.57e-06 ***
t2TRUE         -21.481977    5.800790   -3.703 0.000674 ***
Population       0.017529    0.004947    3.543 0.001066 **

Residual standard error: 17.95 on 38 degrees of freedom
Multiple R-squared: 0.4444,      Adjusted R-squared: 0.4151
F-statistic: 15.19 on 2 and 38 DF,  p-value: 1.416e-05
```

Note that the regression model with t2 and Population has a lower residual standard error (17.95 on 38 d.f.) than the ANOVA from the model suggested by rpart (18.48 on 37 d.f.).

In summary, I prefer the tree function for data inspection, because it shows more detail about the potential interaction structure in the dataframe. On the other hand, rpart is much better at anticipating the results of model simplification. I recommend you use them both, and get the benefit of two perspectives on your data set before embarking on the time-consuming business of carrying out a comprehensive multiple regression exercise.

23.4   Tree models as regressions

To see how a tree model works when there is a single, continuous response variable, it is useful to compare the output with a simple linear regression. Take the relationship between mileage and weight in the car.test.frame data:

```
car.test.frame <- read.table("c:\\temp\\car.test.frame.txt",header=T)
attach(car.test.frame)
names(car.test.frame)

[1] "Price" "Country" "Reliability" "Mileage"
[5] "Type" "Weight" "Disp." "HP"

plot(Weight,Mileage,pch=21,col="brown",bg="green")
```

The heavier cars do fewer miles per gallon, but there is a lot of scatter. The tree model starts by finding the weight that splits the mileage data in a way that explains the maximum deviance. This weight turns out to be 2567.5.

```
a <- mean(Mileage[Weight<2567.5])
b <- mean(Mileage[Weight>=2567.5])
lines(c(1500,2567.5,2567.5,4000),c(a,a,b,b))
```

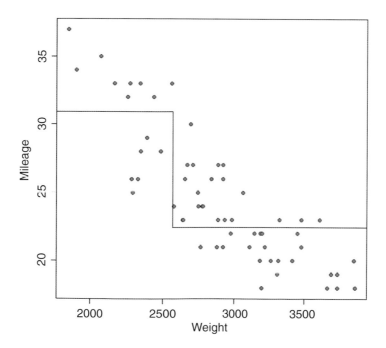

The next thing the tree model does is to work out the threshold weight that would best split the mileage data for the lighter cars: this turns out to be 2280. It then works out the threshold split for the heavier cars: this turns out to be 3087.5. And so the process goes on, until there are too few cars in each split to justify continuation (five or fewer by default). To see the full regression tree as a function plot we can use the `predict` function with the regression tree object `car.model` like this:

```
car.model <- tree(Mileage~Weight)
wt <- seq(1500,4000)
y <- predict(car.model,list(Weight=wt))
plot(Weight,Mileage,pch=21,col="brown",bg="green")
lines(wt,y)
```

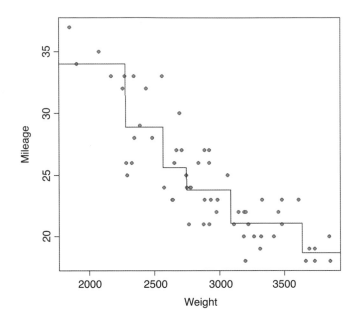

You would not normally do this, of course (and you *could not* do it with more than two explanatory variables) but it is a good way of showing how tree models work with a continuous response variable.

23.5   Model simplification

Model simplification in regression trees is based on a **cost–complexity measure**. This reflects the trade-off between fit and explanatory power (a model with a perfect fit would have as many parameters as there were data points, and would consequently have no explanatory power at all). We return to the air pollution example analysed earlier, where we fitted the tree model object called `model`.

Regression trees can be over-elaborate and can respond to random features of the data (the so-called **training set**). To deal with this, R contains a set of procedures to prune trees on the basis of the cost–complexity measure. The function `prune.tree` determines a nested sequence of sub-trees of the supplied tree by recursively 'snipping' off the least important splits, based upon the cost–complexity measure. The `prune.tree` function returns an object of class `tree.sequence`, which contains the following components:

```
prune.tree(model)

$size
[1] 6 5 4 3 2 1
```

This shows the number of terminal nodes in each tree in the cost–complexity pruning sequence: the most complex model had six terminal nodes (see above)

```
$dev:
[1]  8876.589 9240.484 10019.992 11284.887 14262.750 22037.902
```

This is the total deviance of each tree in the cost–complexity pruning sequence.

```
$k:
[1]  -Inf 363.8942 779.5085 1264.8946 2977.8633 7775.1524
```

It is the value of the cost–complexity pruning parameter of each tree in the sequence. If determined algorithmically (as here, k is not specified as an input), its first value defaults to $-\infty$, its lowest possible bound.

```
plot(prune.tree(model))
```

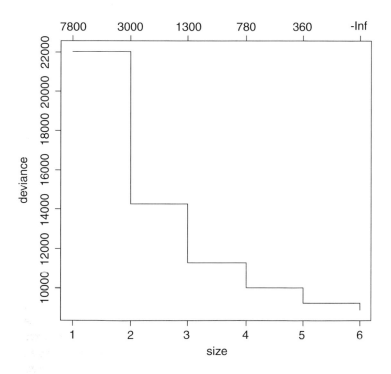

This shows the way that deviance declines as complexity is increased. The total deviance is 22 037.902 ($size = 1$), and this is reduced as the complexity of the tree increases up to six nodes. An alternative is to specify the number of nodes to which you want the tree to be pruned; this uses the "best=" option. Suppose we want the best tree with four nodes:

```
model2 <- prune.tree(model,best=4)
plot(model2)
text(model2)
```

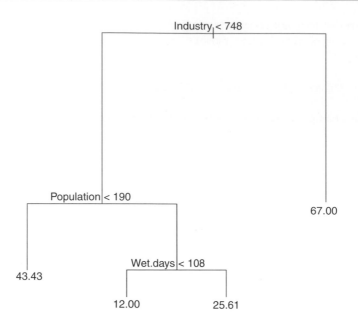

In printed form, this is:

```
print(model2)
```

```
node), split, n, deviance, yval
      * denotes terminal node

 1) root 41 22040 30.05
   2) Industry < 748 36 11260 24.92
     4) Population < 190 7  4096 43.43 *
     5) Population > 190 29  4187 20.45
      10) Wet.days < 108 11    96 12.00 *
      11) Wet.days > 108 18  2826 25.61 *
   3) Industry > 748 5  3002 67.00 *
```

It is straightforward to remove parts of trees, or to select parts of trees, using subscripts. For example, a negative subscript [−3] leaves off everything above node 3, while a positive subscript [3] selects only that part of the tree above node 3.

23.6 Classification trees with categorical explanatory variables

Tree models are a superb tool for helping to write efficient and effective taxonomic keys.

Suppose that all of our explanatory variables are categorical, and that we want to use tree models to write a dichotomous key. There is only one entry for each species, so we want the twigs of the tree to be the individual rows of the dataframe (i.e. we want to fit a tree perfectly to the data). To do this we need to specify two extra arguments: minsize = 2 and mindev = 0. In practice, it is better to specify a very small value for the minimum deviance (say, 10^{-6}) rather than zero (see below).

The following example relates to the nine lowland British species in the genus *Epilobium* (Onagraceae). We have eight categorical explanatory variables and we want to find the optimal dichotomous key. The dataframe looks like this:

```
epilobium <- read.table("c:\\temp\\epilobium.txt",header=T)
attach(epilobium)
epilobium
```

	species	stigma	stem.hairs	glandular.hairs	seeds	pappilose	stolons	petals	base
1	hirsutum	lobed	spreading	absent	none	uniform	absent	>9mm	rounded
2	parviflorum	lobed	spreading	absent	none	uniform	absent	<10mm	rounded
3	montanum	lobed	spreading	present	none	uniform	absent	<10mm	rounded
4	lanceolatum	lobed	spreading	present	none	uniform	absent	<10mm	cuncate
5	tetragonum	clavate	appressed	present	none	uniform	absent	<10mm	rounded
6	obscurum	clavate	appressed	present	none	uniform	stolons	<10mm	rounded
7	roseum	clavate	spreading	present	none	uniform	absent	<10mm	cuneate
8	palustre	clavate	spreading	present	appendage	uniform	absent	<10mm	rounded
9	ciliatum	clavate	spreading	present	appendage	ridged	absent	<10mm	rounded

Producing the key could not be easier:

```
model <- tree(species ~ .,epilobium,mindev=1e-6,minsize=2)
plot(model)
text(model,cex=0.7)
```

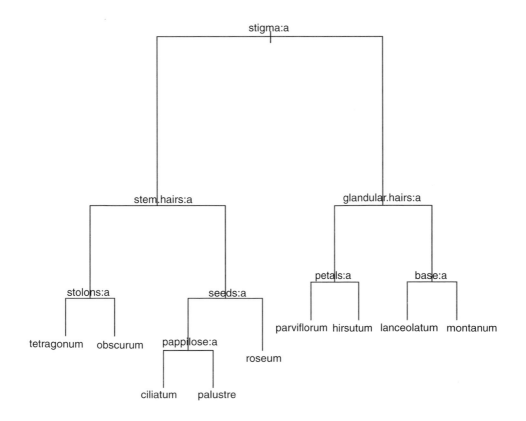

Here is the tree written as a dichotomous key:

1. Stigma entire and club-shaped	2
1. Stigma four lobed	6
2. Stem hairs all appressed	3
2. At least some stem hairs spreading	4
3. Glandular hairs present on hypanthium	*E. obscurum*
3. No glandular hairs on hypanthium	*E. tetragonum*
4. Seeds with a terminal appendage	5
4. Seeds without terminal appendage	*E. roseum*
5. Surface of seed with longitudinal papillose ridges	*E. ciliatum*
5. Surface of seed uniformly papillose	*E. palustre*
6. At least some spreading hairs non-glandular	7
6. Spreading hairs all glandular	8
7. Petals large (>9 mm)	*E. hirsutum*
7. Petals small (<10 mm)	*E. parviflorum*
8. Leaf base cuneate	*E. lanceolatum*
8. Leaf base rounded	*E. montanum*

The computer has produced a working key to a difficult group of plants. The result stands as testimony to the power and usefulness of tree models. The same principle underlies good key-writing as is used in tree models: find the characters that explain most of the variation, and use these to split the cases into roughly equal-sized groups at each dichotomy.

23.7   Classification trees for replicated data

In this next example from plant taxonomy, the response variable is a four-level categorical variable called Taxon (it is a label expressed as Roman numerals I to IV). The aim is to use the measurements from the seven morphological explanatory variables to construct the best key to separate these four taxa (the 'best' key is the one with the lowest error rate – the key that misclassifies the smallest possible number of cases).

```
taxonomy <- read.table("c:\\temp\\taxonomy.txt",header=T)
attach(taxonomy)
names(taxonomy)

[1] "Taxon" "Petals" "Internode" "Sepal" "Bract" "Petiole"
[7] "Leaf" "Fruit"
```

Using the tree model for classification could not be simpler:

```
model1 <- tree(Taxon~., taxonomy)
```

We begin by looking at the plot of the tree:

```
plot(model1)
text(model1)
```

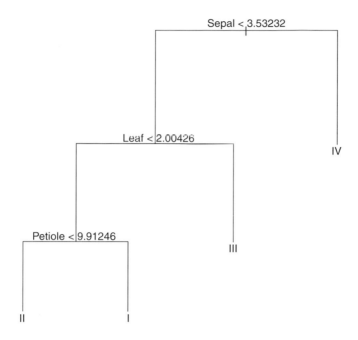

With only a small degree of rounding on the suggested break points, the tree model suggests a simple (and for these 120 plants, completely error-free) key for distinguishing the four taxa:

1. Sepal length >4.0	*Taxon IV*
1. Sepal length ≤4.0	2.
2. Leaf width >2.0	*Taxon III*
2. Leaf width ≤2.0	3.
3. Petiole length <10	*Taxon II*
3. Petiole length ≥10	*Taxon I*

The summary option for classification trees produces the following:

```
summary(model1)

Classification tree:
tree(formula = Taxon ~ ., data = taxonomy)
Variables actually used in tree construction:
[1] "Sepal" "Leaf" Petiole"
Number of terminal nodes: 4
Residual mean deviance: 0 = 0 / 116
Misclassification error rate: 0 = 0 / 120
```

Three of the seven variables were chosen for use (Sepal, Leaf and Petiole); four variables were assessed and rejected (Petals, Internode, Bract and Fruit). The key has four nodes and hence three dichotomies. As you see, the misclassification error rate was an impressive 0 out of 120. It is note-worthy that this classification tree does much better than the multivariate classification methods described in Chapter 25.

For classification trees, the `print` method produces a great deal of information

```
print(model1)

node), split,  n, deviance, yval, (yprob)
      * denotes terminal  node
1) root 120 332.70 I (0.2500 0.2500 0.2500 0.25)
  2) Sepal<3.53232 90 197.80 I (0.3333 0.3333 0.3333 0.00)
    4) Leaf<2.00426 60 83.18 I (0.5000 0.5000 0.0000 0.00)
      8) Petiole<9.91246 30 0.00 II (0.0000 1.0000 0.0000 0.00) *
      9) Petiole>9.91246 30 0.00 I (1.0000 0.0000 0.0000 0.00) *
    5) Leaf>2.00426 30 0.00 III (0.0000 0.0000 1.0000 0.00) *
  3) Sepal>3.53232 30 0.00 IV (0.0000 0.0000 0.0000 1.00) *
```

The **node** number is followed by the **split criterion** (e.g. Sepal < 3.53 at node 2). Then comes the number of cases passed through that node (90 in this case, versus 30 going into node 3, which is the terminal node for taxon IV). The remaining deviance within this node is 197.8 (compared with zero in node 3 where all the individuals are alike; they are all taxon IV). Next is the name of the factor level(s) left in the split (I, II and III in this case, with the convention that the first in the alphabet is listed), then a list of the empirical probabilities (the fractions of all the cases at that node that are associated with each of the levels of the response variable – in this case the 90 cases are equally split between taxa I, II and III and there are no individuals of taxon IV at this node, giving 0.33, 0.33, 0.33 and 0 as the four probabilities).

There is quite a useful plotting function for classification trees called `partition.tree`, but it is only sensible to use it when the model has two explanatory variables. Its use is illustrated here by taking the two most important explanatory variables, Sepal and Leaf:

```
model2 <- tree(Taxon~Sepal+Leaf,taxonomy);
partition.tree(model2)
```

This shows how the phase space defined by sepal length and leaf width has been divided up between the four taxa, but it does not show where the data fall. We could use `points(Sepal,Leaf)` to overlay the points, but for illustration we shall use `text`. We create a vector called `label` that has a for taxon I, b for II, and so on:

```
label <- ifelse(Taxon=="I", "a",
                ifelse(Taxon=="II","b",ifelse(Taxon=="III","c","d")))
```

Then we use these letters as a `text` overlay on the `partition.tree` like this:

```
text(Sepal,Leaf,label,col=1+as.numeric(factor(label)))
```

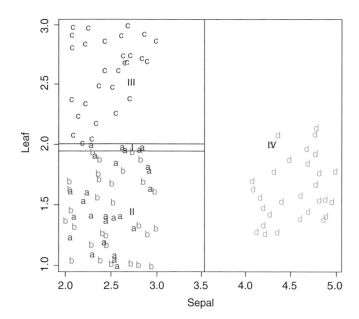

You see that taxa III and IV are beautifully separated on the basis of sepal length and leaf width, but taxa I and II are all jumbled up (recall that they are separated from one another on the basis of petiole length).

23.8 Testing for the existence of humps

Tree models can be useful in assessing whether or not there is a hump in the relationship between y and x. This is difficult to do using other kinds of regression, because linear models seldom distinguish between humps and asymptotes. If a tree model puts a lower section at the right of the graph than in the centre, then this hints at the presence of a hump in the data. Likewise, if it puts an elevated section at the left-hand end of the x axis then that is indicative of a U-shaped function.

Here is a function called hump which extracts information from a tree model to draw the stepped function through a scatterplot:

```
hump <- function(x,y){
library(tree)
model <- tree(y~x)
xs <- grep("[0-9]",model[[1]][[5]])
xv <- as.numeric(substring(model[[1]][[5]][xs],2,10))
xv <- xv[1:(length(xv)/2)]
xv <- c(min(x),sort(xv),max(x))
yv <- model[[1]][[4]][model[[1]][[1]]=="<leaf>"]
plot(x,y,col="red",pch=16,
xlab=deparse(substitute(x)),ylab=deparse(substitute(y)))
i <- 1
j <- 2
k <- 1
b <- 2*length(yv)+1
```

```
for (a in 1:b){
lines(c(xv[i],xv[j]),c(yv[k],yv[i]))
if (a %% 2 == 0){
j <- j+1
k <- k+1}
else{
i <- i+1
}}}
```

We shall test it on the ethanol data which are definitely humped (p. 675):

```
library(lattice)
attach(ethanol)
names(ethanol)
```

```
[1] "NOx" "C" "E"
```

```
hump(E,NOx)
```

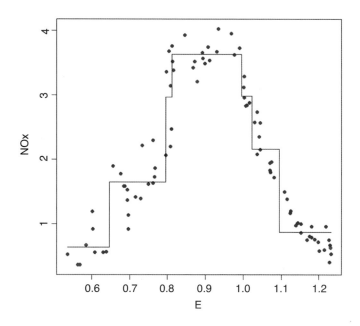

There is a minimum number of points necessary for creating a new step ($n = 5$), and a minimum difference in the mean of one group and the next. To see this, you should contrast these two fits:

```
hump(E[E<1.007],NOx[E<1.007])
```

```
hump(E[E<1.006],NOx[E<1.006])
```

The first data set has evidence of a hump, but the second does not.

24

Time Series Analysis

Time series data are vectors of numbers, typically regularly spaced in time. Yearly counts of animals, daily prices of shares, monthly means of temperature, and minute-by-minute details of blood pressure are all examples of time series, but they are measured on different time scales. Sometimes the interest is in the time series itself (e.g. whether or not it is cyclic, or how well the data fit a particular theoretical model), and sometimes the time series is incidental to a designed experiment (e.g. repeated measures). We cover each of these cases in turn.

The three key concepts in time series analysis are

- trend,

- serial dependence, and

- stationarity.

Most time series analyses assume that the data are untrended. If they do show a consistent upward or downward trend, then they can be detrended before analysis (e.g. by differencing). Serial dependence arises because the values of adjacent members of a time series may well be correlated. Stationarity is a technical concept, but it can be thought of simply as meaning that the time series has the same properties wherever you start looking at it (e.g. white noise is a sequence of mutually independent random variables each with mean zero and variance $\sigma^2 > 0$).

24.1 Nicholson's blowflies

The Australian ecologist, A.J. Nicholson, reared blowfly larvae on pieces of liver in laboratory cultures that his technicians kept running continuously for almost 7 years (361 weeks, to be exact). The time series for numbers of adult flies looks like this:

```
blowfly <- read.table("c:\\temp\\blowfly.txt",header=T)
attach(blowfly)
names(blowfly)

[1] "flies"
```

The R Book, Second Edition. Michael J. Crawley.
© 2013 John Wiley & Sons, Ltd. Published 2013 by John Wiley & Sons, Ltd.

First, make the flies variable into a time series object and plot it:

```
flies <- ts(flies)
plot(flies)
```

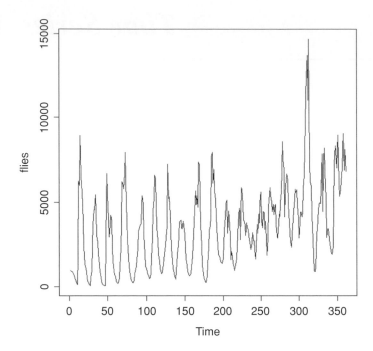

This classic time series has two clear features:

- For the first 200 weeks the system exhibits beautifully regular cycles.

- After week 200 things change (perhaps a genetic mutation had arisen); the cycles become much less clear-cut, and the population begins a pronounced upward trend.

There are two important ideas to understand in time series analysis: **autocorrelation** and **partial autocorrelation**. The first describes how this week's population is related to last week's population. This is the autocorrelation at lag 1. The second describes the relationship between this week's population and the population at lag t once we have controlled for the correlations between all of the successive weeks between this week and week t. This should become clear if we draw the scatterplots from which the first four autocorrelation terms are calculated (lag 1 to lag 4).

There is a snag, however. The vector of flies at lag 1 is shorter (by one) than the original vector because the first element of the lagged vector is the second element of flies. The coordinates of the first data point to be drawn on the scatterplot are (flies[1],flies[2]) and the coordinates of the last plot that can be drawn are (flies[360], flies[361]) because the original vector is 361 element long:

```
length(flies)
```

```
[1] 361
```

Thus, the lengths of the vectors that can be plotted go down by one for every increase in the lag of one. We can produce the four plots for lags 1 to 4 in a function like this:

```
par(mfrow=c(2,2))
sapply(1:4, function(x) plot(flies[-c(361: (361-x+1))], flies[-c(1:x)] ) )
```

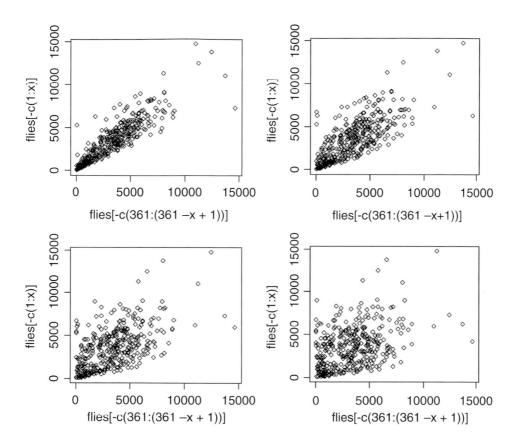

The correlation is very strong at lag 1, but notice how the variance increases with population size: small populations this week are invariably correlated with small populations next week, but large populations this week may be associated with large or small populations next week. The striking pattern here is the way that the correlation fades away as the size of the lag increases. Because the population is cyclic, the correlation goes to zero, then becomes weakly negative and then becomes strongly negative. This occurs at lags that are half the cycle length. Looking back at the time series, the cycles look to be about 20 weeks in length. So let us repeat the exercise by producing scatterplots at lags of 7, 8, 9 and 10 weeks:

```
sapply(7:10, function(x) plot(flies[-c((361-x+1):361)], flies[-c(1:x)] ) )
par(mfrow=c(1,1))
```

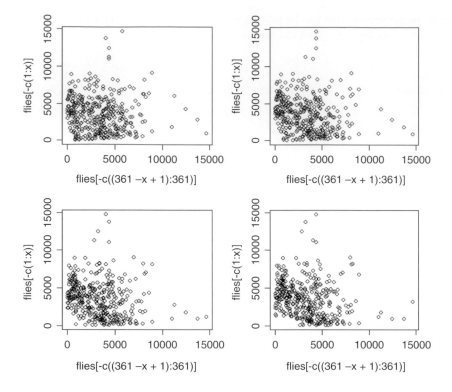

The negative correlation at lag 10 gradually emerges from the fog of no correlation at lag 7.
More formally, the autocorrelation function $\rho(k)$ at lag k is

$$\rho(k) = \frac{\gamma(k)}{\gamma(0)},$$

where $\gamma(k)$ is the autocovariance function at lag k of a stationary random function $\{Y(t)\}$ given by

$$\gamma(k) = \text{cov}\{Y(t), Y(t - k)\}.$$

The most important properties of the autocorrelation coefficient are as follows:

- They are symmetric backwards and forwards, so $\rho(k) = \rho(-k)$.
- The limits are $-1 \leq \rho(k) \leq 1$.
- When $Y(t)$ and $Y(t - k)$ are independent, then $\rho(k) = 0$.
- The converse of this is not true, so that $\rho(k) = 0$ does not imply that $Y(t)$ and $Y(t - k)$ are independent (look at the scatterplot for $k = 7$ in the scatterplots above).

A first-order autoregressive process is written as

$$Y_t = \alpha Y_{t-1} + Z_t.$$

This says that this week's population is α times last week's population plus a random term Z_t. The randomness is **white noise**; the values of Z are **serially independent**, they have a **mean of zero**, and they have **finite variance** σ^2.

In a stationary times series $-1 < \alpha < 1$. In general, then, the autocorrelation function of $\{Y(t)\}$ is

$$\rho_k = \alpha^k, \qquad k = 0, 1, 2, \ldots.$$

Partial autocorrelation is the relationship between this week's population and the population at lag t when we have controlled for the correlations between all of the successive weeks between this week and week t. That is to say, the partial autocorrelation is the correlation between $Y(t)$ and $Y(t + k)$ after regression of $Y(t)$ on $Y(t + 1)$, $Y(t + 2)$, $Y(t + 3)$, $\ldots$, $Y(t + k - 1)$. It is obtained by solving the Yule–Walker equation

$$\rho_k = \sum_{1}^{p} \alpha_i \rho_{k-i}, \qquad k > 0,$$

with the ρ replaced by r (correlation coefficients estimated from the data). Suppose we want the partial autocorrelation between time 1 and time 3. To calculate this, we need the three ordinary correlation coefficients r_{12}, r_{13} and r_{23}. The partial $r_{13,2}$ is then

$$r_{13,2} = \frac{r_{13} - r_{12}r_{23}}{\sqrt{\left(1 - r_{12}^2\right)\left(1 - r_{23}^2\right)}}.$$

For more on partial correlation coefficients, see p. 375.

Let us look at the correlation structure of the blowfly data. The R function for calculating autocorrelations and partial autocorrelations is `acf` (the 'autocorrelation function'). First, we produce the autocorrelation plot to look for evidence of cyclic behaviour:

```
acf(flies,main="",col="red")
```

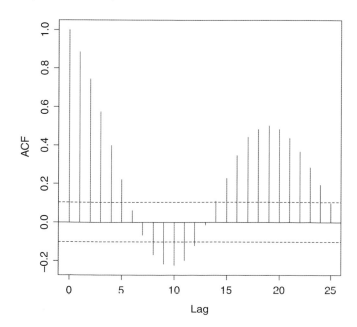

You will not see more convincing evidence of cycles than this. The blowflies exhibit highly significant, regular cycles with a period of 19 weeks. The blue dashed lines indicate the threshold values for significant correlation. What kind of time lags are involved in the generation of these cycles? We use partial autocorrelation (`type="p"`) to find this out:

```
acf(flies,type="p",main="",col="red")
```

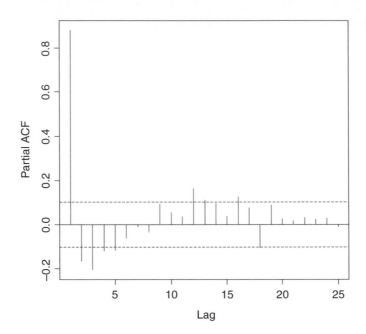

The significant density-dependent effects are manifest at lags of 2 and 3 weeks, with other, marginally significant negative effects at lags of 4 and 5 weeks. These lags reflect the duration of the larval and pupal period (1 and 2 periods, respectively). The cycles are clearly caused by overcompensating density dependence, resulting from intraspecific competition between the larvae for food (what Nicholson christened 'scramble competition'). There is a curious positive feedback at a lag of 12 weeks (12–16 weeks, in fact). Perhaps you can think of a possible cause for this?

We should investigate the behaviour of the second half of the time series separately. Let us say it is from week 201 onwards:

```
second <- flies[201:361]
```

Now test for a linear trend in mean fly numbers against day number, from 1 to `length(second)`:

```
summary(lm(second~I(1:length(second))))
```

Note the use of `I` in the model formula (for 'as is') to tell R that the colon we have used is to generate a sequence of x values for the regression (and not an interaction term as it would otherwise have assumed).

```
Coefficients:
                     Estimate Std. Error t value Pr(>|t|)
(Intercept)          2827.531    336.661   8.399 2.37e-14 ***
I(1:length(second))    21.945      3.605   6.087 8.29e-09 ***
```

```
Residual standard error: 2126 on 159 degrees of freedom
Multiple R-squared: 0.189,        Adjusted R-squared: 0.1839
F-statistic: 37.05 on 1 and 159 DF,    p-value: 8.289e-09
```

This shows that there is a highly significant upward trend of about 22 extra flies on average each week in the second half of time series. We can detrend the data by subtracting the fitted values from the linear regression of second on day number:

```
detrended <- second - predict(lm(second~I(1:length(second))))
par(mfrow=c(2,2))
ts.plot(detrended)
```

There are still cycles there, but they are weaker and less regular. We repeat the correlation analysis on the detrended data:

```
acf(detrended,main="")
```

These look more like damped oscillations than repeated cycles. What about the partials?

```
acf(detrended,type="p",main="")
par(mfrow=c(1,1))
```

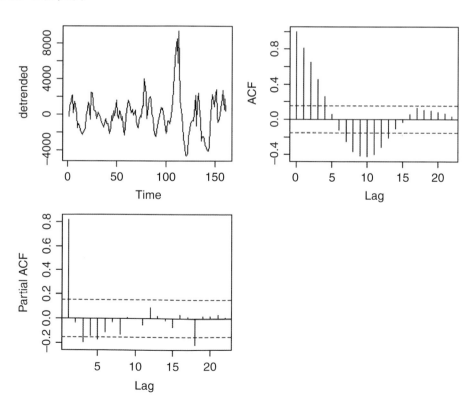

There are still significant negative partial autocorrelations at lags 3 and 5, but now there is a curious extra negative partial at lag 18. It looks, therefore, as if the main features of the ecology are the same (scramble competition for food between the larvae, leading to negative partials at 3 and 5 weeks after 1 and 2 generation lags), but population size is drifting upwards and the cycles are showing a tendency to dampen out.

24.2 Moving average

The simplest way of seeing pattern in time series data is to plot the moving average. A useful summary statistic is the three-point moving average:

$$y_i' = \frac{y_{i-1} + y_i + y_{i+1}}{3}.$$

The function $ma3$ will compute the three-point moving average for any input vector x:

```
ma3 <- function (x) {
     y <- numeric(length(x)-2)
for (i in 2:(length(x)-1)) {
     y[i] <- (x[i-1]+x[i]+x[i+1])/3
}
y }
```

A time series of mean monthly temperatures will illustrate the use of the moving average:

```
temperature <- read.table("c:\\temp\\temp.txt",header=T)
attach(temperature)
tm <- ma3(temps)
plot(temps)
lines(tm[2:158],col="blue")
```

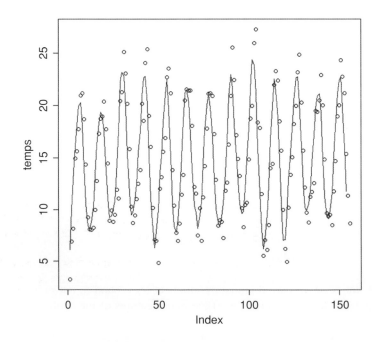

The seasonal pattern of temperature change over the 13 years of the data is clear. Note that a moving average can never capture the maxima or minima of a series (because they are averaged away). Note also that the three-point moving average is undefined for the first and last points in the series.

24.3   Seasonal data

Many time series applications involve data that exhibit seasonal cycles. The commonest applications involve weather data. Here are daily maximum and minimum temperatures from Silwood Park in south-east England over a 19-year period:

```
weather <- read.table("c:\\temp\\SilwoodWeather.txt",header=T)
attach(weather)
names(weather)
```

```
[1] "upper" "lower" "rain" "month" "yr"
```

```
plot(upper,type="l")
```

The seasonal pattern of temperature change is clear, but there is no clear trend (e.g. warming, see p. 791). Note that the *x* axis is labelled by the day number of the time series ('Index').

   We start by modelling the seasonal component. The simplest models for cycles are scaled so that a complete annual cycle is of length 1.0 (rather than 365 days). Our series consists of 6940 days over a 19-year span, so we write:

```
length(upper)
```

```
[1] 6940
```

```
index <- 1:6940
6940/19
```

```
[1] 365.2632
```

```
time <- index/365.2632
```

The equation for the seasonal cycle is:

$$y = \alpha + \beta \sin(2\pi t) + \gamma \cos(2\pi t) + \varepsilon.$$

This is a linear model, so we can estimate its three parameters very simply:

```
model <- lm(upper~sin(time*2*pi)+cos(time*2*pi))
```

To investigate the fit of this model we need to plot the scattergraph using very small symbols (otherwise the fitted line will be completely obscured). The smallest useful plotting symbol is the dot ".".

```
plot(time, upper, pch=".")
lines(time, predict(model),col="red",lwd=2)
```

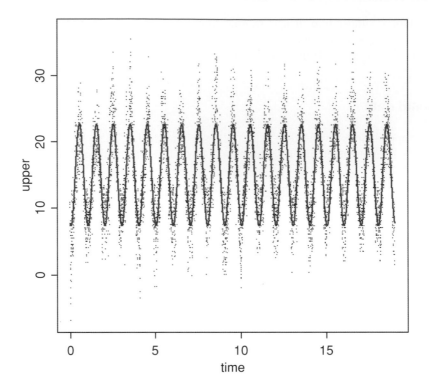

The three parameters of the model are all highly significant:

```
summary(model)

Call:
lm(formula = upper ~ sin(time * 2 * pi) + cos(time * 2 * pi))

Residuals:
     Min       1Q    Median       3Q       Max
-14.1336   -2.4220   -0.1233    2.2162   14.6456

Coefficients:
                    Estimate Std. Error t value Pr(>|t|)
(Intercept)         14.95647    0.04088  365.86   <2e-16 ***
sin(time * 2 * pi)  -2.53883    0.05781  -43.91   <2e-16 ***
cos(time * 2 * pi)  -7.24017    0.05781 -125.23   <2e-16 ***

Residual standard error: 3.406 on 6937 degrees of freedom
Multiple R-squared: 0.7174,      Adjusted R-squared: 0.7173
F-statistic:  8806 on 2 and 6937 DF,  p-value: < 2.2e-16
```

We can investigate the residuals to look for patterns (e.g. trends in the mean, or autocorrelation structure). Remember that the residuals are stored as part of the model object:

```
plot(model$resid,pch=".")
```

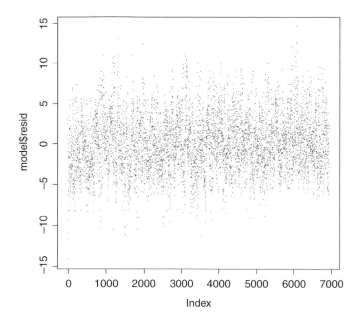

There looks to be some periodicity in the residuals, but no obvious trends. To look for serial correlation in the residuals, we use the `acf` function like this:

```
windows(7,4)
par(mfrow=c(1,2))
acf(model$resid,main="")
acf(model$resid,type="p",main="")
```

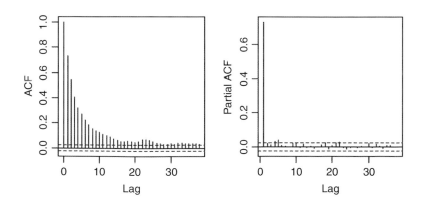

There is very strong serial correlation in the residuals, and this drops off roughly exponentially with increasing lag (left-hand graph). The partial autocorrelation at lag 1 is very large (0.7317), but the correlations at higher lags are much smaller. This suggests that an AR(1) model (autoregressive model with order 1) might be appropriate. This is the statistical justification behind the old joke about the weather forecaster who was asked what tomorrow's weather would be. 'Like today's', he said.

24.3.1   Pattern in the monthly means

The monthly average upper temperatures show a beautiful seasonal pattern when analysed by `acf`:

```
temp <- ts(as.vector(tapply(upper,list(month,yr),mean)))
windows(7,7)
acf(temp,main="")
```

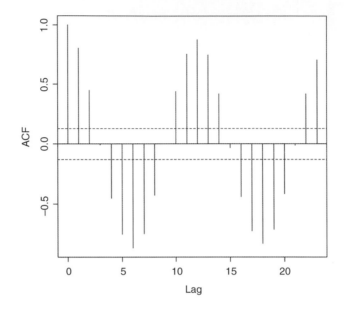

There is a perfect cycle with period 12 (as expected). What about patterns across years?

```
ytemp <- ts(as.vector(tapply(upper,yr,mean)))
acf(ytemp,main="")
```

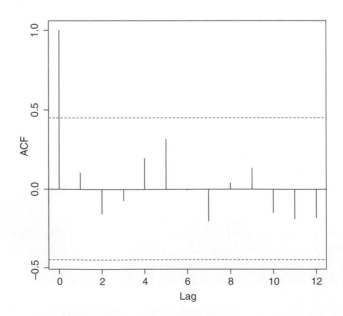

Nothing! The pattern you may (or may not) see depends upon the scale at which you look for it. As for spatial patterns (Chapter 26), so it is with temporal patterns. There is strong pattern between days within months (tomorrow will be like today). There is very strong pattern from month to month within years (January is cold, July is warm). But there is no pattern at all from year to year (there may be progressive global warming, but it is not apparent within this recent time series (see below), and there is absolutely no evidence for untrended serial correlation).

24.4 Built-in time series functions

The analysis is simpler, and the graphics are better labelled, if we convert the temperature data into a regular time series object using `ts`. We need to specify the first date (January 1993) as `start=c(1993,1)`, and the number of data points per year as `frequency=365`.

```
high <- ts(upper,start=c(1993,1),frequency=365)
```

Now use `plot` to see a plot of the time series, correctly labelled by years:

```
plot(high)
```

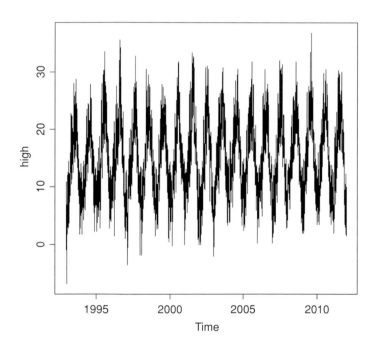

24.5 Decompositions

It is useful to be able to turn a time series into components. The function `stl` (with a lower-case letter L, not numeral one) performs seasonal decomposition of a time series into seasonal, trend and irregular components using `loess`. First, we make a time series object, specifying the start date and the frequency (as in Section 24.4), then use `stl` to decompose the series:

```
up <- stl(high,"periodic")
```

The plot function produces the data series, the seasonal component, the trend and the residuals in a single frame:

```
plot(up)
```

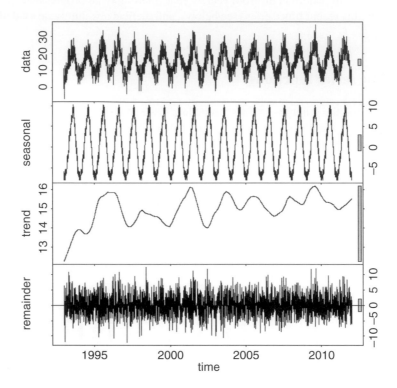

The remainder component is the residuals from the seasonal plus trend fit. The bars at the right-hand side are of equal heights (in user coordinates).

24.6   Testing for a trend in the time series

It is important to know whether these data provide any evidence for global warming. The trend part of the figure indicates a fluctuating increase, but is it significant? The mean temperature in the last 9 years was 0.71°C higher than in the first 10 years:

```
ys <- factor(1+(yr>2002))
tapply(upper,ys,mean)

        1           2
14.62056   15.32978
```

We cannot test for a trend with linear regression because of the massive temporal pseudoreplication. Suppose we tried this:

```
model1 <- lm(upper~index+sin(time*2*pi)+cos(time*2*pi))
summary(model1)
```

```
Coefficients:
                      Estimate Std. Error  t value Pr(>|t|)
(Intercept)          1.433e+01  8.136e-02  176.113   <2e-16 ***
index                1.807e-04  2.031e-05    8.896   <2e-16 ***
sin(time * 2 * pi)  -2.518e+00  5.754e-02  -43.758   <2e-16 ***
cos(time * 2 * pi)  -7.240e+00  5.749e-02 -125.939   <2e-16 ***

Residual standard error: 3.387 on 6936 degrees of freedom
Multiple R-squared: 0.7206,     Adjusted R-squared: 0.7205
F-statistic:  5963 on 3 and 6936 DF,  p-value: < 2.2e-16
```

It would suggest (wrongly, as we shall see) that the warming was highly significant (index p value less than 2×10^{-16} for a slope of 0.000 180 7 degrees of warming per day, leading to a predicted increase in mean temperature of 1.254°C over the 6940 days of the time series).

Since there is so much temporal pseudoreplication we should use a mixed model (lmer, p. 695), and because we intend to compare two models with different fixed effects we use the method of maximum likelihood (REML=FALSE). The explanatory variable for any trend is index, and we fit the model with and without this variable, allowing for different intercepts for the different years as a random effect:

```
model2 <-
  lmer(upper~index+sin(time*2*pi)+cos(time*2*pi)+(1 | factor(yr)),REML=FALSE)
model3 <-
  lmer(upper~sin(time*2*pi)+cos(time*2*pi)+(1 | factor(yr)),REML=FALSE)
anova(model2,model3)

Data:
Models:
model3: upper ~ sin(time * 2 * pi) + cos(time * 2 * pi) + (1 | factor(yr))
model2: upper ~ index + sin(time * 2 * pi) + cos(time * 2 * pi) + (1 |
model2:        factor(yr))
        Df   AIC   BIC logLik Chisq Chi Df Pr(>Chisq)
model3   5 36452 36486 -18221
model2   6 36458 36499 -18223     0      1          1
```

Clearly, the trend is non-significant (chi-squared $= 0$, $p = 1$). If you are prepared to ignore all the variation (from day to day and from month to month), then you can get rid of the pseudoreplication by averaging and test for trend in the yearly mean values: these show a significant trend if the first year (1993) is included, but not if it is omitted:

```
means <- as.vector(tapply(upper,yr,mean))
model <- lm(means~I(1:19))
summary(model)

Coefficients:
             Estimate  Std. Error   t value   Pr(> | t |)
(Intercept)  14.27105     0.32220    44.293       <2e-16 ***
I(1:19)       0.06858     0.02826     2.427       0.0266 *

model <- lm(means[-1]~I(1:18))
summary(model)
```

```
Coefficients:
              Estimate   Std. Error   t value   Pr(> | t |)
(Intercept)   14.59826      0.30901    47.243       <2e-16  ***
I(1:18)        0.04761      0.02855     1.668        0.115
```

Obviously, you need to be circumspect when interpreting trends in time series.

24.7   Spectral analysis

There is an alternative approach to time series analysis, which is based on the analysis of **frequencies** rather than fluctuations of numbers. Frequency is the reciprocal of cycle period. Ten-year cycles would have a frequency 0.1 per year. Here are the famous Canadian lynx data:

```
numbers <- read.table("c:\\temp\\lynx.txt",header=T)
attach(numbers)
names(numbers)

[1]  "Lynx"

plot.ts(Lynx)
```

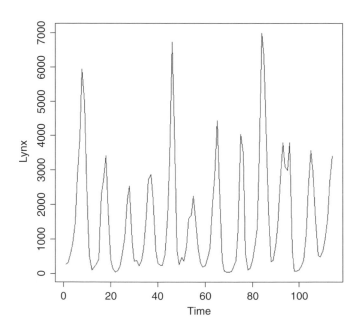

The fundamental tool of spectral analysis is the **periodogram**. This is based on the squared correlation between the time series and sine/cosine waves of frequency ω, and conveys exactly the same information as the autocovariance function. It may (or may not) make the information easier to interpret. Using the function is straightforward; we employ the spectrum function like this:

```
spectrum(Lynx,main="",col="red")
```

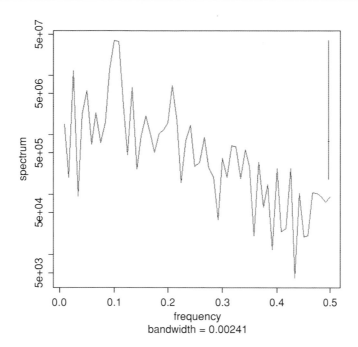

The plot is on a log scale, in units of decibels, and the subtitle on the x axis shows the bandwidth, while the 95% confidence interval in decibels is shown by the vertical blue bar in the top right-hand corner. The figure is interpreted as showing strong cycles with a frequency of about 0.1, where the maximum value of spectrum occurs. That is to say, it indicates cycles with a period of $1/0.1 = 10$ years. There is a hint of longer period cycles (the local peak at frequency 0.033 would produce cycles of length $1/0.033 = 30$ years) but no real suggestion of any shorter-term cycles.

24.8   Multiple time series

When we have two or more time series measured over the same period, the question naturally arises as to whether or not the ups and downs of the different series are correlated. It may be that we suspect that change in one of the variables causes changes in the other (e.g. changes in the number of predators may cause changes in the number of prey, because more predators means more prey eaten). We need to be careful, of course, because it will not always be obvious which way round the causal relationship might work (e.g. predator numbers may go up because prey numbers are higher; ecologists call this a numerical response). Suppose we have the following sets of counts:

```
twoseries <- read.table("c:\\temp\\twoseries.txt",header=T)
attach(twoseries)
names(twoseries)
```

```
[1]  "x"  "y"
```

We start by inspecting the two time series one above the other:

```
plot.ts(cbind(x,y),main="")
```

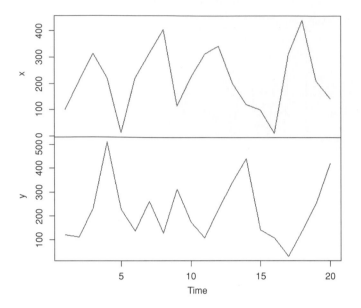

There is some evidence of periodicity (at least in *x*) and it looks as if *y* lags behind *x* by roughly 2 periods (sometimes 1). Now let us carry out straightforward analyses on each time series separately and the cross-correlation between the two series:

```
par(mfrow=c(2,2))
acf(cbind(x,y),type="p",col="red")
```

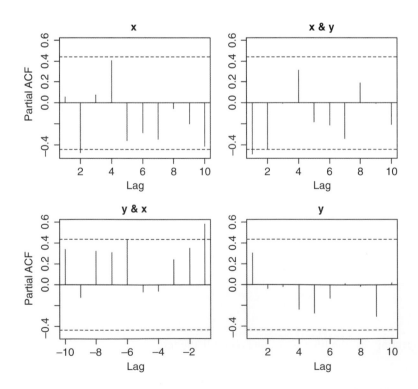

As we suspected, the evidence for periodicity is stronger in x than in y: the partial autocorrelation is significant and negative at lag 2 for x, but not for y. The interesting point is the cross-correlation between x and y which is significant at lags 1 and 2 (top right). Positive changes in x are associated with negative changes in y and vice versa.

24.9 Simulated time series

To see how the correlation structure of an AR(1) depends on the value of α, we can simulate the process over, say, 250 time periods using different values of α. We generate the white noise Z_t using the random number generator `rnorm(n,0,s)` which gives n random numbers with a mean of 0 and a standard deviation of s. To simulate the time series we evaluate

$$Y_t = \alpha Y_{t-1} + Z_t,$$

multiplying last year's population by α then adding the relevant random number from Z_t.

We begin with the special case of $\alpha = 0$ so that $Y_t = Z_t$ and the process is pure white noise:

```
Y <- rnorm(250,0,2)
windows(7,4)
par(mfrow=c(1,2))
plot.ts(Y)
acf(Y,main="")
```

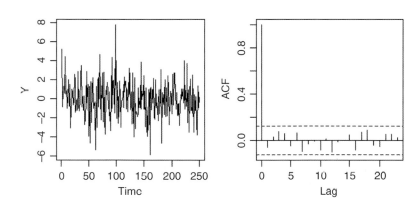

The time series is bound to be stationary because each value of Z is independent of the value before it. The correlation at lag 0 is 1 (of course), but there is absolutely no hint of any correlations at higher lags.

To generate the time series for non-zero values of α we need to use recursion: this year's population is last year's population times α plus the white noise. We begin with a negative value of $\alpha = -0.5$. First we generate all the noise values (by definition, these *do not* depend on population size):

```
Z <- rnorm(250,0,2)
```

Now the initial population at time 0 is set to 0 (remember that the population is stationary, so we can think of the Y values as departures from the long-term mean population size). This means that $Y_1 = Z_1$. Thus, Y_2 will be whatever Y_1 was, times –0.5, plus Z_2. And so on.

```
Y <- numeric(250)
Y[1] <- Z[1]
```

```
for (i in 2:250) Y[i] <-   -0.5*Y[i-1]+Z[i]
plot.ts(Y)
acf(Y,main="")
```

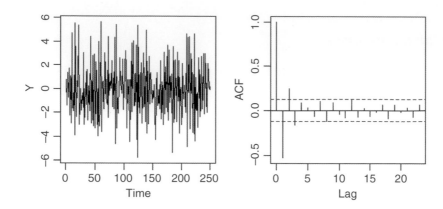

The time series shows rapid return to equilibrium following random departures from it. There is a highly significant negative autocorrelation at lag 1, significant positive autocorrelation at lag 2 and so on, with the size of the correlation gradually damping away.

Let us simulate a time series with a positive value of, say, $\alpha = 0.5$:

```
Z <- rnorm(250,0,2)
Y[1] <- Z[1]
for (i in 2:250) Y[i] <-   0.5*Y[i-1]+Z[i]
plot.ts(Y)
acf(Y, main="")
```

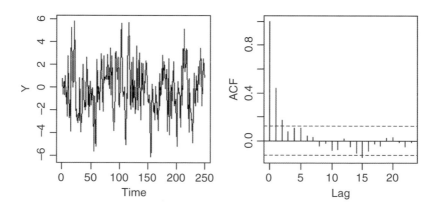

Now the time series plot looks very different, with protracted periods spent drifting away from the long-term average. The autocorrelation plot shows significant positive correlations for the first three lags.

Finally, we look at the special case of $\alpha = 1$. This means that the time series is a classic **random walk**, given by

$$Y_t = Y_{t-1} + Z_t.$$

```
Z <- rnorm(250,0,2)
Y[1] <- Z[1]
for (i in 2:250) Y[i] <-  Y[i-1]+Z[i]
plot.ts(Y)
acf(Y, main="")
```

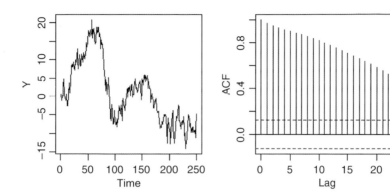

The time series wanders about and strays far away from the long-term average. The `acf` plot shows positive correlations dying away very slowly, and still highly significant at lags of more than 20. Of course, if you do another realization of the process, the time series will look very different, but the autocorrelations will be similar.

24.10 Time series models

Time series models come in three kinds (Box and Jenkins, 1976):

- moving average (MA) models where

$$X_t = \sum_{j=0}^{q} \beta_j \varepsilon_{t-j};$$

- autoregressive (AR) models where

$$X_t = \sum_{i=1}^{p} \alpha_i X_{t-i} + \varepsilon_t;$$

- autoregressive moving average (ARMA) models where

$$X_t = \sum_{i=1}^{p} \alpha_i X_{t-i} + \sum_{j=0}^{q} \beta_j \varepsilon_{t-j}.$$

A moving average of order q averages the random variation over the last q time periods. An autoregressive model of order p computes X_t as a function of the last p values of X, so, for a second-order process, we would use

$$X_t = \alpha_1 X_{t-1} + \alpha_2 X_{t-2} + \varepsilon_t.$$

Typically, we would use the partial autocorrelation plot (above) to determine the order. So, for the lynx data (p. 800) we would use order 2 or 4, depending on taste. Other things being equal, parsimony suggests the use of order 2. The fundamental difference is that a set of random components (ε_{t-j}) influences the current value of a MA process, whereas only the current random effect (ε_t) affects an AR process. Both kinds of effects are at work in an ARMA processes. Ecological models of population dynamics are typically AR models. For instance,

$$N_t = \lambda N_{t-1}$$

(the discrete-time version of exponential growth ($\lambda > 1$) or decay ($\lambda < 1$)) looks just like an first order AR process with the random effects missing. This is somewhat misleading, however, since time series are supposed to be stationary, which would imply a long-term average value of $\lambda = 1$. But, in the absence of density dependence (as here), this is impossible. The α of the AR model is *not* the λ of the population model.

Models are fitted using the `arima` function, and their performances are compared using the AIC (see p. 415). The most important component of the model is `order`. This is a vector of length 3 specifying the order of the autoregressive operators, the number of differences, and the order of moving average operators. Thus `order=c(1,3,2)` is based on a first-order autoregressive process, three differences, and a second-order moving average. The Canadian lynx data are used as an example of `arima` in time series modelling.

Records of the number of skins of predators (lynx) and prey (snowshoe hares) returned by trappers were collected over many years by the Hudson's Bay Company. The lynx numbers are shown on p. 800 and exhibit a clear 10-year cycle. We begin by plotting the autocorrelation and partial autocorrelation functions:

```
windows(7,4)
par(mfrow=c(1,2))
acf(Lynx,main="")
acf(Lynx,type="p",main="")
```

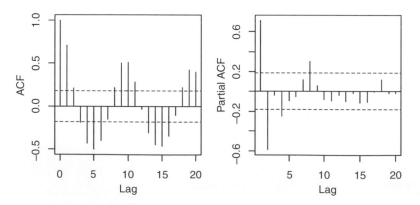

The population is very clearly cyclic, with a period of 10 years. The dynamics appear to be driven by strong, negative density dependence (a partial autocorrelation of –0.588) at lag 2. There are other significant partials at lag 1 and lag 8 (positive) and lag 4 (negative). Of course you cannot infer the mechanism by observing the dynamics, but the lags associated with significant negative and positive feedbacks are extremely interesting and highly suggestive. The main prey species of the lynx is the snowshoe hare and the negative feedback at lag 2 may reflect the timescale of this predator–prey interaction. The hares are known to cause medium-term induced reductions in the quality of their food plants as a result of heavy browsing pressure when the hares are at high density, and this could map through to lynx populations with lag 4.

The order vector specifies the non-seasonal part of the ARIMA model: the three components (p, d, q) are the AR order, the degree of differencing, and the MA order. We start by investigating the effects of AR order with no differencing and no moving average terms, comparing models on the basis of the AIC:

```
model10 <- arima(Lynx,order=c(1,0,0))
model20 <- arima(Lynx,order=c(2,0,0))
model30 <- arima(Lynx,order=c(3,0,0))
model40 <- arima(Lynx,order-c(4,0,0))
model50 <- arima(Lynx,order=c(5,0,0))
model60 <- arima(Lynx,order=c(6,0,0))
AIC(model10,model20,model30,model40,model50,model60)
```

```
        df      AIC
model10  3  1926.991
model20  4  1878.032
model30  5  1879.957
model40  6  1874.222
model50  7  1875.276
model60  8  1876.858
```

On the basis of AR alone, it appears that order 4 is best (AIC = 1874.222). What about MA?

```
model01 <- arima(Lynx,order=c(0,0,1))
model02 <- arima(Lynx,order=c(0,0,2))
model03 <- arima(Lynx,order=c(0,0,3))
model04 <- arima(Lynx,order=c(0,0,4))
model05 <- arima(Lynx,order=c(0,0,5))
model06 <- arima(Lynx,order=c(0,0,6))
AIC(model01,model02,model03,model04,model05,model06)
```

```
        df      AIC
model01  3  1917.947
model02  4  1890.061
model03  5  1887.770
model04  6  1888.279
model05  7  1885.698
model06  8  1885.230
```

The AIC values are generally higher than given by the AR models. Perhaps there is a combination of AR and MA terms that is better than either on their own?

```
model40 <- arima(Lynx,order=c(4,0,0))
model41 <- arima(Lynx,order=c(4,0,1))
model42 <- arima(Lynx,order=c(4,0,2))
model43 <- arima(Lynx,order=c(4,0,3))
AIC(model40,model41,model42,model43)
```

```
           df      AIC
model40   6 1874.222
model41   7 1875.351
model42   8 1862.435
model43   9 1880.432
```

Evidently there is no need for a moving average term (model40 is best). What about the degree of differencing?

```
model400 <- arima(Lynx,order=c(4,0,0))
model401 <- arima(Lynx,order=c(4,1,0))
model402 <- arima(Lynx,order=c(4,2,0))
model403 <- arima(Lynx,order=c(4,3,0))
AIC(model400,model401,model402,model403)
           df      AIC
model400  6 1874.222
model401  5 1890.961
model402  5 1917.882
model403  5 1946.143
```

The model with no differencing performs best. The lowest AIC is 1874.222, which suggests that a model with an AR lag of 4, no differencing and no moving average terms is best. This implies that a rather complex ecological model is required which takes account of both the significant partial correlations at lags of 2 and 4 years, and not just the 2-year lag (i.e. plant–herbivore effects may be necessary to explain the dynamics, in addition to predator–prey effects).

25

Multivariate Statistics

This class of statistical methods is fundamentally different from the others in this book because there is no response variable. Instead of trying to understand variation in a response variable in terms of explanatory variables, in multivariate statistics we look for **structure in the data**. The problem is that structure is rather easy to find, and all too often it is a feature of that particular data set alone. The real challenge is to find *general* structure that will apply to other data sets as well. Unfortunately, there is no guaranteed means of detecting pattern, and a great deal of ingenuity has been shown by statisticians in devising means of pattern recognition in multivariate data sets. The main division is between methods that **assume a given structure** and seek to divide the cases into groups, and methods that seek to **discover structure** from inspection of the dataframe. The really important point is that you need to know exactly what the question is that you are trying to answer. Do not mistake the opaque for the profound.

The multivariate techniques implemented in R include:

- principal components analysis (`prcomp`);
- factor analysis (`factanal`);
- cluster analysis (`hclust, kmeans`);
- discriminant analysis (`lda, qda`);
- neural networks (`nnet`).

These techniques are *not* recommended unless you know exactly what you are doing, and exactly *why* you are doing it. Beginners are sometimes attracted to multivariate techniques because of the complexity of the outputs they produce.

25.1 Principal components analysis

The idea of principal components analysis (PCA) is to find a small number of **linear combinations** of the variables so as to capture most of the variation in the dataframe as a whole. With a large number of variables it may be easier to consider a small number of combinations of the original data rather than the entire dataframe.

The R Book, Second Edition. Michael J. Crawley.
© 2013 John Wiley & Sons, Ltd. Published 2013 by John Wiley & Sons, Ltd.

Suppose, for example, that you had three variables measured on each subject, and you wanted to distil the essence of each individual's performance into a single number. An obvious solution is the arithmetic mean of the three numbers $\frac{1}{3}v_1 + \frac{1}{3}v_2 + \frac{1}{3}v_3$, where v_1, v_2 and v_3 are the three variables (e.g. pupils' exam scores in maths, physics and chemistry). The vector of coefficients $l = (1/3, 1/3, 1/3)$ is called a linear combination. Linear combinations where $\sum l^2 = 1$ are called standardized linear combinations. Principal components analysis finds a set of orthogonal standardized linear combinations which together explain all of the variation in the original data. There are as many principal components as there are variables, but typically it is only the first few of them that explain important amounts of the total variation.

Calculating principal components is easy. Interpreting what the components mean in scientific terms is hard, and potentially equivocal. You need to be more than usually circumspect when evaluating multivariate statistical analyses.

The following dataframe contains mean dry weights (in grams) for 54 plant species on 89 plots, averaged over 10 years; see Crawley et al. (2005) for species names and more background. What are the principal components and what environmental factors are associated with them?

```
pgdata <- read.table("c:\\temp\\pgfull.txt",header=T)
names(pgdata)
```

```
[1]     "AC"      "AE"      "AM"     "AO"    "AP"     "AR"     "AS"
[8]     "AU"      "BH"      "BM"     "CC"    "CF"     "CM"     "CN"
[15]    "CX"      "CY"      "DC"     "DG"    "ER"     "FM"     "FP"
[22]    "FR"      "GV"      "HI"     "HL"    "HP"     "HS"     "HR"
[29]    "KA"      "LA"      "LC"     "LH"    "LM"     "LO"     "LP"
[36]    "OR"      "PL"      "PP"     "PS"    "PT"     "QR"     "RA"
[43]    "RB"      "RC"      "SG"     "SM"    "SO"     "TF"     "TG"
[50]    "TO"      "TP"      "TR"     "VC"    "VK"     "plot"   "lime"
[57]    "species" "hay"     "pH"
```

We need to extract the 54 variables that refer to the species' abundances and leave behind the variables containing the experimental treatments (plot and lime) and the covariates (species richness, hay biomass and soil pH). This creates a smaller dataframe containing all 89 plots (i.e. all the rows) but only columns 1 to 54:

```
pgd <- pgdata[,1:54]
```

There are two functions for carrying out PCA in R. The generally preferred method for numerical accuracy is prcomp (where the calculation is done by a singular value decomposition of the centred and scaled data matrix, not by using eigen on the covariance matrix, as in the alternative function princomp).

The aim is to find linear combinations of a set of variables that maximize the variation contained within them, thereby displaying most of the original variation in fewer dimensions. These principal components have a value for every one of the 89 rows of the dataframe. By contrast, in factor analysis (see p. 813), each factor contains a contribution (which may in some cases be zero) from each variable, so the length of each factor is the total number of variables (54 in the current example). This has practical implications, because you can plot the principal components against other explanatory variables from the dataframe, but you cannot do this for factors because the factors are of length 54 while the covariates are of length 89. You need to think about this until the penny drops.

We shall use the option `scale=TRUE` because the variances are significantly different for the 54 plant species:

```
model <- prcomp(pgd,scale=TRUE)
summary(model)

Importance of components:
                        PC1     PC2     PC3     PC4     PC5     PC6     PC7     PC8
Standard deviation      3.005   2.336   1.9317  1.786   1.7330  1.5119  1.5088  1.3759
Proportion of Variance  0.167   0.101   0.0691  0.059   0.0556  0.0423  0.0422  0.0351
Cumulative Proportion   0.167   0.268   0.3373  0.396   0.4520  0.4943  0.5365  0.5716
...
...
                        PC53    PC54
Standard deviation      0.11255 0.01721
Proportion of Variance  0.00023 0.00001
Cumulative Proportion   0.99999 1.00000
```

You can see that the first principal component (PC1) explains 16.7% of the total variation, and only the next four (PC2–PC5) explain more than 5% of the total variation. Here is the plot of this model, showing the relative importance of PC1.

```
plot(model,main="",col="green")
```

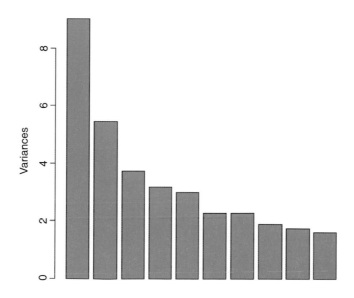

This is called a **scree plot** in PCA because it is supposed to look like a cliff face on a mountainside (on the left), with a scree slope below it (the tail on the right). The standard practice is to assume that you need sufficient principal components to account for 90 % of the total variation (but that would take 24 components in the present case). Principal component loadings show how the original variables (the 54 different species in our example) influence the principal components.

In a biplot, the original variables are shown by arrows (54 of them in this case):

```
biplot(model)
```

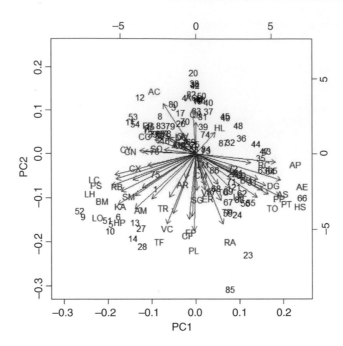

The numbers represent the rows in the original dataframe, and the directions of the arrows show the relative loadings of the species on the first and second principal components. Thus, the species AP, AE and HS have strong positive loadings on PC1 and LC, PS BM and LO have strong negative loadings. PC2 has strong positive loadings from species AO and AC and negative loadings from TF, PL and RA.

If there are explanatory variables available, you can plot these against the principal components to look for patterns. In this example, it looks as if the first principal component is associated with increasing biomass (and hence increasing competition for light) and as if the second principal component is associated with declining soil pH (increasing acidity):

```
yv <- predict(model)[,1]
yv2 <- predict(model)[,2]
windows(7,4)
par(mfrow=c(1,2))
plot(pgdata$hay,yv,pch=16,xlab="biomass",ylab="PC 1",col="red")
plot(pgdata$pH,yv2,pch=16,xlab="soil pH",ylab="PC 2",col="blue")
```

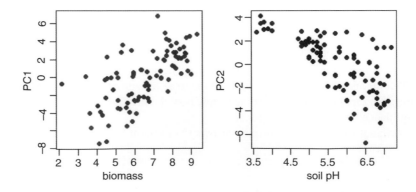

There are, indeed, very strong correlations between biomass and PC1 and between soil pH and PC2. Now the real work would start, because we are interested in the mechanisms that underlie these patterns.

25.2   Factor analysis

With principal components analysis we were fundamentally interested in the variables and their contributions. Factor analysis aims to provide usable numerical values for quantities such as intelligence or social status that are not directly measurable. The idea is to use correlations between observable variables in terms of underlying 'factors'. Note that 'factors' in factor analysis are *not* the same as the categorical explanatory variables we have been calling factors throughout the rest of this book.

Compared with PCA, the variables themselves are of relatively little interest in factor analysis; it is gaining an understanding of the hypothesized underlying factors that is the main aim. The idea is that the correlations amongst the variables are explained by the common factors. The function `factanal` performs maximum likelihood factor analysis on a covariance matrix or data matrix. The `pgd` dataframe is introduced on p. 810. You need to specify the number of factors you want to estimate – we begin with 8:

```
factanal(pgd,8)

Call:
factanal(x = pgd, factors = 8)
Uniquenesses:
    AC     AE     AM     AO     AP     AR     AS     AU     BH     BM
 0.638  0.086  0.641  0.796  0.197  0.938  0.374  0.005  0.852  0.266
    CC     CF     CM     CN     CX     CY     DC     DG     ER     FM
 0.056  0.574  0.786  0.579  0.549  0.733  0.837  0.408  0.072  0.956
    FP     FR     GV     HI     HL     HP     HS     HR     KA     LA
 0.371  0.815  0.971  0.827  0.921  0.218  0.332  0.915  0.319  0.305
    LC     LH     LM     LO     LP     OR     PL     PP     PS     PT
 0.349  0.333  0.927  0.121  0.403  0.005  0.286  0.606  0.336  0.401
    QR     RA     RB     RC     SG     SM     SO     TF     TG     TO
 0.913  0.491  0.005  0.754  0.341  0.212  0.825  0.428  0.476  0.469
    TP     TR     VC     VK
 0.309  0.611  0.651  0.170
Loadings:
    Factor1 Factor2 Factor3 Factor4 Factor5 Factor6 Factor7 Factor8
AC  -0.512  -0.268                           0.121
AE   0.925  -0.107                  -0.146           -0.118
AM  -0.206   0.413   0.213           0.163    0.115   0.153   0.186
AO  -0.312  -0.196  -0.151  -0.105           -0.148  -0.102
AP   0.827  -0.173  -0.195  -0.167           -0.123
AR           0.150           0.111                    0.127
AS   0.778
AU                                                            0.996
BH   0.380
BM  -0.116   0.292           0.695                    0.380
CC  -0.152                   0.159           0.943
CF           0.539                   0.342
CM                   0.434  -0.110
CN  -0.276   0.143                            0.541   0.147
CX                           0.628            0.169   0.146
CY  -0.211          -0.162   0.340                    0.270
```

	Factor1	Factor2	Factor3	Factor4	Factor5	Factor6	Factor7	Factor8
DC		-0.125				0.372		
DG	0.738		-0.127			0.145		
ER					0.960			
FM	-0.108					0.133		
FP	0.245	0.226		0.478	0.493		-0.176	
FR	-0.386		-0.144					
GV	-0.134							
HI	-0.202	-0.129	-0.163	0.182			0.216	
HL		-0.157		-0.127		-0.139		
HP	-0.155	0.832					0.240	
HS	0.746	-0.102	0.257	-0.152				
HR	-0.155	-0.107	-0.122	0.101			0.150	
KA	-0.167	0.774	-0.169	0.139				
LA						0.829		
LC	-0.306	0.378	-0.125	0.529			0.328	
LH	-0.256	0.556	-0.132	0.421		0.223	0.195	
LM					0.112	0.221		
LO	-0.129	0.432		0.781			0.251	
LP	0.115		0.745					
OR							0.996	
PL		0.369	0.675		0.337			
PP	0.527	0.226		-0.167		-0.175		
PS	-0.212	0.301	-0.130	0.681		0.150	0.158	
PT	0.741			-0.100	0.150	-0.105		
QR	-0.194	-0.135						
RA	0.195	0.227	0.578		0.205	-0.166	-0.107	
RB	-0.122	0.158		0.272			0.934	
RC	0.361			-0.198		-0.176	-0.152	
SG					0.806			
SM		0.388					0.787	
SO			-0.100	0.386				
TF		0.702	0.260					
TG	0.141		0.583	-0.110		0.367	0.107	
TO	0.418		0.567	-0.158				
TP			0.818					
TR	0.141	0.306	0.238				0.458	
VC	0.403	0.246	0.309			-0.169		
VK					0.909			

	Factor1	Factor2	Factor3	Factor4	Factor5	Factor6	Factor7	Factor8
SS loadings	5.840	3.991	3.577	3.540	3.028	2.644	2.427	2.198
Proportion Var	0.108	0.074	0.066	0.066	0.056	0.049	0.045	0.041
Cumulative Var	0.108	0.182	0.248	0.314	0.370	0.419	0.464	0.505

Test of the hypothesis that 8 factors are sufficient.
The chi-squared statistic is 1675.57 on 1027 degrees of freedom.
The p-value is 5.92e-34

On factor 1 you see strong positive correlations with AE, AP and AS and negative correlations with AC, AO and FR: this has a natural interpretation as a gradient from tall neutral grassland (positive correlations) to short, acidic grasslands (negative correlations). On factor 2, low-growing species associated with moderate to high soil pH (AM, CF, HP, KA) have large positive values and low-growing acid-loving species (AC and AO) have negative values. Factor 3 picks out the key nitrogen-fixing (legume) species LP and TP with high positive values. And so on.

Note that the loadings are of length 54 (the number of variables) not 89 (the number of rows in the dataframe representing the different plots in the experiment), so we cannot plot the loadings against the covariates as we did with PCA (p. 812). However, we can plot the factor loadings against one another:

```
model <- factanal(pgd,8)
windows(7,7)
par(mfrow=c(2,2))
plot(loadings(model)[,1],loadings(model)[,2],pch=16,xlab="Factor 1",
     ylab="Factor 2",col="blue")
plot(loadings(model)[,1],loadings(model)[,3],pch=16,xlab="Factor 1",
     ylab="Factor 3",col="red")
plot(loadings(model)[,1],loadings(model)[,4],pch=16,xlab="Factor 1",
     ylab="Factor 4",col="green")
plot(loadings(model)[,1],loadings(model)[,5],pch=16,xlab="Factor 1",
     ylab="Factor 5",col="brown")
```

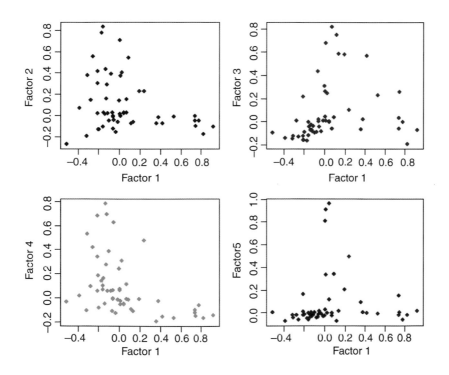

What factanal does would conventionally be described as exploratory, not confirmatory, factor analysis. For the latter, try the sem package:

```
install.packages("sem")
library(sem)
?sem
```

25.3   Cluster analysis

Cluster analysis is a set of techniques that look for groups (clusters) in the data. Objects belonging to the same group resemble each other. Objects belonging to different groups are dissimilar. Sounds simple, doesn't it? The problem is that there is usually a huge amount of redundancy in the explanatory variables. It is not at all obvious which measurements (or combinations of measurements) will turn out to be the ones that are best for allocating individuals to groups. There are three ways of carrying out such allocation:

- partitioning into a number of clusters specified by the user, with functions such as kmeans;
- hierarchical, starting with each individual as a separate entity and ending up with a single aggregation, using functions such as hclust;
- divisive, starting with a single aggregate of all the individuals and splitting up clusters until all the individuals are in different groups.

25.3.1   Partitioning

The kmeans function operates on a dataframe in which the columns are variables and the rows are the individuals. Group membership is determined by calculating the centroid for each group. This is the multidimensional equivalent of the mean. Each individual is assigned to the group with the nearest centroid. The kmeans function fits a *user-specified* number of cluster centres, such that the within-cluster sum of squares from these centres is minimized, based on Euclidian distance. Here are data from six groups with two continuous explanatory variables (*x* and *y*):

```
kmd <- read.table("c:\\temp\\kmeansdata.txt",header=T)
attach(kmd)
names(kmd)

[1] "x" "y" "group"

par(mfrow=c(2,2))
plot(x,y,pch=16)
```

The raw data show hints of clustering (top left panel) but the clustering becomes clear only after the groups have been coloured differently using col=group (top right). kmeans does an excellent job when it is told that there are six clusters (open symbols, bottom left) but, of course, there can be no overlap in assignation (as there was in the original data). When just four clusters are estimated, it is the centre cluster and the south-east cluster that disappear (open symbols, bottom right).

```
plot(x,y,col=group,pch=16)
model <- kmeans(data.frame(x,y),6)
plot(x,y,col=model[[1]])
model <- kmeans(data.frame(x,y),4)
plot(x,y,col=model[[1]])
par(mfrow=c(1,1))
```

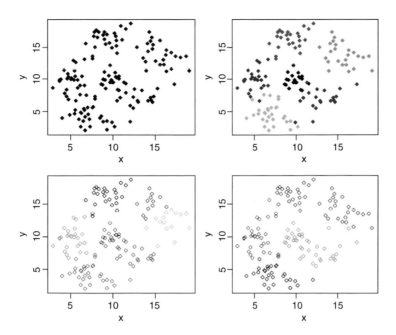

You should compare the bottom two plots carefully to see which points have changed groups.

To see the rate of misclassification we can tabulate the real groups against the groups determined by kmeans:

```
model <- kmeans(data.frame(x,y),6)
table(model[[1]],group)
```

```
  group
    1  2  3  4  5  6
1   0  0 25  1  0  0
2   0  0  0  0  0 25
3   0  0  0  2 10  0
4   0 23  0  0  0  0
5   0  2  0  0 20  0
6  20  0  0 17  0  0
```

The first thing to note is that R has numbered its groups differently (as it would, because it does not know our names for the groups). Group 1 (in the first column) were associated perfectly (20 out of 20 in R's group 6). Group 2 had 23 of its members in R's group 4 but 2 out of 25 were allocated to R's group 5. Group 3 was classified perfectly. Group 4 was less good, with one misapplication to R's group 1 and two to R's group 3. Group 6 was perfectly classified. The only really poor performance was with Group 5, which was split 20 to 10. This is impressive, given that there were several obvious overlaps in the original data.

25.3.2 Taxonomic use of kmeans

In the next example we have measurements of seven variables on 120 individual plants. The question is which of the variables (fruit size, bract length, internode length, petal width, sepal length, petiole length or leaf width) are the most useful taxonomic characters.

```
taxa <- read.table("c:\\temp\\taxon.txt",header=T)
```

```
attach(taxa)
names(taxa)
```

```
[1]  "Petals"  "Internode"  "Sepal"  "Bract"  "Petiole"  "Leaf"
[7]  "Fruit"
```

A simple and sensible way to start is by looking at the dataframe as a whole, using `pairs` to plot every variable against every other:

```
pairs(taxa)
```

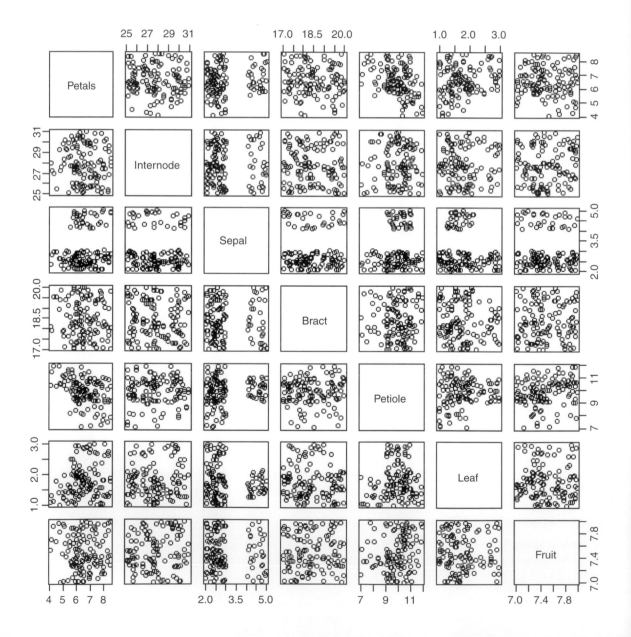

There appears to be excellent data separation on sepal length, and reasonable separation on petiole length and leaf width, but nothing obvious for the other variables.

These data actually come from four taxa (labelled I–IV), so in this contrived case we *know* that there are four groups. In reality, of course, we would not know this, and finding out the number of groups would be one of the central aims of the study. We begin, therefore, by seeing how well kmeans allocates individuals to four groups:

```
kmeans(taxa,4)

K-means clustering with 4 clusters of sizes 22, 19, 36, 43
Cluster means:
     Petals Internode    Sepal    Bract   Petiole     Leaf    Fruit
1 7.167446   29.02460 2.445634 18.41452  8.684403 1.727075 7.475693
2 6.996461   26.56658 4.614698 18.32699 10.059406 1.653439 7.425915
3 6.732706   26.20788 2.486903 18.50806  9.641293 1.992337 7.437037
4 5.834561   29.18766 3.004510 18.45372 10.529315 1.736012 7.558543

Clustering vector:
  1   2   3   4   5   6   7   8   9  10  11  12  13  14  15  16  17  18  19  20  21  22  23  24  25  26  27  28  29  30  31
  4   4   4   3   4   4   4   3   4   4   4   3   4   4   4   4   4   4   3   4   4   4   3   4   4   4   4   4   4   4   3
 32  33  34  35  36  37  38  39  40  41  42  43  44  45  46  47  48  49  50  51  52  53  54  55  56  57  58  59  60  61  62
  3   1   1   1   3   1   3   1   1   4   1   3   1   3   1   1   3   1   3   1  `3   3   3   1   3   1   1   1   3   4   3
 63  64  65  66  67  68  69  70  71  72  73  74  75  76  77  78  79  80  81  82  83  84  85  86  87  88  89  90  91  92  93
  3   3   1   4   3   3   1   4   3   4   3   1   3   3   4   1   4   3   3   3   3   4   1   1   3   3   4   2   4   2
 94  95  96  97  98  99 100 101 102 103 104 105 106 107 108 109 110 111 112 113 114 115 116 117 118 119 120
  4   2   2   4   4   2   2   4   2   2   4   2   2   2   2   2   2   4   4   2   2   2   4   2   4   4

Within cluster sum of squares by cluster:
[1]   75.15813   64.09288 142.36779 186.00036
(between_SS / total_SS =   45.6 %)
```

Not very impressive at all. Of course, the computer was doing its classification blind. But we know the four islands from which the plants were collected. Can we write a key that ascribes taxa to islands better than the blind scheme adopted by kmeans? The answer is definitely yes. When we used a classification tree model on the same data, it discovered a faultless key on its own (see Section 23.6).

25.4   Hierarchical cluster analysis

The idea behind hierarchical cluster analysis is to show which of a (potentially large) set of samples are most similar to one another, and to group these similar samples in the same limb of a tree. Groups of samples that are distinctly different are placed in other limbs. The trick is in defining what we mean by 'most similar'. Each of the samples can be thought of a sitting in an *m*-dimensional space, defined by the *m* variables (columns) in the dataframe. We define similarity on the basis of the **distance** between two samples in this *m*-dimensional space. Several different distance measures could be used, but the default is Euclidean distance (for the other options, see ?dist), and this is used to work out the distance from every sample to every other sample. This quantitative dissimilarity structure of the data is stored in a matrix produced by the dist function. Initially, each sample is assigned to its own cluster, and then the hclust algorithm proceeds iteratively, at each stage joining the two most similar clusters, continuing until there is just a single cluster (see ?hclust for details).

The following data (introduced on p. 810) show the distribution of 54 plant species over 89 plots receiving different experimental treatments. The aim is to see which plots are most similar in their botanical composition, and whether there are reasonably homogeneous groups of plots that might represent distinct plant communities.

```
pgdata <- read.table("c:\\temp\\pgfull.txt",header=T)
attach(pgdata)
labels <- paste(plot,letters[lime],sep="")
```

The first step is to turn the matrix of measurements on individuals into a dissimilarity matrix. In the dataframe, the columns are variables and the rows are the individuals, and we need to calculate the 'distances' between each row in the dataframe and every other using `dist(pgdata[,1:54])`. These distances are then used to carry out hierarchical cluster analysis using the `hclust` function:

```
hpg <- hclust(dist(pgdata[,1:54]))
```

We can plot the object called `hpg`, and we specify that the leaves of the hierarchy are labelled by their plot numbers (pasted together from the plot number and lime treatment):

```
plot(hpg,labels=labels,main= "")
```

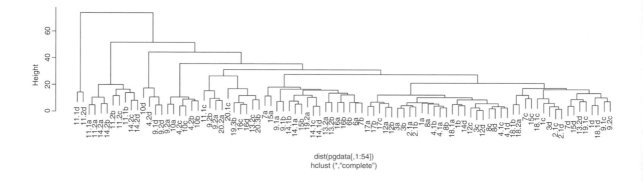

If you view this object in full-screen mode within R you will be able to read all the plot labels, and to work out the groupings. It turns out that the groupings have very natural scientific interpretations. The highest break, for instance, separates the two plots dominated by *Holcus lanatus* (11.1d and 11.2d) from the other 87 plots. The second break distinguishes the high nitrogen plots also receiving phosphorus (plots 11 and 14). The third break takes out the acidified plots (numbers 9, 10 and 4.2). The plots on the right-hand side all have soils that exhibit phosphorus deficiency. The leftmost groups are all from plots receiving high rates of nitrogen and phosphorus input. More subtly, plots 12a and 3a are supposed to be the same (they are replicates of the no-fertilizer, high-lime treatment), yet they are separated at a break at height approximately 15. And so on. The `hclust` function has done an excellent job of recognizing real plant communities over the top seven splits.

Let us try hierarchical clustering on the taxonomic data (p. 817).

```
plot(hclust(dist(taxa)),main="")
```

Because in this artificial example we know that the first 30 rows in the dataframe come from group 1, rows 31–60 from group 2, rows 61–90 from group 3 and rows 91–120 from group 4, we can see than the grouping produced by hclust is pretty woeful. Most of the rows in the leftmost major split are from group 2, but the rightmost split contains members from groups 1, 4 and 3. Neither kmeans nor hclust is up to the job in this case. When we *know* the group identities, then it is easy to use tree models to devise the optimal means of distinguishing and classifying individual cases (see p. 781).

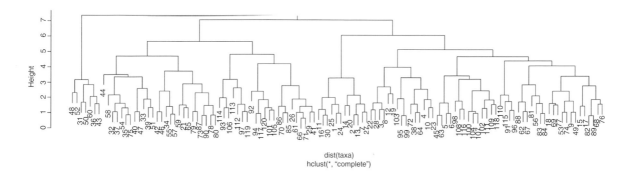

dist(taxa)
hclust(*, "complete")

So there we have it. When we *know* the identity of the species, then tree models are wonderfully efficient at constructing keys to distinguish between the individuals, and at allocating them to the relevant categories. When we do *not* know the identities of the individuals, then the statistical task is much more severe, and inevitably ends up being much more error-prone. The best that cluster analysis could achieve through kmeans was with five groups (one too many, as we know, having constructed the data) and in realistic cases we have no way of knowing whether five groups was too many, too few or just right. Multivariate clustering without a response variable is fundamentally difficult and equivocal.

25.5 Discriminant analysis

In this case, you know the identity of each individual (unlike cluster analysis) and you want to know how the explanatory variables contribute to the correct classification of individuals. The method works by uncovering relationships among the groups' covariance matrices to discriminate between groups. With k groups you will need $k - 1$ discriminators. The functions you will need for discriminant analysis are available in the MASS library. Returning to the taxon dataframe (see p. 817), we will illustrate the use of lda to carry out a linear discriminant analysis. For further relevant information, see ?predict.lda and ?qda in the MASS library.

```
library(MASS)
model <- lda(Taxon~.,taxa)
plot(model,col=rep(1:4,each=30))
```

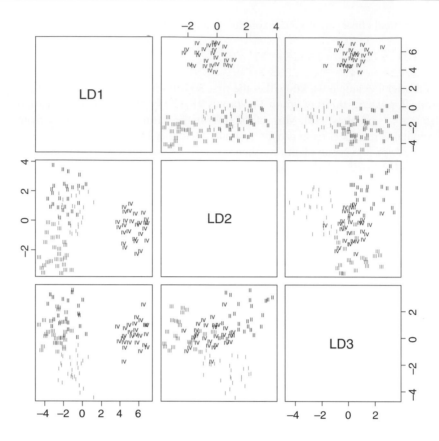

The linear discriminators LD1 and LD2 clearly separate taxon IV without error, but this is easy because there is no overlap in sepal length between this taxon and the others. LD2 and LD3 are quite good at finding taxon II (upper right), and LD1 and LD3 are quite good at getting taxon I (bottom left). Taxon III would be what was left over. Here is the printed model:

```
model

Call:
lda(Taxon ~ ., data = taxa)

Prior probabilities of groups:
   I    II   III    IV
0.25 0.25 0.25 0.25

Group means:
      Petals Internode     Sepal     Bract   Petiole      Leaf     Fruit
I   5.476128  27.91886  2.537955  18.60268 10.864184  1.508029  7.574642
II  7.035078  27.69834  2.490336  18.47557  8.541085  1.450260  7.418702
III 6.849666  27.99308  2.446003  18.26330  9.866983  2.588555  7.482349
IV  6.768464  27.78503  4.532560  18.42953 10.128838  1.645945  7.467917

Coefficients of linear discriminants:
                  LD1          LD2          LD3
Petals    -0.01891137  0.034749952  0.559080267
```

```
Internode    0.03374178    0.009670875    0.008808043
Sepal        3.45605170   -0.500418135    0.401274694
Bract        0.07557480    0.068774714   -0.024930728
Petiole      0.25041949   -0.343892260   -1.249519047
Leaf        -1.13036429   -3.008335468    0.647932763
Fruit        0.18285691   -0.208370808   -0.269924935

Proportion of trace:
   LD1     LD2     LD3
0.7268  0.1419  0.1313
```

So you would base your key on sepal first (3.45) then leaf (–3.008) then petiole (–1.249). Compare this with the key uncovered by the tree model on p. 778. Here are the predictions from the linear discriminant analysis:

```
predict(model)

$class
 [1] I   I   I   I   I   I   I   I   I   I   I   I   I   I   I   I   I   I   I   I   III I   I   I   I   I   I   I   I   I
[31] II  II  II  II  II  II  II  II  II  II  II  II  II  II  II  II  II  II  II  II  II  II  II  II  II  II  II  II  II  II
[61] III III III III III III III III III III III III III III III III III III III III III III III III III III III III III III
[91] IV  IV  IV  IV  IV  IV  IV  IV  IV  IV  IV  IV  IV  IV  IV  IV  IV  IV  IV  IV  IV  IV  IV  IV  IV  IV  IV  IV  IV  IV
Levels: I II III IV
```

One of the members of taxon I is misallocated to taxon III (case 21), but otherwise the discrimination is perfect. You can train the model with a random subset of the data (say, half of it; 60 random cases):

```
train <- sort(sample(1:120,60))
table(Taxon[train])

 I  II  III  IV
13  18   16  13
```

This set has only 13 members of taxon I and IV but reasonable representation of the other two taxa (there would be 15 of each in an even split). Now use this for training:

```
model2 <- lda(Taxon~.,taxa,subset=train)
predict(model2)

$class
 [1] I   I   I   I   I   I   I   I   I   I   I   I   I   I   I   I   I   I   II  II  II  II  II  II  II  II  II  II  II  II
[31] II  III III III III III III III III III III III III III III IV  IV  IV  IV  IV  IV  IV  IV  IV  IV  IV  IV  IV  IV
Levels: I II III IV
```

This is still very good: the first 13 should be I, the next 18 II, and so on. The discrimination is perfect in this randomization. You can use the model based on the training data to predict the unused data:

```
unused <- taxa[-train,]
predict(model,unused)

$class
 [1] I   I   I   I   I   I   I   I   I   I   III I   I   II  II  II  II  II  II  II  II  II  II  II  II  II  II  II  II  II  III
[31] III III III III III III III III III III III III III III III IV  IV  IV  IV  IV  IV  IV  IV  IV  IV  IV  IV  IV  IV
Levels: I II III IV
```

```
table(unused$Taxon)

 I  II  III  IV
17  12   14  17
```

As you can see, one of the first 17 that should have been taxon I was misclassified as taxon III, but all the other predictions were spot on.

25.6   Neural networks

These are computationally intensive methods for finding pattern in data sets that are so large, and contain so many explanatory variables, that standard methods such as multiple regression are impractical (they would simply take too long to plough through). The key feature of neural network models is that they contain a *hidden layer*: each node in the hidden layer receives information from each of many inputs, sums the inputs, adds a constant (the bias) then transforms the result using a fixed function. Neural networks can operate like multiple regressions when the outputs are continuous variables, or like classifications when the outputs are categorical. They are described in detail by Ripley (1996). Facilities for analysing neural networks are in the MASS library.

26

Spatial Statistics

There are three kinds of problems that you might tackle with spatial statistics:

- point processes (locations and spatial patterns of individuals);
- maps of a continuous response variable (kriging);
- spatially explicit responses affected by the identity, size and proximity of neighbours.

26.1   Point processes

There are three broad classes of spatial pattern on a continuum from complete regularity (evenly spaced hexagons where every individual is the same distance from its nearest neighbour) to complete aggregation (all the individuals clustered into a single clump): we call these regular, random and aggregated patterns and they look like this:

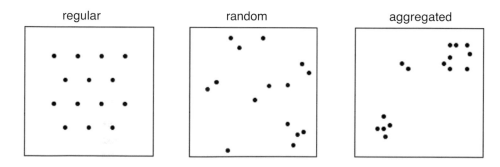

In their simplest form, the data consist of sets of *x* and *y* coordinates within some sampling frame such as a square or a circle in which the individuals have been mapped. The first question is often whether there is any evidence to allow rejection of the null hypothesis of **complete spatial randomness** (CSR). In a **random** pattern the distribution of each individual is completely independent of the distribution of every other. Individuals neither inhibit nor promote one another. In a **regular** pattern individuals are more spaced out than in a random one, presumably because of some mechanism (such as competition) that eliminates individuals that are too close together. In an **aggregated** pattern, individual are more clumped than in a random one, presumably because of some process such as reproduction with limited dispersal, or because of underlying spatial heterogeneity (e.g. good patches and bad patches).

Counts of individuals within sample areas (quadrats) can be analysed by comparing the frequency distribution of counts with a Poisson distribution with the same mean. Aggregated spatial patterns (in which the variance is greater than the mean) are often well described by a negative binomial distribution with aggregation parameter k (see p. 315). The main problem with quadrat-based counts is that they are highly **scale-dependent**. The same spatial pattern could appear to be regular when analysed with small quadrats, aggregated when analysed with medium-sized quadrats, yet random when analysed with large quadrats.

Distance measures are of two broad types: measures from individuals to their nearest neighbours, and measures from random points to the nearest individual. Recall that the nearest individual to a random point is *not* a randomly selected individual: this protocol favours selection of isolated individuals and individuals on the edges of clumps.

In other circumstances, you might be willing to take the existence of patchiness for granted, and to carry out a more sophisticated analysis of the spatial attributes of the patches themselves, their mean size and the variance in size, spatial variation in the spacing of patches of different sizes, and so on.

26.1.1 Random points in a circle

The circle is specified by the x and y coordinates of its centre and by the radius. We can compute the coordinates of the circumference of a circle of radius r, with its centre located at $x = y = 0$ like this

```
x <- r*sin(angle)
y <- r*cos(angle)
```

where `angle` varies between 0 and 2π radians. There are two ways to generate random points within this circle: one is to assume that the circle is a target, and that I am aiming at the centre; and the other is to assume that I am cutting a circular patch out of a sea of spatially independent points. In the first case we might generate a uniformly random angle, then generate a uniformly distributed random distance along this radius. Overall, these points will cluster around the centre of the circle because the random radii are most densely clustered here.

```
point <- function(r) {
angle <- runif(1)*2*pi
length <- runif(1)*r
x <- length*sin(angle)
y <- length*cos(angle)
return (data.frame(x,y))
}
```

The easting and northing of the centre of the circle are `e0` and `n0` respectively, the radius is `r` and we want to plot 1000 random points within the circle:

```
e0 <- 10
n0 <- 10
plot(e0,n0,ylab="",xlab="",ylim=c(0,2*n0),xlim=c(0,2*e0),type="n")
```

```
n <- 1000
r <- 10

for (i in 1:n) {
a <- point(r)
e <- e0+a[1]
n <- n0+a[2]
points(e,n,pch=16,col="blue")
}
```

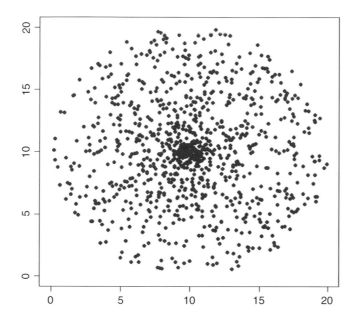

If, instead of plotting the random points, you draw lines from the centre of the circle to the random points, you can see exactly why this algorithm gives random points that are clustered around the centre.

```
e0 <- 10
n0 <- 10
plot(e0,n0,ylab="",xlab="",ylim=c(0,2*n0),xlim=c(0,2*e0),type="n")

n <- 1000
r <- 10

for (i in 1:n) {
a <- point(r)
e <- e0+a[1]
n <- n0+a[2]
lines(c(e0,e),c(n0,n),col= "red")
}
```

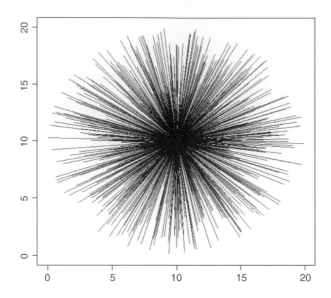

A different question is the 'cookie cutter' case. If I throw a circular quadrat onto a spatially uniform map of random points, what does the distribution of my randomly selected points look like? Here is the pseudo-code:

- Make a square map of *n* random points (uniform eastings and uniform northings).

- Make a polygon to describe the circumference of your circular sampling quadrat.

- Put the quadrat on the square map, and use the function from `maptools` to ask whether or not every point on the map is, or is not, inside your circle (the function is called `point.in.polygon` and returns a 1 for `TRUE` and a zero for `FALSE`).

- Use the output vector called `wanted` to select the points that are in your circle

  Here is the R code for 10 000 random points in a square region whose side is of length 10:

```
n <- 10000
side <- 10

library(maptools)

space <- cbind((runif(n)*side),(runif(n)*side))
plot(space)

circle <- function(e,n,r) {
angle <- seq(0,2*pi,2*pi/360)
x - r*sin(angle)
y - r*cos(angle)
return (cbind((x+e),(y+n)))
}
```

Select the random points in a circle of radius 1 centred at (8, 8):

```
xc - 8
yc <- 8
rc <- 1
```

```
outline <- circle(xc,yc,rc)

wanted <- point.in.polygon(space[,1],space[,2],outline[,1],outline[,2])

points(space[,1][wanted==1],space[,2][wanted==1], col="blue",pch=16)
```

Now add a bigger red circle of points centred at (5, 5):

```
xc <- 5
yc <- 5
rc <- 2

outline<-circle(xc,yc,rc)

wanted<-point.in.polygon(space[,1],space[,2],outline[,1],outline[,2])

points(space[,1][wanted==1],space[,2][wanted==1], col="red",pch=16)
```

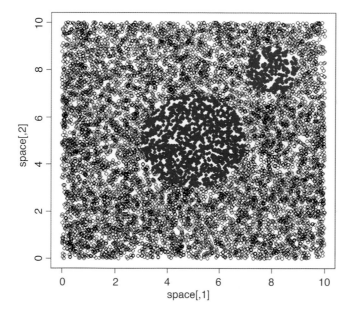

As intended, there is no clustering of these points around the centres of the circles. If the circle represents a small fraction of the total area of the square, then this method is very inefficient.

26.2 Nearest neighbours

Suppose that we have been set the problem of drawing lines to join the nearest neighbour pairs of any given set of points (x, y) that are mapped in two dimensions. There are three steps to the computing: we need to

- compute the distance to every neighbour;
- identify the smallest neighbour distance for each individual;
- use these minimal distances to identify all the nearest neighbours.

We start by generating a random spatial distribution of 100 individuals by simulating their x and y coordinates from a uniform probability distribution:

```
x <- runif(100)
y <- runif(100)
```

The graphics parameter `pty="s"` makes the plotting area square, as we would want for a map like this:

```
par(pty="s")
plot(x,y,pch=21,bg="red")
```

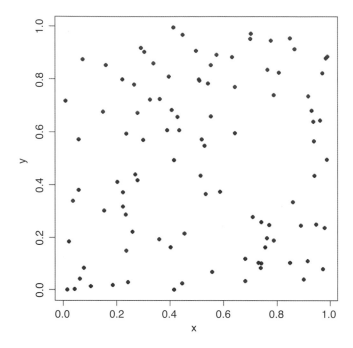

Computing the distances is straightforward: for each individual we use Pythagoras to calculate the distance to every other plant. The distance between two points with coordinates (x_1, y_1) and (x_2, y_2) is d:

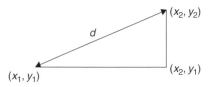

The square on the hypotenuse (d^2) is the sum of the squares on the two adjacent sides: $(x_2 - x_1)^2 + (y_2 - y_1)^2$ so the distance d is given by

$$d = \sqrt{(y_2 - y_1)^2 + (x_2 - x_1)^2}.$$

We write a function for this as follows:

```
distance <- function(x1,y1,x2,y2) sqrt((x2 - x1)^2 + (y2 - y1)^2)
```

Now we loop through each individual i and calculate a vector of distances, d, from every other individual. The nearest neighbour distance is the minimum value of d, and the identity of the nearest neighbour, nn, is found using the which function, which(d==min(d[-i])), which gives the subscript of the minimum value of d (the [-i] is necessary to exclude the distance 0 which results from the ith individual's distance from itself). Here is the complete code to compute nearest neighbour distances, r, and identities, nn, for all 100 individuals on the map:

```
r <- numeric(100)
nn <- numeric(100)
d <- numeric(100)
for (i in 1:100) {
for (k in 1:100) d[k] <- distance(x[i],y[i],x[k],y[k])
r[i] <- min(d[-i])
nn[i] <- which(d==min(d[-i]))
}
```

Now we can fulfil the brief, and draw lines to join each individual to its nearest neighbour, like this:

```
for (i in 1:100) lines(c(x[i],x[nn[i]]),c(y[i],y[nn[i]]),col="green")
```

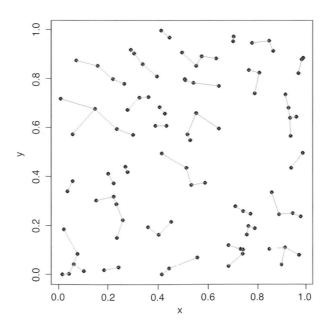

Note that when two points are very close together, and each is the nearest neighbour of the other, it can look as if a single point is not joined to any neighbours.

The next task is to work out how many of the individuals are closer to the edge of the area than they are to their nearest neighbour. Because the bottom and left margins are at $y = 0$ and $x = 0$ respectively, the y coordinate of any point gives the distance from the bottom edge of the area and the x coordinate gives the distance from the left-hand margin. We need only work out the distance of each individual from the top and right-hand margins of the area:

```
topd <- 1-y
rightd <- 1-x
```

Now we use the parallel minimum function `pmin` to work out the distance to the nearest margin for each of the 100 individuals:

```
edge <- pmin(x,y,topd,rightd)
```

Finally, we count the number of cases where the distance to the edge is less than the distance to the nearest neighbour:

```
sum(edge<r)
```

```
[1] 25
```

We identify these points on the map by circling them in red:

```
plot(x,y,pch=16)
id <- which(edge<r)
points(x[id],y[id],col="red",cex=1.5,lwd=2)
```

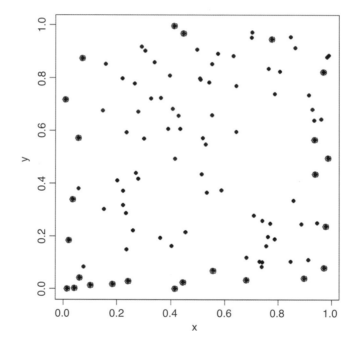

It is the vertical or horizontal distance to the edge that has been used to identify these points, so some of them look suspiciously close to their neighbours (e.g. in the bottom left-hand corner).

Edge effects are potentially very important in spatial point processes, especially when there are few individuals or the mapped area is long and thin (rather than square or circular). Excluding the individuals that are closer to the edge than to their nearest neighbour reduces the mean nearest neighbour distance:

```
mean(r)
```

```
[1] 0.05294168
```

```
mean(r[-id])
```

```
[1] 0.04802602
```

26.2.1 Tessellation

The procedure of splitting a two-dimensional surface into a mosaic by halving the distance between neighbouring pairs of points is called **tessellation**. There is a function to do this in the `tripack` package by Albrecht Gebhardt:

```
install.packages("tripack")
library(tripack)
x<-runif(100)
y<-runif(100)
```

Create a Voronoi object (here called `map`) by applying the function called `voronoi.mosaic` to the vectors of x and y coordinates.

```
map<-voronoi.mosaic(x,y)
```

Start by producing a scatterplot of the random points (in green):

```
plot(x,y,pch=16,col="green")
```

Now add the Voronoi tesselation on top of the scatterplot:

```
plot.voronoi(map,pch=16,add=TRUE)
```

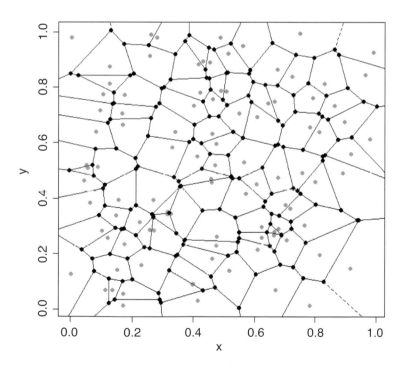

As you can see, it is relatively unusual for the points to be in the 'centre of gravity' of their tessellated patch. Each node (black circle) is a circumcircle centre of some triangle from the Delaunay triangulation.

26.3   Tests for spatial randomness

Clark and Evans (1954) give a very simple test of spatial randomness. Making the strong assumption that you know the population density of the individuals, ρ (generally you do not know this, and would need to estimate it independently), then the expected mean distance to the nearest neighbour is

$$\mathrm{E}(r) = \frac{\sqrt{\rho}}{2}.$$

In our example we have 100 individuals in a unit square, so $\rho = 0.01$ and $\mathrm{E}(r) = 0.05$. The actual mean nearest neighbour distance was

```
mean(r)
```

```
[1] 0.05294168
```

which is very close to expectation: this clearly is a random distribution of individuals (as we constructed it to be). An index of randomness is given by the ratio $\bar{r}/\mathrm{E}(r) = 2\bar{r}/\sqrt{\rho}$. This takes the value 1 for random patterns, more than 1 for regular (spaced-out) patterns, and less than 1 for aggregated patterns.

One problem with such **first-order** estimates of spatial pattern (including measures such as the variance–mean ratio) is that they can give no feel for the way that spatial distribution changes *within* an area.

26.3.1   Ripley's K

The **second-order** properties of a spatial point process describe the way that spatial interactions *change* through space. These are computationally intensive measures that take a range of distances within the area, calculate a pattern measure, then plot a graph of the function against distance, to show how the pattern measure changes with scale. The most widely used second-order measure is the K function, which is defined as

$$K(d) = \frac{1}{\lambda}\mathrm{E}[\text{number of points} \leq \text{distance } d \text{ of an arbitrary point}],$$

where λ is the mean number of points per unit area (the **intensity** of the pattern). If there is *no spatial dependence*, then the expected number of points that are within a distance d of an arbitrary point is πd^2 times the mean density. So, if the mean density is 2 points per square metre ($\lambda = 2$), then the expected number of points within a 5 m radius is $\lambda \pi d^2 = 2 \times \pi \times 5^2 = 50\pi = 157.1$. If there is clustering, then we expect an excess of points at short distances (i.e. $K(d) > \pi d^2$ for small d). Likewise, for a regularly spaced pattern, we expect an excess of long distances, and hence few individuals at short distances (i.e. $K(d) < \pi d^2$). Ripley's K (published in 1976) is calculated as follows:

$$\hat{K}(d) = \frac{1}{n^2}|A|\sum_{i \neq j}\sum \frac{I_d(d_{ij})}{w_{ij}}.$$

Here n is the number of points in region A with area $|A|$, and d_{ij} are the distances between points (the distance between the ith and jth points, to be precise). To account for edge effects, the model includes the term w_{ij} which is the fraction of the area, centred on i and passing through j, that lies within the area A (all the w_{ij} are 1 for points that lie well away from the edges of the area). $I_d(d_{ij})$ is an indicator function to show which points are to be counted as neighbours at this value of d: it takes the value 1 if $d_{ij} \leq d$ and zero otherwise (i.e. points with $d_{ij} > d$ are omitted from the summation). The pattern measure is obtained by plotting $\hat{K}(d)$ against d. This is then compared with the curve that would be observed under complete spatial randomness (namely, a plot of πd^2 against d). When clustering occurs, $K(d) > \pi d^2$ and the curve lies *above* the CSR curve, while regular patterns produce a curve *below* the CSR curve.

You can see *why* you need the edge correction from this simple simulation experiment. For individual number 1, with coordinates (x_1, y_1), calculate the distances to all the other individuals, using the function distance that we wrote earlier (p. 830):

```
distances <- numeric(100)
for(i in 1:100) distances[i] <- distance(x[1],y[1],x[i],y[i])
```

Now find out how many other individuals are within a distance d of this individual. Take as an example $d = 0.1$.

```
sum(distances<0.1)-1
```

```
[1] 4
```

There were four other individuals within a distance $d = 0.1$ of the first individual (the distance 0 from itself is included in the sum, so we have to correct for this by subtracting 1). The next step is to generalize the procedure from this one individual to all the individuals. We make a two-dimensional matrix called dd to contain all the distances from every individual (rows) to every other individual (columns):

```
dd <- numeric(10000)
dd <- matrix(dd,nrow=100)
```

The matrix of distances is computed within loops for both individual (j) and neighbour (i) like this:

```
for (j in 1:100) {for(i in 1:100) dd[j,i] <- distance(x[j],y[j],x[i],y[i])}
```

Alternatively, you could use sapply with an anonymous function like this, which has the advantage that we do not need to prepare the matrix dd in advance:

```
dd <- sapply(1:100,function (i,j=1:100) distance(x[j],y[j],x[i],y[i]))
```

We should check that the number of individuals within 0.1 of individual 1 is still 4 under this new notation. Note the use of blank subscripts [1,] to mean 'all the individuals in row number 1':

```
sum(dd[1,]<0.1)-1
```

```
[1] 4
```

So that's OK. We want to calculate the sum of this quantity over all individuals, not just individual number 1:

```
sum(dd<0.1)-100
```

```
[1] 270
```

This means that there are 270 cases in which other individuals are counted within $d = 0.1$ of focal individuals. Next, create a vector containing a range of different distances, d, over which we want to calculate $K(d)$ by counting the number of individuals within distance d, summed over all individuals:

```
d <- seq(0.01,1,0.01)
```

For each of these distances we need to work out the total number of neighbours of all individuals. So, in place of 0.1 (in the sum, above), we need to put each of the d values in turn. The count of individuals is going to be a vector of length 100 (one for each d):

```
count <- numeric(100)
```

Calculate the count for each distance d:

```
for (i in 1:100) count[i] <- sum(dd<d[i])-100
```

The expected count increases with d as πd^2 so we scale our count by dividing by the square of the total number of individuals $n^2 = 100^2 = 10\,000$:

```
K <- count/10000
```

Finally, plot a graph of K against d:

```
plot(d,K,type="l",col="red")
```

Not surprisingly, when we sample the whole area ($d = 1$), we count all of the individuals in every neighbourhood ($K = 1$). For CSR the graph should follow πd^2 so we add a line to show this:

```
lines(d,pi*d^2,col="blue")
```

Up to about $d = 0.2$ the agreement between the two lines is reasonably good, but for longer distances our algorithm is counting far too few neighbours. This is because much of the area scanned around marginal individuals is invisible, since it lies outside the study area (there may well be individuals out there, but we shall never know). This simple model demonstrates that the *edge correction* is a fundamental part of Ripley's K.

Fortunately, we do not have to write a function to work out a corrected value for K; it is available as Kfn in the built-in spatial library. Here we use it to analyse the pattern of trees in the dataframe called pines. The library function ppinit reads the data from a library file called pines.dat which is stored in the spatial/ppdata directory. It then converts this into a list with names $x, $y and $area. The first row of the file contains the number of trees (71) the second row has the name of the data set (pines), the third row has the four boundaries of the region plus the scaling factor (0, 96, 0, 100, 10 so that the coordinates of lower and upper x are computed as 0 and 9.6, and the coordinates of lower and upper y are 0 and 10). The remaining rows of the data file contain x and y coordinates for each individual tree, and these are converted into a list of x values and a separate list of y values. You need to know these details about the structure of the data files in order to use these library functions with your own data (see p. 845).

```
library(spatial)
pines <- ppinit("pines.dat")
```

First, set up the plotting area with two square frames:

```
windows(7,4)
par(mfrow=c(1,2),pty="s")
```

On the left, make a map using the x and y locations of the trees, and on the right make a plot of $L(t)$ (the pattern measure) against distance:

```
plot(pines,pch=16, col="blue")
plot(Kfn(pines,5),type="s",xlab="distance",ylab="L(t)")
```

Recall that if there was CSR, then the expected value of K would be πd^2; to linearize this, we could divide by π and then take the square root. This is the measure used in the function Kfn, where it is called $L(t) = \sqrt{K(t)/\pi}$. Now for the simulated upper and lower bounds: the first argument in Kenvl (calculating envelopes for K) is the maximum distance (half the length of one side), the second is the number of simulations (100 is usually sufficient), and the third is the number of individuals within the mapped area (71 pine trees in this case).

```
lims <- Kenvl(5,100,Psim(71))
lines(lims$x,lims$lower,lty=2,col="red")
lines(lims$x,lims$upper,lty=2,col="red")
```

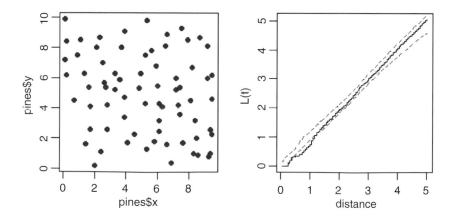

There is a suggestion that at relatively small distances (around 1 or so), the trees are rather regularly distributed (more spaced out than random), because the plot of $L(t)$ against distance falls below the lower envelope of the CSR line (it should lie between the two limits for its whole length if there was CSR). The mechanism underlying this spatial regularity (e.g. non-random recruitment or mortality, competition between growing trees, or underlying non-randomness in the substrate) would need to be investigated in detail. With an aggregated pattern, the line would fall above the upper envelope (see p. 847).

26.3.2 Quadrat-based methods

Another approach to testing for spatial randomness is to count the number of individuals in quadrats of different sizes. Here, the quadrats have an area of 0.01, so the expected number per quadrat is 1. Earlier, we generated 100 random coordinates for x and y:

```
plot(x,y,pch=16,col="red")
grid(10,10,lty=1)
```

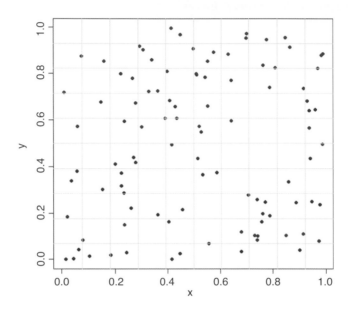

Note that the grid function has not done exactly what we intended (the grids are not exactly on the tick marks). To count the numbers of individuals in each of the cells of the map, the trick is to use cut to convert the *x* and *y* coordinates of the map into bin numbers (between 1 and 10 for the quadrat size we have drawn here). To achieve this, the break points are generated by the sequence (0,1,0.1):

```
xt <- cut(x,seq(0,1,0.1))
yt <- cut(y,seq(0,1,0.1))
```

This creates vectors of integer subscripts between 1 and 10 for xt and yt. Now all we need to do is use table to count up the number of individuals in every cell (i.e. in every combination of xt and yt):

```
count <- as.vector (table(xt,yt))
table(count)
```

```
count
 0  1  2  3  4  5
37 38 16  7  1  1
```

This shows that 37 cells are empty, 1 cell had five individuals, but no cells contained six or more individuals. Now we need to see what this distribution would look like under a particular null hypothesis. For a Poisson process (see p. 314), for example,

$$P(x) = \frac{e^{-\lambda}\lambda^x}{x!}.$$

Note that the mean depends upon the quadrat size we have chosen. With 100 individuals in the whole area, the expected number in any one of our 100 cells, λ, is 1.0. The expected frequencies of counts between 0 and 5 are therefore given by

```
(expected <- 100*exp(-1)/sapply(0:5,factorial))
  [1] 36.7879441 36.7879441 18.3939721  6.1313240  1.5328310  0.3065662
```

The fit between observed and expected is almost perfect (as we should expect, of course, having generated the random pattern ourselves). A test of the significance of the difference between an observed and expected frequency distribution is shown on p. 841.

26.3.3 Aggregated pattern and quadrat count data

Here is an example of a quadrat-based analysis of an aggregated spatial pattern. We begin by producing a map of the trees, then use `abline` rather than `grid` (see above) to make sure that the lines are exactly where we want them to be:

```
trees <- read.table("c:\\temp\\trees.txt",header=T)
attach(trees)
names(trees)

[1]  "x"  "y"

plot(x,y,pch=16,col="blue")
abline(v=seq(0,100,10),col="lightgray",lty=1)
abline(h=seq(0,100,10),col="lightgray",lty=1)
```

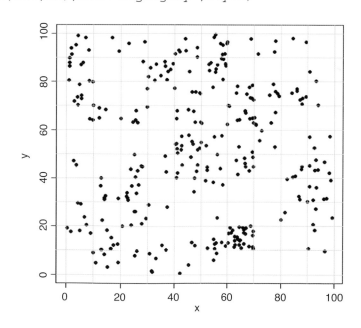

We cut up the data and tabulate the counts:

```
xt <- cut(x,10)
yt <- cut(y,10)
```

```
count <- as.vector(table(xt,yt))
table(count)
```

```
count
 0  1  2  3  4  5  6  7  8  9 11 12 24
27 11 18  7  9  6 10  6  1  1  2  1  1
```

There are quadrats with as many as 24 individuals, and despite the fact that the mean number is greater than 3 individuals per square, there are still 27 completely empty squares. The expected frequencies under the null hypothesis of a random pattern depend only on the mean number per cell,

```
mean(count)
```

```
[1] 3.11
```

and as a preliminary estimate of the departure from randomness we calculate the variance–mean ratio (recall that with the Poisson distribution the variance is equal to the mean):

```
var(count)/mean(count)
```

```
[1] 4.007243
```

These data are distinctly aggregated (the variance–mean ratio is much greater than 1), so we might compare the counts with a negative binomial distribution (p. 315). The expected frequencies are estimated by multiplying our total number of squares (100) by the probability densities from a negative binomial distribution generated by the function dnbinom. This has three arguments: the counts for which we want the probabilities (0:10), the mean (mu=3.11) and the aggregation parameter $k = $ mu^2/(var-mu) = size=1.03417:

```
mean(count)^2/(var(count)- mean(count))
```

```
[1] 1.03417
```

Here are the expected frequencies:

```
(expected <- dnbinom(0:10, size=1.03417, mu=3.11)*100)
```

```
[1] 23.798804 18.470128 14.097756 10.700189  8.098573  6.119123
[7]  4.618259  3.482699  2.624761  1.977235  1.488890
```

These are reasonably close to the observed frequencies (above) but we need to quantify the lack of fit. The plan is to display the observed and expected frequencies as pairs of bars, side by side. We need to make space, therefore, to accommodate, say, 22 bars (11 for each histogram):

```
ht <- numeric(22)
observed <- table(count)
ht[seq(1,21,2)] <- observed
ht[seq(2,22,2)] <- expected
names <- rep("",22)
names[seq(1,21,2)] <- as.character(0:10)
barplot(ht,col=c("darkgray","lightgray"),names=names,
        ylab="frequency",xlab="trees per quadrat")
legend(locator(1),legend=c("observed","expected"),
        fill=c("darkgray","lightgray"))
```

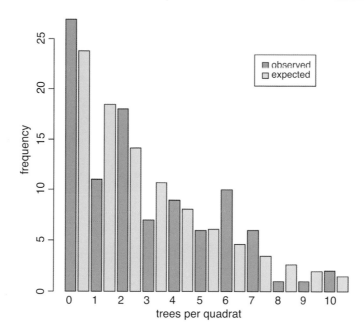

The fit is reasonably good, but we need a quantitative estimate of the lack of agreement between the observed and expected distributions. Pearson's chi-squared is perhaps the simplest (p. 367). We need to trim the observed and expected vectors so that none of the expected frequencies is less than 4. Inspection shows that the lowest expected frequency greater than 4 is in location 7, so we shall accumulate all frequencies in locations 8 and above

```
expected[8] <- sum(expected[8:length(expected)])
expected <- expected[-c(9:length(expected))]
observed[8] <- sum(observed[8:length(observed)])
observed <- observed[-c(9:length(observed))]
```

Now calculate Pearson's chi-squared as $\sum [(O - E)^2/E]$:

```
sum((observed-expected)^2/expected)
```

```
[1] 12.80059
```

The number of degrees of freedom is the number of legitimate comparisons (8) minus the number of parameters estimated from the data (2) minus 1 for contingency (i.e. $8 - 2 - 1 = 5$ d.f.). So the probability of obtaining a chi-squared value of this size (12.8) or greater is

```
1-pchisq(12.8,5)
```

```
[1] 0.02532684
```

We conclude that the negative binomial is an imperfect description of these quadrat data (because $p < 0.05$). The reason for the significant lack of fit is the serious underestimation of quadrats containing just one tree, and the excess of quadrats containing six or seven trees.

26.3.4   Counting things on maps

The convention is that if a point falls exactly on the *x* axis or exactly on the *y* axis, then it is counted as being *inside* the area (green points on the map below), but if it falls on the top axis or on the right hand axis, then it is *outside* the area (red points).

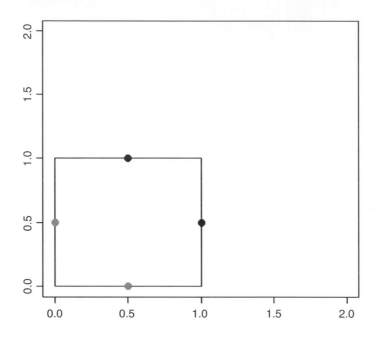

```
plot(c(0,2),c(0,2),type="n",xlab="",ylab="")
lines(c(0,1,1,0,0),c(0,0,1,1,0))
points(c(0.5,0),c(0,0.5),pch=16,col="green",cex=1.5)
points(c(0.5,1),c(1,0.5),pch=16,col="red",cex=1.5)
```

That way, all points on the map have an equal probability of being counted, and there is no double-counting of points. Red points on the top axis will be counted in the next quadrat to the north, and red points on the right hand axis will be counted in the next quadrat to the east.

This counting convention is embodied in the R function called cut, using round brackets and square brackets like this: "(b1, b2]", "(b2, b3]" ... for the default right = TRUE; here, the round bracket means 'greater than' and the square bracket means 'less than or equal to'. For our convention, we need to specify right=FALSE. Now we have "[b1, b2)", "[b2, b3)" ... where the square bracket means 'greater than or equal to' and the round bracket means 'less than'. Here are some test data:

```
data <- read.table("c:\\temp\\countpoints.txt",header=T)
attach(data)
plot(x,y,pch=16,col="green")
points(x[y==1 | x==1],y[y==1 | x==1],pch=16,col="red")
lines(c(0,1,1,0,0),c(0,0,1,1,0),lty=2)
lines(c(0.2,0.2),c(0,1),lty=2)
lines(c(0.4,0.4),c(0,1),lty=2)
lines(c(0.6,0.6),c(0,1),lty=2)
lines(c(0.8,0.8),c(0,1),lty=2)
```

```
lines(c(0,1),c(0.2,0.2),lty=2)
lines(c(0,1),c(0.4,0.4),lty=2)
lines(c(0,1),c(0.6,0.6),lty=2)
lines(c(0,1),c(0.8,0.8),lty=2)
points(0,0.5,cex=2)
points(0.5,0,cex=2)
```

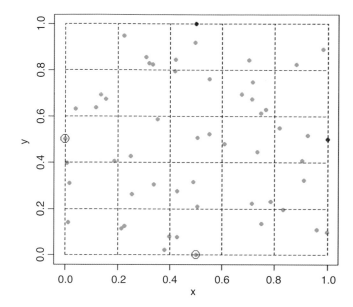

We want to count the number of points in each of 25 quadrats measuring 0.2×0.2. The function to achieve this is cut. We need to specify six values at which to cut the x axis and six values at which to cut the y axis. The function converts the vector of continuous values of x into a factor xc with five levels (and converts the coordinates y into factor levels within yc in the same way):

```
xc <- cut(x,seq(0,1,0.2),right = FALSE)
yc <- cut(y,seq(0,1,0.2),right = FALSE)
```

We count the number of points in each quadrat using table, like this:

```
table(yc,xc)
```

	xc				
yc	[0,0.2)	[0.2,0.4)	[0.4,0.6)	[0.6,0.8)	[0.8,1)
[0,0.2)	1	4	2	1	3
[0.2,0.4)	2	2	3	2	1
[0.4,0.6)	2	2	2	2	3
[0.6,0.8)	4	0	2	5	0
[0.8,1)	0	4	2	1	2

which gives the correct counts, but in a pattern that does not match the map (the y values are upside down). We can fix this by reordering the factor levels of yc using rev to reverse the order of the rows:

```
yc <- factor(yc,rev(levels(yc)))
```

Now, `table` produces a map-like summary of the counts:

```
table(yc,xc)
```

```
            xc
yc          [0,0.2)  [0.2,0.4)  [0.4,0.6)  [0.6,0.8)  [0.8,1)
   [0.8,1)        0          4          2          1        2
   [0.6,0.8)      4          0          2          5        0
   [0.4,0.6)      2          2          2          2        3
   [0.2,0.4)      2          2          3          2        1
   [0,0.2)        1          4          2          1        3
```

There are four points on the edge of the map: the red points are not counted (totals 2 and 3 respectively), but the circled green points are counted (totals 2 and 2, above).

26.4 Packages for spatial statistics

In addition to the built-in library `spatial` there are two substantial contributed packages for analysing spatial data. The `spatstat` library is what you need for the statistical analysis of spatial point patterns, while the `spdep` library is good for the spatial analysis of data from mapped regions.

With point patterns the things you will want to do include

- creation, manipulation and plotting of point patterns,

- exploratory data analysis,

- simulation of point process models,

- parametric model fitting,

- hypothesis tests and diagnostics;

whereas with maps you might

- compute basic spatial statistics such as Moran's I and Geary's C,

- create neighbour objects of class `nb`,

- create weights list objects of class `lw`,

- work out neighbour relations from polygons (outlines of regions),

- colour mapped regions on the basis of derived statistics.

You need to take time to master the different formats of the data objects used by the two packages. You will waste a lot of time if you try to use the functions in these libraries with your own, unreconstructed data files.

Here is the code for installing and reading about `spatstat` and `spdep`:

```
install.packages("spatstat")
library(help=spatstat)
library(spatstat)
demo(spatstat)
```

```
install.packages("spdep")
library(help=spdep)
library(spdep)
```

26.4.1   The spatstat package

You need to use the function ppp to convert your coordinate data into an object of class ppp representing a point pattern data set in the two-dimensional plane. Our next dataframe contains information on the locations and sizes of 3359 ragwort plants in a 30 m × 15 m map:

```
data <- read.table("c:\\temp\\ragwortmap.txt",header=T)
attach(data)
names(data)
```

```
[1] "xcoord" "ycoord" "type"
```

The plants are classified as belonging to one of four types: skeletons are dead stems of plants that flowered the year before, regrowth are skeletons that have live shoots at the base, seedlings are small plants (a few weeks old) and rosettes are larger plants (one or more years old) destined to flower this year. The function ppp requires separate vectors for the *x* and *y* coordinates: these are in our file under the names xcoord and ycoord. The third and fourth arguments to ppp are the boundary coordinates for *x* and *y* respectively (in this example c(0,3000) for *x* and c(0,1500) for *y*). The final argument to ppp contains a vector of what are known as 'marks': these are the factor levels associated with each of the points (in this case, type is either skeleton, regrowth, seedling or rosette). You give a name to the ppp object (ragwort) and define it like this:

```
ragwort <- ppp(xcoord,ycoord,c(0,3000),c(0,1500),marks=type)
```

You can now use the object called ragwort in a host of different functions for plotting and statistical modelling within the spatstat library. For instance, here are maps of the point patterns for the four plant types separately:

```
plot(split(ragwort),main="")
```

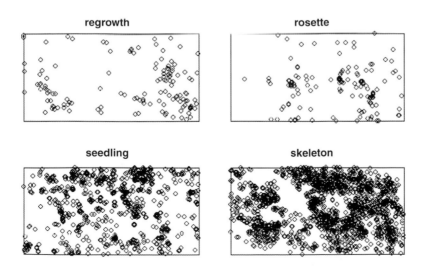

Point patterns are summarized like this:

```
summary(ragwort)
```

```
Marked planar point pattern: 3359 points
Average intensity 0.000746 points per square unit
Multitype:
         frequency proportion intensity
regrowth       135     0.0402  3.00e-05
rosette        146     0.0435  3.24e-05
seedling      1100     0.3270  2.44e-04
skeleton      1980     0.5890  4.40e-04

Window: rectangle = [0, 3000]x[0, 1500]units
Window area =  4500000 square units
```

which computes the frequency and intensity for each mark ('intensity' is the mean density of points per unit area). In this case, where distances are in centimetres, the intensity is the mean number of plants per square centimetre (the highest intensity is skeletons, with $0.000\,44$ cm^{-2}). The function quadratcount produces a useful summary of counts:

```
plot(quadratcount(ragwort),main="")
```

60	147	261	292	76
105	101	188	155	141
64	95	151	228	147
98	129	115	117	154
62	77	88	145	103

This is the default, but you can specify either the numbers of quadrats in the x and y directions (default 5 and 5), or provide numeric vectors giving the x and y coordinates of the boundaries of the quadrats. If we want counts in 0.5 m squares:

```
plot(quadratcount(ragwort,
     xbreaks=c(0,500,1000,1500,2000,2500,3000),
     ybreaks=c(0,500,1000,1500)),main="")
```

100	145	261	388	267	149
100	169	157	261	259	171
83	169	96	190	216	178

There are functions for producing density plots of the point pattern:

```
Z <- density.ppp(ragwort)
plot(Z,main="")
```

The classic graphical description of spatial point patterns is Ripley's K (see p. 834).

```
K <- Kest(ragwort)
plot(K, main = "K function")
```

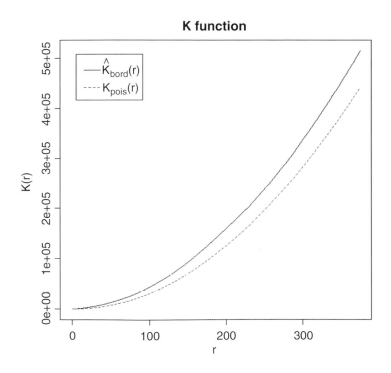

The red dotted line shows the expected number of plants within a radius r of a plant under the assumption of complete spatial randomness. The observed curve (black) lies above this line, indicating strong spatial aggregation at all spatial scales up to more than 300 cm.

The pair correlation function pcf for the ragwort data looks like this:

```
pc <- pcf(ragwort)
plot(pc, main = "Pair correlation function")
```

Pair correlation function

There is strong correlation between pairs of plants at small scales, but much less above $r = 20$ cm. The function `distmap` shows the distance map around individual plants:

```
Z <- distmap(ragwort)
plot(Z,main="")
```

You can use `spatstat` to generate a wide range of patterns of random points, including independent uniform random points, inhomogeneous Poisson point processes, inhibition processes, and Gibbs point processes using Metropolis–Hastings (see `?spatstat` for details). Some useful functions on point-to-point distances in `spatstat` include:

`nndist`	nearest neighbour distances;
`nnwhich`	find nearest neighbours;
`pairdist`	distances between all pairs of points;
`crossdist`	distances between points in two patterns;
`exactdt`	distance from any location to nearest data point;
`distmap`	distance map image;
`density.ppp`	kernel smoothed density.

There are several summary statistics for a multi-type point pattern with a component $marks which is a factor:

`Gcross, Gdot, Gmulti`	multitype nearest neighbour distributions;
`Kcross, Kdot, Kmulti`	multitype *K*-functions;
`Jcross, Jdot, Jmulti`	multitype *J*-functions;
`Alltypes`	estimates of the above for all *i,j* pairs;
`Lest`	multitype *I*-function;
`Kcross.inhom`	inhomogeneous counterpart of `Kcross`;
`Kdot.inhom`	inhomogeneous counterpart of `Kdot`.

Point process models are fitted using the ppm function like this:

```
model <- ppm(ragwort, ~marks + polynom(x, y, 2), Poisson())
plot(model)
```

Fitted trend mark = regrowth

There are eight such maps produced, showing means (four maps) and standard errors (four maps). Typing

```
summary(model)
```

produces a massive table of output, including what the authors refer to as the 'gory details'.

26.4.2   The `spdep` package

The key to using this package is to understand the differences between the various formats in which the spatial data can be held:

- *x* and *y* coordinates (in a two-column matrix, with *x* in column 1 and *y* in 2);

- lists of regions that are neighbours to each region, with (potentially) unequal numbers of neighbours in different cases (this is called a neighbour file and belongs to class `nb`);

- dataframes containing a region, its neighbour and the statistical weight of the association between the two regions on each row (class `data.frame`);

- lists containing the identities of the *k* nearest neighbours (class `knn`);

- a weights list object suitable for computing Moran's *I* or Geary's *C* (class `lw`);

- lists of polygons, defining the outlines of regions on a map (class `polylist`).

Unlike spatstat (Section 26.4.1) where the x and y coordinates were in separate vectors, spdep wants the x and y coordinates in a single two-column matrix. For the ragwort data (p. 845) we need to write:

```
library(spdep)
myco <- cbind(xcoord,ycoord)
myco <- matrix(myco,ncol=2)
```

A raw list of coordinates contains no information about neighbours, but we can use the knearneigh function to convert a matrix of coordinates into an object of class knn. Here we ask for the four nearest neighbours of each plant:

```
myco.knn <- knearneigh(myco, k=4)
```

This list object has the following structure:

```
str(myco.knn)

List of 5
 $ nn       : int [1:3359, 1:4] 2 1 4 3 7 4 8 7 10 9 ...
 $ np       : int 3359
 $ k        : num 4
 $ dimension: int 2
 $ x        : num [1:3359, 1:2] 27 29 20 20 78 25 89 97 253 259 ...
 - attr(*, "class")= chr "knn"
 - attr(*, "call")= language knearneigh(x = myco, k = 4)
```

- $nn contains 3359 lists, each a vector of length 4, containing the identities of the four points that are the nearest neighbours of each of the points from 1 to 3359.

- $np (an integer) is the number of points in the pattern.

- $k is the number of neighbours of each point.

- $dimension is 2.

- $x is the matrix of coordinates of each point (x in the first column, y in the second).

Before you can do much with a knn object you will typically want to convert it to a neighbour object (nb) using the knn2nb function like this:

```
myco.nb <- knn2nb(myco.knn)
```

The essential concept for using the spdep package is the neighbour object (with class nb). For a given location, typically identified by the (x, y) coordinates of its centroid, the neighbour object is a list, with the elements of the list numbered from 1 to the number of locations, and each element of the list contains a vector of integers representing the identities of the locations that share a boundary with that location. The important point is that different vectors are likely to be of different lengths. You can do interesting things with nb objects. Here is a plot with each point joined to its four nearest neighbours – you specify the nb object and the matrix of coordinates:

```
plot(myco.nb,myco)
```

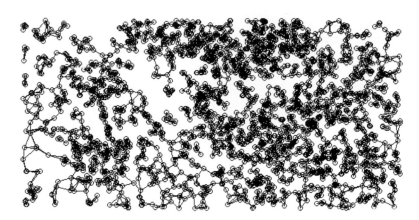

The simplest way to create an `nb` object is to read a text file containing one row for each neighbour relationship, using the special input function `read.gwt2nb`. The header row can take one of two forms. The simplest (called 'old-style GWT') is a single integer giving the number of locations in the file. There will always be many more rows in the data file than this number, because each location will typically have several neighbours. The second form of the header row has four elements: the first is set arbitrarily to zero, the second is the integer number of locations, the third is the name of the shape object and the fourth is the vector of names identifying the locations. An example should make this clear. These are the contents of a text file called naydf.txt:

```
5
1    2    1
1    3    1
2    1    1
2    3    1
2    4    1
3    1    1
3    2    1
3    5    1
4    2    1
4    3    1
4    5    1
5    3    1
5    4    1
```

The 5 in the first row indicates that this file contains information on five locations. On subsequent lines the first number identifies the location, the second number identifies one of its neighbours, and the third number is the weight of that relationship. Thus, location 5 has just two neighbours, and they are locations 3 and 4 (the last two rows of the file). We create a neighbour object for these data with the `read.gwt2nb` function like this:

```
dd <- read.gwt2nb("c:\\temp\\naydf.txt")
```

Here is a summary of the newly-created neighbour object called `dd`:

```
summary(dd)
```

```
Neighbour list object:
```

```
Number of regions: 5
Number of nonzero links: 13
Percentage nonzero weights: 52
Average number of links: 2.6
Non-symmetric neighbours list
Link number distribution:

2 3
2 3
2 least connected regions:
1 5 with 2 links
3 most connected regions:
2 3 4 with 3 links
```

Here are the five vectors of neighbours:

```
dd[[1]]

[1]   2   3

dd[[2]]

[1]   1   3   4

dd[[3]]

[1]   1   2   5

dd[[4]]

[1]   2   3   5

dd[[5]]

[1]   3   4
```

The coordinates of the five locations need to be specified:

```
coox <- c(1,2,3,4,5)
cooy <- c(3,1,2,0,3)
```

and the vectors of coordinates need to be combined into a two-column matrix. Now we can use `plot` with `dd` and the coordinate matrix to indicate the neighbour relations of all five locations like this:

```
plot(dd,matrix(cbind(coox,cooy),ncol=2))
text(coox,cooy,as.character(1:5),pos=rep(3,5))
```

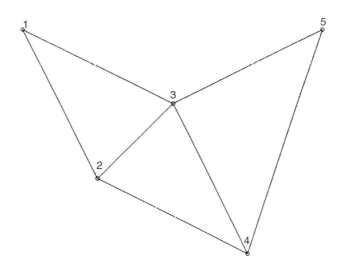

Note the use of pos = 3 to position the location numbers 1 to 5 *above* their points. You can see that locations 1 and 5 are the least connected (two neighbours) and location 3 is the most connected (four neighbours). Note that the specification in the data file was not fully reciprocal, because location 4 was defined as a neighbour of location 3 but not vice versa. There is a comment, Non-symmetric neighbours list, in the output to summary(dd) to draw attention to this. A function make.sym.nb(dd) is available to convert the object dd into a symmetric neighbours list.

For calculating indices much as Moran's *I* and Geary's *C* you need a 'weights list' object. This is created most simply from a neighbour object using the function nb2listw. For the ragwort data, we have already created a neighbour object called myco.nb (p. 850) and we create the weights list object myco.lw like this:

```
myco.lw <- nb2listw(myco.nb, style="W")
myco.lw
```

```
Characteristics of weights list object:
Neighbour list object:
Number of regions: 3359
Number of nonzero links: 13436
Percentage nonzero weights: 0.1190831
Average number of links: 4
Non-symmetric neighbours list

Weights style: W
Weights constants summary:
      n         nn    S0       S1         S2
W  3359  11282881  3359  1458.25  14077.88
```

There are three classic tests based on spatial cross products $C(i, j)$, where $z(i) = (x(i) - \text{mean}(x))/sd(x)$:

- Moran ($C(i, j) = z(i)z(j)$);

- Geary ($C(i, j) = (z(i) - z(j))^2$);

- Sokal ($C(i, j) = |z(i) - z(j)|$).

Here is the Moran *I* test for the ragwort data, using the weights list object `myco.lw`:

```
moran(1:3359,myco.lw,length(myco.nb),Szero(myco.lw))
```

```
$I
[1]  0.9931224
```

```
$K
[1]  1.8
```

Here is Geary's *C* for the same data:

```
geary(1:3359,myco.lw,length(myco.nb),length(myco.nb)-1,Szero(myco.lw))
```

```
$C
[1]  0.004549794
```

```
$K
[1]  1.8
```

Here is Mantel's permutation test:

```
sp.mantel.mc(1:3359,myco.lw,nsim=99)

        Mantel permutation test for moran measure
data:  1:3359
weights: myco.lw
number of simulations + 1: 100

statistic = 3334.905, observed rank = 100, p-value = 0.01
alternative hypothesis: greater
sample estimates:
mean of permutations    sd of permutations
           -5.434839             34.012539
```

In all cases, the first argument is a vector of location numbers (1 to 3359 in the ragwort example), the second argument is the weight list object `myco.lw`. For `moran`, the third argument is the length of the neighbours object, `length(myco.nb)` and the fourth is `Szero(myco.lw)`, the global sum of weights, both of which evaluate to 3359 in this case. The function `geary` has an extra argument, `length(myco.nb)-1`, and `sp.mantel.mc` specifies the number of simulations.

26.4.3   Polygon lists

Perhaps the most complex spatial data handled by `spdep` comprise digitized outlines (sets of *x* and *y* coordinates) defining multiple regions, each of which can be interpreted by R as a polygon. Here is such a list from the built-in `columbus` data set:

```
data(columbus)
polys
```

The `polys` object is of class `polylist` and comprises a list of 49 polygons. Here is the first of them:

```
[[1]]
              [,1]        [,2]
 [1,]  8.624129  14.23698
 [2,]  8.559700  14.74245
 [3,]  8.809452  14.73443
 [4,]  8.808413  14.63652
 [5,]  8.919305  14.63850
 [6,]  9.087138  14.63049
 [7,]  9.099965  14.24483
 [8,]  9.015047  14.24184
 [9,]  9.008951  13.99506
[10,]  8.818140  14.00205
[11,]  8.653305  14.00809
[12,]  8.642902  14.08971
[13,]  8.632592  14.17059
[14,]  8.625826  14.22367
[15,]  8.624129  14.23698
attr(,"bbox")
[1]    8.559700 13.995060   9.099965 14.742450
attr(,"ringDir")
[1] 1
attr(,"after")
[1] NA
attr(,"plotOrder")
[1] 1
attr(,"nParts")
[1] 1
attr(,"pstart")
attr(,"pstart")$from
[1] 1
attr(,"pstart")$to
[1] 15
```

Each element of the list contains a two-column matrix with *x* coordinates in column 1 and *y* coordinates in column 2, with as many rows as there are digitized points on the outline of the polygon in question. After the matrix of coordinates come the boundary box and various plotting options.

There is an extremely useful function poly2nb that takes the list of polygons and works out which regions are neighbours of one another by looking for shared boundaries. The result is an nb object (here called colnbs), and we can get a visual check of how well poly2nb has worked by overlaying the neighbour relations on a map of the polygon outlines:

```
colnbs <- poly2nb(polys)
plot(c(5.5,11.5),c(10.5,15),type="n",xlab="",ylab="")
for (i in 1:49) polygon(polys[[i]][,1],polys[[i]][,2],col="lightgrey")
plot(colnbs,coords,add=T,col="red")
```

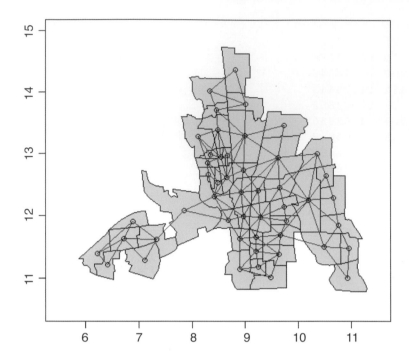

The agreement is perfect. Obviously, creating a polygon list is likely to be a huge amount of work, especially if there are many regions, each with complicated outlines. Before you start making one, you should check that it has not been done already by someone else who might be willing to share it with you. To create polygon lists and bounding boxes from imported shape files you should use one of `read.shapefile` or `Map2poly` (for details, see `?read.shapefile` and `?Map2poly`). Subtleties include the facts that lakes are digitized anti-clockwise and islands are digitized clockwise.

26.5   Geostatistical data

Mapped data commonly show the value of a continuous response variable (e.g. the concentration of a mineral ore) at different spatial locations. The fundamental problem with this kind of data is spatial pseudoreplication. Hot spots tend to generate lots of data, and these data tend to be rather similar because they come from essentially the same place. Cold spots are poorly represented and typically widely separated. Large areas between the cold spots have no data at all.

Spatial statistics takes account of this spatial autocorrelation in various ways. The fundament tool of spatial statistics is the **variogram** (or **semivariogram**). This measures how quickly spatial autocorrelation, $\gamma(h)$, falls off with increasing distance:

$$\gamma(h) = \frac{1}{2\,|N(h)|} \sum_{N(h)} (z_i - z_j)^2.$$

Here $N(h)$ is the set of all pairwise Euclidean distances $i - j = h$, $|N(h)|$ is the number of distinct pairs within $N(h)$, and z_i and z_j are values of the response variable at spatial locations i and j. There are two important rules of thumb: (1) the distance of reliability of the variogram is less than half the maximum distance over the entire field of data; and (2) you should only consider producing an empirical variogram when you have more than 30 data points on the map.

Plots of the empirical variogram against distance are characterized by some quaintly named features which give away its origin in geological prospecting:

- **nugget**, small-scale variation plus measurement error;
- **sill**, the asymptotic value of $\gamma(h)$ as $h \to \infty$, representing the variance of the random field;
- **range**, the threshold distance (if such exists) beyond which the data are no longer autocorrelated.

Variogram plots that do not asymptote may be symptomatic of trended data or a non-stationary stochastic process. The **covariogram** $C(h)$ is the covariance of z values at separation h, for all i and $i + h$ within the maximum distance over the whole field of data:

$$\text{cov}(Z(i + h), Z(i)) = C(h).$$

The **correlogram** is a ratio of covariances:

$$\rho(h) = \frac{C(h)}{C(0)} = 1 - \frac{\gamma(h)}{C(0)}.$$

Here $C(0)$ is the variance of the random field and $\gamma(h)$ is the variogram. Where the variogram increases with distance, the correlogram and covariogram decline with distance.

The variogram assumes that the data are untrended. If there are trends, then one option is median polishing. This involves modelling row and column effects from the map like this:

```
y ~ overall mean + row effect + column effect + residual
```

This two-way model assumes additive effects and would not work if there was an interaction between the rows and columns of the map. An alternative would be to use a generalized additive model (p. 670) with non-parametric smoothers for latitude and longitude.

Anisotropy occurs when spatial autocorrelation changes with direction. If the sill changes with direction, this is called zonal anisotropy. When it is the range that changes with direction, the process is called geometric anisotropy.

Geographers have a wonderful knack of making the simplest ideas sound complicated. **Kriging** is nothing more than linear interpolation through space. Ordinary kriging uses a random function model of spatial correlation to calculate a weighted linear combination of the available samples to predict the response for an unmeasured location. Universal kriging is a modification of ordinary kriging that allows for spatial trends. We say no more about models for spatial prediction here; details can be found in Kaluzny et al. (1998). Our concern is with using spatial information in the interpretation of experimental or observational studies that have a single response variable. The emphasis is on using location-specific measurements to model the spatial autocorrelation structure of the data.

The idea of a variogram is to illustrate the way in which spatial variance increases with spatial scale (or alternatively, how correlation between neighbours falls off with distance). Confusingly, R has two functions with the same name: `variogram` (lower-case 'v') is in the `spatial` library and `Variogram` (upper-case 'V') is in `nlme`. Their usage is contrasted here for the ragwort data (p. 845).

To use `variogram` from the `spatial` library, you need to create a trend surface or a kriging object with columns *x*, *y* and *z*. The first two columns are the spatial coordinates, while the third contains the response variable (basal stem diameter in the case of the ragwort data):

```
library(spatial)
data <- read.table("c:\\temp\\ragwortmap2006.txt",header=T)
attach(data)
names(data)

[1]  "stems"   "diameter"   "xcoord"   "ycoord"

dts <- data.frame(x=xcoord,y=ycoord,z=diameter)
```

Next, you need to create a trend surface using a function such as `surf.ls`:

```
surface <- surf.ls(2,dts)
```

This trend surface object is then the first argument to `variogram`, followed by the number of bins (here 300). The function computes the average squared difference for pairs with separation in each bin, returning results for bins that contain six or more pairs:

```
variogram(surface,300)
```

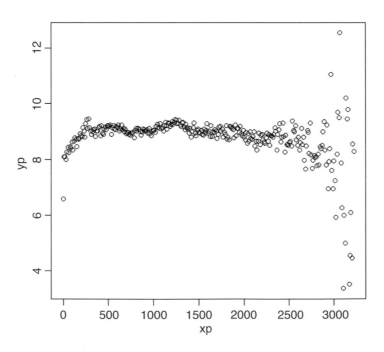

The sister function is `correlogram`, which takes identical arguments:

```
correlogram(surface,300)
```

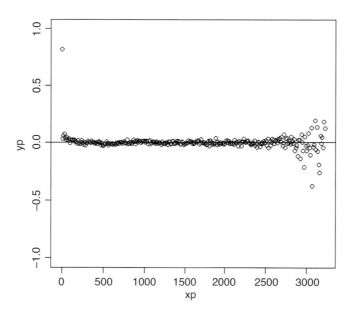

The positive correlations have disappeared by about 100 cm. The correlations at xp = 3000 are spurious edge effects.

For the Variogram function in the nlme library, you need to fit a model (typically using gls or lme), then provide the model object along with a form function in the call:

```
library(nlme)
model <- gls(diameter~xcoord+ycoord)
plot(Variogram(model,form= ~xcoord+ycoord))
```

26.6   Regression models with spatially correlated errors: Generalized least squares

In Chapter 19 we looked at the use of linear mixed-effects models for dealing with random effects and temporal pseudoreplication. Here we illustrate the use of generalized least squares (GLS) for regression modelling where we would expect neighbouring values of the response variable to be correlated. The great advantage of the `gls` function is that the errors are allowed to be correlated and/or to have unequal variances. The `gls` function is part of the `nlme` package:

```
library(nlme)
```

   The following example is a geographic-scale trial to compare the yields of 56 different varieties of wheat. What makes the analysis more challenging is that the farms carrying out the trial were spread out over a wide range of latitudes and longitudes.

```
spatialdata <- read.table("c:\\temp\\spatialdata.txt",header=T)
attach(spatialdata)
names(spatialdata)
```

```
[1]  "Block"  "variety"  "yield"  "latitude"  "longitude"
```

   We begin with graphical data inspection to see the effect of location on yield:

```
windows(7,4)
par(mfrow=c(1,2))
plot(latitude,yield,pch=21,col="blue",bg="red")
plot(longitude,yield,pch=21,col="blue",bg="red")
```

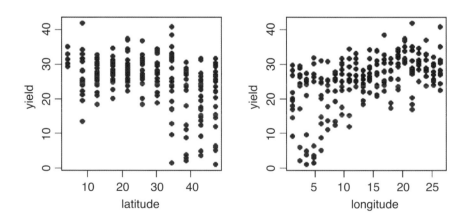

There are clearly big effects of latitude and longitude on both the mean yield and the variance in yield. The latitude effect looks like a threshold effect, with little impact for latitudes less than 30. The longitude effect looks more continuous but there is a hint of non-linearity (perhaps even a hump). The varieties differ substantially in their mean yields:

```
windows(7,7)
barplot(sort(tapply(yield,variety,mean)),col="green")
```

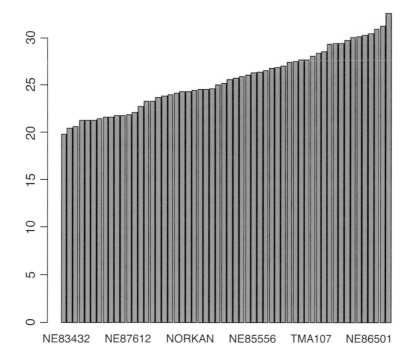

The lowest-yielding varieties are producing about 20 and the highest about 30 kg of grain per unit area. There are also substantial block effects on yield:

```
tapply(yield,Block,mean)

        1         2         3         4
27.57500 28.81091 24.42589 21.42807
```

Here is the simplest possible analysis – a one-way analysis of variance using variety as the only explanatory variable:

```
model1 <- aov(yield~variety)
summary(model1)

            Df Sum Sq Mean Sq F value Pr(>F)
variety     55   2387   43.41    0.73  0.912
Residuals  168   9990   59.47
```

This says that there are no significant differences between the yields of the 56 varieties. We can try a split-plot analysis (see p. 519) using varieties nested within blocks:

```
Block <- factor(Block)
model2 <- aov(yield~Block+variety+Error(Block))
summary(model2)

Error: Block
      Df Sum Sq Mean Sq
Block  3   1854   617.9

Error: Within
```

```
           Df Sum Sq Mean Sq F value Pr(>F)
variety    55   2389   43.43   0.881  0.702
Residuals 165   8135   49.30
```

This has made no difference to our interpretation. We could fit latitude and longitude as covariates:

```
model3 <- aov(yield~Block+variety+latitude+longitude)
summary(model3)
```

```
           Df Sum Sq Mean Sq F value    Pr(>F)
Block       3   1854   617.9  19.858  5.07e-11 ***
variety    55   2389    43.4   1.396    0.0565 .
latitude    1    686   686.1  22.051  5.60e-06 ***
longitude   1   2378  2377.6  76.414  2.69e-15 ***
Residuals 163   5072    31.1
```

This makes an enormous difference. Now the differences between varieties are close to significance ($p = 0.0565$).

Finally, we could use a GLS model to introduce spatial covariance between yields from locations that are close together. We begin by making a grouped data object:

```
space <- groupedData(yield~variety|Block)
```

We use this to fit a model using `gls` which allows the errors to be correlated and to have unequal variances. We shall add these sophistications later:

```
model4 <- gls(yield~variety-1,space)
summary(model4)
```

```
Generalized least squares fit by REML
  Model: yield ~ variety - 1
  Data: space
       AIC      BIC    logLik
  1354.742 1532.808 -620.3709
```

```
Coefficients:
                   Value Std.Error  t-value p-value
varietyARAPAHOE  29.4375  3.855687 7.634827       0
varietyBRULE     26.0750  3.855687 6.762738       0
varietyBUCKSKIN  25.5625  3.855687 6.629818       0
varietyCENTURA   21.6500  3.855687 5.615083       0
```

and so on, for all 56 varieties. The variety means are given, rather than differences between means, because we removed the intercept from the model by using `yield~variety-1` rather than `yield~variety` in the model formula (see p. 398).

Now we want to include the spatial covariance. The `Variogram` function is applied to `model4` like this:

```
plot(Variogram(model4,form=~latitude+longitude))
```

Table 26.1.   Spatial correlation structures. Options for specifying the form and distance dependence of spatial correlation in generalized least squares models. For more detail, see the help on `?corClasses` and on the individual correlation structures (e.g. `?corExp`).

corExp	exponential spatial correlation
corGaus	Gaussian spatial correlation
corLin	linear spatial correlation
corRatio	rational quadratic spatial correlation
corSpher	spherical spatial correlation
corSymm	general correlation matrix, with no additional structure

The sample variogram increases with distance, illustrating the expected spatial correlation. Extrapolating back to zero distance, there appears to be a nugget of about 0.2. There are several assumptions we could make about the spatial correlation in these data. For instance, we could try a spherical correlation structure, using the `corSpher` class (the range of options for spatial correlation structure is shown in Table 26.1). We need to specify the distance at which the semivariogram first reaches 1. Inspection shows this distance to be about 28. We can update `model4` to include this information:

```
model5 <- update(model4,
          corr=corSpher(c(28,0.2),form=~latitude+longitude,nugget=T))
summary(model5)

Generalized least squares fit by REML
  Model: yield ~ variety - 1
  Data: space
      AIC        BIC      logLik
  1185.863  1370.177  -533.9315
```

```
Correlation Structure: Spherical spatial correlation
 Formula: ~latitude + longitude
 Parameter estimate(s):
     range       nugget
27.4574777   0.2093144

Coefficients:
                      Value  Std.Error    t-value  p-value
varietyARAPAHOE     26.65898   3.437352   7.755672        0
varietyBRULE        25.84956   3.441792   7.510496        0
varietyBUCKSKIN     34.84837   3.478290  10.018822        0
varietyCENTURA      25.09472   3.458867   7.255186        0
varietyCENTURK78    26.33425   3.477919   7.571840        0
varietyCHEYENNE     24.67327   3.438914   7.174727        0
```

This is a big improvement, and AIC has dropped from 1354.742 to 1185.863. The range (27.46) and nugget (0.209) are very close to our visual estimates.

There are other kinds of spatial model, of course. We might try a rational quadratic model (corRatio); this needs an estimate of the distance at which the semivariogram is $(1 + \text{nugget})/2 = 1.2/2 = 0.6$, as well as an estimate of the nugget. Inspection gives a distance of about 12.5, so we write:

```
model6 <- update(model4,
corr=corRatio(c(12.5,0.2),form=~latitude+longitude,nugget=T))
```

We can use anova to compare the two spatial models:

```
anova(model5,model6)


        Model df      AIC       BIC      logLik
model5      1 59 1185.863 1370.177 -533.9315
model6      2 59 1183.278 1367.592 -532.6389
```

The rational quadratic model (model6) has the lower AIC and is therefore preferred to the spherical model. To test for the significance of the spatial correlation parameters we need to compare the preferred spatial model6 with the non-spatial model4 (which assumed spatially independent errors):

```
anova(model4,model6)


        Model df      AIC       BIC      logLik   Test L.Ratio p-value
model4      1 57 1354.742 1532.808 -620.3709
model6      2 59 1183.278 1367.592 -532.6389 1 vs 2 175.464  <.0001
```

The two extra degrees of freedom used up in accounting for the spatial structure are clearly justified. We need to check the adequacy of the corRatio model. This is done by inspection of the sample variogram for the normalized residuals of model6:

```
plot(Variogram(model6,resType="n"))
```

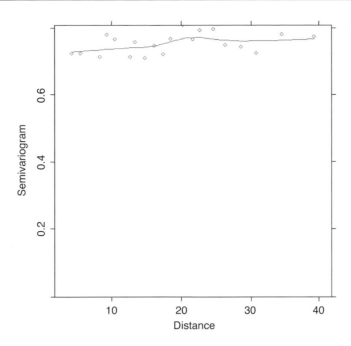

There is no pattern in the plot of the sample variogram, so we conclude that the rational quadratic is adequate. To check for constancy of variance, we can plot the normalized residuals against the fitted values like this:

```
plot(model6,resid( ., type="n")~fitted(.),abline=0)
```

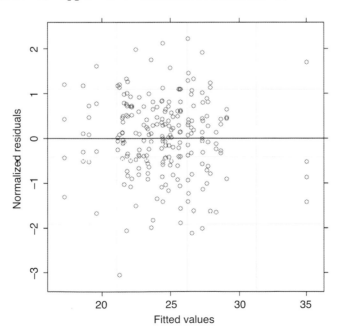

and the normal plot is obtained in the usual way:

```
qqnorm(model6,~resid(.,type="n"))
```

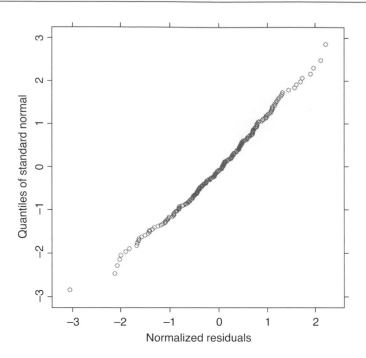

The model looks fine.

The next step is to investigate the significance of any differences between the varieties. Use `update` to change the structure of the model from `yield~variety-1` to `yield~variety`:

```
model7 <- update(model6,model=yield~variety)
anova(model7)
```

```
Denom. DF: 168
              numDF    F-value  p-value
(Intercept)       1  30.399419  <.0001
variety          55   1.850939  0.0015
```

The differences between the varieties now appear to be highly significant (recall that they were only marginally significant with our linear `model3` using analysis of covariance to take account of the latitude and longitude effects). Specific contrasts between varieties can be carried out using the `L` argument to `anova`. Suppose that we want to compare the mean yields of the first and third varieties. To do this, we set up a vector of contrast coefficients `c(-1,0,1)` and apply the contrast like this:

```
anova(model6,L=c(-1,0,1))
```

```
Denom. DF: 168
 F-test for linear combination(s)
varietyARAPAHOE varietyBUCKSKIN
             -1               1
  numDF  F-value p-value
1     1 7.696728  0.0062
```

Note that we use `model6` (with all the variety means), not `model7` (with an intercept and Helmert contrasts). The specified varieties, Arapahoe and Buckskin, exhibit highly significant differences in mean yield.

26.7   Creating a dot-distribution map from a relational database

Here is an example of extracting a relatively small subset of data from a large relational database, and using the information to produce a dot distribution map. The Access database contains two related tables:

- `sites` contains information on 2628 locations;

- `records` contains lists of species found at each site (43 001 in total).

The two tables are related by a variable called `site number`. The task is to extract eastings and northings for each record of a named species, and use these to produce a dot-distribution map, with one dot for each site at which that particular species was recorded.

Instructions on how to make an open database connection are on p. 154. I assume that you have downloaded the Access database called `berks.accdb` from this book's website (see p. iii) and created an ODBC channel called `berks` on your computer. Open the channel to connect R to the Access database, using the function `odbcConnect`:

```
library(RODBC)
channel <- odbcConnect("berks")
```

To complete the map you need to:

- read a file of *x* and *y* coordinates for the outline of the region being mapped (the county of Berkshire in this example);

- draw the outline on a `plot` without any labelling on the axes;

- use `axis` to label using single digits (the grid references have 5 digits);

- add a grid using `abline` with grey lines;

- specify a species to map;

- read the coordinates for this species into R from the Access database;

- use `points` to add the distribution dots to the map.

Here is the outline of the county of Berkshire:

```
data<-read.table("c:\\temp\\vc22outline.txt",header=T)
attach(data)
```

Now plot the outline of the county on blank axes:

```
plot(e,n,type="l",xaxt="n",yaxt="n",xlab="",ylab="")
```

The next task is to label the 100 km squares with a single digit on each of the axes:

```
axis(1,seq(20000,100000,10000),seq(2,10,1))
axis(2,seq(60000,110000,10000),seq(6,11,1))
```

Produce a grid for the 100 km squares in grey:

```
abline(v=10000*(2:10),col="gray")
abline(h=10000*(6:11),col="gray")
```

Use the mouse to stretch or contract the left and bottom margins of the graphics window until the grids make perfect squares.

Next, write a function called `mapit` to specify the name of the species whose data you want to extract from the database, and use this name to construct a query that will extract information for all of the sites at which the named species is found. Finally, execute the query using `sqlQuery` to create a dataframe in R containing the necessary eastings and northings (the *x* and *y* coordinates for the dots to be added to the map):

```
mapit <- function(x) {
mapname <- x
query <- paste(
"SELECT sites.easting, sites.northing
 FROM records
 INNER JOIN sites ON records.site=sites.[site number]
 WHERE records.trimmed ='", mapname, "'", sep='')

mysites <- sqlQuery(channel, query)

with(mysites,points(easting,northing,pch=16,col="red"))
}
```

The two tables in the database are related by shared variables called `site` and `site number`, respectively. The square brackets deal with the gap in the variable name. Run the function by specifying the name of the species you want to map (*Viscum album*):

```
mapit("Viscum album")
```

This will extract the coordinates and add the distribution dots to the map in red:

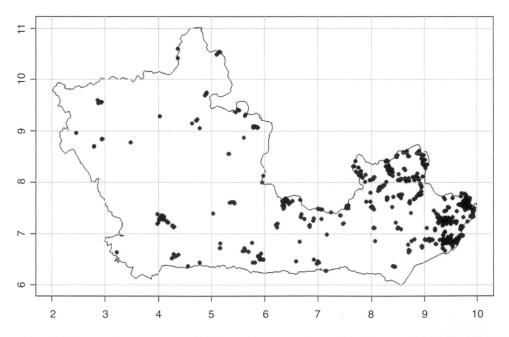

Note how the variable name (`mapname`, which is `"Viscum album"` in this example) has been incorporated into the `WHERE` part of the query using the `paste` function. The way the two tables, `sites` and `records`, are related is specified in the `INNER JOIN` part of the query. The longer of the two tables (`records`) is specified in the `FROM` part of the query. The key here is in dealing correctly with the single quotes that need to appear around the species name inside the `WHERE` part of the query character string.

27

Survival Analysis

A great many studies in statistics deal with deaths or with failures of components: they involve the numbers of deaths, the timing of death, or the risks of death to which different classes of individuals are exposed. The analysis of survival data is a major focus of the statistics business (see Kalbfleisch and Prentice, 1980; Miller, 1981; Fleming and Harrington 1991), for which R supports a wide range of tools. The main theme of this chapter is the analysis of data that take the form of measurements of the **time to death**, or the **time to failure** of a component. Up to now, we have dealt with mortality data by considering the proportion of individuals that were dead *at a given time*. In this chapter each individual is followed until it dies, then the time of death is recorded (this will be the response variable). Individuals that survive to the end of the experiment will die at an unknown time in the future; they are said to be **censored** (as explained below).

27.1   A Monte Carlo experiment

With data on time to death, the most important decision to be made concerns the error distribution. The key point to understand is that the variance in age at death is almost certain to increase with the mean, and hence standard models (assuming constant variance and normal errors) will be inappropriate. You can see this at once with a simple Monte Carlo experiment. Suppose that the per-week probability of failure of a component is 0.1 from one factory but 0.2 from another. We can simulate the fate of an individual component in a given week by generating a uniformly distributed random number between 0 and 1. If the value of the random number is less than or equal to 0.1 (or 0.2 for the second factory), then the component fails during that week and its lifetime can be calculated. If the random number is larger than 0.1 (or 0.2, respectively), then the component survives to the next week. The lifetime of the component is simply the number of the week in which it finally failed. Thus, a component that failed in the first week has an age at failure of 1 (this convention means that there are no zeros in the dataframe).

The simulation is very simple. We create a vector of random numbers, rnos, that is long enough to be certain to contain a value that is less than our failure probabilities of 0.1 and 0.2. Remember that the mean life expectancy is the reciprocal of the failure rate, so our mean lifetimes will be $1/0.1 = 10$ and $1/0.2 = 5$ weeks, respectively. A length of 100 should be more than sufficient:

```
rnos <- runif(100)
```

The trick is to find the week number in which the component failed; this is the lowest subscript for which rnos <=0.1 for factory 1. We can do this very efficiently using the which function: which returns *a vector of subscripts* for which the specified logical condition is true. So for factory 1 we would write:

```
which(rnos<= 0.1)
```

```
[1]  5  8  9 19 29 33 48 51 54 63 68 74 80 83 94 95
```

This means that 16 of my first set of 100 random numbers were less than or equal to 0.1. The important point is that the *first* such number occurred in week 5. So the simulated value of the age of death of this *first* component is 5 and is obtained from the vector of failure ages using the subscript [1]:

```
which(rnos<= 0.1)[1]
```

```
[1]  5
```

All we need to do to simulate the life spans of a sample of 30 components, death1, is to repeat the above procedure 30 times:

```
death1 <- numeric(30)
for (i in 1:30){
rnos <- runif(100)
death1[i] <- which(rnos<= 0.1)[1]
}
death1
```

```
[1]   5  8   7 23 5 4 18 2 6 4 10 12 7 3 5 17 1 3 2 1 12
[22]  8  2 12  6 3 13 16 3 4
```

The fourth component survived for a massive 23 weeks but the 17th component failed during its first week. The simulation has roughly the right average weekly failure rate:

```
1/mean(death1)
```

```
[1]  0.1351351
```

which is as close to 0.1 as we could reasonably expect from a sample of only 30 components.

Now we do the same for the second factory with its failure rate of 0.2:

```
death2 <- numeric(30)
for (i in 1:30) {
rnos <- runif(100)
death2[i] <- which(rnos<= 0.2)[1] }
```

The sample mean is again quite reasonable (the real hazard was 0.20):

```
1/mean(death2)
```

```
[1]   0.2205882
```

We now have the simulated raw data to carry out a comparison in age at death between factories 1 and 2. We combine the two vectors into one, and generate a vector to represent the factory identities:

```
death <- c(death1,death2)
factory <- factor(c(rep(1,30),rep(2,30)))
```

We get a visual assessment of the data as follows:

```
plot(factory,death,xlab="factory",ylab="age at failure",col="wheat3")
```

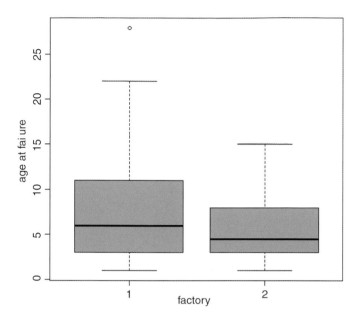

The median age at death for factory 1 is somewhat greater, but the variance in age at death is much higher than from factory 2. For data like this we expect the variance to be proportional to the square of the mean, so an appropriate error structure is the gamma (as explained below). We model the data very simply as a one-way analysis of deviance using glm with `family=Gamma` (note the upper-case 'G'):

```
model1 <- glm(death~factory,Gamma)
summary(model1)


Call:
glm(formula = death ~ factory, family = Gamma)

Coefficients:
              Estimate    Std. Error    t value    Pr(>| t |)
(Intercept)   0.13514     0.02218       6.092      9.6e-08      ***
factory2      0.08545     0.04246       2.013      0.0488       *

(Dispersion parameter for Gamma family taken to be 0.8082631)
      Null deviance: 44.067 on 59 degrees of freedom
Residual deviance: 40.501 on 58 degrees of freedom
AIC: 329.62
Number of Fisher Scoring iterations: 6
```

We conclude that the factories are marginally significantly different in mean age at failure of these components ($p = 0.0488$). So, even with a twofold difference in the true failure rate, it is hard to detect a significant difference in mean age at death with samples of size $n = 30$. The moral is that for data like this on age at death you are going to need really large sample sizes in order to find significant differences.

It is good practice to remove variable names (like `death`) that you intend to use later in the same session (see p. 10):

```
rm(death)
```

27.2 Background

Since everything dies eventually, it is often not possible to analyse the results of survival experiments in terms of the proportion that were killed (as we did in Chapter 16); in due course, they *all* die. Look at the following figure:

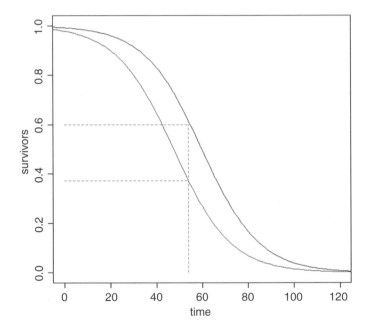

It is clear that the two treatments (red and blue) caused different patterns of mortality, but both start out with 100% survival and both end up with zero. We could pick some arbitrary point in the middle of the distribution at which to compare the percentage survival (say at time 54, shown by the vertical green dashed line, where the red curve shows 38% survival but the blue curve shows 60% survival), but this may be difficult in practice, because one or both of the treatments might have few observations at the same location. Also, the choice of when to measure the difference is entirely subjective and hence open to bias. It is much better to use R's powerful facilities for the analysis of survival data than it is to pick an arbitrary time at which to compare two proportions.

Demographers, actuaries and ecologists use three interchangeable concepts when dealing with data on the timing of death: **survivorship**, **age at death** and **instantaneous risk of death**. There are three broad patterns of survivorship (named by the eminent American animal ecologist Edward Deevey in 1947): Type I, where most of the mortality occurs late in life (e.g. humans); Type II, where mortality occurs at a roughly constant rate throughout life; and Type III, where most of the mortality occurs early in life (e.g. salmonid fishes). On a log scale, the numbers surviving over time would look like this:

27.3 The survivor function

The survivorship curve plots the natural log of the proportion of a cohort of individuals starting out at time 0 that is still alive at time t. For the so-called Type II survivorship curve (the red line above), there is a linear decline in log numbers with time. This means that a constant proportion of the individuals alive at the beginning of a time interval will die during that time interval (i.e. the proportion dying is independent of density and constant for all ages). When the death rate is highest for the younger age classes we get Type III survivorship curve, which descends steeply at first, with the rate of descent casing later on. When it is the oldest animals that have the highest risk of death (as in the case of human populations in affluent societies where there is low infant mortality) we obtain the Type I curve, which has a shallow descent to start, becoming steeper later.

27.4 The density function

The density function describes the fraction of all deaths from our initial cohort that are likely to occur in a given brief instant of time. For the Type II curve this is a negative exponential. Because the fraction of individuals dying is constant with age, the number dying declines exponentially as the number of survivors (the number of individuals at risk of death) declines exponentially with the passage of time. The density function declines more steeply than exponentially for Type III survivorship curves. In the case of Type I curves, however, the density function has a maximum at the time when the product of the risk of death and the number of survivors is greatest (see below).

27.5 The hazard function

The hazard is the instantaneous risk of death. It is the instantaneous rate of change in the log of the number of survivors per unit time. Thus, for the Type II survivorship the hazard function is a horizontal line, because the risk of death is constant with age. Although this sounds highly unrealistic, it is a remarkably robust assumption in many applications. It also has the substantial advantage of parsimony. In some cases, however, it is clear that the risk of death changes substantially with the age of the individuals, and we need to be able to take this into account in carrying out our statistical analysis. In the case of Type III survivorship, the risk of death declines with age, while for Type I survivorship (as in humans) the risk of death increases with age.

27.6 The exponential distribution

This is a one-parameter distribution in which the hazard function is independent of age (i.e. it describes a Type II survivorship curve). The exponential is a special case of the gamma distribution in which the shape parameter α is equal to 1.

27.6.1 Density function

The density function is the probability of a death occurring in the small interval of time between t and $t + dt$, and a plot of the number dying in the interval around time t as a function of t (i.e. the proportion of the original cohort dying at a given age) declines exponentially:

$$f(t) = \frac{e^{-t/\mu}}{\mu},$$

where both μ and t are greater than 0. Note that the density function has an intercept of $1/\mu$ (remember that e^0 is 1). The number from the initial cohort dying per unit time declines exponentially with time, and a fraction $1/\mu$ dies during the first time interval (and, indeed, during every subsequent time interval).

27.6.2 Survivor function

This shows the proportion of individuals from the initial cohort born at time 0 still alive at time t:

$$S(t) = e^{-t/\mu}.$$

The survivor function has an intercept of 1 (i.e. all the cohort is alive at time 0), and shows the probability of surviving at least as long as t.

27.6.3 Hazard function

This is the statistician's equivalent of the ecologist's *instantaneous death rate*. It is defined as the ratio between the density function and the survivor function, and is the conditional density function at time t, given survival up to time t. In the case of Type II curves this has an extremely simple form:

$$h(t) = \frac{f(t)}{S(t)} = \frac{e^{-t/\mu}}{\mu e^{-1/\mu}} = \frac{1}{\mu},$$

because the exponential terms cancel out. Thus, with the exponential distribution *the hazard is the reciprocal of the mean time to death*, and vice versa. For example, if the mean time to death is 25 weeks, then the hazard is 0.04; if the hazard were to increase to 0.05, then the mean time of death would decline to 20 weeks.

The **survivor**, **density** and **hazard** functions of the exponential distribution are as follows (note the changes of scale on the *y* axes):

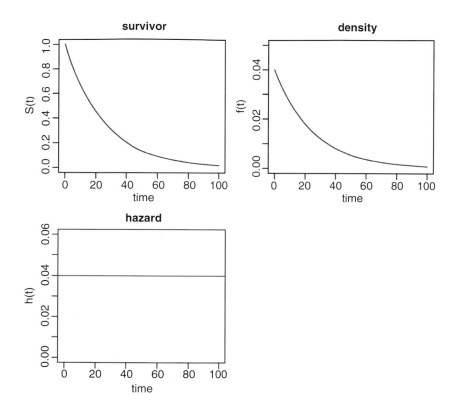

Of course, the death rate may not be constant with age. For example, the death rate may be high for very young as well as for very old animals, in which case the survivorship curve is like an S shape on its side.

27.7 Kaplan–Meier survival distributions

The response in a survival analysis an initially perplexing object called a Kaplan–Meier object. There are two parts to it: the survival part and the status part. The Kaplan–Meier survival distribution is a discrete stepped survivorship curve that accrues information as each death occurs. Suppose we had $n = 5$ individuals and that the times at death were 12, 17, 29, 35 and 42 weeks after the beginning of a trial. The survival curve is horizontal at 1.0 until the first death occurs at time 12. The curve then steps down to 0.8 because 80% of the initial cohort is now alive. It continues at 0.8 until time 17 when the curve steps down to 0.6 (40% of the individuals are now dead). And so on until all of the individuals are dead at time 42.

In general, therefore, we have two variables at any one time: the number of deaths, $d(t_i)$, and the number at risk, $r(t_i)$ (i.e. those that have not yet died: the survivors). The Kaplan–Meier survivor function is

$$\hat{S}_{KM} = \prod_{t_i < t} \frac{r(t_i) - d(t_i)}{r(t_i)}$$

which, as we have seen, produces a step at every time at which one or more deaths occurs.

Sometimes we know when an individual was last seen alive, but not the age at which it died. Patients may leave a study for all sorts of reasons (recovery, emigration, etc.) and it is important that we make the maximum use of the information that we have about them (even though we do not have a value for age at death, which is the response variable). The status of individuals makes up the second component of the response object. It is a vector showing for each individual whether its time value is an age at death or age when last seen alive. Deaths are indicated by 1 and non-deaths by 0. The non-dead individuals are said to be **censored**. These individuals have their ages shown by a + ('plus') on the plot, or after their age in a dataframe (thus 65 means died at time 65, but 65+ means still alive when last seen at age 65). Censored individuals contribute to our understanding of the shape of the survivorship curve but they do not contribute directly to our understanding of the mean age at death.

27.8 Age-specific hazard models

In many circumstances, the risk of death increases with age. There are many models to chose from:

Distribution	Hazard
Exponential	$\text{constant} = \dfrac{1}{\mu}$
Weibull	$\alpha\lambda(\lambda t)^{\alpha-1}$
Gompertz	be^{ct}
Makeham	$a + be^{ct}$
Extreme value	$\dfrac{1}{\sigma}e^{(t-\eta)/\sigma}$
Rayleigh	$a + bt$

The Rayleigh is obviously the simplest model in which hazard increases with time, but the Makeham is widely regarded as the best description of hazard for human subjects. After infancy, there is a constant hazard (a) which is due to age-independent accidents, murder, suicides, etc., with an exponentially increasing hazard in later life. The Gompertz assumption was that 'the average exhaustion of a man's power to avoid death is such that at the end of equal infinitely small intervals of time he has lost equal portions of his remaining power to oppose destruction which he had at the commencement of these intervals'. Note that the Gompertz differs from the Makeham only by the omission of the extra background hazard (a), and this becomes negligible in old age. The Weibull distribution is very flexible because it can deal with hazards that increase with age in an accelerating ($\alpha > 1$) or decelerating ($\alpha < 1$) manner with age.

These plots show how hazard changes with age for four of the hazard functions:

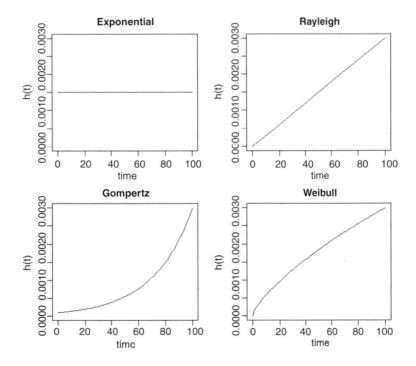

27.9   Survival analysis in R

There are three cases that concern us here:

- constant hazard and no censoring;
- constant hazard with censoring;
- age-specific hazard, with or without censoring.

The first case is dealt with very simply in R by specifying a generalized linear model with gamma errors. The second case involves the use of exponential survival models with a *censoring indicator* (1 indicates that the response is a time at death, 0 indicates that the individual was alive when last seen; see below and p. 883). The third case involves a choice between **parametric** models, based on the **Weibull** distribution, and **non-parametric** techniques, based on the **Cox proportional hazards** model.

27.9.1   Parametric models

We are unlikely to know much about the error distribution in advance of the study, except that it will certainly not be normal. In R we are offered several choices for the analysis of survival data:

- gamma;
- exponential;
- piecewise exponential;
- extreme value;
- log-logistic;
- lognormal;
- Weibull.

In practice, it is often difficult to choose between them. In general, the best solution is to try several distributions and to pick the error structure that produces the minimum error deviance.

27.9.2   Cox proportional hazards model

This is the most widely used regression model for survival data. It assumes that the hazard is of the form

$$\lambda(t;\, Z_i) = \lambda_0(t)\, r_i(t),$$

where $Z_i(t)$ is the set of explanatory variables for individual i at time t. The *risk score* for subject i is

$$r_i(t) = e^{\beta Z_i(t)},$$

in which β is a vector of parameters from the linear predictor; $\lambda_0(t)$ is an *unspecified baseline hazard function* that will cancel out in due course. The antilog guarantees that λ is positive for any regression model $\beta Z_i(t)$.

If a death occurs at time t^*, then, conditional on this death occurring, the likelihood that it is individual i that dies, rather than any other individual at risk, is

$$L_i(\beta) = \frac{\lambda_0(t^*)r_i(t^*)}{\sum_j Y_j(t^*)\lambda_o(t^*)r_j(t^*)} = \frac{r_i(t^*)}{\sum_j Y_j(t^*)r_j(t^*)}.$$

The product of these terms over all times of death, $L(\beta) = \prod L_i(\beta)$, was christened a partial likelihood by Cox (1972). This is clever, because maximizing $\log(L(\beta))$ allows an estimate of β without knowing anything about the baseline hazard function ($\lambda_0(t)$ is a nuisance variable in this context). The proportional hazards model is nonparametric in the sense that it depends only on the **ranks** of the survival times.

27.9.3 Cox's proportional hazard or a parametric model?

In cases where you have censoring, or where you want to use a more complex error structure, you will need to choose between a parametric model, fitted using `survreg`, and a non-parametric model, fitted using `coxph`. If you want to use the model for prediction, then you have no choice: you must use the parametric `survreg` because `coxph` does not extrapolate beyond the last observation. Traditionally, medical studies use `coxph` while engineering studies use `survreg` (so-called accelerated failure-time models), but both disciples could fruitfully use either technique, depending on the nature of the data and the precise question being asked. Here is a typical question addressed with `coxph`: 'How much does the risk of dying decrease if a new drug treatment is given to a patient?' In contrast, parametric techniques are typically used for questions like this: 'What proportion of patients will die in 2 years based on data from an experiment that ran for just 4 months?'

27.10 Parametric analysis

The following example concerns survivorship of two cohorts of seedlings. All the seedlings died eventually, so there is no censoring in this case. There are two questions:

- Was survivorship different in the two cohorts?

- Was survivorship affected by the size of the canopy gap in which germination occurred?

Here are the data:

```
seedlings <- read.table("c:\\temp\\seedlings.txt",header=T)
attach(seedlings)
names(seedlings)
[1] "cohort" "death" "gapsize"
```

We need to load the survival library:

```
library(survival)
```

We begin by creating a variable called `status` to indicate which of the data are censored:

```
status <- 1*(death>0)
```

There are no cases of censoring in this example, so all of the values of status are equal to 1.

The fundamental object in survival analysis is `Surv(death,status)`, the Kaplan–Meier survivorship object (see p. 875 for an introduction to this). We can plot it out using `survfit` like this:

```
plot(survfit(Surv(death,status)~1),ylab="Survivorship",xlab="Weeks",col=4)
```

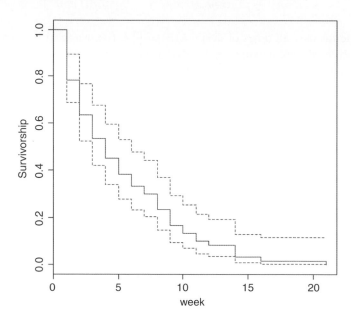

This shows the overall survivorship curve with the confidence intervals. All the seedlings were dead by week 21. Were there any differences in survival between the two cohorts?

```
model <- survfit(Surv(death,status)~cohort)
summary(model)
```

```
Call: survfit(formula = Surv(death, status) ~ cohort)
```

cohort=October

time	n.risk	n.event	survival	std.err	lower 95% CI	upper 95% CI
1	30	7	0.7667	0.0772	0.62932	0.934
2	23	3	0.6667	0.0861	0.51763	0.859
3	20	3	0.5667	0.0905	0.41441	0.775
4	17	2	0.5000	0.0913	0.34959	0.715
5	15	3	0.4000	0.0894	0.25806	0.620
6	12	1	0.3667	0.0880	0.22910	0.587
8	11	2	0.3000	0.0837	0.17367	0.518
9	9	4	0.1667	0.0680	0.07488	0.371
10	5	1	0.1333	0.0621	0.05355	0.332
11	4	1	0.1000	0.0548	0.03418	0.293
14	3	1	0.0667	0.0455	0.01748	0.254
16	2	1	0.0333	0.0328	0.00485	0.229
21	1	1	0.0000	NaN	NA	NA

```
                       cohort=September
time n.risk n.event survival std.err lower 95% CI upper 95% CI
   1     30        6   0.8000  0.0730      0.6689        0.957
   2     24        6   0.6000  0.0894      0.4480        0.804
   3     18        3   0.5000  0.0913      0.3496        0.715
   4     15        3   0.4000  0.0894      0.2581        0.620
   5     12        1   0.3667  0.0880      0.2291        0.587
   6     11        2   0.3000  0.0837      0.1737        0.518
   7      9        2   0.2333  0.0772      0.1220        0.446
   8      7        2   0.1667  0.0680      0.0749        0.371
  10      5        1   0.1333  0.0621      0.0535        0.332
  11      4        1   0.1000  0.0548      0.0342        0.293
  12      3        1   0.0667  0.0455      0.0175        0.254
  14      2        2   0.0000     NaN          NA           NA
```

To plot these figures use plot(model) like this:

```
plot(model,col=c("red","blue"),ylab="Survivorship",xlab="week")
```

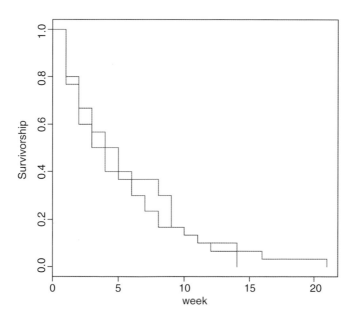

The red line is for the October cohort and the blue line is for September. To see the median times at death for the two cohorts, just type:

```
model
```

```
Call: survfit(formula = Surv(death, status) ~ cohort)
                 records n.max n.start events median 0.95LCL 0.95UCL
cohort=October        30    30      30     30    4.5       3       9
cohort=September      30    30      30     30    3.5       2       7
```

27.11   Cox's proportional hazards

The median age at death was one week later in the October cohort, but look at the width of the confidence intervals: 3 to 9 versus 2 to 7. Clearly there is no significant effect of cohort on time of death. What about gap size? We start with a full analysis of covariance using `coxph` rather than `survfit`.

```
model1 <- coxph(Surv(death,status)~strata(cohort)*gapsize)
summary(model1)

Call:
coxph(formula = Surv(death, status) ~ strata(cohort) * gapsize)

  n= 60, number of events= 60

                                    coef exp(coef) se(coef)      z Pr(>|z|)
gapsize                          -1.1863    0.3054   0.6210 -1.910   0.0561 .
gapsize:strata(cohort)cohort=September  0.5795    1.7852   0.8264  0.701   0.4831

                                 exp(coef) exp(-coef) lower .95 upper .95
gapsize                             0.3054     3.2749   0.09042     1.031
gapsize:strata(cohort)cohort=September  1.7852     0.5602   0.35341     9.018

Concordance= 0.659  (se = 0.077 )
Rsquare= 0.076   (max possible= 0.993 )
Likelihood ratio test= 4.73  on 2 df,    p=0.09372
Wald test         = 4.89  on 2 df,    p=0.08682
Score (logrank) test = 5.04  on 2 df,    p=0.08046
```

There is no evidence of any interaction ($p = 0.483$) and the main effect of gap size is not quite significant in this model ($p = 0.056$). We fit the simpler model with no interaction:

```
model2 <- coxph(Surv(death,status)~strata(cohort)+gapsize)
anova(model1,model2)

Analysis of Deviance Table
 Cox model: response is  Surv(death, status)
 Model 1: ~ strata(cohort) * gapsize
 Model 2: ~ strata(cohort) + gapsize
   loglik   Chisq Df P(>|Chi|)
1 -146.95
2 -147.20 0.4945  1     0.4819
```

There is no significant difference in explanatory power, so we accept the simpler model without an interaction term. Note that removing the interaction makes the main effect of gap size significant ($p = 0.035$):

```
summary(model2)

Call:
coxph(formula = Surv(death, status) ~ strata(cohort) + gapsize)

  n= 60, number of events= 60

           coef exp(coef) se(coef)      z Pr(>|z|)
gapsize -0.8545    0.4255   0.4054 -2.108    0.035 *

        exp(coef) exp(-coef) lower .95 upper .95
gapsize    0.4255       2.35    0.1922    0.9418

Concordance= 0.659  (se = 0.077 )
```

```
Rsquare= 0.068    (max possible= 0.993 )
Likelihood ratio test= 4.24   on 1 df,    p=0.03947
Wald test             = 4.44   on 1 df,    p=0.03504
Score (logrank) test = 4.54   on 1 df,    p=0.03313
```

We conclude that the risk of seedling death is lower in bigger gaps (`coef = -0.855`), but this effect is similar in the September and October germinating cohorts.

You see that the modelling methodology is exactly the same as usual: fit a complicated model and simplify it to find a minimal adequate model. The only difference is the use of `Surv(death,status)` when the response is a Kaplan–Meier object.

27.12   Models with censoring

Censoring occurs when we do not know the time of death for all of the individuals. This comes about principally because some individuals outlive the experiment, while others leave the experiment before they die. We know when we last saw them alive, but we have no way of knowing their age at death. These individuals contribute something to our knowledge of the survivor function, but nothing to our knowledge of the age at death. Another reason for censoring occurs when individuals are lost from the study: they may be killed in accidents, they may emigrate, or they may lose their identity tags.

In general, then, our survival data may be a mixture of times at death and times after which we have no more information on the individual. We deal with this by setting up an extra vector called the **censoring indicator** to distinguish between the two kinds of numbers. If a time really is a time to death, then the censoring indicator takes the value 1. If a time is just the last time we saw an individual alive, then the censoring indicator is set to 0. Thus, if we had the following time data T and censoring indicator W on seven individuals, this would mean that five of the times were times at death while in two cases, one at time 8 and another at time 15, individuals were seen alive but never seen again:

```
T     4      7      8      8      12     15     22
W     1      1      0      1      1      0      1
```

With repeated sampling in survivorship studies, it is usual for the degree of censoring to decline as the study progresses. Early on, many of the individuals are alive at the end of each sampling interval, whereas few if any survive to the end of the last study period. We need to tidy up from the last example:

```
rm(status)
detach(seedlings)
```

The following example comes from a study of cancer patients undergoing one of four drug treatment programmes (drugs A, B and C and a placebo):

```
cancer <- read.table("c:\\temp\\cancer.txt",header=T)
attach(cancer)
names(cancer)

[1] "death" "treatment" "status"

plot(survfit(Surv(death,status)~treatment),
               col=c(1:4),ylab="Surivorship",xlab="Time")
```

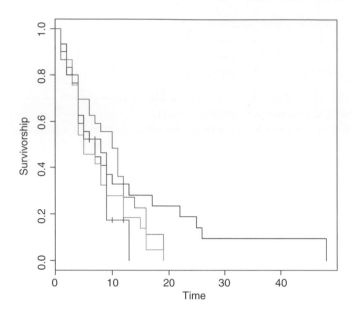

The censored individuals are shown by crosses on the Kaplan–Meier plot, above. Of the non-censored individuals, the mean ages at death were as follows:

```
tapply(death[status==1],treatment[status==1],mean)
```

```
   DrugA    DrugB    DrugC  placebo
9.480000 8.360000 6.800000 5.238095
```

The long tail is for drug A (black line). The latest deaths in the other treatments were at times 14 and 19. The variances in age at death are dramatically different under the various treatments:

```
tapply(death[status==1],treatment[status==1],var)
```

```
   DrugA     DrugB     DrugC   placebo
117.51000  32.65667  27.83333  11.39048
```

27.12.1 Parametric models

The simplest model assumes a constant hazard: `dist="exponential"`.

```
model1 <- survreg(Surv(death,status)~treatment,dist="exponential")
summary(model1)
```

```
Call:
survreg(formula = Surv(death, status) ~ treatment, dist = "exponential")
                 Value Std. Error      z        p
(Intercept)      2.448      0.200 12.238 1.95e-34
treatmentDrugB  -0.125      0.283 -0.443 6.58e-01
treatmentDrugC  -0.430      0.283 -1.520 1.28e-01
treatmentplacebo -0.333     0.296 -1.125 2.61e-01

Scale fixed at 1
```

```
Exponential distribution
Loglik(model)= -310.1   Loglik(intercept only)= -311.5
       Chisq= 2.8 on 3 degrees of freedom, p= 0.42
Number of Newton-Raphson Iterations: 4
n= 120
```

Under the assumption of exponential errors there are no significant effects of drug treatment on survivorship (all $p > 0.1$). How about modelling non-constant hazard using Weibull errors instead (these are the default for survreg)?

```
model2 <- survreg(Surv(death,status)~treatment)
summary(model2)

Call:
survreg(formula = Surv(death, status) ~ treatment)
                 Value Std. Error     z        p
(Intercept)      2.531     0.1572 16.102 2.47e-58
treatmentDrugB  -0.191     0.2193 -0.872 3.83e-01
treatmentDrugC  -0.475     0.2186 -2.174 2.97e-02
treatmentplacebo -0.454    0.2313 -1.963 4.96e-02
Log(scale)      -0.260     0.0797 -3.264 1.10e-03

Scale= 0.771

Weibull distribution
Loglik(model)= -305.4   Loglik(intercept only)= -308.3
       Chisq= 5.8 on 3 degrees of freedom, p= 0.12
Number of Newton-Raphson Iterations: 5
n= 120
```

The scale parameter 0.771, being less than 1, indicates that slope of the hazard *decreases* with age in this study. Drug B is not significantly different from drug A ($p = 0.38$), but drug C and the placebo are significantly poorer ($p < 0.05$). We can use anova to compare model1 and model2:

```
anova(model1,model2)

       Terms Resid. Df    -2*LL Test Df Deviance    Pr(>Chi)
1  treatment       116 620.1856   NA        NA          NA
2  treatment       115 610.7742   =   1   9.4114 0.002156405
```

You can see that model2 with Weibull errors is significant improvement over model1 with exponential errors ($p = 0.002$).

We can try amalgamating the factor levels – the analysis suggests that we begin by grouping A and B together:

```
treat2 <- treatment
levels(treat2)

[1] "DrugA"   "DrugB"   "DrugC"   "placebo"

levels(treat2)[1:2] <- "DrugsAB"
levels(treat2)

[1] "DrugsAB" "DrugC"   "placebo"
```

```
model3 <- survreg(Surv(death,status)~treat2)
anova(model2,model3)
```

```
         Terms Resid. Df     -2*LL     Test Df  Deviance   Pr(>Chi)
1 treatment         115 610.7742         NA        NA         NA
2    treat2         116 611.5190 1 vs. 2 -1 -0.744833 0.3881171
```

That model simplification was justified. What about drug C? Can we lump it together with the placebo?

```
levels(treat2)[2:3] <- "placeboC"
model4 <- survreg(Surv(death,status)~treat2)
anova(model3,model4)
```

```
   Terms Resid. Df     -2*LL Test Df    Deviance   Pr(>Chi)
1 treat2         116 611.5190    NA         NA         NA
2 treat2         117 611.5301   = -1 -0.01101309 0.9164208
```

Yes, we can. That simplification was clearly justified ($p = 0.916$):

```
summary(model4)
```

```
Call:
survreg(formula = Surv(death, status) ~ treat2)
                 Value Std. Error     z         p
(Intercept)      2.439      0.112 21.76 5.37e-105
treat2placeboC -0.374      0.160 -2.33  1.96e-02
Log(scale)     -0.249      0.078 -3.20  1.39e-03

Scale= 0.779

Weibull distribution
Loglik(model)= -305.8   Loglik(intercept only)= -308.3
        Chisq= 5.05 on 1 degrees of freedom, p= 0.025
Number of Newton-Raphson Iterations: 5
n= 120
```

We can summarize the results in terms of the mean age at death, taking account of the censoring:

```
tapply(predict(model4,type="response"),treat2,mean)
```

```
  DrugsAB   placeboC
11.459885   7.887685
```

Here are the uncorrected mean ages at death for those cases where we know the age of death:

```
tapply(death[status==1],treat2[status==1],mean)
```

```
  DrugsAB placeboC
8.920000 6.086957
```

The greater the censoring, the bigger the difference will be.

```
detach(cancer)
rm(death, status)
```

27.12.2 Comparing `coxph` and `survreg` survival analysis

Finally, we shall compare the methods, parametric and non-parametric, by analysing the same data set both ways. It is an analysis of covariance with one continuous explanatory variable (initial weight) and one categorical explanatory variable (group):

```
insects <- read.table("c:\\temp\\roaches.txt",header=T)
attach(insects)
names(insects)
```

```
[1] "death" "status" "weight" "group"
```

First, we plot the survivorship curves of the three groups:

```
plot(survfit(Surv(death,status)~group),col=c(2,3,4),ylab="Survivorship",
xlab="Time")
```

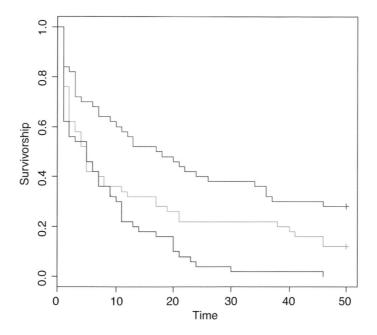

There are clearly big differences between the death rates in the three groups. The crosses $+$ at the end of the survivorship curves for groups A and B indicate that there was censoring in these groups (not all of the individuals were dead at the end of the experiment).

We begin the modelling with parametric methods (`survreg`). We shall compare the default error distribution (Weibull, which allows for non-constant hazard with age) with the simpler exponential (assuming constant hazard):

```
model1 <- survreg(Surv(death,status)~weight*group,dist="exponential")
summary(model1)
```

```
Call:
survreg(formula = Surv(death, status) ~ weight * group, dist = "exponential")
```

```
               Value Std. Error       z        p
(Intercept)    3.8702      0.3854 10.041 1.00e-23
weight        -0.0803      0.0659 -1.219 2.23e-01
groupB        -0.8853      0.4508 -1.964 4.95e-02
groupC        -1.7804      0.4386 -4.059 4.92e-05
weight:groupB  0.0643      0.0674  0.954 3.40e-01
weight:groupC  0.0796      0.0674  1.180 2.38e-01

Scale fixed at 1

Exponential distribution
Loglik(model)= -480.6   Loglik(intercept only)= -502.1
      Chisq= 43.11 on 5 degrees of freedom, p= 3.5e-08
Number of Newton-Raphson Iterations: 5
n= 150
```

For `model2` we employ the default Weibull distribution allowing non-constant hazard:

```
model2 <- survreg(Surv(death,status)~weight*group)
summary(model2)

Call:
survreg(formula = Surv(death, status) ~ weight * group)
                 Value Std. Error       z        p
(Intercept)     3.9506      0.5308  7.443 9.84e-14
weight         -0.0973      0.0909 -1.071 2.84e-01
groupB         -1.1337      0.6207 -1.826 6.78e-02
groupC         -1.9841      0.6040 -3.285 1.02e-03
weight:groupB   0.0826      0.0929  0.889 3.74e-01
weight:groupC   0.0931      0.0930  1.002 3.16e-01
Log(scale)      0.3083      0.0705  4.371 1.24e-05

Scale= 1.36

Weibull distribution
Loglik(model)= -469.6   Loglik(intercept only)= -483.3
      Chisq= 27.42 on 5 degrees of freedom, p= 4.7e-05
Number of Newton-Raphson Iterations: 5
n= 150
```

The fact that the scale parameter is greater than 1 indicates that the risk of death increases with age in this case. We compare the two models in the usual way, using `anova`:

```
anova(model1,model2)

           Terms Resid. Df    -2*LL Test Df  Deviance      Pr(>Chi)
1 weight * group      144 961.1800   NA   NA        NA            NA
2 weight * group      143 939.2261    =   1  21.95388 2.792823e-06
```

The Weibull `model2` is vastly superior to the exponential ($p < 0.000\,01$) so we continue with `model2`. There is no evidence in `model2` (summary above) of any interaction between `weight` and `group` ($p = 0.374$) so we simplify using `step`:

```
model3 <- step(model2)

Start:  AIC=953.23
Surv(death, status) ~ weight * group

                   Df    AIC
- weight:group    2 950.30
<none>               953.23

Step:  AIC=950.3
Surv(death, status) ~ weight + group

              Df    AIC
- weight     1 949.01
<none>          950.30
- group      2 967.75

Step:  AIC=949.01
Surv(death, status) ~ group

            Df    AIC
<none>         949.01
- group     2 970.64
```

After eliminating the non-significant interaction (AIC = 950.30), R has removed the main effect of weight (AIC = 949.01), but has kept the main effect of group (AIC of <none> is less than AIC of -group).

The minimal model with survreg is this:

```
summary(model3)

Call:
survreg(formula = Surv(death, status) ~ group)
                Value Std. Error       z         p
(Intercept)     3.459      0.2283  15.15  7.20e-52
groupB         -0.822      0.3097  -2.65  7.94e-03
groupC         -1.540      0.3016  -5.11  3.28e-07
Log(scale)      0.314      0.0705   4.46  8.15e-06

Scale= 1.37

Weibull distribution
Loglik(model)= -470.5    Loglik(intercept only)= -483.3
        Chisq= 25.63 on 2 degrees of freedom, p= 2.7e-06
Number of Newton-Raphson Iterations: 5
n= 150
```

It is clear that all three groups are required (B and C differ by 0.72, with standard error 0.31), so this is the minimal adequate model. Here are the predicted mean ages at death:

```
tapply(predict(model3),group,mean)

        A         B         C
31.796137 13.972647  6.814384
```

You can compare these with the mean ages of those insects that died

```
tapply(death[status==1],group[status==1],mean)
```

```
        A         B         C
12.611111  9.568182  8.020000
```

and the ages when insects were last seen (dead or alive):

```
tapply(death,group,mean)
```

```
    A     B     C
23.08 14.42  8.02
```

The predicted ages at death are substantially greater than the observed ages at last sighting when there is lots of censoring (e.g. 31.8 vs. 23.08 for group A).

Here are the same data analysed with the Cox proportional hazards model:

```
model10 <- coxph(Surv(death,status)~weight*group)
summary(model10)
```

```
Call:
coxph(formula = Surv(death, status) ~ weight * group)

  n= 150, number of events= 130
                   coef exp(coef) se(coef)      z Pr(>|z|)
weight          0.06330   1.06535  0.06738  0.940  0.34747
groupB          0.79098   2.20555  0.45641  1.733  0.08309 .
groupC          1.28634   3.61953  0.45243  2.843  0.00447 **
weight:groupB -0.05568   0.94585  0.06878 -0.809  0.41824
weight:groupC -0.05869   0.94300  0.06897 -0.851  0.39481

              exp(coef) exp(-coef) lower .95 upper .95
weight           1.0654     0.9387    0.9336     1.216
groupB           2.2056     0.4534    0.9016     5.395
groupC           3.6195     0.2763    1.4912     8.785
weight:groupB    0.9458     1.0573    0.8266     1.082
weight:groupC    0.9430     1.0604    0.8238     1.079

Concordance= 0.608   (se = 0.034 )
Rsquare= 0.135    (max possible= 0.999 )
Likelihood ratio test= 21.83   on 5 df,    p=0.0005645
Wald test              = 20.75   on 5 df,    p=0.000903
Score (logrank) test = 22.05   on 5 df,    p=0.0005132
```

As you see, the interaction terms are not significant ($p > 0.39$), so we simplify using `step` as before:

```
model11 <- step(model10)
```

```
Start:  AIC=1113.54
Surv(death, status) ~ weight * group

                Df    AIC
- weight:group   2 1110.3
<none>             1113.5
```

```
Step:   AIC=1110.27
Surv(death, status) ~ weight + group

          Df    AIC
- weight   1 1108.8
<none>       1110.3
- group    2 1123.7

Step:   AIC=1108.82
Surv(death, status) ~ group

          Df    AIC
<none>       1108.8
- group    2 1125.4
```

Note that the AIC values are different than they were with the parametric model. The interaction term is dropped because this simplification reduces AIC to 1110.3. Then the covariate (`weight`) is dropped because this simplification also reduces AIC (to 1108.8). But removing `group` would *increase* AIC to 1125.4, so this is not done. The minimal model contains a main effect for `group` but no effect of `weight`:

```
summary(model11)

Call:
coxph(formula = Surv(death, status) ~ group)

  n= 150, number of events= 130

           coef exp(coef)  se(coef)      z Pr(>|z|)
groupB 0.5607    1.7520    0.2257 2.485    0.013 *
groupC 1.0084    2.7412    0.2263 4.456 8.33e-06 ***

          exp(coef) exp(-coef) lower .95 upper .95
groupB      1.752     0.5708    1.126     2.727
groupC      2.741     0.3648    1.759     4.271

Concordance= 0.607   (se = 0.032 )
Rsquare= 0.128    (max possible= 0.999 )
Likelihood ratio test= 20.55   on 2 df,    p=3.452e-05
Wald test               = 19.86   on 2 df,   p=4.865e-05
Score (logrank) test = 20.98   on 2 df,   p=2.775e-05
```

To see what these numbers mean, it is a good idea to go back to the raw data on times of death (or last sighting for the censored individuals). Here are the mean values of those that died:

```
tapply(death[status==1],group[status==1],mean)

       A         B         C
12.611111  9.568182  8.020000
```

Evidently, individuals in group A lived a lot longer than those in group C. Compared with group C (the minimum value), the mean age at death for group A expressed as a ratio is

```
12.61/8.02

[1] 1.572319
```

and for group B we have

```
12.61/9.57
```

```
[1]  1.317659
```

These figures are the approximate hazards for an individual in group C or B relative to an individual in group A. In the `coxph` output of `model11` they are labelled `exp(coef)`. The model values are slightly different from the raw means because of the way the model has dealt with censoring (14 censored individuals in group A, six in group B and none in group C).

You should compare the outputs from the two functions `coxph` and `survreg` to make sure you understand their similarities and differences. One fundamental difference is that the parametric Kaplan–Meier survivorship curves refer to the **population**, whereas Cox proportional hazards refer to an **individual** in a particular group.

28

Simulation Models

Simulation modelling is an enormous topic, and all I intend here is to demonstrate a few very simple temporal and spatial simulation techniques that give the flavour of what is possible in R.

Simulation models are used for investigating dynamics in time, in space, or in both space and time together. For temporal dynamics we might be interested in:

- the transient dynamics (the behaviour after the start but before equilibrium is attained – if indeed equilibrium is ever attained);

- equilibrium behaviour (after the transients have damped away);

- chaos (random-looking, but actually deterministic temporal dynamics that are extremely sensitive to initial conditions).

For spatial dynamics, we might use simulation models to study:

- metapopulation dynamics (where local extinction and recolonization of patches characterize the long-term behaviour, with constant turnover of occupied patches);

- neighbour relations (in spatially explicit systems where the performance of individuals is determined by the identity and attributes of their immediate neighbours);

- pattern generation (dynamical processes that lead to the generation of emergent, but more or less coherent patterns).

28.1 Temporal dynamics: Chaotic dynamics in population size

Biological populations typically increase exponentially when they are small, but individuals perform less well as population density rises, because of competition, predation or disease. In aggregate, these effects on birth and death rates are called **density-dependent processes**, and it is the nature of the density-dependent processes that determines the temporal pattern of population dynamics. The simplest density-dependent

The R Book, Second Edition. Michael J. Crawley.
© 2013 John Wiley & Sons, Ltd. Published 2013 by John Wiley & Sons, Ltd.

model of population dynamics is known as the **quadratic map**. It is a first-order non-linear difference equation,

$$N(t+1) = \lambda N(t)[1 - N(t)],$$

where $N(t)$ is the population size at time t, $N(t+1)$ is the population size at time $t+1$, and the single parameter, λ, is known as the per-capita multiplication rate. The population can only increase when the population is small if $\lambda > 1$, the so-called **invasion criterion**. But how does the system behave as λ increases above 1?

We begin by simulating time series of populations for different values of λ and plotting them to see what happens. First, here are the dynamics with $\lambda = 2$:

```
par(mfrow=c(2,2))
lambda <- 2
x <- numeric(40)
x[1] <- 0.6
for (t in 2 : 40) x[t] <- lambda * x[t-1] * (1 - x[t-1])
plot(1:40,x,type="l",ylim=c(0,1),ylab="population",
                                xlab="time",main="lambda = 2.0")
```

The population falls very quickly from its initial value (0.6) to equilibrium ($N* = 0.5$) and stays there; this system has a stable point equilibrium. What if λ were to increase to 3.3?

```
lambda <- 3.3
x <- numeric(40)
x[1] <- 0.6
for (t in 2 : 40) x[t] <- lambda * x[t-1] * (1 - x[t-1])
plot(1:40,x,type="l",ylim=c(0,1),ylab="population",
                            xlab="time",main="lambda = 3.3")
```

Now the dynamics show persistent two-point cycles. What about $\lambda = 3.5$?

```
lambda <- 3.5
x <- numeric(40)
x[1] <- 0.6
for (t in 2 : 40) x[t] <- lambda * x[t-1] * (1 - x[t-1])
plot(1:40,x,type="l",ylim=c(0,1),ylab="population",
                            xlab="time",main="lambda = 3.5")
```

The outcome is qualitatively different. Now we have persistent four-point cycles. Suppose that λ were to increase to 4:

```
lambda <- 4
x <- numeric(40)
x[1] <- 0.6
for (t in 2 : 40) x[t] <- lambda * x[t-1] * (1 - x[t-1])
plot(1:40,x,type="l",ylim=c(0,1),ylab="population",
                            xlab="time",main="lambda = 4.0")
```

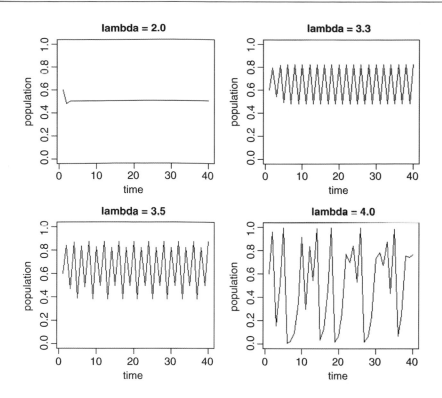

Now this is really interesting. The dynamics do not repeat in any easily described pattern. They are said to be **chaotic** because the pattern shows extreme sensitivity to initial conditions: tiny changes in initial conditions can have huge consequences on numbers at a given time in the future.

28.1.1 Investigating the route to chaos

We have seen four snapshots of the relationship between λ and population dynamics. To investigate this more fully, we should write a function to describe the dynamics as a function of λ, and extract a set of (say, 20) sequential population densities, after any transients have died away (after, say, 380 iterations):

```
numbers <- function (lambda) {
x <- numeric(400)
x[1] <- 0.6
for (t in 2 : 400) x[t] <- lambda * x[t-1] * (1 - x[t-1])
x[381:400] }
```

The idea is to plot these 20 values on the y axis of a new graph, against the value of λ that produced them. A stable point equilibrium will be represented by a single point, because all 20 values of y will be identical. Two-point cycles will show up as two points, four-point cycles as four points, but chaotic systems will appear as many points. Start with a blank graph:

```
par(mfrow=c(1,1))
plot(c(2,4),c(0,1),type="n",xlab="lambda",ylab="population")
```

Now, simulate using a `for` loop a wide range of values for λ between 2 and 4 (the range we investigated earlier), use the function `sapply` to apply our function to the current value of λ, and then use `points` to add these results to the graph:

```
for(lam in seq(2,4,0.01))
points(rep(lam,20),sapply(lam,numbers),pch=16,cex=0.5,col="blue")
```

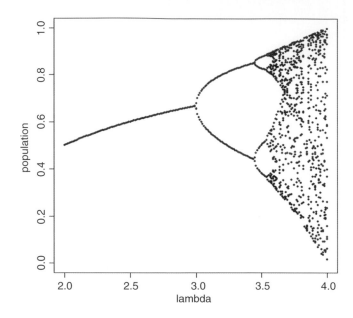

This graph shows what is called 'the period-doubling route to chaos'; see May (1976) for details.

28.2   Temporal and spatial dynamics: A simulated random walk in two dimensions

The idea is to follow an individual as it staggers its way around a two-dimensional random walk, starting at the point (50, 50) and leaving a trail of lines on a square surface which scales from 0 to 100. First, we need to define what we mean by our random walk. Suppose that in the *x* direction the individual could move one step to the left in a given time period, stay exactly where it is for the whole time period, or move one step to the right. We need to specify the probabilities of these three outcomes. Likewise, in the *y* direction the individual could move one step up in a given time period, stay exactly where it is for the whole time period, or move one step down. Again, we need to specify probabilities. In R, the three movement options are `c(1,0,-1)` for each of the types of motion (left, stay or right, and up, stay or down) and we might as well say that each of the three motions is equally likely. We need to select one of the three motions at random independently for the *x* and *y* directions at each time period. In R we use the sample function for this:

```
sample(c(1,0,-1),1)
```

which selects one value (the last argument is 1) with equal probability from the three listed options (+1, 0 or −1). Out of 99 repeats of this procedure, we should expect an average of 33 ups and 33 downs, 33 lefts and 33 rights.

We begin by defining the axes and drawing the start position in red:

```
plot(0:100,0:100,type="n",xlab="",ylab="")
x <- y <- 50
points(50,50,pch=16,col="red",cex=1.5)
```

Now simulate the spatial dynamics of the random walk with up to 10 000 time steps:

```
for (i in 1:10000) {
xi <- sample(c(1,0,-1),1)
yi <- sample(c(1,0,-1),1)
lines(c(x,x+xi),c(y,y+yi),col="blue")
x <- x+xi
y <- y+yi
if (x>100 | x<0 | y>100 | y<0) break
}
```

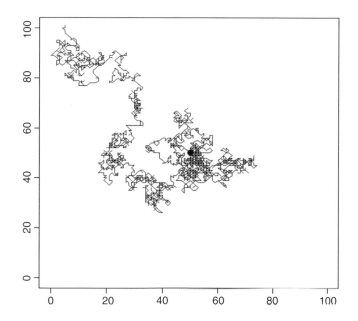

You could make the walk more sophisticated by providing wrap-around margins (see below). On average, of course, the random walk should stay in the middle, where it started, but as you will see by running this model repeatedly, most random walks do nothing of the sort. Instead, they wander off and fall over one of the edges in more or less short order.

28.3   Spatial simulation models

There are two broad types of spatial simulation models:

- spatial models that are divided but not spatially explicit;
- spatially explicit models where neighbours can be important.

Metapopulation dynamics is a classic example of a spatial model that is not spatially explicit. Patches are occupied or not, but the fate of a patch is not related to the fate of its immediate neighbours but rather by the global supply of propagules generated by all the occupied patches in the entire metapopulation.

28.3.1 Metapopulation dynamics

The theory is very simple. The world is divided up into many patches, all of which are potentially habitable. Populations on inhabited patches go extinct with a density-independent probability, e. Occupied patches all contain the same population density, and produce migrants (**propagules**) at a rate m per patch. Empty patches are colonized at a rate proportional to the total density of propagules and the availability of empty patches that are suitable for colonization. The response variable is the proportion of patches that are occupied, p. The dynamics of p, therefore, are just gains minus losses, so

$$\frac{\mathrm{d}p}{\mathrm{d}t} = p(1-p)m - ep.$$

At equilibrium $dp/dt = 0$, and so

$$p(1-p)m = ep,$$

giving the equilibrium proportion of occupied patches p^* as

$$p^* = 1 - \frac{e}{m}.$$

This draws attention to a critical result: there is a threshold migration rate ($m = e$) below which the metapopulation cannot persist, and the proportion of occupied patches will drift inexorably to zero. Above this threshold, the metapopulation persists in dynamic equilibrium with patches continually going extinct (the mean lifetime of a patch is $1/e$) and other patches becoming colonized by immigrant propagules. This model is due to Levins (1969).

The simulation produces a moving cartoon of the occupied (yellow) and empty patches (red). We begin by setting the parameter values

```
m <- 0.15
e <- 0.1
```

We create a square universe of 10 000 patches in a 100×100 array, but this is not a spatially explicit model, and so the map-like aspects of the image should be ignored. The response variable is just the proportion of all patches that are occupied. Here are the initial conditions, placing occupied 100 patches at random in a sea of unoccupied patches:

```
s <- (1-e)
N <- matrix(rep(0,10000),nrow=100)
xs <- sample(1:100)
ys <- sample(1:100)
for (i in 1:100){
N[xs[i],ys[i]] <- 1 }
image(1:100,1:100,N)
```

We want the simulation to run over 1000 generations:

```
for (t in 1:1000){
```

First we model the survival (or not) of occupied patches. Each cell of the universe gets an independent random number from a uniform distribution (a real number between 0 and 1). If the random number is bigger than or equal to the survival rate s (= 1 – e, above) then the patch survives for another generation. If the random number is less than s, then the patch goes extinct and N is set to zero:

```
S   <- matrix(runif(10000),nrow=100)
N <- N*(S<s)
```

Note that this one statement updates the whole matrix of 10 000 patches. Next, we work out the production of propagules, im, by the surviving patches (the rate per patch is m):

```
im <- floor(sum( N*m))
```

We assume that the settlement of the propagules is random, some falling in empty patches but others being 'wasted' by falling in already occupied patches:

```
placed <- matrix(sample(c(rep(1,im) ,rep (0,10000-im))),nrow=100)
N <- N+placed
N <- apply(N,2,function(x) ifelse(x>1,1,x))
```

The last line is necessary to keep the values of N just 0 (empty) or 1 (occupied) because our algorithm gives N = 2 when a propagule falls in an occupied patch. Now we can draw the map of the occupied patches:

```
image(1:100,1:100,N,add=TRUE)
box(col="red")
}
```

Because the migration rate (m = 0.15) exceeds the extinction rate (e = 0.1) the metapopulation is predicted to persist. The analytical solution for the long-term proportion of patches occupied is one-third of patches (1 – 0.1/0.15 = 0.333). At any particular time of stopping the simulation, you can work out the actual proportion occupancy as

```
sum(N)/length(N)
```

```
[1] 0.268
```

because there were 2680 occupied patches in this map at the time we stopped. Here is the code in one block:

```
m <- 0.15
e <- 0.1
s <- (1-e)
N <- matrix(rep(0,10000),nrow=100)
xs <- sample(1:100)
ys <- sample(1:100)
for (i in 1:100){
N[xs[i],ys[i]] <- 1 }
image(1:100,1:100,N)

for (t in 1:1000){
S   <- matrix(runif(10000),nrow=100)
N <- N*(S<s)
```

```
im <- floor(sum( N*m))
placed <- matrix(sample(c(rep(1,im) ,rep (0,10000-im))),nrow=100)
N <- N+placed
N <- apply(N,2,function(x) ifelse(x>1,1,x))
image(1:100,1:100,N,add=TRUE)
box(col="red")
}
```

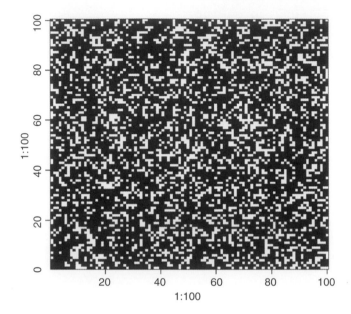

Remember that a metapopulation model is *not* spatially explicit, so you should not read anything into any the apparent neighbour relations in this graph (the occupied patches should be distributed at random over the surface).

28.3.2 Coexistence resulting from spatially explicit (local) density dependence

We have two species which would not coexist in a well-mixed environment because the fecundity of species A is greater than the fecundity of species B, and this would lead, sooner or later, to the competitive exclusion of species B and the persistence of a monoculture of species A. The idea is to see whether the introduction of local neighbourhood density dependence is sufficient to prevent competitive exclusion and allow long-term coexistence of the two species.

The kind of mechanism that might allow such an outcome is the build-up of specialist natural enemies such as insect herbivores or fungal pathogens in the vicinity of groups of adults of species A, that might prevent recruitment by species A when there were more than a threshold number, say *T*, of individuals of species A in a neighbourhood.

The problem with spatially explicit models is that we have to model what happens at the edges of the universe. All locations need to have the same numbers of neighbours in the model, but patches on the edge have fewer neighbours than those in the middle. The simplest solution is to model the universe as having 'wrap-around margins' in which the left-hand edge is assumed to have the right-hand edge as its left-hand neighbour (and vice versa), while the top edge is assumed to have the bottom edge as its neighbour above (and vice versa). The four corners of the universe are assumed to be reciprocal diagonal neighbours.

We need to define who is a neighbour of whom. The simplest method, adopted here, is to assume a square grid in which a central cell has eight neighbours – three above, three below and one to either side:

```
plot(c(0,1),c(0,1),xaxt="n",yaxt="n",type="n",xlab="",ylab="")
abline("v"=c(1/3,2/3))
abline("h"=c(1/3,2/3))
xs <- c(.15,.5,.85,.15,.85,.15,.5,.85)
ys <- c(.85,.85,.85,.5,.5,.15,.15,.15)
for (i in 1:8) text(xs[i],ys[i],as.character(i))
text(.5,.5,"target cell")
```

This code produces a plot showing a target cell in the centre of a matrix, and the numbers in the other cells indicate its 'first-order neighbours':

1	2	3
4	target cell	5
6	7	8

We need to write a function to define the margins for cells on the top, bottom and edge of our universe, N, and which determines all the neighbours of the four corner cells. Our universe is 100×100 cells and so the matrix containing all the neighbours will need to be 102×102. Note the use of subscripts (see p. 36 for revision of this):

```
margins <- function(N){
edges <- matrix(rep(0,10404),nrow=102)
edges[2:101,2:101] <- N
edges[1,2:101] <- N[100,]
edges[102,2:101] <- N[1,]
edges[2:101,1] <- N[,100]
edges[2:101,102] <- N[,1]
edges[1,1] <- N[100,100]
edges[102,102] <- N[1,1]
edges[1,102] <- N[100,1]
edges[102,1] <- N[1,100]
edges}
```

Next, we need to write a function to count the number of species A in the eight neighbouring cells, for any cell i, j:

```
nhood <- function(X,j,i) sum(X[(j-1):(j+1),(i-1):(i+1)]==1)
```

Now we can set the parameter values: the reproductive rates of species A and B, the death rate of adults (which determines the space freed up for recruitment) and the threshold number, T, of species A (out of the eight neighbours) above which recruitment cannot occur:

```
Ra <- 3
Rb <- 2.0
D <- 0.25
s <- (1-D)
T <- 6
```

The initial conditions fill one half of the universe with species A and the other half with species B, so that we can watch any spatial pattern as it emerges:

```
N <- matrix(c(rep(1,5000),rep(2,5000)),nrow=100)
image(1:100,1:100,N)
```

We run the simulation for 1000 time steps:

```
for (t in 1:1000) {
```

First, we need to see if the occupant of a cell survives or dies. For this, we compare a uniformly distributed random number between 0 and 1 with the specified survival rate s = 1 - D. If the random number is less than s the occupant survives, if it is greater than s it dies:

```
S   <- 1*(matrix(runif(10000),nrow=100)<s)
```

We kill the necessary number of cells to open up space for recruitment:

```
N <- N*S
space <- 10000-sum(S)
```

Next, we need to compute the neighbourhood density of A for every cell (using the wrap-around margins):

```
nt <- margins(N)
tots <- matrix(rep(0,10000),nrow=100)
for (a in 2:101) {
for (b in 2:101) {
tots[a-1,b-1] <- nhood(nt,a,b)
}}
```

The survivors produce seeds as follows:

```
seedsA <-   sum(N==1)*Ra
seedsB <-   sum(N==2)*Rb
all.seeds <- seedsA+seedsB
fA <- seedsA/all.seeds
fB <- 1-fA
```

Seeds settle over the universe at random:

```
setA <- ceiling(10000*fA)
placed <- matrix(sample(c(rep(1,setA) ,rep (2,10000-setA))),nrow=100)
```

Seeds only produce recruits in empty cells N[i,j]==0. If the winner of an empty cell (placed) is species B, then species B gets that cell: if (placed[i,j]= = 2) N[i,j] <- 2. If species A is supposed to win a cell, then we need to check that it has fewer than T neighbours of species A. If so, species A gets the cell. If not, the cell is forfeited to species B: `if (tots[i,j]>=T) N[i,j] <- 2`.

```
for (i in 1:100){
for(j in 1:100){
if (N[i,j] == 0 )
if(placed[i,j]== 2) N[i,j] <- 2
else
if (tots[i,j]>=T) N[i,j] <- 2
else N[i,j] <- 1
}}
```

Finally, we can draw the map, showing species A in red and species B in white:

```
image(1:100,1:100,N,add=TRUE)
box(col="red")}
```

You can watch as the initial half-and-half pattern breaks down, and species A increases in frequency at the expense of species B. Eventually, however, species A gets to the point where most of the cells have six or more neighbouring cells containing species A, and its recruitment begins to fail. At equilibrium, species B persists in isolated cells or in small (white) patches, where the cells have six or more occupants that belong to species A. The full code for the model is in the file called janzen.txt on the book's web site.

If you set the threshold $T = 9$, you can watch as species A drives species B to extinction. If you want to turn the tables, and see species B take over the universe, set $T = 0$.

28.4 Pattern generation resulting from dynamic interactions

In this section we look at an example of an ecological interaction between a species and its parasite. The interaction is unstable in a non-spatial model, with increasing oscillations in numbers leading quickly to extinction of the host species and then, in the next generation, its parasite. The non-spatial dynamics look like this:

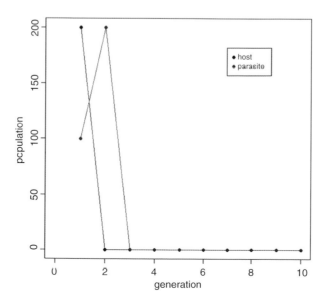

The parasite increases in generation number 1 and drives the host to extinction in generation 2, subsequently going extinct itself in generation 3. The challenge is to see if making the interaction spatially explicit can promote coexistence, and if so, through what pattern of spatial and temporal dynamics.

In a spatial model, we allow that hosts and parasites can move from the location in which they were born to any one of the eight first-order neighbouring cells (p. 901). For the purposes of dispersal, the universe is assumed to have wrap-around margins for both species. The interaction is interesting because it is capable of producing beautiful spatial patterns that fluctuate with host and parasite abundance. We begin by setting the parameter values for the dynamics of the host (r) and the parasite (a) and the migration rates of the host (Hmr=0.1) and parasite (Pmr=0.9). In this case the hosts are relatively sedentary and the parasites are highly mobile:

```
r <- 0.4
a <- 0.1
Hmr <- 0.1
Pmr <- 0.9
```

Next, we set up the matrices of host (N) and parasite (P) abundance. These will form what is termed a coupled map lattice:

```
N <- matrix(rep(0,10000),nrow=100)
P <- matrix(rep(0,10000),nrow=100)
```

The simulation is seeded by introducing 200 hosts and 100 parasites into a single cell at location [33,33]:

```
N[33,33] <- 200
P[33,33] <- 100
image(1:100,1:100,N)
```

We need to define a function called host to calculate the next host population as a function of current numbers of hosts and parasites (N and P), and another function called parasite to calculate the next parasite population as a function of N and P – this is called a Nicholson–Bailey model:

```
host <- function(N,P) N*exp(r-a*P)
parasite <- function(N,P) N*(1-exp(-a*P))
```

Both species need a definition of their wrap-around margins for defining the destinations of migrants from each cell:

```
host.edges <- function(N){
Hedges <- matrix(rep(0,10404),nrow=102)
Hedges[2:101,2:101] <- N
Hedges[1,2:101] <- N[100,]
Hedges[102,2:101] <- N[1,]
Hedges[2:101,1] <- N[,100]
Hedges[2:101,102] <- N[,1]
Hedges[1,1] <- N[100,100]
Hedges[102,102] <- N[1,1]
Hedges[1,102] <- N[100,1]
Hedges[102,1] <- N[1,100]
Hedges}
parasite.edges <- function(P){
```

```
Pedges <- matrix(rep(0,10404),nrow=102)
Pedges[2:101,2:101] <- P
Pedges[1,2:101] <- P[100,]
Pedges[102,2:101] <- P[1,]
Pedges[2:101,1] <- P[,100]
Pedges[2:101,102] <- P[,1]
Pedges[1,1] <- P[100,100]
Pedges[102,102] <- P[1,1]
Pedges[1,102] <- P[100,1]
Pedges[102,1] <- P[1,100]
Pedges}
```

A function is needed to define the eight cells that comprise the neighbourhood of any cell and add up the total number of neighbouring individuals:

```
nhood <- function(X,j,i) sum(X[(j-1):(j+1),(i-1):(i+1)])
```

The number of host migrants arriving in every cell is calculated as follows:

```
h.migration <- function(Hedges){
Hmigs <- matrix(rep(0,10000),nrow=100)
for (a in 2:101) {
for (b in 2:101) {
Hmigs[a-1,b-1] <- nhood(Hedges,a,b)
}}
Hmigs}
```

The number of parasite migrants is given by:

```
p.migration <- function(Pedges){
Pmigs <- matrix(rep(0,10000),nrow=100)
for (a in 2:101) {
for (b in 2:101) {
Pmigs[a-1,b-1] <- nhood(Pedges,a,b)
}}
Pmigs}
```

The simulation begins here, and runs for 600 generations:

```
for (t in 1:600){
he <- host.edges(N)
pe <- parasite.edges(P)
Hmigs <- h.migration(he)
Pmigs <- p.migration(pe)
N <- N-Hmr*N+Hmr*Hmigs/9
P <- P-Pmr*P+Pmr*Pmigs/9
Ni <- host(N,P)
P <- parasite(N,P)
N <- Ni
image(1:100,1:100,N,add=TRUE)
}
```

The full text of the R code is in a file called comins.txt on the book's web site. You can watch as the initial introduction at (33, 33) spreads out and both host and parasite populations pulse in abundance. Eventually, the wave of migration reaches the margin and appears on the right-hand edge. The fun starts when the two waves meet one another. The pattern below is typical of the structure that emerges towards the middle of a simulation run:

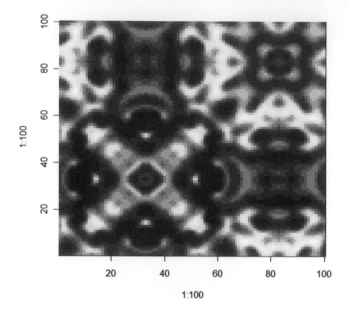

29

Changing the Look of Graphics

Many of the changes that you will want to make to the look of your graphics involve the use of the graphics parameters function, `par`. Other changes, however, can be made through alterations to the arguments to high-level functions such as `plot`, `points`, `lines`, `axis`, `title` and `text` (these options are shown with an asterisk in Table 29.2).

29.1   Graphs for publication

The most likely changes you will be asked to make are to the orientation of the numbers on the tick marks, and to the sizes of the plotting symbols and text labels on the axes. There are four functions involved here:

`las`	determines the orientation of the numbers on the tick marks;
`cex`	determines the size of plotting characters (pch);
`cex.lab`	determines the size of the text labels on the axes;
`cex.axis`	determines the size of the numbers on the tick marks.

Here we show four different combinations of options. You should pick the settings that look best for your particular graph.

```
par(mfrow=c(2,2))
x <- seq(0,150,10)
y <- 16+x*0.4+rnorm(length(x),0,6)

plot(x,y,pch=16,col="blue",xlab="label for x axis",
                    ylab="label for y axis")
plot(x,y,pch=16,col="blue",xlab="label for x axis",
                    ylab="label for y axis",
                    las=1,cex.lab=1.2, cex.axis=1.1)
plot(x,y,pch=16,col="blue",xlab="label for x axis",
                    ylab="label for y axis",las=2,cex=1.5)
plot(x,y,pch=16,col="blue",xlab="label for x axis",
                    ylab="label for y axis",las=3,
                    cex=0.7,cex.lab=1.3, cex.axis=1.3)
```

The R Book, Second Edition. Michael J. Crawley.
© 2013 John Wiley & Sons, Ltd. Published 2013 by John Wiley & Sons, Ltd.

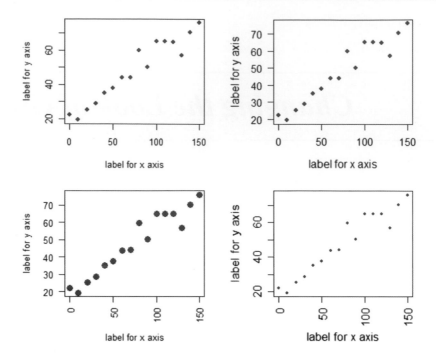

The top left-hand graph uses all the default settings:

```
las = 0, cex = 1, cex.lab = 1
```

In the top right-hand graph the numbers have been rotated so that they are all vertical (las = 1), the label size has been increased by 20% and the numbers by 10%:

```
las = 1, cex = 1, cex.lab = 1.2, cex.axis=1.1
```

In the bottom left-hand graph the plotting symbol (pch = 16) has been increased in size by 50% and the numbers on both axes are parallel with their axes (las = 2):

```
las = 2, cex = 1.5, cex.lab = 1
```

Finally, in the bottom right-hand graph the label size has been increased by 30%, the plotting symbols reduced by 30% and the axes numbers are all at 90 degrees (las = 3):

```
las = 3, cex = 0.7, cex.lab = 1.3, cex.axis = 1.3
```

My favourite is the top right-hand graph with slightly larger text and numbers and with vertical numbering on the *y* axis.

29.2   Colour

Colours are specified in R in one of four ways:

- by colour number (1 to 7; colour 8 is light grey);
- by colour name (e.g. "red" as an element of colors ());

- by a hexadecimal string of the form `#rrggbb`;

- by an integer subscript `i`, on the current `palette()[i]`.

To see all 657 colours available in R (note the American spelling of 'colors' in R), type:

```
colors()
```

```
 [1]  "white"          "aliceblue"      "antiquewhite"   "antiquewhite1"  "antiquewhite2"
 [6]  "antiquewhite3"  "antiquewhite4"  "aquamarine"     "aquamarine1"    "aquamarine2"
[11]  "aquamarine3"    "aquamarine4"    "azure"          "azure1"         "azure2"
[16]  "azure3"         "azure4"         "beige"          "bisque"         "bisque1"
[21]  "bisque2"        "bisque3"        "bisque4"        "black"          "blanchedalmond"
[26]  "blue"           "blue1"          "blue2"          "blue3"          "blue4"
[31]  "blueviolet"     "brown"          "brown1"         "brown2"         "brown3"
...
[641] "violetred"      "violetred1"     "violetred2"     "violetred3"     "violetred4"
[646] "wheat"          "wheat1"         "wheat2"         "wheat3"         "wheat4"
[651] "whitesmoke"     "yellow"         "yellow1"        "yellow2"        "yellow3"
[656] "yellow4"        "yellowgreen"
```

As you can see, there are some pretty odd colour names (`"blanchedalmond"` for instance).

The simplest way to specify a colour is with a character string giving the colour name (e.g. `col = "red"`). Another simple way is to learn the default colour numbers (`col=i`):

```
plot(0:8,0:8,type="n",xlab="",ylab="colour number")
axis(2,at=1:7)
for (i in 1:7) lines(c(0,8),c(i,i),col=i)
```

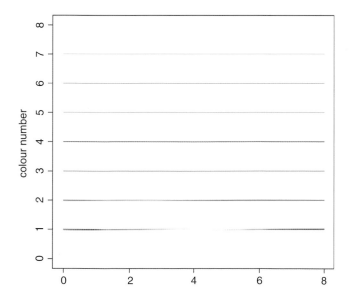

As you can see, 1 = black, 2 = red, 3 = green, 4 = blue, 5 = light blue, 6 = purple and 7 = yellow. Number 8 is light grey, then the pattern repeats itself, so 9 = black, 10 = red and so on.

Alternatively, colours can be specified directly in terms of their red–green–blue (RGB) components with a string of the form `"#RRGGBB"` where each of the pairs RR, GG, BB consists of two hexadecimal digits giving a value in the range 00 to FF. The simplest way to get a colour of your choice is to work out the decimal weighting of red, green and blue (each on a scale of 0 to 1) then use the function rgb like this:

```
par(mfrow=c(2,3))

plot(0:1,0:1,typ="n",xlab="red",ylab="blue",main="green = 0")
for (red in seq(0,1,0.1)) {
green<-0
for (blue in seq(0,1,0.1)) {
points(red,blue,pch=16,col=rgb(red,green,blue)) }}
```

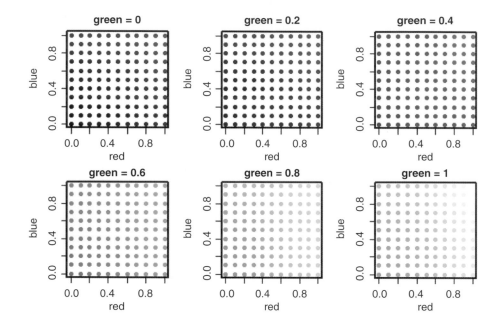

Suppose you wanted some subtle shades of orange. Then you would be working in the high values of red (say, 0.9 or 1), low values of blue (less than 0.4) and intermediate values of green (0.4 to 0.6). Notice that black is (0, 0, 0) and white is (1, 1, 1).

29.2.1 Palettes for groups of colours

Colours can also be specified by giving an index into a small table of colours, known as the **palette**. The functions rgb (red–green–blue) and hsv (hue–saturation–value) provide additional ways of generating colours (see the relevant help ?rgb and ?hsv).

There are several built-in palettes. For instance, the built-in function called rainbow takes the seven colours of the rainbow (red, orange, yellow, green, blue, indigo, violet) and splits them into a specified number of colours on the basis of hue, saturation and value. Here are four examples, with the spectrum split into 7, 14, 28 or 56 segments:

```
par(mfrow=c(2,2))
par(mar=c(1.5,1.5,1.5,1.5))
pie(rep(1, 7),  col = rainbow(7),  radius = 1)
pie(rep(1, 14), col = rainbow(14), radius = 1)
```

```
pie(rep(1, 28), col = rainbow(28), radius = 1)
pie(rep(1, 56), col = rainbow(56), radius = 1)
```

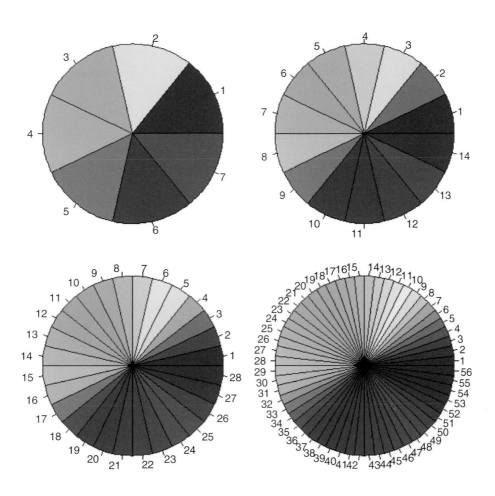

Notice that between them, greens and blues take up more than half of the space, with red, orange, yellow, indigo and violet making up the remainder. Also, note the use of the margin parameter to optimize the size of the pie diagrams, while keeping their labels distinct from each other.

There are four other built-in colour functions that you can use to produce graded hues:

```
pie(rep(1, 14), col = heat.colors(14),
            radius = 0.9,main="heat.colors")
pie(rep(1, 14), col = terrain.colors(14),
            radius = 0.9,main="terrain.colors")
pie(rep(1, 14), col = topo.colors(14),
            radius = 0.9,main="topo.colors")
pie(rep(1, 14), col = cm.colors(14),
            radius = 0.9,main="cm.colors")
par(mfrow=c(1,1))
```

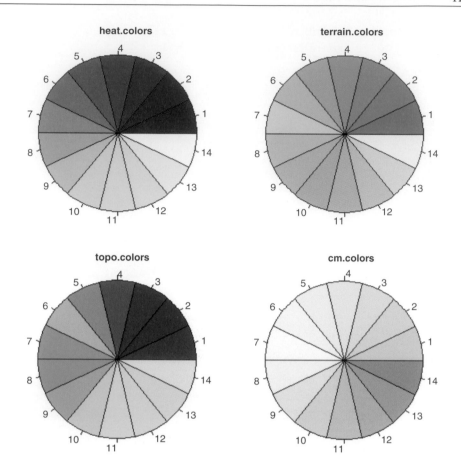

It is simple to create your own customized palettes. He we use the function `rgb` to do it:

```
custom<-c(rgb(0.6,0.8,1),rgb(1,0.8,0.2),rgb(1,0.8,0.4),
          rgb(1,0.8,0.6),rgb(1,0.8,0.8),rgb(1,0.8,1),
          rgb(0.8,0.8,1),rgb(0.7,0.8,1))
pie(rep(1/8,8),col=custom)
```

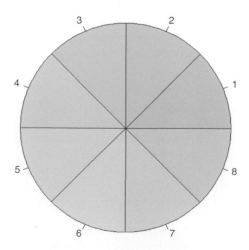

29.2.2 The `RColorBrewer` package

This is a very useful package of tried and tested colour schemes, in which carefully selected colours have been grouped together into a set of palettes (more information on ColorBrewer is available at its website, **http://www.colorbrewer.org**).

```
install.packages("RColorBrewer")
```

These palettes have a minimum of three colours and a maximum of 8–12 depending on the palette. There are three types of palettes – sequential, diverging and qualitative:

- Sequential palettes are suited to ordered data that progress from low to high. Lightness steps dominate the look of these schemes, from light colours for low data values to dark colours for high data values.

- Diverging palettes put equal emphasis on mid-range critical values and extremes at both ends of the data range. The critical class or break in the middle of the legend is emphasized with light colours, and low and high extremes are emphasized with dark colours that have contrasting hues.

- Qualitative palettes do not imply magnitude differences between legend classes, and hues are used to create the primary visual differences between classes. Qualitative schemes are best suited to representing nominal or categorical data.

Here is a demonstration of three palettes from each of the three palette types. You specify how many colours to use in the palette (eight in these examples).

```
library(RColorBrewer)
par(mfrow=c(3,3))
par(mar=c(1,1,1,1))
mypalette   <- brewer.pal(8,"Reds")
pie(rep(1,8), col = mypalette, radius = 0.9,main="Reds")
mypalette   <- brewer.pal(8,"Blues")
pie(rep(1,8), col = mypalette, radius = 0.9,main="Blues")
mypalette   <- brewer.pal(8,"Greens")
pie(rep(1,8), col = mypalette, radius = 0.9,main="Greens")
mypalette   <- brewer.pal(8,"BrBG")
pie(rep(1,8), col = mypalette, radius = 0.9,main="BrBG")
mypalette   <- brewer.pal(8,"PiYG")
pie(rep(1,8), col = mypalette, radius = 0.9,main="PiYG")
mypalette   <- brewer.pal(8,"Spectral")
pie(rep(1,8), col = mypalette, radius = 0.9,main="Spectral")
mypalette   <- brewer.pal(8,"Accent")
pie(rep(1,8), col = mypalette, radius = 0.9,main="Accent")
mypalette   <- brewer.pal(8,"Pastel1")
pie(rep(1,8), col = mypalette, radius = 0.9,main="Pastel1")
mypalette   <- brewer.pal(8,"Set2")
pie(rep(1,0), col = mypalette, radius = 0.9,main="Set2")
par(mfrow=c(1,1))
```

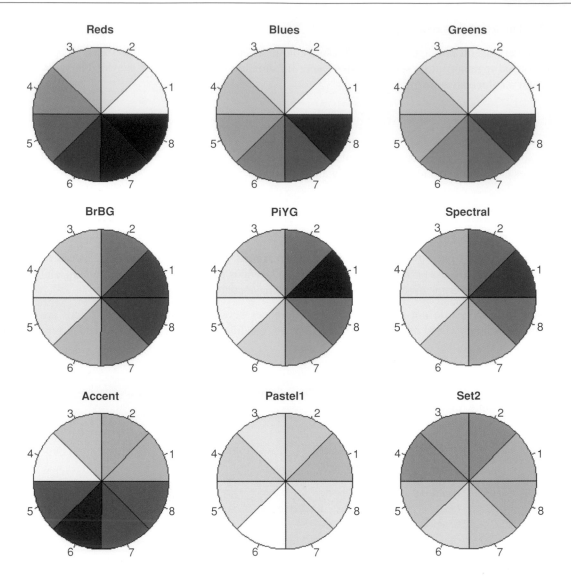

The top row contains three classic sequential palettes, the centre row three different diverging palettes, and the bottom row three quite effective qualitative palettes. Once you have defined a palette, you can refer to colours within it using subscripts in any plotting function that accepts a `col=` argument, for example `col = mypalette[3]`.

To reset the palette back to the default use:

```
palette("default")
```

29.2.3 Coloured plotting symbols with contrasting margins

The five shapes of plotting symbols that allow contrasting margins are circle (`pch=21`), square (`22`), diamond (`23`), triangle up (`24`) and triangle down (`25`). Effective colour combinations are red with a black border, orange with a blue border, yellow with a red border, cyan with a blue border, green with a red border and violet with a blue border. Darker colours like blue and brown are rather less effective with borders.

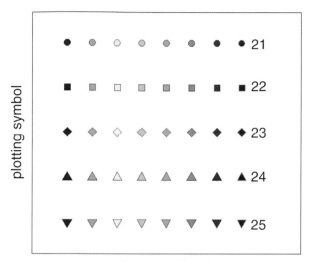

29.2.4 Colour in legends

The issue here is that the background in the plotting symbols 21–25 is called bg, but in figure legends bg controls the background colour of the whole legend box. So when you are creating figure legends for plotting symbols pch=21 to 25 you need to remember to use the argument pt.bg in the legend function in place of the usual bg for the interior colour of the symbol.

```
data<-read.table("c:\\temp\\plotcol.txt",header=T)
attach(data)
names(data)

[1] "x1" "y1" "x2" "y2" "x3" "y3"

plot(c(0,10),c(0,50),type="n",xlab="x",ylab="y",las=1)
points(x1,y1,pch=21,bg="yellow",col="red")
lines(lowess(x1,y1,f=0.8),col="red")
points(x2,y2,pch=21,bg="cyan",col="blue")
lines(lowess(x2,y2,f=0.8),col="blue")
points(x3,y3,pch=21,bg="red",col="black")
lines(lowess(x3,y3,f=0.8),col="black")
```

To emphasize the point, we add a different colour for the background of the legend bg="wheat1", with the pch backgrounds set by pt.bg=c("red","cyan","yellow"):

```
legend(0,50,legend=c("control","heat","dose"),pch=c(21,21,21),
    bg="wheat1",pt.bg=c("red","cyan","yellow"),col=c("black","blue","red"))
```

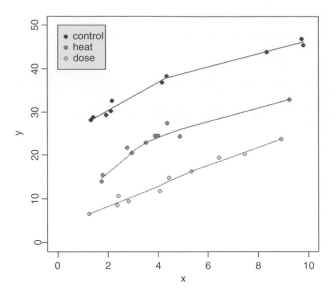

29.2.5   Background colours

It is important to distinguish two contrasting uses of the work 'background' in R graphics. The first and most obvious (using `par(bg=colour)`) refers to the colour of 'paper' on which the graph is produced. The second, somewhat more obscure usage refers to the fill colour of two-tone plotting symbols (`pch=21` to `25`).

```
data<-read.table("c:\\temp\\silwoodweather.txt",header=T)
attach(data)
par(bg="wheat2")
plot(factor(month),lower,col="green4")
```

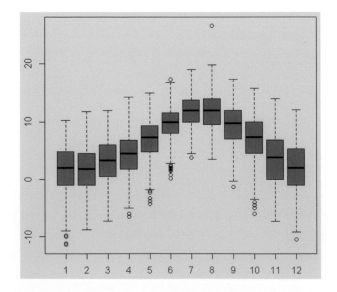

29.2.6 Foreground colours

Changing the colour of such things as axes and boxes around plots uses the 'foreground' parameter, `fg`:

```
par(mar=c(4,4,1,1))
par(mfrow=c(2,2))
plot(1:10,1:10,xlab="x label",ylab="y label")
plot(1:10,1:10,xlab="x label",ylab="y label",fg="blue")
plot(1:10,1:10,xlab="x label",ylab="y label",fg="red")
plot(1:10,1:10,xlab="x label",ylab="y label",fg="green")
```

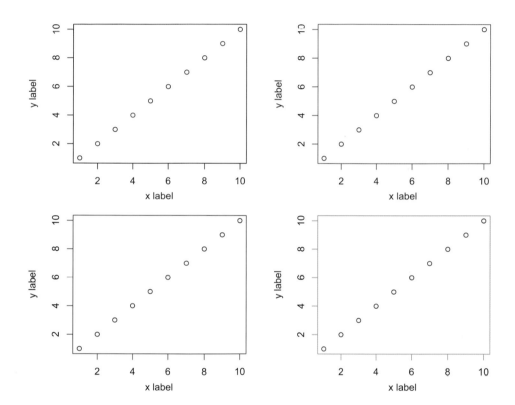

29.2.7 Different colours and font styles for different parts of the graph

The colours for different parts of the graph are specified as follows:

`col.axis`	is the colour to be used for axis annotation;
`col.lab`	is the colour to be used for *x* and *y* labels;
`col.main`	is the colour to be used for plot main titles;
`col.sub`	is the colour to be used for plot subtitles.

The fonts of the various titles are specified in a similar way:

font.axis	is the font to be used for axis annotation;
font.lab	is the font to be used for *x* and *y* labels;
font.main	is the font to be used for plot main titles;
font.sub	is the font to be used for plot subtitles.

This example illustrates the use of expression to allow more complicated formatting of axis labels: for instance, square brackets produce subscripts log[10], caret produces superscripts mday^{-1}, and successive tildes ~~ produce wider spacing between words:

```
plot (1:10,1:10, xlab="x axis label", ylab="y axis label",
pch=16, col="orange", col.lab="green4",col.axis="blue",col.main="red",
col.sub="navy",sub="Subtitle",las=1,font.axis=3,font.lab=2,font.main=4,
font.sub=3,main=expression ("Critical elasticity log"[~10]~~~
    "mday" ^~~-1))
```

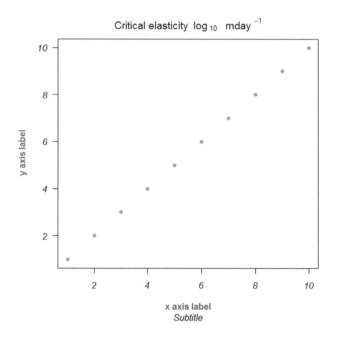

29.2.8 Full control of colours in plots

If you really feel the need, then you can control the colours and line types of all of the components of a plot separately. Here is a box-and-whisker plot in which the following components are specified explicitly:

- the box and its outline (box*);
- the median, its line type, line width and colour (med*);
- the whiskers, their line type, width and colour (whisk*);
- the staples (this is the jargon for the flat ends of the whiskers; staple*);
- the outliers, their plotting symbol, edge and fill (out*).

Box

boxlty	line type
boxlwd	line width
boxcol	line colour
boxfill	fill colour

Median

medlty	line type (medlty="blank" if you want no line, just a point)
medlwd	line width
medpch	plotting symbol (added with the line unless specified otherwise)
medcex	plotting symbol size
medcol	plotting symbol colour
medbg	plotting symbol fill for pch=21 to 25

Whisker

whisklty	line type
whisklwd	line width
whiskcol	colour

Staple

staplelty	line type
staplelwd	line width
staplewex	width expansion
staplecol	colour

Outlier

outlty	line type
outlwd	line width
outwex	width expansion
outpch	plotting symbol
outcex	symbol size
outcol	colour of the outline of the plotting symbol
outbg	colour of the fill of the plotting symbol

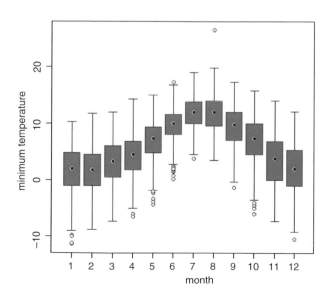

This has most of the options in their non-default settings (but whether you would ever really want to do this is debatable):

```
plot(factor(month),lower,ylab="minimum temperature",xlab="month",
      medlty="blank",medpch=21,medbg="red",medcol="yellow",
      boxcol="red",boxfill="green",outpch=21,outbg="yellow",
      outcol="red",staplecol="blue",whisklty=1,whiskcol="blue")
```

29.3   Cross-hatching

You can control five aspects of shading: the density of the lines, the angle of the shading, the border of the shaded region, the colour of the lines and the line type. Here are their default values:

```
density = NULL
angle = 45
border = NULL
col = NA
lty = par("lty"), ...)
```

Other graphical parameters dealing with lines such as xpd, lend, ljoin and lmitre (Table 29.2 and see p. 947) can be given as arguments. We shall shade each of the bars differently:

```
data <- read.table("c:\\temp\\box.txt",header=T)
attach(data)
names(data)

[1] "fact"      "response"

barplot(tapply(response,fact,mean),density=3:10,angle=seq(30,60,length=8))
```

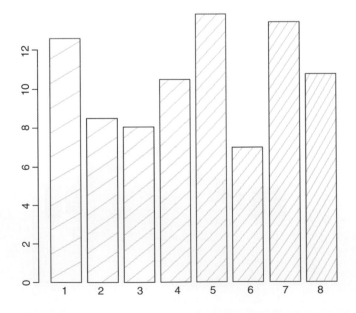

The density and the angle of the shading both increase from left to right (the density from 3 lines per inch to 10 lines per inch, and the angle from 30 to 60 degrees).

29.4   Grey scale

Here is the same example with grey scale instead of shading. Remember that the grey scale goes from 0 to 1, which (counter-intuitively) is from dark to light:

```
barplot(tapply(response,fact,mean),col=grey(seq(0.8,0.2,length=8)))
```

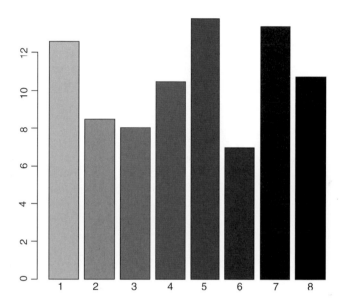

29.5   Coloured convex hulls and other polygons

The following data, which you may recall from Chapter 5, come from a long-term study of the botanical composition of a pasture, where the response variable is the dry mass of a grass species called *Festuca rubra*, and the two explanatory variables are total hay biomass and soil pH:

```
data <- read.table("c:\\temp\\pgr.txt",header=T)
attach(data)
names(data)
```

```
[1]  "FR"  "hay"  "pH"
```

The idea is to draw polygons to represent the convex hulls for the abundance of *Festuca* in the space defined by hay biomass and soil pH. The polygon is to be red for *Festuca* > 5, green for *Festuca* > 10 and cross-hatched in blue for *Festuca* > 20. After all of the solid objects have been drawn, the data are to be overlaid as a scatterplot with pch = 16:

```
plot(hay,pH)
x <- hay[FR>5]
y <- pH[FR>5]
```

```
polygon(x[chull(x,y)],y[chull(x,y)],col="red")
x <- hay[FR>10]
y <- pH[FR>10]
polygon(x[chull(x,y)],y[chull(x,y)],col="green")
x <- hay[FR>20]
y <- pH[FR>20]
polygon(x[chull(x,y)],y[chull(x,y)],density=10,angle=90,col="blue")
polygon(x[chull(x,y)],y[chull(x,y)],density=10,angle=0,col="blue")
points(hay,pH,pch=16)
```

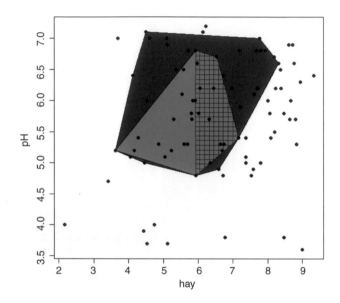

The issue of transparency (i.e. what you can see 'through' what) is described in the help files for ?polygon and ?rgb. If in doubt, use points, lines and polygons in sequence, so that objects ('on top') that you want to be visible in the final image are drawn last. Note that for square cross-hatching you draw the object twice; once with angle=90 then again with angle=0.

29.6   Logarithmic axes

You can transform the variables inside the plot function (e.g. plot(log(y) ~ x)) or you can plot the untransformed variables on logarithmically scaled axes (e.g. log="x").

```
data <- read.table("c:\\temp\\logplots.txt",header=T)
attach(data)
names(data)
```

```
[1]  "x"  "y"
```

```
par(mfrow=c(2,2))
plot(x,y,pch=16, main="untransformed",col="red")
plot(log(x),log(y),pch=16, main="log-log", col="blue")
plot(x,y,pch=16,log="xy", main="both transformed", col="blue")
plot(x,y,pch=16,log="y", main="only y transformed", col="green")
```

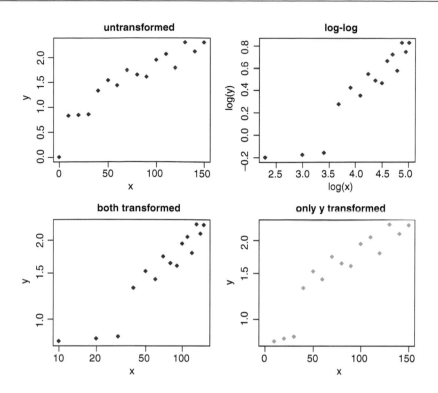

The untransformed data (red) are in the top left-hand graph, and both x and y are transformed to logs before plotting in the upper right (blue). The bottom left-hand plot shows both axes log-transformed (also blue, same pattern, but different axis labelling), while the bottom right shows the data (green) with only the y axis log-transformed. Note that all logs in R are to the base e by default (not base 10). It is important to understand that when R is asked to plot the log of zero it simply omits any such points and issues a warning on the command line (compare the top left-hand graph with a point at $(0, 0)$ with the other three graphs).

29.7 Different font families for text

To change the typeface used for plotted text, change the name of a font family. Standard values are `family="serif"`, `"sans"` (the default font), `"mono"`, and `"symbol"`, and the Hershey font families are also available. Some devices will ignore this setting completely. Text drawn onto the plotting region is controlled using `par` like this:

```
par(mfrow=c(1,1))
plot(1:10,1:10,type="n",xlab="x",ylab="y")
par(family="sans")
text(5,8,"This is the default font")
par(family="serif")
text(5,6,"This is the serif font")
par(family="mono")
text(5,4,"This is the mono font")
par(family="HersheySymbol")
text(5,2,"This is the symbol font")
par(family="sans")
```

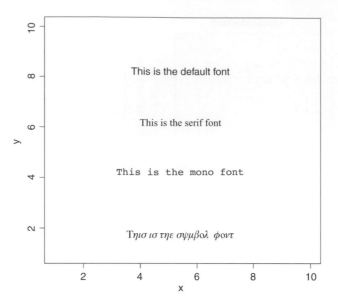

Don't forget to turn the family back to `"sans"`, otherwise you may get some very unexpected symbols in your next text.

29.8 Mathematical and other symbols on plots

To write on plots using more intricate symbols such as mathematical symbols or Greek letters we use `expression` or `substitute` (see Table 29.1 for a list of the available symbols). Here are some examples of their use. First, we produce a plot of $\sin \phi$ against the phase angle ϕ over the range $-\pi$ to $+\pi$ radians:

```
x   <-   seq(-4, 4, len = 101)
plot(x,sin(x),type="l",xaxt="n", col="red",
             xlab=expression(paste("Phase Angle ",phi)),
             ylab=expression("sin "*phi))
axis(1, at = c(-pi, -pi/2, 0, pi/2, pi),
             lab = expression(-pi, -pi/2, 0, pi/2, pi))
```

Note the use of `xaxt="n"` to suppress the default labelling of the x axis, and the use of `expression` in the labels for the x and y axes to obtain mathematical symbols such as phi (ϕ) and pi (π). The more intricate values for the tick marks on the x axis are obtained by the `axis` function, specifying 1 (the x ('bottom') axis is axis no. 1), then using the `at` function to say where the labels and tick marks are to appear, and `lab` with `expression` to say what the labels are to be.

Suppose you wanted to add $\chi^2 = 24.5$ to this graph at location $(-\pi/2, 0.5)$. You use `text` with `substitute`, like this:

```
text(-pi/2,0.5,substitute(chi^2=="24.5"))
```

Note the use of 'double equals' to print a single equals sign, and the use of caret ^ to obtain superscripts. You can write quite complicated formulae on plots using `paste` to join together the elements of an equation. Here is the density function of the normal written on the plot at location $(\pi/2, -0.5)$:

```
text(pi/2, -0.5, expression(paste(frac(1, sigma*sqrt(2*pi)), " ",
     e^{frac(-(x-mu)^2, 2*sigma^2)})))
```

Table 29.1. Drawing mathematical expressions in text. If the `text` argument to one of the text-drawing functions (`text`, `mtext`, `axis`, `legend`) is an expression, the argument is interpreted as a mathematical expression and the output will be formatted according to T$_E$X-like rules (see Knuth, 1998). Expressions can also be used for titles, subtitles and x and y axis labels (but not for axis labels on `persp` plots). It is possible to produce many different mathematical symbols, Greek letters, generate subscripts or superscripts, produce fractions, etc. The available features are listed below.

Syntax	Meaning
`x + y`	x plus y
`x - y`	x minus y
`x*y`	juxtapose x and y
`x/y`	x forwardslash y
`x %+-% y`	x plus or minus y
`x %/% y`	x divided by y
`x %*% y`	x times y
`x %.% y`	x cdot y
`x[i]`	x subscript i
`x^2`	x superscript 2
`paste(x, y, z)`	juxtapose x, y, and z
`sqrt(x)`	square root of x
`sqrt(x, y)`	yth root of x
`x == y`	x is equal to y
`x != y`	x is not equal to y
`x < y`	x is less than y
`x <= y`	x is less than or equal to y
`x > y`	x is greater than y
`x >= y`	x is greater than or equal to y
`x %~~% y`	x is approximately equal to y
`x %=~% y`	x and y are congruent
`x %==% y`	x is defined as y
`x %prop% y`	x is proportional to y
`plain(x)`	draw x in normal font
`bold(x)`	draw x in bold font
`italic(x)`	draw x in italic font
`bolditalic(x)`	draw x in bold italic font
`symbol(x)`	draw x in symbol font
`list(x, y, z)`	comma-separated list
`...`	ellipsis (height varies)
`cdots`	ellipsis (vertically centred)
`ldots`	ellipsis (at baseline)
`x %subset% y`	x is a proper subset of y
`x %subseteq% y`	x is a subset of y
`x %notsubset% y`	x is not a subset of y
`x %supset% y`	x is a proper superset of y
`x %supseteq% y`	x is a superset of y
`x %in% y`	x is an element of y
`x %notin% y`	x is not an element of y
`hat(x)`	x with a circumflex
`tilde(x)`	x with a tilde
`dot(x)`	x with a dot

(*continued*)

Table 29.1. *(Continued)*

Syntax	Meaning
`ring(x)`	*x* with a ring
`bar(x)`	*x* with bar
`widehat(xy)`	*xy* with a wide circumflex
`widetilde(xy)`	*xy* with a wide tilde
`x %<->% y`	*x* double-arrow *y*
`x %->% y`	*x* right-arrow *y*
`x %<-% y`	*x* left-arrow *y*
`x %up% y`	*x* up-arrow *y*
`x %down% y`	*x* down-arrow *y*
`x %<=>% y`	*x* is equivalent to *y*
`x %=>% y`	*x* implies *y*
`x %<=% y`	*y* implies *x*
`x %dblup% y`	*x* double-up-arrow *y*
`x %dbldown% y`	*x* double-down-arrow *y*
`alpha`	
`beta`	
`...`	Greek alphabet (lower case)
`omega`	
`Alpha`	
`Beta`	
`...`	uppercase Greek alphabet
`Omega`	
`theta1, phi1, sigma1, omega1`	cursive Greek symbols
`Upsilon1`	capital upsilon with hook
`aleph`	first letter of Hebrew alphabet
`infinity`	infinity symbol
`partialdiff`	partial differential symbol
`nabla`	nabla, gradient symbol
`32*degree`	32 degrees
`60*minute`	60 minutes of angle
`30*second`	30 seconds of angle
`displaystyle(x)`	draw *x* in normal size (extra spacing)
`textstyle(x)`	draw *x* in normal size
`scriptstyle(x)`	draw *x* in small size
`scriptscriptstyle(x)`	draw *x* in very small size
`underline(x)`	draw *x* underlined
`x ~~ y`	put extra space between *x* and *y*
`x + phantom(0) + y`	leave gap for "0", but do not draw it
`x + over(1, phantom(0))`	leave vertical gap for "0" (do not draw)
`frac(x, y)`	*x* over *y*
`over(x, y)`	*x* over *y*
`atop(x, y)`	*x* over *y* (no horizontal bar)
`sum(x[i], i==1, n)`	sum *x*[*i*] for *i* equals 1 to *n*
`prod(plain(P)(X==x), x)`	product of P(*X* = *x*) for all values of *x*
`integral(f(x)*dx, a, b)`	definite integral of *f*(*x*) with respect to *x*
`union(A[i], i==1, n)`	union of *A*[*i*] for *i* equals 1 to *n*
`intersect(A[i], i==1, n)`	intersection of *A*[*i*]
`lim(f(x), x %->% 0)`	limit of *f*(*x*) as *x* tends to 0

Table 29.1. *(Continued)*

Syntax	Meaning
`min(g(x), x > 0)`	minimum of $g(x)$ for x greater than 0
`inf(S)`	infimum of S
`sup(S)`	supremum of S
`x^y + z`	normal operator precedence
`x^(y + z)`	visible grouping of operands
`x^{y + z}`	invisible grouping of operands
`group("(",list(a, b),"]")`	specify left and right delimiters
`bgroup("(",atop(x,y),")")`	use scalable delimiters
`group(lceil, x, rceil)`	special delimiters

Note the use of `frac` to obtain individual fractions: the first argument is the text for the numerator, the second the text for the denominator. Most of the arithmetic operators have obvious formats (+, -, /, *, ^, etc.); the only non-intuitive symbol that is commonly used is 'plus or minus' $\pm$; this is written as `%+-%` like this:

```
text(pi/2,0,expression(hat(y) %+-% se))
```

To write the results of calculations using text, it is necessary to use `substitute` with `as.expression`. Here, the coefficient of determination (cd) was calculated earlier and we want to write its value on the plot, labelled with '$r^2 =$':

```
cd <-  0.63
text(locator(1),as.expression(substitute(r^2 == cd,list(cd=cd))))
```

Just click when the cursor is where you want the text to appear. Note the use of 'double equals' and the requirement for a list containing the value calculated earlier.

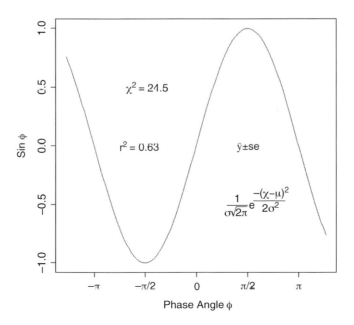

There are several other useful plotting symbols (see `?plotmath`) that you refer to in `text` functions as `symbol` (e.g. the 'universal' character is obtained with `expression(symbol("\042"))`): the full set is universal (`"\\042"`), existential (`"\\044"`), suchthat (`"\\047"`), therefore (`"\\134"`), perpendicular (`"\\136"`), circlemultiply (`"\\304"`), circleplus (`"\\305"`), emptyset (`"\\306"`), angle (`"\\320"`), leftangle (`"\\341"`) and rightangle (`"\\361"`).

29.9   Phase planes

Suppose that we have two competing species (named 1 and 2) and we are interested in modelling the dynamics of the numbers of each species (N_1 and N_2). We want to draw a phase plane showing the behaviour of the system close to equilibrium. Setting the derivatives to zero and dividing both sides by $r_i N_i$, we get

$$0 = 1 - \alpha_{11}N_1 - \alpha_{12}N_2,$$

which is called the isocline for species 1. It is linear in N_1 and N_2 and we want to draw it on a phase plane with N_2 on the y axis and N_1 on the x axis. The intercept on the y axis shows the abundance of N_2 when $N_1 = 0$: this is $1/\alpha_{12}$. Likewise, when $N_2 = 0$ we can see that $N_1 = 1/\alpha_{11}$ (the value of its single-species equilibrium). Similarly,

$$0 = 1 - \alpha_{21}N_1 - \alpha_{22}N_2$$

describes the isocline for species 2. The intercept on the y axis is $1/\alpha_{22}$ and the value of N_1 when $N_2 = 0$ is $1/\alpha_{21}$. Now we draw a phase plane with both isoclines, and label the ends of the lines appropriately. We might as well scale the axes from 0 to 1, but we want to suppress the default tick marks:

```
plot(c(0,1),c(0,1),ylab="",xlab="",xaxt="n",yaxt="n",type="n")
abline(0.8,-1.5,col="blue")
abline(0.6,-0.8, col="red")
```

The solid line shows the isocline for species 1 and the dotted line shows species 2.

Now for the labels. We use `at` to locate the tick marks – first the x axis (`axis = 1`),

```
axis(1, at = 0.805, lab = expression(1/alpha[21]))
axis(1, at = 0.56, lab = expression(1/alpha[11]))
```

and now the y axis (`axis = 2`),

```
axis(2, at = 0.86, lab = expression(1/alpha[12]),las=1)
axis(2, at = 0.63, lab = expression(1/alpha[22]),las=1)
```

Note the use of `las=1` to turn the labels through 90 degrees to the horizontal. Now label the lines to show which species isocline is which. Note the use of the function `frac` to print fractions and square brackets (outside the quotes) for subscripts:

```
text(0.05,0.85, expression(paste(frac("d N"[1],"dt"), " = 0" )))
text(0.78,0.07, expression(paste(frac("d N"[2],"dt"), " = 0" )))
```

We need to draw phase plane trajectories to show the dynamics. Species will increase when they are at low densities (i.e. 'below' their isoclines) and decrease at high densities (i.e. 'above' their isoclines). Species 1 increasing is a horizontal arrow pointing to the right. Species 2 declining is a vertical arrow pointing downwards. The resultant motion shows how both species' abundances change through time from a given point on the phase plane.

```
arrows(-0.02,0.72,0.05,0.72,length=0.1)
arrows(-0.02,0.72,-0.02,0.65,length=0.1)
arrows(-0.02,0.72,0.05,0.65,length=0.1)
arrows(0.65,-0.02,0.65,0.05,length=0.1)
arrows(0.65,-0.02,0.58,-0.02,length=0.1)
arrows(0.65,-0.02,0.58,0.05,length=0.1)
arrows(0.15,0.25,0.15,0.32,length=0.1)
arrows(0.15,0.25,0.22,0.25,length=0.1)
arrows(0.15,0.25,0.22,0.32,length=0.1)
arrows(.42,.53,.42,.46,length=0.1)
arrows(.42,.53,.35,.53,length=0.1)
arrows(.42,.53,.35,.46,length=0.1)
```

All the motions converge, so the point is a stable equilibrium and the two species would coexist. All other configurations of the isoclines lead to competitive exclusion of one of the two species. Finally, label the axes with the species' identities:

```
axis(1, at = 1, lab = expression(N[1]))
axis(2, at = 1, lab = expression(N[2]),las=1)
```

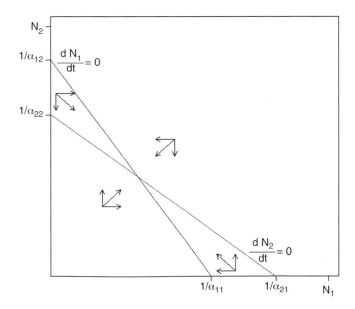

29.10   Fat arrows

You often want to add arrows to plots in order to draw attention to particular features. Here is a function called fat.arrows that uses locator(1) to identify the *bottom* of the point of a vertical fat arrow. You can modify the function to draw the arrow at any specified angle to the clicked point of its arrowhead. The default widths and heights of the arrow are 0.5 scaled *x* or *y* units and the default colour is red:

```
fat.arrow <- function(size.x=0.5,size.y=0.5,ar.col="red"){
size.x <- size.x*(par("usr")[2]-par("usr")[1])*0.1
size.y <- size.y*(par("usr")[4]-par("usr")[3])*0.1
pos <- locator(1)
xc <- c(0,1,0.5,0.5,-0.5,-0.5,-1,0)
yc <- c(0,1,1,6,6,1,1,0)
polygon(pos$x+size.x*xc,pos$y+size.y*yc,col=ar.col)  }
```

Here is the function in use with the defaults (the long, slim, red arrow) and three non-default options:

```
plot(0:10,0:10,type="n",xlab="",ylab="")
fat.arrow()
fat.arrow(ar.col="green")
fat.arrow(ar.col="blue",size.x=0.8)
fat.arrow(ar.col="orange",size.x=0.8,size.y=0.3)
```

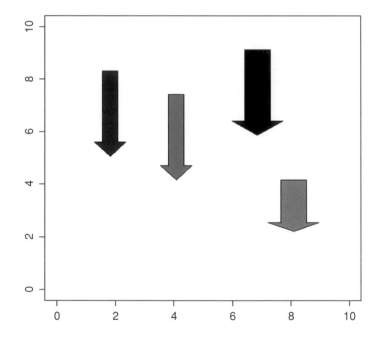

29.11   Three-dimensional plots

When there are two continuous explanatory variables, it is often useful to plot the response as a contour map. In this example, the biomass of one plant species (the response variable) is plotted against soil pH and total community biomass. The species is a grass called *Festuca rubra* that peaks in abundance in communities of intermediate total biomass:

```
data <-  read.table("c:\\temp\\pgr.txt",header=T)
attach(data)
```

```
names(data)
```

```
[1]  "FR"  "hay"  "pH"
```

You need the package called `akima` in order to implement bivariate interpolation onto a grid for irregularly spaced input data like these, using the function `interp`:

```
install.packages("akima")
library(akima)
```

The two explanatory variables are presented first (`hay` and `pH` in this case), with the response variable (the 'height' of the topography), which is `FR` in this case, third:

```
zz <- interp(hay,pH,FR)
```

The list called `zz` can now be used in any of the four functions `contour`, `filled.contour`, `image` or `persp`.

We start by using `contour` and `image` together. Rather than the red and yellows of `heat.colors` we choose the cooler blues and greens of `topo.colors`:

```
image(zz,col = topo.colors(12),xlab="biomass",ylab="pH")
contour(zz,add=T)
```

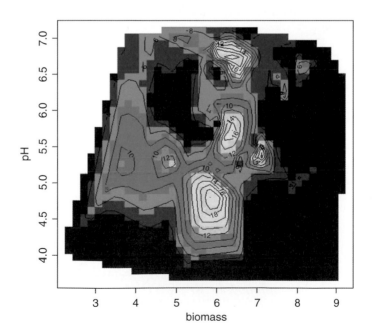

Alternatively, you can use the `filled.contour` function,

```
filled.contour(zz,col = topo.colors(24),xlab="biomass",ylab="pH")
```

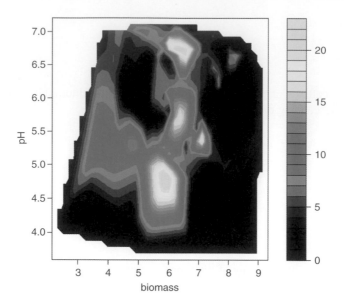

which provides a useful colour key to the abundance of *Festuca*. Evidently the grass peaks in abundance at intermediate biomass, but it also occurs at lower biomasses on soils of intermediate pH (5.0–6.0). It is found in only trace amounts in communities where the biomass is above 7.5 tonnes per hectare, except where soil pH is around 6.6.

The function `persp` allows an angled view of a 3D-like object, rather than the map-like views of `contour` and `image`. The angles `theta` and `phi` define the viewing direction: `theta` gives the azimuthal direction and `phi` gives the colatitude.

```
persp(zz,xlab="biomass",ylab="pH",zlab="Festuca rubra",
                theta = 30, phi = 30,col="lightblue")
```

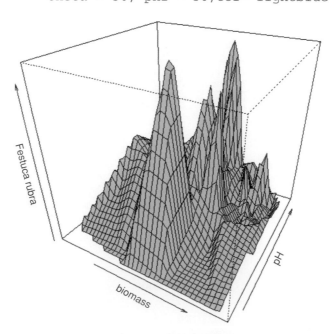

It is straightforward to create 3D images of mathematical functions from regularly spaced grids produced by the `outer` function without using `interp`. First create a series of values for the *x* and *y* axis (the base of the plot):

```
x <-   seq(0,10,0.1)
y <-   seq(0,10,0.1)
```

Now write a function to predict the height of the graph (the response variable, *z*) as a function of the two explanatory variables *x* and *y*:

```
func <-   function(x,y) 3 * x * exp(0.1*x) * sin(y*exp(-0.5*x))
```

Now use the outer function to evaluate the function over the complete grid of points defined by *x* and *y*:

```
image(x,y,outer(x,y,func))
contour(x,y,outer(x,y,func),add=T)
```

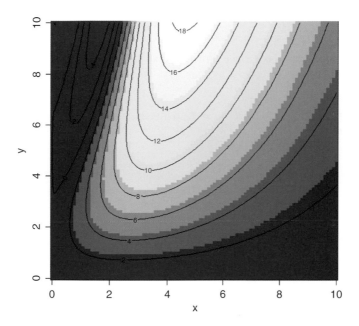

29.12   Complex 3D plots with `wireframe`

If you want to create really fancy 3D graphics you will want to master the `wireframe` function, which allows you to specify the location of the notional light source that illuminates your object (and hence creates the shadows). Here are two examples from `demo(trellis)` that produce pleasing 3D objects. In the first case, the surface is based on data (in the dataframe called `volcano`), whereas in the second case (strips on a globe) the graphic is based on an equation (`z ~ x * y`). It is in `library(lattice)`. This is how `wireframe` is invoked:

```
wireframe(volcano, shade = TRUE, aspect = c(61/87,
         0.4), screen = list(z = -120, x = -45), light.source = c(0,
         0, 10), distance = 0.2, shade.colors = function(irr, ret,
         height, w = 0.5) grey(w * irr + (1 - w) * (1 - (1 - ref)^0.4)))
```

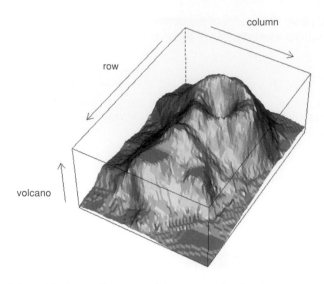

Next, we see a shaded globe with the surface turned into strips by leaving out every other pair of coloured orbits by setting their values to NA.

```
n <- 50
tx <- matrix(seq(-pi, pi, len = 2 * n), 2 * n, n)
ty <- matrix(seq(-pi, pi, len = n)/2, 2 * n, n, byrow = T)
xx <- cos(tx) * cos(ty)
yy <- sin(tx) * cos(ty)
zz <- sin(ty)
zzz <- zz
zzz[, 1:12 * 4] <- NA
```

Now draw the globe and shade the front and back surfaces appropriately:

```
wireframe(zzz ~ xx * yy, shade = TRUE, light.source = c(3,3,3))
```

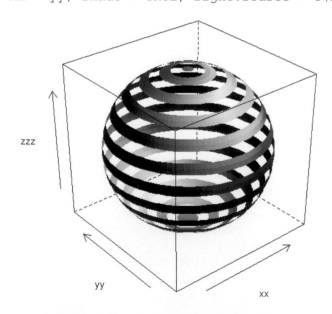

29.13   An alphabetical tour of the graphics parameters

Beginners cannot be expected to know which graphics attributes are changed with the par function, which can be changed inside the plot function, and which stand alone. This section therefore unites all the various kinds of graphics control into a single list (see Table 29.2): properties that are altered by a call to the par function are shown as par(name), while properties that can be altered inside a plot function are shown in that context; other graphics functions that stand alone (such as axis) are not shown in the table.

When writing functions, you need to know things about the current plotting region. For instance, to find out the limits of the current axes, use

```
par("usr")
```

```
[1] -0.05 10.05 -0.05 10.05
```

which shows the minimum *x* value par("usr")[1], the maximum *x* value par("usr")[2], the minimum *y* value par("usr")[3] and the maximum *y* value par("usr")[4] of the current plotting region.

If you need to use par, then the graphics parameters should be altered *before* you use the first plot function. It is a good idea to save a copy of the default parameter settings so that they can be changed back at the end of the session to their default values:

```
default.parameters <- par(no.readonly = TRUE)
...
par(...)
...
par(default.parameters)
```

To inspect the current values of any of the graphics parameters, type the name of the option in double quotes: thus, to see the sizes of the margins (for the gales data on p. 942),

```
par("mar")
```

```
[1] 5.1 4.1 4.1 2.1
```

29.13.1   Text justification, adj

To alter the justification of text strings, run the par function like this:

```
par(adj=0)
```

The parameter adj=0 produces left-justified text, adj=0.5 centred text (the default) and adj=1 right-justified text. For the text function you can vary justification in the *x* and *y* directions independently like this adj=c(1,0).

29.13.2   Annotation of graphs, ann

If you want to switch off the annotation from a plot (i.e. leave the numbers on the tick marks but not write the *x* and *y* axis labels or print any titles on the graph), then set ann = FALSE.

29.13.3   Delay moving on to the next in a series of plots, ask

Setting ask = TRUE means that the user is asked for input before the next figure is drawn.

Table 29.2. Graphical parameters and their default values. Each of the functions is illustrated in detail in the text. The column headed 'In plot?' indicates with an asterisk whether this parameter can be changed as an argument to the `plot`, `points` or `lines` functions.

Parameter	In plot?	Default value	Meaning
adj	*	0.5 (centred)	Justification of text
ann	*	TRUE	Annotate plots with axis and overall titles?
ask		FALSE	Pause before new graph?
bg	*	"transparent"	Background style or colour
bty		full box	Type of box drawn around the graph
cex	*	1	Character expansion: enlarge if > 1, reduce if < 1
cex.axis	*	1	Magnification for axis notation
cex.lab	*	1	Magnification for label notation
cex.main	*	1.2	Main title character size
cex.sub	*	1	Subtitle character size
cin		0.1354167, 0.1875000	Character size (width, height) in inches
col	*	"black"	`colors()` to see range of colours
col.axis		"black"	Colour for graph axes
col.lab	*	"black"	Colour for graph labels
col.main	*	"black"	Colour for main heading
col.sub	*	"black"	Colour for subheading
cra		13, 18	Character size (width, height) in rasters (pixels)
crt		0	Rotation of single characters in degrees (see `srt`)
csi		0.1875	Character height in inches
cxy		0.02255379, 0.03452245	Character size (width, height) in user-defined units
din		7.166666, 7.156249	Size of the graphic device (width, height) in inches (the window is bigger than this)
family	*	"sans"	Font style: from "serif", "sans", "mono" and "symbol" (and see `font`, below)
fg		"black"	Colour for objects such as axes and boxes in the foreground
fig		0, 1, 0, 1	Coordinates of the figure region within the display region: `c(x1, x2, y1, y2)`
fin		7.166666, 7.156249	Dimensions of the figure region (width, height) in inches
font	*	1	Font (regular = 1, bold = 2 or italics = 3) in which text is written (and see family, above)
font.axis	*	1	Font in which axis is numbered
font.lab	*	1	Font in which labels are written
font.main	*	1	Font for main heading
font.sub	*	1	Font for subheading
gamma		1	Correction for hsv colours
hsv		1, 1, 1	Values (range [0, 1]) for hue, saturation and value of colour
lab		5, 5, 7	Number of tick marks on the x axis, y axis and size of labels
las		0	Orientation of axis numbers: use `las=1` for publication
lend		"round"	Style for the ends of lines; could be "square" or "butt"
lheight		1	Height of a line of text used to vertically space multi-line text
ljoin		"round"	Style for joining two lines; could be "mitre" or "bevel"

Table 29.2. *(Continued)*

Parameter	In plot?	Default value	Meaning
lmitre		10	Controls when mitred line joins are automatically converted into bevelled line joins
log	*	Neither	Which axes to log: `"log=x"`, `"log=y"` or `"log=xy"`
lty	*	`"solid"`	Line type (e.g. dashed: `lty=2`)
lwd	*	1	Width of lines on a graph
mai		0.95625, 0.76875, 0.76875, 0.39375	Margin sizes in inches for c(`bottom`, `left`, `top`, `right`)
mar		5.1, 4.1, 4.1, 2.1	Margin sizes in numbers of lines for c(`bottom`, `left`, `top`, `right`)
mex		1	Margin expansion specifies the size of font used to convert between `"mar"` and `"mai"`, and between `"oma"` and `"omi"`
mfcol		1, 1	Multiple frames per page (same layout as `mfrow` (see below), but graphs produced columnwise)
mfg		1, 1, 1, 1	Which figure in an array of figures is to be drawn next (if setting) or is being drawn (if enquiring); the array must already have been set by `mfcol` or `mfrow`
mfrow		1, 1	Multiple frames per page (first number = rows, second number = columns): `mfrow = c(2,3)` gives graphs in two rows each with three columns, drawn row-wise
mgp		3, 1, 0	Margin line (in `mex` units) for the axis title, axis labels and axis line
new		FALSE	To draw another plot on top of the existing plot, set `new=TRUE` so that plot does not wipe the slate clean
oma		0, 0, 0, 0	Size of the outer margins in lines of text c(`bottom`, `left`, `top`, `right`)
omd		0, 1, 0, 1	Size of the outer margins in normalized device coordinate (NDC) units, expressed as a fraction (in [0,1]) of the device region c(`bottom`, `left`, `top`, `right`)
omi		0, 0, 0, 0	Size of the outer margins in inches c(`bottom`, `left`, `top`, `right`)
pch	*	1	Plotting symbol; e.g. `pch=16`
pin		6.004166, 5.431249	Current plot dimensions (width, height), in inches
plt		0.1072675, 0.9450581, 0.1336245, 0.8925764	Coordinates of the plot region as fractions of the current figure region c(`x1`, `x2`, `y1`, `y2`)
ps		12	Point size of text and symbols
pty		`"m"`	Type of plot region to be used: `pty="s"` generates a square plotting region, `"m"` stands for maximal
srt	*	0	String rotation in degrees
tck		`tck = -0.5`	Big tick marks (grid-lines); to use this set `tcl=NA`
tcl		−0.5	Tick marks outside the frame
tmag		1.2	Enlargement of text of the main title relative to the other annotating text of the plot

(continued)

Table 29.2. (*Continued*)

Parameter	In plot?	Default value	Meaning
type	*	"p"	Plot type: e.g. type="n" to produce blank axes
usr		set by the most recent plot function	Extremes of the user-defined coordinates of the plotting region c(xmin, xmax, ymin, ymax)
xaxp		0, 1, 5	Tick marks for log axes: xmin, xmax and number of intervals
xaxs		"r"	Pretty x axis intervals
xaxt		"s"	x axis type: use xaxt="n" to set up the axis but not plot it
xlab	*	label for the x axis	xlab="label for x axis"
xlim	*	pretty	User control of x axis scaling: xlim=c(0,1)
xlog		FALSE	Is the x axis on a log scale? If TRUE, a logarithmic scale is in use; e.g. following plot(y~x, log ="x")
xpd		FALSE	The way plotting is clipped: if FALSE, all plotting is clipped to the plot region; if TRUE, all plotting is clipped to the figure region; and if NA, all plotting is clipped to the device region
yaxp		0, 1, 5	Tick marks for log axes: ymin, ymax and number of intervals
yaxs		"r"	Pretty y axis intervals
yaxt		"s"	y axis type: use yaxt="n" to set up the axis but not plot it
ylab	*	label for the y axis	ylab="label for y axis"
ylim	*	pretty	User control of y axis scaling: ylim=c(0,100)
ylog		FALSE	Is the y axis on a log scale? If TRUE, a logarithmic scale is in use; e.g. following plot(y ~ x, log ="xy")

29.13.4 Control over the axes, axis

The attributes of four sides of the graph (1 = bottom (the x axis); 2 = left (the y axis); 3 = above and 4 = right) are controlled by the axis function.

When you want to put two graphs with different y scales on the same plot, you are likely to want to scale the right axis (axis=4) differently from the usual y axis on the left (see below).

Again, you may want to label the tick marks on the axis with letters (rather than the usual numbers) and this, too, is controlled by the axis function.

First, draw the graph with no axes at all using plot with the axes=FALSE option:

```
plot(1:10, 10:1, type="n", axes=FALSE,xlab="",ylab="")
```

For the purposes of illustration only, we use different styles on each of the four axes.

```
axis(1, 1:10, LETTERS[1:10], col.axis = "blue")
axis(2, 1:10, letters[10:1], col.axis = "red")
axis(3, lwd=3, col.axis = "green")
axis(4, at=c(2,5,8), labels=c("one","two","three"))
```

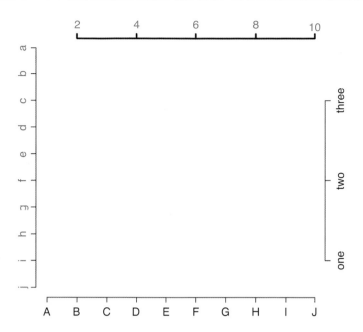

On axis 1 there are upper-case letters in place of the default numbers 1 to 10 with blue rather than black lettering. On axis 2 there are lower-case letters in reverse sequence in red on each of the 10 tick marks (note the order of the *y* values `10:1` in the original plot function). On axis 3 (the top of the graph) there is green lettering for the default numbers (2 to 10 in steps of 2) and an extra thick black line for the axis itself (`lwd = 3`). On axis 4 we have overwritten the default number and location of the tick marks using `at`, and provided our own labels for each tick mark (note that the vectors of `at` locations and `labels` must be the same length).

Because we did not use `box()` there are gaps between the ends of each of the four axes.

29.13.5   Background colour for plots, `bg`

The colour to be used for the background of plots is set by the `bg` function like this:

```
par(bg="cornsilk")
```

There is an example on p. 916. The default setting is `par(bg="transparent")`.

29.13.6   Boxes around plots, `bty`

Boxes are altered with the `bty` parameter, and `bty="n"` suppresses the box. If the character is one of `"o"`, `"1"`, (lower-case L, not numeral 1), `"7"`, `"c"`, `"u"`, or `"]"` the resulting box resembles the corresponding *upper*-case letter. Here are six options:

```
par(mfrow=c(3,2))
plot(1:10,10:1,type="n",main="default complete box")
```

```
plot(1:10,10:1,type="n",bty="n",main="no box")
plot(1:10,10:1,type="n",bty="]",main="open on the left")
plot(1:10,10:1,type="n",bty="c",main="open on the right")
plot(1:10,10:1,type="n",bty="u",main="open on the top")
plot(1:10,10:1,type="n",bty="7",main="top and right only")
```

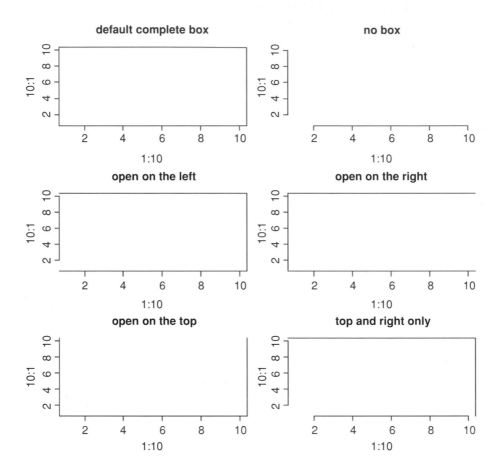

29.13.7 Size of plotting symbols using the character expansion function, cex

You can use points with cex to create 'bubbles' of different sizes. You need to specify the x, y coordinates of the centre of the bubble, then use cex = value to alter the diameter of the bubble (in multiples of the default character size: cex stands for character expansion).

```
plot(0:10,0:10,type="n",xlab="",ylab="")
for (i in 1:10) points(2,i,cex=i)
for (i in 1:10) points(6,i,cex=(10+(2*i)))
```

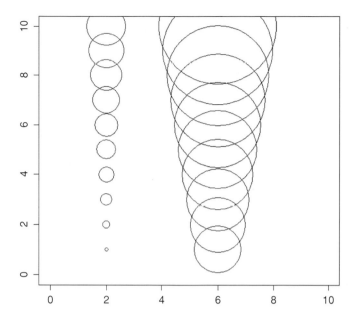

The left column shows points of size 1, 2, 3, 4, etc. (`cex = i`) and the big circles on the right are in sequence `cex = 12, 14, 16`, etc. (`cex=(10+(2*i))`).

29.13.8 Changing the shape of the plotting region, `plt`

Suppose that you wanted to draw a map that was 30 m along the *x* axis and 15 m along the *y* axis. The standard plot would have roughly twice the scale on the *y* axis as the *x*. What you want to do is reduce the height of the plotting region by half while retaining the full width of the *x* axis so that the scales on the two axes are the same. You achieve this with the `plt` option, which allows you to specify the coordinates of the plot region as fractions of the current figure region. Here we are using the full screen for one figure so we want to use only the central 40% of the region (from $y = 0.3$ to 0.7):

```
par(plt=c(0.15,0.94,0.3,0.7))
plot(c(0,3000),c(0,1500),type-"n",ylab="y",xlab="x")
```

29.13.9 Locating multiple graphs in non-standard layouts using `fig`

Generally, you would use `mfrow` to get multiple plots on the same graphic screen (see p. 209); for instance, `mfrow=c(3,2)` would give six plots in three rows of two columns each. Sometimes, however, you want a non-standard layout, and `fig` is the function to use in this case. Suppose we want to have two graphs, one in the bottom left-hand corner of the screen and one in the top right-hand corner. What you need to know is that `fig` considers that the whole plotting region is scaled from (0,0) in the bottom left-hand corner to (1,1) in the top right-hand corner. So we want our bottom left-hand plot to lie within the space $x = c(0,0.5)$ and $y = (0,0.5)$, while our top right-hand plot is to lie within the space $x = c(0.5,1)$ and $y = (0.5,1)$. Here is how to plot the two graphs: `fig` is like a new `plot` function and the second use of `fig` would normally wipe the slate clean, so we need to specify that `new=TRUE` in the second `par` function to stop this from happening:

```
par(fig=c(0.5,1,0.5,1))
plot(0:10,25*exp(-0.1*(0:10)),col="blue",type="l",ylab="remaining",xlab="time")
par(fig=c(0,0.5,0,0.5),new=T)
plot(0:100,0.5*(0:100)^0.5, col="red",type="l",xlab="amount",ylab="rate")
```

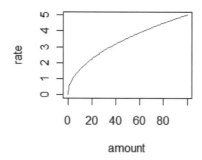

29.13.10 Two graphs with a common *x* scale but different *y* scales using `fig`

The idea here is to draw to graphs with the same *x* axis, one directly above the other, but with different scales on the two *y* axes (see also `plot.ts` on p. 800). Here are the data:

```
data <- read.table("c:\\temp\\gales.txt",header=T)
attach(data)
names(data)
```

```
[1]  "year"  "number"  "February"
```

We use `fig` to split the plotting area into an upper figure (where `number` will be drawn first) and a lower figure (for February gales, to be drawn second but on the same page, so `new=T`). The whole plotting area scales from (0,0) in the bottom left-hand corner to (1,1) in the top right-hand corner, so

```
par(fig=c(0,1,0.5,1))
```

Now think about the margins for the top graph. We want to label the *y* axis, and we want a normal border above the graph and to the right, but we want the plot to sit right on top of the lower graph, so we set the bottom margin to zero (the first argument):

```
par(mar=c(0,5,2,2))
```

Now we plot the top graph, leaving off the *x* axis label and the *x* axis tick marks:

```
plot(year,number,xlab="",xaxt="n",type="b",pch=16,col="blue",
            ylim=c(0,2000),ylab="Population")
```

Next, we define the lower plotting region and declare that new=T:

```
par(fig=c(0,1,0,0.5),new=T)
```

For this graph we *do* want a bottom margin, because we want to label the common *x* axes (Year), but we want the top of the second graph to be flush with the bottom of the first graph, so we set the upper margin to zero (argument 3):

```
par(mar=c(5,5,0,2))
plot(year,February,xlab="Year",type="h",col="red",ylab="February gales")
```

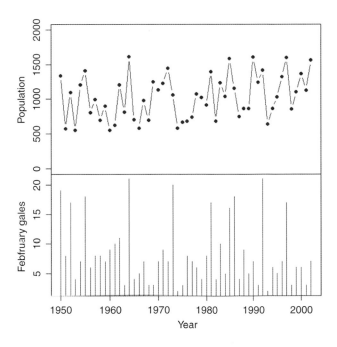

Contrast this with the overlaid plots on p. 952.

29.13.11 The layout function

If you do not want to use mfrow (p. 209) or fig (p. 942) to configure your multiple plots, then layout might be the function you need. This function allows you to alter both the location and shape of multiple plotting regions independently. The layout function is used like this:

```
layout(matrix, widths = ws, heights = hs, respect = FALSE)
```

where matrix is a `matrix` object specifying the location of the next n figures on the output device (see below), ws is a vector of column widths (with `length=ncol(matrix)`) and hs is a vector of row heights (with `length=nrow(matrix)`). Each value in the matrix must be 0 or a positive integer. If n is the largest positive integer in the matrix, then the integers $\{1, \ldots, n-1\}$ must also appear at least once in the matrix. Use 0 to indicate locations where you do not want to put a graph. The `respect` argument controls whether a unit column width is the same physical measurement on the device as a unit row height and is either a logical value or a matrix object. If it is a matrix, then it must have the same dimensions as `matrix` and each value in the matrix must be either 0 or 1. Each figure is allocated a region composed from a subset of these rows and columns, based on the rows and columns in which the figure number occurs in `matrix`. The function `layout.show(n)` plots the outlines of the next *n* figures.

Here is an example of the kind of task for which `layout` might be used. We want to produce a scatterplot with histograms on the upper and right-hand axes indicating the frequency of points within vertical and horizontal strips of the scatterplot (see the result below). This is example was written by Paul R. Murrell. Here are the data:

```
x <- pmin(3, pmax(-3, rnorm(50)))
y <- pmin(3, pmax(-3, rnorm(50)))
xhist <- hist(x, breaks=seq(-3,3,0.5), plot=FALSE)
yhist <- hist(y, breaks=seq(-3,3,0.5), plot=FALSE)
```

We need to find the ranges of values within *x* and *y* where the two histograms will lie:

```
top <- max(c(xhist$counts, yhist$counts))
xrange <- c(-3,3)
yrange <- c(-3,3)
```

Now the `layout` function defines the location of the three figures: Fig. 1 is the scatterplot which we want to locate in the lower left of four boxes, Fig. 2 is the top histogram which is to be in the upper left box, and Fig. 3 is the side histogram which is to be drawn in the lower right location (the top right location is empty), Thus, the matrix is specified as

```
matrix(c(2,0,1,3),2,2,byrow=TRUE)
```

like this:

```
nf <- layout(matrix(c(2,0,1,3),2,2,byrow=TRUE), c(3,1), c(1,3), TRUE)
layout.show(nf)
```

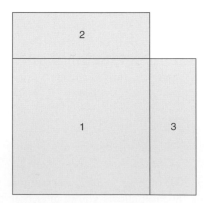

The figures in the first (left) column of the matrix (Figs 1 and 2) are of width 3 while the figure in the second column (Fig. 3) is of width 1, hence `c(3,1)` is the second argument. The heights of the figures in the first column of the matrix (Figs 2 and 1) are 1 and 3 respectively, hence `c(1,3)` is the third argument. The missing figure is 1 by 1 (top right).

```
par(mar=c(3,3,1,1))
plot(x, y, xlim=xrange, ylim=yrange, pch=21,col="blue",bg="red",
                        xlab="", ylab="")

par(mar=c(0,3,1,1))
barplot(xhist$counts, axes=FALSE, col="green",ylim=c(0, top), space=0)

par(mar=c(3,0,1,1))
barplot(yhist$counts, axes=FALSE, col="green",
                        xlim=c(0, top), space=0, horiz=TRUE)
```

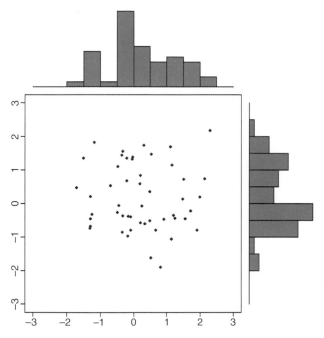

Note the way that the margins for the three figures are controlled, and how the `horiz=TRUE` option is specified for the histogram on the right-hand margin of the plot.

29.13.12   Creating and controlling multiple screens on a single device

The function `split.screen` defines a number of regions within the current device which can be treated as if they were separate graphics devices. It is useful for generating multiple plots on a single device (see also `mfrow` and `layout`). Screens can themselves be split, allowing for quite complex arrangements of plots. The function `screen` is used to select which screen to draw in, and `erase.screen` is used to clear a single screen, which it does by filling with the background colour, while `close.screen` removes the specified screen definition(s) and split-screen mode is exited by `close.screen(all = TRUE)`. You should complete each graph before moving on to the graph in the next screen, because returning to a screen can create problems.

You can create a matrix in which each row describes a screen with values for the left, right, bottom, and top of the screen (in that order) in normalized device coordinate (NDC) units, that is, 0 at the lower left-hand corner of the device surface, and 1 at the upper right-hand corner (see `fig`, above).

First, set up the matrix to define the corners of each of the plots. We want a long, narrow plot on the top of the screen as Fig. 1, a tall rectangular plot on the bottom left as Fig. 2, then two small square plots on the bottom right as Figs 3 and 4. The dataframe called `gales` is read on p. 942. Here is the matrix:

```
fig.mat <- c(0,0,.5,.5,1,.5,1,1,.7,0,.35,0,1,.7,.7,.35)
fig.mat <- matrix(fig.mat,nrow=4)
fig.mat
```

```
       [,1]  [,2]  [,3]  [,4]
[1,]    0.0   1.0  0.70  1.00
[2,]    0.0   0.5  0.00  0.70
[3,]    0.5   1.0  0.35  0.70
[4,]    0.5   1.0  0.00  0.35
```

Now we can draw the four graphs:

```
split.screen(fig.mat)
```

```
[1]  1  2  3  4
```

```
screen(1)
plot(year,number,type="l",col="blue")
```

```
screen(2)
plot(year,February,type="h", col="red")
```

```
screen(3)
plot(1:10,0.5*(1:10)^0.5,xlab="concentration",
                 ylab="rate",type="l", col="green4")
```

```
screen(4)
plot(1:10,600*exp(-0.5*(1:10)),xlab="time",
                 ylab="residue",type="l", col="green4")
```

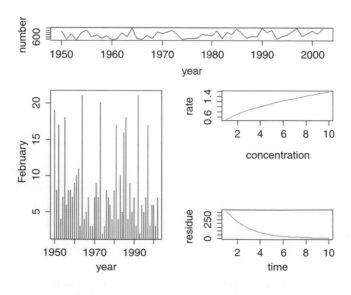

29.13.13 Orientation of numbers on the tick marks, las

Many journals require that the numbers used to label the y axis must be horizontal. To change from the default, use las:

las=0	always parallel to the axis (the default);
las=1	always horizontal (preferred by many journals);
las=2	always perpendicular to the axis;
las=3	always vertical.

Note that you cannot use character or string rotation for this. Examples are shown on p. 907.

29.13.14 Shapes for the ends and joins of lines, lend and ljoin

The default is that the bare ends of lines should be rounded (see also arrows if you want pointed ends). You can change this to "butt" or "square". This example shows the use of overwriting with successive colours to achieve special effects:

```
plot(0:10,0:10,type="n",xlab="",ylab="")
lines(c(2,5,8),c(8,2,8),lwd=50,lend="square",ljoin="mitre")
lines(c(2,5,8),c(8,2,8),col="green",lwd=50,lend="round",ljoin="round")
lines(c(2,5,8),c(8,2,8),col="red",lwd=50,lend="butt",ljoin="bevel")
```

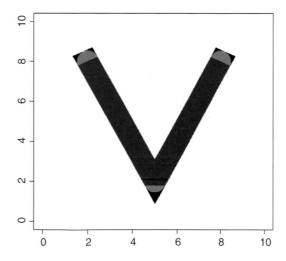

A black V was drawn first with 'square' ends and 'mitre' joins, then the green V with 'round' ends and 'round' joins, then finally the red V with 'butt' ends and 'bevel' joins.

To get the effect of bordered lines (e.g. to produce roads on a map) first draw wide lines in black (or whatever colour you want the border to be), then draw the colour for the body of the line using a slightly smaller value for the line width. Here is a two-dimensional random walk drawn as a smoothly snaking road in red with black margins. Here is the random walk:

```
x<-numeric(100)
y<-numeric(100)
```

```
x[1]<-1
y[1]<-1
for (i in 2:100)  {
a<-runif(1)*2*pi
d<-runif(1)*1
x[i]<-x[i-1]+d*sin(a)
y[i]<-y[i-1]+d*cos(a)
}
```

Now we plot blank axes with an intact box around the outside of the map:

```
plot(0:10,0:10,type="n",xaxt="n",yaxt="n",xlab="",ylab="")
```

The red road with black margins and smooth curves is added like this:

```
lines(x,y,lwd=13,lend="round",ljoin="round")
lines(x,y,lwd=10,col="red",lend="round",ljoin="round")
```

29.13.15 Line types, `lty`

Line types (like solid, dotted or dashed) are changed with the line-type parameter `lty`:

`lty = 1`	solid (the default);
`lty = 2`	dashed;
`lty = 3`	dotted;
`lty = 4`	dot-dash;
`lty = 5`	long-dash;
`lty = 6`	two-dash.

Invisible lines are drawn if `lty=0` (i.e. the line is not drawn). Alternatively, you can use text to specify the line types with one of the following character strings: `blank`, `solid`, `dashed`, `dotted`, `dotdash`, `longdash` or `twodash` (see below).

29.13.16 Line widths, `lwd`

To increase the widths of the plotted lines use `lwd = 2` (or greater; the default is `lwd=1`). The interpretation is device specific, and some devices do not implement line widths less than 1. The function `abline` is so called because it has two arguments – the first is the intercept (a) and the second is the slope (b) of a linear relationship $y = a + bx$ (see p. 191 for background):

```
plot(1:10,1:10,xlim=c(0,10),ylim=c(0,10),xlab="",ylab="",type="n")

for(i in 1:7) {
abline((-6+2*i),1,lty=i,col=i)
}

for(i in 2:7) {
y <- (-6+2*i)+2
x <- 2
text(x,y,i)
}

abline(-6,1,lty=1,lwd=4)
abline(-8,1,lty=1,lwd=8)

points(5,1,pch=16,cex=3,col="white")
points(7,1,pch=16,cex=3,col="white")
points(9,1,pch=16,cex=3,col="white")
text(5,1,1,col="red")
text(7,1,4,col="red")
text(9,1,8,col="red")
```

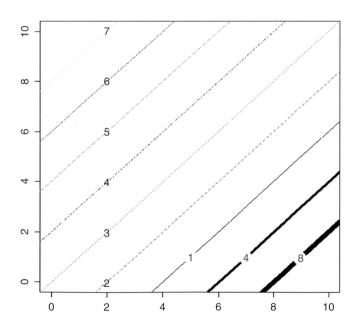

The numerals in a vertical line above $x = 2$ indicate the line types 2 to 7 in colours 2 to 7. In the bottom right-hand corner are three solid lines `lty=1` of widths `lwd=1`, 4 and 8. Note the use of the large white `pch=16` across the lines to make a gap in which the red labels indicating the line widths can be printed clearly.

29.13.17   Several graphs on the same page, `mfrow` and `mfcol`

The way to remember the names of these functions is to think of them as standing for 'multiple frames in rows' (`mfrow`) or 'multiple frames in columns' (`mfcol`). You can obtain multiple graph panels on the same graphics device by `par(mfrow)`, `par(mfcol)`, `par(fig)`, `par(split.screen)` and `par(layout)`, but `par(mfrow)` is much the most frequently used. You specify the number of rows of graphs (first argument) and number of columns of graphs per row (second argument) like this:

`par(mfrow=c(1,1))`	the default of one plot per screen;
`par(mfrow=c(1,2))`	one row of two columns of plots;
`par(mfrow=c(2,1))`	two rows of one column of plots;
`par(mfrow=c(2,2))`	four plots in two rows of two columns each;
`par(mfrow=c(3,2))`	six plots in three rows of two columns each;
`par(mfrow=c(3,3))`	nine plots in three rows of three columns each.

The graphs will be produced row-wise, starting in the top left-hand corner. You need to complete each graph (add all `points`, `lines` and `text`) before going on to the next by issuing a new `plot` command.

In a layout with exactly two rows and columns the base value of `cex` is reduced by a factor of 0.83; if there are three or more of either rows or columns, the reduction factor is 0.66. Remember to set `par` back to `par(mfrow=c(1,1))` when you have finished with multiple plots.

29.13.18   Margins around the plotting area, `mar`

You need to control the size of the margins when you intend to use large symbols or long labels for your axes, or when you want to position multiple plots closer together. The four margins of the plot are defined by integers 1 to 4 as follows:

1 = bottom (the x axis);
2 = left (the y axis);
3 = top;
4 = right.

The sizes of the margins of the plot are measured *in lines of text*. The four arguments to the `mar` function are given in the sequence bottom, left, top, right. The default is

`par(mar=(c(5, 4, 4, 2) + 0.1))`

with more spaces on the bottom (5.1) than on the top (4.1) to make room for a subtitle (if you should want one), and more space on the left (4.1) than on the right (2) on the assumption that you will not want to label the right-hand axis. Suppose that you *do* want to put a label on the right-hand axis, then you would need to increase the size of the fourth number, for instance like this: `par(mar=(c(5, 4, 4, 4) + 0.1))`.

To get rid of margins altogether, use `par(mar=(c(0, 0, 0, 0))`, but bear in mind that there will be no space for any labels under this format.

29.13.19 Plotting more than one graph on the same axes, `new`

The `new` parameter is a logical variable, defaulting to `new=FALSE`. If it is set to `new=TRUE`, the next high-level plotting command (like `plot(y~x)`) does *not* wipe the slate clean in the default way. This allows one plot to be placed on top of another.

29.13.20 Two graphs on the same plot with different scales for their *y* axes

```
gales <- read.table("c:\\temp\\gales.txt",header=T)
attach(gales)
names(gales)
```

```
[1]  "year"  "number"  "February"
```

In this example we want to plot the number of animals in a wild population as a time series over the years 1950–2000 with the scale of animal numbers on the left-hand axis (numbers fluctuate between about 600 and 1600). Then, on top of this, we want to overlay the number of gales in February each year. This number varies between 1 and 22, and we want to put a scale for this on the right-hand axis (`axis=4`).

First we need to make room in the right-hand margin for labelling the axis with the information on February gales:

```
par(mar=c(5,4,4,4)+0.1)
```

Now draw the time series using a thicker line than usual (`lwd=2`) for emphasis:

```
plot(year,number,type="l",lwd=2,las=1,col="blue")
```

Next, indicate that the next graph will be overlaid on the present one:

```
par(new=T)
```

Now plot the graph of gales against years. This is to be displayed as vertical (`type="h"`) dashed lines (`lty=2`) in red:

```
plot(year,February,type="h",axes=F,ylab="",lty=2,col="red")
```

and it is drawn with its own scale (with ticks from 5 to 20, as we shall see). The right-hand axis is ticked and labelled as follows. First use `axis(4)` to create the tick marks and scaling information, then use the `mtext` function to produce the axis label (the name stands for 'margin text').

```
axis(4,las=1)
mtext(side=4,line=2.5,"February gales")
```

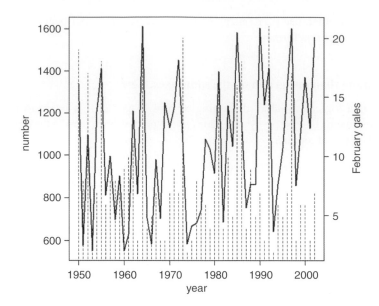

It looks as if unusually severe February gales are associated with the steepest population crashes (contrast this with the separate plots on p. 943).

29.13.21   Outer margins, oma

There is an area outside the margins of the plotting area called the **outer margin**. Its default size is zero, oma=c(0,0,0,0), but if you want to create an outer margin you use the function oma.

Here is the function to produce an outer margin big enough to accommodate five lines of text on the bottom and left-hand sides of the plotting region (see p. 956):

```
par(oma=c(5,5,0,0))
```

You will want to combine large outer margins with reduced inner margins when you want to produce your own multiple-panel plots.

When using par(mfrow=c(2,2)) to get a panel of plots, you will probably use main to get a unique title for each of the plots, but you may want an overall title (the equivalent of main but for the entire set of panel plots. You use mtext ('margin text') for this with along with outer=T. Here are Anscombe's infamous four plots; he contrived the data so that they all have exactly the same regression models and *p* values, but they are obviously very different once you plot them. Moral: always plot the data first, then do the modelling, once you know what the data look like. The data are built into R:

```
attach(anscombe)
par(mfrow=c(2,2))
```

You can see the contrasting patterns in the four data sets

```
plot(x1,y1,main="set 1",col="red", pch=21, bg =
"orange",xlim=c(0,20),ylim=c(0,16))
abline(lm(y1~x1),col="navy")

plot(x2,y2,main="set 2",col="red", pch=21, bg =
"orange",xlim=c(0,20),ylim=c(0,16))
```

```
abline(lm(y2~x2),col="navy")
```

```
plot(x3,y3,main="set 3",col="red", pch=21, bg =
"orange",xlim=c(0,20),ylim=c(0,16))
abline(lm(y3~x3),col="navy")
```

```
plot(x4,y4,main="set 4",col="red", pch=21, bg =
"orange",xlim=c(0,20),ylim=c(0,16))
abline(lm(y4~x4),col="navy")
```

but there is no room for an overall title.

You need to plan ahead, and to make space for at least one line of (potentially large) text at the top of the page. This is the *third margin*, and the space you want to create is *outside* the space of the existing plot. Using oma, you specify the width of the margin (in units of text lines) for each of the four margins in a vector (in our case, c(0,0,2,0), to leave two lines at the top (third) margin):

```
par(mfrow=c(2,2), oma= c(0,0,2,0))
```

Now redraw the four plots (as above) then add the graph title using mtext like this:

```
mtext("Anscombe's 4 regression data sets", outer = TRUE, cex=1.5)
```

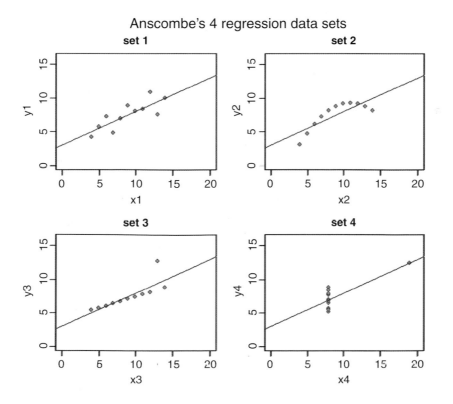

29.13.22   Packing graphs closer together

In this example we want to create nine closely spaced plots in a 3 × 3 pattern without any tick marks, and to label only the outer central plot on the *x* and *y* axes. We need to take care of four things:

- `mfrow=c(3,3)` to get the nine plots in a 3 × 3 pattern;
- `mar=c(0.2,0.2,0.2,0.2)` to leave a narrow strip (0.2 lines looks best for tightly packed plots) between each graph;
- `oma=c(5,5,0,0)` to create an outer margin on the bottom and left for labels;
- `outer = T` in title to write the titles in the outer margin.

The plots consist of 100 pairs of ranked uniform random numbers `sort(runif(100))`, and we shall plot the nine graphs (see p. 209) with a `for` loop:

```
par(mfrow=c(3,3))
par(mar=c(0.2,0.2,0.2,0.2))
par(oma=c(5,5,0,0))
for (i in 1:9) plot(sort(runif(100)),sort(runif(100)),
                        xaxt="n",yaxt="n",pch=21, bg="green")
title(xlab="time",ylab="distance",outer=T,cex.lab=2)
```

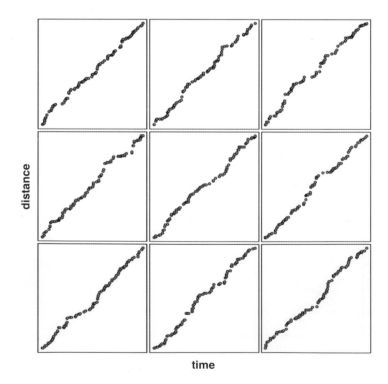

29.13.23   Square plotting region, `pty`

If you want to have a square plotting region (e.g. when producing a map or a grid with true squares on it), then use the `pty="s"` option. The option `pty="m"` generates the maximal plotting region which is not square on most devices. See the example on p. 829.

29.13.24   Character rotation, `srt`

To rotate characters in the plotting plane, use `srt` (which stands for 'string rotation'). The argument to the function is in degrees of counter-clockwise rotation:

```
plot(1:10,1:10,type="n",xlab="",ylab="")
for (i in 1:10)  text (i,i,LETTERS[i],srt=(20*i),col="red")
for (i in 1:10)  text (10-i+1,i,letters[i],srt=(20*i),col="blue")
```

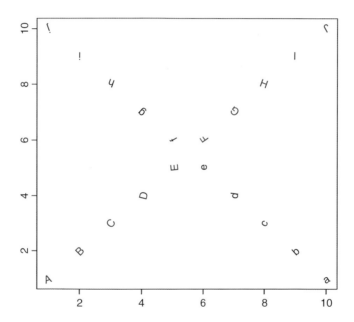

Observe how the letters i and I have been turned upside down (`srt=180`).

29.13.25   Rotating the axis labels

When you have long text labels (e.g. for bars on a `barplot`) it is a good idea to rotate them through 45 degrees so that all the labels are printed, and all are easy to read.

```
spending <- read.csv("c:\\temp\\spending.csv")
attach(spending)
names(spending)
```

```
[1] "spend"    "country"
```

There are three steps involved:

- Make the bottom margin big enough to take the long labels (`mar`).

- Find the *x* coordinates of the centres of the bars (`xvals`) with `usr`.

- Use `text` with `srt = 45` to rotate the labels.

```
par(mar = c(7, 4, 4, 2) + 0.1)
xvals <- barplot(spend,ylab="spending",col="wheat2")

text(xvals, par("usr")[3]-0.25, srt = 45,
                  adj = 1,labels = country, xpd = TRUE)
```

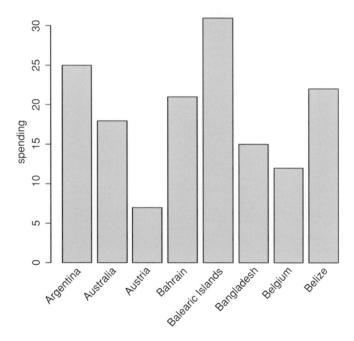

Note the use of `xpd=TRUE` to allow for text outside the plotting region, and `adj=1` to place the right-hand end of text at the centre of the bars. The vertical location of the labels is set by `par("usr")[3]-0.25` and you can adjust the value of the offset (here 0.25) as required to move the axis labels up or down relative to the *x* axis.

29.13.26   Tick marks on the axes

The functions `tck` and `tcl` control the length and location of the tick marks. Negative values put the tick marks *outside* the box (`tcl=-0.5` is the default setting in R); `tcl` gives the length of tick marks as a fraction of the height of a line of text.

The default setting for `tck` is `tck=NA` but you can use this for drawing grid lines: `tck=0` means no tick marks, while `tck=1` means fill the whole frame (i.e. the tick marks make a grid). The tick is given as a fraction of the frame width (they are $+0.03$ in the bottom right-hand graph, so are internal to the plotting region).

```
par(mfrow=c(2,2))
plot(1:10,1:10,xlab="",ylab="",type="n",main="default ticks")
```

```
plot(1:10,1:10,xlab="",ylab="",type="n",main="maximum ticks",tck=1)
plot(1:10,1:10,xlab="",ylab="",type="n",main="no ticks",tck=0)
plot(1:10,1:10,xlab="",ylab="",type="n",main="interior ticks",tck=0.03)
```

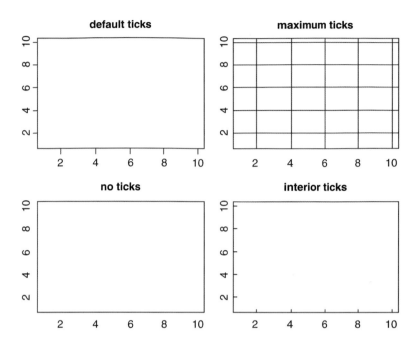

29.13.27 Axis styles

There are three functions that you need to distinguish:

axis	select one of the four sides of the plot to work with;
xaxs, yaxs	intervals for the tick marks;
xaxt, yaxt	suppress production of the axis with xaxt="n".

The axis function is described on pp. 909 and 924.

The xaxs function is used infrequently: style "r" (regular) first extends the data range by 4% and then finds an axis with pretty labels that fits within the range; style "i" (internal) just finds an axis with pretty labels that fits within the original data range.

Finally, xaxt and yaxt are often used when you want to specify your own kind of axes with different locations for the tick marks and/or different labelling. If you do not want any tick marks or numbers on the axes, then suppress the tick marks and value labels using xaxt="n" and/or yaxt="n" (see pp. 924 and 928 for examples).

29.14 Trellis graphics

The main purpose of trellis graphics is to produce multiple plots per page and multi-page plots, particularly in the context of mixed-effects modelling (see p. 696). The plots are produced in adjacent panels, typically with one plot for each level of a categorical variable (called the **conditioning variable**). For instance, you might plot weight against age for each of two genders (males and females). The response variable is weight, the continuous explanatory variable is age (also called the **primary covariate** in documentation on trellis graphics) and the categorical explanatory variable is gender (a factor with two levels). In a case like this, the default would produce two panels side by side in one row, with the panel for females on the left (simply because 'f' comes before 'm' in the alphabet). In the jargon of trellis graphics, gender is a **grouping factor** that divides the observations into distinct groups. Here are the data:

```
data <- read.table("c:\\temp\\panels.txt",header=T)
attach(data)
names(data)

[1]  "age"  "weight"  "gender"
```

The package for producing trellis graphics in R is called `lattice` (not trellis as you might have guessed, because that name was pre-empted by a commercial package):

```
library(lattice)
```

The panel plots are created by the `xyplot` function, using a formula to indicate the grouping structure: `weight ~ age | gender`. This is read as 'weight is plotted as a function of age, given gender' (the vertical bar | is the 'given' symbol).

```
xyplot(weight ~ age | gender)
```

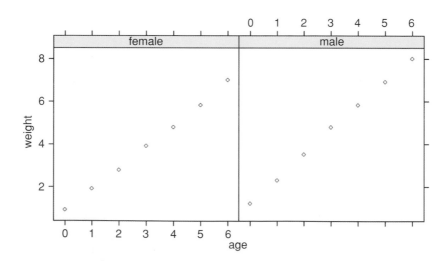

Trellis graphics is a framework for data visualization developed at Bell Laboratories by Rick Becker, Bill Cleveland and others, extending the ideas about what makes for an effective graph (layout, colour, style, symbol sizes and so forth) presented in Cleveland (1993). The interface is based on the implementation in S-PLUS, but there are several differences, and code produced for S-PLUS might not work in R. The R version was written by Deepayan Sarkar, and the plots created by `lattice` are rendered by the Grid Graphics engine for R (developed by Paul Murrell). Most of the high-level trellis functions in S-PLUS are implemented in R, with the exception of the pie chart:

- `barchart` for barplots;
- `bwplot` for box-and-whisker plots;
- `densityplot` for kernel density plots;
- `dotplot` for dot plots;
- `histogram` for panels of histograms;
- `qqmath` for quantile plots against mathematical distributions;
- `stripplot` for a one-dimensional scatterplot;
- `qq` for a quantile–quantile plot for comparing two distributions;
- `xyplot` for a scatterplot;
- `levelplot` for creating level plots (similar to image plots);
- `contourplot` for contour plots;
- `cloud` for three-dimensional scatterplots;
- `wireframe` for 3D surfaces (similar to `persp` plots);
- `splom` for a scatterplot matrix;
- `parallel` for creating parallel coordinate plots;
- `rfs` to produce a residual and fitted value plot (see also `oneway`);
- `tmd` for a Tukey mean–difference plot.

Lattice plots are highly customizable via user-modifiable settings, but these are completely unrelated to base graphics settings. In particular, changing `par()` settings usually has **no** effect on lattice plots. To read more about the background and capabilities of the lattice package, type

```
help(package = lattice)
```

29.14.1   Panel box-and-whisker plots

Here is an example trellis plot for the interpretation of a designed experiment where all the explanatory variables are categorical. It uses `bwplot` to illustrate the results of a three-way analysis of variance (p. 528).

```
data <- read.table("c:\\temp\\daphnia.txt",header=T)
attach(data)
names(data)
```

```
[1]   "Growth.rate"   "Water"   "Detergent"   "Daphnia"
```

```
bwplot(Growth.rate~Water+Daphnia|Detergent)
```

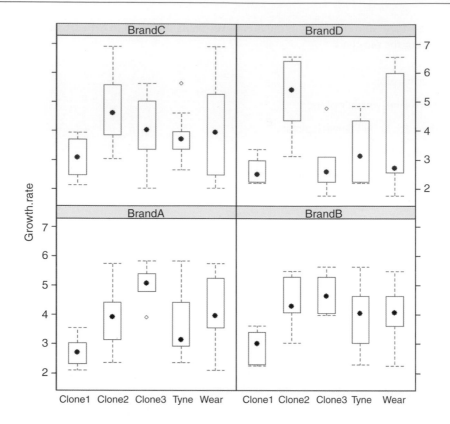

29.14.2   Panel scatterplots

Panels are by default drawn starting from the bottom left-hand corner, going right and then up, unless `as.table = TRUE`, in which case panels are drawn from the top left-hand corner, going right and then down. Both of these orders can be modified using the `index.cond` and `perm.cond` arguments.

There are some grid-compatible replacements for commonly used base R graphics functions: for example, `lines` can be replaced by `llines` (or equivalently, `panel.lines`). Note that base R graphics functions like `lines` simply will not work in a lattice `panel` function. The following example is concerned with root growth measured over time, as repeated measures on 12 individual plants:

```
results <- read.table("c:\\temp\\fertilizer.txt",header=T)
attach(results)
names(results)

[1]  "root"   "week"   "plant"   "fertilizer"
```

Here is a set of 12 scatterplots, showing `root ~ week` with one panel for each plant. The syntax uses the "given" bar | like this:

```
xyplot(root ~ week | plant)
```

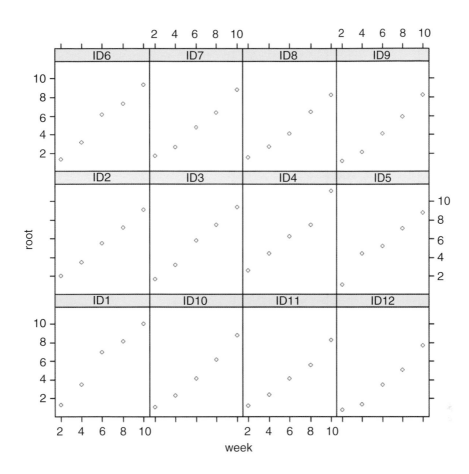

By default, the panels are shown in alphabetical order by plant name from bottom left (ID1) to top right (ID9). If you want to change things like the plotting symbol you can do this within the `xyplot` function,

```
xyplot(root ~ week | plant,pch=16)
```

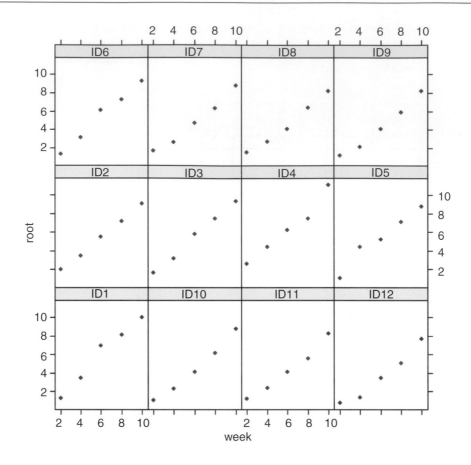

but if you want to make more involved changes, you should use a `panel` function.

Suppose we want to fit a separate linear regression for each individual plant. We write

```
xyplot(root ~ week | plant ,
          panel = function(x, y) {
          panel.xyplot(x, y, pch=16)
          panel.abline(lm(y ~ x))
     })
```

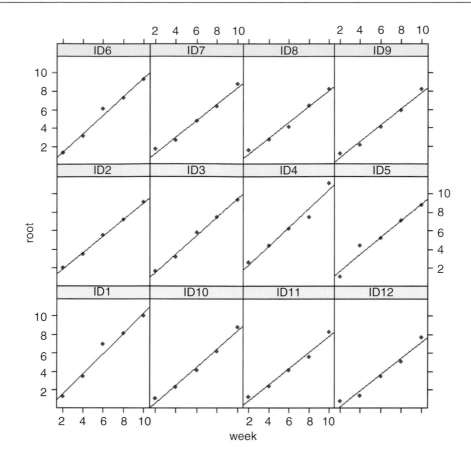

You might want to do different things in different panels. Here, we draw a horizontal red dashed line, highlighting the location of the fourth data point in each panel using subscripts [4]:

```
xyplot(root ~ week | plant ,
panel = function(x, y) {
panel.xyplot(x, y, pch=16)
panel.abline(lm(y ~ x))
panel.abline(h=y[4],col="red",lty=3)
})
```

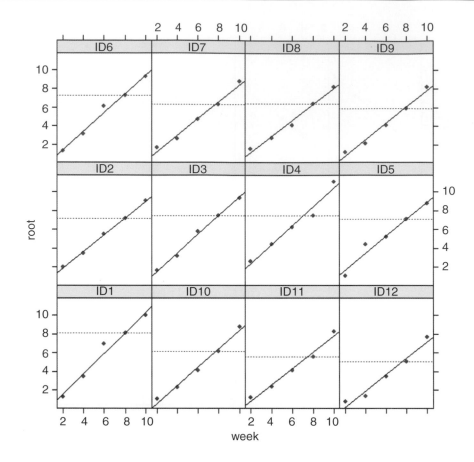

The panels are numbered by default from lower left to upper right. Here we use `panel.number()` to illustrate this, by adding a text label to each panel showing the panel number in green:

```
xyplot(root ~ week | plant ,
panel = function(x, y) {
panel.xyplot(x, y, pch=16)
panel.abline(lm(y ~ x))
panel.abline(h=y[4],col="red",lty=3)
panel.text(8,2,panel.number(),col="green",cex=0.7)
})
```

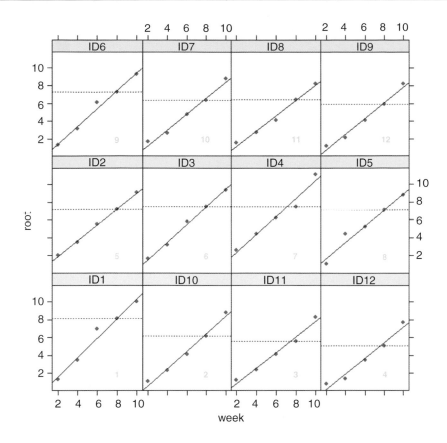

You can add extra points and extra lines to each panel using `panel.lines` and `panel.points` with `panel.number()` as a subscript like this:

```
panel.points(xnew[panel.number()],ynew[panel.number()])
```

29.14.3   Panel barplots

The following example shows the use of the trellis version of the barchart with the built-in barley data. The data are shown separately for each year (`groups=year`) and the bars are stacked for each year (`stack=TRUE`) in different shades of blue (`col=c("cornflowerblue","blue")`):

The barcharts are produced in three rows of two plots each (`layout = c(2,3)`). Note the use of `scales` to rotate the long labels on the *x* axis through 45 degrees:

```
barchart(yield ~ variety | site, data = barley,
         groups = year, layout = c(2,3), stack = TRUE,
         col=c("cornflowerblue","blue"),
         ylab = "Barley Yield (bushels/acre)",
         scales = list(x = list(rot = 45)))
```

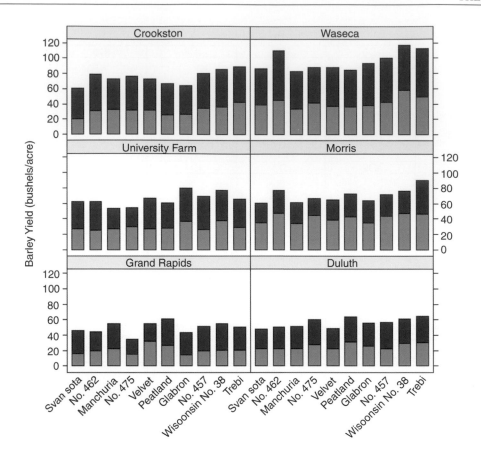

29.14.4 Panels for conditioning plots

In this example we put each of the panels side by side (layout=c(9,1)) on the basis of an equal-count split of the variable called E within the built-in ethanol dataframe:

```
EE <- equal.count(ethanol$E, number=9, overlap=1/4)
```

Within each panel defined by EE we draw a grid (panel.grid(h=-1, v=2)), create a scatterplot of NOx against C (panel.xyplot(x, y)) and draw an individual linear regression (panel.abline(lm(y ~ x))):

```
xyplot(NOx ~ C | EE, data = ethanol,layout=c(9,1),
          panel = function(x, y) {
              panel.grid(h=-1, v= 2)
              panel.xyplot(x, y)
              panel.abline(lm(y ~ x))
          })
```

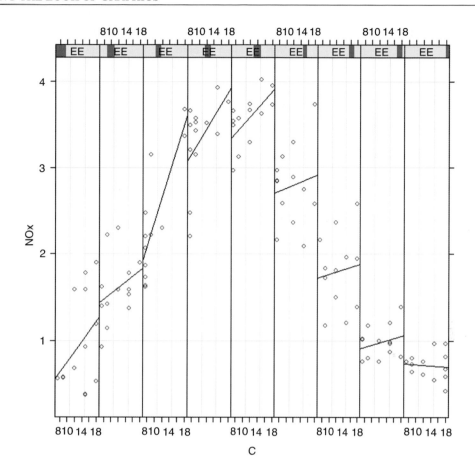

This is an excellent way of illustrating that the correlation between NOx and C is positive for all levels of EE except the highest one, and that the relationship is steepest for values of EE just below the median (i.e. in the third panel from the left).

29.14.5 Panel histograms

The task is to use the Silwood weather data to draw a panel of histograms, one for each month of the year, showing the number of days per month during the period 1987–2005 with particular minimum temperatures.

```
data <- read.table("c:\\temp\\SilwoodWeather.txt",header=T)
attach(data)
names(data)
```

```
[1]  "upper"  "lower"  "rain"  "month"  "yr"
```

```
histogram( ~ lower| month,type="count",
          xlab="mimimum temerature",ylab="frequency",
          breaks=seq(-12,28,2),strip=strip.custom(factor.levels=month))
```

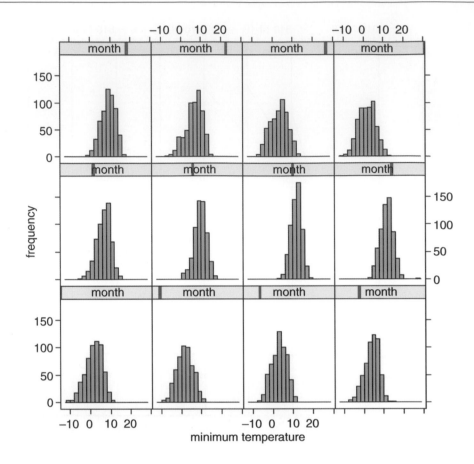

The panel histogram is drawn using the histogram function which takes a model formula *without a response variable ~* lower|month as its first argument.

29.14.6   Effect sizes

An alternative is to use the effects package which takes a model object (a linear model or a generalized linear model) and provides trellis plots of specified effects.

```
install.packages("effects")
library(effects)
model <- lm(Growth.rate~Water*Detergent*Daphnia)
```

First calculate all the effects using the allEffects function, then plot this object, specifying the interaction you want to see, using double quotes:

```
daph.effects <- allEffects(model)
plot(daph.effects,"Water:Detergent:Daphnia")
```

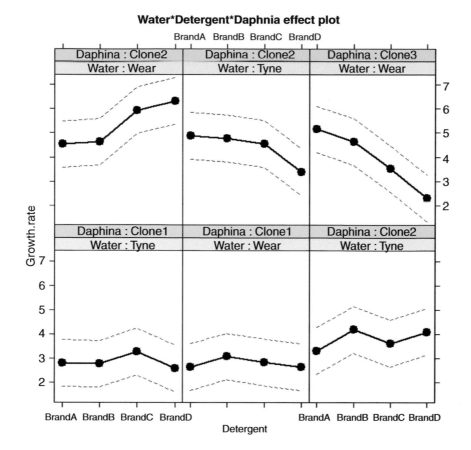

Water*Detergent*Daphnia effect plot

29.14.7 More panel functions

Plots can be transformed by specifying the grouping (`groups=rowpos`), indicating that each group should be drawn in a different colour (`panel="panel.superpose"`), or by specifying that the dots should be joined by lines for each member of the group (`panel.groups="panel.linejoin"`). Here are the orchard spray data with each row shown in a different colour and the treatment means joined together by lines. This example also shows how to use `auto.key` to locate a key to the groups on the right of the plot, showing `lines` rather than `points`:

```
xyplot(decrease ~ treatment, OrchardSprays, groups = rowpos,
         type="a",
         auto.key =
         list(space = "right", points = FALSE, lines = TRUE))
```

You can find more examples of lattice graphics in Chapter 19 on mixed-effects models.

References and Further Reading

Agresti, A. (1990) *Categorical Data Analysis*. New York: John Wiley & Sons, Inc.

Aitkin, M., Francis, B., Hinde, J. and Darnell, R. (2009) *Statistical Modelling in R*. Oxford: Clarendon Press.

Atkinson, A.C. (1985) *Plots, Transformations, and Regression*. Oxford: Clarendon Press.

Baddeley, A. and Turner, R. (2012) *spatstat: Spatial Point Pattern analysis, model-fitting, simulation, tests*. Maintainer: Adrian Baddeley, Adrian.Baddeley@csiro.au.

Bates, D., Maechler, M. and Bolker, B. (2012) *lme4: Linear mixed-effects models using S4 classes*. Maintainer: lme4-author@R-forge.wu-wien.ac.at.

Bishop, Y.M.M., Fienberg, S.J. and Holland, P.W. (1980) *Discrete Multivariate Analysis: Theory and Practice*. New York: John Wiley & Sons, Inc.

Bivand, R. et al. (2012) *spdep: Spatial dependence: weighting schemes, statistics and models*. Maintainer: Roger Bivand, Roger.Bivand@nhh.no.

Borenstein, M., Hedges, L.V., Higgins, J.P.T. and Rothstein, H.R. (2009) *Introduction to Meta-analysis*. Chichester: John Wiley & Sons, Ltd.

Box, G.E.P. and Cox, D.R. (1964) An analysis of transformations. *Journal of the Royal Statistical Society, Series B*, **26**, 211–246.

Box, G.E.P. and Jenkins, G.M. (1976) *Time Series Analysis: Forecasting and Control*. Oakland, CA: Holden-Day.

Box, G.E.P., Hunter, W.G. and Hunter, J.S. (1978) *Statistics for Experimenters: An Introduction to Design, Data Analysis and Model Building*. New York: John Wiley & Sons, Inc.

Breiman, L., Friedman, L.H., Olshen, R.A. and Stone, C.J. (1984) *Classification and Regression Trees*. Belmont, CA: Wadsworth International Group.

Canty, A. and Ripley, B. (2012). *boot: Bootstrap R (S-Plus) Functions*. R package version 1.3-4.

Caroll, R.J. and Ruppert, D. (1988) *Transformation and Weighting in Regression*. New York: Chapman & Hall.

Casella, G. and Berger, R.L. (1990) *Statistical Inference*. Pacific Grove, CA: Wadsworth and Brooks/Cole.

Chambers, J.M. and Hastie, T.J. (1992) *Statistical Models in S*. Pacific Grove, CA: Wadsworth and Brooks Cole.

Chambers, J.M., Cleveland, W.S., Kleiner, B. and Tukey, P.A. (1983) *Graphical Methods for Data Analysis*. Belmont, CA: Wadsworth.

Chatfield, C. (1989) *The Analysis of Time Series: An Introduction*. London: Chapman & Hall.

Clark, P.J. and Evans, F.C. (1954) Distance to nearest neighbour as a measure of spatial relationships in populations. *Ecology*, **35**, 445–453.

Cleveland, W.S. (1993) *Visualizing Data*. Summit, NJ: Hobart Press.

Cochran, W.G. and Cox, G.M. (1957) *Experimental Designs*. New York: John Wiley & Sons, Inc.

Collett, D. (1991) *Modelling Binary Data*. London: Chapman & Hall.

Conover, W.J. (1980) *Practical Nonparametric Statistics*. New York: John Wiley & Sons, Inc.

Conover, W.J., Johnson, M.E. and Johnson, M.M. (1981) A comparative study of tests for homogeneity of variances, with applications to the outer continental shelf bidding data. *Technometrics*, **23**, 351–361.

The R Book, Second Edition. Michael J. Crawley.
© 2013 John Wiley & Sons, Ltd. Published 2013 by John Wiley & Sons, Ltd.

Cook, R.D. and Weisberg, S. (1982) *Residuals and Influence in Regression*. New York: Chapman & Hall.

Cox, D.R. (1972) Regression models and life-tables (with discussion). *Journal of the Royal Statistical Society, Series B*, **34**, 187–220.

Cox, D.R. and Hinkley, D.V. (1974) *Theoretical Statistics*. London: Chapman & Hall.

Cox, D.R. and Oakes, D. (1984) *Analysis of Survival Data*. London: Chapman & Hall.

Cox, D.R. and Snell, E.J. (1989) *Analysis of Binary Data*. London: Chapman & Hall.

Crawley, M.J. (2002) *Statistical Computing: An Introduction to Data Analysis using S-PLUS*. Chichester: John Wiley & Sons, Inc.

Crawley, M.J., Johnston, A.E., Silvertown, J., Dodd, M., de Mazancourt, C., Heard, M.S., Henman, D.F. and Edwards, G.R. (2005) Determinants of species richness in the Park Grass Experiment. *American Naturalist*, **165**, 348–362.

Cressie, N.A.C. (1991) *Statistics for Spatial Data*. New York: John Wiley & Sons, Inc.

Crowder, M.J. and Hand, D.J. (1990) *Analysis of Repeated Measures*. London: Chapman & Hall.

Dalgaard, P. (2002) *Introductory Statistics with R*. New York: Springer-Verlag.

Davidian, M. and Giltinan, D.M. (1995) *Nonlinear Models for Repeated Measurement Data*. London: Chapman & Hall.

Davison, A.C. and Hinkley, D.V. (1997) *Bootstrap Methods and Their Application*. Cambridge: Cambridge University Press.

Deevey, E.S. Jr (1947) Life tables for natural populations of animals. *Quarterly Review of Biology*, **22**, 283–314. Reprinted in W.E. Hazen (ed.) (1964) *Readings in Population and Community Ecology*. Philadelphia: Saunders.

DerSimonian, R. and Laird, N. (1986) Meta-analysis in clinical trials. *Controlled Clinical Trials*, **7**, 177–188.

Diggle, P.J. (1983) *Statistical Analysis of Spatial Point Patterns*. London: Academic Press.

Diggle, P.J., Liang, K.-Y. and Zeger, S.L. (1994) *Analysis of Longitudinal Data*. Oxford: Clarendon Press.

Dobson, A.J. (1990) *An Introduction to Generalized Linear Models*. London: Chapman & Hall.

Draper, N.R. and Smith, H. (1981) *Applied Regression Analysis*. New York: John Wiley & Sons, Inc.

Edwards, A.W.F. (1972) *Likelihood*. Cambridge: Cambridge University Press.

Efron, B. and Tibshirani, R.J. (1993) *An Introduction to the Bootstrap*. New York: Chapman & Hall.

Eisenhart, C. (1947) The assumptions underlying the analysis of variance. *Biometrics*, **3**, 1–21.

Everitt, B.S. (1994) *Handbook of Statistical Analyses Using S-PLUS*. Boca Raton, FL: Chapman & Hall/CRC.

Ferguson, T.S. (1996) *A Course in Large Sample Theory*. London: Chapman & Hall.

Fisher, L.D. and Van Belle, G. (1993) *Biostatistics*. New York, John Wiley & Sons, Inc.

Fisher, R.A. (1954) *Design of Experiments*. Edinburgh: Oliver and Boyd.

Fleming, T. and Harrington, D. (1991) *Counting Processes and Survival Analysis*. New York: John Wiley & Sons, Inc.

Fox, J. (2002) *An R and S-Plus Companion to Applied Regression*. Thousand Oaks, CA: Sage.

Fox, J. and Weisberg, S. (2011) *An R Companion to Applied Regression*, 2nd edn. Thousand Oaks CA: Sage. http://socserv.socsci.mcmaster.ca/jfox/Books/Companion.

Fox, J. and Weisberg, S. (2012) *car: Companion to Applied Regression*, with contributions by Douglas Bates, David Firth, Michael Friendly, Gregor Gorjanc, Spencer Graves, Richard Heiberger, Rafael Laboissiere, Georges Monette, Henric Nilsson, Derek Ogle, Brian Ripley and Achim Zeileis. Maintainer: John Fox, jfox@mcmaster.ca.

Gebhardt, A. (2012) *akima: Interpolation of irregularly spaced data*. Fortran code by H. Akima. R port by Albrecht Gebhardt. aspline function by Thomas Petzoldt. Eenhancements and corrections by Martin Maechler. Maintainer: Albrecht Gebhardt, albrecht.gebhardt@uni-klu.ac.at.

Gelman, A., Carlin, J.B., Stern, H.S. and Rubin, D.B. (2004) *Bayesian Data Analysis*, 2nd edn. Boca Raton, FL: Chapman & Hall/CRC.

Gordon, A.E. (1981) *Classification: Methods for the Exploratory Analysis of Multivariate Data*. New York: Chapman & Hall.

Gosset, W.S. ('Student') (1908) The probable error of a mean. *Biometrika*, **6**, 1–25.

Grimmett, G.R. and Stirzaker, D.R. (1992) *Probability and Random Processes*. Oxford: Clarendon Press.

Hairston, N.G. (1989) *Ecological Experiments: Purpose, Design and Execution*. Cambridge: Cambridge University Press.

Hampel, F.R., Ronchetti, E.M., Rousseeuw, P.J. and Stahel, W.A. (1986) *Robust Statistics: The Approach Based on Influence Functions*. New York: John Wiley & Sons, Inc.

Harman, H.H. (1976) *Modern Factor Analysis*. Chicago: University of Chicago Press.

Hastie, T. and Tibshirani, R. (1990) *Generalized Additive Models*. London: Chapman & Hall.

Hicks, C.R. (1973) *Fundamental Concepts in the Design of Experiments*. New York: Holt, Rinehart and Winston.

Hoaglin, D.C., Mosteller, F. and Tukey, J.W. (1983) *Understanding Robust and Exploratory Data Analysis*. New York: John Wiley & Sons, Inc.

Hochberg, Y. and Tamhane, A.C. (1987) *Multiple Comparison Procedures*. New York: John Wiley & Sons, Inc.

Hosmer, D.W. and Lemeshow, S. (2000) *Applied Logistic Regression*, 2nd edn. New York: John Wiley & Sons, Inc.

Hsu, J.C. (1996) *Multiple Comparisons: Theory and Methods*. London: Chapman & Hall.

Huber, P.J. (1981) *Robust Statistics*. New York: John Wiley & Sons, Inc.

Huitema, B.E. (1980) *The Analysis of Covariance and Alternatives*. New York: John Wiley & Sons, Inc.

Hurlbert, S.H. (1984) Pseudoreplication and the design of ecological field experiments. *Ecological Monographs*, **54**, 187–211.

Johnson, N.L. and Kotz, S. (1970) *Continuous Univariate Distributions*, Volume 2. New York: John Wiley & Sons, Inc.

Kalbfleisch, J. and Prentice, R.L. (1980) *The Statistical Analysis of Failure Time Data*. New York: John Wiley & Sons, Inc.

Kaluzny, S.P., Vega, S.C., Cardoso, T.P. and Shelly, A.A. (1998) *S + Spatial Stats*. New York: Springer-Verlag.

Kauffman, J., Matsik, B. and Spencer, K. (2001) *Beginning SQL Programming*. Birmingham: Wrox Press.

Kendall, M.G. and Stewart, A. (1979) *The Advanced Theory of Statistics*. Oxford: Oxford University Press.

Keppel, G. (1991) *Design and Analysis: A Researcher's Handbook*. Upper Saddle River, NJ: Prentice Hall.

Khuri, A.I., Mathew, T. and Sinha, B.K. (1998) *Statistical Tests for Mixed Linear Models*. New York: John Wiley & Sons, Inc.

Knuth, D.E. (1998) *The $T_EX$book*. Reading, MA: Addison Wesley.

Krause, A. and Olson, M. (2000) *The Basics of S and S-PLUS*. New York: Springer-Verlag.

Lee, P.M. (2012) *Bayesian Statistics: An Introduction*, 4th edn. Chichester: John Wiley & Sons, Ltd.

Lehmann, E.L. (1986) *Testing Statistical Hypotheses*. New York: John Wiley & Sons, Inc.

Levins, R. (1969) Some demographic and genetic consequences of environmental heterogeneity for biological control. *Bulletin of the Entomological Society of America*, **15**, 237–240

Lunn, D., Spiegelhalter, D., Thomas, A. and Best, N. (2009) The BUGS project: Evolution, critique and future directions. *Statistics in Medicine*, **28**, 3049–3082.

Mandelbrot, B.B. (1977) *Fractals, Form, Chance and Dimension*. San Francisco: Freeman.

Mardia, K.V., Kent, J.T. and Bibby, J.M. (1979) *Multivariate Statistics*. London: Academic Press.

May, R.M. (1976) Simple mathematical models with very complicated dynamics. *Nature*, **261**, 459–467.

McCullagh, P. and Nelder, J. A. (1989) *Generalized Linear Models*, 2nd edn. London: Chapman & Hall.

McCulloch, C.E. and Searle, S.R. (2001) *Generalized, Linear and Mixed Models*. New York: John Wiley & Sons, Inc.

Michelson, A.A. (1880) Experimental determination of the velocity of light made at the U.S. Naval Academy, Annapolis. *Astronomical Papers*, **1**, 109–145

Millard, S.P. and Krause, A. (2001) *Using S-PLUS in the Pharmaceutical Industry*. New York: Springer-Verlag.

Miller, R.G. (1981) *Survival Analysis*. New York: John Wiley & Sons, Inc.

Miller, R.G. (1997) *Beyond ANOVA: Basics of Applied Statistics*. London: Chapman & Hall.

Mosteller, F. and Tukey, J.W. (1977) *Data Analysis and Regression*. Reading, MA: Addison-Wesley.

Murrell, P. (2006) *R Graphics*. Boca Raton, FL: Chapman & Hall.

Nelder, J.A. and Wedderburn, R.W.M. (1972) Generalized linear models. *Journal of the Royal Statistical Society, Series A*, **135**, 370–384.

Neter, J., Wasserman, W. and Kutner, M.H. (1985) *Applied Linear Regression Models*. Homewood, IL: Irwin.

Neter, J., Kutner, M., Nachsteim, C. and Wasserman, W. (1996) *Applied Linear Statistical Models*. New York: McGraw-Hill.

Neuwirth, E. (2012) *RColorBrewer: ColorBrewer palettes*. Maintainer: Erich Neuwirth, erich.neuwirth@univie.ac.at.

OED (2004) *Oxford English Dictionary*. Oxford: Oxford University Press.

O'Hagen, A. (1988) *Probability: Methods and Measurement*. London: Chapman & Hall.

Perry, J.N., Rothery, P., Clark, S.J., Heard, M.S. and Hawes, C. (2003) Design, analysis and statistical power of the farm-scale evaluations of genetically modified herbicide-tolerant crops. *Journal of Applied Ecology*, **40**, 17–31.

Pinheiro, J. and Bates, D. (2012) *nlme: Linear and nonlinear mixed effects models*, with Saikat DebRoy, Deepayan Sarkar and the R Core team. Maintainer: R-core, R-core@R-project.org.

Pinheiro, J.C. and Bates, D.M. (2000) *Mixed-Effects Models in S and S-PLUS*. New York: Springer-Verlag.

Platt, J.R. (1964) Strong inference. *Science*, **146**, 347–353.

Plummer, M. (2012) *JAGS: A program for analysis of Bayesian graphical models using Gibbs sampling.* http://mcmc-jags.sourceforge.net/.

Priestley, M.B. (1981) *Spectral Analysis and Time Series*. London: Academic Press.

Rao, P.S.R.S. (1997) *Variance Components Estimation: Mixed Models, Methodologies and Applications*. London: Chapman & Hall.

Riordan, J. (1978) *An Introduction to Combinatorial Analysis*. Princeton, NJ: Princeton University Press.

Ripley, B. (2012) *tree: Classification and regression trees*. Maintainer: Brian Ripley, ripley@stats.ox.ac.uk.

Ripley, B. and Lapsley, M. (2012) *RODBC: ODBC database access*. Maintainer: Brian Ripley, ripley@stats.ox.ac.uk.

Ripley, B.D. (1996) *Pattern Recognition and Neural Networks*. Cambridge: Cambridge University Press.

Robert, C.P. and Casella, G. (1999) *Monte Carlo Statistical Methods*. New York: Springer-Verlag.

Rosner, B. (1990) *Fundamentals of Biostatistics*. Boston: PWS-Kent.

Ross, G.J.S. (1990) *Nonlinear Estimation*. New York: Springer-Verlag.

Santer, T.J. and Duffy, D.E. (1990) *The Statistical Analysis of Discrete Data*. New York: Springer-Verlag.

Schwarzer, G. (2012) *meta: Meta-analysis with R*. Maintainer: Guido Schwarzer, sc@imbi.uni-freiburg.de.

Scott, D.W. (1992) *Multivariate Density Estimation. Theory, Practice and Visualization*. New York: John Wiley & Sons, Inc.

Searle, S.R., Casella, G. and McCulloch, C.E. (1992) *Variance Components*. New York: John Wiley & Sons, Inc.

Shao, J. and Tu, D. (1995) *The Jackknife and Bootstrap*. New York: Springer-Verlag.

Shumway, R.H. (1988) *Applied Statistical Time Series Analysis*. Englewood Cliffs, NJ: Prentice Hall.

Silverman, B.W. (1986) *Density Estimation*. London: Chapman & Hall.

Silvey, S.D. (1970) *Statistical Inference*. London, Chapman & Hall.

Snedecor, G.W. and Cochran, W.G. (1980) *Statistical Methods*. Ames: Iowa State University Press.

Soetaert, K., Petzoldt, T. and Setzer, R.W. (20102) *deSolve: General solvers for initial value problems of ordinary differential equations (ODE), partial differential equations (PDE), differential algebraic equations (DAE), and delay differential equations (DDE)*. Maintainer: Thomas Petzoldt, thomas.petzoldt@tu-dresden.de.

Sokal, R.R. and Rohlf, F.J. (1995) *Biometry: The Principles and Practice of Statistics in Biological Research*. San Francisco, W.H. Freeman.

Spiegelhalter, D. Thomas, A., Best, N. and Lunn, D. (2003) *WinBUGS User Manual*. Version 1.4. http://www.mrc-bsu.cam.ac.uk/bugs/winbugs/manual14.pdf.

Sprent, P. (1989) *Applied Nonparametric Statistical Methods*. London: Chapman & Hall.

Taylor, L.R. (1961) Aggregation, variance and the mean. *Nature*, **189**, 732–735.

Therneau, T.M. and Atkinson, B. (2012) *rpart: Recursive partitioning*. Maintainer: Brian Ripley, ripley@stats.ox.ac.uk.

Upton, G. and Fingleton, B. (1985) *Spatial Data Analysis by Example*. Chichester: John Wiley & Sons, Ltd.

Venables, W.N. and Ripley, B.D. (2002) *Modern Applied Statistics with S-PLUS*, 4th edn. New York: Springer-Verlag.

Venables, W.N., Smith, D.M. and the R Development Core Team (2012) *An Introduction to R. Notes on R: A Programming Environment for Data Analysis and Graphics*. Version 2.15.0. ISBN 3-900051-12-7. http://cran.r-project.org/.

Wedderburn, R.W.M. (1974) Quasi-likelihood functions, generalized linear models and the Gauss-Newton method. *Biometrika*, **61**, 439–447.

Weisberg, S. (1985) *Applied Linear Regression*. New York: John Wiley & Sons, Inc.

Wetherill, G.B., Duncombe, P., Kenward, M., Kollerstrom, J., Paul, S.R. and Vowden, B.J. (1986) *Regression Analysis with Applications*. London: Chapman & Hall.

Winer, B.J., Brown, D.R. and Michels, K.M. (1991) *Statistical Principles in Experimental Design*. New York: McGraw-Hill.

Wood, S.N. (2000) Modelling and smoothing parameter estimation with multiple quadratic penalties. *Journal of the Royal Statistical Society, Series B*, **62**, 413–428.

Wood, S.N. (2003) Thin plate regression splines. *Journal of the Royal Statistical Society, Series B*, **65**, 95–114.

Wood, S.N. (2004) Stable and efficient multiple smoothing parameter estimation for generalized additive models. *Journal of the American Statistical Association*, **99**, 673–686.

Wood, S.N. (2006) *Generalized Additive Models: An Introduction with R*. Boca Raton, FL: CRC Press.

Wood, S. (2012) *mgcv: Mixed GAM computation vehicle with GCV/AIC/REML smoothness estimation*. Maintainer: Simon Wood, simon.wood@r-project.org.

Yu-Sung Su and Masanao Yajima (2012) *R2jags: A package for running jags from R*. Maintainer: Yu-Sung Su, suyusung@tsinghua.edu.cn.

Zar, J.H. (1999) *Biostatistical Analysis*. Englewood Cliffs, NJ: Prentice Hall.

REFERENCES AND FURTHER READING

Index

Text in **bold** represents R functions or arguments to R functions. Greek letters appear at the end of the index.